T0270936

SHARKS AND THEIR RELATIVES II

BIODIVERSITY, ADAPTIVE PHYSIOLOGY, AND CONSERVATION

CRC
MARINE BIOLOGY
SERIES

The late Peter L. Lutz, Founding Editor
David H. Evans, Series Editor

PUBLISHED TITLES

Biology of Marine Birds
E.A. Schreiber and Joanna Burger

Biology of the Spotted Seatrout
Stephen A. Bortone

The Biology of Sea Turtles, Volume II
Peter L. Lutz, John A. Musick, and Jeanette Wyneken

Biology of Sharks and Their Relatives
Jeffrey C. Carrier, John A. Musick,
and Michael R. Heithaus

*Early Stages of Atlantic Fishes: An Identification Guide
for the Western Central North Atlantic*
William Richards

The Physiology of Fishes, Third Edition
David H. Evans

Biology of the Southern Ocean, Second Edition
George A. Knox

Biology of the Three-Spined Stickleback
Sara Östlund-Nilsson, Ian Mayer, and Felicity Anne
Huntingford

*Biology and Management of the World Tarpon
and Bonefish Fisheries*
Jerald S. Ault

*Methods in Reproductive Aquaculture: Marine and
Freshwater Species*
Elsa Cabrita, Vanesa Robles, and Paz Herráez

SHARKS AND THEIR RELATIVES II

BIODIVERSITY, ADAPTIVE PHYSIOLOGY, AND CONSERVATION

EDITED BY

JEFFREY C. CARRIER

JOHN A. MUSICK

MICHAEL R. HEITHAUS

CRC Press
Taylor & Francis Group
Boca Raton   London   New York

CRC Press is an imprint of the
Taylor & Francis Group, an **informa** business

CRC Press
Taylor & Francis Group
6000 Broken Sound Parkway NW, Suite 300
Boca Raton, FL 33487-2742

© 2010 by Taylor and Francis Group, LLC
CRC Press is an imprint of Taylor & Francis Group, an Informa business

No claim to original U.S. Government works

International Standard Book Number: 978-1-4200-8047-6 (Hardback)

Library of Congress Cataloging-in-Publication Data

Sharks and their relatives II : biodiversity, adaptive physiology, and conservation / editors Jeffrey C. Carrier, John A. Musick, and Michael R. Heithaus. -- 1st ed.
    p. cm. -- (Marine biology series ; 11)
  Includes bibliographical references and index.
  ISBN 978-1-4200-8047-6 (alk. paper)
    1. Chondrichthyes--Physiology. 2. Chondrichthyes--Conservation. 3. Sharks--Physiology. 4. Sharks--Conservation. 5. Animal diversity. I. Carrier, Jeffrey C. II. Musick, John A. III. Heithaus, Michael R. IV. Biology of sharks and their relatives.

QL638.6.B56 2010
597.3--dc22
                                             2009044922

Visit the Taylor & Francis Web site at
http://www.taylorandfrancis.com

and the CRC Press Web site at
http://www.crcpress.com

To our students, past, present, and future, whose work has inspired our curiosity,

challenged us to continue to grow intellectually, and kept us forever young.

Contents

Section I Chondrichthyan Biodiversity: Ecosystems and Distribution of Fauna

Section II Adaptive Physiology

Preface

The publication of *The Biology of Sharks and Their Relatives* in 2004 provided a comprehensive review of research spanning the long interval since the publication of Perry Gilbert's classic works, *Sharks and Survival* and *Sharks, Skates, and Rays*. We were satisfied with our range of coverage and with the expertise provided by the authors who contributed to this work. Yet we realized that the volume would necessarily be incomplete because of the breadth of new research being undertaken, the rapid advances in technology that seemed to be opening new avenues for investigation, and the emergence of new investigators who were beginning to explore issues of biodiversity and distribution, physiology, and ecology in ways that have eluded more traditional studies.

To address subject areas and subdisciplines where our coverage was absent or superficial in volume one, we have assembled in the current volume a collection of works that reveal patterns of biodiversity, the physiological attributes that contribute to elasmobranchs' successful exploitation of oceanic and freshwater realms, and the unique issues associated with the interaction between elasmobranchs and humans, all of this with overarching attention to issues of conservation. We begin with chapters examining biodiversity. We have chosen to approach this discussion by presenting elasmobranchs as inhabitants of the range of zoogeographic provinces, realizing that significant overlap may occur for more pelagic species. This realization was reflected in the dialogue that occurred during preparation of the book between our chapter authors, and the recognition that many species simply cannot be confined to a specific habitat or range of habitats. We then continue by examining some of the unique physiological adaptations that allow these animals to exploit the range of habitats where they are found, from unique sensory modalities to compensatory mechanisms for physiological and environmental stress. Our concluding section presents some of the challenges faced by members of these groups. We have asked our authors to consider human interactions and anthropogenic effects on worldwide populations and the potential extinction risks posed from survival under increasing threats from changes in habitat, changes in water chemistry, and increasing commercial exploitation. Conservation of species under threat remains a theme throughout the book.

Our authors represent an international group of investigators including established scientists whose work has been widely published and respected, and emerging younger scientists who have exploited recent advances in technology to ask and answer new questions as well as offering new insights and interpretations to enduring problems in the fields of ecology and physiology. We have asked them to be speculative and challenging, and we have asked them to predict future areas for investigation in hopes that their work will both inspire and provoke additional studies of these fascinating animals.

The Editors

Jeffrey C. Carrier, Ph.D., is a professor of biology at Albion College, Albion, Michigan, where he has been a faculty member since 1979. He earned his B.S. in biology in 1970 from the University of Miami, and completed a Ph.D. in biology from the University of Miami in 1974. While at Albion College, Dr. Carrier has received multiple awards for teaching and scholarship and has held the A. Merton Chickering and W.W. Diehl Endowed Professorships in biology. His primary research interests have centered on various aspects of the physiology and ecology of nurse sharks in the Florida Keys. His most recent work has centered on the reproductive biology and mating behaviors of this species in a long-term study from an isolated region of the Florida Keys. Dr. Carrier has been a member of the American Elasmobranch Society, the American Society of Ichthyologists and Herpetologists, Sigma Xi, and the Council on Undergraduate Research. He served as secretary, editor, and two terms as president of the American Elasmobranch Society, and has received that society's Distinguished Service Award.

John A. (Jack) Musick, Ph.D., is the Marshall Acuff Professor Emeritus in Marine Science at the Virginia Institute of Marine Science (VIMS), College of William and Mary, where he has served on the faculty since 1967. He earned his B.A. in biology from Rutgers University in 1962 and his M.A. and Ph.D. in biology from Harvard University in 1964 and 1969, respectively. While at VIMS, he has successfully mentored 37 masters and 49 Ph.D. students. Dr. Musick has been awarded the Thomas Ashley Graves Award for Sustained Excellence in Teaching from the College of William and Mary, the Outstanding Faculty Award from the State Council on Higher Education in Virginia, and the Excellence in Fisheries Education Award from the American Fisheries Society. He has published more than 150 scientific papers and co-authored or edited 16 books focused on the ecology and conservation of sharks, other marine fishes, and sea turtles. In 1985, he was elected a fellow by the American Association for the Advancement of Science. He has received Distinguished Service Awards from both the American Fisheries Society and the American Elasmobranch Society, for which he has served as president, and been recognized as a distinguished fellow. Dr. Musick also has served as president of the Annual Sea Turtle Symposium (now the International Sea Turtle Society), and as a member of the World Conservation Union (IUCN) Marine Turtle Specialist Group. Dr. Musick served as co-chair of the IUCN Shark Specialist Group for nine years, and is currently the vice chair for science. He also has served on three national, five regional, and five state scientific advisory committees concerned with marine resource management and conservation. In 2008, Dr. Musick was awarded The Lifetime Achievement Award in Science by the State of Virginia.

Michael R. Heithaus, Ph.D., is an assistant professor of marine biology at Florida International University in Miami. He received his B.A. in biology from Oberlin College in Ohio and his Ph.D. from Simon Fraser University in British Columbia, Canada. He was a postdoctoral scientist and staff scientist at the Center for Shark Research at Mote Marine Laboratory in Sarasota, Florida (2001–2003) and also served as a research fellow at the National Geographic Society (2002–2003). Dr. Heithaus' main research interests are in predator–prey interactions and the factors influencing behavioral decisions, especially of large marine taxa including marine mammals, sharks and rays, and sea turtles. Currently, he is investigating how behavioral decisions, especially anti-predator behavior, may influence behavioral decisions of other individuals and community dynamics. The majority of Dr. Heithaus' previous field work has focused on tiger sharks and their prey species in Western Australia. Dr. Heithaus is a member of the Ecological Society for America, Animal Behavior Society, International Society for Behavioral Ecology, Society for Marine Mammalogy, and Sigma Xi.

Contributors

Diego Bernal
Department of Biology
University of Massachusetts
Dartmouth, Massachusetts

Robert H. Buch
Florida Program for Shark Research
Florida Museum of Natural History
University of Florida
Gainesville, Florida

George H. Burgess
Florida Program for Shark Research
Florida Museum of Natural History
University of Florida
Gainesville, Florida

Felipe Carvalho
Florida Program for Shark Research
Florida Museum of Natural History
University of Florida
Gainesville, Florida

Patricia Charvet-Almeida
Research Collaborator
Museu Paraense Emílio Goeldi (MPEG)
SENAI-PR and Projeto Trygon
Curitiba, PR, Brazil

Geremy Cliff
KwaZulu-Natal Sharks Board
Umhlanga, South Africa *and* Biomedical
    Resource Unit
University of KwaZulu-Natal
Durban, South Africa

Anthony D. Cornett
Department of Mathematics, Science, and
    Psychology
Valencia Community College
Kissimmee, Florida

Sheldon F.J. Dudley
KwaZulu-Natal Sharks Board
Umhlanga, South Africa *and* Biomedical
    Resource Unit
University of KwaZulu-Natal
Durban, South Africa

Nicholas K. Dulvy
Department of Biological Sciences
Simon Fraser University
Burnaby, British Columbia, Canada

David A. Ebert
Pacific Shark Research Center
Moss Landing Marine Laboratories
Moss Landing, California

Robyn E. Forrest
Fisheries Centre University of British
    Columbia
Vancouver, British Columbia, Canada

Alejandro Frid
Vancouver Aquarium
Vancouver, British Columbia, Canada

Michael G. Frisk
School of Marine and Atmospheric
    Sciences
Stony Brook University
Stony Brook, New York

Brittany A. Garner
Florida Program for Shark Research
Florida Museum of Natural History
University of Florida
Gainesville, Florida

James Gelsleichter
Department of Biology
University of North Florida
Jacksonville, Florida

R. Dean Grubbs
Florida State University Coastal and
    Marine Laboratory
St. Teresa, Florida

Michael R. Heithaus
Marine Science Program
Department of Biological Sciences
Florida International University
Miami, Florida

Stephen M. Kajiura
Department of Biological Sciences
Florida Atlantic University
Boca Raton, Florida

Peter M. Kyne
Tropical Rivers and Coastal Knowledge
Charles Darwin University
Darwin, Northern Territory, Australia

David S. Portnoy
Department of Wildlife and Fisheries
    Science
Texas A&M University
College Station, Texas

Carla Christie Diban Quijada
Faculdade de Ciências Médicas
Campina Grande, PB, Brazil

Ricardo S. Rosa
Departamento de Sistemática e Ecologia
Universidade Federal da Paraíba
João Pessoa, PB, Brazil

Mahmood S. Shivji
The Guy Harvey Research Institute
and Save Our Seas Shark Center
Oceanographic Center
Nova Southeastern University
Fort Lauderdale, Florida

Colin A. Simpfendorfer
Fishing and Fisheries Research Centre
School of Earth and Environmental
    Sciences
James Cook University
Townsville, Queensland, Australia

David W. Sims
Marine Biological Association of the
    United Kingdom
The Laboratory
Citadel Hill
Plymouth, United Kingdom

Gregory Skomal
Massachusetts Shark Research Program
Massachusetts Marine Fisheries
Vineyard Haven, Massachusetts

Emma Sommerville
Centre for Fish and Fisheries Research
Murdoch University
Murdoch, Western Australia, Australia

John D. Stevens
CSIRO Marine and Atmospheric Research
Hobart, Tasmania, Australia

Jeremy J. Vaudo
Marine Science Program
Department of Biological Sciences
Florida International University
Miami, Florida

Christina J. Walker
Department of Biology
University of North Florida
Jacksonville, Florida

William T. White
CSIRO Wealth from Oceans Flagship
Hobart, Tasmania, Australia

Megan V. Winton
Pacific Shark Research Center
Moss Landing Marine Laboratories
Moss Landing, California

Aaron J. Wirsing
College of Forest Resources
University of Washington
Seattle, Washington

Boris Worm
Department of Biology
Dalhousie University
Halifax, Nova Scotia, Canada

Kara E. Yopak
Center for Scientific Computation in
    Imaging
University of California San Diego
San Diego, California

Section I

Chondrichthyan Biodiversity: Ecosystems and Distribution of Fauna

1

Epipelagic Oceanic Elasmobranchs

John D. Stevens

CONTENTS

1.1 Introduction: The Epipelagic Oceanic Ecosystem

The oceanic zone is generally defined as those waters beyond the 200 m isobath at the edge of the continental shelf extending out and encompassing the ocean basins. This contrasts with the neritic zone that extends from the shore and is the water above the continental shelf. The pelagic zone includes all open waters that are not close to the bottom and is divided into the epipelagic, mesopelagic, bathypelagic, abyssopelagic, and hadopelagic zones that are distinguished by their depth and ecology (Bone, Marshall, and Blaxter 1995). The epipelagic zone extends from the surface down to 200 m and has an abundance of light that allows for photosynthesis. Nearly all primary production in the ocean occurs here and this is where most organisms are concentrated (Pinet 2006). About 80% of the ocean's surface is above water greater than 200 m in depth. In general, species diversity tends to decrease as one leaves the shore and the food web becomes supported by the planktonic production. Extending down from the epipelagic zone is the mesopelagic zone, or twilight zone, that reaches to a depth of 1000 m and has a little light but not enough for photosynthesis to occur. Together, the epipelagic and mesopelagic zones, where light penetrates the water, are known as the photic zone. The pelagic zone occupies about 1370 million cubic km (330 million cubic miles) and has a vertical range up to 11 km (6.8 miles). The diversity and abundance of pelagic life decrease with increasing depth. It is affected by light levels, pressure, temperature, salinity, the supply of dissolved oxygen and nutrients, and the submarine topography.

Epipelagic oceanic ecosystems occur in all major oceans, and in the context of this book there is some potential overlap with Chapters 2 (deepwater chondrichthyans), 3 (high latitude seas and chondrichthyans), and 4 (chondrichthyans of tropical marine ecosystems). While most epipelagic oceanic elasmobranchs migrate vertically into deeper zones, the distinction for this book is the zone in which they spend the majority of their time, although for some species this may not be well known. The majority of epipelagic oceanic elasmobranchs are tropical but, to avoid overlap, Chapter 9 will largely ignore these species. Some epipelagic oceanic elasmobranchs occur at high latitudes and these will be dealt with in this chapter. However, at the extremes of high latitude in the Arctic and Southern Oceans epipelagic elasmobranchs are absent.

1.2 Biodiversity and Biogeography

1.2.1 Biodiversity and Systematics

As with any division of species into specific ecosystems, the inclusion or exclusion of some will be debatable. Epipelagic is taken here to refer to highly mobile species that occur in the top 200 m of the water column; oceanic species primarily inhabit ocean basins away from the shelf edge of land masses. Some oceanic species may at times come into more coastal waters for a variety of reasons. Epipelagic oceanic elasmobranchs dealt with in this chapter are listed in Table 1.1 and comprise about 2% of the global extant chondrichthyan fauna (estimated at 1200 species). However, when examined by order about 73% of the Lamniformes are epipelagic and oceanic although the Carcharhiniformes probably dominate in terms of biomass. A number of other species that could be argued for inclusion (Table 1.2) have been omitted because either little is known about them (they may be based on only a few specimens), they are primarily associated with land masses and are only semi-oceanic, or if they have oceanic components to their populations it is not clear they are epipelagic. Including these species would take the total to about 33 or 3% of the extant chondrichthyan fauna. However, categorizing some species is still problematic. For example, it could be argued that the white shark (*Carcharodon carcharias*) is primarily coastal. However, recent tracking data have revealed a considerable oceanic life-history component that warrants its inclusion in this chapter. Of the 21 species in Table 1.1, 19 are sharks and two are batoids, with the most speciose family being the Lamnidae (five species) (Table 1.3). A further four families and six species are lamniform sharks. Four species are squaliform sharks (from two families). There is one epipelagic oceanic orectolobiform shark (family Rhincodontidae) and the two batoids are from the families Dasyatidae and Mobulidae (Table 1.3).

As noted by Compagno (2008), pelagic elasmobranchs exhibit three basic body plans or ecomorphotypes specialized for life in this environment, macroceanic, microceanic, and dorso-ventrally flattened. Compagno (1990) further examined sharks in terms of ecomorphotypes which can include diverse taxa that may or may not be related but which are grouped together by similarities in morphology, habitat, and behavior. He considered several oceanic ecomorphotypes: a high-speed or tachypelagic body form that parallels the tunas and is represented by the lamnids *Isurus oxyrinchus* (shortfin mako) and *Lamna* spp. (porbeagle and salmon sharks); the archipelagic (modified tachypelagic) superpredator represented by the white shark; the macroceanic morphotype with long pectoral fins such as the blue shark (*Prionace glauca*), oceanic whitetip shark (*Carcharhinus longimanus*),

TABLE 1.1

Epipelagic Oceanic Elasmobranchs

Family	Species	Distribution/Depth	Size Range (TL)
Somniosidae	*Scymnodalatias albicauda*	Southern Ocean 150–510 m	>20–111 cm
Dalatiidae	*Euprotomicrops bispinatus*	Circumglobal 0–400 or 1800 m	8–27 cm
	Isistius brasiliensis	Circumglobal 0–>1000 m	14–50 cm
	Squaliolus laticaudus	Nearly circumglobal 200–500 m	Attains 28 cm
Rhincodontidae	*Rhincodon typus*	Circumglobal 0–1000 m	50–1200 cm
Pseudocarchariidae	*Pseudocarcharias kamoharai*	Circumglobal 0–600 m	40–118 cm
Megachasmidae	*Megachasma pelagios*	Circumglobal 150–1000 m	Attains 550 cm
Alopiidae	*Alopias pelagicus*	Indo-Pacific 0–150 m	130–390 cm
	Alopias superciliosus	Circumglobal 0–700 m	100–484 cm
	Alopias vulpinus	Circumglobal 0–650 m	115–570 cm
Cetorhinidae	*Cetorhinus maximus*	Atlantic and Pacific 0–1000 m	150–1000 cm
Lamnidae	*Carcharodon carcharias*	Circumglobal 0–1280 m	130–600 cm
	Isurus oxyrinchus	Circumglobal 0–650 m	60–394 cm
	Isurus paucus	Circumglobal Depth range uncertain	97–417 cm
	Lamna nasus	North and South Atlantic, South Pacific, and South Indian Oceans 0–370 m	70–324 cm
	Lamna ditropis	North Pacific 0–225 m	40–300 cm
Carcharhinidae	*Carcharhinus falciformis*	Circumglobal 0–500 m	70–330 cm
	Carcharhinus longimanus	Circumglobal 0–150 m	60–350 cm
	Prionace glauca	Circumglobal 0–1000 m	35–383 cm
Dasyatidae	*Pteroplatytrygon violacea*	Circumglobal 0–>100 m	16–80 cm disc width
Mobulidae	*Manta birostris*	Circumglobal Near surface (0–? m)	122–670 cm disc width

TABLE 1.2

Secondary* Epipelagic Oceanic Elasmobranchs

Family	Species
Squalidae	*Squalus acanthias*
Somniosidae	*Scymnodalatias oligodon*
	Zameus squamulosus
Etmopteridae	*Etmopterus pusillus*
	Etmopterus bigelowi
Dalatiidae	*Isistius plutodus*
	Squaliolus aliae
Triakidae	*Galeorhinus galeus*
Carcharhinidae	*Carcharhinus signatus*
Sphyrinidae	*Sphyrna lewini*
Mobulidae	*Mobula japonica*
	Mobula thurstoni

* Species that might be argued to be epipelagic and oceanic, but not included in this chapter.

TABLE 1.3

Number and Percentage of Oceanic Species Relative to All Species in That Order

Order	No. of Oceanic Species	% of Oceanic Species
Squaliformes	4	3.1
Orectolobiformes	1	3.0
Lamniformes	11	73.3
Carcharhiniformes	3	1.3
Myliobatiformes	2?	
Total	21	2.8

silky shark (*C. falciformis*), thresher sharks (*Alopias* spp.), longfin mako (*Isurus paucus*), and megamouth shark (*Megachasma pelagios*); the microceanic morphotype of small- or moderate-sized sharks with long fusiform bodies and small pectoral fins such as dwarf members of the squaliformes and the crocodile shark (*Pseudocarcharias kamoharai*); the rhomboidal form of the pelagic stingray (*Pteroplatytrygon violacea*); and the aquilopelagic morphotype of the manta ray (*Mobula brevirostris*).

Compagno (2008) suggests that the relatively low diversity of extant pelagic chondrichthyans, as in freshwater, is evidence that the open ocean is a marginal habitat for this group when compared to the high diversity of oceanic teleosts and cephalopods. Pelagic sharks are best known from the epipelagic zone, which contains the highest known diversity of species and the largest biomass. Indeed, the cosmopolitan blue shark is (or was) arguably the most wide-ranging and abundant chondrichthyan.

1.2.2 Biogeography

All species in Table 1.1 are widespread, with most having circumglobal distributions. Nine species have primarily tropical distributions, eight are found in tropical and temperate waters (five of these are more tropical and three are more temperate), and four have temperate distributions. In tropical reef fish, vagility or dispersal ability is lowest in small benthic species that lack pelagic eggs or larvae (Rosenblatt 1963; Rosenblatt and Waples 1986). Vagility is also inversely correlated with speciation, intrataxon diversity, and endemism. Musick, Harbin, and Compagno (2004), in their account of shark zoogeography, found that vagility increased with body size and was lowest in benthic species, higher in benthpelagic species, and highest in pelagic species. Coastal sharks also tend to have lower vagility than bathyal or oceanic species. These authors found a strong relationship between species diversity and body size, with about eight times the number of small species (<100 cm) than very large species (>300 cm). They also found that, regardless of size, benthic species had distributional ranges more than five times smaller than pelagic sharks. Musick, Harbin, and Compagno (2004) suggested that smaller benthic species that have reduced distributional ranges become more easily isolated, leading to higher rates of speciation and greater diversity than large, wide-ranging species. Of the species considered to be primarily epipelagic and oceanic in Table 1.1, 14 (67%) are very large (>300 cm). However, it is interesting that 19% are small (<100 cm), of which three are diminutive squaliform sharks in the family Dalatiidae. These species make diel migrations between near surface waters and depths of 500 to more than 1000 m and may be associated with specific water masses (Musick, Harbin, and Compagno 2004). It is possible they are able to take advantage of currents at different depths to aid in their dispersal.

The antitropical distributions of the porbeagle (*Lamna nasus*) and basking shark (*Cetorhinus maximus*) probably arose during glaciation periods when the tropical zone was more constricted, allowing these species to cross the tropics by remaining at depth in the equatorial zone. The salmon shark (*Lamna ditropis*) probably diverged from the porbeagle after closure of the Arctic seaway by an ice sheet in the late Cenozoic (Reif and Saure 1987). The absence of the pelagic thresher (*Alopias pelagicus*) from the Atlantic is more enigmatic. It is tempting to presume that this species evolved after the formation of the Isthmus of Panama and that Cape Horn and the Cape of Good Hope acted as barriers to the distribution of this tropical species into the Atlantic. However, this hypothesis is contradicted by the fossil record and it may be that the pelagic thresher was initially present in the Atlantic and subsequently died out (Musick, Harbin, and Compagno 2004).

1.3 Life-Histories

1.3.1 Reproductive Biology and Strategies

It is no great surprise that all epipelagic oceanic elasmobranchs considered in this chapter are viviparous. All known oviparous chondrichthyans lay, and usually attach, their eggs on the substrate that is not really an option for the species in question, and pelagic eggs do not appear to have evolved in this group (Musick and Ellis 2005). As discussed in Wourms, Grove, and Lombardi (1988), the relationship between viviparity and the ecology of viviparous species is poorly understood and attempts to explain viviparity using general life-history strategy models are inadequate. Viviparity confers a number of obvious advantages that facilitate protection and development of the young, dispersal and parental care with few of the constraints on mobility. Specializations for the transfer of nutrients to the developing embryo may further enhance survival of the offspring. While viviparity is found in a range of different habitats in elasmobranchs, specific advantages for the epipelagic lifestyle would seem to be associated with freedom from the substrate (for egg laying) and production of well-advanced and self-sufficient young. Viviparous teleosts also have diverse lifestyles and occur in a range of habitats suggesting viviparity is a flexible reproductive strategy in teleosts (Wourms, Grove, and Lombardi 1988). However, the supply method of nutrients to the embryos in teleosts may be restricted by environmental conditions. In one group of teleosts (Poeciliidae), it has been suggested that lecithotrophy is successful in unpredictable environments but that matrotrophy requires a predictable food supply (Thibault and Schultz 1978). Lecithotrophic embryos are nourished by yolk reserves stored in the egg and are not dependent on food availability once vitellogenesis is complete. Conversely, the growth of embryos that rely on transfer of nutrients from the mother may be affected by changes in food availability. The epipelagic lifestyle does not appear to have restricted the methods of embryonic nutrition in elasmobranchs, with the dominant method, oophagy (52%), reflecting the dominance of lamnoid sharks in this habitat. Three (14%) of the species in Table 1.1 are placentotrophic and seven (33%) are either lecithotrophic or histotrophic. This dominance of matroprophic methods would suggest, if the teleostean argument applied, that the oceanic epipelagic habitat provided a predictable food supply; however, this seems counterintuitive. The method of embryonic nutrition even varies among the four giant plankton feeders. The temperate basking shark and mainly tropical megamouth shark are oophagous, the tropical manta ray is histotrophic, and in the mainly tropical whale shark (*Rhincodon typus*) the method of nutrient transfer to the embryos is uncertain. Whale sharks were once thought to be oviparous, but they are now known to show a more primitive form of viviparity where the egg cases are retained inside the female until hatching.

Snelson et al. (2006) state that pelagic elasmobranchs have slightly larger litters of smaller young than coastal elasmobranchs and suggest this is due to the challenges of their respective habitats. These authors suggest that food and predators are more abundant in coastal habitats and that selective pressures have consequently produced larger, faster growing young to take advantage of more food and to combat higher predation. However, I suggest that this relationship is not clear and that, if anything, the reverse applies. Indeed, most authors have concluded that open waters are more dangerous than complex coastal habitats due to a lack of spatial refuges (Branstetter 1990; Heithaus 2001, 2007). So, it could be argued that large, well-developed young that are less vulnerable to predation might be selected for in the pelagic realm where protected nurseries or cryptic avoidance is less possible. Where information is available, birth size in the species from Table 1.1 varies from 3.8% (whale shark) to 38% (crocodile shark) of maximum size. However, the majority of species (60%) had a birth size between 15% and 25% of maximum size and 30% had a birth size greater than 25% of maximum size. In only 10% of species was the birth size less than 15% of maximum size. The only recorded litter size in the whale shark was 300 (Joung et al. 1996) while the blue shark, which has an average litter size of 30 to 40, can have up to 135 pups. In these two species it seems that a higher natural mortality associated with a small birth size (3.8% to 11.2% of maximum size) is traded off against larger litter sizes. Such trade-offs between offspring number and size are common in many taxa.

It is interesting that relatively few pregnant females of several of these epipelagic elasmobranchs have been recorded. While for some this is due to relatively few records for the species in general, for others like the basking shark, whale shark, white shark, shortfin mako, and manta ray this is not the case and they are common and frequently caught (even targeted commercially) species. In the case of the white shark this may partly be explained by escapement of these large, powerful fish but it is also possible that pregnant females of these species are occupying a habitat or behaving in a way that makes them less likely to be captured. Epipelagic elasmobranchs show a diversity of gestation periods (4 to 18 months) and breeding frequencies that suggest this habitat does not impose any stringent restrictions on these parameters. The oceanic environment does, however, pose potential challenges, including encountering mates and a lack of protected nursery areas for neonates and small juveniles. Little is known about mechanisms associated with mating areas in these species; but in the relatively well-studied blue shark, females are able to store sperm for long periods after mating and there are complex movement patterns that bring the normally spatially segregated sexes together for mating (see Section 1.3.4). Again, little is known about nursery areas for most of these species but some (shortfin mako, thresher shark, silky shark) utilize coastal areas and others like the blue shark have spatially segregated areas in oceanic waters that are usually in more productive zones at higher latitudes. In salmon sharks, pupping and nursery grounds have been proposed both along the transition boundary of the subarctic and central Pacific currents (Nakano and Nagasawa 1996) and nearshore from the Alaska-Canada border to the northern end of Baja California, Mexico (Goldman and Musick 2008).

1.3.2 Age and Growth

It is not obvious that the epipelagic environment imposes any specific selective forces on growth rates or longevity of elasmobranchs. One probable exception is that the neonates of the whale shark and blue shark that are born at a relatively small size would need to grow quickly through the predation window. Of the 21 species listed in Table 1.1, age and growth are reasonably well studied in five species and there is some information for another eight. Age and growth parameters for these 13 species are shown in Table 1.4. Nothing is known about age and growth in eight species.

1.3.3 Feeding Ecology and Behavior

Three main feeding strategies are employed by epipelagic oceanic elasmobranchs; there are the huge planktivorous species, vertically migrating diminutive dalatiids and a diverse group of mostly lamnid and carcharhinid species that feed mainly on fish and cephalopods. The plankton feeders (those considered here) comprise the whale, basking and megamouth shark, and the manta ray. These species exploit different temperature regimes and depth strata and have different morphological adaptions to capture their prey.

The whale shark occurs in tropical and warm temperate waters around the world, both in the open ocean but also close to the coast where it takes advantage of seasonal pulses of productivity. The prey taken varies considerably in size from coral and teleost spawn, krill, copepods, and jellyfish to small cephalopods and schooling fishes such as anchovies, mackerel, and even tuna (see reviews by Compagno 2001; Stevens 2007). Taylor, Compagno, and Struhsaker (1983) reviewed the feeding biology and filter apparatus of whale sharks in relation to the basking and megamouth shark. They concluded that the dense filter screens of whale sharks act as more efficient filters for short suction intakes and confer more versatile feeding behaviors, in contrast to the flow-through system of the other two species. The filter apparatus of the whale shark comprises parallel plates that transversely bridge the internal gill openings and connect adjacent holobranchs (Taylor,

TABLE 1.4

Age and Growth Parameters for Epipelagic Oceanic Elasmobranchs

Species	Area	L_∞	K	T_0	Age at Maturity	Maximum Age	References
R. typus	South Africa				F-22 M-20	F-19–27 M-20–31	Wintner 2000
A. pelagicus	Taiwan	F-197.2 M-182.2	F-0.085 M-0.118	F-7.67 M-5.48	F-9 M-7–8	F-16 M-14	Liu et al. 1999
A. superciliosus	Taiwan	F-224.6 M-218.8	F-0.092 M-0.088	F-4.21 M-4.22	F-12–13 M-9–10	F-20 M-14	Liu et al. 1998
A. vulpinus	California	F-464.3 M-416.2	F-0.124 M-0.189	F-3.35 M-2.08	F-5 M-5	F-22 M-19	Smith et al. 2008
C. maximus	Worldwide	Not currently possible to age from vertebrae					Natanson et al. 2008
C. carcharias	South Africa	544-PCL	0.065	-4.4	8–13	13(35?)	Wintner and Cliff 1999
I. oxyrinchus	NW Atlantic	366-FL 253-FL	F-0.087 M-0.125		F-18 M-8	F-32 M-29	Natanson et al. 2006
L. nasus	NW Atlantic	F-369.8 M-257.7	F-0.061 M-0.03	F-5.9 M-5.87	F-13 M-8	F-24 M-25	Natanson et al. 2002
L. ditropis	NW Pacific	F-203.8-PCL M-180.3-PCL	F-0.136 M-0.171	F-3.95 M-3.63	F-8–10 M-5	F-17 M-25	Tanaka 1980
	NE Pacific	F-207.4-PCL M-182.8-PCL	F-0.17 M-0.23	F-2.3 M-1	F-6–9 M-3–5	F-20 M-17	Goldman 2002

C. falciformis	Gulf of Mexico	311.0	0.101	-2.72	F-12 M-10	F-22 M-20	Bonfil et al. 1993
	Taiwan	F-341.0 M-315.0	F-0.077 M-0.091	F-3.03 M-2.32	F-8-10 M-9	F-11 M-14	Joung et al. 2008
C. longimanus	Central Pacific	244.6-PCL	0.103	-2.70	4-5	11	Seki et al. 1998
	South Atlantic	284.9	0.100	-3.39	6-7	F-17 M-13	Lessa et al. 1999
P. glauca	North Pacific	F-304.0 M-369.0	F-0.16 M-0.1	F-1.01 M-1.38			Tanaka et al. 1990
	Central Pacific	F-243.3-PCL M-289.7-PCL	F-0.144 M-0.129	F-0.85 M-0.76	F-5-6 M-4-5	F-15-16 M-15-16	Nakano 1994
	NW Atlantic	F-310.0-FL M-282.0-FL	F-0.13 M-0.18	F-1.77 M-1.35	F-5 M-5	F-15 M-16	Skomal and Natanson 2003
P. violacea	California	F-103 M-67	F-0.32 M-0.8	F-8.2 M-5.6	3	8.6-12	Mollet et al. 2002

F = Female; M = Male; PCL = Precaudal Length; FL = Fork Length; L_∞ is cm total length (unless otherwise noted); age is in years.

Compagno, and Struhsaker 1983). Whale sharks have been observed feeding on copepods by swimming through the patches, lifting their heads partly out of the water and gulping them in. They are also reported to feed almost vertically in the water, sucking in prey near the surface. Whale sharks can capture larger, more active nektonic prey such as small fishes and squid, but are not so well adapted for concentrating diffuse planktonic food, probably making them more dependent on dense aggregations of prey. Certainly, whale sharks arrive predictably at annual mass spawning events in various parts of the world, and individuals return in following years to the same site. Satellite tracking has revealed extensive oceanic movements (see review in Stevens 2007) and documented dives to at least 1000 m depth (Graham, Roberts, and Smart 2005; Wilson et al. 2006). Like most pelagic sharks and large teleosts, whale sharks show diel behavior, generally diving deeper during the day and remaining closer to the surface at night, probably associated with vertical migration of their prey. However, this pattern may vary when the sharks are feeding on dense prey concentrations when in coastal waters, probably due to differences in the behavior of their prey (Wilson et al. 2006).

The basking shark exploits the plankton-rich waters of the temperate zones, having an antitropical distribution. They feed on microscopic plankton such as copepods, trapping them on their unique gill rakers with the help of mucus secreted in the pharynx. The filter apparatus, with its enormous gill cavities and streamlined gill raker denticles, is adapted for high rates of water flow generated by swimming and is a more dynamic process than in the whale or megamouth sharks. When feeding they usually cruise near the surface with their mouth open and gill slits distended, occasionally closing their mouth to swallow their prey. Assuming a swimming speed of 3.7 km/h, an average adult basking shark may filter 2000 t of water per hour (Compagno 2001). Their large livers (up to 25% of body weight) are high in squalene oil, a low-density hydrocarbon, giving them near-neutral buoyancy. They actively locate plankton concentrations at the surface at tidal fronts or at boundaries of water masses, which they probably detect from chemical cues, and may occur in aggregations in coastal waters during spring and autumn plankton blooms. Basking sharks have been reported to loose their gill rakers in winter and along with the low plankton concentrations at this time of year it was hypothesized that they hibernated on the bottom. Certainly, these sharks have been caught in trawls near the bottom in winter in New Zealand (Francis and Duffy 2002). However, recent satellite tracking work in the North Atlantic has shown that they do not hibernate during winter but instead make extensive horizontal (up to 3400 km) and vertical (to more than 750 m depth) movements to take advantage of productive continental shelf and slope habitats during summer, autumn, and winter (Sims et al. 2003; Skomal, Wood, and Caloyianis 2004). Basking sharks probably also exploit the mesopelagic realm for plankton and usually show diel behavior, occurring deeper during the day than at night. However, when in productive shelf waters, like whale sharks, they may at times reverse this pattern in response to prey behavior (Sims et al. 2005).

The megamouth shark probably exploits both epipelagic and mesopelagic habitats where it feeds on euphausid shrimps, copepods, and jellyfish in mainly tropical and subtropical waters. This rarely reported (it was only discovered in 1976) but probably widespread species has a soft, flabby body, heterocercal tail, small gill slits, and low-flow filter apparatus that suggests it is a less active species than the whale or basking shark. Taylor, Compagno, and Struhsaker (1983) suggested it might swim slowly through prey schools with its jaws wide open occasionally closing its mouth and contracting its pharynx to expel water and concentrate its prey before swallowing it. Tissue in the mouth may be bioluminescent, acting to attract prey (Taylor, Compagno, and Struhsaker 1983), and Compagno (2001) has also suggested that it may use its mouth as a bellows to suck in prey. However, Nakaya, Matsumoto, and Suda (2008) consider that the megamouth shark has a unique feeding mechanism among sharks, which they call engulfment feeding, that is typically seen in balaenopterid whales. As noted by these authors, this species has a terminal mouth, large gape, small gill openings, long bucco-pharyngeal cavity, and unique elastic skin and loose connective tissue around the pharyngeal region. When feeding, the head is raised, opening the mouth and allowing water to flow in by suction. As the shark swims, it gulps water and the forward motion forces the jaws fully open and fills the fully expanded bucco-pharyngeal cavity stretching the skin around the pharyngeal region. The mouth is then closed forcing the water out through the gill slits, sieving the prey on the gill rakers. The weak body structure relative to whale and basking sharks may be an adaption to a more nutrient-poor deepwater habitat, as seen in various mesopelagic teleosts (Taylor, Compagno, and Struhsaker 1983). However, Compagno (2001) noted that the coloration, liver oil composition, and catch records were more suggestive of an epipelagic rather than a deepwater habitat; he may also have been influenced by the two-day acoustic track of a megamouth shark that showed strong diel behavior, swimming at 12 to 25 m at night and descending to 120 to 166 m during the day. However, given that most epipelagic sharks regularly dive to at least 600 m, the usual depth preferences of the megamouth shark are still uncertain.

The manta ray has a similar, mainly tropical, distribution to the whale shark and probably exploits similar resources with the two species often found in the same area. They employ a dynamic filtering process, swimming slowly and channeling the plankton into their terminal mouths with their cephalic scoops. However, little is known about the depth behavior of manta rays or of the specific adaptions of their filtering plates or their preferred planktonic prey.

Of the pelagic and oceanic squaloids, the whitetail dogfish (*Symnodalatias albicauda*) is known from only a few specimens and nothing is known about its diet and feeding behavior. It is probably mainly mesopelagic or bathypelagic, migrating into the epipelagic zone at night, feeding mainly on small teleosts and cephalopods. The biology of the diminutive dalatiids the pygmy shark (*Euprotomicrus bispinatus*) and the spined pygmy shark (*Squaliolus laticaudus*) is also poorly known. Both are vertical migrators, ascending at night into the epipelagic zone from meso- or even bathypelagic depths. Both feed on small teleosts, cephalopods, and crustaceans. It is not known whether their luminous organs play any role in feeding behavior. The cookie-cutter shark (*Isistius brasiliensis*) is well known for its "cookie-cutting" behavior and is ectoparasitic on large fish and marine mammals to which it attaches itself with its suctorial lips and modified pharynx. It then spins, boring out a plug of flesh with its large lower teeth and leaving a crater-shaped wound on its victim. It has a large, oily liver that makes it neutrally buoyant and able to hang motionless in the water; victims may be lured by the shark's strong luminescence or by the patch of skin lacking luminescence that occurs under its head and that may resemble the silhouette of prey for the cookie-cutter sharks' hosts. It is probably bathypelagic during the day and rising to near the surface at night. These little sharks have even attacked nuclear submarines, leaving crater-marks on their rubber sonar domes. The diet also includes whole prey, particularly squid, some of which are nearly as large as their captors.

Within the lamnoid sharks, the medium-sized crocodile shark is poorly known. Its grasping dentition suggests it feeds mainly on small fish, cephalopods, and crustaceans and its large eyes imply that it is a visual hunter, living mainly in mesopelagic depths but migrating to and feeding in the epipelgic zone at night. Its squalene-rich liver oil is thought to aid in buoyancy control or vertical behavior.

The alopiid sharks have developed a highly specialized body form with their enormously elongated tails that are used to round up and then stun the small fishes on which they feed (Compagno 2001). In the thresher shark (*Alopias vulpinus*) (and maybe the two other species) the vertebrae near the tail-tip are broadened and strengthened to give it rigidity when striking their prey. While the three thresher shark species all appear to feed on similar prey, they exploit slightly different habitats. The pelagic thresher is predominantly tropical in the Indo-Pacific with a known depth distribution down to 150 m, although it probably goes deeper than that. The bigeye thresher (*Alopias superciliosus*) occurs in all tropical and warm temperate seas and satellite tracking of one individual in Australia (Figure 1.1) showed strong diel movements, with most of the day spent below the thermocline at 300 to 500 m (6°C to 12°C) and the night between 10 and 100 m (20°C to 26°C). Short-term tracking in the eastern Pacific showed similar movements from deeper (200 to 550 m) and cooler (6°C to 11°C) water during the day into the mixed layer (50 to 130 m) and warmer temperatures (15°C to 26°C) at night (H. Nakano reported in Smith et al. 2008). It may be able to maintain body temperatures above that of the surrounding sea water to conserve heat in its brain and eyes during its periods at depth (Carey et al. 1971). Its large eye suggests it is a visual predator, feeding in the mesopelagic layer during the day and in the epipelagic zone at night. Compagno (2001) suggested the keyhole-shaped orbits extending onto the dorsal surface of the head may help it to strike prey from below with its tail. Temperate waters of all oceans are the main habitat of the thresher shark, although it also occurs in tropical seas. Like the lamnids, this powerful swimmer (that often jumps clear of the water) has a well-developed heat-exchanging circulatory system, enabling it to maintain body temperatures higher than that of the surrounding water, which probably enables it to exploit cooler water than the other alopiids (Bone and Chubb 1983). Satellite tracking of one individual in Australia (Figure 1.1) showed daily vertical migrations, with most of the night spent in the top 50 m while during the day most time was spent at 300 to 400 m. The feeding behavior of all the thresher sharks results in them often being tail hooked in longline fisheries presumably as a result of them trying to stun the bait.

The lamnid sharks are highly specialized for a pelagic lifestyle and show parallel evolution in many of their morphological and physiological adaptions with the tunas. They have a thunniform body shape, cardiovascular and muscular systems that allow prolonged aerobic swimming speeds, and they are endothermic (Compagno 2001). Endothermy is best developed in the salmon shark and porbeagle, which have consequently been able to exploit subarctic and subantarctic waters where temperatures may be only a few degrees above zero. In the salmon shark, body temperatures may be as much as 15.6°C and stomach temperatures 21.2°C above the surrounding water, and increased amounts of certain proteins in the heart muscle allow it to maintain cardiac contractility in the cold (Goldman et al. 2004; Weng et al. 2005). This species occurs from the surface to about 370 m in subarctic waters in the Gulf of Alaska during winter, with 98% of the time of satellite-tracked individuals spent above 150 m. During this season they are probably feeding mainly on herring, while during the summer in the same area they feed on salmon. Some sharks migrated into subtropical waters during winter where they showed bimodal diving behavior, with one area of occupancy at 100 to 200 m and another below the thermocline at 300 to 500 m, going as deep as 830 m (Weng et al. 2005). Porbeagles tracked in the northeast Atlantic showed considerable plasticity in behavior, probably associated with feeding. In summer, they occupied shelf habitats utilizing the whole water column and probably feeding on both pelagic and demersal fish. In autumn, they occupied cooler shelf-edge habitats diving down to 550 m, possibly exploiting prey not available to ectothermic predators such as the blue shark (Pade et al. 2009).

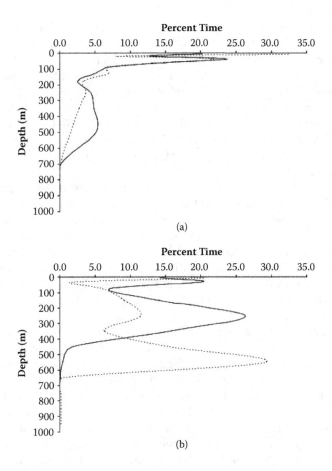

FIGURE 1.1
Percentage time-at-depth recorded by pop-up satellite archival tags from four species of epipelagic oceanic sharks. (a) Solid line is blue shark, dotted line is shortfin mako. (b) Solid line is thresher shark, dotted line is bigeye thresher.

At the other extreme, the longfin mako is a tropical species and has a poorly developed heat-exchanger system. *Isurus* and *Lamna,* with their pointed, grasping teeth, feed predominantly on teleost fishes and cephalopods. In the shortfin mako, probably the fastest of all sharks, large individuals (over 3 m total length [TL]) have broader, cutting teeth and can take billfish and even small cetaceans. Satellite tracking of one individual in Australia showed that it spent 82% of its time at less than 100 m and 4% of its time at greater than 300 m, diving as deep as 620 m (Figure 1.1).

The white shark is described by Compagno (2008) as a superpredator and is one of the few sharks that, once it reaches a subadult size, regularly feeds on marine mammals. Its large cutting teeth and powerful jaws equip it for this purpose and it has a range of hunting strategies depending on the prey being targeted (Bruce, Stevens, and Malcolm 2006). Acoustic and satellite tracking studies have led to a great increase in our knowledge of the movements and behavior of this shark in the last 10 years. Around seal colonies, they patrol particularly the entry and exit points hunting mainly during the day, swimming close to the bottom and attacking their prey from underneath. When attacking large prey such as elephant seals they may first immobilize them, withdraw a safe distance, and then only move in to feed once the seal is dead or severely weakened. Some seasonal movement patterns demonstrated from satellite tracking studies in Australasia and California have been linked to whale migrations (Weng et al. 2007) and it has been hypothesized that these sharks may prey upon the calves; however, this has yet to be substantiated. It is certainly intriguing why individuals would leave what appear to be productive feeding areas to travel into open ocean areas and oceanic island locations (see Section 1.3.4). When in open ocean areas they may dive regularly to >300 m and occasionally to >700 m and experience temperatures of 5°C to 26°C (Boustany et al. 2002). White sharks also prey on a variety of teleost and chondrichthyan fishes, as well as marine birds, turtles, cephalopods, and other molluscs and crustaceans.

The oceanic pelagic carcharhinids feed mainly on teleost fishes and cephalopods but the patchy nature of food resources leads to them opportunistically taking other prey. The blue shark, with its slender body and long pectoral fins, is well adapted to use ocean currents and is an inquisitive and persistent (if not immediately aggressive) predator that will also take birds and may consume cetaceans, although probably mainly through scavenging. Juvenile blue sharks will feed on aggregations of large planktonic crustaceans and have gill rakers that may aid in trapping these small prey items. Like most large pelagic predators, blue sharks show strong diel behavior, diving deeper during the day and remaining nearer the surface at night, probably following the deep-scattering layer and associated prey resources. Pop-off archival tag data from three sharks off the east coast of Australia showed that they spent between 35% and 58% of their time in depths of less than 50 m, between 52% and 78% of their time in less than 100 m, and between 10% and 16% in depths greater than 300 m (Figure 1.1). They dived as deep as 1000 m (the limit of the depth sensor). Blue sharks have been reported to show tropical submergence but there was little evidence of this in these data, with the sharks spending much time at or near the surface in 26°C to 27°C water. While most of their fish prey is pelagic, bottom fishes also feature in the diet in coastal waters. They are known to feed throughout the day and night but have been reported to be more active at night, with highest activity in the early evening (Sciarrotta and Nelson 1977).

The silky shark is often associated with tuna schools and will also follow groups of cetaceans; it is most likely these sharks are feeding on the same prey rather than on the tuna or cetaceans. It has been suggested that the mottled white tips on the fins of the oceanic whitetip shark may mimic a school of baitfish, attracting such prey as tunas and mackerels (Myrberg 1991). Compagno (1984) also reported an instance where several of these sharks cruised erratically among a school of small tuna that were feeding on baitfish. They did not chase the tuna but apparently just waited for them to swim into their jaws! Like the blue shark, the oceanic whitetip shark is a persistent predator, although it is more aggressive than the former species.

With the exception of the plankton-feeding mobulid rays, the pelagic stingray is the only batoid to exploit the epipelagic oceanic realm where its diet consists mainly of jellyfish, squid, crustaceans, and fishes. While it is relatively abundant and successful in this environment, little is known about the feeding behavior of this species.

1.3.4 Spatial Dynamics, Population Structure, and Migrations

The challenges of exploiting often patchy food resources and of finding mates and productive areas for giving birth in the oceanic environment have resulted in many of the species having complex population structures and migrations that we are still far from understanding. Our knowledge is greatest for a few of the species that are commonly taken by fisheries and, in terms of movements, that have also been the subject of large tagging programs and recent studies using electronic tags. Without question we know most about the blue shark, which provides an excellent example of specialized adaptions to the oceanic lifestyle. While sex and size segregation is widespread among chondrichthyans, few species demonstrate it better than the blue shark. Recreational and commercial fisheries data provide a complex picture of sex and size structuring by both latitude and longitude in most oceans, although less is known from the Indian Ocean. Segregation is thought to reduce, through habitat partitioning, competition for food resources and to protect subadult females from the dangers associated with male mating behavior and pups from adult predation (Nakano 1994). Off the east coast of Australia, there is a decrease in body size and an increase in the proportion and abundance of females with increasing latitude so that at high latitudes (40°S) the population mainly comprises juvenile females (Stevens and Wayte 2008). Juveniles also predominate in higher latitudes of the North Pacific, North Atlantic, and Indian Ocean (Suda 1953; Gubanov and Grigor'yev 1975; Nakano 1994; Kohler and Turner 2008). The blue shark is highly migratory, with complex movement patterns related to water temperature, reproduction, and the distribution of prey. A seasonal shift in population abundance to higher latitudes is associated with high-productivity oceanic convergence or boundary zones. Tagging studies of blue sharks have demonstrated extensive movements in the North Atlantic, suggesting a single stock with numerous trans-Atlantic migrations (Kohler, Casey, and Turner 1998; Kohler and Turner 2008), which are probably accomplished by swimming slowly, but assisted by the major current systems (Stevens 1976, 1990; Casey 1985). More limited tagging in the Pacific has also shown extensive movements of up to 9200 km (P. Saul, NIWA, Wellington, personal communication).

A mainly tagging-based movement model has been developed for blue sharks in the North Atlantic (Casey 1985; Stevens 1990; Kohler and Turner 2008). In spring and summer the western Atlantic population consists mainly of juveniles, subadult females, and adult males that move inshore from the Gulf Stream. During summer, they extend northward in large numbers along the continental shelf from southern New England to the Grand Banks, where they feed and mate (Casey 1985). During late summer, autumn, and winter, subadult females and adult males move offshore into the Gulf Stream or south, with some traveling as far as the Caribbean and South America. Some subadult females, most of which have recently mated, move offshore and travel the current systems to the eastern Atlantic. During winter in the eastern Atlantic, adult females occur off the Canary Islands and African coast at about 27°N to 32°N (Muñoz-Chápuli 1984); many of these are pregnant (Casey 1985). Adult males are found farther north off Portugal, as are juvenile and subadult females that have moved south from northern Europe. Immature males are not caught in this region and may be offshore. Some mating of these subadult females probably occurs during winter. In spring and summer, adults of both sexes are found from 32°N to 35°N, where they mate. Immature males also occur in this area. Adult females seem to have a seasonal reproductive cycle, while males and subadult females are sexually active throughout the year (Pratt 1979; Stevens 1984). In summer, the immature females migrate north to northern Europe where they are common off the coast of southwestern England (Stevens 1976). Birth probably occurs in early spring. Nursery areas are found in the Mediterranean and off the Iberian peninsula, particularly off Portugal and near the Azores (Aires-da-Silva, Ferreira, and Pereira 2008), but extend as far north as the Bay of Biscay. Juvenile sharks remain in the nursery areas and do not take part in the extensive migrations of the adults until they reach a length of about 130 cm (Stevens 1976; Muñoz-Chapuli 1984). In the eastern Atlantic, mature females, pregnant sharks, and newborn young are common during certain seasons, and it seems that a large proportion of the

North Atlantic breeding population occurs in this region (Casey 1985). A similar movement cycle associated with reproduction appears to occur in the South Atlantic, although the picture is only slowly starting to be pieced together from fisheries data.

In the North Pacific, Nakano (1994) suggested that mating takes place in early summer at 20°N to 30°N, and that the pregnant females migrate north to the parturition grounds by the next summer. Birth occurs in early summer in pupping grounds that are located at 35°N to 45°N. The pupping and nursery areas are located in the subarctic boundary where there is a large prey biomass for the juveniles, who remain there for 5 to 6 years prior to maturity (Nakano and Nagasawa 1996). Adults occur mainly from equatorial waters to the south of the nursery grounds.

The picture is less clear for the shortfin mako. Although sex and size segregation occur, for example, small juveniles are found mainly in coastal waters, large males occur in the northeast Atlantic, and large females in the northwest Atlantic, there is no evident pattern of changes in sex ratio and size with latitude. Results from a large tagging study in the northwest Atlantic show that the shortfin mako makes extensive movements of up to 4543 km, with 36% of recaptures caught at greater than 556 km from their tagging site (Casey and Kohler 1992; Kohler, Casey, and Turner 1998). However, only one fish crossed the Mid-Atlantic Ridge, suggesting that trans-Atlantic migrations are not as common as in the blue shark. Casey and Kohler (1992) proposed the following hypothesis for migrations of the shorfin mako in the western North Atlantic. From January to April they are found along the western margin of the Gulf Stream north to Cape Hatteras. In April and May, as inshore shelf waters start to warm and the Gulf Stream strengthens, they move onto the shelf between Cape Hatteras and Georges Bank. From June to October they occur on the shelf between Cape Hatteras and the southern Gulf of Maine, as well as offshore to the Gulf Stream. They suggest that this area may be the main feeding grounds for a large part of the juvenile and subadult population in the western North Atlantic. During autumn and winter, they move offshore and south to the Gulf Stream and Sargasso Sea, with some also entering the Caribbean and Gulf of Mexico. The core distribution in the western North Atlantic seems to be between 20°N and 40°N and bordered by the Gulf Stream in the west and the Mid-Atlantic Ridge in the east (Casey and Kohler 1992). More limited data from the Pacific also show large movements of up to 5500 km, although most tag returns from New Zealand and southeast Australia are restricted to the southwest Pacific (see summary in Stevens 2008).

Tagging and genetic data suggest there is only one population of the blue shark in the North Atlantic, although there is some evidence that the Mediterranean stock may be separate; stock structure in other oceans is poorly understood although there is some exchange across ocean basins. Based on tagging and genetic data, northeast and northwest Atlantic populations of the shortfin mako appear to be separate, at least for management purposes, with little exchange between them. Several fisheries stocks of this species probably occur across the other oceans (Heist 2008).

In the salmon shark, the western side of the North Pacific is male dominated and the eastern side female dominated; dominance increases with latitude as does size (Goldman and Musick 2008). Satellite tracking of salmon sharks has demonstrated seasonal migrations from subarctic to temperate and subtropical regions of up to 18,220 km that are associated with feeding or reproduction. In the summer and autumn, the mainly female sharks were feeding in the Gulf of Alaska but in winter some sharks moved as far south as Hawaii and California, while others remained in Alaskan waters (Weng et al. 2005). Conventional tagging of the porbeagle in the North Atlantic has demonstrated movements from the English Channel to northern Norway (2370 km) and northern Spain and in the northwest

Atlantic from New England north to Newfoundland and offshore 1861 km into oceanic waters (Stevens 1990; Kohler, Casey, and Turner 1998). Tagging data suggest that northeast and northwest Atlantic stocks of porbeagle are essentially separate. There appears to be no exchange between northern and southern hemisphere populations, and the number of stocks in the southern hemisphere is unknown.

Observations on white sharks at some viewing sites such as the Neptune Islands off South Australia suggest relatively nomadic habits with only limited time spent at these sites, although they may be revisited periodically and in successive years. However, tagging at Guadalupe Island, Mexico, shows that at this site the sharks remain there for 5 to 8 months each year (Domeier and Nasby-Lucas 2008). The white shark makes seasonal movements and in Australia satellite-tracked juveniles and subadults move northward up both coasts in autumn as far as about 22°S where they spend the winter before returning to southern Australian waters in spring. Some of these small juveniles show site specificity to certain beaches on the New South Wales coast, where they appear to be feeding on schooling fishes (Bruce, Stevens, and Malcolm 2006). Subadults tracked in New Zealand have shown movements to the tropical waters of New Caledonia (M. Francis, NIWA, Wellington, personal communication). Trans-Tasman Sea migrations between Australia and New Zealand also occur. A shark tagged off South Africa traveled to northwest Australia in 99 days and returned to the tagging location in just under nine months (Bonfil et al. 2005). A tracking study off California showed that following periods of decreased pinniped abundance at the Farallon Island, offshore migrations of subadult and adult white sharks occurred during November to March. The sharks followed a migration corridor to a focal area 2500 km to the west in the eastern Pacific, with some sharks moving as far as Hawaii. The sharks remained in the eastern Pacific focal area for up to 167 days during spring and summer, occupying depths from the surface to more than 700 m (Boustany et al. 2002; Weng et al. 2007). Interestingly, sharks tagged off Guadalupe Island, Mexico, made annual migrations between December and May to the same area in the eastern Pacific visited by the Farallon Island sharks, and also as far as Hawaii. Both sexes travelled at the same time, but males returned to Guadulupe earlier than females. These long-distance movements are more likely to be associated with feeding than reproduction, but at this stage the targeted prey species are unknown (Domeier and Nasby-Lucas 2008). While some exchange between continents or across ocean basins is suggested by tagging and genetics data, global stock structure is still poorly known (Pardini et al. 2001; Bonfil et al. 2005).

The movements of some whale shark populations are linked to predictable seasonal food pulses in certain areas such as Ningaloo Reef in Western Australia, Gladden Spit in Belize, Holbox, Mexico and Donsol, Philippines, where ecotourism has developed around their presence. Satellite tracking has shown that after leaving Ningaloo Reef, the sharks move northward into the Indian Ocean and sometimes into the waters of Christmas Island and Indonesia (Wilson et al. 2006). Photo-identification studies have shown that individuals may return to Ningaloo in successive years (Meekan et al. 2006). Satellite tracking in other areas has shown both relatively localized and long-distance movements. However, because of difficulties with keeping tags attached to these animals, few regular migratory routes have been identified and interpretation of some long-distance movements is difficult because the tags have remained submerged (consequently giving no positions) for long periods. Sharks tagged in the Seychelles have moved toward the African coast and also to south of Sri Lanka (3380 km; Rowat and Gore 2007). Juveniles tagged off Taiwan moved offshore into the Taiwan Strait and northwestern Pacific to north off Okinawa (5900 km) where they appeared to be related to boundary currents (Hsu et al. 2007). Tagging in the Gulf of California showed mainly localized movements but one individual apparently travelled 13,000 km into the Pacific Ocean (Eckert and Stewart 2001). Stock structure in the whale shark is poorly understood.

The basking shark appears to be highly migratory and as Compagno (2001) notes it is well known for its seasonal appearance during spring and autumn in large numbers in northern coastal waters of the North Atlantic and North Pacific and its subsequent disappearance in winter. These sharks have been seen in deep water above the continental slopes and in the ocean basins, and it is thought they move into coastal waters to take advantage of seasonal plankton blooms. Satellite tracking has shown long-distance movements (up to 3400 km) mainly associated with the shelf edge (Sims et al. 2003; Skomal, Wood, and Caloyianis 2004). However, one shark tagged off the British Isles migrated nearly 10,000 km across the North Atlantic to off Newfoundland (Gore et al. 2008), providing evidence that they use epi- and mesopelagic oceanic waters. The stock structure of basking sharks is poorly known.

For the remaining oceanic species considered in this chapter, we have very limited information on population structure or migration patterns. In most of the oceanic squalids we know that they make vertical migrations between the epipelagic and meso- or bathypelagic zones but the details of these movements are lacking. There are some limited tag-recapture data for the thresher sharks, longfin mako, silky and oceanic whitetip sharks from a large cooperative study that had most tagging effort on the Atlantic coast of the United States (Kohler, Casey, and Turner 1998). A few returns from the bigeye thresher showed movements from the New England or central Atlantic coast to Cuba, the Gulf of Mexico, and out into the central Atlantic (2767 km). Longfin mako returns ($n = 4$) showed some movement from shelf to oceanic waters (>1590 km) while the silky shark showed mainly coastal movements of up to 1339 km and the oceanic whitetip shark ($n = 6$) of up to 2270 km. Acoustic tagging studies show that the manta ray has strong site fidelity and limited movements in some areas such as Hawaii (Clark 2007). Little information is available on stock structure of any of these pelagic shark species.

1.4 Exploitation, Population Status, Management, and Conservation

1.4.1 Exploitation

Traditional subsistence fisheries for a variety of pelagic sharks have existed in undeveloped countries for hundreds of years. In the 1930s to 1980s, a number of target fisheries for the porbeagle and basking shark operated in Europe and the Americas for meat and liver oil. In more recent times, increasing demand and prices in the international shark fin trade have led to huge increases in catches of most pelagic species to supply this market. However, the only global database of reported catches maintained by the United Nations Food and Agriculture Organization (FAO) grossly underestimates the magnitude of catches (Camhi et al. 2008). An analysis of trade data from Hong Kong shark fin auctions estimated that pelagic sharks represented about a third of the fins traded (Clarke et al. 2006). A number of Regional Fishery Management Organisations (RFMOs) and tuna commissions are now improving their data collection of pelagic sharks (Camhi et al. 2008).

Targeted basking shark fisheries use nets to deliberately entangle the fish or harpoon guns to take basking sharks swimming or feeding on the surface. Targeted fisheries have been recorded from Norway, Ireland, Scotland, Iceland, California, China, Japan, Peru, and Ecuador. There are a few well-documented fisheries for basking sharks, particularly in the northeast Atlantic, and these suggest stock reductions of 50% to 90% over a few decades or less. These declines have persisted into the long term, with no apparent recovery several decades after exploitation has ceased. However, factors other than exploitation, such as market and economic changes, food supply, and oceanographic changes may also be involved in these declines. The basking shark was traditionally targeted in the eighteenth to twentieth centuries for its liver oil, which was initially used for tanning leather, for lamp oil, and as a source of vitamin A; more recently it has been used as a rich source of squalene. The liver comprises 17% to 25% of the body weight and yields 60% to 75% oil. The meat and fins are also valuable products, while the cartilage and skin are of secondary importance. The meat has been used, either fresh or dried, for food or fishmeal since early fisheries. It was the secondary product of most traditional fisheries after oil, but is still valuable in some areas. Fins are recorded as a byproduct of the Monterey fishery in the 1940s, and were an important product of the Irish fishery by 1960. The increased value of fins during the past decade means they are probably the major incentive for continued directed basking shark fisheries in some areas. According to Compagno (2001), the huge pectoral and dorsal fins sold in 1999 for US$10,000 to $20,000 each. The cartilage has been used to produce fishmeal, and more recently for medical research and the health market. The thick skin can produce high quality leather. The high value of basking shark fins in international trade is reportedly the reason why the northeast Atlantic fishery for this species is still viable, now that liver oil prices have fallen.

Historically, targeted whale shark fisheries for meat, liver oil, and fins occurred in locations such as India, Pakistan, the Maldives, China, Taiwan, Japan, the Philippines, Indonesia, Malaysia, and Senegal using harpoons or gaffs, fish traps, and set nets. Liver oil was traditionally one of the most important products, being used to waterproof artisanal wooden fishing boats in the Maldives, India, and elsewhere. Flesh was traditionally utilized locally in fresh, dried, and salted form, and traded locally for food as in the Philippines where most other parts such as skin, gills, and intestines were used for food or medicinal purposes. Meat is the main traded product stimulated by increased demand in Taiwan over the past two decades. Rising prices and declining catches off Taiwan stimulated whale shark fishing in the Philippines and India turning incidental and traditional subsistence fisheries into targeted fisheries supplying the international market. Whale shark meat is reputed to be the world's most expensive shark meat. However, there is little good data in the primary literature from existing fisheries. From 1988 to 1991, some 647 whale sharks were caught off Verval, India (Vivekanandan and Zala 1994). Joung et al. (1996) note that in the 1970s and early 1980s, 30 to 100 whale sharks were caught per season in southwest Taiwan, but by the late 1980s some seasons produced less than ten sharks. Chang, Leu, and Fang (1997) provide some fishery information that differs from that of Joung et al. (1996), stating that up to 100 sharks per year have been taken off the east coast and about 60 per year off the west coast of Taiwan. Alava et al. (2002) describe whale shark catches in the Philippines as ranging from about 20 to 150 individuals per year between 1990 and 1997. Chen, Liu, and Joung (2002) noted that the Taiwan whale shark fishery captured an annual average of 158 individuals from set nets and 114 individuals from harpoon fisheries. Catches were higher in the mid-1980s and lower in the mid-1990s. In addition to being a target species in certain areas, whale sharks are also taken as bycatch, notably in gillnet and purse seine fisheries (Silas 1986; Romanov 2002).

Manta rays are targeted in harpoon fisheries in parts of southeast Asia, where they are also a retained bycatch of gillnet fisheries. It has been estimated that 1575 mobulid rays are landed annually at one fishing port in Lombok, Indonesia, or about 320 t (White et al. 2006) as drift gillnet bycatch in the skipjack tuna fishery. Five species of mobulids were recorded, including the manta ray (14% of the mobulid catch). The meat is consumed domestically and the skin is also deep-fried as *kerupuk* (similar to prawn crackers). However, the gill arches are traded internationally and are much more valuable. A buyer in Lombok was sold three adult manta rays for US$545 and said he would receive US$490 for the filter plates but only $109 for skins and cartilage (White et al. 2006).

Porbeagles have been fished in the northeast Atlantic principally by Denmark, France, Norway, and Spain. Norway began a target longline fishery for this species in the 1920s. Landings reached their first peak of 3884 t in 1933; however, about 6000 t were taken in 1947, when the fishery restarted after the Second World War. From 1953 to 1960, there was a progressive drop in landings to between 1200 and 1900 t, which then fell to a 20 t mean over the past decade. Average Danish landings fell from above 1500 t in the early 1950s to a recent mean of ~50 t. Reported landings from the historically important UK and adjacent waters fishery have decreased to very low levels during the past 30 to 40 years. French and Spanish longliners have operated directed porbeagle fisheries since the 1970s, but in the last few years there have been only 8 to 11 French vessels targeting this species. In 2008, the quota was set at 580 t, but the European Commission has now recommended a zero take. Porbeagles have virtually disappeared from Mediterranean records.

Targeted porbeagle fishing started in the northwest Atlantic in 1961 when Norway switched its operations to the coast of New England and Newfoundland, following depletion of the northeast Atlantic stock. Catches increased from about 1900 t in 1961 to more than 9000 t in 1964. By 1965, many vessels had switched to other species or moved to other grounds because of depleted stocks. The fishery collapsed after only six years, landing less than 1000 t in 1970, and it took 25 years for only very limited recovery to occur. Norwegian and Faroese fleets were excluded from Canadian waters following the establishment of Canada's Exclusive Economic Zone (EEZ) in 1995. Three offshore and several inshore Canadian vessels entered the targeted northwest Atlantic fishery during the 1990s. Catches of 1000 to 2000 t/year throughout much of this decade reduced population levels to a new low in less than 10 years. Commercial catch rates are now only 10% to 30% of those in the early 1990s (Campana et al. 2008). Porbeagle landings from the southern hemisphere are poorly known, although the longline fleet of Uruguay and pelagic and bottom longliners and trawl fisheries in New Zealand are known to take this species. The porbeagle is also an important target species for recreational fishing in Ireland and the United Kingdom; the recreational fishery in Canada and the United States is very small.

Some small target fisheries for shortfin mako and thresher sharks exist, for example, in California and Spain. Although a relatively productive species, the U.S. west coast fishery for thresher shark showed signs of declining less than 10 years after a target fishery was initiated in the late 1970s (Smith et al. 2008). However, the majority of the catch of makos and threshers is taken incidentally by longlines and gillnets directed at tuna and billfish (Holts et al. 1998). Consequently the magnitude of the catch and mortality is not reflected in catch statistics. Stevens (2000) estimated that 12,500 metric tons (mt) of shortfin mako were caught by longline fleets in the Pacific in 1994, and Babcock and Nakano (2008) reported that about 10,000 mt were caught by tuna fleets in the Atlantic in 1995. Other annual catches from smaller areas or more specific fisheries are generally between 20 and 800 mt (Mejuto 1985; Muñoz-Chápuli et al. 1993; Bonfil 1994; Francis, Griggs, and Baird 2001; Stevens and Wayte 2008). In general, shortfin mako catches tend to be about 3% to 13% of blue shark catches in the same longline or gillnet fishery. For anglers, shortfin mako is the most desirable and commonly retained big-game shark because it puts up a good fight and has high-quality meat (Babcock 2008). Casey and Hoey (1985) stated that the recreational catch of shortfin mako along the U.S. Atlantic Coast and in the Gulf of Mexico in 1978 was 17,973 fish weighing some 1223 mt. From 1987 to 1989, this annual catch was about 1000 mt (Casey and Kohler 1992). However, only 2882 makos (about 200 mt) were reported in 2001 (Babcock 2008).

Pelagic sharks are a major bycatch of longline and gillnet fisheries, particularly from nations with high-seas fleets such as Japan, Taiwan, South Korea, China, and Spain. Some species, notably the silky shark, are also a large bycatch of tuna purse seine fisheries. In the past, only a few species were targeted commercially, particularly porbeagle and short-fin mako, for their high-value meat. However, the increasing price of shark fins, rapidly growing Asian economies, and increasing management restrictions on coastal populations are leading to greater pressure on high-seas stocks of pelagic sharks. In the Pacific, high-seas fish catches from pelagic longlining are increasing. Because there is usually no requirement for these fisheries to record their shark catch, the magnitude of the catch is not reflected in catch statistics. Commercial catch data are poor because individuals are often finned and their carcasses discarded at sea. Of the species identified in the Hong Kong fin market, some 70% were pelagic sharks (Clarke et al. 2006). The median number and biomass of sharks entering the shark fin trade have been estimated at 38 million individuals and 1.7 million mt, respectively (Clarke et al. 2006). These figures suggest that official (reported) landings in the FAO database may underestimate real catches by three to four times (Clarke et al. 2006). In the light of international concern over shark stocks (FAO, International Union for Conservation of Nature [IUCN]) there are questions over the sustainability of this shark catch, and the resulting effects on the ecosystem (see Chapter 17).

The blue shark is the most frequently captured of the pelagic sharks in high-seas longline fisheries and is taken from the world's oceans in greater quantities than any other single species of chondrichthyan. It is also the most common pelagic shark taken by sport fishermen, particularly in the United States, Europe, and Australia (Babcock 2008). While reliable catch data for this species from longline and gillnet fishing are sparse, it is clear that very large quantities are being taken globally. The high-seas catch of blue sharks from North Pacific fisheries in 1988 was estimated at 5 million individuals, or 100,000 mt at an average weight of 20 kg (Nakano and Watanabe 1992), and the catch from longline fleets in the Pacific in 1994 was about 137,000 mt (Stevens 2000). Bonfil (1994) estimated that 6.2 to 6.5 million blue sharks were taken annually by high-seas fisheries around the world. Although these figures are only rough estimates, they give some idea of the magnitude of the exploitation. Blue shark fins are the most common in the Hong Kong fin market, comprising at least 17% of the total, and Clarke et al. (2006) estimated 10.7 million individuals (0.36 million tones) are killed for the global fin trade annually.

1.4.2 Demography and Population Status

Fishing is a major source of mortality for oceanic sharks and so it is important to understand their life-history traits and the constraints these impose on the species' ability to withstand exploitation. As with most other sharks and rays, epipelagic oceanic elasmobranchs have life-history traits that result in generally slow intrinsic rates of population increase. Au, Smith, and Show (2008) calculated rebound potentials (rates of population increase at maximum sustainable yield) for four pelagic sharks at 3.8% to 6.9% compared to 8.9% to 18.2% for large tunas and billfish. Within the oceanic elasmobranch group, however, there is considerable variation in demographic parameters, and this is illustrated in Table 1.5. However, these parameters are reliably known for few species. On average, these species mature at about 11 years (range 2 to 21 years) and live for between 8 and 65 years. Typically, they have gestation periods of 9 to 18 months and reproductive cycles of 1 to 3

TABLE 1.5

Demographic Parameters for Epipelagic Oceanic Elasmobranchs

Species	Population	Annual Survivorship	Generation Time (years)	Annual Rate of Increase	References
A. pelagicus	NW Pacific	0.77–0.90	13	0.033	Liu et al. 1999, 2006; Otake and Mizue 1981
A. superciliosus	NW Pacific	0.77–0.89	17	0.002	Chen et al. 1997; Liu et al. 1998
A. vulpinus	NE Pacific	0.56–0.93	8	0.254	Cailliet and Bedford 1983; Hixon 1979
C. carcharias	Global	0.71–0.96	22	0.051	Dulvy et al. 2008
I. oxyrinchus	NW Atlantic	0.79–0.94	24	0.047	Campana et al. 2005; Mollet et al. 2000; Natanson et al. 2006; Pratt and Casey 1983; Wood et al. 2007
	SW Pacific	0.79–0.93	23	0.034	Bishop et al. 2006; Francis 2007; Francis and Duffy 2005
L. nasus	North Atlantic	0.82–0.93	18	0.081	Campana et al. 2002; Natanson et al. 2002
	SW Pacific	0.78–0.94	26	0.086	Francis et al. 2007; Francis and Duffy 2005; Francis and Stevens 2000
	Central Pacific	0.64–0.90	10	0.058	Oshitani et al. 2003
L. ditropis	NE Pacific	0.67–0.91	13	0.081	Goldman 2007; Goldman and Musick 2006; Nagasawa 1998; Tanaka 1980
C. falciformis	Gulf of Mexico	0.75–0.90	16	0.067	Bonfil 1990; Bonfil et al. 1993; Branstetter 1987
C. longimanus	Pacific and Atlantic	0.72–0.92	11	0.110	Lessa et al. 1999; Seki et al. 1998
P. glauca	North Atlantic	0.65–0.91	10	0.287	Skomal and Natanson 2003
P. violacea	NE Pacific	0.68–0.88	6	0.311	Mollet et al. 2002; Mollet 2002; Neer 2008

years. Smith et al. (2008) calculated intrinsic rebound rates for 11 pelagic species and found they mostly fell in the middle range of the productivity spectrum. However, some species lay near both the low and high ends of the spectrum.

Of the species for which information is available, the pelagic stingray and the blue shark have the highest annual rates of population increase (31% and 29%, respectively), which is more than three times greater than the other species. In contrast, the bigeye thresher and pelagic thresher have annual rates of population increase of only 2% and 3.3%, respectively (Dulvy et al. 2008). The relatively productive pelagic stingray matures at 2 to 3 years, has a gestation period of 2 to 3 months, and a reproductive cycle of about 6 months, endowing the population with a fairly quick turnover rate. Toward the other end of the spectrum, in the shortfin mako females do not mature until 18 to 21 years, they have a gestation period of 15 to 18 months, a 3-year reproductive cycle, and an annual rate of population increase of 3.4% to 4.7%. Rates of population increase calculated by different authors can vary depending on the assumptions made, the life-history parameters chosen, and the degree to which uncertainty is incorporated. So, for example, estimated rates of population increase for the pelagic stingray vary from 6% to 31% in the studies of Smith et al. (2008) and Dulvey et al. (2008). However, the relative ranking of species in these two studies is generally similar.

Cortés (2008) used principal component analysis to examine the difference in life-history traits between eight species of pelagic sharks. He noted that the early maturity and large litter size of the blue shark resulted in a high rate of population increase despite the small size of its pups, which would be subject to higher natural mortality rates. The silky shark, oceanic whitetip, and porbeagle formed a group that had moderate rates of population increase and shared similar-sized pups (70 to 76 cm TL), had similar annual fecundities (three to five pups per year), slow overall growth rates (k values of 0.06 to 0.10), and longevities of 17 to 25 years. The shortfin mako, bigeye and pelagic threshers formed a group that had very low productivities and shared large adult size, low annual fecundities of two to four, and low k values (0.08 to 0.09). As for sharks in general, the population growth rates of these pelagic species were more sensitive to survival of juveniles and adults than to survival of age 0 neonates or to fecundity (Cortés 2008). Cortés (2008) postulated that these pelagic species had maximum sustainable yields at or above 50% of their carrying capacity.

Comprehensive stock assessments for chondrichthyan fishes are very limited, and the situation is no different for oceanic elasmobranchs. A number of studies have attempted to assess the status of blue shark stocks, and to a lesser extent those of shortfin makos and porbeagles. A few studies have focused on other species such as the thresher and silky shark, and the pelagic stingray. West, Stevens, and Basson (2004) reviewed assessments carried out for blue sharks that included demographic methods, age structured models, and food web and ecosystem models, as well as various forms of catch rate analysis; they also used a yield analysis in the southwest Pacific. The International Commission for the Conservation of Atlantic Tunas (ICCAT) has also carried out assessments on this species (as well as on shortfin mako) and a fishery-independent approach using tag-recapture data has been carried out in the North Atlantic. These models have shown a conflicting picture of blue shark sustainability in the North Atlantic and central Pacific. Different catch rate analyses generate diverging trends even in the same ocean (Nakano 1998; Matsunaga, Nakano, and Minami 2001; Simpfendorfer et al. 2002; Baum et al. 2003; Aires-da-Silva, Ferreira, and Pereira 2008). The poor quality of catch data and problems of catch rate standardization have hampered analyses of both blue shark and shortfin mako population status by ICCAT. These data problems mean that any assessment will have high uncertainty.

For porbeagles, there is a recent stock assessment for the northwestern Atlantic (Campana et al. 2008). Based on tag-recapture and an age and sex structured population model, it is estimated that the total population biomass of this stock is 1572 to 7695 t (11% to 17% of virgin biomass) and 2612 to 13,847 mature females (4% to 14% of virgin abundance). Because of the very low numbers of mature females now present in the stock, it is unlikely that the strict quota management and area closures will allow quick rebuilding of the population (Campana et al. 2008). There is no stock assessment for the more heavily fished, unmanaged and possibly more seriously depleted northeast Atlantic and Mediterranean population, or for southern stocks.

For other species, evaluation of population status is limited to fisheries catch rate data, which show apparent declines in abundance of all three thresher sharks, white sharks, shortfin mako, silky, and oceanic whitetip. Salmon shark populations seem to be stable and few data are available for pelagic stingrays, although one study suggests numbers may be increasing in the Pacific (Baum et al. 2003; Baum and Myers 2004; Ward and Myers 2005; Ferretti et al. 2008; Camhi et al. 2008).

Because of the paucity of data for many species (teleosts as well as chondrichthyans) another technique for assessing relative risk from fishing to population status is ecological risk assessment (ERA). This can operate at different levels, from purely qualitative to fully quantitative depending on the amount of data available; the general methodology has been described by Hobday et al. (2007). Risk is considered on two axis, productivity (the ability of the species to withstand or recover from fishing) and susceptibility or the level at which a species is likely to be affected by fishing.

The global threatened status of oceanic epipelagic elasmobranchs has also been assessed using the IUCN Red List categories and criteria (Mace 1995; IUCN 2004). Dulvy et al. (2008) considered 21 species in their analysis, which showed strong overlap with the species considered in this chapter (Table 1.1). Eleven of the species they considered were assessed as globally threatened (one Endangered and ten Vulnerable), five as Near Threatened, two as Least Concern, and three were Data Deficient. The proportion of epipelagic oceanic elasmobranchs considered to be threatened, 52% from Dulvy et al. (2008) or 48% from Table 1.1 (Figure 1.2a) is considerably higher than the 22% of all chondrichthyans that are threatened (Figure 1.2b). The giant devilray (*Mobula mobular*) was the only globally Endangered oceanic pelagic elasmobranch considered by Dulvy et al. (2008), although this species was not included in Table 1.1 of this chapter.

An integrated approach to assessing the risk faced by Atlantic pelagic sharks to overexploitation was used by Simpfendorfer et al. (2008). This incorporated ERA, the IUCN Red List assessments, and an approximation of the biomass at maximum sustainable yield

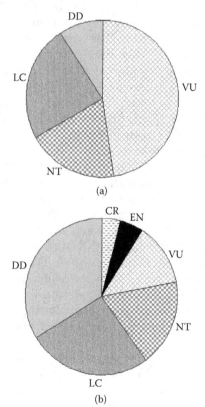

(a)

(b)

FIGURE 1.2
Percentage of (a) epipelagic oceanic elasmobranchs (n = 21 from Table 1.1) and (b) globally assessed chondrichthyan fishes (n = 591) within each IUCN Red List category. CR = Critically Endangered, EN = Endangered, VU = Vulnerable, NT = Near Threatened, LC = Least Concern, DD = Data Deficient.

(the inflection point on the population growth curve). Multivariate statistics were used to produce the integrated results. According to this analysis, the species at highest risk from longline fisheries were makos (both shortfin and longfin), bigeye thresher, and to a lesser extent silky sharks. The porbeagle, oceanic whitetip, and thresher shark formed a middle group and the pelagic stingray and blue shark had the lowest level of risk.

1.4.3 Management and Conservation

A growing concern over the global status of shark populations led to the FAO International Plan of Action for the Conservation and Management of Sharks (IPOA-Sharks) in 1999. This requested all countries engaged in shark fishing, as well as RFMOs, to assess their shark resources and prepare National and Regional Shark Plans by 2001. However, IPOA-Sharks is voluntary and to date few countries and no RFMOs have produced Shark Plans. Few countries have comprehensive management plans for any of their chondrichthyan resources within their EEZs. For the high-seas component of pelagic elasmobranchs there is virtually no management because of the difficulty of establishing international agreements to take responsibility for these resources beyond national EEZs.

This lack of management, in particular of pelagic elasmobranchs, is due to the low priority afforded to them relative to high-value target species such as tunas. This is reflected in the poor quality of the data that is usually not species-specific and may not be reported accurately, if at all. While the larger RFMOs have now generally accepted that pelagic elasmobranchs fall under their mandate, few have attempted assessments for any of these species. Where they have, as in the case of ICCAT for blue shark and shortfin mako, the assessments are severely limited by the data quality and reporting problems. Even for the relatively data-rich blue shark, catch rate standardization, stock and population structure problems lead to major uncertainties in the assessment and to date there would be little confidence in the outputs of these models.

There is management of some pelagic elasmobranchs within national EEZs. Whale, basking, and white sharks are protected in a number of countries. In New Zealand, commercial target fishing for basking sharks was banned in 1991, although they are allowed to be taken as bycatch. Norway agreed to an annual quota on catches of 800 t liver weight in 1982. This was progressively reduced and has been at 100 mt (or about 200 to 300 sharks per year) since 1994. Porbeagles are managed by quota in parts of the North Atlantic and in New Zealand. Quota management based on stock assessment and scientific advice has been in place in the Canadian EEZ since 2002. This has maintained a relatively stable population, but with a slight decline in mature females; there is also a U.S. quota. Fisheries in the northeast Atlantic are effectively unrestricted (quotas greatly exceed landings). Scientific advice in 2005 that no fishery should be permitted in the northeast Atlantic was not adopted. New Zealand introduced quota management in 2004. A number of countries (and some RFMOs) have implemented finning bans that prohibit the retention of fins without the corresponding carcasses, or bycatch limits on pelagic sharks. However, in many cases these operate by allowing a certain ratio of fin to carcass weight. Some fleets argue for unrealistically high ratios based on cutting practices that may allow a loophole for finning (Dulvy et al. 2008). Better controls are achieved by requiring carcasses to be landed with fins attached, as in certain Australian fisheries. See Dulvy et al. (2008) for more details of countries and RFMOs that currently apply finning bans or other management measures.

On the high seas, no catch limits have yet been imposed by RFMOs. However, there are international treaties that include some pelagic sharks. White, basking, and whale sharks are listed on both Appendix II of the Convention on International Trade in Endangered Species (CITES) and the Convention on Migratory Species (CMS). Shortfin makos are also included on Appendix II of CMS. Appendix II of CITES allows trade only if the take of that species can be demonstrated to be sustainable. Two regional treaties, the Barcelona and Bern Conventions, include some pelagic sharks (white and basking sharks, shortfin mako, porbeagle, and blue shark) that would give various levels of protection or permit certain levels of exploitation depending on stock status (Dulvy et al. 2008). However, the effectiveness of any of these listings has yet to be demonstrated.

1.5 Summary

Epipelagic oceanic elasmobranchs show a wide range of morphological, physiological, and behavioral specializations for exploiting their environment. However, although this zone comprises over 70% of the oceans, the number of species regularly living there is relatively low. The adaptive diversity of these species encompasses the giant plankton feeders, diminutive squaloid sharks, bizarre body form of the alopiids, ectoparasitic feeding strategies of *Isistius*, and the highly active and powerful endothermic lamnids. Most of these species are widely distributed, highly migratory, and show complex spatial population structuring, all of which help them exploit the often patchy food resources and meet the challenges of reproduction over the vast distances of the open ocean. However, these oceanic elasmobranchs face an uncertain future. Growing Asian economies are fueling increases in the international shark fin trade, and high-seas exploitation from longlining, gillnetting, and purse-seining activities are resulting in unsustainable fishing pressure. Three-quarters of these species are listed as Threatened or Near Threatened on the IUCN Red List. Fisheries catch and effort data for most of these species is poor leading to high uncertainty in those few species where population assessments have been attempted.

References

Aires-da-Silva A, Ferreira RL, Pereira JG (2008) Case study: Blue shark catch-rate patterns from the Portuguese swordfish longline fishery in the Azores. In: Camhi MD, Pikitch EK (eds) Sharks of the Open Ocean: Biology, Fisheries and Conservation. Blackwell, Oxford, pp 230–235.

Alava MNR, Dolumbalo ERZ, Yaptinchay AA, Trono RB (2002) Fishery and trade of whale sharks and manta rays in the Bohol Sea, Philippines. In: Fowler SL, Reed TM, Dipper FA (eds) Elasmobranch Biodiversity, Conservation and Management: Proceedings of the International Seminar and Workshop, Sabah, Malaysia, July 1997. IUCN SSC Shark Specialist Group. IUCN, Gland, Switzerland, pp 132–148.

Au DW, Smith SE, Show C (2008) Shark productivity and reproductive protection, and a comparison with teleosts. In: Camhi MD, Pikitch EK (eds) Sharks of the Open Ocean: Biology, Fisheries and Conservation. Blackwell, Oxford, pp 298–308.

Babcock EA (2008) Recreational fishing for pelagic sharks worldwide. In: Camhi MD, Pikitch EK (eds) Sharks of the Open Ocean: Biology, Fisheries and Conservation. Blackwell, Oxford, pp 193–204.

Babcock EA, Nakano H (2008) Data collection, research, and assessment efforts for pelagic sharks by the International Commission for the Conservation of Atlantic Tuna. In: Camhi MD, Pikitch EK (eds) Sharks of the Open Ocean: Biology, Fisheries and Conservation. Blackwell, Oxford, pp 472–477.

Baum JK, Myers RA (2004) Shifting baselines and the decline of pelagic sharks in the Gulf of Mexico. Ecol Lett 7:135–145.

Baum JK, Myers RA, Kehler DG, Worm B, Harley SJ, Doherty PA (2003) Collapse and conservation of shark populations in the northwest Atlantic. Science 299:389–392.

Bishop SDH, Francis MP, Duffy C, Montgomery JC (2006) Age, growth, maturity, longevity and natural mortality of the shortfin mako (*Isurus oxyrinchus*) in New Zealand waters. Mar Freshw Res 57:143–154.

Bone Q, Chubb AD (1983) The retial system of the locomotor muscles in the thresher shark. J Mar Biol Assoc UK 63:239–241.

Bone Q, Marshall NB, Blaxter JHS (1995) Biology of fishes. Second edition. Blackie Academic & Professional, Glasgow.

Bonfil R (1990) Contribution to the fisheries biology of the silky shark *Carcharhinus falciformis* (Bibron, 1839) from Yucatan, Mexico. MSc thesis, University of Wales.

Bonfil R (1994) Overview of world elasmobranch fisheries. FAO Fisheries Technical Paper 341.

Bonfil R, Mena R, de Anda D (1993) Biological parameters of commercially exploited silky sharks, *Carcharhinus falciformis*, from the Campeche bank, Mexico. Report No. 115. NOAA National Marine Fisheries Service, Silver Spring, MD.

Bonfil R, Meyer M, Scholl MC, Johnson R, O'Brien S, Oosthuizen H, Swanson S, Kotze D, Paterson M (2005) Transoceanic migration, spatial dynamics, and population linkages of white sharks. Science 310:100–103.

Boustany AM, Davis SF, Pyle P, Anderson SD, Le Boeuf BJ, Block BA (2002) Expanded niche for white sharks. Nature 415:35–36.

Branstetter S (1987) Age, growth and reproductive biology of the silky shark, *Carcharhinus falciformis*, and the scalloped hammerhead, *Sphyrna lewini*, from the northwestern Gulf of Mexico. Environ Biol Fish 19:161–173.

Branstetter S (1990) Early life-history implications of selected carcharhinoid and lamnoid sharks of the Northwest Atlantic. In: Pratt H, Gruber S, Taniuchi T (eds) Elasmobranchs as Living Resources: Advances in the Biology, Ecology, Systematics, and the Status of the Fisheries. NOAA Tech Report NMFS 90. NOAA/WMFS, Silver Spring, MD, pp 17–28.

Bruce BD, Stevens JD, Malcolm H (2006) Movements and swimming behaviour of white sharks (*Carcharodon carcharias*) in Australian waters. Mar Biol 150:161–172.

Cailliet GM, Bedford DW (1983) The biology of three pelagic sharks from California waters, and their emerging fisheries: a review. Report No. 24. California Cooperative Oceanic Fisheries Investigations.

Camhi MD, Lauck E, Pikitch EK, Babcock EA (2008) A global overview of commercial fisheries for open ocean sharks. In: Camhi MD, Pikitch EK (eds) Sharks of the Open Ocean: Biology, Fisheries and Conservation. Blackwell, Oxford, pp 166–192.

Campana SE, Joyce W, Marks L, Natanson LJ, Kohler NE, Jensen CF, Mello JJ, Pratt HL, Myklevoll S (2002) Population dynamics of the porbeagle in the northwest Atlantic Ocean. N Am J of Fish Manage 22:106–121.

Campana SE, Marks L, Joyce W (2005) The biology and fishery of shortfin mako sharks (*Isurus oxyrinchus*) in Atlantic Canadian waters. Fish Res 73:341–352.

Campana SE, Joyce W, Marks L, Hurley P, Natanson LJ, Kohler NE, Jensen CF, Mello JJ, Pratt HL, Myklevoll S, Harley S (2008) The rise and fall (again) of the porbeagle shark population in the Northwest Atlantic. Fish and Aquatic Resources Series 13:445–461.

Carey FG, Teal JM, Kanwisher JW, Lawson KD, Beckett JS (1971) Warm-bodied fish. Am Zool 11: 137–145.

Casey JG (1985) Transatlantic migrations of the blue shark: A case history of cooperative shark tagging. In: Proceedings of the First World Angling Conference, Cap d'Agde, France, 12–18 September, 1984 (ed. R. H. Stroud). International Game Fish Association, Dania Beach, FL, pp 253–267.

Casey JG, Hoey JJ (1985) Estimated catches of large sharks by U.S. recreational fishermen in the Atlantic and Gulf of Mexico. In: Shark Catches from Selected Fisheries off the U.S. East Coast. NOAA Technical Report NMFS SSRF No. 31. NOAA/NMFS, Silver Spring, MD, pp 15–19.

Casey JG, Kohler NE (1992) Tagging studies on the shortfin mako (*Isurus oxyrinchus*) in the western North Atlantic. In: Pepperell JG (ed) Sharks: Biology and Fisheries. Aust J Mar Fresh Res 43 (special volume):45–60.

Chang WB, Leu MY, Fang LS (1997) Embryos of the whale shark, *Rhincodon typus*: early growth and size distribution. Copeia 2:444–446.

Chen CT, Liu KM, Chang YC (1997) Reproductive biology of the bigeye thresher shark, *Alopias superciliosus* (Lowe, 1839) (Chondrichthyes: Alopiidae), in the northwestern Pacific. Ichthyol Res 44:227–235.

Chen CT, Liu KM, Joung SJ (2002) Preliminary report on Taiwan's whale shark fishery. In: Fowler S, Reed TM, Dipper FA (eds) Elasmobranch Biodiversity, Conservation and Management: Proceedings of the International Seminar and Workshop, Sabah, Malaysia, July 1997. IUCN SSC Shark Specialist Group. IUCN, Gland, Switzerland and Cambridge, UK, pp 162–167.

Clark T (2007) Remote tracking of the manta ray (*Manta birostris*) in Hawaii. Pac Sci 59(1):113.

Clarke SC, McAllister MK, Milner-Gulland EJ, Kirkwood GP, Michielsens CGJ, Agnew DJ, Pikitch EK, Nakano H, Shivji MS (2006) Global estimates of shark catches using trade records from commercial markets. Ecol Lett 9:1115–1126.

Clarke S, Milner-Gulland EJ, Bjørndal T (2007) Social, economic and regulatory drivers of the shark fin trade. Mar Res Econ 22:305–327.

Compagno LJV (1984) FAO Species Catalogue. Vol 4. Sharks of the World: An Annotated and Illustrated Catalogue of Shark Species Known to Date. Part 2. Carcharhiniformes. FAO Fish Synop (125) FAO, Rome, pp 251–655.

Compagno LJV (1990) Alternate life history styles of cartilaginous fishes in time and space. Environ. Biol Fish 28(1-4):33–75.

Compagno LJV (2001) Sharks of the world. An annotated and illustrated catalogue of shark species known to date. Volume 2. Bullhead, mackerel and carpet sharks (Heterodontiformes, Lamniformes and Orectolobiformes). FAO Species Catalogue for Fishery Purposes 1:1–269.

Compagno LJV (2008) Pelagic elasmobranch diversity. In: Camhi MD, Pikitch EK (eds) Sharks of the Open Ocean: Biology, Fisheries and Conservation. Blackwell, Oxford, pp 14–23.

Cortes E (2008) Comparative life history and demography of pelagic sharks. In: Camhi MD, Pikitch EK (eds) Sharks of the Open Ocean: Biology, Fisheries and Conservation. Blackwell, Oxford, pp 309–322.

Domeier ML, Nasby-Lucas N (2008) Migration patterns of white sharks *Carcharodon carcharias* tagged at Guadalupe Island, Mexico, and identification of an eastern Pacific shared offshore foraging area. Mar Ecol Prog Ser 370:221–237.

Dulvy NK, Baum JK, Clarke S, Compagno LJV, Cortés E, Domingo A, Fordham S, Fowler S, Francis MP, Gibson C, Martínez J, Musick JA, Soldo A, Stevens JD, Valenti S (2008) You can swim but you can't hide: the global status and conservation of oceanic pelagic sharks and rays. Aquatic Conserv: Mar Freshw Ecosyst DOI: 10.1002/aqc.975.

Eckert SA, Stewart BS (2001) Telemetry and satellite tracking of whale sharks, *Rhincodon typus*, in the Sea of Cortez, Mexico, and the North Pacific Ocean. Environ Biol Fish 60:299–308.

Ferretti F, Myers RA, Serena F, Lotze HK (2008) Loss of large predatory sharks from the Mediterranean Sea. Conserv Biol 22:952–964.

Francis MP (2007) Unpublished data cited in Dulvy NK, Baum JK, Clarke S, Compagno LJV, Cortés E, Domingo A, Fordham S, Fowler S, Francis MP, Gibson C, Martínez J, Musick JA, Soldo A, Stevens JD, Valenti S (2008) You can swim but you can't hide: the global status and conservation of oceanic pelagic sharks and rays. Aquatic Conserv: Mar Freshw Ecosyst Doi: 10.1002/aqc.975.

Francis MP, Duffy C (2002) Distribution, seasonal abundance and bycatch of basking sharks (*Cetorhinus maximus*) in New Zealand, with observations on their winter habitat. Mar Biol 140:831–842.

Francis MP, Duffy C (2005) Length at maturity in three pelagic sharks (*Lamna nasus, Isurus oxyrinchus,* and *Prionace glauca*) from New Zealand. Fish Bull 103:489–500.

Francis MP, Stevens JD (2000) Reproduction, embryonic development and growth of the porbeagle shark, *Lamna nasus,* in the south-west Pacific Ocean. Fish Bull 98:41–63.

Francis MP, Griggs LH, Baird SJ (2001) Pelagic shark bycatch in the New Zealand tuna longline fishery. Mar Freshw Res 52:165–178.

Francis MP, Campana SE, Jones CM (2007) Age under-estimation in New Zealand porbeagle sharks (*Lamna nasus*): is there an upper limit to ages that can be determined from shark vertebrae? Mar Freshw Res 58:10–23.

Goldman KJ (2002) Aspects of age, growth, demographics and thermal biology of two lamniform shark species. PhD dissertation, College of William and Mary. School of Marine Science, Virginia Institute of Marine Science, Williamsburg, VA.

Goldman KJ (2007) personal communication, cited in Dulvy NK, Baum JK, Clarke S, Compagno LJV, Cortés E, Domingo A, Fordham S, Fowler S, Francis MP, Gibson C, Martínez J, Musick JA, Soldo A, Stevens JD, Valenti S (2008) You can swim but you can't hide: the global status and conservation of oceanic pelagic sharks and rays. Aquatic Conserv: Mar Freshw Ecosyst Doi: 10.1002/aqc.975.

Goldman KJ, Musick JA (2006) Growth and maturity of salmon sharks (*Lamna ditropis*) in the eastern and western North Pacific, and comments on back-calculation methods. Fish Bull 104:278–292.

Goldman KJ, Musick JA (2008) The biology and ecology of the salmon shark, *Lamna ditropis*. In: Camhi MD, Pikitch EK (eds) Sharks of the Open Ocean: Biology, Fisheries and Conservation. Blackwell, Oxford, pp 95–104.

Goldman KJ, Anderson SD, Latour RJ, Musick JA (2004) Homeothermy in adult salmon sharks, *Lamna ditropis*. Env Biol Fish 71:403-411.

Gore MA, Rowat D, Hall J, Gell FR, Ormond RF (2008) Transatlantic migration and mid-ocean diving by basking shark. Biol Lett Doi:10.1098/rsbl.2008.0147 pp 1–4.

Graham RT, Roberts CM, Smart CR (2005) Diving behaviour of whale sharks in relation to a predictable food pulse. J R Soc Interface doi:10.1098/rsif.2005.0082.

Gubanov YeP, Grigor'yev VN (1975) Observation on the distribution and biology of the blue shark *Prionace glauca* (Carcharhinidae) of the Indian Ocean. J Ichthyol 15:37–43.

Heist EJ (2008) Molecular markers and genetic population structure of pelagic sharks. In: Camhi MD, Pikitch EK (eds) Sharks of the Open Ocean: Biology, Fisheries and Conservation. Blackwell, Oxford, pp 323–333.

Heithaus MR (2001) Predator-prey and competitive interactions between sharks (order Selachii) and dolphins (suborder Odontoceti): a review. J Zool 253:53–68.

Heithaus MR (2007) Nursery areas as essential shark habitats: a theoretical perspective. American Fisheries Society Symposium 50:3–13.

Hixon MA (1979) Term fetuses from a large common thresher shark, *Alopias vulpinus*. Calif Fish Game 65:191–192.

Hobday AJ, Smith A, Webb H, Daley R, Wayte S, Bulman C, Dowdney J, Williams A, Sporcic M, Dambacher J, Fuller M, Walker T (2007) Ecological Risk Assessment for the Effects of Fishing Methodology. Report R04/1072 Australian Fisheries Management Authority, Canberra.

Holts DB, Julian A, Sosa-Nishizaki O, Bartoo NW (1998) Pelagic shark fisheries along the west coast of the United States and Baja California, Mexico. In: Hueter RE (ed) Proceedings of an International Symposium Held at the 125th Annual Meeting of the American Fisheries Society (Tampa, Florida, 30 August, 1995). Fish Res 39:115–125.

Hsu HH, Joung SJ, Liao YY, Liu KM (2007) Satellite tracking of juvenile whale sharks, *Rhincodon typus*, in the Northwestern Pacific. Proceedings of the International Whale Shark Conference in Perth, Western Australia. Special issue: Fish Res 84:25–31.

IUCN (International Union for Conservation of Nature) (2004) Guidelines for using the IUCN Red List Categories and Criteria. IUCN Species Survival Commission, Gland, Switzerland.

Joung SJ, Chen CS, Clark E, Uchida S, Huang YP (1996) The whale shark, *Rhincodon typus*, is a live-bearer: 300 embryos found in one 'megamamma' supreme. Environ Biol Fish 46:219–223.

Joung SJ, Chen CT, Lee HH, Liu KM (2008) Age, growth, and reproduction of silky sharks, *Carcharhinus falciformis*, in northeastern Taiwan waters. Fish Res 90:78–85.

Kohler NE, Turner PA (2008) Stock structure of the blue shark, *Prionace glauca*, in the North Atlantic Ocean based on tagging data. In: Camhi MD, Pikitch EK (eds) Sharks of the Open Ocean: Biology, Fisheries and Conservation. Blackwell, Oxford, pp 339–350.

Kohler NE, Casey JG, Turner PA (1998) NMFS Cooperative shark tagging program, 1962–93: An atlas of shark tag and recapture data. Mar Fish Rev 60(2):1–87.

Lessa R, Santana FM, Paglerani R (1999) Age, growth and stock structure of the oceanic whitetip shark, *Carcharhinus longimanus*, from the southwestern equatorial Atlantic. Fish Res 42(1–2):21–30.

Liu KM, Chiang PJ, Chen CT (1998) Age and growth estimates of the bigeye thresher shark, *Alopias superciliosus*, in northeastern Taiwan waters. Fish B-NOAA 96(3):482–491.

Liu KM, Chen CT, Liao TH, Joung SJ (1999) Age, growth, and reproduction of the pelagic thresher shark, *Alopias pelagicus* in the northwestern Pacific. Copeia (1):68–74.

Liu KM, Changa YT, Ni IH, Jin CB (2006) Spawning per recruit analysis of the pelagic thresher shark, *Alopias pelagicus*, in the eastern Taiwan waters. Fish Res 82:52–64.

Mace GM (1995) Classification of threatened species and its role in conservation planning. In: Lawton JH, May RM (eds) Extinction Rates. Oxford University Press, Oxford, pp 197–213.

Matsunaga H, Nakano H, Minami H (2001) Standardised CPUE and catch for the main pelagic sharks species dominated in the SBT fishery. Report No. 011/66. CCSBT-ERS.

Meekan MG, Bradshaw CJA, Press M, McLean C, Richards A, Quasnichka S, Taylor JG (2006) Population size and structure of whale sharks *Rhincodon typus* at Ningaloo reef, Western Australia. Mar Ecol Prog Ser 319:275–285.

Mejuto J (1985) Associated Catches of Sharks, *Prionace glauca, Isurus oxyrinchus* and *Lamna nasus*, with NW and N Spanish Longline Fishery, in 1984. ICES C.M. 1985/H:42. International Council for the Exploration of the Sea, Pelagic Fish Committee, Copenhagen.

Mollet HF (2002) Distribution of the pelagic stingray, *Dasyatis violacea* (Bonaparte, 1832), off California, Central America, and worldwide. Mar Freshw Rev 53:525–530.

Mollet H, Cliff G, Pratt HL, Stevens JD (2000) Reproductive biology of the female shortfin mako, *Isurus oxyrinchus* Rafinesque, 1810, with comments on the embryonic development of lamnoids. Fish Bull 98:299–318.

Mollet H, Ezcurra JM, O'Sullivan JB (2002) Captive biology of the pelagic stingray, *Dasyatis violacea* (Bonaparte, 1832). Mar Freshw Res 53:531–541.

Muñoz-Chápuli R (1984) Ethologie de la reproduction chez quelques requins de l'Atlantique nord-est. Cybium 8(4):1–14.

Muñoz-Chápuli R, Notarbartolo di Sciara G, Seret B, Stehmann M (1993) The Status of the Elasmobranch Fisheries in Europe. Report of the Northeast Atlantic subgroup of the IUCN Shark Specialist Group.

Musick JA, Ellis JK (2005) Reproductive evolution of chondrichthyans. In: Hamlett WC, Jamieson BGM (eds) Reproductive Biology and Phylogeny. Science Publishers, Enfield, pp 45–79.

Musick JA, Harbin MM, Compagno LJV (2004) Historical zoogeography of the selachii. In: Carrier JC, Musick JA, Heithaus MR (eds) Biology of Sharks and Their Relatives. CRC Press, Boca Raton, FL, pp 33–78.

Myrberg AA (1991) Distinctive markings of sharks: ethological considerations of visual function. J Exp Zool 5:156–166.

Nagasawa K (1998) Predation by salmon sharks (*Lamna ditropis*) on Pacific salmon (*Oncorhynchus* spp.) in the North Pacific Ocean. NPAFC Bulletin 1:419–433.

Nakano H (1994) Age, reproduction and migration of blue shark in the North Pacific Ocean. Bulletin National Research Institute of Far Seas Fisheries No. 31:141–256.

Nakano H (1998) Standardized CPUE for pelagic shark caught by the Japanese longline fishery in the Atlantic Ocean. Collective Volume of Scientific Papers of the International Commission for the Conservation of Atlantic Tunas (ICCAT).

Nakano H, Nagasawa K (1996) Distribution of pelagic elasmobranchs caught by salmon research gillnets in the North Pacific. Fisheries Science 62 (5):860–865.

Nakano H, Watanabe Y (1992) Effect of high seas driftnet fisheries on blue shark stock in the North Pacific. Compendium of documents submitted to the scientific review of North Pacific highseas driftnet fisheries. Sidney, B.C., Canada, June 11–14, 1991.

Nakaya K, Matsumoto R, Suda K (2008) Feeding strategy of the megamouth shark *Megachasma pelagios* (Lamniformes: Megachasmidae). J Fish Biol 73:17–34.

Natanson LJ, Mello JJ, Campana SE (2002) Validated age and growth of the porbeagle shark (*Lamna nasus*) in the western North Atlantic Ocean. Fish Bull 100:266–278.

Natanson LJ, Kohler NE, Ardizzone D, Cailliet GM, Wintner SP, Mollet HF (2006) Validated age and growth estimates for the shortfin mako, *Isurus oxyrinchus*, in the North Atlantic Ocean. Environ Biol Fish 77:367–383.

Natanson LJ, Wintner, SP, Johansson F, Piercy A, Campbell P, De Maddalena A, Gulak, SJB, Human B, Fulgosi FC, Ebert DA, Hemida F, Mollen FH, Vanni S, Burgess, GH, Compagno LJV, Wedderburn-Maxwell A (2008) Ontogenetic vertebral growth patterns in the basking shark *Cetorhinus maximus*. Mar Ecol Prog Ser 361:267–278.

Neer JA (2008) The biology and ecology of the pelagic stingray, *Pteroplatytrygon violacea* (Bonaparte, 1832). In: Camhi MD, Pikitch EK, Babcock EA (eds) Sharks of the Open Ocean: Biology, Fisheries and Conservation. Blackwell, Oxford, pp 152–159.

Oshitani S, Nakano H, Tanaka S (2003) Age and growth of the silky shark *Carcharhinus falciformis* from the Pacific Ocean. Fisheries Sci 69:456–464.

Otake T, Mizue K (1981) Direct evidence for oophagy in thresher shark, *Alopias pelagicus*. Jpn J Ichthyol 28:171–172.

Pade NG, Queiroz N, Humphries NE, Witt MJ, Jones CS, Noble LR, Sims DW (2009) First results from satellite-linked archival tagging of porbeagle shark, *Lamna nasus*: area fidelity, wider-scale movements and plasticity in diel depth changes. J Exp Mar Biol Ecol 370:64–74.

Pardini AT, Jones CS, Noble LR, Kreiser B, Malcolm H, Bruce BD, Stevens JD, Cliff G, Scholl MC, Francis M, Duffy CAJ, Martin AP (2001) Sex-biased dispersal of great white sharks. Nature 412:139–140.

Pinet PR (2006) Invitation to Oceanography, fourth edition. Jones & Bartlett, Sudbury.

Pratt HL (1979) Reproduction in the blue shark, *Prionace glauca*. Fish Bull 77:445–470.

Pratt HL, Casey JG (1983) Age and growth of the shortfin mako, *Isurus oxyrinchus*, using four methods. Can J Fish Aquat Sci 40:1944–1957.

Reif WE, Saure C (1987) Zoogeography of shore fishes of the Indo-Pacific region. Zool Stud 37(4):227–268.

Romanov EV (2002) Bycatch in the tuna purse-seine fisheries of the western Indian Ocean. Fish Bull 100:90–105.

Rosenblatt RH (1963) Some aspects of marine speciation in marine shore fishes. In: Harding JP, Tebble N (eds) Speciation in the Sea, a symposium. Systematics Association, London, pp 117–180.

Rosenblatt RH, Waples RS (1986) A genetic comparison of allopatric populations of shore fish species from the eastern and central Pacific Ocean, dispersal or vicariance? Copeia 1986:275–284.

Rowat D, Gore M (2007) Regional scale horizontal and local scale vertical movements of whale sharks in the Indian Ocean off Seychelles. Proceedings of the International Whale Shark Conference in Perth, Western Australia. Special issue: Fish Res 84:4–9.

Sciarrotta TC, Nelson DR (1977) Diel behaviour of the blue shark, *Prionace glauca*, near Santa Catalina Island, California. Fish Bull 75(3):519–528.

Seki T, Taniuchi T, Nakano H, Shimizu M (1998) Age, growth and reproduction of the oceanic white-tip shark from the Pacific Ocean. Fish Sci 64(1):14–20.

Silas EG (1986) The whale shark (*Rhiniodon typus* Smith) in Indian coastal waters: is the species endangered or vulnerable? Indian Counc Agric Res Tech Extension Ser.

Simpfendorfer CA, Hueter RE, Bergman U, Connett SMH (2002) Results of a fishery-independent survey for pelagic sharks in the western North Atlantic, 1977–1994. Fish Res 55:175–192.

Simpfendorfer C, Cortes E, Heupel M, Brooks E, Babcock E, Baum J, McAuley R, Dudley S, Stevens J, Fordham S, Soldo A (2008) An integrated approach to determining the risk of over-exploitation for data-poor pelagic Atlantic sharks. Lenfest Ocean Program SCRS/2008/140.

Sims DW, Southall EJ, Richardson AJ, Reid PC, Metcalfe JD (2003) Seasonal movements and behaviour of basking sharks from archival tagging: no evidence of winter hibernation. Mar Ecol Prog Ser 248:187–196.

Sims DW, Southall EJ, Tarling GA, Metcalfe JD (2005) Habitat-specific normal and reverse diel vertical migration in the plankton-feeding basking shark. J Anim Ecol 74:755–761.

Skomal GB, Natanson LJ (2003) Age and growth of the blue shark (*Prionace glauca*) in the North Atlantic Ocean. Fish Bull 101:627–639.

Skomal GB, Wood G, Caloyianis N (2004) Archival tagging of a basking shark, *Cetorhinus maximus*, in the western North Atlantic. J Mar Biol Assoc UK 84:795–799.

Smith SE, Rasmussen RC, Ramon DA, Cailliet GM (2008) The biology and ecology of thresher sharks (Alopiidae). In: Camhi MD, Pikitch EK (eds) Sharks of the Open Ocean: Biology, Fisheries and Conservation. Blackwell, Oxford, pp 60–68.

Snelson FF, Roman BL, Burgess GH (2006) The reproductive biology of pelagic elasmobranchs. In: Camhi MD, Pikitch EK (eds) Sharks of the Open Ocean: Biology, Fisheries and Conservation. Blackwell, Oxford, pp 24–53.

Stevens JD (1976) First results of shark tagging in the north-east Atlantic. J Mar Biol Assoc UK 56:929–937.

Stevens JD (1984) Biological observations on sharks caught by sport fishermen off New South Wales. Aust J Mar Fresh Res 35:573–590.

Stevens JD (1990) Further results from a tagging study of pelagic sharks in the north-east Atlantic. J Mar Biol Assoc UK 70:707–720.

Stevens JD (2000) The population status of highly migratory oceanic sharks. In: K Hinman (ed) Getting Ahead of the Curve: Conserving the Pacific Ocean's Tunas, Swordfish, Billfishes and Sharks. Marine Fisheries Symposium No. 16. National Coalition for Marine Conservation, Leesburg, VA.

Stevens JD (2007) Whale shark (*Rhincodon typus*) biology and ecology: a review of the primary literature. Proceedings of the International Whale Shark Conference in Perth, Western Australia. Special issue: Fish Res 84:4–9.

Stevens JD (2008) The biology and ecology of the shortfin mako shark, *Isurus oxyrinchus*. In: Camhi MD, Pikitch EK (eds) Sharks of the Open Ocean: Biology, Fisheries and Conservation. Blackwell, Oxford, pp 87–94.

Stevens JD, Wayte SS (2008) Case study: the bycatch of pelagic sharks in Australia's tuna longline fisheries. In: Camhi MD, Pikitch EK (eds) Sharks of the Open Ocean: Biology, Fisheries and Conservation. Blackwell, Oxford, pp 260–267.

Suda A (1953) Ecological study of blue shark (*Prionace glauca* Linné). Bull Nankai Fish Res Lab 1(26):1–11.

Tanaka S (1980) Biological survey of salmon shark, *Lamna ditropis*, in the western North Pacific Ocean. Report of new shark resource exploitation survey in the fiscal year 1979 (North Pacific Ocean). Japan Marine Fisheries Resource Research Centre, Tokyo, pp 59–84.

Tanaka S, Cailliet GM, Yudin KG (1990) Differences in growth of the blue shark, *Prionace glauca*: technique or population? In: Pratt HL, Gruber SH, Taniuchi T (eds) Elasmobranchs as Living Resources: Advances in the Biology, Ecology, Systematics, and the Status of the Fisheries. NOAA Tech Rep NMFS 90:177–187.

Taylor LR, Compagno LJV, Struhsaker PJ (1983) Megamouth: a new species, genus and family of lamnoid shark (*Megachasma pelagios*, Family Megachasmidae) from the Hawaiian Islands. Proc Calif Acad Sci 43(8):87–110.

Thibault RE, Schultz RJ (1978) Reproductive adaptions among viviparous fishes (Cyprinodontiformes: Poeciliidae) Evolution (Lawrence, Kans) 32:320–333.

Vivekanandan E, Zala MS (1994) Whale shark fishery off Veraval. Indian J Fish 41:37–40.

Ward P, Myers RA (2005) Shifts in open-ocean fish communities coinciding with the commencement of commercial fishing. Ecology 86:835–847.

Weng KC, Castilho PC, Morrissette JM, Landeira-Fernandez AM, Holts DB, Schallert RJ, Goldman KJ, Block BA (2005) Satellite tagging and cardiac physiology reveal niche expansion in salmon sharks. Science 310(5745):104–106.

Weng KC, Boustany AM, Pyle P, Anderson SD, Brown A, Block BA (2007) Migration and habitat of white sharks (*Carcharodon carcharias*) in the eastern Pacific Ocean. Mar Biol 152:877–894.

West G, Stevens J, Basson M (2004) Assessment of blue shark population status in the western South Pacific. AFMA Project R01/1157. CSIRO Marine Research, Hobart, Tasmania, Australia.

White WT, Giles J, Dharmadi, Potter IC (2006) Data on the bycatch fishery and reproductive biology of mobulid rays (Myliobatiformes) in Indonesia. Fish Res 82:65–73.

Wilson SG, Polovina JJ, Stewart BS, Meekan MG (2006) Movements of whale sharks (*Rhincodon typus*) tagged at Ningaloo Reef, Western Australia. Mar Biol 148:1157–1166.

Wintner SP (2000) Preliminary study of vertebral growth rings in the whale shark, *Rhincodon typus*, from the east coast of South Africa. Env Biol Fish 59:441–451.

Wintner SP, Cliff G (1999) Age and growth determinations of the white shark *Carcharodon carcharias*, from the east coast of South Africa. Fish Bull 97:153–169.

Wood AD, Collie JS, Kohler NE (2007) Estimating survival of the shortfin mako *Isurus oxyrinchus* (Rafinesque) in the north-west Atlantic from tag-recapture data. J Fish Biol 71:1679–1695.

Wourms JP, Grove BD, Lombardi J (1988) The maternal-embryonic relationship in viviparous fishes. In: Hoar WS, Randall DJ (eds) Fish Physiology. Volume 11. The Physiology of Developing Fish. Part B: Viviparity and Posthatching Juveniles. Academic Press, San Diego, pp 1–134.

2

Deepwater Chondrichthyans

Peter M. Kyne and Colin A. Simpfendorfer

CONTENTS

2.1 Introduction

The deep sea is a relatively stable environment, characterized by cold temperatures and poor or absent light. Relative to inshore shelf habitats, the ocean's deepwater environments remain poorly known. The continued expansion of global fishing into the deep ocean has raised new concerns about the ability of deepwater organisms to sustain the pressures of exploitation (Morato et al. 2006). General knowledge on the deep sea lags behind the expansion of fisheries (Haedrich, Merrett, and O'Dea 2001) and as such management is often further behind. The intrinsic vulnerability of the chondrichthyan fishes given their life history characteristics (Hoenig and Gruber 1990; Cahmi et al. 1998; Musick 1999) is widely acknowledged and often cited. This vulnerability may be heightened in the deep sea, where conditions result in slower growth rates and reduced recruitment to populations. The vast majority of available life history data on the sharks, batoids, and chimaeras comes from the shallow water. Logistical, biological, and geographical difficulties with sampling (i.e., scattered distributions, deep occurrence, taxonomic uncertainty, and limited material) in the deep sea have limited the present state of knowledge. The amount of available information has, however, increased in recent years, and demonstrates that deepwater species are among the most unproductive of the chondrichthyans.

Here we review the present state of knowledge concerning this diverse group of chondrichthyans, which represents some 46% of the global shark, ray, and chimaera fauna. We focus mainly on their life history, the essential information required to effectively direct management. We define the deep sea to be that region of the ocean that lies beyond the 200 m isobath, the depth generally recognized as the continental and insular shelf edge (Merrett and Haedrich 1997). Hence, deepwater chondrichthyans are those sharks, rays, and holocephalans whose distribution is confined to (or predominantly at) depths below 200 m, or those that spend a considerable part of their lifecycle below this depth. This encompasses the continental and insular slopes and beyond, including the abyssal plains and oceanic seamounts. In the broadest sense, the deepwater fauna can be divided into pelagic species that occupy the water column and demersal species that occur on (benthic) or just above (epibenthic or benthopelagic) the ocean floor (Haedrich 1996). We discuss both of these groups, although the demersal species feature more prominently. With regards to the pelagic fauna we have limited our discussion of those species that occur in the mesopelagic (200 to 1000 m) and bathypelagic (1000 to 4000 m) zones if they also readily occupy the epipelagic (0 to 200 m) zone. Epipelagic species are treated in Chapter 1.

Excluded from our discussion are many chondrichthyans that have been recorded at depths of >200 m but that are predominantly species of the shelf. These species are recorded far less commonly or irregularly in the deep sea and include such examples as spiny dogfish *Squalus acanthias* (recorded exceptionally to 1446 m) and many skates whose distribution extends to the upper slope (including sandy skate *Leucoraja circularis*, blonde skate *Raja brachyura*, thornback skate *R. clavata*, clearnose skate *R. eglanteria*, brown skate *R. miraletus*, and rough skate *R. radula*). Other species that are abundant on the shelf in parts of their range are included here if they are also widely occurring on the slope. Examples include shortnose spurdog *Squalus megalops*, longnose skate *R. rhina*, big skate *R. binoculata*, barndoor skate *Dipturus laevis*, and spotted ratfish *Hydrolagus colliei*, and for these species, much of the available information on life history comes from the shelf (where the species may be subject to strong seasonal light and temperature signals).

2.2 Biodiversity and Biogeography

2.2.1 Biodiversity and Systematics

The total number of extant, formally described chondrichthyan species currently stands at 1144; this comprises 482 sharks, 671 batoids, and 45 holocephalans (W.T. White, personal communication). Of the global fauna, 530 chondrichthyans can be considered to be deepwater species (according to our definition) representing 46.3% of the global total (Table 2.1). The deepwater fauna is divided between 254 sharks (52.7% of global), 236 batoids (38.2% of global), and 40 holocephalans (88.9% of global). All nine orders of elasmobranchs and the single holocephalan order are represented in the deep sea (see the appendix to this chapter).

In addition to the formally described species mentioned above, there are many more new or recently identified sharks, batoids, and chimaeras that are known to researchers, but have not yet been formally described. At present this includes about 70 deepwater species (calculated from the literature and available checklists), and more are sure to be discovered and described. The number of undescribed species not only highlights the overall lack of knowledge of the deep sea fauna at even the most basic (i.e., taxonomic) level, but also that the deep sea chondrichthyan fauna is far from fully documented and species will continue to be added as deepwater surveys continue (e.g., the 2003 NORFANZ cruise surveying the seamounts and abyssal plains around the Lord Howe and Norfolk Ridges in the Western Pacific; Last 2007). In addition, the systematics and interrelationships of several groups of deep sea chondrichthyans remain unresolved, and much work is still needed on such groups as the spurdogs (*Squalus*), gulper sharks (*Centrophorus*), catsharks of the genus *Apristurus*, and some of the skate assemblages.

The bulk of the deepwater chondrichthyan fauna is attributable to four main groups: (1) squaloid dogfishes (order Squaliformes), which represent 46.1% of the deepwater shark fauna; (2) scyliorhinid catsharks (order Carcharhiniformes, family Scyliorhinidae) (40.2% of the deepwater shark fauna); (3) skates (order Rajiformes, families Arhynchobatidae, Rajidae and Anacanthobatidae) (89.8% of the deepwater batoid fauna); and (4) holocephalans (order Chimaeriformes, families Rhinochimaeridae and Chimaeridae).

The squaloid dogfishes are mostly benthic or benthopelagic, although the kitefin sharks (Dalatiidae) are primarily pelagic and include many species that undergo daily vertical migrations (there are also a few pelagic species within the Somniosidae; Compagno, Dando, and Fowler 2005). This group includes what is probably the world's smallest shark, the smalleye pygmy shark *Squaliolus aliae*, which reaches a maximum size of 22 cm total length (TL) (Last and Stevens 2009), and some of the largest shark species, that is, sleeper sharks (*Somniosus*), some of which can reach >600 cm TL (Compagno, Dando, and Fowler 2005). The lanternsharks (Etmopteridae) and some kitefin sharks possess photophores that produce a bioluminescence, aiding in counter-illumination (emitted light eliminates the silhouette of the fish formed when it is illuminated from above; Claes and Mallefet 2008).

The catsharks are the largest shark family, with about two-thirds of the known fauna occurring in the deep sea. These are generally small species (<100 cm TL), which are benthic and relatively poor swimmers (Compagno, Dando, and Fowler 2005). The skates are a morphologically conservative, yet highly diverse group of benthic batoids (McEachran and Dunn 1998). Species range from small (<0.3 m TL) to some of the largest batoids (>2.5 m TL) and diversity is greatest on the outer continental shelves and upper slopes (Ebert and Bizzarro 2007). The holocephalans—among the most poorly known cartilaginous

TABLE 2.1

Diversity of Deepwater Chondrichthyan Fishes by Order and Family

Order	Family	Common Name	No. of Species
Sharks			
Hexanchiformes	Chlamydoselachidae	Frilled sharks	1
	Hexanchidae	Sixgill and sevengill sharks	3
Squaliformes	Echinorhinidae	Bramble sharks	2
	Squalidae	Dogfish sharks	25
	Centrophoridae	Gulper sharks	18
	Etmopteridae	Lanternsharks	42
	Somniosidae	Sleeper sharks	16
	Oxynotidae	Roughsharks	5
	Dalatiidae	Kitefin sharks	9
Squatiniformes	Squatinidae	Angelsharks	7
Pristiophoriformes	Pristiophoridae	Sawsharks	3
Heterodontiformes	Heterodontidae	Bullhead sharks	1
Orectolobiformes	Parascylliidae	Collared carpetsharks	2
Lamniformes	Odontaspididae	Sand tiger sharks	2
	Pseudocarchariidae	Crocodile sharks	1
	Mitsukurinidae	Goblin sharks	1
	Alopiidae	Thresher sharks	1
	Cetorhinidae	Basking sharks	1
Carcharhiniformes	Scyliorhinidae	Catsharks	102
	Proscylliidae	Finback catsharks	3
	Pseudotriakidae	False catsharks	2
	Triakidae	Houndsharks	6
	Carcharhinidae	Requiem sharks	1
		Subtotal sharks	254
Batoids			
Rajiformes	Rhinobatidae	Guitarfishes	1
	Narcinidae	Numbfishes	7
	Narkidae	Sleeper rays	4
	Torpedinidae	Torpedo rays	7
	Arhynchobatidae	Softnose skates	75
	Rajidae	Hardnose skates	116
	Anacanthobatidae	Legskates	21
	Plesiobatidae	Giant stingarees	1
	Urolophidae	Stingarees	2
	Hexatrygonidae	Sixgill stingrays	1
	Dasyatidae	Whiptail stingrays	1
		Subtotal batoids	236
Holocephalans			
Chimaeriformes	Rhinochimaeridae	Longnose chimaeras	8
	Chimaeridae	Shortnose chimaeras	32
		Subtotal holocephalans	40
		Total	**530**

fishes—are a group of mostly deepwater benthic soft-bodied chondrichthyans with body lengths reaching >1.5 m.

2.2.2 Biogeography and Bathymetry

The deepwater chondrichthyan fauna reaches its highest diversity in the Indo-West Pacific (Table 2.2). The Western and Eastern Atlantic have similar numbers of species, but diversity is lower in the Eastern Pacific. The Arctic and Antarctic regions are depauperate in terms of deep sea fauna. These biogeographical patterns follow general trends of chondrichthyan biogeography and diversity (Musick, Harbin, and Compagno 2004). The Indo-West Pacific is a large ocean region with a high level of endemism and thus it is not surprising that the highest diversity is recorded there. Lower diversity in the Eastern Pacific is attributable to a general lack of squaloid sharks (Musick, Harbin, and Compagno 2004). In contrast to sharks and holocephalans, skate diversity is great at high latitudes (Ebert and Bizzarro 2007; see Chapter 3 in this volume).

Deepwater chondrichthyans include some very wide-ranging species, for example the smooth lanternshark *Etmopterus pusillus* and the longnose velvet dogfish *Centroselachus crepidater*, which both occur widely in the Atlantic and Indo-West Pacific. However, the vast majority of deepwater chondrichthyans are geographically and bathymetrically restricted. Levels of endemism are high in the lanternsharks, catsharks, and the skates, and many species have localized regional distributions (Ebert and Bizzarro 2007). Endemism is often associated with seamounts and deep sea ridges. Localized species are often poorly known; many species are known from only one specimen (e.g., pocket shark *Mollisquama parini* and Aguja skate *Bathyraja aguja* from the Southeast Pacific).

The deepest recorded chondrichthyan fishes are great lanternshark *Etmopterus princeps* (to 4500 m; Compagno, Dando, and Fowler 2005) and Bigelow's skate *Rajella bigelowi* (to 4156 m; Stehmann 1990). These are the only species to be recorded below 4000 m. However, the Portuguese dogfish *Centroscymnus coelolepis* occurs to 3675 m, leafscale gulper shark *Centrophorus squamosus* to 3280 m (Priede et al. 2006), and among the skates, pallid skate *Bathyraja pallida* reaches 3280 m (Priede et al. 2006), several other *Bathyraja* reach ~2900 m, while thickbody skate *Amblyraja frerichsi* and gray skate *Dipturus batis* have been documented to ~2600 m (Priede et al. 2006). The deepest records for holocephalans are for a *Harriotta* species at 3010 m and Atlantic chimaera *Hydrolagus affinis* at 2909 m (Priede et al. 2006).

TABLE 2.2

Diversity of Deepwater Chondrichthyan Fishes by Major Ocean Region

Ocean Region	Number of Species			
	Sharks	Batoids	Holocephalans	Total
Arctic Sea	1	0	0	1
Indo–West Pacific	181	116	25	322
Eastern Pacific	45	44	6	95
Western Atlantic	71	68	8	147
Eastern Atlantic	67	48	10	125
Antarctic seas	1	7	1	9

Note: The sum of totals exceeds the known number of deepwater species (530) as wider-ranging species occur in more than one region.

Priede et al. (2006) note that the chondrichthyans "have generally failed to colonize the oceans deeper than 3,000 m and it is unlikely that major new populations will be discovered in abyssal regions." For many of the deepest recorded chondrichthyans, their core bathymetric range is at much shallower depths, including *E. princeps*, *C. coelolepis*, and *R. bigelowi* (Priede et al. 2006). Some of the *Bathyraja* skates, however, have very deep minimum depths and are thus more specialized to the deeper habitats (i.e., fine-spined skate *B. microtrachys* with a minimum depth of ~2000 m, and *B. pallida* with a minimum depth of 2200 m). The general absence of chondrichthyans from the abyssal plains and their complete absence from the hadal and hadopelagic zones (demersal and pelagic habitats >6000 m, respectively) are not related to a lack of surveying at depth, but rather may be a result of an inability to meet their high energy demands, and support their metabolism and large, lipid-rich livers (Priede et al. 2006).

2.3 Life History

For the vast majority of deepwater chondrichthyans, details of their life history characteristics are lacking. There is a reasonable amount of information on the biology of some species of dogfish sharks (Squalidae), gulper sharks (Centrophoridae), lanternsharks (Etmopteridae), sleeper sharks (Somniosidae), catsharks (Scyliorhinidae), softnose skates (Arhynchobatidae), and hardnose skates (Rajidae), with much of the information on the biology of deepwater sharks coming from the Northeast Atlantic and southeastern Australia.

2.3.1 Reproductive Biology

Many coastal and shelf chondrichthyans display seasonal reproductive cycles, but the majority of deepwater species exhibit aseasonal reproductive cycles, with asynchronicity among the population. This aseasonality—which may be related to the relative stability of the deep sea environment (Wetherbee 1996)—makes it difficult to determine the gestation period and for most species, reproductive periodicity remains unknown. Follicle and embryonic development rates and sizes, however, indicate a long reproductive cycle. Additionally, many deepwater squaloid sharks exhibit a resting period between parturition and the next ovulation (e.g., ovarian follicles do not develop while gestation proceeds), extending the reproductive cycle (Irvine 2004; Irvine, Stevens, and Laurenson 2006b).

For oviparous species, estimates of fecundity are difficult because egg-laying periods and rates are mostly unknown. Ovarian fecundity (counts of the number of developing or developed follicles) may provide a proxy. However, the relationship between the number of follicles and actual reproductive output is not clear. For example, among the viviparous squaloid sharks, Irvine (2004) noted that mean litter size was three to four less than mean ovarian fecundity in New Zealand lanternshark *Etmopterus baxteri*, with atretic follicles observed in some early pregnancies. Yano (1995) found a similar disparity between ovarian and uterine fecundity in black dogfish *Centroscyllium fabricii*. Very large numbers of follicles have been reported for *Somniosus* species, 372+ for Pacific sleeper shark *S. pacificus* (Ebert, Compagno, and Natanson 1987) and up to 2931 in Greenland shark *S. microcephalus* (Yano et al. 2007), although the few observed *Somniosus* litter sizes are in the range of only 8 to 10 young (Barrull and Mate 2001; Compagno, Dando, and Fowler 2005).

2.3.1.1 Sharks

Of the major deepwater shark groups, the squaloid dogfishes are viviparous, while the catsharks are either oviparous (the dominant reproductive mode) or viviparous. Deepwater squaloid dogfishes generally do not have well-defined reproductive seasons (Chen, Taniuchi, and Nose 1981; Yano 1995; Daley, Stevens, and Graham 2002; Irvine 2004; Graham 2005; Hazin et al. 2006) and it is difficult to elucidate their reproductive cycle. Gestation periods are generally unknown and many species have a resting period after parturition (e.g., kitefin shark *Dalatias licha* [Daley, Stevens, and Graham 2002]; etmopterids [Yano 1995; Daley, Stevens, and Graham 2002; Irvine 2004]; somniosids [Daley, Stevens, and Graham 2002; Irvine 2004]). The large size of preovulatory follicles suggests lengthy vitellogenesis because the energy demands to develop such large oocytes are high. It seems plausible then, that many reproductive cycles are biennial or triennial. For example, Tanaka et al. (1990) suggested a 3.5 year gestation period for frilled shark *Chlamydoselachus anguineus* based on growth rates of embryos held in artificial conditions, and Braccini, Gillanders, and Walker (2006) demonstrated that *S. megalops* off Australia has an ovarian cycle and gestation period of two years. The results of Braccini, Gillanders, and Walker (2006), as well as data from the shelf-occurring *S. acanthias* (Holden 1977; Jones and Geen 1977), suggest that many other squalids exhibit biennial reproductive cycles.

Reproductive output in deepwater hexanchoid and squaloid sharks is generally limited, with small litter sizes the normal condition for many species. Among the hexanchoids, reported litter sizes include 2 to 10 (average 6) for *C. anguineus* (Tanaka et al. 1990) and 9 to 20 in sharpnose sevengill shark *Heptranchias perlo* (Bigelow and Schroeder 1948), but up to 108 in bluntnose sixgill shark *Hexanchus griseus* (Vaillant 1901). Table 2.3 provides a summary of reported litter sizes for deepwater squaloid sharks. In *Squalus* (litter size range 1 to 15; see Table 2.3), smaller litter sizes are more common, for example, in *S. megalops*, Braccini, Gillanders, and Walker (2006) found that 69.3% of gravid females examined had a litter size of two, 30.0% a litter size of three, and only 0.7% a litter size of four. For some *Squalus* species larger females are able to carry more embryos (Watson and Smale 1998; Braccini, Gillanders, and Walker 2006). Similarly, Irvine (2004) found strong correlations between maternal size and the number of follicles and the number of embryos within *E. baxteri* from southern Australia, Yano and Tanaka (1988) found that fecundity increased with maternal size in *C. coelolepis*, and Yano (1995) found similar results for *C. fabricii*.

Gulper sharks of the genus *Centrophorus* are among, if not the most, unproductive of chondrichthyan fishes. Fecundity is low, with litter sizes of one or two for most species examined, with the exception of recorded litter sizes of six and seven in *C. squamosus* (Bañón, Piñeiro, and Casas 2006; Figueiredo et al. 2008). Figueiredo et al. (2008) hypothesized that old *C. squamosus* may undergo senescence, which would further restrict the reproductive output of the species. Daley, Stevens, and Graham (2002) found that litter size in southern dogfish *C. zeehaani* was invariably one in 37 gravid females examined, and only a single embryo or ovum has been noted in gulper shark *C. granulosus* (Golani and Pisanty 2000; Guallart and Vicent 2001; Megalofonou and Chatzispyrou 2006). *Centrophorus* have a continuous reproductive cycle with follicles continuing to develop as gestation ensues and at the time of ovulation the follicles are very large (Guallart and Vicent 2001; Daley, Stevens, and Graham 2002; McLaughlin and Morrissey 2005). Irvine (2004) suggested that to allow for maturation of oocytes to these large ovulatory sizes in species with continuous reproductive cycles, a long gestation period is required. Gestation is two years in *C. granulosus* (Guallart and Vincent 2001) and *C.* cf. *uyato* may have a three-year gestation period (McLaughlin and Morrissey 2005). *Deania* also probably has a two- or three-year

TABLE 2.3

Summary of Litter Sizes for Deepwater Squaloid Sharks

Family	Species	Location	Litter Size (Average)	Reference
Echinorhinidae	*Echinorhinus brucus*	NE Brazil, SW Atlantic	15–26	Cadenat and Blanche (1981)
	Echinorhinus cookei	Hawaii, Eastern Central Pacific	114	Crow et al. (1996)
Squalidae	*Cirrhigaleus asper*	NE Brazil, SW Atlantic	12–19	Fischer et al. (2006)
	Cirrhigaleus barbifer	New Zealand, SW Pacific	6–10	Duffy et al. (2003)
	Squalus blainville	Mediterranean	2–6	Sion et al. (2003)
	Squalus grahami	SE Australia	2–7	Graham (2005)
	Squalus japonicus	Choshi, Japan, NW Pacific	2–8 (5.3)	Chen et al. (1981)
		Nagasaki, Japan, NW Pacific	2–8 (3.9)	Chen et al. (1981)
	Squalus megalops	Agulhas Bank, South Africa, SE Atlantic	2–4	Watson and Smale (1998)
		Andaman Is., Eastern Indian	5–7	Soundararajan and Dam Roy (2004)
		SE Australia	1–3 (2.1)	Graham (2005)
		SE Australia	2–4	Braccini et al. (2006)
		NE Brazil, SW Atlantic	1–8	Hazin et al. (2006)
	Squalus mitsukurii	Choshi, Japan, NW Pacific	4–15 (8.8)	Taniuchi et al. (1993)
		Masseiba, Japan, NW Pacific	6–9 (7.1)	Taniuchi et al. (1993)
		Ogasawara Is., Japan, NW Pacific	2–9 (4.5)	Taniuchi et al. (1993)
		Hancock Seamount, NW Pacific	1–6	Wilson and Seki (1994)
		NE Brazil, SW Atlantic	3–11	Fischer et al. (2006)
	Squalus montalbani	SE Australia	4–10	Daley et al. (2002); Graham (2005)
Centrophoridae	*Centrophorus acus*	Andaman Is., Eastern Indian	1–2	Soundararajan and Dam Roy (2004)
	Centrophorus granulosus	Mediterranean	1	Golani and Pisanty (2000); Guallart and Vicent (2001); Megalofonou and Chatzispyrou (2006)
	Centrophorus harrissoni	SE Australia	1–2	Daley et al. (2002)
	Centrophorus moluccensis	SE Australia	1–2	Daley et al. (2002)
	Centrophorus squamosus	Galicia, Spain, NE Atlantic	7	Bañón et al. (2006)
		Portugal, NE Atlantic	1–6	Figueiredo et al. (2008)
	Centrophorus cf. *uyato*	Cayman Trench, Western Central Atlantic	1–2	McLaughlin and Morrissey (2005)

TABLE 2.3 (*Continued*)

Summary of Litter Sizes for Deepwater Squaloid Sharks

Family	Species	Location	Litter Size (Average)	Reference
	Centrophorus zeehaani	SE Australia	1	Daley et al. (2002)
	Deania calcea	Rockall Trough and Porcupine Bank, NE Atlantic	8–14	Clarke et al. (2002b)
		SE Australia	1–17 (7)	Daley et al. (2002)
		SE Australia	5–10	Irvine (2004)
	Deania quadrispinosum	SE Australia	8–17	Daley et al. (2002)
Etmopteridae	*Aculeola nigra*	Chile, SE Pacific	(10)	Acuña et al. (2003)
	Centroscyllium fabricii	Greenland, NE Atlantic	4–40 (16.4)	Yano (1995)
	Centroscyllium kamoharai	SE Australia	3–22 (12)	Daley et al. (2002)
	Etmopterus baxteri	New Zealand, SW Pacific	9–15 (12.7)	Wetherbee (1996)[a]
		SE Australia	6–16 (10)	Daley et al. (2002)[a]
		SE Australia	1–16 (8.8)	Irvine (2004)
	Etmopterus spinax	Portugal, NE Atlantic	1–16 (7.6)	Coelho and Erzini (2008)
		Mediterranean	6–18	Serena et al. (2006)
Somniosidae	*Centroscymnus coelolepis*	Suruga Bay, Japan, NW Pacific	15–29	Yano and Tanaka (1988)
		West of British Isles, NE Atlantic	8–19 (14)	Girard and Du Buit (1999)
		West of British Isles, NE Atlantic	8–21 (13.8)	Clarke et al. (2001)
		SE Australia	8–19 (12)	Daley et al. (2002)
		Portugal, NE Atlantic	1–25 (9.9)	Veríssimo et al. (2003)
		Galicia, Spain, NE Atlantic	5–22 (14)	Bañón et al. (2006)
		Portugal, NE Atlantic	(11.3)	Figueiredo et al. (2008)
	Centroscymnus owstoni	Suruga Bay, Japan, NW Pacific	16–31	Yano and Tanaka (1988)
		SE Australia	5–13	Daley et al. (2002)
	Centroselachus crepidater	SE Australia	3–9 (6)	Daley et al. (2002)
		NE Atlantic	1–9	Nolan and Hogan (2003)
		SE Australia	4–8 (6)	Irvine (2004)

Continued

TABLE 2.3 (*Continued*)

Summary of Litter Sizes for Deepwater Squaloid Sharks

Family	Species	Location	Litter Size (Average)	Reference
	Scymnodalatias albicauda	SE Atlantic	59	Nakaya and Nakano (1995)
	Somniosus rostratus	Mediterranean	8	Barrull and Mate (2001)
Oxynotidae	*Oxynotus bruniensis*	SE Australia	7	Last and Stevens (2009)
	Oxynotus centrina	Mediterranean	10–15	Capapè et al. (1999); Megalofonou and Damalas (2004)
Dalatiidae	*Dalatias licha*	SE Australia; ?	7–16	Daley et al. (2002); Compagno et al. (2005)
	Isistius brasiliensis	Brazil, SW Atlantic	9	Gadig and Gomes (2002)
	Squaliolus laticaudus	Brazil, SW Atlantic	4	Cunha and Gonzalez (2006)

[a] Referred to in Wetherbee (1996) and Daley et al. (2002) as *E. granulosus*, this is correctly *E. baxteri*.

reproductive cycle that is noncontinuous (i.e., with a resting period between parturition and the development of new oocytes) (Daley, Stevens, and Graham 2002).

Specific information on the reproductive biology of the deepwater mackerel sharks (Lamniformes) is limited. While oophagy has been confirmed in several lamnoids (see Compagno 2001), gravid females have never been observed in some deepwater species, such as the rare goblin shark *Mitsukurina owstoni*, and the smalltooth sand tiger *Odontaspis ferox* (Compagno 2001; Yano et al. 2007; Fergusson, Graham, and Compagno 2008). Sund (1943) reported a litter size of six from a single basking shark *Cetorhinus maximus*, and the litter size of the deepwater odontaspidids is likely two (with biennial reproduction) based on the mode of reproduction of the inshore sand tiger shark *Carcharias taurus* (Compagno 2001; Lucifora, Menni, and Escalante 2002). In the crocodile shark *Pseudocarcharias kamoharai*, two embryos develop in each uterus by feeding on ova in uterine egg capsules (Fujita 1981; Compagno 2001).

The false catsharks (Pseudotriakidae) are the only carcharhinoid sharks, and indeed the only nonlamnoid sharks, to display oophagy (Yano 1992, 1993; Musick and Ellis 2005). Consequently, fecundity is low. Yano (1993) found the slender smoothhound *Gollum attenuatus* to generally possess two embryos, one in each uterus, although a small number of specimens contained only a single embryo (1.8% of gravid females examined) and two females contained three embryos (in both cases, one of these had failed to develop). In the false catshark *Pseudotriakis microdon*, only litter sizes of two have been observed (Saemundsson 1922; Taniuchi, Kobayashi, and Otake 1984; Yano 1992; Stewart 2000). The gestation period of both species is unknown, but for *P. microdon* at least, it is likely extended, presumably more than year and possibly two or three years (K. Yano, unpublished data, cited in Kyne, Yano, and White 2004).

Of the species of catsharks (Scyliorhinidae) for which reproductive mode has been confirmed, the majority are oviparous (Ebert, Compagno, and Cowley 2006). Viviparity is known only from a small number of species, and has been confirmed from broadhead catshark *Bythaelurus clevai*, mud catshark *B. lutarius*, lollipop catshark *Cephalurus cephalus*, and African sawtail catshark *Galeus polli* (Springer 1979; Séret 1987; Balart, Gonzalez-Garcia, and Villavicencio-Garayzar 2000; Compagno, Dando, and Fowler 2005; Ebert, Compagno, and Cowley 2006; Francis 2006). Despite earlier reports of viviparity in Dawson's catshark

B. dawsoni and roughtail catshark *G. arae*, these species have since been shown to be oviparous (Konstantinou, McEachran, and Woolley 2000; Francis 2006). Furthermore, a report of viviparity in the Saldanha catshark *Apristurus saldanha* by Myagkov and Kondyurin (1978) is in error, and that manuscript is in fact referring to *G. polli* (Ebert, Compagno, and Cowley 2006).

Litter size is small in the viviparous catsharks: two (one embryo per uterus) in both *B. clevai* and *B. lutarius* (Compagno, Dando, and Fowler 2005), one or two in *C. cephalus* (Balart, Gonzalez-Garcia, and Villavicencio-Garayzar 2000) and 5 to 13 in *G. polli* (Ebert, Compagno, and Cowley 2006). Both single and multiple oviparity can occur among the *Galeus* species, but only single oviparity is known in *Apristurus*, *Parmaturus*, and *Scyliorhinus* (Cross 1988; Ebert, Compagno, and Cowley 2006). Of the multiple oviparous sawtail sharks, blackmouth catshark *G. melastomus* can possess up to 13 egg cases in the uteri at one time, while Atlantic sawtail catshark *G. atlanticus* has been observed to carry up to nine egg cases in the uteri at one time (Muñoz-Chápuli and Perez Ortega 1985; Iglésias, Du Buit, and Nakaya 2002). For the oviparous *Bythaelurus*, only single oviparity has been observed, while *Halaelurus* are multiple oviparous (Compagno, Dando, and Fowler 2005; Francis 2006).

Catsharks generally reproduce throughout the year, and while most species lack significant patterns of reproductive seasonality, there are often seasonal peaks in egg production (Ebert, Compagno, and Cowley 2006). Richardson et al. (2000) suggested year-round reproduction in Izak catshark *Holohalaelurus regani* with no significant difference observed in the proportion of mature females carrying egg cases between seasons, similar to *B. dawsoni* (Francis 2006). Cross (1988) suggested that brown catshark *A. brunneus* and filetail catshark *P. xaniurus* may be reproductively active throughout the year, although in *A. brunneus* more females carried egg cases in December to May than June to November. Both gecko catshark *G. eastmani* and broadfin sawtail catshark *G. nipponensis* from Suruga Bay, Japan, reproduce throughout the year, but *G. nipponensis* has a higher incidence of carrying egg cases in December and January (Horie and Tanaka 2000). Off southern Portugal, *G. melastomus* is reproductively active year-round but with bimodal peaks in summer and winter (Costa, Erzini, and Borges 2005).

Chain catshark *Scyliorhinus retifer* held in captivity laid pairs of eggs at intervals of 14.1 to 16.7 days (average ~15.3 days; Castro, Bubucis, and Overstrom 1998). If egg-laying continued at this level all year, an individual could produce 46 egg cases annually. Capapé et al. (2008) estimated a maximum annual egg case production of 97 in *G. melastomus* off Mediterranean France, and Richardson et al. (2000) suggested that fecundity is high in *H. regani*, based on the proportion of mature females carrying egg cases and a continuous reproductive cycle. Like many other oviparous chondrichthyans (see Hoff 2008), development time of embryos is long in *S. retifer* (mean = 256 days; Castro, Bubucis, and Overstrom 1988). Little is known about egg-laying sites in the deep sea, although underwater video footage analyzed by Flammang, Ebert, and Cailliet et al. (2007) revealed that *A. brunneus* and *P. xaniurus* entangle egg case tendrils on sessile invertebrates on rocky outcrops at depths of 300 to 400 m.

2.3.1.2 Skates

Despite the high diversity of the softnose skates (Arhynchobatidae), there is little detailed information on their biology. There is considerably more information available on the hardnose skates (Rajidae) than most other batoid families, particularly for shelf species. The biology of the entirely deepwater family Anacanthobatidae is very poorly known.

Skates are oviparous and many species exhibit year-round or protracted egg-laying seasons, although inshore species may have more defined laying seasons (Sulikowski et al.

2005b, 2007; Ruocco et al. 2006). Henderson, Arkhipkin, and Chtcherbich (2004) observed year-round egg-laying in white-dotted skate *Bathyraja albomaculata* from the Falkland/ Malvinas Islands, with peaks in deposition in autumn and winter. *Bathyraja albomaculata* from off continental South America (Uruguay, Argentina, and southwest Chile) were carrying egg cases in April, September, and October, also suggesting year-round laying (Ruocco et al. 2006). An examination of a small (~2 km²) Alaska skate *B. parmifera* nursery area on the shelf-slope edge in the eastern Bering Sea showed a peak in reproductive activity in June and July, although egg-laying occurred year-round (Hoff 2008). Egg case densities were as high as 549,843 eggs/km² in this nursery ground (with 53% to 84% viability) (Hoff 2008).

Ebert (2005) provided a detailed account of the reproductive biology of seven *Bathyraja* and one *Rhinoraja* skate from the eastern Bering Sea in the North Pacific, noting that for these skates there is an extended juvenile stage, a brief adolescent stage, and little growth after maturity. In Aleutian skate *B. aleutica*, Ebert (2005) recognized that ovarian fecundity increased in females up to 1450 mm TL, only to decline in the larger animals. In addition, he observed that some of the largest *B. aleutica*, commander skate *B. lindbergi*, and small-thorn skate *B. minispinosa* were reproductively inactive with atrophied ovaries, suggesting either a period of reproductive inactivity or senescence.

Estimates of annual fecundity of rajid skates are highly variable between species. Two species of rajids are known to carry multiple embryos per egg case: mottled skate *Raja pulchra*, an inshore coastal species from the Northwest Pacific, and *R. binoculata* from the eastern North Pacific (Ishiyama 1958; Hitz 1964; Ebert and Davis 2007). *Raja binoculata* can carry two to seven embryos per egg case, with an average of three to four (Hitz 1964). Annual fecundity has been estimated as 90 for cuckoo skate *Leucoraja naevus* (Du Buit 1976) and 10 to 20 for thorny skate *Amblyraja radiata* (Berestovskii 1994) (40.5 in captivity; Parent et al. 2008). Parent et al. (2008) provided annual fecundities of 69 to 115 for an individual *D. laevis* held in captivity, with very long incubation periods of 342 to 494 days (mean = 421 days). Ebert and Davis (2007) calculated that annual fecundity in *R. binoculata* may reach 1260 based on an assumed average of 3.5 embryos per egg case, and egg-laying rates of 360 egg cases per year in captivity (K. Lewand, personal communication in Ebert and Davis 2007). Temperature, however, affects egg-laying rates (Holden, Rout, and Humphreys 1971) and will alter estimates of fecundity. Hoff (2008) showed that the protracted incubation period of oviparous chondrichthyans was correlated with the rearing temperature; development is longest in *B. parmifera* in the Bering Sea (~3.5 years). It is reasonable then to suggest that incubation times could be very protracted for deepwater oviparous skates, catsharks, and holocephalans in the cold waters of the continental slope and beyond.

2.3.1.3 Holocephalans

Holocephalans are oviparous but, as with many other deepwater groups, their biology is poorly known. Chimaeras may be reproductively active throughout the year, without an apparent well-defined egg-laying season (ninespot chimaera *Hydrolagus barbouri*; Kokuho, Nakaya, and Kitagawa 2003) or with a seasonal peak in activity (*H. colliei*; see Sathyanesan 1966). Barnett et al. (2009) hypothesized that *H. colliei* has a six- to eight-month parturition season and Malagrino, Takemura, and Mizue (1981) suggested that rabbitfish *Chimaera monstrosa* has a reproductive season lasting six to seven months, with a peak in activity in the northern winter. In captivity, *H. colliei* egg cases are laid in pairs every 10 to 14 days over a period of several months (Didier and Rosenberger 2002). Barnett et al. (2009)

combined captive egg-laying rates with the duration of the parturition season in *H. colliei* to arrive at an estimated annual fecundity of 19.5 to 28.9 for that species, the only estimate of annual fecundity for a deepwater holocephalan.

2.3.2 Age and Growth

Determining age at maturity, longevity, and growth rates is crucial to provide management advice through stock assessment and population models, and to accurately assess productivity and vulnerability of a species to fisheries (Cailliet and Goldman 2004). Sharks and rays have traditionally been aged by examining seasonal changes in the deposition of growth bands in their vertebrae (Cailliet and Goldman 2004). Band counts can be correlated to age where the pattern of deposition has been shown to be annual (processes termed verification and validation; see Cailliet and Goldman 2004). For most sharks and rays examined, growth bands are laid down annually, although there are exceptions (i.e., angel sharks *Squatina* spp; Natanson and Cailliet 1990). Vertebral ageing of deepwater chondrichthyans has generally been unsuccessful and alternative methods have had to be developed, many still in their infancy.

The internal and external examination of the dorsal spines of dogfishes (Squaliformes) and holocephalans has proved useful for estimating age (Sullivan 1977; Tanaka 1990; Irvine 2004; Irvine, Stevens, and Laurenson 2006b), while skates can be aged using caudal thorns (Gallagher and Nolan 1999). Neural arches have shown potential for estimating age in sixgill sharks (Hexanchiformes) (McFarlane, King, and Saunders 2002) and radiometric ageing has been tested on some dogfishes (Fenton 2001). Gennari and Scacco (2007) successfully enhanced growth bands on the small, poorly calcified, fragile, deep-coned vertebrae of *E. spinax*, and the use of their technique may prove useful for other deepwater species with similar vertebral structure. In other deepwater groups, attempts at ageing have proved unsuccessful; for sleeper sharks *Somniosus* spp., which lack dorsal spines, an examination of vertebrae and neural arches could not identify any bands that may represent growth (S. Irvine, personal communication). There is a complete lack of age and growth estimates for the deepwater catsharks (Scyliorhinidae); the vertebrae of many species may be too poorly calcified to yield age estimates and attempts to age various scyliorhinids, including *Apristurus* spp. and *Parmaturus* spp., have proved unsuccessful (B. Flammang, personal communication). The continued development of ageing methods for deepwater chondrichthyans (particularly those lacking dorsal spines) is required.

Table 2.4 provides a summary of all available age and growth studies for deepwater chondrichthyan species. Published estimates of age and growth for deepwater chondrichthyans are available for only 34 of the 530 species (13 dogfishes, one thresher shark, 19 skates, and one chimaera) (Table 2.4). Many of these aging estimates are unvalidated and there is a clear need for research into suitable validation techniques for age determination in the deepwater sharks. A correlation between external dorsal spine band counts and the results of radiometric age estimates (Fenton 2001) in *C. crepidater* allowed Irvine, King, and Saunders (2006b) to suggest that external bands were laid down annually.

The oldest age estimates of a deepwater shark are 70 years for female and 71 years for male *C. squamosus* from the Northeast Atlantic (Clarke, Connolly, and Bracken 2002a). Maturity was suggested to be reached at 35 years (Clarke, Connolly, and Bracken 2002a). Fenton (2001) provided a preliminary age estimate of 46 years for *C. zeehaani* from Australia (although this included only immature individuals). Maximum age estimates for needle dogfish *C. acus* from Japan were considerably lower at 18 years for females and 17 years for males (Tanaka 1990), while *C. granulosus* in the western Mediterranean has been estimated

TABLE 2.4

Summary of Age and Growth Studies for Deepwater Chondrichthyan Species

Family	Species	Species Depth Range (m)	Location	Method	von Bertalanffy Growth Parameters			Max Age (t_{max})	Age at Maturity (t_{mat})	Reference
					L_∞	k	t_0			
Squalidae	*Squalus blainville*	16–>440	Western Mediterranean	Vertebrae	♀ 1179 mm TL ♂ 960 mm TL	0.102 0.135	-1.380 -1.397	8 8	5.1 3.3	Cannizzaro et al. (1995)
	Squalus megalops	0–732	Agulhas Bank, South Africa, SE Atlantic	External DS	♀ 932 mm TL ♂ 526 mm TL	0.03 0.09	-8.1 -6.9	32 29	15 9	Watson and Smale (1999)
			SE Australia	External DS	♀ 756 mm TL ♂ 455 mm TL	0.042 0.158	-9.77 -4.86	28 15	— —	Braccini et al. (2007)
	Squalus mitsukurii	4–954	Sala y Gomez Seamount, SE Pacific	External DS	♀ — ♂ —	— —	— —	16 14	— —	Litvinov (1990)
			Hancock Seamount, NW Pacific	External DS	♀ 1070 mm TL ♂ 660 mm TL	0.041 0.155	-10.09 -4.64	27 18	15 4	Wilson and Seki (1994)
			Hancock Seamount, NW Pacific	External DS	♀ 831 mm TL ♂ 645 mm TL	0.103 0.252	-2.94 -0.430	17 12	14–16 6–7	Taniuchi and Tachikawa (1999)
			Choshi, Japan, NW Pacific	External DS	♀ 1628 mm TL ♂ 1093 mm TL	0.039 0.066	-5.21 -5.03	21 20	9 10–11	Taniuchi and Tachikawa (1999)
			Ogasawara Is., Japan, NW Pacific	External DS	♀ 1112 mm TL ♂ 880 mm TL	0.051 0.060	-5.12 -5.51	27 21	9 5	Taniuchi and Tachikawa (1999)
Centrophoridae	*Centrophorus acus*	150–950	Suruga Bay, Japan, NW Pacific	Internal DS	♀ 1262 mm TL ♂ 1172 mm TL	0.155 0.173	-0.485 -1.403	18 17	— 10	Tanaka (1990)
	Centrophorus granulosus	50–1440	Western Mediterranean	Internal DS	♀ 1094 mm TL ♂ 917 mm TL	0.096 0.107	-5.48 -9.78	39 25	16.5 8.5	Guallart (1998)
	Centrophorus squamosus	230–2400	Rockall Trough and Porcupine Bank, NE Atlantic	Internal DS	♀ — ♂ —	— —	— —	70 71	35 30	Clarke et al. (2002a)

Centrophorus zeehaani	208–701	Australia	Radiometric	♀+♂ —	—	—	46[a]	—	Fenton (2001)
Deania calcea	70–1470	Rockall Trough and Porcupine Bank, NE Atlantic	Internal DS	♀ 1193 mm TL / ♂ 935 mm TL	0.077 / 0.135	-0.933 / -0.165	35 / 32	25 / —	Clarke et al. (2002b)
Etmopteridae									
Etmopterus baxteri	250–1500	SE Australia	External DS	♀ 1225 mm TL / ♂ —	0.051 / —	-5.11 / —	37 / 33	21.5 / 13.5	Irvine (2004)
		SE Australia	External DS	♀ 681 mm TL / ♂ 606 mm TL	0.040 / 0.082	-4.51 / -1.43	57 / 48	30 / 20	Irvine et al. (2006a)
		SE Australia	Internal DS	♀ 693 mm TL / ♂ 596 mm TL	0.163 / 0.13	-2.00 / —	22 / 17	10.5 / 10	Irvine et al. (2006a)
Etmopterus pusillus	274–1000	Portugal, NE Atlantic	Internal DS	♀ 540 mm TL / ♂ 490 mm TL	0.13 / 0.17	—	17 / 13	10 / 7	Coelho and Erzini (2007)
Etmopterus spinax	70–2000	Western Mediterranean	Vertebrae	♀ 450 mm TL / ♂ 394 mm TL	0.16 / 0.19	-1.09 / -1.41	9 / 7	—	Gennari and Scacco (2007)
		Western Mediterranean	Vertebrae	—	—	—	7	5	Sion et al. (2002)
		Portugal, NE Atlantic	Internal DS	♀ 558 mm TL / ♂ 580 mm TL	0.12 / 0.09	—	11 / 8	4.67 / 3.97	Coelho and Erzini (2008)
Somniosidae									
Centroselachus crepidater	270–2080	SE Australia	External DS	♀ 961 mm TL / ♂ 732 mm TL	0.072 / 0.141	-6.13 / -2.99	54 / 34	20 / ~9	Irvine et al. (2006b)
		SE Australia	Internal DS	♀ 932 mm TL / ♂ 706 mm TL	0.163 / 0.362	-1.92 / -1.51	27 / 22	—	Irvine (2004)
Proscymnodon plunketi	219–1427	SE Australia	Radiometric	♀+♂	—	—	43	29	Fenton (2001)
		SE Australia	External DS	♀ — / ♂ —	—	—	39 / 32	18	Irvine (2004)
Alopiidae									
Alopias superciliosus	0–732	Taiwan, NW Pacific	Vertebrae	♀ 2246 mm TL / ♂ 2188 mm TL	0.092 / 0.088	-4.21 / -4.24	20 / 19	12.3–13.4 / 9–10	Liu et al. (1998)
Arhynchobatidae									
Bathyraja albomaculata	55–861	Falkland/Malvinas Is., SW Atlantic	Caudal thorn	♀ 649 mm DW / ♂ 706 mm DW	0.09 / 0.07	-1.94 / -2.39	17 / 17	10 / 11	Henderson et al. (2004)
Bathyraja brachyurops	28–604	Falkland/Malvinas Is., SW Atlantic	Caudal thorn	♀ 1204 mm TL / ♂ 1311 mm TL	0.07 / 0.05	-1.95 / -2.55	20 / 20	8.2 / 10	Arkhipkin et al. (2008)

Continued

TABLE 2.4 (Continued)

Summary of Age and Growth Studies for Deepwater Chondrichthyan Species

Family	Species	Species Depth Range (m)	Location	Method	von Bertalanffy Growth Parameters			Max Age (t_{max})	Age at Maturity (t_{mat})	Reference
					L_∞	k	t_0			
	Bathyraja griseocauda	82–941	Falkland/ Malvinas Is., SW Atlantic	Caudal thorn	♀ 3652 mm TL[b] ♂ 3589 mm TL[b]	0.02 0.02	−3.46 −3.04	28 23	17.8 14	Arkhipkin et al. (2008)
	Bathyraja parmifera	20–1425	Eastern Bering Sea, North Pacific	Vertebrae	♀ 1446 mm TL ♂ 1263 mm TL	0.087 0.12	−1.75 −1.39	17 15	10 9	Matta and Gunderson (2007)
	Bathyraja trachura	213–2550	United States, NE Pacific	Vertebrae	♀+♂ 1013 mm TL	0.09	—	♀ 17 ♂ 20	—	Davis et al. (2007)
Rajidae	*Amblyraja georgiana*	20–800	Ross Sea, Antarctica	Caudal thorn	♀ 692 mm PL ♂ 799 mm PL	0.402 0.163	−0.73 −2.41	14 12	8–11 6–7	Francis and Ó Maolagáin (2005)[c]
	Amblyraja radiata	18–1400	Gulf of Maine, NW Atlantic	Vertebrae	♀ 1270 mm TL ♂ 1200 mm TL	0.12 0.11	−0.4 −0.37	16 16	11 10.9	Sulikowski et al. (2005a, 2006)
	Dipturus batis	100–1000	Celtic Sea, NE Atlantic	Vertebrae	♀+♂ 2537 mm TL	0.057	−1.629	50	11	Du Buit (1972)
	Dipturus innominatus	15–1310	New Zealand, SW Pacific	Vertebrae	♀+♂ 1505 mm PL	0.095	−1.06	♀ 24 ♂ 15	13.0 8.2	Francis et al. (2001)
	Dipturus laevis	0–430	Georges Bank, NW Atlantic	Vertebrae	♀+♂ 1663 mm TL	0.141	−1.291	— —	♀ 6.5–7.2 ♂ 5.8–6.1	Gedamke et al. (2005)
	Dipturus pullopunctata	15–457	South Africa, SE Atlantic	Vertebrae	♀ 1327 mm DW ♂ 771 mm DW	0.05 0.10	−2.20 −2.37	14 18	9	Walmsley-Hart et al. (1999)
	Dipturus trachydermus	20–450	Chile, SE Pacific	Vertebrae	♀ 2650 mm TL ♂ 2465 mm TL	0.079 0.087	−1.438 −1.157	26 25	17 15	Licandeo et al. (2007)
	Leucoraja naevus	20–500	Celtic Sea, NE Atlantic	Vertebrae	♀+♂ 916 mm TL	0.1085	−0.465	14	9	Du Buit (1972)
			Irish Sea, NE Atlantic	Vertebrae	♀ 839 mm TL ♂ 746 mm TL	0.197 0.294	−0.151 −0.997	8 6	4.25 4.17	Gallagher et al. (2004)

Species		Location	Structure	Size					Reference
Leucoraja wallacei	70–500	South Africa, SE Atlantic	Vertebrae	♀ 435 mm DW ♂ 405 mm DW	0.26 0.27	−0.21 −0.8	15 12	7	Walmsley-Hart et al. (1999)
Malacoraja senta	46–914	Gulf of Maine, NW Atlantic	Vertebrae	♀ 696 mm TL ♂ 754 mm TL	0.12 0.12	— —	14 15	— —	Natanson et al. (2007)
Raja binoculata	3–800	Canada, NE Pacific	Vertebrae	♀ 2935 mm TL ♂ 2330 mm TL	0.04 0.05	−1.60 −2.10	26 25	8 6	McFarlane and King (2006)
		California, Eastern Central Pacific	Vertebrae	♀ - ♂	— —	— —	12 11	10–12 7–8	Zeiner and Wolf (1993)
		Gulf of Alaska, NE Pacific	Vertebrae	♀ 2475 mm TL ♂ 1533 mm TL	0.0796 0.1524	−1.075 −0.632	14 15	— —	Gburski et al. (2007)
Raja rhina	9–1069	Canada, NE Pacific	Vertebrae	♀ 1372 mm TL ♂ 1315 mm TL	0.06 0.07	−1.80 −2.17	26 23	10 7	McFarlane and King (2006)
		California, Eastern Central Pacific	Vertebrae	♀ 1067 mm TL ♂ 967 mm TL	0.16 0.25	−0.3 0.73	13 13	10 7	Zeiner and Wolf (1993)
		Gulf of Alaska, NE Pacific	Vertebrae	♀ 2341 mm TL ♂ 1688 mm TL	0.0368 0.0561	−1.993 −1.671	24 25	— —	Gburski et al. (2007)
Zearaja chilensis	28–500	Valdivia, Chile, SE Pacific	Vertebrae	♀ 1283 mm TL ♂ 1078 mm TL	0.112 0.134	−0.514 −0.862	21 18	14 11	Licandeo et al. (2006)
		Southern fjords, Chile, SE Pacific	Vertebrae	♀ 1364 mm TL ♂ 1179 mm TL	0.104 0.116	−0.669 −1.056	21 17	13.5 10.7	Licandeo and Cerna (2007)
		Patagonia fjords, Chile, SE Pacific	Vertebrae	♀ 1496 mm TL ♂ 1220 mm TL	0.087 0.110	−1.266 −1.263	22 19	12.8 10.3	Licandeo and Cerna (2007)
Zearaja nasutus	10–1500	New Zealand, SW Pacific	Vertebrae	♀+♂ 913 mm PL	0.16	−1.20	♀ 9 ♂ 7	5.7 4.3	Francis et al. (2001)
Chimaeridae *Chimaera monstrosa*	50–1000	Portugal, NE Atlantic	Internal DS	♀ 636 mm PSCFL ♂ 533 mm PSCFL	0.10 0.14	−1.08 −0.50	17 15	— —	Moura et al. (2004)

a No mature specimens were aged.

b The estimates of L_∞ for *B. griseocauda* are inflated (maximum size of species ~2000 mm TL) (Arkhipkin et al. 2008).

c Francis and Ó Maolagáin (2005) advise that their aging estimates are preliminary and should be utilized with caution.

DS, dorsal spine; PL, pelvic length; PSCFL, pre-supracaudal fin length; DW, disc width.

to reach 39 years (Guallart 1998). Female birdbeak dogfish *Deania calcea* were estimated to reach similar maximum ages of 37 and 35 years off southern Australia (Irvine 2004) and in the Northeast Atlantic (Clarke, Connolly, and Bracken 2002b), respectively. This species also exhibited late maturity, at 21.5 years for Australia and 25 years for the Northeast Atlantic (Clarke, Connolly, and Bracken 2002b; Irvine 2004).

Estimating the age of squaloid sharks using dorsal spines has been undertaken using counts of either internal or external bands. Irvine, Stevens, and Laurenson (2006a, 2006b) showed that there were considerable differences in age estimations between these techniques for *E. baxteri* (maximum age 57 years using external band counts; 26 years using internal band counts) and *C. crepidater* (54 years external; 27 years internal). Irvine, Stevens, and Laurenson (2006b) suggested that internal bands were inaccurate and underestimate age, and Irvine, Stevens, and Laurenson (2006a) noted that internal bands become unreadable as internal dentine appears to stop forming in adult fish. As such, count estimates from external bands likely are more reliable (Irvine, Stevens, and Laurenson 2006a, 2006b). *Etmopterus spinax* reaches a far smaller size (rarely >450 mm TL, although it can reach 600 mm TL) than *E. baxteri* (880 mm TL) (Compagno, Dando, and Fowler 2005), but even this small species is relatively slow-growing (Coelho and Erzini 2008) and long-lived (extrapolated longevity 22 years; Gennari and Scacco 2007).

Until recently, age and growth data for the softnose skates (Arhynchobatidae) were very limited; there have been several aging studies of hardnose species (Rajidae). For many species here included as deepwater fauna, aging studies were conducted on the shelf end of their bathymetrical range, for example broadnose skate *Bathyraja brachyurops* from the Falkland/Malvinas Islands (Arkhipkin et al. 2008), and *R. binoculata* and *R. rhina* from the Northeast Pacific (McFarlane and King 2006; Gburski, Gaichas, and Kimura 2007; Ebert, Smith, and Cailliet 2008), among others. Maximum age estimates for arhynchobatid skates (*Bathyraja*) are 17 to 28 years (Henderson, Arkhipkin, and Chtcherbich 2004; Davis, Cailliet, and Ebert 2007; Matta and Gunderson 2007; Arkhipkin et al. 2008), and for rajid species 8 to 26 years (see Table 2.4), with the exception of *D. batis*, for which Du Buit (1972) estimated a maximum age of 50 years. Longevity of at least 20 years has been confirmed for *A. radiata* from skates tagged off Newfoundland (Templeman 1984). Age and growth estimates follow the general trend that larger skates are slower growing and longer lived than the smaller species, which are faster growing (Sulikowski et al. 2005b).

Age and growth estimates are available for a single deepwater holocephalan *C. monstrosa* from the Northeast Atlantic (Moura et al. 2004; Calis et al. 2005), with males aged to 30 years and females to 26 years off the west of Ireland. Age estimates off Portugal were considerably lower, up to 17 years for females and 15 years for males. Variation in these estimates is likely due to the size range of fish sampled, with Calis et al. (2005) sampling individuals to a far greater maximum size than Moura et al. (2004) (i.e., 740 mm vs. 571 mm PSCFL). Johnson and Horton (1972) attempted several methods to age *H. colliei*; however, none proved successful.

2.3.3 Demography and Population Dynamics

Combining some of the information on reproductive biology and age for deepwater chondrichthyan species can provide estimates of the number of reproductive years, and possible lifetime reproductive output (lifetime fecundity). The number of reproductive years for some deepwater chondrichthyans may be very small, particularly for the skates, which may also undergo senescence at large sizes (Ebert 2005). Table 2.5 gives the estimated number

TABLE 2.5

Estimates of the Number of Reproductive Years and Lifetime Fecundities for a Selection of Deepwater Squaloid Sharks, Assuming Different Reproductive Periodicities (i.e., Biennial or Triennial)

Species	Reproductive Periodicity	No. of Reproductive Years	Lifetime Fecundity	References
Centrophorus granulosus	Biennial	23.5	12	Guallart (1998); Irvine (2004)
Deania calcea	Biennial	15.5	62	Irvine (2004)
	Triennial		41	
Etmopterus baxteri	Biennial	28	128[a]	Irvine (2004)
	Triennial		81[a]	
Centroselachus crepidater	Biennial	34	102	Irvine (2004)
	Triennial		68	
Proscymnodon plunketi	Biennial	10	85	Irvine (2004)
	Triennial		56	

Source: Irvine SB (2004) Ph.D. Thesis, Deakin University, Australia.

[a] These calculations consider the maternal–litter size relationship (Irvine 2004).

of reproductive years and lifetime fecundities of several deepwater squaloid sharks, based largely on Irvine (2004), and Table 2.6 gives those for a selection of deepwater rajid skates, based largely on Ebert (2005).

The lowest calculated lifetime fecundity is for *C. granulosus*; a single female will produce a maximum of 12 pups throughout its lifetime (assuming continuous breeding from maturity to maximum age with no senescence) (Table 2.5). Among the skates, lifetime fecundity in *A. radiata* may be low if one applies the upper annual egg case production of 20 from

TABLE 2.6

Estimates of the Number of Reproductive Years and Lifetime Fecundities for a Selection of Deepwater Rajid Skates

Species	No. of Reproductive Years	Lifetime Fecundity	References
Amblyraja georgiana	4–7	—	Francis and Ó Maolagáin (2005)
Amblyraja radiata	6–10	120–200	Berestovskii (1994); Templeman (1984); Sulikowski et al. (2005a; 2006)
Dipturus innominatus	13	—	Francis et al. (2001); Ebert (2005)
Dipturus pullopunctata	6	—	Walmsley-Hart et al. (1999); Ebert (2005)
Leucoraja naevus	6	540	Du Buit (1972, 1976); Ebert (2005)
	4	360	Du Buit (1976); Gallagher et al. (2004)
Leucoraja wallacei	9	—	Walmsley-Hart et al. (1999); Ebert (2005)
Raja binoculata	1–3	—	Zeiner and Wolf (1993); Ebert (2005)
	17	21,420	McFarlane and King (2006); Ebert and Davis (2007)
Raja rhina	4	—	Zeiner and Wolf (1993); Ebert (2005)
	19	—	McFarlane and King (2006)
Zearaja chilensis	8	—	Licandeo et al. (2006)
Zearaja nasutus	4	—	Francis et al. (2001); Ebert (2005)

Berestovskii (1994), together with maximum ages of 16 to 20 (Templeman 1984; Sulikowski et al. 2005a) and a female age at maturity of 11 years (Sulikowski et al. 2006). This results in an estimated lifetime fecundity of 120 to 200 egg cases. At the other end of the scale, if the annual fecundity estimate of *R. binoculata* from Ebert and Davis (2007) (1260 young per year) is combined with the aging results of McFarlane and King (2006; 17 reproductive years), the big skate in the eastern North Pacific may have a lifetime fecundity of 21,420 young. This calculation assumes no senescence and continuous breeding throughout the adult life of the species, which likely overestimate lifetime reproductive output.

García, Lucifora, and Myers (2008) showed that when compared to continental shelf and oceanic chondrichthyans, deepwater species had a later age at maturity, higher longevity, and a lower growth completion rate. They estimated that the average fishing mortality required to drive a deepwater species to extinction was 38% of that of an oceanic species, and 58% of that of a continental shelf species (García, Lucifora, and Myers 2008).

2.3.4 Diet and Feeding Habits

Many sharks are upper trophic level predators (Cortés 1999; Wetherbee and Cortés 2004), but an overall understanding of their diets, food consumption patterns, and feeding habits lags behind our knowledge of other marine groups, including the teleost fishes (Wetherbee and Cortés 2004). That lag is even more apparent in relation to the deepwater chondrichthyans. Stomach content analysis has been conducted for a number of species, allowing standardized estimates of trophic levels for deepwater sharks, which are tertiary consumers (trophic levels 4.2 to 4.3 for hexanchids, 3.8 to 4.4 for squaloids, 3.6 to 4.2 for scyliorhinids, and 4.1 to 4.3 for pseudotriakids; Cortés 1999), and for skates (mean trophic levels 3.8 for Rajidae, 3.9 for Arhynchobatidae, and 3.5 for Anacanthobatidae; Ebert and Bizzarro 2007). Among the skates, many of the larger deepwater species showed the highest trophic level estimates (Ebert and Bizzarro 2007).

Many deepwater sharks show fairly opportunistic feeding, although teleosts are often primary prey. In addition, deepwater sharks scavenge on cetacean falls at depth (Smith and Baco 2003) while the cookiecutter sharks (*Isistius* spp.) are parasitic on large marine fauna (see Chapter 1). Jakobsdóttir (2001) found that teleosts, cephalopods, and crustaceans occurred frequently in the diet of *C. fabricii*. Teleosts were the most frequent prey class in the stomachs of *E. princeps*, but cephalopods and crustaceans were also important (Jakobsdóttir 2001). Similarly, *E. spinax* exhibits a generalized benthopelagic diet (Neiva, Coelho, and Erzini 2006) as does the catshark *G. melastomus* (Carrassón, Stefanescu, and Cartes 1992). In contrast, the diet of *C. coelolepis* in the Mediterranean is highly specialized on cephalopods (Carrassón, Stefanescu, and Cartes 1992). Blaber and Bulman (1987) classified *D. calcea* as an epibenthic piscivore with a diet consisting almost entirely of teleosts (these authors observed little dietary overlap with a sympatric teleost epibenthic piscivore, suggesting resource partitioning between these upper slope species).

Ontogenetic dietary shifts are evident in some deepwater sharks. Larger *C. fabricii* specialized more on teleosts (Jakobsdóttir 2001), and *E. spinax* showed a shift from a diet dominated by crustaceans in smaller size classes to a more diverse diet in larger size classes, with a higher relative importance of teleosts and cephalopods (Neiva, Coelho, and Erzini 2006). Such shifts may be related to changes in habitat with size (Neiva, Coelho, and Erzini 2006). Carrassón, Stefanescu, and Cartes (1992) showed that *G. melastomus* diet varied between the upper and mid slope, probably due to a change in resource availability. Some geographic variation in the diet of *H. perlo* is apparent, as would be expected from such a wide-ranging species. This shark is a teleost specialist off southern Australia and off

Tunisia, while on the Great Meteor Seamount in the Eastern Atlantic, teleosts and cephalopods are of similar importance (see Braccini 2008).

Skates feed primarily on decapod crustaceans and secondarily teleosts, with polychaetes and amphipods also important prey groups (see Ebert and Bizzarro 2007 for a review). There is relatively little information on the diets of the benthic-feeding holocephalans. The diet of *C. monstrosa* is dominated by crustaceans, with a shift evident from smaller fish that fed mostly on amphipods, to larger individuals that ate mainly decapods (Moura et al. 2005). MacPherson (1980) recorded rather different prey preferences for *C. monstrosa* than Moura et al. (2005), with ophiuroid echinoderms a significant prey group in all but the largest size class of fish. Indeed, Moura et al. (2005) observed geographical and bathymetric differences in diet, suggesting opportunistic foraging.

2.3.5 Population Segregation, Movements, and Migration

Deepwater chondrichthyans are often segregated bathymetrically by size, sex, or maturity stage. Segregation has been demonstrated by changes in the catch composition with depth, including for such species as *E. baxteri*, *C. fabricii*, *E. princeps*, roughskin dogfish *Centroscymnus owstoni*, *C. coelolepis*, and *H. colliei* (Yano and Tanaka 1988; Wetherbee 1996; Jakobsdóttir 2001; Didier and Rosenberger 2002). Several studies have noted a lack of gravid females, and it has been suggested that these may make movements into, or occur in, deeper water (possibly occupying nursery areas; although this is uncertain and nursery areas have not been identified for any deepwater viviparous shark species) or that they may be bathypelagic (and thus less susceptible to capture in benthic sampling/fishing gear; Yano and Tanaka 1988; Wetherbee 1996; Jakobsdóttir 2001; Figueiredo et al. 2008).

A number of deepwater squaloid sharks, particularly small pelagic species in the family Dalatiidae but also sleeper sharks, undertake daily migrations from deep water toward the surface at night, returning to depth during the day. Daily vertical movements appear to be linked to the diel migrations of prey. Hulbert, Sigler, and Lunsford (2006) showed that diel vertical migrations in *S. pacificus* occurred only 25% of the time (i.e., 177 out of 726 days) and as such movement patterns are more complex than repeated daily vertical movements. Sharks were also seen to undertake "systematic vertical oscillations" ("methodical ascents and descents with little pause in transition") and "irregular vertical movements" ("small-amplitude movements with random frequency") with the most time spent at depths of 150 to 450 m (Hulbert, Sigler, and Lunsford 2006).

The basking shark, a highly migratory species occurring in coastal, pelagic, and deepwater habitats, was previously thought to migrate to deep water to undertake a winter hibernation (Compagno 2001; Francis and Duffy 2002). However, basking sharks tracked over extended periods (up to 6.5 months) have shown that they do not hibernate, but undertake horizontal and vertical (>750 m) movements to exploit prey concentrations (Sims et al. 2003). Around New Zealand, basking sharks overwinter, probably on or near the bottom, in deep slope waters (to 904 m; Francis and Duffy 2002).

The short- and long-term (including seasonal) movement and migration patterns of deepwater chondrichthyans are poorly known. The problems associated with tagging animals caught from depth and ensuring their survival once returned to the water, along with logistical constraints, has limited the tagging and tracking of deepwater species. Bagley, Smith, and Priede (1994) used acoustic transmitters imbedded in baits employed at 1517 to 1650 m to briefly track three *C. coelolepis*, the deepest tracked chondrichthyan. All sharks moved outside the range of the recording equipment (500 m) within six hours of bait deployment and, although Bagley, Smith, and Priede (1994) suggested that this indicated no site fidelity, the spatial and temporal scales of the study were limited. A single *C. acus* tracked acoustically for nearly 21 hours in Japan generally swam parallel to the 500 m depth contour, mostly remaining close to the seafloor although making some vertical movements to between 10 and 50 m into the water column (Yano and Tanaka 1986). Short-term acoustic tracking of two *H. griseus* (2 to 4 days) revealed that the sharks generally remained between depths of 600 and 1100 m, swimming back and forth within limited areas, although one took a deep excursion to 1500 m (Carey and Clark 1995).

2.4 Exploitation, Management, and Conservation

2.4.1 Exploitation and Threats

As traditional marine resource stocks are depleted, global demand for fish products increases, and fishing technology advances, fisheries are moving into deeper water and new commercial deepwater fisheries are continuing to develop (Gordon 1999; Haedrich, Merrett, and O'Dea 2001; Morato et al. 2006). Deepwater chondrichthyans are taken as either targeted species or as bycatch in these deep-sea fishing operations, although there is a general lack of available trade and landings data for deepwater sharks, rays, and chimaeras (Cavanagh and Kyne 2005). An assessment of the global catch of deepwater chondrichthyans is made difficult because many species are taken as bycatch, are often discarded, or landed under generic species-codes such as "shark" or "other." Additionally, catches are under-reported globally, and poor taxonomic resolution and species identification limits the availability of species-specific data. As chondrichthyans are generally absent from the deepest oceans, they do not have a refuge at depth; all species are within the depths exploited by commercial fisheries (Priede et al. 2006).

There are few time-series data available for catches of deepwater chondrichthyans, but what data are available show considerable declines for a variety of species (Graham, Andrew, and Hodgson 2001; Devine, Baker, and Haedrich 2006). There are also several examples of the collapse of deepwater stocks due to direct or indirect fishing. Below are four case studies of the effects of commercial fishing activities on deepwater chondrichthyan stocks. These are among the few such instances where sufficient information is available to discuss the impacts of fishing. Also of concern for the deep sea are the impacts of trawling on habitat, including seamounts (Koslow et al. 2000). For example, Hoff (2008) noted that skate nursery areas are in highly productive areas, and thus vulnerable to the impacts of benthic trawling.

2.4.1.1 Case Study 1: Australian Southern and Eastern Scalefish and Shark Fishery

In the Australian Southern and Eastern Scalefish and Shark Fishery (SESSF), which extends across southern and southeastern Australia over an extensive area of the Australian Fishing Zone, intensive fishing has depleted upper slope chondrichthyan species. The SESSF is a complex multispecies, multigear fishery that comprises several sectors. Little information is available on the catch of deepwater sharks from the earlier years of slope fishing, which developed off the east coast (the state of New South Wales, NSW) in the 1970s and off the southern states of Victoria and Tasmania in the 1980s. Initial catch levels off NSW were reportedly high and as there was no market for deepwater shark carcasses, discarding levels were also very high (Daley, Stevens, and Graham 2002). Logbook data showed that substantial numbers of deepwater shark were taken off NSW in the 1980s and fisheries-independent surveys on the NSW upper slope have shown the depletion of the upper slope species (Graham, Andrew, and Hodgson 2001).

Regulations in the State of Victoria on the mercury content of shark flesh restricted any targeting of large deepwater sharks off Victoria and Tasmania before the regulation was eased in 1995. Daley, Stevens, and Graham (2002) believe that there would have been significant discarding prior to that time. A number of market and management changes led to an increased interest in the deepwater dogfish resource, and the targeting of slope species. These included the development of a liver oil market (considerably higher value than flesh), relaxation of the Victorian mercury content regulations, and the introduction to the fishery of a quota management system for target teleost fishes (Daley, Stevens, and Graham 2002). Once quotas were introduced, fishers sought alternative species that were not under quota, including deepwater sharks.

Daley, Stevens, and Graham (2002) reported a 75% decline in catch rates of upper slope dogfish from 1986 to 1999 in a large trawl sector of the SESSF, and a 59% decline in the logbook reported catch of *Centrophorus* spp. off NSW. *Centrophorus* spp. and other squaloids were essentially depleted on the upper slope before the late 1990s (Graham, Andrew, and Hodgson 2001) with the largest declines apparent from the 200 to 399 m and 400 to 649 m depth ranges off NSW (Daley, Stevens, and Graham 2002). The majority of the dogfish catch for 1999 and 2000 was taken from the mid slope (650 to 1200 m; Daley, Stevens, and Graham 2002).

With a lack of species-specific reporting of catches it cannot be determined from available industry data presented by Daley, Stevens, and Graham (2002) which species have been most affected. This information, however, can be drawn from fisheries-independent surveys undertaken on the NSW upper slope. Graham, Andrew, and Hodgson (2001) compared catches of chondrichthyans in three areas of the upper slope off NSW at depths of 200 to 605 m from standardized surveys undertaken in 1976–1977 and 1996–1997. A summary of changes in chondrichthyan populations is given in Table 2.7. The largest declines occurred among the *Centrophorus* species; longnose gulper shark *C. harrissoni*

TABLE 2.7

Changes in the Relative Abundance of Chondrichthyan Species on the NSW Upper Slope between 1976–1977 and 1996–1997

Family	Species	Trend	Extent of Change
Hexanchidae	*Heptranchias perlo*	Decline	−91.4%
Squalidae	*Squalus megalops*	Increase	+18.0%
	Squalus montalbani and *Squalus grahami*	Decline	−97.3%
Centrophoridae	*Centrophorus harrissoni*	Decline	−99.7%
	Centrophorus moluccensis	Decline	−98.4%
	Centrophorus zeehaani	Decline	−99.7%
	Deania quadrispinosum	Decline	−87.3%
Squatinidae	*Squatina albipunctata*	Decline	−96.0%
Pristiophoridae	*Pristiophorus cirratus*	Decline	−47.9%
Scyliorhinidae	*Cephaloscyllium albipinnum*	Decline	−31.9%
Rajidae	*Dipturus* spp	Decline	−83.2%
Chimaeridae	*Hydrolagus ogilbyi*	Decline	−96.4%

Source: Data from Graham KJ et al. (2001) Mar Freshw Res 52:549–561.

and *C. zeehaani* were taken in 1996–1997 at 0.3% of the 1976–1977 catch level, and smallfin gulper shark *C. moluccensis* at 1.6% of the 1976–1977 level. Changes in relative abundance of greater than −90% were also demonstrated for *H. perlo*, Philippine spurdog *Squalus montalbani*, eastern longnose spurdog *S. grahami*, eastern angelshark *Squatina albipunctata*, and Ogilby's ghostshark *Hydrolagus ogilbyi*. Longsnout dogfish *Deania quadrispinosum* and skates (*Dipturus* spp.) declined by >80%. The only chondrichthyan to show an increase in relative abundance was *S. megalops*, increasing by 18% across the three study areas. This is an abundant small dogfish, also common on the outer shelf and likely to be recruiting from there to the upper slope (Graham, Andrew, and Hodgson 2001). It may also have benefited from the decline of other species. It is clear that two decades of intensive trawl fishing caused the collapse of chondrichthyan populations on the upper slope off NSW. Continued exploitation after 1996 and targeting of deepwater sharks on the mid slope (650 to 1200 m) added to the pressures already faced by these species. When considering catch data, localized extirpations for many mid slope species may have gone unnoticed due to vessels moving from area to area during fishing operations.

2.4.1.2 Case Study 2: Northeast Atlantic Deepwater Fisheries

Fisheries research in the Northeast Atlantic is coordinated through ICES (the International Council for the Exploration of the Sea) using data reported from a system of regional sub-areas. The majority of deepwater shark fishing activities occur in northern ICES Sub-areas (Sub-areas V, VI, VII, XII), particularly around the Rockall Trough and on the Porcupine Bank slopes, which lie to the west of the British Isles (Clarke et al. 2003). Species are taken in multispecies trawl, and multispecies and directed longline fisheries. France, Iceland, Ireland, Norway, Portugal, Spain, and the United Kingdom are the main countries in the region landing deepwater sharks (Walker et al. 2005). The most commonly landed species are *C. squamosus* and *C. coelolepis*. It has only been since 1988 that *C. squamosus* and *C. coelolepis* have begun to be landed, initially by the French, then the Spanish and later by other nations (Heessen 2003; ICES 2006). The short time period of the fisheries has generally limited assessments of trend data, until recently when ICES presented a series of CPUE (catch-per-unit-effort) trends for these species to advise that the stock had been depleted (ICES 2005).

Prior to the ICES advice, Basson et al. (2002) reported a declining trend for French trawl *C. squamosus* and *C. coelolepis* CPUE over a short time series of 1990 to 1998 in ICES Sub-areas VI and VII and Division Vb (the area to the West of the British Isles). They used production models to undertake biomass estimates, which suggested that the exploitable biomass of these two species in those regions was below 50% virgin biomass. However, the short time series and the effects of fishing at different depths constrained their results. They also made the observation that the French trawl fleet was moving into deeper water to exploit sharks as well as other fish (Basson et al. 2002).

As part of the European-wide DELASS project (Development of Elasmobranch Assessments) (Heessen 2003) stock assessments were attempted for these two species for the Northeast Atlantic as a whole. Although Heessen (2003) reports evidence of a decline in abundance (from CPUE data) for the two species combined in ICES Sub-areas V, VI, and VII, the stock assessment was unsuccessful due to a short time series, a lack of species-specific data, and a poor understanding of the stock structure of the species in the region.

In 2005, ICES advised a zero catch limit for *C. squamosus* and *C. coelolepis* in ICES areas (i.e., the Northeast Atlantic; ICES 2005). ICES (2005) noted that exploitation increased significantly from the commencement of fishing at the end of the 1980s while CPUE declined considerably for French, Irish, Norwegian, Portuguese, and Scottish trawlers and longliners from 1994 to 2005 in the northern region (ICES Sub-areas VI, VII, and XII). Longline CPUE for ICES Sub-area IX appears to be stable, although the time series is short (2001 to 2006; ICES 2006).

Jones et al. (2005) provided further evidence of considerable declines for these two species utilizing fisheries-independent data. Trawl surveys undertaken to the west of Scotland between 1998 and 2004 were compared with historical trawl survey data from 1970 to 1978, which is prior to exploitation of deepwater marine resources in the region. There were considerable declines in CPUE between the 1970s surveys and the recent surveys for *C. squamosus* and *C. coelolepis* as well as for *D. licha* and *D. calcea* (1998 to 2004 survey catches 62% to 99% lower than 1970s surveys; Jones et al. 2005). CPUE increased considerably for *C. crepidater* as the result of two unusually large catches in 1998. CPUE also increased for *G. melastomus*, and was comparable between the surveys for *E. spinax* and *Apristurus* spp. (Jones et al. 2005). Separate preliminary analysis of the survey data from 1998 to 2004, although a short time series, showed that CPUE continued to decline for *C. squamosus* and *C. coelolepis* (Jones et al. 2005).

2.4.1.3 Case Study 3: Azores Kitefin Shark Fishery

A considerable biomass decline is apparent from nearly 30 years of fishing pressure for the Azores kitefin shark stock, which was likely exploited beyond sustainable levels. The cessation of fishing should allow the stock to begin to recover, but 30 years of exploitation appears to have depleted the local population (Heessen 2003). *Dalatias licha* was targeted in an artisanal longline (handline) fishery in the Azores from 1972 to 2001. It was also taken in an industrial benthic gillnet fishery from 1980 onward (Heessen 2003). The catch was dominated by the benthic gillnet fishery that showed three peaks, in its second year (1981) at 667 t, in 1984–1985 at 814 to 842 t, and in 1991 at 794 t. Catch then declined rapidly to 148 t in 1996 and <1 t in 2000 and 2001. CPUE in the gillnet fishery peaked in the late 1980s to early 1990s before declining to initial levels by 1998. Catch in the handline fishery peaked in 1979–1980 at 227 to 232 t before declining to <100 t where it remained relatively stable (with some fluctuations) until the early 1990s. Catch then gradually declined to ~11 t in 2001. CPUE fluctuated throughout the period of the fishery, with a rapid increase in 1996, the last year for which CPUE data are presented in Heessen (2003).

The poor market value of kitefin livers apparently underlies the decline in landings, rather than the collapse of the stock. The species is no longer targeted around the Azores but taken as bycatch in other demersal fisheries (Heessen 2003). Heessen (2003) undertook a stock assessment for *D. licha* around the Azores and modeled biomass time series trends based on the catch and CPUE data from the longline and gillnet fisheries. A continuous decline in the predicted biomass was demonstrated from the commencement of fishing in 1972, through the 1970s and 1980s. The severity of the decline (i.e., the slope of the predicted biomass time series) was greater after the gillnet fishery commenced in 1980. The model suggested that the stock began to rebuild as the fishery came to its end. The model also gave estimated probabilities of 51% to 85% that the stock was overexploited (based on comparing estimated biomass in 2001 with predictions of maximum sustainable yield; Heessen 2003).

2.4.1.4 Case Study 4: Maldives Gulper Shark Fishery

The exploitation of gulper sharks (*Centrophorus* spp.) in the Maldives commenced in 1980 using multihook handlines (vertical longlines). The fishery was short-lived, with a complete collapse of the gulper shark fishery (targeted for liver oil that was exported to the Japanese market) after only ~20 years of exploitation. Anderson and Ahmed (1993) surveyed shark fishing activities around the Maldives, reporting 31 full-time and 274 part-time vessels, operating from 37 islands involved in the gulper shark fishery. There are no time-series data of catches available for the fishery (M.S. Adam, personal communication), only export figures for shark liver oil. Exports peaked rapidly in 1982 (87,400 liters) before showing a general downward trend until 1989. Exports then increased again in 1990 and 1991 but decreased to very low levels up to 2002 (Anderson and Ahmed 1993; MOFA 1996, 1997; MOFAMR 2001, 2002). This marked the effective end of the fishery due to stock collapse from overfishing of the resource (R.C. Anderson, personal communication). The present status of the stock is unknown, but fishing for gulper sharks has ceased. The resource was probably fished beyond sustainable levels in its early years and depleted to very low levels by continued fishing.

2.4.2 Management

In general, management of deep-sea fisheries is insufficient to ensure ecological sustainability of the resource. In some instances management has been implemented after the collapse of stocks. In many instances, fisheries continue to exploit or impact upon stocks even after considerable declines have been demonstrated. In Australia's SESSF, trip limits were first imposed for landings of *Centrophorus* (upper slope species) in 2003, well after the collapse of the population. The landing of mid-slope deepwater dogfishes was largely unregulated until 2005. However, from 2003 the landing of dogfishes (including mid-slope species) required the landing of both the liver and carcass so that accurate landing information could be obtained. Due to the difficulty in accurately identifying many deepwater dogfish species, a "basket" quota management system was introduced in 2005. This "basket" included various species of mid-slope dogfishes (Centrophoridae, Etmopteridae, Somniosidae, and Dalatiidae), many considered "discard" species, and the quota system set a Transferable Allowable Catch (TAC) for 2005 and 2006 at around half the reported 2004 catch (200 t). A bycatch-only TAC of 22 t was introduced for 2007 as a precautionary management measure, while the quota for 2008 increased to 50 t in each of the east and west components of the fishery. The quota does not include *Squalus* dogfishes (outer shelf and upper slope species), known locally as "greeneye" sharks and catches of these remain unregulated (for some species significant declines are apparent). *Centrophorus* spp. (upper slope species) remain managed under individual trip limits. As these sharks are taken in a multispecies fishery, these trip limits affect discard rate, rather than capture rate (R. Daley, personal communication).

A recent ban on commercial trawling at depths >700 m in Australia's SESSF (>750 m in the Great Australian Bight) should have significant conservation benefits for deepwater chondrichthyans. However, while a bycatch-only TAC is in place for "deepwater sharks" (mid-slope species), ongoing fishing pressure on the upper slope will continue to impact some of the most unproductive shark species (e.g., *Centrophorus*) through bycatch. Pressure is currently on the Australian government to improve management of *Centrophorus* within the SESSF.

The Namibian government has taken a precautionary approach to the management of deepwater shark fishing, attempting to balance the exploitation of a potentially valuable resource with an understanding of the biological vulnerability of the group (NATMIRC 2003). Two fishing companies were granted exploratory rights in 2000 to target deepwater dogfishes (primarily *C. squamosus*, *D. quadrispinosum*, and *C. coelolepis*) off Namibia utilizing set nets. The exploratory licenses were governed by rigid management measures, including monitoring and logbook recording of all individuals caught by species and sex (NATMIRC 2003). The exploratory directed deepwater shark licenses expired in January 2006 and insignificant abundance and biomass data to set effective Total Allowable Catches meant that there was no immediate decision to grant commercial rights to the fishery (H. Holtzhausen, personal communication). Three deep-sea surveys were conducted and subsequent analysis of catch data indicated that the long-term sustainability of the resource could not be guaranteed.

Marine protected areas will be important in the conservation and management of the deepwater fauna and its habitats (Davies, Roberts, and Hall-Spencer 2007). Relative to the global area of the deep sea, existing reserves are small, although some are significant regionally. Australia's South-East Commonwealth Marine Reserve Network includes representative examples of temperate benthic and seamount habitats. The world's largest marine park, the Phoenix Islands Protected Area in the central Pacific (administered by the Republic of Kiribati), also encompasses deep-sea habitats. While several other deep-sea protected areas exist around the world (see Davies, Roberts, and Hall-Spencer 2007), the Darwin Mounds closed area, which protects a relatively small portion of the Northeast Atlantic (De Santo and Jones 2007), is the only internationally agreed upon area (Davies, Roberts, and Hall-Spencer 2007). In addition to existing protected areas, the ban on benthic trawling below 1000 m in the Mediterranean, which was declared in 2005, provides an important deep-sea refuge. Overcoming the challenges of implementing and enforcing high-seas protected areas under international management will be a significant future conservation goal (Davies, Roberts, and Hall-Spencer 2007).

2.4.3 Conservation Status

The IUCN (International Union for Conservation of Nature) *Red List of Threatened Species*™ (the IUCN Red List) provides a robust instrument to assess an individual species' risk of extinction, based on its life history, distribution, abundance, and the threats that it faces. The IUCN Red List is widely considered the most objective and authoritative system for assessing the global conservation status of the world's plant and animal species (IUCN 2001).

One half of the deepwater chondrichthyan fauna (265 species) has been assessed against the IUCN Red List Categories and Criteria, and published on the Red List (IUCN 2008). The remaining species are considered Not Evaluated, although these are presently in preparation for publication (there is an ongoing *Global Shark Assessment* program that aims to assess all chondrichthyan species; S.V. Valenti, personal communication). Approximately 8% of deepwater chondrichthyans are globally threatened (i.e., they are Critically Endangered, Endangered, or Vulnerable), ~12% are Near Threatened, 26% are Least Concern, and ~54% are Data Deficient (Figure 2.1). The situation is similar if only the sharks are considered: 6% are threatened, ~11% Near Threatened, ~25% Least Concern, and ~58% are Data Deficient. The proportion of threatened batoids is considerably higher (18%), while ~13% are Near Threatened, ~28% are Least Concern, and ~40% are Data Deficient. The holocephalans are largely Data Deficient (~63%), with no threatened species, although ~9% are Near Threatened, and ~28% are Least Concern.

The considerable number of Data Deficient species highlights the overall lack of information on much of the fauna. Cavanagh and Kyne (2005) note that this category has often been unavoidably assigned to deepwater species, even when concerns of a species' vulnerability exist, simply because there is insufficient data to support any other listing. It is not only the threatened species that require urgent conservation actions; research funding

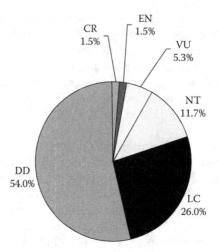

FIGURE 2.1

A color version of this figure follows page 336. The conservation status of deepwater chondrichthyans accord-ing to the IUCN *Red List of Threatened Species*™. Percentages of species within each Red List category are shown: CR, Critically Endangered; EN, Endangered; VU, Vulnerable; NT, Near Threatened; LC, Least Concern; DD, Data Deficient.

and effort is needed for Data Deficient species in order to provide the necessary informa-tion to accurately assess their conservation status (Cavanagh et al. 2003).

The four globally Critically Endangered deepwater chondrichthyans are *Centrophorus harrissoni*, an Australian endemic upper to mid-slope shark (depth range 220 to 790 m), which has suffered dramatic declines (>99%) from an intensive multispecies trawl fishery; sawback angelshark *Squatina aculeata*, a shelf and upper slope shark (depth range 30 to 500 m), which is apparently extirpated from large parts of its Mediterranean and West African range where industrial and artisanal fishing pressure is intense; *Dipturus batis*, an outer shelf, upper to deep slope and deep-sea rise batoid (depth range 100 to 2619 m), which has been depleted across much of its range while fishing activities continue; and Maltese skate *Leucoraja melitensis*, a Mediterranean endemic shelf and upper to mid-slope batoid (depth range 60 to 800 m), which now appears to be restricted to one small loca-tion within its former range, an area subject to heavy trawling (Brander 1981; Graham, Andrew, and Hodgson 2001; IUCN 2008). For these species, together with the Endangered and Vulnerable fauna, management and recovery plans are urgently required.

2.5 Conclusions

The chondrichthyan fauna of the deep sea remains poorly known, despite increas-ing research effort in recent years. A review of available biological data emphasizes the unproductive nature of many species, and their subsequent vulnerability to the effects of commercial fishing activities. There are several well-documented cases of the collapse of deepwater stocks, and these only serve to reinforce the need for a precautionary approach to the management of deep-sea resources. As Haedrich, Merrett, and O'Dea (2001) and Clarke et al. (2003) note, management of deep-sea fisheries needs to take an alternative approach to that of traditional shelf fisheries; rather it requires an approach based on the life history of the deepwater fauna. It is our hope that this chapter will provide a baseline from which to prioritize research, conservation, and management of deep-sea sharks, batoids, and holocephalans. Knowledge gaps are considerable, many species groups remain poorly known, and reliable catch-and-trade data are all but lacking. The specific data needed to inform management requires a further input of research effort; however, the absence of data should not restrict the complete implementation of precautionary management.

Acknowledgments

Funding support for an original compilation and summarization of available information on deepwater chondrichthyans was provided by the Marine Conservation Biology Institute (Bellevue, Washington). The authors thank Lance Morgan for project assistance. Claudine Gibson, Sarah Fowler, Catherine McCormack, and Sarah Valenti of the IUCN Shark Specialist Group, and Michael Bennett provided administrative and technical assistance for that project. We are also thankful to Sarah Irvine for access to life history data and invaluable discussions on the biology, fisheries, and conservation of deepwater chondrichthyans. Thanks are also extended to Ross Daley, Dave Ebert, Malcolm Francis, and William White for providing access to resources and information. Micha Jackson assisted in the compilation of this chapter.

References

Acuña E, Villarroel JC, Catalán R et al. (2003) Reproduction and feeding habits of two deep-sea sharks from Central-Northern Chile: the etmopterid *Aculeola nigra* De Buen, 1959 and the scylliorinid *Bythalaelurus canescens* (Günther, 1878). Abstract. Conservation and Management of Deepsea Chondrichthyan Fishes. Joint FAO and IUCN Shark Specialist Group Pre-conference Meeting in conjunction with Deepsea 2003. University of Otago, Portobello Marine Laboratory, Portobello, New Zealand, 27–29 November 2003.

Anderson RC, Ahmed H (1993) The shark fisheries of the Maldives. Ministry of Fisheries and Agriculture, Republic of Maldives, Malé and FAO, Rome.

Arkhipkin AI, Baumgartner N, Brickle P et al. (2008) Biology of the skates *Bathyraja brachyurops* and *B. griseocauda* in waters around the Falkland Islands, Southwest Atlantic. ICES J Mar Sci 65(4):560–570.

Bagley PM, Smith A, Priede IG (1994) Tracking movements of deep demersal fishes in the Porcupine Seabight, North-East Atlantic Ocean. J Mar Biol Assoc UK 74:473–480.

Balart EF, Gonzalez-Garcia J, Villavicencio-Garayzar C (2000) Notes on the biology of *Cephalurus cephalus* and *Parmaturus xaniurus* (Chondrichthyes: Scyliorhinidae) from the west coast of Baja California Sur, Mexico. Fish Bull 98(1):219–221.

Bañón R, Piñeiro C, Casas M (2006) Biological aspects of deep-water sharks *Centroscymnus coelolepis* and *Centrophorus squamosus* in Galician waters (north-western Spain). J Mar Biol Assoc UK 86(4):843–846.

Barnett LAK, Earley RL, Ebert DA et al. (2009) Maturity, fecundity, and reproductive cycle of the spotted ratfish, *Hydrolagus colliei*. Mar Biol 156:301–316.

Barrull J, Mate I (2001) First record of a pregnant female little sleeper shark *Somniosus rostratus* (Risso, 1826) on the Spanish Mediterranean coast. Bol Inst Esp Oceanogr 17(3–4):323–325.

Basson M, Gordon JDM, Large P et al. (2002) The effects of fishing on deep-water fish species to the west of Britain. Joint Nature Conservation Committee Report No. 324. JNCC, Petersborough.

Berestovskii EG (1994) Reproductive biology of skates of the family Rajidae in the seas of the far north. J Ichthyol 34(6):26–37.

Bigelow HB, Schroeder WC (1948) Sharks. In: Fishes of the Western North Atlantic. Mem Sears Found Mar Res 1(1):59–546.

Blaber SJM, Bulman CM (1987) Diets of fishes of the upper continental slope of eastern Tasmania: content, calorific values, dietary overlap and trophic relationships. Mar Biol 95:345–356.

Braccini JM (2008) Feeding ecology of two high-order predators from south-eastern Australia: the coastal broadnose and the deepwater sharpnose sevengill sharks. Mar Ecol Prog Ser 371:273–284.

Braccini JM, Gillanders BM, Walker TI (2006) Determining reproductive parameters for population assessments of chondrichthyan species with asynchronous ovulation and parturition: piked spurdog (*Squalus megalops*) as a case study. Mar Freshw Res 57(1):105–119.

Braccini JM, Gillanders BM, Walker TI et al. (2007) Comparison of deterministic growth models fitted to length-at-age data of the piked spurdog (*Squalus megalops*) in south-eastern Australia. Mar Freshw Res 58:24–33.

Brander K (1981) Disappearance of Common skate *Raia batis* from Irish Sea. Nature 290:48–49.

Cadenat J, Blanche J (1981) Requins de Méditerranée et d'Atlantique (plus particulièrement de la côte occidentale d'Afrique). Faune tropicale, OSTROM 21:1–330.

Camhi M, Fowler S, Musick J et al. (1998) Sharks and their relatives: ecology and conservation. Occasional Paper of the IUCN Species Survival Commission No. 20.

Cailliet GM, Goldman KJ (2004) Age determination and validation in chondrichthyan fishes. In: Carrier JC, Musick JA, Heithaus MR (eds) Biology of sharks and their relatives. CRC Press, Boca Raton, FL, pp 399–447.

Calis E, Jackson EH, Nolan CP et al. (2005) Preliminary age and growth estimates of the rabbitfish, *Chimaera monstrosa*, with implications for future resource management. e-J Northwest Atl Fish Sci 35, art. 21.

Cannizzaro L, Rizzo P, Levi D et al. (1995) Age determination and growth of *Squalus blainvillei* (Risso, 1826). Fish Res 23(1–2):113–125.

Capapé C, Seck AA, Quignard J-P (1999) Observations on the reproductive biology of the angular rough shark, *Oxynotus centrina* (Oxynotidae). Cybium 23(3):259–271.

Capapé C, Guélorget O, Vergne Y et al. (2008) Reproductive biology of the blackmouth catshark, *Galeus melastomus* (Chondrichthyes: Scyliorhinidae) off the Languedocian coast (southern France, northern Mediterranean). J Mar Biol Assoc UK 88(2):415–421.

Carey FG, Clark E (1995) Depth telemetry from the sixgill shark, *Hexanchus griseus*, at Bermuda. Environ Biol Fishes 42:7–14.

Carrassón M, Stefanescu C, Cartes JE (1992) Diets and bathymetric distributions of two bathyal sharks of the Catalan deep sea (western Mediterranean). Mar Ecol Prog Ser 82:21–30.

Castro JI, Bubucis PM, Overstrom NA (1988) The reproductive biology of the chain dogfish, *Scyliorhinus retifer*. Copeia 1988(3):740–746.

Cavanagh RD, Kyne PM (2005) The conservation status of deep-sea chondrichthyan fishes. In: Shotton R (ed) Deep Sea 2003: Conference on the Governance and Management of Deep-Sea Fisheries. Part 2: Conference poster papers and workshop papers. Queenstown, New Zealand, 1–5 December 2003 and 27–29 November 2003, Dunedin, New Zealand. FAO Fish Proc No. 3/2. Rome, FAO, pp 366–380.

Cavanagh RD, Kyne PM, Fowler SL et al. (2003) The conservation status of Australasian chondrichthyans: Report of the IUCN Shark Specialist Group Australia and Oceania Regional Red List Workshop, Queensland, Australia, 7–9 March, 2003. School of Biomedical Sciences, The University of Queensland, Brisbane.

Chen C, Taniuchi T, Nose Y (1981) Some aspects of reproduction in the pointed-snout dogfish *Squalus japonicus* taken off Nagasaki and Choshi. Bull Jpn Soc Sci Fish 47(9):1157–1164.

Claes JM, Mallefet J (2008) Early development of bioluminescence suggests camouflage by counter-illumination in the velvet belly lantern shark *Etmopterus spinax* (Squaloidea: Etmopteridae). J Fish Biol 73:1337–1350.

Clarke MW, Connolly PL, Bracken JJ (2001) Aspects of reproduction of the deep water sharks *Centroscymnus coelolepis* and *Centrophorus squamosus* from west of Ireland and Scotland. J Mar Biol Assoc UK 81(6):1019–1029.

Clarke MW, Connolly PL, Bracken JJ (2002a) Age estimation of the exploited deepwater shark *Centrophorus squamosus* from the continental slopes of the Rockall Trough and Porcupine Bank. J Fish Biol 60(3):501–514.

Clarke MW, Connolly PL, Bracken JJ (2002b) Catch, discarding, age estimation, growth and maturity of the squalid shark *Deania calceus* west and north of Ireland. Fish Res 56(2):139–153.

Clarke M, Girard M, Hareide N-R et al. (2003) Approaches to assessment of deepwater sharks in the Northeast Atlantic. NAFO Scientific Council Research Document 02/136.

Coelho R, Erzini K (2007) Population parameters of the smooth lantern shark, *Etmopterus pusillus*, in southern Portugal (NE Atlantic). Fish Res 86:42–57.

Coelho R, Erzini K (2008) Life history of a wide-ranging deepwater lantern shark in the north-east Atlantic, *Etmopterus spinax* (Chondrichthyes: Etmopteridae), with implications for conservation. J Fish Biol 73:1419–1443.

Compagno LJV (2001) Sharks of the world. An annotated and illustrated catalogue of shark species known to date. Volume 2. Bullhead, mackerel and carpet sharks (Heterodontiformes, Lamniformes and Orectolobiformes). FAO Species Catalogue for Fishery Purposes. No. 1, Vol. 2. FAO, Rome.

Compagno L, Dando M, Fowler S (2005) A field guide to sharks of the world. Collins, London.

Cortés E (1999) Standardized diet compositions and trophic levels of sharks. ICES J Mar Sci 56:707–717.

Costa ME, Erzini K, Borges TC (2005) Reproductive biology of the blackmouth catshark, *Galeus melastomus* (Chondrichthyes: Scyliorhinidae) off the south coast of Portugal. J Mar Biol Assoc UK 85(5):1173–1183.

Cross JN (1988) Aspects of the biology of two scyliorhinid sharks, *Apristurus brunneus* and *Parmaturus xaniurus*, from the upper continental slope off southern California. Fish Bull 86(4):691–702.

Crow GL, Lowe CG, Wetherbee BM (1996) Shark records from longline fishing programs in Hawai'i with comments on Pacific Ocean distributions. Pac Sci 50(4):382–392.

Cunha CM, Gonzalez MB (2006) Pregnancy in *Squaliolus laticaudus* (Elasmobranch: Dalatiidae) from Brazil. Environ Biol Fishes 75(4):465–469.

Daley R, Stevens J, Graham K (2002) Catch analysis and productivity of the deepwater dogfish resource in southern Australia. FRDC Final Report, 1998/108. Fisheries Research and Development Corporation, Canberra.

Davies AJ, Roberts JM, Hall-Spencer J (2007) Preserving deep-sea natural heritage: emerging issues in offshore conservation and management. Biol Conserv 138:299–312.

Davis CD, Cailliet GM, Ebert DA (2007) Age and growth of the roughtail skate *Bathyraja trachura* (Gilbert 1892) from the eastern North Pacific. Environ Biol Fishes 80:325–336.

De Santo EM, Jones PJS (2007) The Darwin Mounds: from undiscovered coral to the development of an offshore marine protected area regime. In: George RY, Cairns SD (eds) Conservation and adaptive management of seamount and deep-sea coral ecosystems. Rosenstiel School of Marine and Atmospheric Science, University of Miami, pp 147–156.

Devine JA, Baker KD, Haedrich RL (2006) Deep-sea fishes qualify as endangered. Nature 439:29.

Didier DA, Rosenberger LJ (2002) The spotted ratfish, *Hydrolagus colliei*: notes on its biology with a redescription of the species (Holocephali: Chimaeridae). Calif Fish Game 88(3):112–125.

Du Buit MH (1972) Age et croissance de *Raja batis* et de *Raja naevus* en Mer Celtique. J Cons Int Explor Mer 37:261–265.

Du Buit MH (1976) The ovarian cycle of the cuckoo ray, *Raja naevus* (Müller and Henle), in the Celtic Sea. J Fish Biol 8:199–207.

Duffy C, Bishop S, Yopak K (2003) Distribution and biology of *Cirrhigaleus barbifer* (Squalidae) in New Zealand waters. Abstract. Conservation and management of deepsea chondrichthyan fishes. Joint FAO and IUCN Shark Specialist Group Pre-conference Meeting in conjunction with Deepsea 2003. University of Otago, Portobello Marine Laboratory, Portobello, New Zealand, 27–29 November, 2003.

Ebert DA (2005) Reproductive biology of skates, *Bathyraja* (Ishiyama), along the eastern Bering Sea continental slope. J Fish Biol 66(3):618–649.

Ebert DA, Bizzarro JJ (2007) Standardized diet compositions and trophic levels of skates (Chondrichthyes: Rajiformes: Rajoidei). Environ Biol Fishes 80:221–237.

Ebert DA, Davis CD (2007) Descriptions of skate egg cases (Chondrichthyes: Rajiformes: Rajoidei) from the eastern North Pacific. Zootaxa 1393:1–18.

Ebert DA, Compagno LJV, Cowley PD (2006) Reproductive biology of catsharks (Chondrichthyes: Scyliorhinidae) off the west coast of southern Africa. ICES J Mar Sci 63:1053–1065.

Ebert DA, Compagno LJV, Natanson LJ (1987) Biological notes on the Pacific sleeper shark, *Somniosus pacificus* (Chondrichthyes: Squalidae). Calif Fish Game 73(2):117–123.

Ebert DA, Smith WD, Cailliet GM (2008) Reproductive biology of two commercially exploited skates, *Raja binoculata* and *R. rhina*, in the western Gulf of Alaska. Fish Res 94:48–57.

Fenton GE (2001) Radiometric ageing of sharks. FRDC Final Report, 1994/021. Fisheries Research and Development Corporation, Canberra.

Fergusson IK, Graham KJ, Compagno LJV (2008) Distribution, abundance and biology of the small-tooth sandtiger shark *Odontaspis ferox* (Risso, 1810) (Lamniformes: Odontaspididae). Environ Biol Fishes 81:207–228.

Figueiredo I, Moura T, Neves A et al. (2008) Reproductive strategy of leafscale gulper shark *Centrophorus squamosus* and the Portuguese dogfish *Centroscymnus coelolepis* on the Portuguese continental slope. J Fish Biol 73:206–225.

Fischer AF, Veras DP, Hazin FHV et al. (2006) Maturation of *Squalus mitsukurii* and *Cirrhigaleus asper* (Squalidae, Squaliformes) in the southwestern equatorial Atlantic Ocean. J Appl Ichthyol 22:495–501.

Flammang BE, Ebert DA, Cailliet GM (2007) Egg cases of the genus *Apristurus* (Chondrichthyes: Scyliorhinidae): Phylogenetic and ecological implications. Zoology 110:308–317.

Francis MP (2006) Distribution and biology of the New Zealand endemic catshark, *Halaelurus dawsoni*. Environ Biol Fishes 75(3):295–306.

Francis MP, Duffy C (2002) Distribution, seasonal abundance and bycatch of basking sharks (*Cetorhinus maximus*) in New Zealand, with observations on their winter habitat. Mar Biol 140(4):831–842.

Francis MP, Ó Maolagáin C (2005) Age and growth of the Antarctic skate (*Amblyraja georgiana*) in the Ross Sea. CCAMLR Science 12:183–194.

Francis MP, Ó Maolagáin C, Stevens D (2001) Age, growth, and sexual maturity of two New Zealand endemic skates, *Dipturus nasutus* and *D. innominatus*. NZ J Mar Freshw Res 35(4):831–842.

Fujita K (1981) Oviphagous embryos of the pseudochariid shark, *Pseudocarcharias kamoharai*, from the central Pacific. Jpn J Ichthyol 28(1):37–44.

Gadig OBF, Gomes UL (2002) First report on embryos of *Isistius brasiliensis*. J Fish Biol 60(5):1322–1325.

Gallagher MJ, Nolan CP (1999) A novel method for the estimation of age and growth in rajids using caudal thorns. Can J Fish Aquat Sci 56(9):1590–1599.

Gallagher MJ, Nolan CP, Jeal F (2004) Age, growth and maturity of the commercial ray species from the Irish Sea. J Northwest Atl Fish Sci 35:47–66.

García VB, Lucifora LO, Myers RA (2008) The importance of habitat and life history to extinction risk in sharks, skates, rays and chimaeras. Proc R Soc B 275:83–89.

Gburski CM, Gaichas SK, Kimura DK (2007) Age and growth of the big skate (*Raja binoculata*) and longnose skate (*R. rhina*) in the Gulf of Alaska. Environ Biol Fishes 80:337–349.

Gedamke T, DuPaul WD, Musick JA (2005) Observations on the life history of the barndoor skate, *Dipturus laevis*, on Georges Bank (Western North Atlantic). J Northwest Atl Fish Sci 35:67–78.

Gennari E, Scacco U (2007) First age and growth estimates in the deep water shark, *Etmopterus spinax* (Linnaeus, 1758), by deep coned vertebral analysis. Mar Biol 152:1207–1214.

Girard M, Du Buit M-H (1999) Reproductive biology of two deep-water sharks from the British Isles, *Centroscymnus coelolepis* and *Centrophorus squamosus* (Chondrichthyes: Squalidae). J Mar Biol Assoc UK 79(5):923–931.

Golani D, Pisanty S (2000) Biological aspects of the gulper shark, *Centrophorus granulosus* (Bloch & Schneider, 1801), from the Mediterranean coast of Israel. Acta Adriat 41(2):71–77.

Gordon JDM (1999) Management considerations of deep-water shark fisheries. FAO Fish Tech Pap 378/2:774–818.

Graham KJ (2005) Distribution, population structure and biological aspects of *Squalus* spp. (Chondrichthyes: Squaliformes) from New South Wales and adjacent Australian waters. Mar Freshw Res 56(4):405–416.

Graham KJ, Andrew NL, Hodgson KE (2001) Changes in relative abundance of sharks and rays on Australian South East Fishery trawl grounds after twenty years of fishing. Mar Freshw Res 52:549–561.

Guallart J (1998) Contribución al conocimientto de la taxonomía y la biología del tiburón batial *Centrophorus granulosus* (Bloch & Schneider, 1801) (Elasmobranchii, Squalidae) en el Mar Balear (Mediterráneo occidental). Ph.D. Thesis, Universitat de Valencia, Valencia.

Guallart J, Vicent JJ (2001) Changes in composition during embryo development of the gulper shark, *Centrophorus granulosus* (Elasmobranchii, Centrophoridae): an assessment of maternal-embryonic nutritional relationships. Environ Biol Fishes 61:135–150.

Haedrich RL (1996) Deep-water fishes: evolution and adaptation in the earth's largest living spaces. J Fish Biol 49(Suppl A):40–53.

Haedrich RL, Merrett NR, O'Dea NR (2001) Can ecological knowledge catch up with deep-water fishing? A North Atlantic perspective. Fish Res 51:113–122.

Hazin FHV, Fischer AF, Broadhurst MK et al. (2006) Notes on the reproduction of *Squalus megalops* off northeastern Brazil. Fish Res 79:251–257.

Heessen, HJL (ed) (2003) Development of elasmobranch assessments DELASS. Final Report of DG Fish Study Contract 99/055.

Henderson AC, Arkhipkin AI, Chtcherbich JN (2004) Distribution, growth and reproduction of the white-spotted skate *Bathyraja albomaculata* (Norman, 1937) around the Falkland Islands. J Northwest Atl Fish Sci 35:79–87.

Hitz CR (1964) Observations on egg cases of the big skate (*Raja binoculata* Girard) found in Oregon coastal waters. J Fish Res Board Can 21(4):851–854.

Hoenig JM, Gruber SH (1990) Life-history patterns in the elasmobranchs: implications for fisheries management. NOAA Tech Rep NMFS 90:1–16.

Hoff GR (2008) A nursery site of the Alaska skate (*Bathyraja parmifera*) in the eastern Bering Sea. Fish Bull 106:233–244.

Holden MJ (1977) Elasmobranchs. In: Gulland A (ed) Fish population dynamics. J. Wiley and Sons, New York, pp 187–215.

Holden MJ, Rout DW, Humphreys CN (1971) The rate of egg laying by three species of ray. J Cons Int Explor Mer 33(3):335–339.

Horie T, Tanaka S (2000) Reproduction and food habits of two species of sawtail catsharks, *Galeus eastmani* and *G. nipponensis*, in Suruga Bay, Japan. Fish Sci 66(5):812–825.

Hulbert LB, Sigler MF, Lunsford CR (2006) Depth and movement behaviour of the Pacific sleeper shark in the north-east Pacific Ocean. J Fish Biol 69:406–425.

ICES (2005) Report of the ICES Advisory Committee on Fishery Management, Advisory Committee on the Marine Environment and Advisory Committee on Ecosystems, 2005. Volume 1–11.

ICES (2006) Report of the Working Group on Elasmobranch Fishes (WGEF), 14–21 June, 2006, ICES Headquarters. ICES CM 2006/ACFM:31.

Iglésias SP, Du Buit M-H, Nakaya K (2002) Egg capsules of deep-sea catsharks from eastern North Atlantic, with first descriptions of the capsule of *Galeus murinus* and *Apristurus aphyodes* (Chondrichthyes: Scyliorhinidae). Cybium 26(1):59–63.

Irvine SB (2004) Age, growth and reproduction of deepwater dogfishes from southeastern Australia. PhD Thesis, Deakin University, Australia.

Irvine SB, Stevens JD, Laurenson LJB (2006a) Comparing external and internal dorsal-spine bands to interpret the age and growth of the giant lantern shark, *Etmopterus baxteri* (Squaliformes: Etmopteridae). Environ Biol Fishes 77:253–264.

Irvine SB, Stevens JD, Laurenson LJB (2006b) Surface bands on deepwater squalid dorsal-fin spines: an alternative method for ageing *Centroselachus crepidater*. Can J Fish Aquat Sci 63(3):617–627.

Ishiyama R (1958) Studies on the Rajid fishes (Rajidae) found in the waters around Japan. J Shimonoseki College Fish 7(2–3):193–394.

IUCN (2001) IUCN Red List Categories and Criteria: Version 3.1. IUCN Species Survival Commission. IUCN, Gland, Switzerland and Cambridge, UK.

IUCN (2008) 2008 IUCN Red List of Threatened Species. Available via http://www.redlist.org. Accessed December 15, 2008.

Jabobsdóttir KB (2001) Biological aspects of two deep-water squalid sharks: *Centroscyllium fabricii* (Reinhardt, 1825) and *Etmopterus princeps* (Collett, 1904) in Icelandic waters. Fish Res 51(2–3):247–265.

Johnson AG, Horton HF (1972) Length-weight relationship, food habits, parasites, and sex and age determination of the ratfish, *Hydrolagus colliei* (Lay and Bennett). Fish Bull 70(2):421–429.

Jones BC, Geen GH (1977) Reproduction and embryonic development of spiny dogfish (*Squalus acanthias*) in the Strait of Georgia, British Columbia. J Fish Res Board Can 34:1286–1292.

Jones E, Beare D, Dobby H et al. (2005) The potential impact of commercial fishing activity on the ecology of deepwater chondrichthyans from the west of Scotland. ICES CM 2005/N:16.

Kokuho T, Nakaya K, Kitagawa D (2003) Distribution and reproductive biology of the nine-spot ratfish *Hydrolagus barbouri* (Holocephali; Chimaeridae). Mem Grad School Fish Sci Hokkaido Univ 50(2):63–87.

Konstantinou H, McEachran JD, Woolley JB (2000) The systematics and reproductive biology of the *Galeus arae* subspecific complex (Chondrichthyes: Scyliorhinidae). Environ Biol Fishes 57(2):117–129.

Koslow JA, Boehlert GW, Gordon JDM et al. (2000) Continental slope and deep-sea fisheries: implications for a fragile ecosystem. ICES J Mar Sci 57:548–557.

Kyne PM, Yano K, White WT (2004) *Pseudotriakis microdon*. In: 2008 IUCN Red List of Threatened Species. IUCN. Available via http://www.iucnredlist.org. Accessed December 15, 2008.

Last PR (2007) The state of chondrichthyan taxonomy and systematics. Mar Freshw Res 58:7–9.

Last PR, Stevens JD (2009) Sharks and rays of Australia. CSIRO Publishing, Collingwood.

Licandeo R, Cerna FT (2007) Geographic variation in life-history traits of the endemic kite skate *Dipturus chilensis* (Batoidea: Rajidae), along its distribution in the fjords and channels of southern Chile. J Fish Biol 71:421–440.

Licandeo R, Cerna F, Céspedes R (2007) Age, growth, and reproduction of the roughskin skate, *Dipturus trachyderma*, from the southeastern Pacific. ICES J Mar Sci 64:141–148.

Licandeo RR, Lamilla JG, Rubilar PG et al. (2006) Age, growth, and sexual maturity of the yellownose skate *Dipturus chilensis* in the south-eastern Pacific. J Fish Biol 68(2):488–506.

Litvinov FF (1990) Ecological characteristics of the spiny dogfish, *Squalus mitsukurii* from the Nazca and Sala-y-Gomez submarine ridges. J Ichthyol 30:104–115.

Liu KM, Chiang P-J, Chen C-T (1998) Age and growth estimates of the bigeye thresher shark, *Alopias superciliosus*, in northeastern Taiwan waters. Fish Bull 96(3):482–491.

Lucifora LO, Menni RC, Escalante AH (2002) Reproductive ecology and abundance of the sand tiger shark, *Carcharias taurus*, from the southwestern Atlantic. ICES J Mar Sci 59:553–561.

Lucifora LO, Valero JL, García VB (1999) Length at maturity of the greeneye spurdog shark, *Squalus mitsukurii* (Elasmobranchii: Squalidae), from the SW Atlantic, with comparisons with other regions. Mar Freshw Res 50(7):629–632.

MacPherson E (1980) Food and feeding of *Chimaera monstrosa*, Linnaeus, 1758, in the western Mediterranean. J Cons Int Explor Mer 39(1):26–29.

Malagrino G, Takemura A, Mizue K (1981) Studies on Holocephali. 2. On the reproduction of *Chimaera phantasma* Jordan et Snyder caught in the coastal waters of Nagasaki. Bull Fac Fish Nagasaki Univ 51:1–7.

Matta ME, Gunderson DR (2007) Age, growth, maturity, and mortality of the Alaska skate, *Bathyraja parmifera*, in the eastern Bering Sea. Environ Biol Fishes 80:309–323.

McEachran JD, Dunn KA (1998) Phylogenetic analysis of skates, a morphologically conservative clade of elasmobranchs (Chondrichthyes: Rajidae). Copeia 1998:271–290.

McFarlane GA, King JR (2006) Age and growth of big skate (*Raja binoculata*) and longnose skate (*Raja rhina*) in British Columbia waters. Fish Res 78(2–3):169–178.

McFarlane GA, King JR, Saunders MW (2002) Preliminary study on the use of neural arches in the age determination of bluntnose sixgill sharks (*Hexanchus griseus*). Fish Bull 100(4):861–864.

McLaughlin DM, Morrissey JF (2005) Reproductive biology of *Centrophorus* cf. *uyato* from the Cayman Trench, Jamaica. J Mar Biol Assoc UK 85(5):1185–1192.

Megalofonou P, Chatzispyrou A (2006) Sexual maturity and feeding of the gulper shark, *Centrophorus granulosus*, from the eastern Mediterranean Sea. Cybium 30(4) (Suppl):67–74.

Megalofonou P, Damalas D (2004) Morphological and biological characteristics of a gravid angular rough shark (*Oxynotus centrina*) and its embryos from the Eastern Mediterranean Sea. Cybium 28(2):105–110.

Merrett NR, Haedrich RL (1997) Deep-sea demersal fish and fisheries. Chapman and Hall, London.

MOFA (1996) Basic fisheries statistics Jan–Dec 1996. Economic Planning and Coordination Section, Ministry of Fisheries and Agriculture, Malé.

MOFA (1997) Basic fisheries statistics Jan–Dec 1997. Economic Planning and Coordination Section, Ministry of Fisheries and Agriculture, Malé.

MOFAMR (2001) Basic fisheries statistics Jan–Dec 2001. Economic Planning and Coordination Section, Ministry of Fisheries, Agriculture and Marine Resources, Malé.

MOFAMR (2002) Basic fisheries statistics Jan–Dec 2002. Economic Planning and Coordination Section, Ministry of Fisheries, Agriculture and Marine Resources, Malé.

Morato T, Watson R, Pitcher TJ et al. (2006) Fishing down the deep. Fish Fish 7:24–34.

Moura T, Figueiredo I, Bordalo Machado P et al. (2004) Growth pattern and reproductive strategy of the holocephalan *Chimaera monstrosa* along the Portuguese continental slope. J Mar Biol Assoc UK 84(4):801–804.

Moura T, Figueiredo I, Bordalo-Machado P et al. (2005) Feeding habits of *Chimaera monstrosa* L. (Chimaeridae) in relation to its ontogenetic development on the southern Portuguese continental slope. Mar Biol Res 1(2):118–126.

Muñoz-Chápuli R, Perez Ortega A (1985) Resurrection of *Galeus atlanticus* (Vaillant, 1888), as a valid species from the NE-Atlantic Ocean and the Mediterranean Sea. Bull Mus Hist Natl Paris, 4e série, 7, section A, 1:219–233.

Musick JA (1999) Ecology and conservation of long-lived marine animals. In: Musick JA (ed) Life in the slow lane: ecology and conservation of long-lived marine animals. Am Fish Soc Symp 23, pp 1–10.

Musick JA, Ellis JK (2005) Reproductive evolution of chondrichthyans. In: Hamlett, WC (ed) Reproductive biology and phylogeny of Chondrichthyes. Sharks, batoids and chimaeras. Science Publishers, Inc., Enfield, pp 45–79.

Musick JA, Harbin MM, Compagno LJV (2004) Historical zoogeography of the Selachii. In: Carrier JC, Musick JA, Heithaus MR (eds) Biology of sharks and their relatives. CRC Press, Boca Raton, FL, pp 33–78.

Myagkov NA, Kondyurin VV (1978) Reproduction of the cat shark *Apristurus saldanha* (Barnard, 1925) (Scyliorhinidae, Lamniformes). Biol Morya 1978(2):86–87.

Nakaya K, Nakano H (1995) *Scymnodalatias albicauda* (Elasmobranchii, Squalidae) is a prolific shark. Jpn J Ichthyol 42(3–4):325–328.

Natanson LJ, Cailliet GM (1990) Vertebral growth zone deposition in angel sharks. Copeia 1990:1133–1145.

Natanson LJ, Sulikowski JA, Kneebone JR et al. (2007) Age and growth estimates for the smooth skate, *Malacoraja senta*, in the Gulf of Maine. Environ Biol Fishes 80: 293–308.

NATMIRC (2003) The Namibian Plan of Action for the Conservation and Management of Sharks (NPOA). NATMIRC, Swakopmund, Namibia.

Neiva J, Coelho R, Erzini K (2006) Feeding habits of the velvet belly lanternshark *Etmopterus spinax* (Chondrichthyes: Etmopteridae) off the Algarve, southern Portugal. J Mar Biol Assoc UK 86:835–841.

Nolan CP, Hogan F (2003) The reproductive biology of the longnose velvet dogfish, *Centroscymnus crepidater*, in the Northeast Atlantic. Poster Presentation. Deep Sea 2003: Conference on the Governance and Management of Deep-Sea Fisheries, Queenstown, New Zealand, December 1–5, 2003.

Parent S, Pépin S, Genet J-P et al. (2008) Captive breeding of the barndoor skate (*Dipturus laevis*) at the Montreal Biodome, with comparison notes on two other captive-bred skate species. Zoo Biol 27:145–153.

Priede IG, Froese R, Bailey DM et al. (2006) The absence of sharks from the abyssal regions of the world's oceans. Proc R Soc B 273:1435–1441.

Richardson AJ, Maharaj G, Compagno LJV et al. (2000) Abundance, distribution, morphometrics, reproduction and diet of the Izak catshark. J Fish Biol 56(3):552–576.

Ruocco NL, Lucifora LO, Díaz de Astarloa JM et al. (2006) Reproductive biology and abundance of the white-dotted skate, *Bathyraja albomaculata*, in the Southwest Atlantic. ICES J Mar Sci 63(1):105–116.

Saemundsson B (1922) Zoologiske Meddelelser fra Island. XIV. Vidensk Medd Dan Nat.hist Foren 74:159–201.

Sathyanesan AG (1966) Egg laying of the chimaeroid fish *Hydrolagus colliei*. Copeia 1966(1):132–134.

Serena F, Cecchi E, Mancusi C et al. (2006) Contribution to the knowledge of the biology of *Etmopterus spinax* (Linnaeus 1758) (Chondrichthyes, Etmopteridae). FAO Fish Proc 3(2):388–394.

Séret B (1987) *Halaelurus clevai*, sp. n., a new species of catshark (Scyliorhinidae) from off Madagascar, with remarks on the taxonomic status of the genera *Halaelurus* Gill and *Galeus* Rafinesque. Spec Publ JLB Smith Inst Ichthyol 44:1–27.

Sims DW, Southall EJ, Richardson AJ et al. (2003) Seasonal movements and behaviour of basking sharks from archival tagging: no evidence of winter hibernation. Mar Ecol Prog Ser 248:187–196.

Sion L, D'Onghia G, Carlucci R (2002) A simple technique for ageing the velvet belly, *Etmopterus spinax* (Squalidae). In: Vacchi M, La Mesa G, Serena F et al. (eds) Proceedings of the 4th European Elasmobranch Association Meeting, Livorno, Italy, pp 135–139.

Sion L, D'Onghia G, Tursi A et al. (2003) First data on distribution and biology of *Squalus blainvillei* (Risso, 1826) from the eastern Mediterranean Sea. J Northwest Atl Fish Sci 31:213–219.

Smith CR, Baco AR (2003) Ecology of whale falls at the deep-sea floor. Oceanogr Mar Biol Annu Rev 41:311–354.

Soundararajan R, Dam Roy S (2004) Distributional record and biological notes on two deep-sea sharks, *Centrophorus acus* Garman and *Squalus megalops* (Macleay), from Andaman waters. J Mar Biol Assoc India 46(2):178–184.

Springer S (1979) A revision of the catsharks, family Scyliorhinidae. NOAA Tech Rep NMFS Circ 422.

Stehmann MFW (1990) Rajidae. In: Quero J-C, Hureau J-C, Karrer C (eds) Check-list of the fishes of the eastern tropical Atlantic, Vol. 1. Junta Nacional de Investigaçao Cientifica e Tecnológica, Lisbon pp 29–50.

Stewart A (2000) False catshark: a real rarity. Seaf NZ 8(1):74–76.

Sulikowski JA, Kneebone J, Elzey S et al. (2005a) Age and growth estimates of the thorny skate (*Amblyraja radiata*) in the western Gulf of Maine. Fish Bull 103(1):161–168.

Sulikowski JA, Kneebone J, Elzey S et al. (2005b) The reproductive cycle of the thorny skate (*Amblyraja radiata*) in the western Gulf of Maine. Fish Bull 103(3):536–543.

Sulikowski JA, Kneebone J, Elzey S et al. (2006) Using the composite variables of reproductive morphology, histology and steroid hormones to determine age and size at sexual maturity for the thorny skate *Amblyraja radiata* in the western Gulf of Maine. J Fish Biol 69:1449–1465.

Sulikowski JA, Elzey S, Kneebone J et al. (2007) The reproductive cycle of the smooth skate, *Malacoraja senta*, in the Gulf of Maine. Mar Freshw Res 58(1):98–103.

Sullivan KJ (1977) Age and growth of the elephantfish *Callorhinchus milii* (Elasmobranchii: Callorhynchidae). NZ J Mar Freshw Res 11(4):745–753.

Sund O (1943) Et brugdebarsel. Naturen 67:285–286.

Tanaka S (1990) The structure of the dorsal spine of the deep sea squaloid shark *Centrophorus acus* and its utility for age determination. Bull Jpn Soc Sci Fish 56(6):903–909.

Tanaka S, Shiobara Y, Hioki S et al. (1990) The reproductive biology of the frilled shark, *Chlamydoselachus anguineus*, from Suruga Bay, Japan. Jpn J Ichthyol 37(3):273–291.

Taniuchi T, Tachikawa H (1999) Geographical variation in age and growth of *Squalus mitsukurii* (Elasmobranchii: Squalidae) in the North Pacific. In: Séret B, Sire J-Y (eds) Proceedings 5th Indo-Pacific Fish Conference, Noumea, New Caledonia, November 3–8, 1997. Société Française d'Ichtyologie, Paris, pp 321–328.

Taniuchi T, Kobayashi H, Otake T (1984) Occurrence and reproductive mode of the false cat shark, *Pseudotriakis microdon*, in Japan. Jpn J Ichthyol 31(1):88–92.

Taniuchi T, Tachikawa H, Shimizu M et al. (1993) Geographical variations in reproductive parameters of shortspine spurdog in the North Pacific. Bull Jpn Soc Sci Fish 59(1):45–51.

Templeman W (1984) Migrations of thorny skate, Raja radiata, tagged in the Newfoundland area. J Northwest Atl Fish Sci 5(1):55–63.

Vaillant L (1901) Sur un griset (*Hexanchus griseus* L. Gm.) du golfe de Gascogne. Bull Mus Hist Natl Paris 7:202–204.

Veríssimo A, Gordo L, Figueiredo I (2003) Reproductive biology and embryonic development of *Centroscymnus coelolepis* in Portuguese mainland waters. ICES J Mar Sci 60(6):1335–1341.

Walker P, Cavanagh RD, Ducrocq M et al. (2005) Northeast Atlantic (including Mediterranean and Black Sea). In: Fowler SL, Cavanagh, RD, Camhi M et al. (eds) Sharks, rays and chimaeras: the status of the chondrichthyan fishes. IUCN SSC Shark Specialist Group. IUCN, Gland, Switzerland and Cambridge, UK, pp 71–95.

Walmsley-Hart SA, Sauer WHH, Buxton CD (1999) The biology of the skates *Raja wallacei* and *R. pullopunctata* (Batoidea: Rajidae) on the Agulhas Bank, South Africa. S Afr J Mar Sci 21:165–179.

Watson G, Smale MJ (1998) Reproductive biology of shortnose spiny dogfish, *Squalus megalops*, from the Agulhas Bank, South Africa. Mar Freshw Res 49(7):695–703.

Watson G, Smale MJ (1999) Age and growth of the shortnose spiny dogfish *Squalus megalops* from the Agulhas Bank, South Africa. S Afr J Mar Sci 21:9–18.

Wetherbee BM (1996) Distribution and reproduction of the southern lantern shark from New Zealand. J Fish Biol 49(6):1186–1196.

Wetherbee BM, Cortés E (2004) Food consumption and feeding habits. In: Carrier JC, Musick JA, Heithaus MR (eds) Biology of sharks and their relatives. CRC Press, Boca Raton, FL, pp 225–246.

Wilson CD, Seki MP (1994) Biology and population characteristics of *Squalus mitsukurii* from a seamount in the central North Pacific Ocean. Fish Bull 92(4):851–864.

Yano K (1992) Comments on the reproductive mode of the false cat shark *Pseudotriakis microdon*. Copeia 1992(2):460–468.

Yano K (1993) Reproductive biology of the slender smoothhound, *Gollum attenuatus*, collected from New Zealand waters. Environ Biol Fishes 38(1–3):59–71.

Yano K (1995) Reproductive biology of the black dogfish, *Centroscyllium fabricii*, collected from waters off western Greenland. J Mar Biol Assoc UK 75(2):285–310.

Yano K, Tanaka S (1986) A telemetric study on the movements of the deep sea squaloid shark, *Centrophorus acus*. In: Uyeno T, Arai R, Taniuchi T et al. (eds) Indo-Pacific fish biology. Proceedings of the Second International Conference on Indo-Pacific Fishes. Ichthyological Society of Japan, Tokyo, pp 372–380.

Yano K, Tanaka S (1988) Size at maturity, reproductive cycle, fecundity and depth segregation of the deep sea squaloid sharks *Centroscymnus owstoni* and *C. coelolepis* in Suruga Bay, Japan. Bull Jpn Soc Sci Fish 54(2):167–174.

Yano K, Miya M, Aizawa M et al. (2007) Some aspects of the biology of the goblin shark, *Mitsukurina owstoni*, collected from the Tokyo Submarine Canyon and adjacent waters, Japan. Ichthyol Res 54:388–398.

Zeiner SJ, Wolf PG (1993) Growth characteristics and estimates of age at maturity of two species of skates (*Raja binoculata* and *Raja rhina*) from Monterey Bay, California. NOAA Tech Rep NMFS 115:87–90.

Appendix 2.1

Annotated Checklist of Extant Deepwater Chondrichthyans

The following checklist contains all known described deepwater species of the class Chondrichthyes. Deepwater chondrichthyans are defined as those sharks, rays, and holocephalans whose distribution is predominantly at, are restricted to, or spend the majority of their lifecycle at, depths below 200 m. This depth is generally recognized as the continental and insular shelf edge and, therefore, deepwater species are those occurring on or over the continental and insular slopes and beyond, including the abyssal plains and oceanic seamounts. Incorporated into this definition are benthic and epibenthic species, those occurring on, or associated with, the bottom of the ocean floor; and pelagic species, those occurring in the water column.

The checklist is based largely on Compagno and Duffy (2003), *A Checklist of Deep Water Chondrichthyans of the World*, and Compagno (2005), "Global Checklist of Living Chondrichthyan Fishes." The present checklist contains recent species descriptions, synonymy, and taxonomic changes and updates since preparation of the above-mentioned checklists, that is, up until early 2009. Primary literature (in the form of species descriptions and reviews) and consultation with researchers possessing expert knowledge of taxonomic groups were used to update the checklist.

Phylogeny and systematic arrangement of the checklist follows Compagno (2001, 2005) with the Holocephali here presented first, followed by the squalomorph sharks (incorporating the batoids) and finally the galeomorph sharks. The checklist contains only valid described species, and does not include undescribed species (new or recently identified species yet to be formally treated by science and as such do not yet possess a binomial scientific name).

Where a ? follows a generic name, the placement of this species in that particular genera is questionable and thus tentative. Further research may result in placement in another genus, evaluation of a subgenera to generic level, or designation of an altogether new genus. Where a ? follows a specific name, the validity of this species is questionable or the use of that specific name may be invalid.

Each species entry incorporates a summary of its known distribution, habitat zones, depth range, and, where relevant, taxonomic notes, including recent or commonly used synonyms.

Information on distribution, habitat, and depth was collated from Compagno and Duffy (2003), global and regional field guides, specifically Last and Stevens (2009), Carpenter and Niem (1998, 1999), Carpenter (2002), Ebert (2003), Compagno, Dando, and Fowler (2005), and White, Last, and Stevens et al. (2006), as well as the primary literature, gray literature sources, and consultation with experts.

Habitat zones broadly follow Compagno, Dando, and Fowler (2005). Benthic habitat zones are defined as follows:

Shelf: continental and insular shelves, 0 to 200 m depth; outer shelf is defined as >90 m

Slope: continental and insular slopes, 200 to ~2250 m depth, divided into upper slope (200 to 750 m), mid slope (750 to 1500 m), and deep slope (1500 to 2250 m)

Abyssal plains: deep oceanic plains at ~2000–2250 to 6000 m depth

Hadal: benthic zone of deepsea trenches at 6000 to 11000 m depth

Where available, more specific habitat information may be provided, including occurrences on submarine ridges and rises, troughs, plateaus, seamounts, and deepsea reefs and shoals.

Pelagic habitat zones are defined as follows:

Epipelagic: 0 to 200 m

Mesopelagic: 200 to 1000 m

Bathypelagic: 1000 to 6000 m

Hadopelagic: 6000 to 11000 m

No chondrichthyans have ever been observed or recorded in the hadal or hadopelagic zones (Compagno, Dando, and Fowler 2005; Priede et al. 2006). Priede et al. (2006) hypothesized that the high energy demands of chondrichthyans exclude them from the deepest habitat zones.

References

Carpenter KE (ed) (2002) The living marine resources of the Western Central Atlantic. Volume 1. Introduction, molluscs, crustaceans, hagfishes, sharks, batoid fishes and chimaeras. FAO Species Identification Guide for Fishery Purposes. FAO, Rome.

Carpenter KE, Niem VH (eds) (1998) The living marine resources of the Western Central Pacific. Volume 2. Cephalopods, crustaceans, holothurians and sharks. FAO Species Identification Guide for Fishery Purposes. FAO, Rome.

Carpenter KE, Niem VH (eds) (1999) The living marine resources of the Western Central Pacific. Volume 3. Batoid fishes, chimaeras and bony fishes, Part 1: (Elopidae to Linophrynidae). FAO Species Identification Guide for Fishery Purposes. FAO, Rome.

Compagno LJV (2001) Sharks of the world. An annotated and illustrated catalogue of shark species known to date. Volume 2. Bullhead, mackerel and carpet sharks (Heterodontiformes, Lamniformes and Orectolobiformes). FAO Species Catalogue for Fishery Purposes. Number 1, Volume 2. FAO, Rome.

Compagno LJV (2005) Global checklist of living chondrichthyan fishes. In: Fowler SL, Cavanagh RD, Camhi M et al. (eds) Sharks, rays and chimaeras: the status of chondrichthyan fishes. IUCN/SSC Shark Specialist Group. IUCN, Gland, Switzerland, and Cambridge, UK, pp 401–423.

Compagno LJV, Duffy CAJ (2003) A checklist of deep water chondrichthyans of the world. IUCN SSG Deepwater Shark Workshop, Portobello, New Zealand, November 29, 2003.

Compagno L, Dando M, Fowler S (2005) A field guide to sharks of the world. Collins, London.

Ebert DA (2003) Sharks, rays and chimaeras of California. University of California Press, Berkeley.

Last PR, Stevens JD (2009) Sharks and rays of Australia. CSIRO Publishing, Collingwood.

Priede IG, Froese R, Bailey DM et al (2006) The absence of sharks from abyssal regions of the world's oceans. Proc R Soc B 273(1592):1435–1441.

White WT, Last PR, Stevens JD et al. (2006) Economically important sharks and rays of Indonesia. Australian Centre for International Agricultural Research, Canberra.

Class Chondrichthyes

Subclass Holocephali.

Order Chimaeriformes. Modern Chimaeras.

Family Rhinochimaeridae. Longnose Chimaeras.

Harriotta haeckeli Karrer, 1972. Smallspine spookfish

Patchy in the Southern Ocean and Atlantic. Abyssal plains and deepsea troughs. 1114–2603 m.

Harriotta raleighana Goode & Bean, 1895. Narrownose or longnose chimaera, bent-nose rabbitfish or bigspine spookfish

Wide-ranging but patchy in the Indian, Pacific, and Atlantic. Upper to deep slope, abyssal plains, and seamounts. 380–2600 m.

Neoharriotta carri Bullis & Carpenter, 1966. Dwarf sicklefin chimaera

Western Central Pacific: southern Caribbean. Upper slope. 240–600 m.

Neoharriotta pinnata (Schnakenbeck, 1931). Sicklefin chimaera

Eastern Central and Southeast Atlantic: West Africa. Outermost shelf and upper slope. 200–470 m.

Neoharriotta pumila Didier & Stehmann, 1996. Arabian sicklefin chimaera

Western Indian: Arabian Sea and Gulf of Aden. Outer shelf and upper to mid slope. 100–1120 m.

Rhinochimaera africana Compagno, Stehmann & Ebert, 1990. Paddlenose chimaera or spookfish

Patchy in the Indo-West Pacific. Upper to mid slope and seamounts. 550–1450 m.

Rhinochimaera atlantica Holt & Byrne, 1909. Spearnose chimaera or straightnose rabbitfish

Wide-ranging in the Atlantic. Upper to mid slope. 500–1500 m.

Rhinochimaera pacifica (Mitsukuri, 1895). Pacific spookfish or knifenose chimaera

Patchy in the Indo-West Pacific and the Southeast Pacific. Outermost shelf, upper to mid slope, deepsea troughs, deepsea plateaus, and seamounts. 191–1290 m (mostly >700 m).

Family Chimaeridae. Shortnose Chimaeras.

Chimaera argiloba Last, White & Pogonoski, 2008. Whitefin chimaera

Eastern Indian: western Australia. Upper slope. 370–520 m.

Chimaera cubana Howell-Rivero, 1936. Cuban chimaera

Western Central Atlantic: Caribbean. Upper slope. 234–450 m.

Chimaera fulva Didier, Last & White, 2008. Southern chimaera

Eastern Indian and Southwest Pacific: southern Australia. Mid slope. 780–1095 m.

Chimaera jordani Tanaka, 1905. Jordan's chimaera

Confirmed from the Northwest Pacific (Japan) and Western Indian, but probably more wide-ranging. Upper to deep slope. 383–1600 m.

Chimaera lignaria Didier, 2002. Giant, purple or carpenter's chimaera

Primarily Southwest Pacific: southern Australia and New Zealand. Upper to deep slope, deepsea plateaus, and seamounts. 400–1800 m (mostly >800 m).

Chimaera macrospina Didier, Last & White, 2008. Longspine chimaera

Eastern Indian and Western Pacific: Australia. Upper to mid slope. 435–1300 m (mostly >800 m).

Chimaera monstrosa Linnaeus, 1758. Rabbitfish

Wide-ranging in the Northeast Atlantic including the Mediterranean. Shelf and upper to mid slope. 50–1000 m (mostly 300–500 m).

Chimaera obscura Didier, Last & White, 2008. Shortspine chimaera

Western Central and Southwest Pacific: eastern Australia. Upper to mid slope. 450–1000 m.

Chimaera owstoni Tanaka, 1905. Owston's chimaera

Northwest Pacific: Japan. Upper to mid slope. 500–1200 m.

Chimaera panthera Didier, 1998. Leopard or roundfin chimaera

Southwest Pacific: New Zealand. Upper to mid slope, deepsea rises, submarine ridges, and seamounts. 327–1020 m.

Chimaera phantasma Jordan & Snyder, 1900. Silver chimaera

Northwest Pacific: Japan and Taiwan. Shelf and upper to mid slope. 20–962 m. Includes the junior synonym *Chimaera pseudomonstrosa* Fang & Wang, 1932.

Hydrolagus affinis (Capello, 1867). Atlantic chimaera or smalleyed rabbitfish

Wide-ranging in the North Atlantic. Upper to deep slope, abyssal plains, and seamounts. 300–2909 m (mostly >1000 m).

Hydrolagus africanus (Gilchrist, 1922). African chimaera

Western Indian and probably Southeast Atlantic: southern Africa. Upper to mid slope. 300–1300 m (mostly 421–750 m).

Hydrolagus alberti Bigelow & Schroeder, 1951. Gulf chimaera

Western Central Atlantic: Caribbean and the Gulf of Mexico. Upper to mid slope. 348–1100 m.

Hydrolagus alphus Quaranta, Didier, Long & Ebert, 2006. Whitespot ghostshark

Southeast Pacific: Galapagos Islands. Upper to mid slope. 600–900 m.

Hydrolagus barbouri (Garman, 1908). Ninespot chimaera

Northwest Pacific: Japan. Upper to mid slope. 250–1100 m (most common 600–800 m).

Hydrolagus bemisi Didier, 2002. Pale ghostshark

Southwest Pacific: New Zealand. Upper to mid slope, deepsea plateaus, and rises. 400–1100 m (mostly 500–700 m).

Hydrolagus colliei (Lay & Bennett, 1839). Spotted ratfish

Northeast and Eastern Central Pacific: Alaska to Mexico. Shelf and upper to mid slope. 0–971 m.

Hydrolagus homonycteris Didier, 2008. Black ghostshark

Primarily Southwest Pacific: southern Australia and New Zealand. Mid slope and seamounts. 866–1447 m.

Hydrolagus lemures (Whitley, 1939). Blackfin ghostshark

Eastern Indian and Western Central and Southwest Pacific: Australia. Outer shelf and upper slope. 146–700 m.

Hydrolagus lusitanicus Moura, Figueiredo, Bordalo-Machado, Almeida & Gordo, 2005.

Northeast Atlantic: Portugal. Deep slope. 1600 m.

Hydrolagus macrophthalmus de Buen, 1959. Bigeye chimaera

Southeast Pacific: Chile. Habitat data not available.

Hydrolagus marmoratus Didier, 2008. Marbled ghostshark

Western Central and Southwest Pacific: eastern Australia. Upper to mid slope. 548–995 m.

Hydrolagus matallanasi Soto & Vooren, 2004. Striped Rabbitfish

Southwest Atlantic: southern Brazil. Upper slope. 416–736 m.

Hydrolagus mccoskeri Barnett, Didier, Long & Ebert, 2006. Galapagos ghostshark

Southeast Pacific: Galapagos Islands. Upper slope. 396–506 m.

Hydrolagus mirabilis (Collett, 1904). Large-eyed rabbitfish or spectral chimaera

Wide-ranging in the Eastern Atlantic and also in the Western Central Atlantic. Upper to deep slope. 450–1933 m (mostly >800 m).

Hydrolagus mitsukurii (Dean, 1904). Mitsukurii's chimaera

Northwest and Western Central Pacific: Japan to Philippines. Upper to mid slope. 325–770 m. Includes the junior synonym *Hydrolagus deani* (Smith & Radcliffe, 1912).

Hydrolagus novaezealandiae (Fowler, 1910). Dark or New Zealand ghostshark

Southwest Pacific: New Zealand. Shelf and upper to mid slope. 25–950 m (most common 150–500 m).

Hydrolagus ogilbyi (Waite, 1898). Ogilby's ghostshark

Eastern Indian and Southwest Pacific: southeastern Australia. Outer shelf and upper slope. 120–350 m.

Hydrolagus pallidus Hardy & Stehmann, 1990. Pale chimaera

Scattered records in the Northeast Atlantic. Mid to deep slope and deepsea troughs. 1200–2650 m.

Hydrolagus purpurescens (Gilbert, 1905). Purple chimaera

Northwest Pacific (Japan) and Eastern Central Pacific (Hawaii). Mid to deep slope, deepsea troughs, and seamounts. 920–1130 m (Japan), 1750–1951 m (Hawaii). Includes the probable junior synonym *Hydrolagus eidolon* (Jordan & Hubbs, 1925).

Hydrolagus trolli Didier & Séret, 2002. Pointy-nosed blue chimaera

Patchy in the Southwest and Western Central Pacific. Upper to mid slope and seamounts. 610–2000 m (mostly >1000 m).

Subclass Elasmobranchii.

Superorder Squalomorphii. Squalomorph Sharks.
  Order Hexanchiformes. Cow and Frilled Sharks.

Family Chlamydoselachidae. Frilled Sharks.
  Chlamydoselachus anguineus Garman, 1884. Frilled shark

Wide-ranging but patchy in temperate and tropical waters of the Atlantic and Indo-Pacific. Shelf and upper to mid slope. 50–1500 m.

Family Hexanchidae. Sixgill and Sevengill Sharks.

Heptranchias perlo (Bonnaterre, 1788). Sharpnose sevengill shark or perlon

Wide-ranging but patchy in temperate and tropical waters of the Atlantic and Indo-Pacific. Shelf (occassional) and upper to mid slope. 27–1000 m.

Hexanchus griseus (Bonnaterre, 1788). Bluntnose sixgill shark

Wide-ranging but patchy in temperate and tropical waters of the Atlantic and Indo-Pacific. Shelf, upper to deep slope, submarine ridges, and sea-mounts. Mainly deepwater, young inshore in cold water. 0–2490 m (500–1100 m usual).

Hexanchus nakamurai Teng, 1962. Bigeye sixgill shark

Wide-ranging but patchy in warm-temperate and tropical waters of the Atlantic and Indo-Pacific. Shelf (occasional) and upper slope. 90–621 m.

Order Squaliformes. Dogfish Sharks.

Family Echinorhinidae. Bramble Sharks.

Echinorhinus brucus (Bonnaterre, 1788). Bramble shark

Wide-ranging but patchy in the Atlantic and Indo-Pacific. Shelf (occasional) and upper to mid slope. 18–900 m.

Echinorhinus cookei (Pietschmann, 1928). Prickly shark

Patchy in the Eastern, Central, and Western Pacific. Shelf and upper to mid slope and seamounts. 11–1100 m.

Family Squalidae. Dogfish Sharks.

Cirrhigaleus asper (Merrett, 1973). Roughskin spurdog

Wide-ranging but patchy in warm-temperate and tropical waters of the Atlantic and Indo-Pacific. Shelf and upper slope. 73–600 m.

Cirrhigaleus australis White, Last & Stevens, 2007. Southern mandarin dogfish

Eastern Indian and Southwest Pacific: southern Australia and New Zealand. Upper slope. 360–640 m.

Cirrhigaleus barbifer Tanaka, 1912. Mandarin dogfish

Patchy in the Western Pacific. Outer shelf and upper slope. 146–640 m. May represent a species complex.

Squalus albifrons Last, White & Stevens, 2007. Eastern highfin spurdog

Western Central and Southwest Pacific: eastern Australia. Outer shelf and upper slope. 131–450 m.

Squalus altipinnis Last, White & Stevens, 2007. Western highfin spurdog

Eastern Indian: western Australia. Upper slope. ~300 m.

Squalus blainville (Risso, 1826). Longnose spurdog

Nominally from the Eastern Atlantic and the Mediterranean, but not well defined due to confusion with other species and taxonomic issues. Shelf and upper slope. 16–>440 m. Considerable taxonomic issues (records from elsewhere in the Atlantic and the Indo-Pacific based in part on *S. mitsukurii* or close relatives).

Squalus brevirostris Tanaka, 1917. Japanese shortnose spurdog

Northwest Pacific: Japan and Taiwan. Habitat information not available. Tentatively placed on checklist as distinct from *S. megalops*.

Squalus bucephalus Last, Séret & Pogonoski, 2007. Bighead spurdog

Western Pacific: Norfolk Ridge and New Caledonia. Upper to mid slope and submarine ridges. 448–880 m.

Squalus chloroculus Last, White & Motomura, 2007. Greeneye spurdog

Southwest Pacific and Eastern Indian: southern Australia. Upper to mid slope. 216–1360 m.

Squalus crassispinus Last, Edmunds & Yearsley, 2007. Fatspine spurdog

Eastern Indian: northwestern Australia. Outermost shelf and upper slope. 187–262 m.

Squalus cubensis Howell-Rivero, 1936. Cuban dogfish

Warm-temperate waters of the Western Atlantic. Shelf and upper slope. 60–380 m.

Squalus edmundsi White, Last & Stevens, 2007. Edmund's spurdog

Eastern Indian: western Australia and Indonesia. Upper to mid slope. 204–850 m (mostly 300–500 m).

Squalus grahami White, Last & Stevens, 2007. Eastern longnose spurdog

Western Central and Southwest Pacific: eastern Australia. Outer shelf and upper slope. 148–504 m (mainly 220–450 m).

Squalus griffini Phillipps, 1931. Northern spiny dogfish

Southwest Pacific: New Zealand. Shelf and upper slope. 37–616 m.

Squalus hemipinnis White, Last & Yearsley, 2007. Indonesian shortsnout spurdog

Eastern Indian: Indonesia. Upper slope.

Squalus japonicus Ishikawa, 1908. Japanese spurdog

Northwest and Western Central Pacific: East Asia. Outer shelf and upper slope. 120–340 m.

Squalus lalannei Baranes, 2003. Seychelles spurdog

Western Indian: Seychelles. Mid slope. 1000 m.

Squalus megalops (Macleay, 1881). Shortnose spurdog

Widespread in temperate and warm-temperate waters of the Indo-West Pacific and Eastern Atlantic. Shelf and upper slope. 0–732 m. May represent a species complex.

Squalus melanurus Fourmanoir, 1979. Blacktail spurdog

Western Central Pacific: New Caledonia. Shelf and upper slope. 34–480 m.

Squalus mitsukurii Jordan & Snyder, in Jordan & Fowler, 1903. Shortspine spurdog

Wide-ranging, but patchy in temperate and tropical waters of the Atlantic and Indo-Pacific, but not well defined due to taxonomic issues. Shelf and upper to mid slope, submarine ridges, and seamounts. 4–954 m (100–500 m usual). Resolution of considerable taxonomic issues is ongoing. Tentatively includes the junior synonym *Squalus probatovi* Myagkov & Kondyurin, 1986.

Squalus montalbani Whitley, 1931. Philippine spurdog

Patchy in the Eastern Indian and Western Pacific. Upper slope. 295–670 m.

Squalus nasutus Last, Marshall & White, 2007. Western longnose spurdog

Patchy in the Eastern Indian and Western Pacific. Upper to mid slope. 300–850 m (mainly 300–400 m).

Squalus notocaudatus Last, White & Stevens, 2007. Bartail spurdog

Western Central Pacific: northeastern Australia. Upper slope. 225–454 m.

Squalus rancureli Fourmanoir, 1978. Cyrano spurdog

Western Central Pacific: Vanuatu. Upper slope. 320–400 m.

Squalus raoulensis Duffy & Last, 2007. Kermadec spiny dogfish

Southwest Pacific: Kermadec Ridge. Submarine ridges. 320 m.

Family Centrophoridae. Gulper Sharks.

Centrophorus acus Garman, 1906. Needle dogfish

Patchy in the Eastern Indian and Western Pacific. Outer shelf and upper to mid slope. 150–950 m (mostly >200 m, possibly to 1786 m).

Centrophorus atromarginatus Garman, 1913. Dwarf gulper shark

Patchy in the Indo-West Pacific, but not well defined. Outer shelf and upper slope. 150–450 m.

Centrophorus granulosus (Bloch & Schneider, 1801). Gulper shark

Wide-ranging but patchy in the Atlantic and Indo-Pacific, but some records may represent additional species. Shelf and upper to mid slope. 50–1440 m (mostly 200–600 m). Resolution of taxonomic issues ongoing and as such occurrence and distribution not well defined.

Centrophorus harrissoni McCulloch, 1915. Longnose gulper shark

Patchy in the Western Central and Southwest Pacific. Upper to mid slope and submarine ridges. 220–1050 m.

Centrophorus isodon (Chu, Meng & Liu, 1981). Blackfin gulper shark

Patchy in the Indo-West Pacific. Mid slope. 760–770 m.

Centrophorus lusitanicus Bocage & Capello, 1864. Lowfin gulper shark

Patchy in the Eastern Atlantic and Indo-West Pacific. Upper to mid slope. 300–1400 m (mostly 300–600 m). Indo-West Pacific form may represent a separate species (*Centrophorus* cf. *lusitanicus* in White et al., 2006).

Centrophorus moluccensis Bleeker, 1860. Smallfin gulper shark

Patchy in the Indo-West Pacific. Outer shelf and upper to mid slope. 125–820 m. Resolution of taxonomic issues ongoing. May represent a species complex.

Centrophorus niaukang Teng, 1959. Taiwan gulper shark

Patchy in the Eastern Indian and Western Pacific. Outer shelf and upper to mid slope. 98–~1000 m.

Centrophorus seychellorum Baranes, 2003. Seychelles gulper shark

Western Indian: Seychelles. Mid slope. 1000 m.

Centrophorus squamosus (Bonnaterre, 1788). Leafscale gulper shark

Wide-ranging, but patchy in the Eastern Atlantic and Indo-West Pacific. Upper to deep slope and abyssal plains. 230–3280 m.

Centrophorus tessellatus Garman, 1906. Mosaic gulper shark

Scattered locations in the Western Atlantic and Pacific, but some records provisional. Upper slope. 260–730 m. Taxonomic issues mean that occurrence and distribution are not well defined and as such many records are provisional.

Centrophorus westraliensis White, Ebert & Compagno, 2008. Western gulper shark

Eastern Indian: western Australia. Upper to mid slope. 616–750 m.

Centrophorus zeehaani White, Ebert & Compagno, 2008. Southern dogfish

Eastern Indian and Southwest Pacific: southern Australia. Upper slope. 208–701 m (usually >400 m).

Deania calcea (Lowe, 1839). Birdbeak or shovelnose dogfish

Wide-ranging in the Eastern Atlantic and Indo-West Pacific. Shelf and upper to mid slope. 70–1470 m (usually 400–900 m). Indonesian form may represent a distinct species (*Deania* cf. *calcea* in White, Last & Stevens 2006).

Deania hystricosum (Garman, 1906). Rough longnose dogfish

Scattered locations in the Eastern Atlantic and Western Pacific. Upper to mid slope. 470–1300 m.

Deania profundorum (Smith & Radcliffe, 1912). Arrowhead dogfish

Scattered locations in the Atlantic and Indo-West Pacific. Upper to deep slope. 275–1785 m.

Deania quadrispinosum (McCulloch, 1915). Longsnout dogfish

Southern regions of the Eastern Atlantic and Indo-West Pacific. Outer shelf and upper to mid slope. 150–1360 m (usually <400 m).

Family Etmopteridae. Lanternsharks.

Aculeola nigra de Buen, 1959. Hooktooth dogfish

Southeast Pacific: Peru to Chile. Outer shelf and upper slope. 110–735 m.

Centroscyllium excelsum Shirai & Nakaya, 1990. Highfin dogfish

Northwest Pacific: Emperor Seamount Chain. Seamounts. 800–1000 m.

Centroscyllium fabricii (Reinhardt, 1825). Black dogfish

Wide-ranging in temperate waters of the Atlantic. Outer shelf and upper to deep slope. 180–1600 m (usually >275 m, probably to 2250 m).

Centroscyllium granulatum Günther, 1887. Granular dogfish

Southeast Pacific: Chile. Upper slope. 300–500 m.

Centroscyllium kamoharai Abe, 1966. Bareskin dogfish

Indo-West Pacific: Japan, Australia, and possibly the Philippines. Upper to mid slope. 500–1200 m (mostly >900 m).

Centroscyllium nigrum Garman, 1899. Combtooth dogfish

Patchy in the Central and Eastern Pacific. Upper to mid slope. 400–1143 m.

Centroscyllium ornatum (Alcock, 1889). Ornate dogfish

Northern Indian: Arabian Sea and the Bay of Bengal. Upper to mid slope. 521–1262 m.

Centroscyllium ritteri Jordan & Fowler, 1903. Whitefin dogfish

Northwest Pacific: Japan. Upper to mid slope and seamounts. 320–1100 m.

Etmopterus baxteri Garrick, 1957. New Zealand lanternshark

Southern regions of the Eastern Atlantic and Indo-West Pacific. Upper to mid slope. 250–1500 m. Occurrence and distribution not well defined in the Southeast Atlantic and Western Indian.

Etmopterus bigelowi Shirai & Tachikawa, 1993. Blurred smooth lanternshark

Wide-ranging but patchy in the Atlantic and Indo-Pacific. Outer shelf, upper to mid slope, submarine ridges, and seamounts. 163–1000 m.

Etmopterus brachyurus Smith & Radcliffe, 1912. Shorttail lanternshark

Scattered in the Indo-West Pacific: Japan, the Philippines, and Australia. Upper slope. 400–610 m. References to the species off southern Africa refer to the as yet undescribed sculpted lanternshark.

Etmopterus bullisi Bigelow & Schroeder, 1957. Lined lanternshark

Western Central Atlantic: Caribbean. Upper to mid slope. 275–824 m (mostly >350 m).

Etmopterus burgessi Schaaf-da Silva & Ebert, 2006. Broad-snout lanternshark

Northwest Pacific: Taiwan. Slope. >300 m.

Etmopterus carteri Springer & Burgess, 1985. Cylindrical lanternshark

Western Central Atlantic: Caribbean coast of Colombia. Upper slope. 283–356 m.

Etmopterus caudistigmus Last, Burgess & Séret, 2002. New Caledonia tailspot lanternshark

Western Central Pacific: New Caledonia. Upper to mid slope. 638–793 m.

Etmopterus decacuspidatus Chan, 1966. Combtooth lanternshark

Northwest Pacific: South China Sea. Upper slope. 512–692 m.

Etmopterus dianthus Last, Burgess & Séret, 2002. Pink lanternshark

Western Central Pacific: Australia and New Caledonia. Upper to mid slope. 700–880 m.

Etmopterus dislineatus Last, Burgess & Séret, 2002. Lined lanternshark

Western Central Pacific: Australia. Upper slope. 590–700 m.

Etmopterus evansi Last, Burgess & Séret, 2002. Blackmouth lanternshark

Eastern Indian: Australia and Arafura Sea. Shoals and reefs on the upper slope. 430–550 m.

Etmopterus fusus Last, Burgess & Séret, 2002. Pygmy lanternshark

Eastern Indian: Australia. Upper slope. 430–550 m.

Etmopterus gracilispinis Krefft, 1968. Broadband lanternshark

Wide-ranging but patchy in the Western Atlantic and off southern Africa. Outer shelf and upper to mid slope. 70–1000 m.

Etmopterus granulosus (Günther, 1880). Southern lanternshark

Southwest Atlantic and Southeast Pacific: southern South America. Upper slope. 220–637 m.

Etmopterus hillianus (Poey, 1861). Caribbean lanternshark

Northwest and Western Central Atlantic including the Caribbean. Upper slope. 311–695 m.

Etmopterus litvinovi Parin & Kotlyar, in Kotlyar, 1990. Smalleye lanternshark

Southeast Pacific: Nazca and Sala y Gomez Submarine Ridges. Submarine ridges. 630–1100 m.

Etmopterus lucifer Jordan & Snyder, 1902. Blackbelly lanternshark

Patchy in the Western Pacific, but provisional records from elsewhere require confirmation. Outer shelf, upper to mid slope. 158–1357 m.

Etmopterus molleri (Whitley, 1939). Slendertail lanternshark

Patchy in the Western Pacific and possibly also the Western Indian (Mozambique). Upper slope. 238–655 m. Includes the probable junior synonym *Etmopterus schmidti* Dolganov, 1986.

Etmopterus perryi Springer & Burgess, 1985. Dwarf lanternshark

Western Central Atlantic: Caribbean off Colombia. Upper slope. 283–375 m.

Etmopterus polli Bigelow, Schroeder & Springer, 1953. African lanternshark

Eastern Atlantic: West Africa. Possibly Caribbean off Venezuela. Upper to mid slope. 300–1000 m.

Etmopterus princeps Collett, 1904. Great lanternshark

Wide-ranging in the North and Central Atlantic. Upper to deep slope, deepsea rises, deepsea plateaus, and abyssal plains. 567–4500 m.

Etmopterus pseudosqualiolus Last, Burgess & Séret, 2002. False pygmy lanternshark

Western Central Pacific: Norfolk and Lord Howe Ridges. Submarine ridges. 1043–1102 m.

Etmopterus pusillus (Lowe, 1839). Smooth lanternshark

Wide-ranging in temperate and warm-temperate waters of the Atlantic and Indo-West and Central Pacific. Upper to mid (possibly to deep) slope. 274–1000 m (possibly to 1998 m).

Etmopterus pycnolepis Kotlyar, 1990. Densescale lanternshark

Southeast Pacific: Nazca and Sala y Gomez Submarine Ridges. Submarine ridges. 330–763 m.

Etmopterus robinsi Schofield & Burgess, 1997. West Indian laternshark

Western Central Atlantic: Caribbean. Upper to mid slope. 412–787 m.

Etmopterus schultzi Bigelow, Schroeder & Springer, 1953. Fringefin lanternshark

Western Central Atlantic: Gulf of Mexico. Upper to mid slope. 220–915 m (mostly >350 m).

Etmopterus sentosus Bass, D'Aubrey & Kistnasamy, 1976. Thorny lanternshark

Western Indian: East Africa. Upper slope. 200–500 m.

Etmopterus spinax (Linnaeus, 1758). Velvet belly

Widespread in the Eastern Atlantic and Mediterranean. Shelf, upper to deep slope, and deepsea rises. 70–2490 m (mostly 200–500 m).

Etmopterus splendidus Yano, 1988. Splendid lanternshark

Northwest Pacific: Japan and Taiwan. Uppermost slope. 210 m.

Etmopterus unicolor (Engelhardt, 1912). Brown lanternshark

Patchy in the Eastern Atlantic and Indo-West Pacific. Upper to mid slope and seamounts. 750–1500 m.

Etmopterus villosus Gilbert, 1905. Hawaiian lanternshark

Eastern Central Pacific: Hawaii. Upper to mid slope. 406–911 m.

Etmopterus virens Bigelow, Schroeder & Springer, 1953. Green lanternshark

Western Central Atlantic: Gulf of Mexico and Caribbean. Upper to mid slope. 196–915 m (mostly >350 m).

Miroscyllium sheikoi (Dolganov, 1986). Rasptooth dogfish

Northwest Pacific: off Japan. Upper slope of submarine ridges. 340–370 m.

Trigonognathus kabeyai Mochizuki & Ohe, 1990. Viper dogfish

North and Central Pacific: Japan and Hawaii. Upper slope and seamounts. 270–360 m.

Family Somniosidae. Sleeper Sharks.

Centroscymnus coelolepis Bocage & Capello, 1864. Portuguese dogfish

Wide-ranging in the Atlantic and Indo-Pacific. Outer shelf, upper to deep slope and abyssal plains. 128–3675 m (mostly >400 m).

Centroscymnus owstoni Garman, 1906. Roughskin dogfish

Wide-ranging, but patchy in the Atlantic, Pacific, and Eastern Indian. Upper to mid slope and submarine ridges. 426–1459 m (mostly >600 m). Includes the junior synonym *Centroscymnus cryptacanthus* Regan, 1906.

Centroselachus crepidater (Bocage & Capello, 1864). Longnose velvet dogfish

Wide-ranging, but patchy in the Western Atlantic, Pacific, and Western Indian. Upper to deep slope. 270–2080 m (mostly >500 m).

Proscymnodon macracanthus (Regan, 1906). Largespine velvet dogfish

Southeast Pacific: Straits of Magellan (Chile). Habitat and depth unrecorded.

Proscymnodon plunketi (Waite, 1909). Plunket shark

Patchy in southern regions of the Indo-West Pacific. Upper to mid slope. 219–1427 m (most common 550–732 m).

Scymnodalatias albicauda Taniuchi & Garrick, 1986. Whitetail dogfish

Patchy in the Southern Ocean. Outer shelf, upper slope, and submarine ridges. 150–500 m.

Scymnodalatias garricki Kukuyev & Konovalenko, 1988. Azores dogfish

Northeast Atlantic: North Atlantic Ridge. Mesopelagic over seamounts. 300 m.

Scymnodalatias sherwoodi (Archey, 1921). Sherwood's dogfish

Patchy in the Southern Ocean. Upper slope. 400–500 m.

Scymnodon ringens Bocage & Capello, 1864. Knifetooth dogfish

Northeast and Eastern Central Atlantic. Uncertain from New Zealand in the Southwest Pacific. Upper to deep slope. 200–1600 m.

Somniosus antarcticus Whitley, 1939. Southern sleeper shark.

Patchy in the Southern Ocean. Outer shelf and upper to mid slope. 145–1200 m.

Somniosus longus (Tanaka, 1912). Frog shark

Western Pacific: Japan and New Zealand. Upper to mid slope. 250–1160 m.

Somniosus microcephalus (Bloch & Schneider, 1801). Greenland shark

Cool temperate and boreal waters of the North Atlantic. Shelf (inshore in Arctic winter) and upper to mid slope. 0–1200 m.

Somniosus pacificus Bigelow & Schroeder, 1944. Pacific sleeper shark

Wide-ranging in the North Pacific. Shelf and upper to deep slope (shallower in north, deeper in south of range). 0–2000 m.

Somniosus rostratus (Risso, 1810). Little sleeper shark

Northeast Atlantic and the Mediterranean. Uncertain from Cuba in the Western Central Atlantic. Outermost shelf and upper to mid slope. 200–2200 m.

Zameus ichiharai (Yano & Tanaka, 1984). Japanese velvet dogfish

Northwest Pacific: Japan. Upper to mid slope. 450–830 m.

Zameus squamulosus (Günther, 1877). Velvet dogfish

Wide-ranging but patchy in the Atlantic, Indian, and Pacific. Upper to mid slope. Also epipelagic and mesopelagic. 550–1450 m (when benthic); 0–580 m in water up to 6000 m deep (when pelagic).

Family Oxynotidae. Roughsharks.

Oxynotus bruniensis (Ogilby, 1893). Prickly dogfish

Indo-West Pacific: New Zealand and Australia. Shelf and upper to mid slope. 45–1067 m (most common 350–650 m).

Oxynotus caribbaeus Cervigon, 1961. Caribbean roughshark

Western Central Atlantic: Gulf of Mexico and Caribbean (Venezuela). Upper slope. 402–457 m.

Oxynotus centrina (Linnaeus, 1758). Angular roughshark

Wide-ranging in the Eastern Atlantic and the Mediterranean. Shelf and upper slope. 50–660 m (mostly >100 m).

Oxynotus japonicus Yano & Murofushi, 1985. Japanese roughshark

Northwest Pacific: Japan. Uppermost slope. 225–270 m.

Oxynotus paradoxus Frade, 1929. Sailfin roughshark

Northeast and Eastern Central Atlantic: Scotland to West Africa. Upper slope. 265–720 m.

Family Dalatiidae. Kitefin Sharks.

Dalatias licha (Bonnaterre, 1788). Kitefin shark

Wide-ranging but patchy in the Atlantic, Indo-West, and Central Pacific. Shelf and upper to deep slope. 37–1800 m (mainly >200 m).

Euprotomicroides zantedeschia Hulley & Penrith, 1966. Taillight shark

Patchy in the South Atlantic: off Brazil and southern Africa. Upper slope. Also epipelagic. 458–641 m (when benthic), 0–25 m (when pelagic).

Euprotomicrus bispinatus (Quoy & Gaimard, 1824). Pygmy shark

Wide-ranging but scattered records in the Atlantic and Indo-Pacific. Epipelagic, mesopelagic, and bathypelagic. 0–>1500 m in water up to 9938 m deep.

Heteroscymnoides marleyi Fowler, 1934. Longnose pygmy shark

Scattered records in the southern Atlantic, Southeast Pacific, and Southwest Indian. Epipelagic and mesopelagic. 0–502 m in water 830–>4000 m deep.

Isistius brasiliensis (Quoy & Gaimard, 1824). Cookiecutter or cigar shark

Circumglobal in tropical and warm temperate waters. Epipelagic, mesopelagic, and bathypelagic. 0–3500 m.

Isistius labialis Meng, Chu & Li, 1985. South China cookiecutter shark

Northwest Pacific: South China Sea. Mesopelagic over slope. 520 m.

Mollisquama parini Dolganov, 1984. Pocket shark

Southeast Pacific: Nazca Submarine Ridge. Submarine ridges. 330 m.

Squaliolus aliae Teng, 1959. Smalleye pygmy shark

Patchy in the Western Pacific and Eastern Indian. Epipelagic, mesopelagic, and bathypelagic near land. 200–2000 m.

Squaliolus laticaudus Smith & Radcliffe, 1912. Spined pygmy shark

Wide-ranging but scattered records in the Atlantic, Western Pacific, and Western Indian. Epipelagic and mesopelagic. 200–500 m.

Order Squatiniformes. Angelsharks.

Family Squatinidae. Angelsharks.

Squatina aculeata Dumeril, in Cuvier, 1817. Sawback angelshark

Eastern Atlantic off West Africa, and the Western Mediterranean. Shelf and upper slope. 30–500 m.

Squatina africana Regan, 1908. African angelshark

Western Indian: East Africa. Shelf and upper slope. 0–494 m (mainly 60–300 m).

Squatina albipunctata Last & White, 2008. Eastern angelshark

Western Pacific and (marginally) Eastern Indian: eastern Australia. Shelf and upper slope. 37–415 m.

Squatina argentina (Marini, 1930). Argentine angelshark

Southwest Atlantic: Brazil to Argentina. Shelf and upper slope. 51–320 m (mostly 120–320 m).

Squatina formosa Shen & Ting, 1972. Taiwan angelshark

Western Pacific: Taiwan and Philippines. Outer shelf and upper slope. 183–385 m.

Squatina pseudocellata Last & White, 2008. Western angelshark

Eastern Indian: western Australia. Outer shelf and upper slope. 150–312 m.

Squatina tergocellata McCulloch, 1914. Ornate angelshark

Eastern Indian: Australia. Outer shelf and upper slope. 130–400 m (most common ~300 m).

Order Pristiophoriformes. Sawsharks.

Family Pristiophoridae. Sawsharks.

Pliotrema warreni Regan, 1906. Sixgill sawshark

Southeast Atlantic and Western Indian: southern Africa. Shelf and upper slope. 37–500 m.

Pristiophorus delicatus Yearsley, Last & White, 2008. Tropical sawshark

Western Central Pacific: northeastern Australia. Upper slope. 246–405 m.

Pristiophorus schroederi Springer & Bullis, 1960. Bahamas sawshark

Western Central Atlantic: between Cuba, Florida (United States), and the Bahamas. Upper to mid slope. 438–952 m.

Order Rajiformes. Batoids.

Family Rhinobatidae. Guitarfishes.

Rhinobatos variegatus Nair & Lal Mohan, 1973. Stripenose guitarfish

Eastern Indian: Gulf of Mannar, India. Upper slope. 366 m.

Family Narcinidae. Numbfishes.

Benthobatis kreffti Rincon, Stehmann & Vooren, 2001. Brazilian blind torpedo

Southwest Atlantic: Brazil. Upper slope. 400–600 m.

Benthobatis marcida Bean & Weed, 1909. Pale or blind torpedo

Western Central Atlantic: South Carolina (United States) to northern Cuba. Upper to mid slope. 274–923 m.

Benthobatis moresbyi Alcock, 1898. Dark blindray

Western Indian: Arabian Sea. Mid slope. 787–1071 m.

Benthobatis yangi Carvalho, Compagno & Ebert, 2003. Narrow blindray

Northwest and Western Central Pacific: Taiwan. Upper slope (possibly also outer shelf). <300 m.

Narcine lasti Carvalho & Séret, 2002. Western numbfish

Eastern Indian: western Australia. Outermost shelf and upper slope. 180–320 m.

Narcine nelsoni Carvalho, 2008. Eastern numbfish

Western Central Pacific: northeastern Australia. Outer shelf and upper slope. 140–540 m.

Narcine tasmaniensis Richardson, 1840. Tasmanian numbfish

Eastern Indian and Southwest Pacific: southeastern Australia. Shelf (south of range) and upper slope (north of range). 5–640 m (in north of range mainly 200–640 m).

Family Narkidae. Sleeper Rays.

Heteronarce garmani Regan, 1921. Natal sleeper ray

Western India: Mozambique and South Africa. Shelf and upper slope. 73–329 m.

Heteronarce mollis (Lloyd, 1907). Soft sleeper ray

Western Indian: Gulf of Aden. Shelf and upper slope. 73–346 m.

Typhlonarke aysoni (Hamilton, 1902). Blind electric ray

Southwest Pacific: New Zealand. Shelf and upper to mid slope. 46–900 m (most common 300–400 m).

Typhlonarke tarakea Phillipps, 1929. Oval electric ray

Southwest Pacific: New Zealand. Shelf and upper to mid slope. 46–900 m (most common 300–400 m).

Family Torpedinidae. Torpedo Rays.

Torpedo fairchildi Hutton, 1872. New Zealand torpedo ray

Southwest Pacific: New Zealand. Shelf and upper to mid slope. 5–1135 m (most common 100–300 m).

Torpedo fuscomaculata Peters, 1855. Blackspotted torpedo

Western Indian: Mozambique and South Africa. Reports from Western Indian Ocean islands likely refer to similar undescribed species. Shelf and upper slope. 0–439 m.

Torpedo macneilli (Whitley, 1932). Short-tail torpedo ray

Eastern Indian and Western Pacific: Australia. Outer shelf and upper slope. 90–825 m.

Torpedo microdiscus Parin & Kotlyar, 1985. Smalldisk torpedo

Southeast Pacific: Nazca and Sala y Gomez Submarine Ridges. Submarine ridges.

Torpedo nobiliana Bonaparte, 1835. Great, Atlantic, or black torpedo

Wide-ranging in the Eastern and Western Atlantic. Shelf and upper slope. 2–530 m.

Torpedo puelcha Lahille, 1928. Argentine torpedo

Southwest Atlantic: Brazil to Argentina. Shelf and upper slope. Inshore–600 m.

Torpedo tremens de Buen, 1959. Chilean torpedo

Patchy in the Eastern Central and Southeast Pacific. Shelf and upper slope. Inshore–700 m.

Family Arhynchobatidae. Softnose Skates.

Arhynchobatis asperrimus Waite, 1909. Longtailed skate

Southwest Pacific: New Zealand. Outer shelf and upper to mid slope. 90–1070 m.

Bathyraja abyssicola (Gilbert, 1896). Deepsea skate

Wide-ranging in the North Pacific. Upper to deep slope and abyssal plains. 362–2906 m.

Bathyraja aguja (Kendall & Radcliffe, 1912). Aguja skate

Southeast Pacific: Peru. Mid slope. 980 m.

Bathyraja albomaculata (Norman, 1937). White-dotted skate

Southwest Atlantic and Southeast Pacific: southern South America. Shelf and upper to mid slope. 55–861 m.

Bathyraja aleutica (Gilbert, 1895). Aleutian skate

Wide-ranging in the North Pacific. Outer shelf and upper slope. 91–700 m.

Bathyraja andriashevi Dolganov, 1985. Little-eyed skate

Northwest Pacific: Japan and Russia. Mid to deep slope. 1400–2000 m.

Bathyraja bergi Dolganov, 1985. Bottom skate

Northwest Pacific: Japan and Russia. Outer shelf and upper to mid slope. 70–900 m. Includes the junior synonym *Bathyraja pseudoisotrachys* Ishihara & Ishiyama, 1985.

Bathyraja brachyurops (Fowler, 1910). Broadnose skate

Southwest Atlantic and Southeast Pacific: southern South America. Shelf and upper slope. 28–604 m.

Bathyraja caeluronigricans Ishiyama & Ishihara, 1977. Purpleblack skate

Northwest Pacific: Japan and Russia. Upper slope. 200–400 m. Possible synonym of *Bathyraja matsubarai* (Ishiyama, 1952).

Bathyraja cousseauae Diaz de Astarloa & Mabragaña, 2004. Cousseau's skate

Southwest Atlantic: Argentina and the Falkland/Malvinas Islands. Outer shelf and upper to mid slope. 119–1011 m.

Bathyraja diplotaenia (Ishiyama, 1950). Duskypink skate

Northwest Pacific: Japan. Outer shelf and upper slope. 100–700 m.

Bathyraja eatonii (Günther, 1876). Eaton's skate

Circum-Antarctic. Shelf, upper to mid slope, deepsea plateaus, and submarine ridges. 15–800 m.

Bathyraja fedorovi Dolganov, 1985. Cinnamon skate

Northwest Pacific: Japan and Russia. Upper to deep slope and abyssal plains. 447–2025 m.

Bathyraja griseocauda (Norman, 1937). Graytail skate

Southwest Atlantic and Southeast Pacific: southern South America. Shelf and upper to mid slope. 82–941 m.

Bathyraja hesperafricana Stehmann, 1995. West African skate

Eastern Central Atlantic: West Africa. Mid to deep slope. 750–2000 m.

Bathyraja interrupta (Gill & Townsend, 1897). Bering skate

Wide-ranging in the North and Eastern Central Pacific, although not well defined due to possible misidentifications. Shelf and upper to mid slope. 23–1500 m.

Bathyraja irrasa Hureau & Ozouf-Costaz, 1980. Kerguelen sandpaper skate

Antarctic Indian: Kerguelen Islands. Upper to mid slope. 300–1200 m.

Bathyraja ishiharai Stehmann, 2005. Abyssal skate

Eastern Indian: western Australia. Abyssal plains. ~2300 m.

Bathyraja isotrachys (Günther, 1877). Raspback skate

Northwest Pacific: Russia, Japan, and Korea. Upper to deep slope. 370–2000 m.

Bathyraja kincaidii (Garman, 1908). Sandpaper skate

Eastern Central and Northeast Pacific: Baja California north to Gulf of Alaska. Shelf and upper to mid slope. 55–1372 m (most common 200–500 m).

Bathyraja lindbergi Ishiyama & Ishihara, 1977. Commander skate

Northwest Pacific: Japan, Russia, and the Bering Sea. Outer shelf and upper to mid slope. 120–950 m.

Bathyraja longicauda (de Buen, 1959). Slimtail skate

Southeast Pacific: Peru and Chile. Upper slope. 580–735 m.

Bathyraja maccaini Springer, 1972. McCain's skate

Southeast Pacific and Antarctic Atlantic: Chile and Antarctica (Orkney and South Shetland islands to the Antarctic Peninsula). Shelf and upper slope. Inshore–500 m.

Bathyraja macloviana (Norman, 1937). Patagonian skate

Southwest Atlantic and Southeast Pacific: southern South America. Shelf and upper slope. 53–514 m.

Bathyraja maculata Ishiyama & Ishihara, 1977. Whiteblotched skate

Northern Pacific: Aleutian Islands and Bering Sea westward to Japan. Shelf and upper to mid slope. 73–1110 m.

Bathyraja magellanica (Philippi, 1902 or Steindachner, 1903). Magellan skate

Southwest Atlantic and Southeast Pacific: southern South America. Shelf and upper slope. 51–600 m (mostly <70 m in Falklands).

Bathyraja matsubarai (Ishiyama, 1952). Duskypurple skate

Northwest Pacific: Japan and Russia. Occurrence in the Northeast Pacific requires confirmation. Outer shelf and upper to deep slope. 120–2000 m.

Bathyraja meridionalis Stehmann, 1987. Darkbelly skate

Primarily Sub-Antarctic in the Atlantic, possibly circum-Antarctic. Upper to deep slope, abyssal plains and seamounts. 300–2240 m.

Bathyraja microtrachys (Osburn & Nichols, 1916). Fine-spined skate

Eastern Central Pacific: western United States. Deep slope and abyssal plains. 1995–2900 m. Validity follows Ebert (2003) as separate from *Bathyraja trachura* (Gilbert, 1892).

Bathyraja minispinosa Ishiyama & Ishihara, 1977. Smallthorn skate

Northern Pacific: Aleutian Islands and Bering Sea westward to Japan. Outer shelf and upper to mid slope. 150–1420 m.

Bathyraja multispinis (Norman, 1937). Multispine skate

Southwest Atlantic and Southeast Pacific: southern South America. Outer shelf and upper slope. 82–740 m. A record at 1900 m off Uruguay requires confirmation.

Bathyraja murrayi (Günther, 1880). Murray's skate

Antarctic Indian: Kerguelen and Heard Islands, but possibly circum-Antarctic. Shelf and upper slope. 30–650 m.

Bathyraja notoroensis Ishiyama & Ishihara, 1977. Notoro skate

Northwest Pacific: Japan. Upper slope. ~600 m. Possible synonym of *Bathyraja matsubarai* (Ishiyama, 1952).

Bathyraja pallida (Forster, 1967). Pallid skate

Northeast Atlantic: Bay of Biscay. Deep slope and abyssal plains. 2200–3280 m.

Bathyraja papilionifera Stehmann, 1985. Butterfly skate

Southwest Atlantic: Argentina and the Falkland/Malvinas Islands. Upper to deep slope and seamounts. 637–2000 m.

Bathyraja parmifera (Bean, 1881). Alaska or flathead skate

Wide-ranging in the North Pacific: Gulf of Alaska to Japan. Shelf and upper to mid slope. 20–1425 m. Includes the synonym *Bathyraja simoterus* (Ishiyama, 1967).

Bathyraja peruana McEachran & Miyake, 1984. Peruvian skate

Southeast Pacific: Ecuador to Chile. Upper to mid slope. 600–1060 m.

Bathyraja richardsoni (Garrick, 1961). Richardson's skate

Southwest Pacific: Australia and New Zealand (possibly also in the North Atlantic). Mid to deep slope, abyssal plains, and seamounts. 1370–2909 m.

Bathyraja scaphiops (Norman, 1937). Cuphead skate

Southwest Atlantic: Argentina and the Falkland/Malvinas Islands. Outer shelf and upper slope. 104–509 m (most common 104–159 m).

Bathyraja schroederi (Krefft, 1968). Whitemouth skate

Southwest Atlantic and Southeast Pacific: southern South America. Mid to deep slope and abyssal plains. 800–2380 m.

Bathyraja shuntovi Dolganov, 1985. Narrownose skate

Southwest Pacific: New Zealand. Upper to mid slope. 300–1470 m.

Bathyraja smirnovi (Soldatov & Lindberg, 1913). Golden skate

Northwest Pacific: Bering Sea to Japan. Outer shelf and upper to mid slope. 100–1000 m.

Bathyraja smithii (Müller & Henle, 1841). African softnose skate

Southeast Atlantic: Namibia and South Africa. Upper to mid slope. 400–1020 m.

Bathyraja spinicauda (Jensen, 1914). Spinetail or spinytail skate

Wide-ranging in the Northeast and Northwest Atlantic. Outer shelf and upper to mid slope. 140–1209 m (mainly >400 m).

Bathyraja spinosissima (Beebe & Tee-Van, 1941). White skate

Eastern Central and Southeast Pacific. Records from the Northwest Pacific may be a separate species. Mid to deep slope and abyssal plains. 800–2938 m.

Bathyraja trachouros (Ishiyama, 1958). Eremo skate

Northwest Pacific: Japan. Upper slope. 300–500 m.

Bathyraja trachura (Gilbert, 1892). Roughtail skate

Wide-ranging in the North Pacific. Upper to deep slope and abyssal plains. 213–2550 m.

Bathyraja tzinovskii Dolganov, 1985. Creamback skate

Northwest Pacific: Japan and Russia. Deep slope and abyssal plains. 1776–2500 m.

Bathyraja violacea (Suvorov, 1935). Okhotsk skate

North Pacific: Bering Sea to Japan. Shelf and upper to mid slope. 23–1100 m.

Brochiraja aenigma Last & McEachran, 2006.

Southwest Pacific: New Zealand. Upper slope. 660–665 m.

Brochiraja albilabiata Last & McEachran, 2006.

Southwest Pacific: New Zealand. Mid slope. 785–1180 m.

Brochiraja asperula (Garrick & Paul, 1974). Prickly deepsea skate

Southwest Pacific: New Zealand. Mid slope. 350–1010 m.

Brochiraja leviveneta Last & McEachran, 2006.

Southwest Pacific: New Zealand. Mid slope. 960–1015 m.

Brochiraja microspinifera Last & McEachran, 2006.

Southwest Pacific: New Zealand. Upper to mid slope. 510–900 m.

Brochiraja spinifera (Garrick & Paul, 1974). Spiny deepsea skate

Southwest Pacific: New Zealand. Upper slope. 500–750 m.

Insentiraja laxipella (Yearsley & Last, 1992). Eastern looseskin skate

Western Central Pacific: northeastern Australia. Mid slope. 800–880 m.

Insentiraja subtilispinosa (Stehmann, 1985). Velvet or western looseskin skate

Patchy in the Eastern Indian and Western Central Pacific. Upper to mid slope. 320–1460 m.

Notoraja azurea McEachran & Last, 2008. Blue skate

Eastern Indian: western and southern Australia. Mid slope. 765–1440 m.

Notoraja hirticauda Last & McEachran, 2006. Ghost skate

Eastern Indian: western Australia. Upper to mid slope. 500–760 m.

Notoraja lira McEachran & Last, 2008. Broken Ridge skate

Eastern Indian: Broken Ridge. Submarine ridges. 1050 m.

Notoraja ochroderma McEachran & Last, 1994. Pale skate

Western Central Pacific: northeastern Australia. Upper slope. 400–455 m.

Notoraja sticta McEachran & Last, 2008. Blotched skate

Eastern Indian: southern Australia. Mid slope. 820–1200 m.

Notoraja tobitukai (Hiyama, 1940). Leadhued skate

Northwest Pacific: Japan and Taiwan. Shelf and upper to mid slope. 60–900 m.

Pavoraja alleni McEachran & Fechhelm, 1982. Allen's skate

Eastern Indian: western Australia. Upper slope. 304–458 m.

Pavoraja arenaria Last, Mallick & Yearsley, 2008. Sandy peacock skate

Eastern Indian: southern Australia. Outermost shelf and upper slope. 192–712 m (mainly 300–400 m).

Pavoraja mosaica Last, Mallick & Yearsley, 2008. Mosaic skate

Western Central Pacific: northeastern Australia. Upper slope. 300–405 m.

Pavoraja nitida (Günther, 1880). Peacock skate

Eastern Indian and Southwest Pacific: southern Australia. Shelf and upper slope. 75–432 m.

Pavoraja pseudonitida Last, Mallick & Yearsley, 2008. False peacock skate

Western Central Pacific: northeastern Australia. Upper slope. 212–512 m.

Pavoraja umbrosa Last, Mallick & Yearsley, 2008. Dusky skate

Western Pacific: eastern Australia. Upper slope. 360–731 m.

Psammobatis scobina (Philippi, 1857). Raspthorn sandskate

Southeast Pacific: Chile. Shelf and upper slope. 40–450 m.

Pseudoraja fischeri Bigelow & Schroeder, 1954. Fanfin skate

Scattered records in the Western Central Atlantic. Upper slope. 412–576 m.

Rhinoraja kujiensis (Tanaka, 1916). Dapple-bellied softnose skate

Northwest Pacific: Japan. Upper to mid slope. 450–~1000 m.

Rhinoraja longi Raschi & McEachran, 1991. Aleutian dotted skate

North Pacific: Aleutian Islands. Upper slope. 390–435 m.

Rhinoraja longicauda Ishiyama, 1952. White-bellied softnose skate

Northwest Pacific: Japan and Russia. Upper to mid slope. 300–1000 m.

Rhinoraja obtusa (Gill & Townsend, 1897). Blunt skate

North Pacific: Bering Sea. Habitat and depth information not available.

Rhinoraja odai Ishiyama, 1952. Oda's skate

Northwest Pacific: Japan. Upper to mid slope. 300–870 m.

Rhinoraja taranetzi Dolganov, 1985. Mudskate

Reasonably wide-ranging in the North Pacific. Shelf and upper to mid slope. 81–1000 m. Includes the junior synonym *Bathyraja hubbsi* Ishihara & Ishiyama, 1985.

Family Rajidae. Hardnose Skates.

Amblyraja badia Garman, 1899. Broad skate

Patchy in the North Pacific. Mid to deep slope and abyssal plains. 846–2324 m.

Amblyraja doellojuradoi (Pozzi, 1935). Southern thorny skate

Southwest Atlantic and Southeast Pacific: southern South America. Shelf and upper slope. 51–642 m.

Amblyraja frerichsi Krefft, 1968. Thickbody skate

Southwest Atlantic and Southeast Pacific: southern South America. Mid to deep slope and abyssal plains. 720–2609 m.

Amblyraja georgiana (Norman, 1938). Antarctic starry skate

Circum-Antarctic, including off Chile and the Falkland/Malvinas Islands. Shelf and upper to mid slope. 20–800 m.

Amblyraja hyperborea (Collett, 1879). Arctic or boreal skate

Wide-ranging but patchy in deep temperate waters. Upper to deep slope and abyssal plains. 300–2500 m. Taxonomic issues unresolved, may represent a species complex.

Amblyraja jenseni (Bigelow & Schroeder, 1950). Jensen's skate

North Atlantic: Nova Scotia (Canada) to southern New England (United States) and Iceland. Upper to deep slope. 366–2196 m.

Amblyraja radiata (Donovan, 1808). Thorny skate

Wide-ranging in the North Atlantic and also off South Africa. Shelf and upper to mid slope. 18–1400 m (most common 27–439 m).

Amblyraja reversa (Lloyd, 1906). Reversed skate

Western Indian: Arabian Sea. Deep slope. 1500 m.

Amblyraja robertsi (Hulley, 1970). Bigmouth skate

Southeast Atlantic: South Africa. Mid slope. 1350 m.

Amblyraja taaf (Meisner, 1987). Whiteleg skate

Antarctic Indian: Crozet Islands. Outer shelf and upper slope. 150–600 m.

Breviraja claramaculata McEachran & Matheson, 1985. Brightspot skate

Western Central Atlantic: South Carolina to Florida (United States). Upper to mid slope. 293–896 m.

Breviraja colesi Bigelow & Schroeder, 1948. Lightnose skate

Western Central Atlantic: Florida (United States), the Bahamas, and Cuba. Upper slope. 220–415 m.

Breviraja marklei McEachran & Miyake, 1987. Nova Scotia skate

Northwest Atlantic: Canada. Upper to mid slope. 443–988 m.

Breviraja mouldi McEachran & Matheson, 1995. Blacknose skate

Western Central Atlantic: Honduras to Panama. Upper to mid slope. 353–776 m.

Breviraja nigriventralis McEachran & Matheson, 1985. Blackbelly skate

Western Central Atlantic: Panama and the northern coast of South America. Upper to mid slope. 549–776 m.

Breviraja spinosa Bigelow & Schroeder, 1950. Spinose skate

Western Central Atlantic: North Carolina to Florida (United States). Upper slope. 366–671 m.

Dactylobatus armatus Bean & Weed, 1909. Skilletskate

Patchy in the Western Central Atlantic. Upper slope. 338–685 m.

Dactylobatus clarki (Bigelow & Schroeder, 1958). Hookskate

Patchy in the Western Central Atlantic. Upper to mid slope. 366–915 m.

Dipturus acrobelus Last, White & Pogonoski, 2008. Deepwater skate

Eastern Indian and Southwest Pacific: southern Australia. Upper to mid slope. 446–1328 m (mainly 800–1000 m).

Dipturus apricus Last, White & Pogonoski, 2008. Pale tropical skate

Western Central Pacific: northeastern Australia. Outermost shelf and upper slope. 196–606 m (mainly 300–500 m).

Dipturus batis (Linnaeus, 1758). Gray or blue skate

Formerly wide-ranging in the Northeast and Eastern Central Atlantic, including the Mediterranean, but now extirpated from or reduced in much of its historical range. Outer shelf, upper to deep slope and deepsea rises. 100–2619 m.

Dipturus bullisi (Bigelow & Schroeder, 1962). Tortugas skate

Patchy in the Western Central Atlantic. Upper slope. 183–549 m.

Dipturus campbelli (Wallace, 1967). Blackspot skate

Western Indian: patchy off South Africa and Mozambique. Outer shelf and upper slope. 137–403 m.

Dipturus canutus Last, 2008. Grey skate

Eastern Indian and Southwest Pacific: southeastern Australia. Upper slope. 330–730 m (mainly 400–600 m).

Dipturus cerva (Whitley, 1939). White-spotted skate

Eastern Indian and Southwest Pacific: southern Australia. Shelf and upper slope. 20–470 m.

Dipturus crosnieri (Séret, 1989). Madagascar skate

Western Indian: Madagascar. Upper to mid slope. 300–850 m.

Dipturus doutrei (Cadenat, 1960). Violet or javelin skate

Eastern Central and Southeast Atlantic: West Africa. Outer shelf and upper to mid slope. 163–800 m (mostly >400 m).

Dipturus endeavouri Last, 2008. Endeavour skate

Western Central and Southwest Pacific: eastern Australia. Outer shelf and upper slope. 125–500 m.

Dipturus garricki (Bigelow & Schroeder, 1958). San Blas skate

Western Central Atlantic: northern Gulf of Mexico and Nicaragua. Upper slope. 275–476 m.

Dipturus gigas (Ishiyama, 1958). Giant skate

Northwest and Western Central Pacific: Japan to Philippines. Upper to mid slope. 300–1000 m.

Dipturus grahami Last, 2008. Graham's skate

Western Central and Southwest Pacific: eastern Australia. Outer shelf and upper slope. 146–490 m (mainly 250–450 m).

Dipturus gudgeri (Whitley, 1940). Bight skate

Eastern Indian and Southwest Pacific: southern Australia. Outer shelf and upper slope. 160–765 m (most common 400–550 m).

Dipturus healdi Last, White & Pogonoski, 2008. Heald's skate

Eastern Indian: western Australia. Upper slope. 304–520 m.

Dipturus innominatus (Garrick & Paul, 1974). New Zealand smooth skate

   Southwest Pacific: New Zealand. Shelf and upper to mid slope. 15–1310 m.

Dipturus johannisdavesi (Alcock, 1899). Travancore skate

   Western Indian: Gulf of Aden and India. Upper slope. 457–549 m.

Dipturus laevis (Mitchell, 1817). Barndoor skate

   Northwest Atlantic: Canada and the United States. Shelf and upper slope.
   0–430 m.

Dipturus lanceorostratus (Wallace, 1967). Rattail skate

   Western Indian: Mozambique. Upper slope. 430–439 m.

Dipturus leptocaudus (Krefft & Stehmann, 1975). Thintail skate

   Southwest Atlantic: Brazil. Upper slope. 400–550 m.

Dipturus linteus (Fries, 1838). Sailskate

   North Atlantic, primarily in the Northeast Atlantic. Upper to mid slope. 150–
   1200+ m.

Dipturus macrocaudus (Ishiyama, 1955). Bigtail skate

   Northwest Pacific: Japan, Korea, and Taiwan. Upper to mid slope. 300–800 m.

Dipturus melanospilus Last, White & Pogonoski, 2008. Blacktip skate

   Western Central and Southwest Pacific: eastern Australia. Upper slope. 239–
   695 m.

Dipturus mennii Gomes & Paragó, 2001. South Brazilian skate

   Southwest Atlantic: Brazil. Outer shelf and upper slope. 133–500 m. Includes
   the likely synonym *Dipturus diehli* Soto & Mincarone, 2001.

Dipturus nidarosiensis (Collett, 1880). Norwegian skate

   Northeast Atlantic: Norway, Iceland, and the Rockall Trough. Upper to mid
   slope and deepsea rises. 200–1000 m.

Dipturus oculus Last, 2008. Ocellate skate

   Eastern Indian: western Australia. Upper slope. 200–389 m.

Dipturus oregoni (Bigelow & Schroeder, 1958). Hooktail skate

   Western Central Atlantic: northern Gulf of Mexico. Upper to mid slope. 475–
   1079 m.

Dipturus oxyrhynchus (Linnaeus, 1758). Sharpnose skate

   Wide-ranging in the Northeast and Eastern Central Atlantic, including the
   Mediterranean. Shelf and upper to mid slope. 15–900 m.

Dipturus polyommata (Ogilby, 1910). Argus skate

   Western Central Pacific: northeastern Australia. Outer shelf and upper slope.
   135–320 m.

Dipturus pullopunctata (Smith, 1964). Slime skate

   Southeast Atlantic: Namibia and South Africa. Shelf and upper slope. 15–457 m
   (most common 100–300 m).

Dipturus queenslandicus Last, White & Pogonoski, 2008. Queensland deepwater skate

   Western Central Pacific: northeastern Australia. Upper slope. 399–606 m.

Dipturus springeri (Wallace, 1967). Roughbelly skate

Southeast Atlantic and Western Indian: southern Africa. Upper slope. 400–740 m (mostly 400–500 m).

Dipturus stenorhynchus (Wallace, 1967). Prownose skate

Southeast Atlantic and Western Indian: South Africa and Mozambique. Upper to mid slope. 253–761 m.

Dipturus teevani (Bigelow & Schroeder, 1951). Caribbean skate

Patchy throughout the Western Central Atlantic. Upper slope. 311–732 m.

Dipturus tengu (Jordan & Fowler, 1903). Acutenose or tengu skate

Northwest and Western Central Pacific: Japan to the Philippines. Shelf and upper slope. 45–300 m.

Dipturus trachydermus (Krefft & Stehmann, 1974). Roughskin skate

Southwest Atlantic and Southeast Pacific: southern South America. Shelf and upper slope. 20–450 m.

Dipturus wengi Séret & Last, 2008. Weng's skate

Eastern Indian and Western Pacific: Australia and Indonesia. Upper to mid slope. 486–1165 m.

Fenestraja atripinna (Bigelow & Schroeder, 1950). Blackfin pygmy skate

Western Central Atlantic: North Carolina (United States) to Cuba. Upper to mid slope. 366–951 m.

Fenestraja cubensis (Bigelow & Schroeder, 1950). Cuban pygmy skate

Western Central Atlantic: Florida (United States), the Bahamas, and Cuba. Upper to mid slope. 440–869 m.

Fenestraja ishiyamai (Bigelow & Schroeder, 1962). Plain pygmy skate

Scattered records in the Western Central Atlantic. Upper to mid slope. 503–950 m.

Fenestraja maceachrani (Séret, 1989). Madagascar pygmy skate

Western Indian: Madagascar. Upper to mid slope. 600–765 m.

Fenestraja mamillidens (Alcock, 1889). Prickly skate

Eastern Indian: Bay of Bengal. Mid slope. 1093 m.

Fenestraja plutonia (Garman, 1881). Pluto skate

Patchy in the Western Central Atlantic. Upper to mid slope. 293–1024 m.

Fenestraja sibogae (Weber, 1913). Siboga pygmy skate

Western Central Pacific: Bali Sea, Indonesia. Upper slope. 290 m.

Fenestraja sinusmexicanus (Bigelow & Schroeder, 1950). Gulf of Mexico pygmy skate

Patchy in the Western Central Atlantic. Shelf and upper to mid slope. 56–1096 m.

Gurgesiella atlantica (Bigelow & Schroeder, 1962). Atlantic pygmy skate

Western Atlantic: Central and South America. Upper to mid slope. 247–960 m.

Gurgesiella dorsalifera McEachran & Compagno, 1980. Onefin skate

Southwest Atlantic: Brazil. Upper to mid slope. 400–800 m.

Gurgesiella furvescens de Buen, 1959. Dusky finless skate.

Southeast Pacific: Galapagos Islands and Peru to Chile. Upper to mid slope. 300–960 m.

Leucoraja compagnoi Stehmann, 1995. Tigertail skate

Southeast Atlantic and Western Indian: South Africa. Upper slope. 497–625 m.

Leucoraja fullonica Linnaeus, 1758. Shagreen skate

Northeast (and marginally into the Eastern Central) Atlantic, including the western Mediterranean. Shelf and upper to mid slope. 30–800 m.

Leucoraja garmani (Whitley, 1939). Rosette skate

Northwest and Western Central Atlantic: Cape Cod to Florida (United States). Shelf and upper slope. 37–530 m.

Leucoraja lentiginosa (Bigelow & Schroeder, 1951). Freckled skate

Western Central Atlantic: northern Gulf of Mexico. Shelf and upper slope. 53–588 m.

Leucoraja leucosticta Stehmann, 1971. Whitedappled skate

Eastern Central and Southeast Atlantic: West Africa. Shelf and upper slope. 70–600 m.

Leucoraja melitensis Clark, 1926. Maltese skate

Southwestern and south-central Mediterranean. Shelf and upper to mid slope. 60–800 m (more common 400–800 m).

Leucoraja naevus Müller & Henle, 1841. Cuckoo skate

Reasonably wide-ranging in the Northeast and Eastern Central Atlantic, including the Mediterranean. Shelf and upper slope. 20–500 m.

Leucoraja pristispina Last, Stehmann & Séret, 2008. Sawback skate

Eastern Indian: western Australia. Upper slope. 202–504 m.

Leucoraja wallacei Hulley, 1970. Yellowspot or blancmange skate

Southeast Atlantic and Western Indian: southern Africa. Shelf and upper slope. 70–500 m (most common 150–300 m).

Leucoraja yucatanensis (Bigelow & Schroeder, 1950). Yucatan skate

Western Central Atlantic: Central America. Outermost shelf and upper slope. 192–457 m.

Malacoraja kreffti (Stehmann, 1978). Krefft's skate

Northeast Atlantic: Rockall Trough and Iceland. Mid slope and deepsea rises. 1200 m.

Malacoraja obscura Carvalho, Gomes & Gadig, 2006. Brazilian soft skate

Southwest Atlantic: Brazil. Mid slope. 808–1105 m.

Malacoraja senta (Garman, 1885). Smooth skate

Northwest Altlantic: Canada and the United States. Shelf and upper to mid slope. 46–914 m.

Malacoraja spinacidermis (Barnard, 1923). Prickled or roughskin skate

Wide-ranging but patchy in the Eastern Atlantic. Upper to deep slope. 450–1568 m.

Neoraja africana Stehmann & Séret, 1983. West African pygmy skate

Eastern Central Atlantic: Gabon and Western Sahara. Mid to deep slope. 900–1550 m.

Neoraja caerulea (Stehmann, 1976). Blue pygmy skate

Northeast Atlantic: Rockall Trough. Upper to mid slope and deepsea rises. 600–1262 m.

Neoraja carolinensis McEachran & Stehmann, 1984. Carolina pygmy skate

Western Central Atlantic: North Carolina to Florida (United States). Upper to mid slope. 695–1010 m.

Neoraja iberica Stehmann, Séret, Costa & Baro, 2008. Iberian pygmy skate

Northeast Atlantic: Iberian Peninsula. Upper slope. 270–670 m.

Neoraja stehmanni (Hulley, 1972). South African pygmy skate

Southeast Atlantic: South Africa. Upper to mid slope. 292–1025 m (most common >600 m).

Okamejei arafurensis Last & Gledhill, 2008. Arafura skate

Western Central Pacific and Eastern Indian: northern Australia. Outermost shelf and upper slope. 179–298 m.

Okamejei heemstrai McEachran & Fechhelm, 1982. Narrow or East African skate

Western Indian: East Africa. Upper slope. 200–500 m.

Okamejei leptoura Last & Gledhill, 2008. Thintail skate

Eastern Indian: western Australia. Upper slope. 202–735 m.

Raja africana? Capape, 1977. African skate

Disjunct range in the Mediterranean (Tunisia) and Eastern Central Atlantic (Mauritania). Shelf and upper slope. 50–400 m. Homonym, requires a replacement name.

Raja bahamensis Bigelow & Schroeder, 1965. Bahama skate

Western Central Atlantic: Florida (United States) and the Bahamas. Upper slope. 366–411 m.

Raja binoculata Girard, 1854. Big skate

Wide-ranging in the Northeast and Eastern Central Pacific (and marinally into the Northwest Pacific). Shelf and upper to mid slope. 3–800 m.

Raja inornata Jordan & Gilbert, 1880. California skate

Eastern Central Pacific: western United States and Mexico. Shelf and upper to mid slope. 17–671 m.

Raja maderensis Lowe, 1841. Madeira skate

Eastern Central Atlantic: Madeira. Shelf and upper slope. ?–500 m.

Raja polystigma Regan, 1923. Speckled skate

Western Mediterranean. Outer shelf and upper slope. 100–400 m.

Raja rhina Jordan & Gilbert, 1880. Longnose skate

Wide-ranging in the Northeast and Eastern Central Pacific. Shelf and upper to mid slope. 9–1069 m.

Raja stellulata Jordan & Gilbert, 1880. Pacific starry skate

Eastern Central Pacific: United States and Mexico. Shelf and upper slope. 18–732 m (usually <100 m). Records from Alaska and Bering Sea likely other species (i.e., *B. parmifera*).

Raja straeleni Poll, 1951. Biscuit skate

Wide-ranging in the Eastern Central and Southeast Atlantic: West Africa; Western Indian: southern Africa. Shelf and upper slope. 80–690 m (mostly 200–300 m).

Rajella annandalei (Weber, 1913). Indonesian round or Annandale's skate

Western Central Pacific: eastern Indonesia. Upper to mid slope. 400–830 m.

Rajella barnardi (Norman, 1935). Bigthorn skate

Wide-ranging in the Eastern Atlantic: West Africa. Outer shelf and upper to deep slope. 102–1700 m. Includes the junior synonym *Raja confundens* Hulley, 1970.

Rajella bathyphila (Holt & Byrne, 1908). Deepwater skate

Wide-ranging in the North Atlantic. Upper to deep slope and abyssal plains. 600–2050 m (mostly >1000 m).

Rajella bigelowi (Stehmann, 1978). Bigelow's skate

Wide-ranging but patchy in the North and Central Atlantic. Upper to deep slope, deepsea rises, and abyssal plains. 650–4156 m (mostly >1500 m).

Rajella caudaspinosa (von Bonde & Swart, 1923). Munchkin skate

Southeast Atlantic and Western Indian: Namibia and South Africa. Upper slope. 310–520 m.

Rajella challengeri Last & Stehmann, 2008. Challenger skate

Eastern Indian and Southwest Pacific: southern Australia. Mid slope. 860–1500 m.

Rajella dissimilis (Hulley, 1970). Ghost skate

Patchy in the Eastern Atlantic. Upper to deep slope. 400–1570 m.

Rajella eisenhardti Long & McCosker, 1999. Galapagos gray skate

Southeast Pacific: Galapagos Islands. Mid slope. 757–907 m.

Rajella fuliginea (Bigelow & Schroeder, 1954). Sooty skate

Patchy in the Western Central Atlantic. Upper to mid slope. 731–1280 m.

Rajella fyllae (Luetken, 1888). Round skate

Wide-ranging in the Northeast and Northwest Atlantic. Outer shelf and upper to deep slope. 170–2050 m (average depth range 400–800 m).

Rajella kukujevi (Dolganov, 1985). Mid-Atlantic skate

Patchy in the Northeast Atlantic. Mid slope. 750–1341 m.

Rajella leopardus (von Bonde & Swart, 1923). Leopard skate

Patchy in the Eastern Central and Southeast Atlantic and possibly Western Indian (South Africa). Outer shelf and upper to deep slope. 130–1920 m.

Rajella nigerrima (de Buen, 1960). Blackish skate

Southeast Pacific: Ecuador to Chile. Upper to mid slope. 590–1000 m.

Rajella purpuriventralis (Bigelow & Schroeder, 1962). Purplebelly skate

Patchy in the Western Central Atlantic. Upper to deep slope. 732–2010 m.

Rajella ravidula (Hulley, 1970). Smoothback skate

Eastern Central and Southeast Atlantic: Morocco and South Africa. Mid slope. 1000–1250 m.

Rajella sadowskyii (Krefft & Stehmann, 1974). Brazilian skate

Southwest Atlantic and Southeast Pacific: southern South America. Mid slope. 1200 m.

Rostroraja alba (Lacepede, 1803). White, bottlenose, or spearnose skate

Wide-ranging in the Eastern Atlantic, including the western Mediterranean, and the Western Indian (to Mozambique). Shelf and upper slope. 30–600 m.

Zearaja chilensis (Guichenot, 1848). Yellownose skate

Southwest Atlantic and Southeast Pacific: southern South America. Shelf and upper slope. 28–500 m. Includes the junior synonym *Dipturus flavirostris* (Philippi, 1892).

Zearaja nasutus Banks, in Müller & Henle, 1841. Rough skate

Southwest Pacific: New Zealand. Shelf and upper to mid slope. 10–1500 m.

Family Anacanthobatidae. Legskates.

Anacanthobatis americanus Bigelow & Schroeder, 1962. American legskate

Western Atlantic: Caribbean and the northern coast of South America. Outermost shelf and upper to mid slope. 183–915 m.

Anacanthobatis donghaiensis (Deng, Xiong & Zhan, 1983). East China legskate

Northwest Pacific: East China Sea. Upper to mid slope. 200–1000 m.

Anacanthobatis folirostris (Bigelow & Schroeder, 1951). Leafnose legskate

Western Central Atlantic: northern Gulf of Mexico. Upper slope. 300–512 m.

Anacanthobatis longirostris Bigelow & Schroeder, 1962. Longnose legskate

Western Central Atlantic: northern Gulf of Mexico and Caribbean. Upper to mid slope. 520–1052 m.

Anacanthobatis marmoratus (von Bonde & Swart, 1924). Spotted legskate

Western Indian: South Africa and Mozambique. Upper slope. 230–322 m.

Anacanthobatis nanhaiensis (Meng & Li, 1981). South China legskate

South China Sea. Validity uncertain.

Anacanthobatis ori (Wallace, 1967). Black legskate

Western Indian: Mozambique and Madagascar. Mid to deep slope. 1000–1725 m.

Anacanthobatis stenosoma (Li & Hu, 1982). Narrow legskate

South China Sea. Validity uncertain.

Cruriraja andamanica (Lloyd, 1909). Andaman legskate

Patchy in the Indian Ocean: Tanzania and the Andaman Sea. Upper slope. 274–511 m.

Cruriraja atlantis Bigelow & Schroeder, 1948. Atlantic legskate

Western Central Atlantic: Florida (United States) to Cuba. Upper to mid slope. 512–778 m.

Cruriraja cadenati Bigelow & Schroeder, 1962. Broadfoot legskate

Western Central Atlantic: Florida (United States) and Puerto Rico. Upper to mid slope. 457–896 m.

Cruriraja durbanensis (von Bonde & Swart, 1924). Smoothnose legskate

Southeast Atlantic: South Africa. Mid slope. 860 m. Note types collected off Northern Cape Province of South Africa, not off KwaZulu-Natal as suggested by the specific name.

Cruriraja parcomaculata (von Bonde & Swart, 1924). Roughnose legskate

Southeast Atlantic and Western Indian: Namibia and South Africa. Outer shelf and upper slope. 150–620 m.

Cruriraja poeyi Bigelow & Schroeder, 1948. Cuban legskate

Scattered records throughout the Western Central Atlantic. Upper to mid slope. 366–870 m.

Cruriraja rugosa Bigelow & Schroeder, 1958. Rough legskate

Western Central Atlantic: Gulf of Mexico and Caribbean. Upper to mid slope. 366–1007 m.

Cruriraja triangularis Smith, 1964. Triangular legskate

Western Indian: South Africa and Mozambique. Upper slope. 220–675 m.

Sinobatis borneensis Chan, 1965. Borneo legskate

Patchy in the Northwest and Western Central Pacific. Upper to mid slope. 475–835 m.

Sinobatis bulbicauda Last & Séret, 2008. Western Australian legskate

Eastern Indian: western Australia and Indonesia. Outer shelf and upper to mid slope. 150–1125 m (mostly 400–800 m).

Sinobatis caerulea Last & Séret, 2008. Blue legskate

Eastern Indian: western Australia. Upper to mid slope. 482–1168 m.

Sinobatis filicauda Last & Séret, 2008. Eastern Australian legskate

Western Central Pacific: northeastern Australia. Upper to mid slope. 606–880 m.

Sinobatis melanosoma (Chan, 1965). Blackbodied legskate

Northwest and Western Central Pacific: South China Sea. Mid slope. 900–1100 m.

Family Plesiobatidae. Giant Stingarees.

Plesiobatis daviesi (Wallace, 1967). Deepwater stingray or giant stingaree

Patchy in the Indo-West and Central Pacific. Shelf (single record at 44 m off Mozambique) and upper slope. 44–680 m (mostly 275–680 m).

Family Urolophidae. Stingarees.

Urolophus expansus McCulloch, 1916. Wide stingaree

Eastern Indian: southern Australia. Outer shelf and upper slope. 130–420 m (mainly 200–300 m).

Urolophus piperatus Séret & Last, 2003. Coral Sea stingaree

Western Central Pacific: northeastern Australia. Upper slope and deep reefs. 170–370 m.

Family Hexatrygonidae. Sixgill Stingrays.

Hexatrygon bickelli Heemstra & Smith, 1980. Sixgill stingray

Wide-ranging but patchy in the Indo-West and Central Pacific. Upper to mid slope (although some shallow water and beach stranding records). 362–1120 m.

Family Dasyatidae. Whiptail Stingrays.

Dasyatis brevicaudata (Hutton, 1875). Shorttail or smooth stingray

Patchy in the Indo-West Pacific (New Zealand, Australia, and southern Africa). Inshore and shelf in Australia and New Zealand, outermost shelf and upper slope off South Africa. Intertidal–160 m (NZ, Australia), 180–480 m (southern Africa).

Superorder Galeomorphii. Galeomorph Sharks.
Order Heterodontiformes. Bullhead Sharks.

Family Heterodontidae. Bullhead Sharks.

Heterodontus ramalheira (Smith, 1949). Whitespotted bullhead shark

Western Indian: East Africa and Arabian Sea. Outer shelf and uppermost slope. 40–275 m (mostly >100 m).

Order Orectolobiformes. Carpetsharks.

Family Parascylliidae. Collared Carpetsharks.

Cirrhoscyllium japonicum Kamohara, 1943. Saddled carpetshark

Northwest Pacific: Japan. Uppermost slope. 250–290 m.

Parascyllium sparsimaculatum Goto & Last, 2002. Ginger carpetshark

Eastern Indian: Australia. Upper slope. 245–435 m.

Order Lamniformes. Mackerel Sharks.

Family Odontaspididae. Sand Tiger Sharks.

Odontaspis ferox (Risso, 1810). Smalltooth sand tiger or bumpytail raggedtooth

Wide-ranging but patchy in the Atlantic, including the Mediterranean, and the Indo-Pacific. Possibly circumglobal in warm-temperate and tropical waters. Shelf and upper slope. 13–880 m.

Odontaspis noronhai (Maul, 1955). Bigeye sand tiger

Scattered records in the Central and Southwest Atlantic and the Central Pacific. Possibly circumglobal in deep tropical seas. Upper to mid slope as well as mesopelagic. 600–1000 m.

Family Pseudocarchariidae. Crocodile Sharks.

Pseudocarcharias kamoharai (Matsubara, 1936). Crocodile shark

Cosmopolitan in tropical and warm temperate waters. Epipelagic and mesopelagic. 0–590 m.

Family Mitsukurinidae. Goblin Sharks.

Mitsukurina owstoni Jordan, 1898. Goblin shark

Wide-ranging but patchy in the Atlantic and the Indo-Pacific. Shelf, upper to mid slope and seamounts. 0–1300 m (mainly 270–960 m).

Family Alopiidae. Thresher Sharks.

Alopias superciliosus (Lowe, 1839). Bigeye thresher

Cosmopolitan in tropical and warm temperate waters. Shelf, epipelagic, and mesopelagic. 0–732 m (mostly >100 m).

Family Cetorhinidae. Basking Sharks.

Cetorhinus maximus (Gunnerus, 1765). Basking shark

Cosmopolitan in cold to warm-temperate waters. Shelf and upper to mid slope, epipelagic, and mesopelagic. 0–904 m.

Order Carcharhiniformes. Ground Sharks.

Family Scyliorhinidae. Catsharks.

Apristurus albisoma Nakaya & Séret, 1999. White catshark

Western Central Pacific: Norfolk and Lord Howe Ridges. Submarine ridges. 935–1564 m.

Apristurus ampliceps Sasahara, Sato & Nakaya, 2008. Roughskin catshark

Eastern Indian and Southwest Pacific: Australia and New Zealand. Mid to deep slope and submarine ridges. 800–1503 m.

Apristurus aphyodes Nakaya & Stehmann, 1998. White ghost catshark

Northeast Atlantic: Atlantic Slope. Mid to deep slope. 1014–1800 m.

Apristurus australis Sato, Nakaya & Yorozu, 2008. Pinocchio catshark

Eastern Indian and Western Pacific: Australia. Upper to mid slope and seamounts. 486–1035 m.

Apristurus brunneus (Gilbert, 1892). Brown catshark

Eastern Pacific: Alaska to Mexico, and possibly Southeast Pacific. Shelf and upper to mid slope. 33–1298 m.

Apristurus bucephalus White, Last & Pogonoski, 2008. Bighead catshark

Eastern Indian: western Australia. Mid slope. 920–1140 m.

Apristurus canutus Springer & Heemstra, in Springer, 1979. Hoary catshark

Western Central Pacific: Caribbean Islands and Venezuela. Upper to mid slope. 521–915 m.

Apristurus exsanguis Sato, Nakaya & Stewart, 1999. Pale catshark

Southwest Pacific: New Zealand. Upper to mid slope and submarine ridges. 573–1200 m.

Apristurus fedorovi Dolganov, 1985. Stout catshark

Northwest Pacific: Japan. Mid slope. 810–1430 m.

Apristurus gibbosus Meng, Chu & Li, 1985. Humpback catshark

Northwest Pacific: South China Sea off southern China. Mid slope. 913 m.

Apristurus herklotsi (Fowler, 1934). Longfin catshark

Northwest and Western Central Pacific: Japan, China, and Philippines. Upper to mid slope. 520–910 m.

Apristurus indicus (Brauer, 1906). Smallbelly catshark

Western Indian: Arabian Sea. Records off southern Africa probably erroneous. Mid to deep slope. 1289–1840 m.

Apristurus internatus Deng, Xiong & Zhan, 1988. Shortnose demon catshark

Northwest Pacific: East China Sea off China. Upper slope. 670 m.

Apristurus investigatoris (Misra, 1962). Broadnose catshark

Eastern Indian: Andaman Sea. Mid slope. 1040 m.

Apristurus japonicus Nakaya, 1975. Japanese catshark

Northwest Pacific: Japan. Mid slope. 820–915 m.

Apristurus kampae Taylor, 1972. Longnose catshark

Eastern Pacific: United States and Mexico. Provisional from the Galapagos Islands. Outer shelf and upper to deep slope. 180–1888 m.

Apristurus laurussoni (Saemundsson, 1922). Iceland catshark

Wide-ranging but patchy in the North and Central Atlantic. Upper to deep slope. 560–2060 m. Includes the junior synonyms *Apristurus atlanticus* (Koefoed, 1932) and *Apristurus maderensis* Cadenat & Maul, 1966.

Apristurus longicephalus Nakaya, 1975. Longhead catshark

Patchy in the Indo-West Pacific. Upper to mid slope and submarine ridges. 500–1140 m.

Apristurus macrorhynchus (Tanaka, 1909). Flathead catshark

Northwest Pacific: Japan, China, and Taiwan. Upper to mid slope. 220–1140 m.

Apristurus macrostomus Meng, Chu & Li, 1985. Broadmouth catshark

Northwest Pacific: South China Sea off China. Mid slope. 913 m.

Apristurus manis (Springer, 1979). Ghost catshark

Patchy records in the North and Southeast Atlantic. Upper to deep slope. 658–1740 m.

Apristurus melanoasper Iglésias, Nakaya & Stehmann, 2004. Black roughscale catshark

Patchy in the Atlantic and Indo-West Pacific. Upper to deep slope and submarine ridges. 512–1520 m.

Apristurus microps (Gilchrist, 1922). Smalleye catshark

Patchy in the North and Southeast Atlantic. Upper to deep slope. 700–2200 m.

Apristurus micropterygeus Meng, Chu & Li, in Chu, Meng & Li, 1986. Smalldorsal catshark.

Northwest Pacific: South China Sea off China. Mid slope. 913 m.

Apristurus nasutus de Buen, 1959. Largenose catshark

Eastern Pacific: patchy off Central and South America. Eastern Central Atlantic record probably erroneous. Upper to mid slope. 400–925 m.

Apristurus parvipinnis Springer & Heemstra, in Springer, 1979. Smallfin catshark

Western Central Atlantic: Gulf of Mexico, Caribbean and northern South America. Upper to mid slope. 636–1115 m.

Apristurus pinguis Deng, Xiong & Zhan, 1983. Fat catshark

Patchy in the Eastern Indian and Western Pacific. Mid to deep slope and submarine ridges. 996–2057 m.

Apristurus platyrhynchus (Tanaka, 1909). Spatulasnout catshark

Patchy in the Eastern Indian and Western Pacific. Upper to mid slope and submarine ridges. 400–1080 m. Includes the junior synonyms *Apristurus acanutus* Chu, Meng & Li, in Meng, Chu & Li, 1985 and *Apristurus verweyi* (Fowler, 1934).

Apristurus profundorum (Goode & Bean, 1896). Deepwater catshark

Northwest Atlantic off the United States, possibly Eastern Central Atlantic off West Africa. Deep slope. ~1500 m. Records from the Indian Ocean require confirmation.

Apristurus riveri Bigelow & Schroeder, 1944. Broadgill catshark

Western Central Atlantic: Gulf of Mexico, Caribbean, and northern South America. Upper to mid slope. 732–1461 m.

Apristurus saldanha (Barnard, 1925). Saldanha catshark

Southeast Atlantic: southern Africa. Upper to mid slope. 344–1009 m.

Apristurus sibogae (Weber, 1913). Pale catshark

Western Central Pacific: Indonesia. Upper slope. 655 m.

Apristurus sinensis Chu & Hu, in Chu, Meng, Hu & Li, 1981. South China catshark

Eastern Indian and Western Pacific: Australia and South China Sea. Upper to mid slope. 940–1290 m.

Apristurus spongiceps (Gilbert, 1895). Spongehead catshark

Patchy in the Western and Central Pacific: Indonesia and Hawaii. Upper to mid slope. 572–1482 m.

Apristurus stenseni (Springer, 1979). Panama ghost catshark

Eastern Central Pacific: Panama. Mid slope. 915–975 m.

Asymbolus galacticus Séret & Last, 2008. New Caledonia spotted catshark

Western Central Pacific: New Caledonia. Upper slope. 235–550 m.

Asymbolus pallidus Last, Gomon & Gledhill, in Last, 1999. Pale spotted catshark

Western Central Pacific: northeastern Australia. Upper slope. 225–400 m.

Bythaelurus? alcocki (Garman, 1913). Arabian catshark

Eastern Indian: Arabian Sea. Mid slope. 1134–1262 m. Original and only specimen has been lost, and thus placement of the species in this genus is tentative.

Bythaelurus canescens (Günther, 1878). Dusky catshark

Southeast Pacific: Peru and Chile. Upper slope. 250–700 m.

Bythaelurus clevai (Séret, 1987). Broadhead catshark

Eastern Indian: Madagascar. Upper slope. 400–500 m.

Bythaelurus dawsoni (Springer, 1971). Dawson's catshark

Southwest Pacific: New Zealand. Shelf, upper to mid slope. 50–790 m (most common 300–700 m).

Bythaelurus hispidus (Alcock, 1891). Bristly catshark

Eastern Indian: India and the Andaman Islands. Upper to mid slope. 293–766 m.

Bythaelurus immaculatus (Chu & Meng, in Chu, Meng, Hu & Li, 1982). Spotless catshark

Northwest Pacific: South China Sea. Upper to mid slope. 534–1020 m.

Bythaelurus incanus Last & Stevens, 2008. Dusky catshark

Eastern Indian: northwestern Australia. Mid slope. 900–1000 m.

Bythaelurus lutarius (Springer & D'Aubrey, 1972). Mud catshark

Western Indian: patchy off East Africa. Upper to mid slope. 338–766 m.

Cephaloscyllium albipinnum Last, Motomura & White, 2008. Whitefin swellshark

Eastern Indian and Southwest Pacific: southern Australia. Outer shelf and upper slope. 126–554 m.

Cephaloscyllium cooki Last, Séret & White, 2008. Cook's swellshark

Eastern Indian: northwestern Australia and Indonesia. Upper slope. 223–300 m.

Cephaloscyllium fasciatum Chan, 1966. Reticulated swellshark

Patchy in the Western Pacific. Upper slope. 219–450 m.

Cephaloscyllium hiscosellum White & Ebert, 2008. Australian reticulate swellshark

Eastern Indian: northwestern Australia. Upper slope. 294–420 m.

Cephaloscyllium isabellum (Bonnaterre, 1788). Draughtsboard shark

Southwest Pacific: New Zealand. Shelf and upper slope. Inshore–690 m.

Cephaloscyllium signourum Last, Séret & White, 2008. Flagtail swellshark

Western Central Pacific: northeastern Australia (and possibly elsewhere in Oceania). Upper slope and submarine ridges. 480–700 m.

Cephaloscyllium silasi (Talwar, 1974). Indian swellshark

Western Indian: India. Upper slope. ~300 m.

Cephaloscyllium speccum Last, Séret & White, 2008. Speckled swellshark

Eastern Indian: northwestern Australia. Outer shelf and upper slope. 150–455 m.

Cephaloscyllium sufflans (Regan, 1921). Balloon shark

Western Indian: South Africa and Mozambique. Shelf and upper slope. 40–440 m.

Cephaloscyllium variegatum Last & White, 2008. Saddled swellshark

Western Central and Southwest Pacific: eastern Australia. Outer shelf and upper slope. 114–606 m.

Cephaloscyllium zebrum Last & White, 2008. Narrowbar swellshark

Western Central Pacific: northeastern Australia. Upper slope. 444–454 m.

Cephalurus cephalus (Gilbert, 1892). Lollipop catshark

Eastern Central Pacific: Mexico. Outer shelf and upper to mid slope. 155–927 m.

Figaro boardmani (Whitley, 1928). Australian sawtail catshark

Eastern Indian and Western Pacific: Australia. Outer shelf and upper slope. 150–640 m.

Figaro striatus Gledhill, Last & White, 2008. Northern sawtail catshark

Western Central Pacific: northeastern Australia. Upper slope. 300–419 m.

Galeus antillensis Springer, 1979. Antilles catshark

Western Central Atlantic: Straits of Florida and the Caribbean. Upper slope. 293–658 m.

Galeus arae (Nichols, 1927). Roughtail catshark

Northwest and Western Central Atlantic: United States, Gulf of Mexico, Caribbean, and Central America. Upper slope. 292–732 m.

Galeus atlanticus (Vaillant, 1888). Atlantic sawtail catshark

Patchy in the Northeast Atlantic and Mediterranean. Upper slope. 400–600 m.

Galeus cadenati Springer, 1966. Longfin sawtail catshark

Western Central Atlantic: off Panama and Colombia. Upper slope. 439–548 m.

Galeus eastmani (Jordan & Snyder, 1904). Gecko catshark

Patchy in the Northwest Pacific. Habitat information not available.

Galeus gracilis Compagno & Stevens, 1993. Slender sawtail catshark

Eastern Indian and Western Central Pacific: patchy across northern Australia. Upper slope. 290–470 m.

Galeus longirostris Tachikawa & Taniuchi, 1987. Longnose sawtail catshark

Northwest Pacific: Japan. Upper slope. 350–550 m.

Galeus melastomus Rafinesque, 1810. Blackmouth catshark

Wide-ranging in the Northeast and Western Central Atlantic and the Mediterranean. Shelf and upper to mid slope. 55–1440 m (mainly 200–500 m).

Galeus mincaronei Soto, 2001. Brazillian sawtail catshark

Southwest Atlantic: southern Brazil. Deep reefs on upper slope. 236–600 m.

Galeus murinus (Collett, 1904). Mouse catshark

Northeast Atlantic: Iceland and the Faeroe Islands. Upper to mid slope. 380–1250 m.

Galeus nipponensis Nakaya, 1975. Broadfin sawtail catshark

Northwest Pacific: Japan. Upper slope and submarine ridges. 362–540 m.

Galeus piperatus Springer & Wagner, 1966. Peppered catshark

Eastern Central Pacific: Gulf of California. Upper to mid slope. 275–1326 m.

Galeus polli Cadenat, 1959. African sawtail catshark.

Wide-ranging in the Eastern Atlantic off West Africa. Outer shelf and upper slope. 159–720 m.

Galeus priapus Séret & Last, 2008.

Western Central Pacific: New Caledonia and Vanuatu. Upper to mid slope. 262–830 m.

Galeus schultzi Springer, 1979. Dwarf sawtail catshark

Western Central Pacific: Philippines. Shelf (one record) and upper slope. 50–431 m (single record at 50 m, usually 329–431 m).

Galeus springeri Konstantinou & Cozzi, 1998. Springer's sawtail catshark

Western Central Atlantic: Greater and Lesser Antilles. Upper slope. 457–699 m.

Holohalaelurus favus Human, 2006. Honeycomb or Natal izak

Western Indian: southern Africa. Upper to mid slope. 200–1000 m.

Holohalaelurus grennian Human, 2006. East African spotted catshark

Scattered records in the Western Indian off East Africa. Upper slope. 238–300 m.

Holohalaelurus melanostigma (Norman, 1939). Tropical izak catshark

Western Indian: off East Africa. Upper slope. 607–658 m.

Holohalaelurus punctatus (Gilchrist & Thompson, 1914). African spotted catshark

Southeast Atlantic and Western Indian: southern Africa. Upper slope. 220–420 m.

Holohalaelurus regani (Gilchrist, 1922). Izak catshark

Southeast Atlantic and Western Indian: southern Africa. Shelf and upper to mid slope. 40–910 m (mainly ~100–300 m).

Parmaturus albimarginatus Séret & Last, 2007. Whitetip catshark

Western Central Pacific: New Caledonia. Upper slope. 590–732 m.

Parmaturus albipenis Séret & Last, 2007. White-clasper catshark

Western Central Pacific: New Caledonia. Upper slope. 688–732 m.

Parmaturus bigus Séret & Last, 2007. Beige catshark

Western Central Pacific: northeastern Australia. Upper slope. 590–606 m.

Parmaturus campechiensis Springer, 1979. Campeche catshark

Western Central Atlantic: Bay of Campeche in the Gulf of Mexico. Mid slope. 1097 m.

Parmaturus lanatus Séret & Last, 2007. Velvet catshark

Eastern Indian: Indonesia. Mid slope. 840–855 m.

Parmaturus macmillani Hardy, 1985. New Zealand filetail

Southwest Pacific: New Zealand; Western Indian: south of Madagascar. Submarine ridges. 950–1003 m (may also occur deeper).

Parmaturus melanobranchius (Chan, 1966). Blackgill catshark

Western Pacific: East and South China Seas. Upper to mid slope. 540–835 m.

Parmaturus pilosus Garman, 1906. Salamander shark

Northwest Pacific: Japan and East China Sea. Upper to mid slope. 358–895 m.

Parmaturus xaniurus (Gilbert, 1892). Filetail catshark

Eastern Central Pacific: southern United States and Mexico. Outer shelf and upper to mid slope. Juveniles mesopelagic. 91–1251 m.

Pentanchus profundicolus Smith & Radcliffe, 1912. Onefin catshark

Western Central Pacific: Philippines. Upper to mid slope. 673–1070 m.

Schroederichthys maculatus Springer, 1966. Narrowtail catshark

Western Central Atlantic: patchy off Central and northern South America. Outermost shelf and upper slope. 190–410 m.

Schroederichthys saurisqualus Soto, 2001. Lizard catshark

Southwest Atlantic: southwestern Brazil. Outer shelf (sporadic) and deep reefs on the upper slope. 122–500 m.

Schroederichthys tenuis Springer, 1966. Slender catshark

Restricted range in the Western Atlantic off Suriname and northern Brazil. Shelf and upper slope. 72–450 m.

Scyliorhinus boa Goode & Bean, 1896. Boa catshark

Western Central Atlantic: Caribbean and northern South America. Upper slope. 229–676 m.

Scyliorhinus capensis (Smith, in Müller & Henle, 1838). Yellowspotted catshark

Southeast Atlantic and Western Indian: southern Africa. Shelf and upper slope. 26–695 m (mostly 200–400 m).

Scyliorhinus cervigoni Maurin & Bonnet, 1970. West African catshark

Eastern Atlantic off West Africa. Shelf and upper slope. 45–500 m.

Scyliorhinus comoroensis Compagno, 1989. Comoro catshark

Western Indian: Comoro Islands. Upper slope. 200–400 m.

Scyliorhinus haeckelii (Ribeiro, 1907). Freckled catshark

Western Atlantic: off South America. Deep reefs on the shelf and upper slope. 37–439 m (mostly >250 m).

Scyliorhinus hesperius Springer, 1966. Whitesaddled catshark

Western Central Atlantic: Central America and Colombia. Upper slope. 274–457 m.

Scyliorhinus meadi Springer, 1966. Blotched catshark

Western Central Atlantic: the United States, Gulf of Mexico, Cuba, and the Bahamas. Upper slope. 329–548 m.

Scyliorhinus retifer (Garman, 1881). Chain catshark

Northwest and Western Central Atlantic: the United States, Gulf of Mexico, and Caribbean. Shelf and upper to mid slope. 75–754 m.

Scyliorhinus torrei Howell-Rivero, 1936. Dwarf catshark

Western Central Atlantic: Florida (United States), the Bahamas, Cuba, and the Lesser Antilles. Upper slope. 229–550 m (most >366 m).

Family Proscylliidae. Finback Catsharks.

Eridacnis barbouri (Bigelow & Schroeder, 1944). Cuban ribbontail catshark

Western Central Atlantic: Florida Straits and Cuba. Upper slope. 430–613 m.

Eridacnis radcliffei Smith, 1913. Pygmy ribbontail catshark

Wide-ranging but patchy in the Indo-West Pacific. Shelf, upper to mid slope. 71–766 m.

Eridacnis sinuans (Smith, 1957). African ribbontail catshark

Western Indian off East Africa. Outermost shelf and upper slope. 180–480 m.

Family Pseudotriakidae. False Catsharks.

Gollum attenuatus (Garrick, 1954). Slender smoothhound

Southwest Pacific: New Zealand. Outer shelf, upper slope, and seamounts. 129–724 m (most common 300–600 m).

Pseudotriakis microdon Capello, 1868. False catshark

Wide-ranging but patchy in the Atlantic, Indo-West, and Central Pacific. Shelf (occasional) and upper to deep slope. 200–1900 m.

Family Triakidae. Houndsharks.

Hemitriakis abdita Compagno & Stevens, 1993. Darksnout or deepwater sicklefin houndshark

Western Central Pacific: northeastern Australia. Upper slope. 225–400 m.

Iago garricki Fourmanoir, 1979. Longnose houndshark

Patchy in the Eastern Indian and Western Central Pacific. Upper slope. 250–475 m.

Iago omanensis (Norman, 1939). Bigeye houndshark

Northern Indian including the Red Sea; possibly in the Bay of Bengal (although this may represent a distinct species). Shelf and upper to mid (possibly to deep) slope. <110–1000+ m (possibly to 2195 m in Red Sea).

Mustelus canis insularis (Mitchell, 1815). Dusky smoothhound (island subspecies)

Western Central Atlantic: Caribbean and northern South America. Outer shelf and upper to mid slope. 137–808 m (mostly >200 m). Shelf subspecies (*M. c. canis*) prefers shallow waters (generally <18 m).

Mustelus stevensi White & Last, 2008. Western spotted gummy shark

Eastern Indian: northwestern Australia. Outer shelf and upper slope. 121–735 m.

Mustelus walkeri White & Last, 2008. Eastern spotted gummy shark

Western Central Pacific: northeastern Australia. Shelf and upper slope. 52–403 m.

Family Carcharhinidae. Requiem sharks.

Carcharhinus altimus (Springer, 1950). Bignose shark

Wide-ranging but patchy in tropical and warm-temperate waters of the Atlantic and Indo-Pacific. Shelf and upper slope. 25–430 m (mostly >90 m, young shallower, to 25 m).

3

Chondrichthyans of High Latitude Seas

David A. Ebert and Megan V. Winton

CONTENTS

3.1 Introduction

Chondrichthyans are one of the most successful groups of fishes and have penetrated most marine ecosystems, including high latitude seas. Worldwide, most chondrichthyans (~55%) occur on continental shelves from the intertidal zone down to 200 m depth and to a lesser extent on insular shelves (Compagno 1990). The diversity of shelf species is greatest in the tropics and lower in high latitude seas. However, studies on the biodiversity and life history of high latitude seas chondrichthyans, those occurring in cold-temperate latitudes and higher, are relatively few compared to those in lower latitude seas. The paucity of data from these habitats may be due to the inaccessibility of the environment and a general lack of directed chondrichthyan fisheries in lieu of more economically important bony fish species. Several major teleost fisheries (e.g., groundfish and Patagonian toothfish) in high latitude seas take chondrichthyans as bycatch, but they are discarded, and little information is gathered regarding species complexes or abundance. One exception is for larger pelagic shark species (e.g., *Lamna* spp.) that have been intensely targeted by directed fisheries in some regions.

The diversity of high latitude seas chondrichthyans has likely been underestimated, as species complexes, especially among the skates (Order: Rajiformes) and catsharks (Family: Scyliorhinidae), are poorly known in several regions. Only within the past decade has significant progress been made toward understanding and identifying the species-rich skate faunas that occur in the Bering Sea and southwestern Atlantic, while those of the Southeastern Australian Continental Shelf and upper slope remain relatively unknown. Other groups, such as the catsharks and chimaeroids (Order: Chimaeriformes) are still poorly known taxonomically and virtually nothing is known about the life histories of either group in high latitude seas.

The purpose of this chapter is to highlight the species richness of chondrichthyans occurring in high latitude seas, characterize their life histories, briefly review several high latitude fisheries, especially those in which chondrichthyans are commonly taken as bycatch, and to summarize the conservation status of chondrichthyans occurring in these regions.

3.2 High Latitude Seas Ecosystems

3.2.1 High Latitude Seas Ecosystems

The distribution of marine organisms is influenced by a variety of factors, with temperature the principal factor determining many species' geographic ranges. Most marine species exhibit horizontally stratified distribution patterns, with species diversity decreasing from warm equatorial to cool high latitude seas (Barnes and Hughes 1982; Luning and Asmus 1991; Briggs 1995). Species distribution patterns and the latitudinal thermal gradient can be used to subdivide the ocean into four zones: tropical, warm-temperate, cold-temperate, and polar zones (Briggs 1995).

The high latitude seas of the world's oceans are comprised of three major zones, the cold-temperate, Arctic Polar, and Antarctic Polar zones (Figure 3.1). The cold-temperate zone is subdivided by ocean basins and their adjacent land masses. Although the exact latitudinal range varies with location, in general cold-temperate seas are found at latitudes between

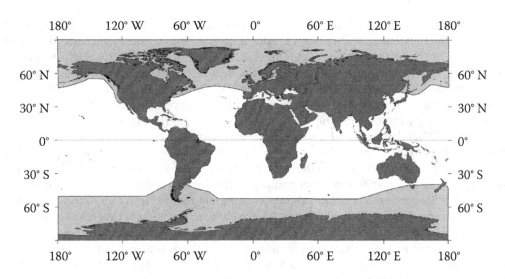

FIGURE 3.1
World map highlighting high latitude seas in gray. (Map by Joseph J. Bizzarro. Used with permission.)

30° and 40° and higher, with polar waters occurring at latitudes near or greater than the Arctic and Antarctic circles, at approximately 66°N and S, respectively. Cold-temperate and cold polar ecosystems differ in temperature, ice cover, light, and productivity regimes. Winter surface temperatures range from a low of –1.9°C (the freezing point of seawater) to 2°C in the polar zones and from 2°C to 12°C in cold-temperate zones (Briggs 1995). Unlike the nearly constant temperatures of polar waters, which exhibit average temperature changes of less than 2°C annually, surface temperatures of cold-temperate waters vary seasonally, warming by up to 7°C during the summer months. The range in temperature varies in each region with depth, decreasing to negligible levels below the thermocline at around 200 m (Hela and Laevastu 1962). Productivity levels of high latitude ecosystems span the range from low to highly productive, with most regions experiencing a high degree of seasonality. At high latitudes, productivity is limited primarily by light availability (Briggs 1995).

3.2.2 Northern Oceans Ecosystems

The North Pacific Boreal includes cold-temperate shelf, upper slope, and open ocean regions, and is subdivided into the eastern and western North Pacific regions. The eastern North Pacific (ENP) cold-temperate region extends along the North American Pacific coast from Point Conception, California, northward to the eastern Bering Sea. The extent of the cold-temperate zone in this region is determined primarily by the eastward flowing North Pacific Current, which splits to form the California Current to the south and the Alaska Current to the north (Briggs 1995). The western North Pacific (WNP) cold-temperate region extends along the eastern coast of the Asian continent from approximately Japan in the south to the northeast coast of Russia. The southern extent of cold-temperate conditions is determined largely by the position of the Oyashio Current, which transports cold, nutrient-rich waters of Arctic Polar Zone origin.

The North Atlantic Boreal encompasses the cold-temperate continental shelves, upper slopes, and open ocean region of the northern Atlantic Ocean, extending from the warm-temperate boundary in the south to approximately the Arctic Circle. Like the North Pacific, the North Atlantic cold-temperate zone is divided into western and eastern regions. The eastern North Atlantic (ENA) cold-temperate region extends from the English Channel north to the Murmansk Peninsula, Russia. In the mid-North Atlantic, the Gulf Stream continues into the North Atlantic Current, forming the North Atlantic Gyre. The Norwegian Current branches northward off of the gyre, carrying relatively warm, saline water of Gulf Stream origin into the seas off the western coast of Europe (Briggs 1995). As a result, the border of the warm and cold-temperate regions occurs farther north than in the western North Atlantic (WNA). The western North Atlantic cold-temperate region extends from Cape Hatteras, North Carolina, to the northern entrance of the Gulf of St. Lawrence. The waters along the Atlantic North American coast are bounded to the east by the northward-flowing Gulf Stream, which isolates the cold-temperate region from the warmer waters of the Sargasso Sea (Briggs 1974). Along most of the cold-temperate coastline, the Gulf Stream courses far out to sea, veering into shallow waters near the region's northern boundary (Barnes and Hughes 1982). The cold Labrador Current flows southward between the Gulf Stream and the coast, extending cold-temperate conditions farther south than they occur in the eastern Atlantic (Briggs 1995). As a result, the environment along the western North American coastline transitions rapidly from warm-temperate to polar regimes, completing the shift within 20° of latitude (Barnes and Hughes 1982).

The Arctic Ocean (ARC) region occupies all areas north of the cold-temperate boundary, primarily encompassing the polar waters within the Arctic Circle but also extending into the northwest Atlantic and the Bering Sea (Briggs 1995). The shallowest of the high latitude seas, the Arctic Ocean, features a broad continental shelf and has an average depth of 1050 m (Pidwirny 2006). Separated from the relatively warm, saline waters of the cold-temperate zone by boundary currents and ridges, the water column over the wide shelf areas is well stratified, with dense Arctic bottom water ranging in temperature from −0.8°C to −0.9°C filling the depths below approximately 800 m. Overall productivity levels are low due to seasonal depletion of nutrients in the surface layer and limited light availability; however, the region does experience bursts of production in the summer season when melting sea ice releases nutrients and sunlight is continuous (Hempel 1990).

3.2.3 Southern Oceans Ecosystems

The cold-temperate and polar seas of the southern hemisphere, commonly referred to as the Southern Oceans, cover more surface area and have been colder for a longer period of time than their northern counterparts. The Southern Oceans region is dominated and defined primarily by the Antarctic Circumpolar Current (ACC), also known as the West Wind Drift. The largest (in some places covering up to 20° of latitude) and most important current in the Southern Ocean, the ACC flows clockwise around the Antarctic continent and contributes to the formation of the Antarctic Convergence, which provides a natural boundary isolating cold polar waters from warmer waters to the north. The ACC and its various branches delineate not only the border between cold-temperate and polar waters but also subdivide the cold-temperate zone into the southern South American (SSA), Tasmanian (TAS), and southern New Zealand (SNZ) regions (Briggs 1995). The SSA is primarily influenced by the ACC as it travels along the South American coastline and is deflected northward, contributing to the formation of the Peru and Falkland currents, which extend cold-temperate conditions along the coasts of Chile and Argentina, respectively (Briggs 1995). As the ACC also flows past the southern coasts of Australia and New Zealand, it contributes to the formation of cold coastal currents. The TAS region is defined by the northern boundary of the ACC (Briggs 1995). This cold-temperate zone comprises the waters surrounding Tasmania and the province of Victoria. Though the waters off southeast Tasmania are highly productive, the waters along the shelf are less productive than typically expected due to the intrusion of low nutrient, tropical waters via the East Australian and Leeuwin Currents. Off the coast of New Zealand, the continental shelf extends from tens to hundreds of kilometers, with all but the northernmost waters in the cold-temperate zone. New Zealand's cold-temperate region includes an area extending from its coastline to several sub-Antarctic islands southeast of the mainland (Briggs 1995). Habitat availability is similar to that of the Tasmanian cold-temperate region, featuring numerous estuaries, mudflats, kelp beds, sea mounts, and deep sea trenches.

Within the boundary of the ACC lies the cold Antarctic (ANT) circumpolar zone, the northward extent of which is determined by the fluctuating position of the Antarctic Convergence. The Antarctic Convergence is a zone of the ACC in which cold surface waters sink, creating large differences in surface temperature with latitude. The resulting temperature gradient provides a natural boundary isolating cold polar waters from warmer waters to the north (Briggs 1995). The position of the Antarctic Convergence oscillates between 48° and 60°S based on the strength and direction of winds and currents. Water in contact with sea ice cools and sinks, flowing northward and eastward as it descends the Antarctic shelf. The resulting bottom water is the coldest of the deep ocean (Cushing and Walsh 1976). The displacement of surface waters due to the formation and presence of sea ice results in a system of rapid upwelling along the Antarctic coastline. As a result, surface waters are often nutrient rich but oxygen poor, with the rest of the water column fairly consistent in terms of temperature and depth (Hela and Laevastu 1962; Briggs 1995). Even with the presence of a strong upwelling system, the polar region's overall productivity is moderate due to iron limitation, ice cover, and extreme weather conditions (Briggs 1995).

3.3 Classification and Zoogeography

3.3.1 Classification

The class Chondrichthyes is generally divided into two major groups, the large subclass Neoselachii, which is subdivided into two cohorts, the Selachii (sharks) and the Batoidea (rays), and the smaller subclass Holocephalii, which includes all of the chimaeras. The Selachii is further subdivided into two superorders, the Squalomorphii and Galeomorphii. The superorder Squalomorphii includes the orders Hexanchiformes, Squaliformes, Squatiniformes, and Pristiophoriformes, while the superorder Galeomorphii includes the Heterodontiformes, Orectolobiformes, Lamniformes, and Carcharhiniformes (Table 3.1). The cohort Batoidea as defined here includes all of the batoids and recognizes four orders: Torpediniformes, Pristiformes, Rajiformes, and Myliobatiformes. The ordinal classification used here follows Compagno (2001, 2005), McEachran and Aschliman (2004), and Nelson (2006), but recognizes that there is still considerable disagreement among taxonomists (Shirai 1992; de Carvalho 1996; Naylor et al. 2005) regarding the relationships of these groups. There are currently 13 recognized orders of "sharks," a term that in the broad sense includes the batoids (flat-sharks) and chimaeras (ghost- or silver-sharks). Eight of these orders comprise the sharks, while the batoids include four orders and the chimaeras a single order.

TABLE 3.1

Biodiversity of High Latitude Seas Chondrichthyans by Order, Family, Genera, and Species

Order	Family	High Latitude Seas Chondrichthyans	
		Number Genera	Number Species
Sharks			
Hexanchiformes	Chlamydoselachidae	1 (1)	1 (2)
	Hexanchidae	3 (3)	4 (4)
Squaliformes	Echinorhinidae	1 (1)	2 (2)
	Squalidae	2 (2)	9 (26)
	Centrophoridae	2 (2)	10 (17)
	Etmopteridae	2 (5)	16 (46)
	Somniosidae	7 (7)	14 (17)
	Oxynotidae	1 (1)	3 (6)
	Dalatiidae	4 (7)	4 (10)
Pristiophoriformes	Pristiophoridae	1 (2)	3 (6)
Squatiniformes	Squatinidae	1 (1)	9 (23)
Heterodontiformes	Heterodontidae	1 (1)	2 (9)
Orectolobiformes	Parascyllidae	1 (2)	2 (9)
	Brachaeluridae	0 (1)	0 (1)
	Orectolobidae	1 (3)	1 (11)
	Hemiscyllidae	0 (2)	0 (17)
	Ginglymostomatidae	0 (3)	0 (3)
	Stegostomatidae	0 (1)	0 (1)
	Rhincodontidae	1 (1)	1 (1)
Lamniformes	Mitsukurinidae	1 (1)	1 (1)
	Odontaspididae	2 (2)	2 (3)
	Pseudocarchariidae	1 (1)	1 (1)
	Megachasmidae	0 (1)	0 (1)
	Alopiidae	1 (1)	2 (3)
	Cetorhinidae	1 (1)	1 (1)
	Lamnidae	3 (3)	5 (5)
Carcharhiniformes	Scyliorhinidae	9 (17)	33 (146)
	Proscyllidae	0 (3)	0 (6)
	Pseudotriakidae	2 (2)	2 (2)
	Leptochariidae	0 (1)	0 (1)
	Triakidae	5 (9)	18 (46)
	Hemigaleidae	0 (4)	0 (8)
	Carcharhinidae	4 (12)	10 (51)
	Sphyrnidae	1 (2)	2 (8)
Batoids			
Torpediniformes	Narcinidae	2 (4)	3 (36)
	Narkidae	1 (6)	2 (12)
	Hypnidae	0 (1)	0 (1)
	Torpedinidae	1 (1)	7 (24)
Pristiformes	Pristidae	0 (2)	0 (7)

TABLE 3.1 (*Continued*)

Biodiversity of High Latitude Seas Chondrichthyans by Order, Family, Genera, and Species

		High Latitude Seas Chondrichthyans	
Order	Family	Number Genera	Number Species
Rajiformes	Anacanthobatidae	0 (3)	0 (21)
	Arhynchobatidae	10 (12)	68 (102)
	Rajidae	13 (18)	58 (164)
	Rhinidae	0 (1)	0 (1)
	Rhynchobatidae	0 (1)	0 (5)
	Rhinobatidae	3 (4)	5 (46)
Myliobatiformes	Platyrhinidae	1 (2)	1 (3)
	Zanobatidae	0 (1)	0 (2)
	Plesiobatidae	0 (1)	0 (1)
	Hexatrygonidae	0 (1)	0 (1)
	Urolophidae	2 (2)	6 (28)
	Urotrygonidae	1 (2)	1 (16)
	Potamotrygonidae	0 (3)	0 (20)
	Dasyatidae	3 (7)	8 (87)
	Gymnuridae	1 (2)	1 (13)
	Myliobatidae	2 (4)	6 (20)
	Rhinopteridae	1 (1)	1 (11)
	Mobulidae	2 (2)	3 (10)
Chimaeras			
Chimaeriformes	Callorhinchidae	1 (1)	2 (3)
	Chimaeridae	2 (2)	19 (33)
	Rhinochimaeridae	2 (3)	5 (8)

Note: Global numbers (in parenthesis) as of May 1, 2009, from database maintained by D.A. Ebert.

The elasmobranchs are the dominant chondrichthyan group, with approximately 56 families representing 96% of extant species. The remaining 4% includes the three chimaera families. In higher taxonomic groupings (genera and above) the sharks are more diverse than the batoids, but among all elasmobranch species there are more batoids (54%) than sharks (42%). Worldwide approximately 1169 species of cartilaginous fishes (D.A. Ebert, personal database, May 1, 2009) have been described, with approximately 179 new species named within the past decade (Figure 3.2). This is only slightly less than the total number of chondrichthyans (*n* = 199) named over the previous three decades (1970 to 1999). In addition, it is estimated that at least 100 or more species are still awaiting formal description by researchers.

Twelve of 13 chondrichthyan orders are found in high latitude seas. The Carharhiniformes, Rajiformes, and Squaliformes are the most dominant groups in terms of abundance and species richness within these regions. The occurrence of two other groups, the chimaeroids and lamnoids, is notable, but for different reasons. The chimaeroids are well represented in terms of species richness, with 26 of 44 known species occurring in cold temperate and polar seas. The lamnoids are notable because both species in the genus *Lamna* are largely confined to cold temperate and polar seas, are well known for their voracious appetite on

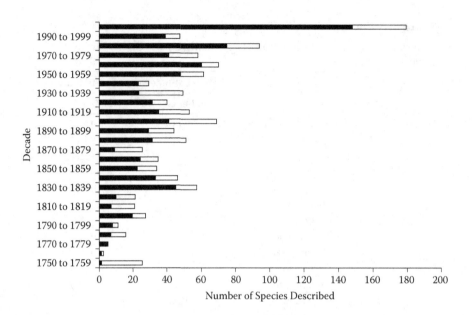

FIGURE 3.2

Total number of chondrichthyans described by decade. Open bars represent the number of high latitude species. Total chondrichthyan numbers from database maintained by D.A. Ebert.

economically important species, and are themselves the subject of directed fisheries in these regions. Although the number of families and genera tends to decrease with increasing latitude, a relatively high percentage of families (72.9%) and genera (55.7%) are represented in high latitude seas. Approximately 354 (30.4%) of the estimated 1169 described chondrichthyan species have been recorded in high latitude seas. Of the sharks, 26 of 34 families, 59 of 106 genera, and 158 of 494 described species have been recorded in high latitude seas. The batoids are represented by three of four orders, 14 of 22 families, 43 of 80 genera, and 170 of 631 species. The Pristiformes (sawfishes) are the only chondrichthyan order not represented in cold temperate or polar seas. The chimaeras are represented by all three families, with five of six genera occurring in high latitude seas. In recent years, the number of species reported to occur in high latitude seas has increased as more species have been discovered in these regions ($n = 31$) over the past decade than in any previous decade (Figure 3.2). Nearly 20% of new species described over the past decade are known to occur in high latitude seas.

3.3.2 Zoogeographic Sources

The geographic distribution of high latitude chondrichthyans was compiled from individual species checklists and regional guides, and through the generous assistance of knowledgeable colleagues who shared their personal insights into regional species accounts. The North Pacific species were compiled from Mecklenberg, Mecklenburg, and Thorsteinson (2002), Ebert (2003), Stevenson et al. (2007), Nakabo (2002), Nakaya and Shirai (1992), Lindberg and Legeza (1959), and Ishihara and Ishiyama (1985, 1986). The North Atlantic species list was compiled from Bigelow and Schroeder (1948, 1953, 1957), Collette and Klein-MacPhee (2002), Moore et al. (2003), Scott and Scott (1988), Hartel et al. (2008), Whitehead et al. (1984), and Gibson et al. (2008). There are no definitive references on Arctic Ocean chondrichthyans since this region has been very poorly studied; therefore, this checklist was compiled from the aforementioned references for the North Pacific and North Atlantic regions. The Southern Ocean regions included regional checklists from southern South America (Bovcon and Cochia 2007; Cousseau et al. 2007; Menni and Lucifora 2007; Menni and Stehmann 2000; Lamilla and Saez 2003; Pequeño and Lamilla 1993), Tasmania (Last and Stevens 2009; Last, White, and Pogonoski 2007; Last, White, and Pogonoski 2008; Last et al. 2008), and New Zealand (Cox and Francis 1997). The Antarctic region has also been very poorly studied with regards to the chondrichthyan fauna and additional species, mostly skates, will be added as improved taxonomic resolution clarifies the status of several species complexes (Gon and Heemstra 1990; Smith et al. 2008).

3.3.3 Zoogeography

The overall species diversity of high latitude seas chondrichthyans is comparable to that of other marine ecosystems, for example, tropical and deep-sea (see Chapters 2 and 4), but differs in terms of composition of species groups (Figure 3.3). Of these two ecosystems, the deep-sea fauna is closest in terms of major species groups, whereas the tropical chondrichthyan fauna is largely composed of groups whose diversity is relatively low or absent at higher latitudes. A total of 158 shark, 170 batoid, and 26 chimaera species are known to occur in one or more of the high latitude regions covered in this chapter. This compares to the 199 shark and 251 batoid species known to occur in tropical regions; no chimaeras are known to occur on shallow tropical reefs (see Chapter 4). The most specious tropical groups are the scyliorhinids, triakids, carcharhinids, and dasyatids. Although the scyliorhinids and triakids are also major groups in high latitudes, the genera and species are quite different. The major groups comprising the deep-sea environment (rajoids, squaloids, scyliorhinids, and chimaeroids) are most similar to the major groups occurring at high latitudes and is likely due to the similarities in the environment of both regions, for example, cold and with limited light and productivity. A total of 254 shark, 236 batoid, and 40 chimaera species are known to occur in the deep-sea marine ecosystem (see Chapter 2).

High latitude regions with the highest diversity of families, genera, and overall species numbers are the TAS, SNZ, and ENA. The SNZ has the highest number of families with 32, followed by the TAS and ENA with 31 and 29, respectively (Table 3.2). The TAS has the highest number of genera and species with 62 and 98, respectively. The SNZ has a slightly higher number of genera (61) than the ENA (55); however, the latter has a slightly greater diversity in number of species (95) than does the SNZ (91). The regions with the least diversity of families, genera, and species are the ANT and ARC (Table 3.2).

The North Pacific region, with the exception of the ANT and ARC regions, has the least diverse chondrichthyan fauna, being represented by 26 families, 44 genera, and 84 species (Table 3.2). Most of the families, genera, and species are in the order Rajiformes, with the skates being the most dominant group with 36 species. The genus *Bathyraja* has its greatest diversity in this region, with 22 of 46 known species occurring in the North Pacific. The ENP has 21 families, 30 genera, and 47 species, as compared to the WNP, which has 19 families, 31 genera, and 57 species represented. Twenty species, eight sharks, and 12 batoids, occur on both sides of the Pacific. Of the shark species known to occur on both sides of the Pacific at high latitudes, three, *Squalus suckleyi*, *Somniosus pacificus*, and *Lamna ditropis*, are endemic to this region. Most of the skate species that occur on both sides of the North Pacific are endemic to the Bering Sea and Aleutian Islands region. There are no high latitude seas chimaera species that occur in both the ENP and WNP.

The North Atlantic region has one of the highest diversities, with 31 families, 61 genera, and 119 species. Of these totals, the ENA has a higher diversity in terms of families, genera,

FIGURE 3.3
A color version of this figure follows page 336. Representatives of the five major groups inhabiting high latitude seas going clockwise starting in the upper left-hand corner. Rajiformes: Rajidae (*Amblyraja badia*); Carcharhiniformes: Scyliorhinidae (*Apristurus kampae*); Chimaeriformes: Chimaeridae (*Hydrolagus colliei*); Squaliformes: Somniosidae (*Somniosus pacificus*); and Carcharhiniformes: Triakidae (*Triakis semifasciata*). (Photos © Monterey Bay Aquarium Research Institute (catshark, skate, sleeper shark); Joseph J. Bizzarro (chimaera); Aaron B. Carlisle (leopard shark). All rights reserved.

and species than the WNA (Table 3.2). In terms of overall species numbers, most of the difference between the two regions occurs in the Squaliformes, from which seven families and 23 species are reported from the ENA as compared to only six families and 10 species in the WNA. Additionally, the number of skate species is higher in the ENA (27) than the WNA (16). A total of 25 shark species (most in the orders Squaliformes, Lamniformes, and Carcharhiniformes), 13 batoids (including 10 skate species), and five chimaeriformes occur in both the ENA and WNA.

The ARC, although poorly known, has relatively low diversity, with only 12 families, 16 genera, and 29 species known to occur in this region (Table 3.2). Most species (23) found to occur in this region belong to the Rajiformes, Squaliformes, or Chimaeriformes, and are also found in the North Atlantic, with one exception. Excluding wide-ranging pelagic species, the only North Pacific species known to extend its range into the ARC is *Somniosus pacificus* (Benz et al. 2004).

The Southern Ocean regions have a relatively diverse fauna, with 39 families, 87 genera, and 208 species reported between the ANT, SSA, SNZ, and TAS (Table 3.2). Of the total

TABLE 3.2

The Families and Species Number Represented within Each High Latitude Region

Family	ENP	WNP	ENA	WNA	ARC	SSA	SNZ	TAS	ANT
Chlamydoselachidae			1	1	1		1	1	
Hexanchidae	2		3	2		2	3	3	
Echinorhinidae	1		1	1		1	2		
Squalidae	1	1	3	1	1	2	4	3	
Centrophoridae			4	1			5	5	
Etmopteridae		3	4	3	2	8	4	6	1
Somniosidae	1	1	7	2	2	2	10	8	1
Dalatidae			2	2		1	3	3	
Oxynotidae			2				1	1	
Squatinidae	1	2	1	1		3		1	
Pristiophoridae		1						2	
Heterodontidae		1					1	1	
Rhincodontidae	1	1				1			
Parascyllidae								2	
Orectolobidae								1	
Odontaspididae				1			1		
Mitsukurinidae			1				1		
Pseudocarchariidae							1		
Alopiidae	1		2	1		1	2	1	
Cetorhinidae	1	1	1	1	1	1	1	1	
Lamnidae	3	3	4	4	1	2	3	3	1
Scyliorhinidae	3	3	9	5	2	2	9	11	
Pseudotriakidae			1	1	1		2		
Triakidae	3	3	3	1		6	2	4	
Carcharhinidae	2	1	2	6		2	7	3	
Sphyrnidae		1	1	2		1	1	1	
Narcinidae						2		1	
Narkidae							2		
Torpedinidae	1		3	1		1	1	1	
Rhinobatidae	1		2					1	
Platyrhinidae	1					1			
Arhynchobatidae	15	23	3	2	2	27	9	2	5
Rajidae	5	4	24	14	11	7	3	11	2
Urolophidae	1							6	
Urotrygonidae									
Dasyatidae	1	2	2	2	2		3	3	
Gymnuridae			1	1	1		1		
Myliobatidae	1		2				1	1	
Rhinopteridae			0	1	1				
Mobulidae		1	1	1	1		2		
Callorhinchidae							1	1	

Continued

TABLE 3.2 (*Continued*)

The Families and Species Number Represented within Each High Latitude Region

Family	ENP	WNP	ENA	WNA	ARC	SSA	SNZ	TAS	ANT
Chimaeridae	1	3	3	4	2	2	3	6	
Rhinochimaeridae		2	2	3	3	3	1	4	
Number Families	21	19	29	27	16	22	32	31	5
Number Species	47	57	95	65	34	78	91	98	10

Note: Abbreviations for regions as given in chapter.

number of species found in the Southern Oceans, only three shark species are known to occur in all four regions. Fourteen shark and one chimaera species are known to occur in three of these regions, the SSA, SNZ, and TAS. Species overlap between regions is highest between the SNZ and TAS, with 35 species restricted to these two regions. This includes 25 shark, four batoid, and six chimaeroid species. The SSA in comparison has relatively little overlap between the SNZ and TAS regions. Only two shark species, *Echinorhinus brucus* and *Proscymnodon macracanthus*, are known to occur in both the SSA and SNZ, and only *Etmopterus bigelowi* and *Carcharhinus brachyurus* are known to occur exclusively in both the SSA and TAS. No batoid or chimaeroid species are known to occur in both the SSA and the SNZ or TAS. Only one species, *Rhinoraja murrayi*, appears to be endemic to the ANT region. Five skate species are known to occur in both the ANT and SSA regions, but not in the other regions.

3.4 Biodiversity

3.4.1 Hexanchiformes

The Hexanchiformes (cow and frilled sharks) comprises two families, four genera, and six species of moderate-sized to very large sharks. These sharks are unique in that they are the only species to possess six or seven paired gill slits, a single dorsal fin, and an anal fin. The hexanchoids are usually considered to be one of the more primitive groups of modern day sharks. Members of this group occupy a benthopelagic habitat and range from tropical to cold temperate seas. Despite the low number of species within this group their wide-ranging distribution is comparable to that of more species-rich groups as evidenced by the occurrence of hexanchoids in both boreal and southern high latitude seas. Representatives of both families and all four genera, and five of six known species occur in high latitude seas (Tables 3.1 and 3.2). All members of this order exhibit yolksac viviparity, with litter sizes ranging from less than a dozen to more than 100 depending on the species (Ebert 1990). The Chlamydoselachidae are sporadically distributed worldwide and are found in the North Atlantic, TAS, and SNZ regions. There is one record of *Chlamydoselachus anguineus* occurring in the ARC region of northern Norway. The species feeds primarily on other sharks and cephalopods and as such has a relatively high trophic level (TL) 4.2 (Ebert 1990; Cortés 1999). The Hexanchidae are also wide ranging, occurring in all regions except for polar seas. The hexanchoids occur mostly along upper continental slopes and the outer continental shelf, with the exception of *Notorynchus cepedianus*, a primarily coastal continental shelf species. One species, *Hexanchus griseus*, is perhaps one of the most abundant and widest ranging shark species, with a distribution comparable to *Squalus acanthias* and *Prionace glauca*. The hexanchoids are apex predators in those environments in which they occur. The TL of this shark group is one of the highest of any group, ranging between 4.2 and 4.7 (Cortés 1999), with members feeding on cephalopods, large teleosts, including billfish, pinnipeds, and even small cetaceans on occasion (Ebert 1990, 1991, 1994).

3.4.2 Squaliformes

The Squaliformes (dogfish sharks) are the third largest group of chondrichthyans, with seven families, 24 genera, and at least 124 described species. The squaloids are a morphologically diverse group that contains some of the smallest and largest known shark species. This order is characterized by two small to moderately large dorsal fins, which may be preceded by a fin-spine in some species, the absence of an anal fin, eyes without a nictitating membrane, and five paired gill slits located anterior to the pectoral fin. In cold temperate and polar seas, members of this group inhabit both shallow and deep water environs, but are generally replaced by requiem (Carcharhinidae) and hammerhead (Sphyrnidae) sharks in warm temperate and tropical seas. All seven families, 19 genera, and at least 58 species occur in high latitude seas (Tables 3.1 and 3.2). The number of species in this group is expected to increase with continued exploration of the deep sea and improved taxonomic resolution incorporating molecular tools. The members of this group occupy a wide variety of habitats and are found from inshore to over 6000 m. Individual species may be demersal or epipelagic, with some species exhibiting pronounced ontogenetic changes in habitat during their lifetime. This is one of the most species-rich groups in cold temperate and polar seas, surpassed only by the Rajiformes and Carcharhiniformes. Most of the species in this group occur in four families, the Centrophoridae, Etmoperidae, Squalidae, and Somniosidae, the latter three of which have species representatives occurring in polar waters, with one genus (*Somniosus* spp.) containing three very large species considered to be truly polar sharks. The other nine species represented in this group are distributed among three relatively small families, the Echinorhinidae, Dalatiidae, and Oxynotidae whose members have scattered but wide-ranging distributions and occur in almost all cold temperate seas. All members of this shark group exhibit yolksac viviparity or limited histotrophy (Musick and Ellis 2005), with litter sizes as small as one for some species (e.g., *Centrophorus* spp.) but upward of 300 or more in some of the large *Somniosus* species. The overall mean TL for this group is 4.1, but ranges widely from 3.5 to 4.4 depending on the individual species (Cortés 1999).

The Squalidae are a wide-ranging, benthopelagic family with two genera and 26 species recognized globally. Both genera comprising nine species are reported to occur in cold-temperate and polar seas. The SNZ region has four representatives, the TAS and ENA regions three each, and the SSA region two. A single species has been reported from the ENP and WNP, WNA, and ARC regions. The family shows a high degree of endemism, with only one species (*Squalus acanthias*) known to occur in multiple geographic regions. *Squalus acanthias* is considered one of the most wide ranging of all shark species, occurring in all high latitude seas except for ANT and the North Pacific; recent taxonomic and molecular studies (Ebert et al. in review) comparing *S. acanthias* forms globally have revealed the North Pacific form to be a distinctly different species (*S. suckleyi*). The members of this group have relatively small litters of between 2 and 12 young. The mean TL of this group is 4.1; the diets of these sharks consist largely of cephalopods and teleosts (Cortés 1999).

The Centrophoridae are a taxonomically challenging group with two genera and 17 species currently being recognized worldwide. Both genera and ten species from this family have been recorded in cold temperate seas, but no species are known to occur in polar seas. This group of benthopelagic sharks is most diverse in the SNZ and TAS regions, with each reporting five species, three of which (*Centrophorus harrissoni*, *C. squamosus*, and *Deania calcea*) occur in both regions. The ENA region has four species, one of which (*C. granulosus*) is also the lone representative of the family in the WNA. Two wide-ranging species, *C. squamosus* and *D. calcea*, occur in the SNZ, TAS, and ENA regions. No members of this group are reported to occur in the ENP, WNP, SSA, or either polar region. Members of this family are extremely vulnerable to overexploitation as litter sizes are typically small, ranging from as little as a single pup per reproductive cycle in some species to 12 per litter. One species, *Centrophorus zeehaani*, has reportedly declined by up to 99% in some portions of its range (Graham, Andrew, and Hodgson 2001). Available dietary information reveals that the members of this group are voracious predators, consuming mostly teleosts and cephalopods, with a relatively high mean TL at 4.2 (Cortés 1999).

The Etmopteridae are the largest family of squaloid sharks, with five genera and 46 or more species occurring worldwide. Members are benthopelagic deepwater sharks that mostly occur between 200 and 1500 m deep though some species may range down to 4500 m (Compagno et al. 2005). Two genera (*Centroscyllium* and *Etmopterus*) and 16 species occur in high latitude seas, with the highest diversity occurring in the SSA ($n = 8$) and TAS ($n = 6$) regions. The ENA and SNZ regions each have four species, and the WNA and WNP regions three each. The family has a single representative in the ARC and ANT regions, with the ENP the only high latitude region not represented by this group. Most species in this family are regional endemics occurring in only one or two regions, but three (*Centroscyllium fabricii*, *Etmopterus lucifer*, and *E. pusillus*) are known to occur across several wide-ranging regions. All three may eventually prove to be less widely distributed as systematic revisions within these genera reveal them to be species complexes. Very little information is available on the litter size of individual species within this family, except that they appear to be small (usually less than ten per litter), with some species possibly producing only one or two pups per litter. The TL range is somewhat broad in this group, with some species feeding primarily on euphausiids and crustaceans (TL = 3.5 to 3.8), and others on teleosts and cephalopods (TL = 4.0 to 4.2) (Cortés 1999).

The Somniosidae are one of the widest-ranging groups, occurring in most seas from the tropics to polar seas (Figure 3.3). Members of this group occupy a benthopelagic or pelagic habitat. The group consists of seven genera and 17 described species, containing some of the smallest and largest known shark species. The group has representatives from all seven genera, comprising 14 of 17 known species, in all high latitude seas and includes one genus (the large *Somniosus* spp.) that are considered to be primarily polar, though members may occur in very deep water in tropical regions. In cold temperate and polar seas large *Somnoisus* spp. may occur in shallow water, even in the intertidal, but move into progressively deeper water in warm temperate and tropical seas. The SNZ and TAS regions are the most diverse, with 10 and 8 species, respectively. Considerable overlap occurs between these two regions, with all eight species found in the TAS region also found in the SNZ region. The ENA is the next most diverse region, containing seven, mostly wide-ranging species. The other regions (ENP and WNP, WNA, SSA, ARC, and ANT) have only one or two species representatives each. Litter size for members of this group is variable, with some species having as few as two pups per litter to 300 or more in some of the large *Somniosus* species (Ebert 2003). The members of this family for which dietary information is available feed primarily on cephalopods, teleosts, other chondrichthyans, and in some instances pinnipeds and cetaceans. It is therefore not surprising that they have an overall mean TL of 4.2 (Cortés 1999).

The Echinorhinidae are a small group consisting of a single genus with two wide-ranging benthopelagic species (*Echinorhinus brucus* and *E. cookei*) and a scattered worldwide distribution. Representatives of one or both species have been reported from the ENP, ENA, WNA, SSA, and SNZ. Although they have not been reported from the WNP and TAS regions it would not be surprising if they are eventually documented in either region. The difference in litter size between these two species is quite striking as *E. brucus* has a relatively small litter size at 15 to 26 young per litter (Compagno, Ebert, and Smale 1989), while a maximum litter size of 114 has been reported for *E. cookei* (Ebert 2003). The mean TL for this family is among the highest of any squaloid at 4.4 as these sharks tend to feed on other chondrichthyans and teleosts (Cortés 1999; Ebert 2003).

The Oxynotidae are a morphologically distinct group of six small to medium-sized sharks within a single genus. Members are benthic dwelling sharks that occur between 50 and 650 m deep. Three species are found in high latitude seas; two (*Oxynotus centrina* and *O. paradoxus*) in the ENA and a single species (*O. bruniensis*) found in each of the SNZ and TAS regions. There are no records of this genus from the other cold-temperate, ARC, or ANT regions. Information on litter sizes for this group is sparse, but reportedly ranges from 7 to 22 (Compagno 1984; Capape, Seck, and Quignard 1999). Virtually nothing is known about the diet of these sharks except that they consume polychaetes and other small benthic invertebrates.

The Dalatiidae are a small group of seven genera and 10 species of dwarf to medium-sized, wide-ranging sharks. Members of this group are mostly pelagic or benthopelagic and occur from near surface waters to over 1500 m deep. Four genera each with a single species are represented in high latitude seas. Three species (*Dalatias licha*, *Euprotomicrus bispinatus*, and *Isistius brasiliensis*), all of which co-occur in both regions, are found in the SNZ and TAS regions. The ENA and WNA regions each contain two species and the SSA region a single species. *Dalatias licha* is the widest-ranging species in this group occurring in four regions, with *E. bispinatus* found in three regions. No representatives have been reported for the ENP and WNP, ARC, and ANT regions. Litter sizes range from 6 to 16 young (Compagno 1984; Ebert 2003). The mean TL for this group is high at 4.2 reflecting the voracious appetite of this group's members, which includes *Isistius brasiliensis*, a species known to parasitize large teleosts and marine mammals and consume cephalopods (Cortés 1999).

3.4.3 Squatiniformes

The Squatiniformes (angel sharks) comprise a single family (Squatinidae) and genus (*Squatina*) of small, rather morphologically undiverse dorso-ventrally flattened sharks. The members of this group are all very similar morphologically and are often misidentified. Recent systematic work has helped clarify the regional status of the group and has led to the description of six new species in recent years. Members of this group are benthic inhabitants and are often seen partially buried in the sediment. There are approximately 23 described species worldwide, mostly in temperate to subtropical seas, with nine species known to extend their range into cold temperate seas (Tables 3.1 and 3.2). Of the nine species occurring in cold temperate seas, the North Pacific and SSA regions have the greatest number, with three species each, the North Atlantic with two species, and the TAS region a single species. All nine species are regional endemics, with none occurring in more than a single region. No species occurs in the SNZ or either polar region. Members of this group exhibit yolk-sac viviparity. Litter sizes for individual species in this group are poorly known, but may be as high as 25 per reproductive cycle, with an average range between 6 and 12 (Ebert 2003; unpublished data). These sharks are voracious benthic predators, feeding on crustaceans, cephalopods, and teleosts, and have a mean TL of 4.1 (Cortés 1999).

3.4.4 Pristiophoriformes

The Pristiophoriformes (saw sharks) are a small, little known, morphologically distinct, shark group with a single family (Pristiophoridae), two genera, and six described species (Tables 3.1 and 3.2). These slender-bodied sharks have a long, flattened, saw-like snout, lateral rostral teeth, a pair of long string-like barbels in front of the nostrils, two dorsal fins, and no anal fin. Most species have five paired gill slits, though one species (*Pliotrema warreni*) endemic to southern Africa has six. These benthopelagic sharks primarily occur on continental shelves in temperate areas and in deeper seas, usually along the upper continental slope, in tropical regions. The distributional range of three species, all in the genus *Pristiophorus*, extends into cold temperate seas. One species (*Pristiophorus japonicus*) occurs in the WNP primarily as a vagrant in cold temperate seas, with two species (*P. cirratus* and *P. nudipinnis*) commonly occurring in TAS waters. No species occurs in polar seas although fossil records show that some early forms did occur in Antarctic waters. Members of this group exhibit yolk-sac viviparity. Litter size in these sharks ranges from 3 to 22 depending on the species. These sharks feed primarily on demersal fishes and have a mean TL of 4.1 (Cortés 1999).

3.4.5 Heterodontiformes

The Heterodontiformes (bullhead or horn sharks) are a small shark group with one family (Heterodontidae) and a single genus (*Heterodontus*) of nine similar-looking species (Tables 3.1 and 3.2). These are the only living sharks with a dorsal spine preceding each dorsal fin and an anal fin. Members of this group are primarily benthic-dwelling, nocturnal sharks that tend to rest in caves and rocky crevices during the day, but become quite active at night. They are mostly near-shore sharks, occurring from the intertidal down to about 100 m, but at least one species is known to occur down to 275 m. The members of this group are mostly warm-temperate to tropical sharks, although two species do occur in cold-temperate waters. *Heterodontus portjacksoni* occurs in both the SNZ and TAS regions and *H. japonica* occurs in the WNP. No other *Heterodontus* species are known to occur in any of the other cold-temperate or polar seas. Members of this group all have an oviparous reproductive mode. Although fecundity is not well known for this group, estimates for those species where some information is available suggest that they may deposit between 12 and 24 egg cases per season (Compagno 2001; Ebert 2003). These sharks occupy a relatively low mean TL at 3.2 (Cortés 1999) because they feed mostly on gastropods, clams, crabs, shrimps, sea urchins, polychaete worms, and small fishes.

3.4.6 Orectolobiformes

The Orectolobiformes (carpet sharks) are composed of seven families, 14 genera, and 44 species globally (Tables 3.1 and 3.2). Members of this group can be distinguished by the presence of nasal barbels, two spineless dorsal fins, and an anal fin. Primarily benthic tropical to warm-temperate sharks found from intertidal waters to about 200 m deep and usually on rocky or coral reefs, but one notable species (*Rhincodon typus*) occupies an oceanic habitat. Three families (Orectolobidae, Parascyllidae, and Rhincodontidae), three genera (*Rhincodon, Parascyllium,* and *Orectolobus*), and four species have been reported from cold-temperate seas; three of four species occur exclusively in the TAS region. On rare occasion *Rhincodon typus* has been found to stray into boreal regions of the North Pacific (Yano, Sugimoto, and Nomura 2003; Ebert et al. 2004). There are no records of these sharks occurring in polar seas. The families Brachaeluridae, Hemiscyllidae, Ginglymostoma, and Stegostomatidae are not represented in high latitude seas. Members of this group may exhibit either an oviparous or yolksac viviparous mode of reproduction. Fecundity estimates for most species in this group are poorly known except for *R. typus*, which has one of the largest litters known among modern chondrichthyans, producing at least 300 young per reproductive cycle (Joung et al. 1996). The mean TL for this group is 3.6, though it ranges from 3.1 to 4.0 between the various families (Cortés 1999).

3.4.7 Lamniformes

The Lamniformes (mackerel sharks) are a small but diverse group containing seven very distinctive families, 10 genera, and 15 species (Tables 3.1 and 3.2). Although each of these families appears to be morphologically distinct, they are all united by a number of unique features, including a pointed snout, lack of dorsal fin spines, a similar style of dentition, and oophagous reproduction. These are mostly large, actively swimming, voracious sharks, and are among the most media-publicized species. These sharks occupy a wide range of nearshore, coastal, deep-sea, and pelagic habitats from polar and cold-temperate to tropical seas. Six of seven families, nine genera, and 12 species have been recorded from high latitude seas. Two families, the Cetorhinidae and Lamnidae, are large, wide-ranging groups that include three species (*Cetorhinus maximus, Lamna ditropis,* and *L. nasus*) that occur primarily in cold-temperate and polar seas. *Lamna nasus* is one of the few known shark species that occurs in both the ARC and ANT regions. In fact, it occurs in all other cold-temperate regions except the North Pacific where it is replaced by the endemic *Lamna ditropis. Cetorhinus maximus* is also wide ranging in all cold-temperate seas and the ARC, but has not been reported from the ANT region. The lamnids, *Carcharodon carcharias* and *Isurus oxyrinchus*, are wide-ranging, oceanic species occurring in most cold-temperate seas, but are absent in polar waters. *Alopias vulpinus* (Alopidae) is widely distributed in most cold-temperate seas, though it does not occur in either the WNP or polar waters. The Odontaspididae, Mitsukurinidae, and Pseudocarcharhinidae are small families, the latter two of which are monotypic. Most records of their occurrence in high latitude seas are anomalous accounts. The only lamniform shark family not recorded from any of the cold-temperate or polar regions is the Megachasmidae. Reproduction in these sharks, for which information is available, reveals that they share a unique form of viviparity in which embryonic sharks feed on unfertilized eggs (oophagy) and in one species their fellow embryos within the uterus (intrauterine cannibalism). Information on fecundity in this group of sharks is sparse, but is generally low, with estimates of only one or two young per cycle to perhaps 16 in some species. TL estimates for this group are high, ranging from 4.2 to 4.5, with the exception of the filter-feeding *Cetorhinus maximus* (TL = 3.2). Depending on the species, diets may consist of marine mammals, other chondrichthyans, large teleosts, and cephalopods (Cortés 1999).

3.4.8 Carcharhiniformes

The Carcharhiniformes (ground sharks) are the most diverse and largest group of shark-like cartilaginous fishes, comprising eight families, 50 genera, and at least 268 described species. Five families (Scyliorhinidae, Pseudotriakidae, Triakidae, Carcharhinidae, and Sphyrnidae), 21 genera, and 65 species are known to occur in cold-temperate seas (Tables 3.1 and 3.2). Most of these species (*n* = 51) are members of two families, the Scyliorhinidae and Triakidae, and may be very common in those regions in which they occur. No members of the relatively small families Proscyllidae, Leptochariidae, and Hemigaleidae have been found in cold-temperate seas.

The Scyliorhinidae (catsharks) are the largest family of sharks, comprising 17 genera and at least 146 species worldwide; nine genera and 33 species have been reported from high latitude seas (Figure 3.3). The number of catshark species is likely to increase with increased exploration. The TAS and SNZ regions are the most diverse in the number of genera with five and four, respectively. The ENA has three genera and the WNA, ARC, ENP and WNP, and SSA regions each have two genera. The reproductive mode of this group is mostly oviparous, although a few species exhibit yolksac viviparity. Few fecundity estimates for scyliorhinids exist, but depending on the species, fecundity may range from 29 to 190, with an average of 60 (Musick and Ellis 2005). The mean TL of these generally small demersal sharks is 3.9 (Cortés 1999). Diets include crustaceans, cephalopods, and small bony fishes.

The TAS and SNZ regions are the most speciose in terms of total number of catshark species, with 11 and 9, respectively. Five of these species occur in both regions. The North Atlantic region has nine species, with three of these also occurring in the WNA, which has five catshark species overall. Two species are known to occur in the ARC region (*Apristurus laurussoni* and *Galeus murinus*), both of which also occur in the ENA. The ENP and WNP each have three endemic species. The SSA region has two catshark species, whereas the ANT has no catshark species.

The genus *Apristurus*, a highly diverse group of largely endemic deepwater sharks, is the most speciose genera of catsharks, with 15 species found in high latitude seas. No other catshark genus has more than four species represented in either the cold-temperate or polar regions. The SNZ and TAS regions are the most diverse, with six species occurring in each region, five of which occur in both regions. The ENA has five *Apristurus* species and the WNA four. Three of the North Atlantic species occur in both the ENA and WNA. *Apristurus laurussoni* occurs in the ARC and is the only polar representative of this genus. The ENP has two species and the WNP one.

The Triakidae (houndsharks) are a large family consisting of nine genera and 46 species worldwide, of which five genera and 18 species (approximately 40% of all triakids) are found in cold-temperate and polar seas (Figure 3.3). The TAS region has four genera represented, with two or three representative genera reported from other cold-temperate regions. The ARC and ANT regions have no genera or species represented. The SSA region has the most species with six, five of which are in the genus *Mustelus*. The ENP and TAS regions each have four species represented while the ENA and WNP have three species each. The SNZ region has two species and the WNA one. The group mostly consists of endemic species, with the exception of *Galeorhinus galeus*, which occurs in all high latitude regions except for the WNA, WNP, ARC, and ANT regions. The triakids exhibit a viviparous reproductive mode and are either mucoid histotrophs or placental (Musick and Ellis 2004), with litter sizes ranging from 2 to over 50 (Ebert 2003). The members of this group have a mean TL of 3.8 (Cortés 1999) and are voracious predators on the local crustacean fauna, although some species (e.g., *G. galeus*) predominantly prey upon teleosts.

The Pseudotriakidae (false catsharks) are a small group of only two species, one of which, *Pseudotraikis microdon*, is fairly wide ranging, occurring in the ENA, WNA, ARC, and in the SNZ regions. A rarely seen deep-sea species, it may eventually be found to occur in other high latitude seas. Both false catshark species exhibit a special form of oophagous reproduction. The Carcharhinidae (requiem sharks) are one of the larger shark families, comprising 12 genera and more than 50 species worldwide. The family is most diverse in tropical to warm-temperate seas, but one species, *Prionace glauca*, is perhaps one of the most wide-ranging pelagic shark species occurring in all cold-temperate seas, but not in polar waters. The Sphyrnidae (hammerhead sharks) are a small group with two genera and eight species, of mostly tropical to warm-temperate species, although *Sphyrna zygaena* is found in most cold-temperate seas, including the ENA, WNA, WNP, SNZ, TAS, and SSA regions. It is absent from the ENP cold-temperate, ARC, and ANT regions. A second species, *Sphyrna tiburo*, occasionally ranges into the WNA region, but no other sphyrnids are known from cold-temperate seas. The cold-temperate members of these families are live bearers exhibiting placental viviparity, and depending on the species litters may range from 2 to over 80. The mean TL for these shark groups is 4.0 with a range of 3.9 to 4.3 (Cortés 1999). Most species feed on cephalopods, teleost fishes, and in some instances other chondrichthyans.

3.4.9 Rajiformes

Worldwide, the batoids are the largest cartilaginous fish group comprised of four orders, 22 families, 80 genera, and at least 631 species, a number likely to increase as new species are described (Tables 3.1 and 3.2). The members of this group are morphologically diverse in appearance, but are united by their dorso-ventrally flattened bodies and ventrally located gill slits. Fourteen batoid families have representative species occurring in high latitude seas, with the two skate families, Arhynchobatidae and Rajidae, comprising approximately three-quarters (*n* = 126) of all high latitude batoid species (*n* = 168). Approximately 27% of all known batoids occur in high latitude seas, with about 44% of all skate species occurring in these regions. Members of the order Myliobatiformes (stingray rays) have eight families, 13 genera, and 27 species represented, but most records of these species occurring in cold-temperate seas are of vagrants. The Myliobatiformes generally replace the skates in warm temperate and tropical seas where they are quite abundant and specious. The order Torpediniformes (electric rays or numbfish) has three families, four genera, and 12 widely scattered species represented in cold-temperate seas, with none occurring in polar waters. Since skates are the dominant batoid group in high latitude seas, and with individual species from most other groups being anomalous or vagrant records, only the skates will be discussed further.

The Rajiformes (skates) are the dominant batoid group in cold-temperate and polar seas, which also are the most abundant group representing nearly one-fourth of all known chondrichthyan species (Figure 3.3). The skates are comprised of six families (Rhinidae, Rhynchobatidae, Rhinobatidae, Anacanthobatidae, Arhynchobatidae, and Rajidae), two of which are extremely abundant in cold-temperate and polar seas, 39 genera, and at least 339 described species (D.A. Ebert, personal database, May 1, 2009). The suborder Rajoidei consists of the "true" skates (Anacanthobatidae, Arhynchobatidae, and Rajidae), which are the most specious group of chondrichthyans, with 33 genera and at least 287 species (Ebert and Compagno 2007; Last et al. 2008; D.A. Ebert, personal database, May 1, 2009). The arhynchobatids and rajids, the two dominant chondrichthyan families, are composed of 23 genera, with at least 126 species recorded from cold-temperate and polar seas. The anacanthobatids, comprising three genera and 21 species, are a comparatively small skate group with a scattered distribution and are largely found on continental and insular slopes in tropical and warm-temperate seas. There are no anacanthobatids species reported to occur in high latitude seas. Most skates are primarily benthic species occurring from the near shore to depths of over 4000 m. All skate species are oviparous, with highly variable estimates of litter size ranging from about 18 to 350 egg cases per year for those species for which limited information is available (Lucifora and Garcia 2004; Musick and Ellis 2005; Ebert, Smith, and Cailliet 2008). The mean TL for the Arhynchobatidae and Rajidae is 3.9 and 3.8, respectively (Ebert and Bizzarro 2007). The relationship of the families Rhinidae, Rhynchobatidae, and Rhinobatidae within the batoids remains uncertain (McEachran and Aschliman 2004). Two of these families (Rhinidae and Rhynchobatidae) have no representatives in high latitudes, while the third (Rhinobatidae) has three genera and five species known to occur in high latitude seas. All five species are uncommon or vagrants in high latitude seas and therefore will not be discussed here further.

The Arhynchobatidae (softnose skates) have their greatest diversity and abundance in cold-temperate and polar seas, with 10 of 12 genera, and 68 of 102 described species having been reported from these regions. The softnose skates can be distinguished from other skate groups by their soft flexible rostrum. In terms of genera, most of the diversity is found in the SSA region where four of six genera are largely endemic to this region. The SNZ region has three genera, two of which are small endemic groups. Other less diverse regions have either one or two genera.

The dominant genus of the Arhynchobatidae is the *Bathyraja*, which has 35 species in total represented in high latitude seas of which 22 are represented in the North Pacific and 10 in the SSA regions. Nine of the *Bathyraja* species found in the North Pacific occur in both the ENP and WNP regions. The North Atlantic, ARC, SNZ, and ANT regions have between two and four representative species. *Bathyraja richardsoni* and *B. shuntovi* are the only softnose skate species found to occur in both hemispheres. The only region not occupied by this genus is the TAS.

The SSA and North Pacific regions are the most diverse in total species numbers, with 26 and 27 species, respectively. The North Atlantic region by comparison is rather depauperate with only three species, two of which (*Bathyraja richardsoni* and *B. spinicauda*) also occur in the ARC region. In the southern hemisphere, the SNZ region has nine representative species while the ANT and TAS regions have five and two species, respectively. No softnose skate species overlap in their distribution between the SSA, SNZ, and TAS regions, but three species (*Bathyraja eatoni*, *B. mccaini*, and *B. meridionalis*) occur in both the SSA and ANT regions.

The Rajidae (hardnose skates) are those skate species with a firm, inflexible rostrum. Members of this skate family exhibit their highest degree of diversity in high latitude seas, with 13 genera and 58 of 164 described species occurring in cold-temperate and polar seas. The ENA and TAS regions have the highest diversity, with seven and six genera, respectively. The WNA and ARC regions each have five genera represented. The other regions are less diverse in terms of genera, having between one (ANT) and four (WNP) genera. The North Pacific genus "A" consisting of five species is at present undescribed.

The most speciose genus of hardnose skates is *Dipturus* with 14 species, of which six, all endemics, occur in the TAS region. Five species occur in the North Atlantic region, four in the ENA, and two in the WNA. The ARC region has four species, all of which also occur in the ENA. *Dipturus linteus* is the only species that occurs on both sides of the North Atlantic and in Arctic seas. The WNP, SSA, and SNZ regions each have a single representative species.

The Rajidae are most abundant in the North Atlantic, with a total of 30 species of which 24 occur in the ENA and 14 in the WNA. Ten of these species occur on both sides of the North Atlantic Ocean. Eleven species have been reported from the ARC region, of which eight also occur in both the ENA and WNA, four are known only from the ENA and one from the WNA. Interestingly, of the 11 known skate species reported from Arctic waters all occur in the North Atlantic, but no North Pacific skates are known to occur in the Arctic.

In the Southern Ocean the Rajidae are much less common, with only the TAS region exhibiting any degree of diversity with 11 species. The other regions of the southern hemisphere, the SSA, SNZ, and ANT, have only seven, three, and two species, respectively. There is no distributional overlap of species between the TAS, SNZ, and SSA regions, but two Antarctic species (*Amblyraja georgiana* and *A. taaf*) also occur in the SSA region. One species, *Amblyraja hyperborea*, is the only hardnose skate that occurs in both hemispheres. It has been reported from the ENA, WNA, and ARC, and from the SNZ and TAS regions.

3.4.10 Chimaeriformes

The Chimaeriformes (chimaeras, ratfish, or silver sharks) are a small, primitive group of chondrichthyans comprised of three families, six genera, and 44 described species (Figure 3.3). The chimaeras are readily distinguished by their elongated tapering body, filamentous whip-like tails, smooth scales, large venomous dorsal fin spine, broad wing-like pectoral fins, and open lateral line canals, which appear as grooves on the head and along the trunk. Chimaeras are sluggish swimming species usually found along outer continental shelves and upper slopes, although some species, especially members of the Callorhinchidae, are common in near-shore coastal waters. All three families, five genera, and 26 species occur in high latitude seas (Tables 3.1 and 3.2). At least five species, all deep-sea forms, are known to occur in the ARC region, but there are no records of any chimaeroid species reported from the ANT region. Two of three members of the Callorhinchidae, both regional endemics, occur in cold-temperate seas; *Callorhinchus callorhynchus* around SSA and *C. milli* in the SNZ and TAS regions. The Chimaeridae, the most species-rich family, have 19 of 33 described species occurring in cold-temperate and polar seas. Most members of the Chimaeridae are regional endemics except for *Hydrolagus pallidus*, which is found in the North Atlantic and in the seas around the SSA region. The SNZ and TAS regions have the most species with six and five, respectively. The ENA has four species, two of which also occur in the ARC region. The ENP has three species, the SSA and WNA regions two species each, and the ENP one species. No species are known to occur in the ANT region. The family Rhinochimaeridae is a wide-ranging group with representative species in all high latitude seas except for the ENP and ANT. Two species, *Harriotta haeckelli* and *H. raleighana*, co-occur in five regions, the ENA, WNA, ARC, SNZ, and TAS, while *Rhinochimaera pacificus* occurs in four regions, WNP, SSA, SNZ, and TAS. All members of this group are oviparous, but estimates of fecundity, ranging from 6 to 24 depending on the species, are sketchy (Barnett et al. 2009).

3.5 Life Histories

3.5.1 Reproduction

Chondrichthyans exhibit two main reproductive modes, oviparity and viviparity, with variations that have evolved within each primary mode. Oviparous species are those that deposit egg cases, usually a leathery external shell that protects developing embryos, on the seafloor bed. These egg cases are useful taxonomic tools because most egg cases have morphological characteristics that are unique to each species (Ebert 2005; Treloar, Laurenson, and Stevenson 2006; Ebert and Davis 2007). This reproductive mode is exhibited by all members of the order Chimaeriformes and the families Arhynchobatidae, Rajidae, and most of the Scyliorhinidae. Viviparous species, in which the developing embryos are retained within the mother's uterus, exhibit two primary developmental modes, lecithotrophy and matrotrophy, which in itself has a variety of reproductive modes (Musick and Ellis 2005). Lecithotrophic, or yolk-sac, viviparous species are those chondrichthyans whereby the young are nourished by a yolk sac and receive no maternal nourishment during development. Embryos of matrotophic species receive maternal nourishment through ingestion of lipids or mucous (histotroph) produced by the uterine wall or of extra unfertilized eggs produced by the mother (Musick and Ellis 2005).

Of the 354 species identified as occurring in high latitude seas, approximately 53% exhibit an oviparous reproductive mode. This is in sharp contrast to the 43% of all chondrichthyans worldwide that exhibit this reproductive mode. However, among those groups that actually reproduce in high latitude regions (e.g., mating, nursery grounds), the percentage of species exhibiting oviparity increases to 70%. This is likely due to the high number of skate, scyliorhinid, and chimaeroid species occurring in these regions. Of the 30% of high latitude seas chondrichthyans exhibiting a viviparous reproductive mode most (70%) are members of the Squaliformes, a group in which nearly 50% of all known species occur in high latitude seas. Members of the Squaliformes exhibit yolk-sac viviparity or mucoid histotrophy. Most of the remaining high latitude species are members of the Triakidae, a group that exhibits either yolk-sac viviparity, or histotrophy or placental viviparity, depending on the species.

Virtually nothing is known about the reproductive cycle of high latitude chondrichthyans. Generally speaking, oviparous chondrichthyans display three types of reproductive cycles. They are reproductively active throughout the year, exhibit a partially defined annual cycle with one or two peaks, or have a well-defined annual or biennial cycle. It has been suggested shallow water oviparous species may exhibit a more defined egg-laying season than deeper water species. However, the reproductive cycle of most high latitude seas chondrichthyans are unknown or poorly known at best. Of those species for which information is available, most appear to be reproductively active year-round or exhibit a partially defined annual cycle (Colonello, Garcia, and Lasta 2007; Ebert et al. 2007; Ebert, Smith, and Cailliet 2008; Kneebone et al. 2007). Viviparous species may exhibit similar reproductive cycles, but most cold-temperate species for which information is available appear to have an annual cycle.

It is a commonly held belief that the females of most chondrichthyan species mature at larger sizes than males. However, this axiom does not hold true for most chondrichthyans that exhibit an oviparous reproductive mode. Contrary to this outdated paradigm, the sexes of most oviparous chondrichthyans mature at a similar size and in many species the males may mature at a slightly larger size (Ebert 2005; Ebert, Compagno, and Cowley 2006; Ebert et al. 2007). A notable exception among oviparous chondrichthyans, however, can be found among those skate species that attain a maximum total length exceeding 1.5 m. In these species, the females tend to grow to a notably larger size than males (Ebert, Compagno, and Cowley 2008). Reasons for this are unclear, but may include (1) no advantage for oviparous females to attain a larger size to produce larger young, as is commonly found in viviparous species, or (2) the time an egg case might remain *in utero* is relatively minimal (as little as every 1 to 6 days) when compared to most viviparous species, which may carry their young *in utero* for up to two years or more (Holden, Rout, and Humphreys 1971; Ishihara et al. 2002; Ebert 2003; Carrier, Pratt, and Castro 2004; Ebert, Compagno, and Cowley 2008).

Fecundity estimates for most oviparous chondrichthyans are poorly known except for captive studies that suggest fecundity in some species may be relatively high both annually and throughout the entire life span of an individual skate (Holden, Rout, and Humphreys

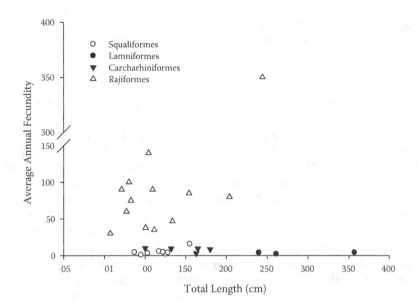

FIGURE 3.4
Relationship between total length and average annual fecundity for high latitude chondrichthyan species. Note the high estimated fecundity of the Rajiformes.

1971; Ishihara et al. 2002; Ebert 2005). Although data are limited, it appears that though oviparous species possess a reproductive strategy that produces smaller offspring, they may have a notably higher fecundity than observed in their viviparous relatives (Lucifora and Garcia 2004; Musick and Ellis 2005; Ebert, Smith, and Cailliet 2008). Fecundity estimates in skates range from a low of 18 to 350 egg cases per year, depending on the species (Figure 3.4). Of particular interest is the potential fecundity of two skate species, *Raja binoculata* and *R. pulchra*, each of which has multiple embryos per egg case; a unique mode even among skates. The fecundity of *R. binoculata* is known to exceed 350 egg cases per year and has been estimated to produce upward of 48,000 embryos during its lifetime, a number far exceeding that of any other chondrichthyan species (Ebert, Smith, and Cailliet 2008). Ishihara et al. (2002) reported that *Okamejei kenojei* matured within three years and produced from 300 to 600 egg cases during its life span. These authors also found that these skates grew very little after attaining sexual maturity, putting most of their energy into reproduction.

It has been suggested that chondrichthyans may become less fecund and may even senesce with age. This has been observed in *Galeorhinus galeus* and in several eastern Bering Sea and southwestern Pacific skate species (Peres and Vooren 1991; Ebert 2005). Ebert (2005) found evidence of senescence in three species of Bering Sea skates and found that the number of ovarian eggs appeared to decline slightly in larger, and presumably older, individuals in at least one species, *Bathyraja aleutica*. Frisk (2004) found that the net reproductive rate of *Leucoraja erinacea* and *L. ocellata* peaked at 7 and 13 years, respectively, but declined in fecundity beyond these ages. These observations combined with those of Ebert (2005) indicate that younger mature fish are more fecund than older, and in some instances larger, fish. This is in contrast to several species of viviparous elasmobranchs, especially those in the family Triakidae, in which size appears to be a more important indicator of fecundity than age (Ebert and Ebert, 2005). These life-history traits have both ecological and fisheries management implications. The removal of older, less fecund oviparous individuals that tend to consume more resources may in effect provide more resources and habitat for younger, more virile fish. This may create a much healthier, more robust population of these important demersal elasmobranchs.

3.5.2 Age and Growth

Age estimates are used in calculations of a species' growth, mortality, and fecundity rates, and therefore are fundamental to the assessment of a population's status and vulnerability to exploitation. As with other fishes, age estimates for chondrichthyans are produced primarily through counts of growth bands in calcified structures. To date, most high latitude chondrichthyan aging studies have relied on counts of pairs of opaque and translucent vertebral bands, which are usually interpreted as representing one year of growth (Cailliet and Goldman 2004). Of the chondrichthyans occurring in high latitude seas for which age estimates are available, most are pelagic migrants or inshore, relatively shallow-dwelling species of commercial interest. Few published age estimates exist for most deepwater high latitude species or for species predominantly taken as bycatch (Table 3.3).

A common problem with aging many deep-dwelling high latitude species is that banding patterns are difficult to discern due to poor vertebral calcification. In past studies, a variety of staining techniques have been used in attempts to improve band clarity, with differential success (Cailliet and Goldman 2004). Methods involving the decalcification of vertebral centra may prove particularly useful in interpreting banding patterns in deepwater species, as has been demonstrated for *Malacoraja senta* and *Galeus melastomus* (Correia and Figueiredo 1997; Natanson et al. 2007). Several studies have used alternative structures to obtain age estimates for species in which vertebral banding patterns are unclear. Counts of growth bands in the dorsal fin spines of squaloid sharks and chimaeras and in the neural arches of hexanchiformes have proven useful, though they have yet to be validated for most species (Cailliet and Goldman 2004). Gallagher and Nolan (1999) investigated the suitability of caudal thorns as an aging structure for four bathyrajid species and found a high degree of agreement between age estimates based on counts of thorn and vertebral bands. The use of this alternative method, although successful in some skate species (Matta and Gunderson 2007), has been found to result in a high degree of imprecision between thorn and vertebral band counts (Perez 2005; Davis, Cailliet, and Ebert 2007). Therefore, although caudal thorns may provide a nonlethal approach to aging, their use may not be appropriate for some skate species and should be investigated further before being widely applied.

Of the chondrichthyan species occurring in high latitude seas, several have proven difficult or impossible to age using conventional methods, even for species in which vertebral banding is evident. Natanson and Cailliet (1990) found that vertebral growth zones in *Squatina californica* reflected somatic growth but were not deposited annually. This was also found to be the case for *Cetorhinus maximus* (Natanson et al. 2008). Age estimates for deepwater catsharks (Scyliorhinidae) are also lacking, though a modified decalcification technique developed for *Galeus melastomus* may have application to other deepwater scyliorhinid species (Correia and Figueiredo 1997).

Although chondrichthyans are widely believed to be a slow growing, late maturing group exhibiting typically low fecundities, available estimates of growth coefficients (k) for species occurring predominantly in high latitude seas vary widely (from 0.036 for *Squalus suckleyi* to 0.370 for *Raja binoculata*), reflecting the diversity of life history modes present (Table 3.3). For high latitude sharks, k values range from 0.036 to 0.12 in the Squaliformes, from 0.061 to 0.17 in the Lamniformes, and from 0.073 to 0.369 in the Carcharhiniformes. Available growth data for the latter are primarily from members of the family Triakidae.

TABLE 3.3

Summary of Available Age, Growth, and Demography Studies for Nonvagrant High Latitude Chondrichthyan Species

Species	Region	TL_{max} (cm)	t_{mat}	t_{max}	$L_∞$ (cm)	k	Fecundity** (offspring/year)	(Z)	R_0	G	r	e'	t_{x2} (years)	References
Squaliformes														
Squalus acanthias	NWA	125.0	7	14	157.00	0.120	2	0.176	3.463	11.406	0.118	1.125	5.874	Smith, Au, and Show 1998; Avsar 2001; Chen and Yuan 2006
Squalus mitsukurii	NP	99.9	15	27	107.00	0.041								Cailliet and Goldman 2004; Kyne and Simpfendorfer 2007
Squalus suckleyi	NEP	128	25	70	128.50	0.036	3.57	0.094			0.071	1.074	9.763	Jones and Geen 1977; Smith, Au, and Show 1998
Centrophorus granulosus	NEA	95	16.5	39	109.40	0.096	0.5							Kyne and Simpfendorfer 2007
Centrophorus squamosus	NEA	145	45	70										Cailliet and Goldman 2004; Kyne and Simpfendorfer 2007
Deania calcea	NEA	117	25	35	119.30	0.077	4–7							Kyne and Simpfendorfer 2007
	SWP	122	21.5	37	122.50	0.051	.5–8.5							Kyne and Simpfendorfer 2007
Etmopterus baxteri	SWP	87	30	57	68.10	0.040	.5–8							Kyne and Simpfendorfer 2007
Centroselachus crepidator	SWP	103	20	54	96.10	0.072	1.5–4.5							Kyne and Simpfendorfer 2007
Proscymnodon plunketi	SWP	155	29	39			9.5–21.5							Kyne and Simpfendorfer 2007
Lamniformes														
Lamna ditropis	NEP	261	6	20	207.40	0.17	2			13	0.081	1.084	8.557	Cailliet and Goldman 2004; Dulvy et al. 2008
Lamna nasus	NWA	357	13	24	369.80	0.061	4			18	0.081	1.084	8.557	Cailliet and Goldman 2004; Dulvy et al. 2008
	SWP	240	15	65			3.75			26	0.086	1.090	8.060	Dulvy et al. 2008

Continued

TABLE 3.3 (Continued)

Summary of Available Age, Growth, and Demography Studies for Nonvagrant High Latitude Chondrichthyan Species

Species	Region	TL_{max} (cm)	t_{mat}	t_{max}	L_∞ (cm)	k	Fecundity** (offspring/year)	(Z)	R_0	G	r	e^r	t_{x2} (years)	References
Carcharhiniformes														
Galeus melastomus	NEA	67		8										Cailliet and Goldman 2004
Scyliorhinus canicula	NEA	100	7.9	12	75.14	0.15								Ivory et al. 2005
Furgaleus macki	SWP	150	6.5	11.5	120.70	0.369	9.5							Simpfendorfer and Unsworth 1998; Simpfendorfer et al. 2000
Galeorhinus galeus	NEP	165	12	40			9.3							Smith et al. 1998; Lucifora et al. 2004
Mustelus californicus	NEP	163	5	17			3	0.253	2.431	7.97	0.118	1.125	5.874	Yudin and Cailliet 1990; Smith et al. 1998; Chen and Yuan 2006
Mustelus canis	NWA	132	4	16	123.50	0.292	9.53							Conrath et al. 2002; Cailliet and Goldman 2004
Mustelus henlei	NEP	100	3	11	97.70	0.244	10	0.383	1.929	4.847	0.143	1.154	4.847	Natanson and Cailliet 1986; Yudin and Cailliet 1990; Chen and Yuan 2006
Mustelus lenticulatus	SWP	151	5.5		147.2	0.119								Francis 2003; Cailliet and Goldman 2004
Triakis semifasciata	NEP	180	10	24	160.20	0.073	8.44	0.140	4.467	22.348	0.067	1.069	10.345	Cailliet 1992; Au and Smith 1997; Cailliet and Goldman 2004
Rajiformes														
Bathyraja aleutica	NEP	154	9	17	172.58	0.110	50–85	0.16–0.26	23.36	13.24	0.225	1.252	3.140	Ebert et al. 2007; Kyne and Simpfendorfer 2007
Bathyraja trachura	NEP	91	5	20	99.38	0.090								Davis et al. 2007

Bathyraja parmifera	NEP	111	10	17	135.39	0.100								Kyne and Simpfendorfer 2007; Matta and Gunderson 2007
Rhinoraja interrupta	NEP	82.5	7.5	13	126.40	0.080	30–75	0.10–0.33	30.29	10.32	0.307	1.360	2.350	Ebert et al. 2007; Kyne and Simpfendorfer 2007
Amblyraja radiata	NEA	100.5	5		68.70	0.125	38	0.7*			0.035*	1.036*	19.804	Walker and Hislop 1998; Kyne and Simpfendorfer 2007; McPhie and Campana 2009
Dipturus batis	NEA	250	11	23	253.73	0.057								Cailliet and Goldman 2004; Dulvy et al. 2006
Dipturus innominata	SWP	240	13	24	150.50	0.095								Cailliet and Goldman 2004; Dulvy and Reynolds 2002
Dipturus laevis	NWA	133.5	12	50			47	0.090			0.2	1.221	3.466	Frisk et al. 2002; Kyne and Simpfendorfer 2007
Raja brachyura	NEA	109	5.5	8	154.70	0.129	90							Gallagher et al. 2005
Raja clavata	NEA	104	10	9	109.00	0.155	60–140	0.588*			−0.073*	0.93*		Walker and Hislop 1998; Gallagher et al. 2005
Raja microocellata	NEA	80			137.00	0.086	100							Ryland and Ajayi 1984
Raja montagui	NEA	77	4.1	7	78.43	0.256	60							Gallagher et al. 2005; Ellis et al. 2007
Raja binoculata	NEP	244	5	26	167.90	0.370	300–350	0.13–0.18	48.69	12.44	0.288	1.334	2.450	Dulvy and Reynolds 2002; McFarlane and King 2006; Ebert et al. 2007
Raja rhina	NEP	203.9	9	26	106.90	0.160	50–85	0.06–0.21	36.47	18.66	0.184	1.202	3.820	McFarlane and King 2006; Ebert et al. 2007
Leucoraja erinacea	NWA	57	4	8			30	0.450			0.21	1.234	3.301	Frisk et al. 2002; Frisk and Miller 2006
Leucoraja naevus	NEA	71	4.25	8	83.92	0.197	90							Gallagher et al. 2005; Kyne and Simpfendorfer 2007
Leucoraja ocellata	NWA	111	9	20.8	137.40	0.059	21–35	0.210			0.13	1.139	5.332	Frisk et al. 2002; Cailliet and Goldman 2004; Frisk and Miller 2006

Continued

TABLE 3.3 (*Continued*)

Summary of Available Age, Growth, and Demography Studies for Nonvagrant High Latitude Chondrichthyan Species

Species	Region	TL_{max} (cm)	t_{mat}	t_{max}	L_∞ (cm)	k	Fecundity** (offspring/ year)	(Z)	R_0	G	r	e^r	t_{x2} (years)	References
Chimaeriformes														
Callorhynchus milii	SWP	125	5.6	9	94.10	0.224								Cailliet and Goldman 2004
Chimaera monstrosa	NEA	57.1	13	29	78.87	0.067								Cailliet and Goldman 2004; Kyne and Simpfendorfer 2007

Note: Fecundity estimates have also been included if available. Due to latitudinal differences in age and demographic parameters, only studies conducted in whole or in part within high latitude waters are presented here. If age estimates are available for both sexes, only estimates for females are included in the summary table.

Legend: TL_{max} = maximum observed total length; t_{mat} = estimated age at maturity; t_{max} = maximum estimated age; L_∞ = asymptotic length; k = von Bertalanffy growth coefficient; Z = mortality rate used for demographic analysis; R_0 = net reproductive rate; G = generation time; r = intrinsic rate of population increase; e^r = finite rate of population increase; t_{x2} = population doubling time.

* Includes fishing mortality

** Fecundity incorporates reproductive periodicity

Available estimates of growth coefficients for the Rajiformes cover the broadest range, varying from 0.057 for *Dipturus batis* to 0.370 for *Raja binoculata*. Estimates are only available for two species of chimaera occurring in high latitude waters, with *k* values of 0.067 for *Chimaera monstrosa* and 0.224 for *Callorhynchus milii*. Though estimates for all groups vary widely, differences in *k* values may also reflect differences in aging methodologies, growth models, or insufficient sample sizes (Cailliet and Goldman 2004).

As would be expected given the diversity of high latitude chondrichthyans, estimates of age at maturity and longevity vary widely within and among chondrichthyan groups (Table 3.3). Ages at maturity range from a low of three for *Mustelus henlei* to a high of 45 for the deepwater gulper shark *Centrophorus squamosus*. Higher ages at maturity generally correspond to higher estimates of longevity, though there are some exceptions. For sharks, longevity estimates range from 14 to 70 years in the Squaliformes, from 20 to 65 years in the Lamniformes, and from 8 to 40 years in the Carcharhiniformes. Maximum age estimates for skates range from 7 to 50 years, with the two available estimates for chimaera species of 9 and 29 years. It is important to note that age estimates produced from different regions of a species' range may provide different estimates of age at maturity and longevity, whether due to latitudinal differences in population parameters or differences in sampling methodology (Lombardi-Carlson et al. 2003; Cailliet and Goldman 2004; Frisk and Miller 2006).

Few studies reporting on the age and growth of high latitude species have verified or validated their results; most simply assume annual patterns of vertebral band deposition based on the few validated studies to date. Validation of the temporal periodicity of vertebral banding is essential in order to ensure the accuracy of age estimates since several studies have demonstrated that band deposition is not directly related to time in all chondrichthyan species (Natanson and Cailliet 1990; Cailliet and Goldman 2004; Natanson et al. 2008). Age estimates for several shark and skate species inhabiting high latitude seas have been validated. Kusher, Smith, and Cailliet (1992) and Smith, Mitchell, and Fuller (2003) both validated annual band deposition patterns in *Triakis semifasciata* through tag-recapture of oxytetracycline (OTC)-injected individuals, with Smith, Mitchell, and Fuller (2003) reporting on a recapture of an individual after 20 years at large. Age estimates for *Lamna nasus* have also been validated by OTC tag recapture and bomb radiocarbon in the western North Atlantic (Campana et al. 2002; Natanson, Mello, and Campana 2002). While rates of band deposition in dorsal spines have only been validated for *Squalus acanthias* using OTC injection and bomb radiocarbon (Beamish and McFarlane 1985; Campana et al. 2006), banding patterns in the spines of *Centroselachus crepidater* have been verified by correlation between band counts and the results of radiometric analysis (Irvine, Stevens, and Laurenson 2006). McPhie and Campana (2009) reported the only direct validation for a skate species using bomb radiocarbon to provide evidence of annual band-pair formation in *Amblyraja radiata*. Gallagher and Nolan (1999) tagged 550 bathyrajids in the waters off the Falkland Islands but only reported one return for an undescribed species not investigated in their study. Though the individual was only at liberty for 10 months, OTC was incorporated into an opaque band with a subsequent translucent band forming during that time period, suggesting an annual banding pattern.

3.5.3 Demography

Chondrichthyans generally exhibit long life spans, late ages at maturity, and relatively low fecundities, characteristics that make age structured models, such as life history tables and matrix population models, more appropriate for analyzing population dynamics than biomass dynamic models traditionally applied to teleost stocks (Cortés 1998). These types of models can provide insight into relationships among life history traits, population productivity, and a species' consequent ability to withstand exploitation. However, as these types of analyses are dependent on detailed, age-specific life history data, it is not surprising that little exists in the way of published demographic studies for high latitude chondrichthyans, given the relative lack of life history information available.

Several studies have recognized body size as an indicator of a species' vulnerability to exploitation in skate assemblages, based on the slower growth, later maturity, and lower productivity of larger species (Walker and Hislop 1998; Dulvy and Reynolds 2002; Frisk, Miller, and Fogarty 2002; Frisk this volume). Though this observation may hold true for some species assemblages, body size should not be considered a reliable indicator in all cases. For example, *Raja binoculata* is the largest skate in the eastern North Pacific skate complex but has among the highest growth and reproductive rates estimated for a skate species to date (Table 3.3). Smith, Au, and Show (1998) provided estimates of intrinsic rebound potentials based on reported life history traits for 26 Pacific Ocean shark species and predicted one of the smallest species included in their analysis, *Squalus suckleyi* (= *S. acanthias* in Smith, Au, and Show 1998), to have the highest vulnerability to exploitation. Cortés (2002) found no relationship between body size and population growth rates in a study investigating 38 shark species. Although body size is a convenient measure that is easily obtainable, in reality other traits, such as growth rate, age at maturity, and fecundity are driving this observation. Therefore, although it may correlate with other demographic parameters in many species (Frisk, this volume), body size alone should not be prioritized as an indicator of vulnerability; many relatively small deepwater species are among the most unproductive and consequently most vulnerable chondrichthyan species (Table 3.3; Smith, Au, and Show 1998; Kyne and Simpfendorfer 2007; Garcia, Lucifora, and Myers 2008).

Estimates of population growth rates have only been published for 16 species in high latitude areas (Table 3.3). Demographic analyses have calculated intrinsic rates of population increase (r) ranging from 0.067 for *Triakis semifasciata* to 0.307 for *Rhinoraja interrupta*, with corresponding population doubling times of 10.345 and 2.350 years in the absence of fishing pressure, respectively. Intrinsic rates of increase for squaloid sharks range from 0.071 for *Squalus suckleyi* to 0.118 for *S. acanthias*, for lamnids from 0.081 to 0.086, and for the carcharhinids from 0.067 for *T. semifasciata* to 0.143 for *Mustelus henlei*. Predicted r values for the Rajiformes were at the higher end of the range, varying from 0.13 for *Leucoraja ocellata* to 0.307 for *R. interrupta*. For two species of skate, the only estimates of population growth incorporated both natural and fishing mortality, reporting low or declining r values for *Amblyraja radiata* and *Raja clavata*, respectively (Table 3.3; Walker and Hislop 1998). To date, no studies investigating the demography of high latitude chimaeras have been published. Gedamke et al. (2007) highlighted problems with the methodology and interpretation of the population models used to produce such estimates. As advances are made in this field, the above estimates and the management recommendations based on them will doubtlessly be revised.

Stock assessments are only as good as the data used in their production; the better the life history estimates input, the more reliable the management strategy based on the results of the assessment. Because latitudinal variation in growth rates is often reported in marine species, with members of the same species from higher latitude seas attaining greater ages and ages at maturity, it is important to obtain life history information and conduct demographic analyses in all regions of a species' range, particularly in areas where the population is subject to exploitation (Lombardi-Carlson et al. 2003; Frisk and Miller 2006; Frisk, this volume). Since most studies to date have been conducted at lower latitudes, the available data may not provide an appropriate basis for the management of high latitude populations. As such, there is an obvious need to collect detailed life history information on chondrichthyans inhabiting high latitude seas so that population growth rates can be calculated and vulnerability to exploitation assessed.

3.5.4 Diet and Feeding Ecology

Trophically chondrichthyans occupy a similar role to that of other high level predators (e.g., birds and marine mammals) and as such may influence and shape those marine communities in which they occur (Table 3.4; Chapter 16). However, the diet and feeding ecology of most high latitude seas chondrichthyans, and their influence in shaping those ecosystems in which they occur, are largely unknown. Limited studies have been conducted on less than half of the species known to occur in these regions, with the TL being determined for only 129 of these species (Cortes 1999; Ebert and Bizzarro 2007). Furthermore, in many instances the diet study was carried out on populations occurring in lower latitudes that may not be reflective of high latitude ecosystems. Studies on the trophic interactions between chondrichthyans and traditionally targeted groundfish species have been carried out only to a limited degree on the Northeast Shelf in the WNA (Link, Garrison, and Almeida 2002). This is unfortunate because in many ecosystems it is likely that as once commercially valuable groundfish populations have been fished down they have subsequently been replaced by demersal chondrichthyans (Link, Garrison, and Almeida 2002; Link 2007). Furthermore, in some high latitude seas ecosystems an increase in chondrichthyan biomass has been documented possibly indicating niche expansion into those left vacant by the depletion of other competitors (Link, Garrison, and Almeida 2002; Link 2007).

The major demersal groups, the squaloids, skates, scyliorhinids, and triakids, can be broadly separated by TL, with mean averages of 4.2, 3.9, 3.8, and 3.8, respectively. However, many individual species estimates of TL were based on a single regional study with few data points, thus not taking into account that TLs may vary ontogenetically, spatially, or temporally. Furthermore, most of these studies do not take into account possible shifts in habitat preference due to changes in life stage. In addition, the diets of many larger, potentially trophically higher, demersal species, for example, skates, are virtually unknown. The feeding ecology and, hence, trophic role in marine communities of the chimaeroids is by and large unknown.

The high latitude seas region with the most species for which the TL has been calculated is the ENA with 49 species (Table 3.4). This is followed by the WNA and TAS regions, where estimates of TL are available for 30 and 29 species, respectively. These regions contrast with the ANT region, where the TL of only three species has been calculated. Interestingly, estimates of TL have been calculated for approximately 50% of the species occurring in four of the five northern hemisphere regions, while estimates have been produced for 30% or less of regional chondrichthyan species in the southern hemisphere. Given that most southern hemisphere regions, with the exception of the ANT region, have higher species diversity compared to those in the northern hemisphere it is striking how poorly these regions are known.

3.5.5 Habitat Association

The majority of high latitude seas chondrichthyans (94%) occupy a benthic or benthopelagic habitat, with pelagic species representing only about 6% of the fauna. This compares

TABLE 3.4

Mean Standardized Trophic Levels of Select Dominant High Latitude Seas Demersal Chondrichthyan Families by Region

	n	ENP	n	WNP	n	ENA	n	WNA	n	ARC	n	SSA	n	SNZ	n	TAS	n	ANT
Squaliformes																		
Echinorhinidae	1	4.40	1		1	4.40	0		0		1	4.40	2	4.40	0		0	
Squalidae	1	4.20	3	4.20	1	4.03	1	3.90	1	3.90	2	4.05	1	3.90	2	4.05	0	
Centrophoridae	0		3		1	4.17	0		0		0		2	4.20	3	4.20	0	
Etmopteridae	0		4	3.80	2	4.00	2	4.00	2	4.00	4	4.05	3	4.17	4	4.13	1	4.10
Somniosidae	1	4.20	5	4.20	2	4.14	2	4.20	2	4.20	1	4.20	4	4.13	3	4.20	1	4.20
Dalatidae	0		1		1	4.10	0		0		1	4.10	3	4.20	3	4.20	0	
Carcharhiniformes																		
Scyliorhinidae	2	3.65	5	0.00	2	3.82	1	3.90	1	4.00	1	3.80	1	4.20	0		0	
Triakidae	3	3.77	3	3.80	1	3.90	0	3.60	0		4	3.90	2	3.85	2	4.20	0	
Rajiformes																		
Arhynchobatidae	9	3.85	1	3.87	1	4.02	1	4.02	1	4.02	6	3.86	0		0		0	
Rajidae	3	3.82	17	3.86	13	3.86	8	3.75	8	3.91	1	3.90	1	3.98	7	3.94	0	

Note: The number of species represented in each family by region is followed by the mean trophic level. Trophic level values are from Cortés (1999); Ebert and Bizzarro (2007); and Treloar et al. (2007).

with global numbers, which reveal that about 90% of chondrichthyans occupy a benthic or benthopelagic habitat (Compagno 1990). Most benthic and benthopelagic species belong to one of the five major chondrichthyan groups, including the skates, squaloids, scyliorhinids, triakids, and chimaeroids, occurring in high latitude seas. Approximately 70% of the benthic and benthopelagic high latitude chondrichthyan species occur along the outer continental shelf and upper slopes, while 30% are considered shelf and near-shore species.

Although chondrichthyans may be characterized by broad assumptions about their habitat preference or type very little is known regarding ontogenetic, spatial, and temporal changes for most species. For example, several species of squaloids (e.g., *Somniosus* and some *Squalus* spp.; Ebert 2003) are known to occupy a midwater habitat as neonates, but develop a more benthic lifestyle as they grow in size and mature. Skates may also change habitat as they grow and mature, as demonstrated by adult *Bathyraja aleutica*, which tend to inhabit the area along the shelf-slope break where their nurseries are found, while neonates of the species migrate down the slope and occupy a deeper habitat. As this species grows in size, it migrates back up the slope and eventually occupies the slope shelf break (Ebert et al. 2006).

3.6 Fisheries and Conservation

3.6.1 Fisheries

Several commercially targeted high latitude chondrichthyan species have suffered declines in abundance over the past century as the result of unsustainable fishing practices. Intensive shark fisheries began during the 1940s in order to fill market demand for vitamin A and continue today primarily for flesh, oil, and fins (Stevens et al. 2000, 2006). As a result of a directed longline fishery, the WNA population of *Lamna nasus* collapsed to approximately 11% of its virgin biomass in the 1990s, while ENA stocks have also been severely depleted (Stevens et al. 2000, 2006). Though the recent implementation of more conservative management schemes in the WNA should aid in the population's recovery, unregulated fishing of this wide-ranging species in the ENA will certainly hamper rebound rates. In the southern hemisphere, *L. nasus* is taken primarily as valuable bycatch in tuna and Patagonian toothfish (*Dissostichus eleginoides*) longline fisheries, though catch records are often unreported (Stevens et al. 2006). In the North Pacific, the endemic congener of *Lamna nasus*, *L. ditropis*, is taken mostly indirectly in longline fisheries, though a small recreational fishery has developed in recent years in Alaskan waters. Based on the slow growth, late age at maturity, and low productivity of *L. nasus*, *L. ditropis* may be similarly susceptible to exploitation (Goldman and Human 2000; Goldman and Musick 2008).

Galeorhinus galeus is another wide-ranging high latitude species of concern. Due to the species' longevity and low biological productivity, populations off the coast of Argentina, Australia, and New Zealand have collapsed or been depleted as the result of overexploitation (Stevens et al. 2000; Walker, Cavanagh, and Stevens 2006). According to calculations by Smith, Au, and Show (1998), the species has among the lowest intrinsic rebound potential of 26 species of Pacific Ocean sharks surveyed. While populations in the ENP have rebounded and management measures have been in place for Australian and New Zealand stocks since the late 1980s, the species' abundance off Argentina continues to decline due to lack of restrictions and intense targeting of gravid females at pupping grounds (Walker, Cavanagh, and Stevens 2006).

Populations of several commercially targeted squaloid sharks have also undergone marked alterations in abundance in the past century. Based on their typically late onset of maturity, low reproductive output (e.g., only one pup per litter and two year gestation period in the deepwater gulper shark, *Centrophorus granulosus*), and high longevity (up to 70 years in certain populations of the spiny dogfish, *Squalus acanthias*), populations are unable to sustain even low levels of exploitation (Smith, Au, and Show 1998; Fordham et al. 2006; Guallart et al. 2006; Kyne and Simpfendorfer 2007). Both Smith, Au, and Show (1998) and Kyne and Simpfendorfer (2007) predicted squaloids to have the lowest intrinsic rebound potentials and highest population doubling times of all chondrichthyan species assessed. Though stocks of *Squalus acanthias* and several deepwater gulper sharks (*Centrophorus granulosus*, *C. harrissoni*, and *C. squamosus*) have suffered declines in biomass in excess of 80% in various portions of their range, little in the way of management is currently in place (Pogonoski and Pollard 2003; White 2003; Fordham et al. 2006; Guallart et al. 2006). Even if zero catch limits are enacted, bycatch will remain an issue for these extremely unproductive species (Kyne and Simpfendorfer 2007).

Although the majority of high latitude chondrichthyan fisheries research has been conducted on species of commercial interest, most chondrichthyan catches in high latitude fisheries are largely the result of indirect fishing efforts. Benthic or benthopelagic groups, such as skates and dogfish sharks, are particularly vulnerable to bottom trawling and longline gear and are often taken in fisheries targeting bony fish species such as hake, Patagonian toothfish, cod, haddock, halibut, shrimps, and other groundfish species complexes. Several high latitude species primarily taken as bycatch in commercial fisheries have also suffered dramatic declines in population size in the past century, of which the rajids *Dipturus batis*, *D. laevis*, and *Rostroraja alba* are the most oft-cited examples (Brander 1981; Casey and Myers 1998; Dulvy et al. 2000; Stevens et al. 2000; Frisk, Miller, and Fogarty 2002). Both *D. batis* and *R. alba* have been driven close to local extirpation by trawling in the Irish Sea (Brander 1981; Casey and Myers 1998; Dulvy et al. 2000; Frisk, Miller, and Fogarty 2002). *Dipturus laevis* was also thought to be on the brink of local extirpation (Casey and Myers 1998) but has since been shown to have a much wider and deeper distribution than originally estimated (Gedamke, DuPaul, and Musick 2005). Since skates often have limited distributions and vary widely in life history traits and productivity, the impact of exploitation on population structure and abundance may be significant (Dulvy and Reynolds 2002; Ebert et al. 2007). The increasing exploitation of this group is especially alarming because their typical demographic traits may severely restrict their ability to sustain fishing pressure or recover from overexploitation (Holden 1974; Walker and Hislop 1998; Stevens et al. 2000; Cailliet and Goldman 2004).

Along with species for which declines have been documented, there is concern for unevaluated rajids that may become the target of developing fisheries in the ENP and elsewhere, as targeting new species and utilizing bycatch have become an alternative strategy employed by fishers and fishery managers to supplement income lost from global population declines of traditional fish stocks (Ebert et al. 2007). Skates are often taken as part of a species complex; however, since most species are reported as "skate unidentified" and a large portion of skate bycatch is discarded, historical population sizes and species assemblages are unknown (Ebert et al. 2007). Fisheries in which skate bycatch is not sorted may mask declines of larger, later-maturing species (Agnew et al. 2000; Dulvy et al. 2000). For example, over the course of a 10-year fishery in the Falkland Islands, catch records indicate that the composition of the skate community shifted, with two smaller, faster-growing species replacing a larger member of the assemblage (Agnew et al. 2000). Along with the high degree of skate bycatch in demersal trawls and its documented impact, the recent development of directed skate fisheries in the Gulf of Alaska and Prince William Sound (Gburski, Gaichas, and Kimura 2007; Kenneth J. Goldman, Alaska Department of Fish and Game, personal communication) emphasizes the necessity of documenting current species-specific baseline biological information.

Although many non-rajid chondrichthyans are also taken as bycatch, the lack of appropriate life history information makes population assessment impossible for many species. Despite the fact that the catsharks (Scyliorhinidae) are the largest shark family (Ebert 2003) and comprise almost 10% of all high latitude chondrichthyans, most species have not been evaluated for risk of overexploitation. Of those that have been assessed, only one high latitude catshark, *Scyliorhinus stellaris*, has been listed as Near Threatened by the International Union for Conservation of Nature (IUCN) to date (Gibson et al. 2008). The biology and habits of most scyliorhinids are poorly known, with those of deepwater species even less well known. However, based on low estimates of intrinsic rebound potential for deepwater chondrichthyans (Kyne and Simpfendorfer 2007), most catsharks may have limited productivity and consequently may be susceptible to overexploitation. Therefore, a precautionary management approach should be taken until species-specific biological parameters are better assessed.

The Chimaeriformes are yet another poorly known chondrichthyan group that may prove of future concern. Catches of chimaeras are perhaps the least reported among chondrichthyan groups. There are currently directed fisheries for this group in the SSA, SNZ, and TAS regions. Chimaera bycatch is often discarded at sea due to its relatively low economic value, which is unfortunate since several specimens retained from the Patagonian toothfish fishery have proven to be undescribed species. The order faces many of the same obstacles to assessment as other primarily deepwater groups. Though biological information is severely lacking, three species are currently listed as Near Threatened by the IUCN (*Chimaera monstrosa, Hydrolagus mirabilus,* and *Hydrolagus ogilbyi*) due to high rates of bycatch, depth range overlap with current deepwater fisheries, or highly restricted ranges (Dagit, Compagno, and Clarke 2007; Dagit, Hareide, and Clò 2007; Dagit and Kyne 2006). The only ageing studies conducted on deepwater chimaeras published to date provided maximum age estimates of 9 and 29 years for *Callorhynchus milii* and *Chimaera monstrosa*, respectively, indicating that chimaeras are likely as vulnerable to overexploitation as other deepwater chondrichthyans (Dagit, Hareide, and Clò 2007; Kyne and Simpfendorfer 2007).

3.6.2 Conservation

Fourteen percent of chondrichthyans known to inhabit high latitude seas are currently classified as threatened (Critically Endangered, Endangered, or Vulnerable) by the IUCN. Nine species are listed as Critically Endangered, 14 as Endangered, and 27 as Vulnerable (Table 3.5), with a further 45 species listed as Near Threatened (IUCN 2008). High latitude regions featuring the highest proportion of threatened species are the ENA, WNA, and SSA regions, which have historically been centers of intensive fishing activity. Threatened species include commercially important chondrichthyans, endemics, deep sea inhabitants, and species prone to trawl bycatch, all of which are linked by their vulnerability to exploitation due to restricted ranges or unproductive life history characteristics (Smith, Au, and Show 1998; Kyne and Simpfendorfer 2007). The status of these 95 near-threatened and threatened species should be closely monitored and appropriate management and recovery plans developed and implemented to ensure their continued survival.

TABLE 3.5

IUCN Red List Status of High Latitude Chondrichthyans

| IUCN Redlist Category | Northern Hemisphere | | | | | Southern Hemisphere | | | | Total All Areas |
| | North Pacific | | North Atlantic | | | Southern Ocean | | | | |
	ENP	WNP	ENA	WNA	ARC	SSA	SNZ	TAS	ANT	
Critically Endangered	0	0	7	0	1	1	1	1	0	9
Endangered	1	0	6	1	0	5	0	1	0	14
Vulnerable	4	3	11	10	3	13	10	8	0	27
Near Threatened	9	4	21	14	9	12	17	12	1	45
Least Concern	5	1	29	22	10	14	24	36	0	80
Conservation Dependent	0	0	0	1	0	0	0	1	0	1
Data Deficient	5	7	23	9	5	21	22	13	1	70
Not Evaluated	22	42	0	5	1	13	19	26	8	108
Total	46	57	97	62	29	79	93	98	10	354

Note: The number of chondrichthyan species assigned to each Red List category within and among all high latitude regions.

Although half of all high latitude chondrichthyans were listed as Data Deficient or were not evaluated by the IUCN, this does not imply that they are not of concern; rather that the lack of species-specific catch and biological data has hindered the assessment of chondrichthyans predominantly taken as bycatch as well as deep-dwelling species (Kyne and Simpfendorfer 2007). As global population declines in traditionally targeted, shallow near-shore stocks continue, fisheries have begun to exploit populations in deeper and more remote waters (Roberts 2002). Given the low productivity of many deepwater high latitude chondrichthyans (Cavanagh et al. 2003; Kyne and Simpfendorfer 2007; Gibson et al. 2008; IUCN 2008), the importance of directing funding and research effort toward these unevaluated, as well as threatened, species becomes apparent.

3.7 Summary

The chondrichthyan fauna occurring in high latitude seas is quite diverse, with approximately 30% of all known species spending all or a portion of their life cycle in this environment. Although not as diverse as that of tropical or deep-sea environs the high latitude seas chondrichthyan fauna has penetrated and occupies most marine habitats within their environment. However, despite the faunal diversity the vast majority of biological information that has been documented is primarily on those species of commercial interest or pelagic vagrants that are usually found in tropical or warm-temperate climates. The available biological information indicates that high latitude chondrichthyans exhibit a wide variety of life history characteristics and subsequent variability in vulnerability to exploitation, with several species among the least productive and most susceptible to exploitation of all chondrichthyans. Problems with taxonomic resolution remain for many of the deep-dwelling, less economically valuable and hence less well known species complexes taken primarily as bycatch. In fact, far more high latitude seas chondrichthyans are likely taken by indirect fisheries, which tend to underestimate their biomass, than those that are the focus of targeted fisheries. Historically, high latitude fisheries targeting chondrichthyans have proven unsustainable. As fisheries begin to exploit populations in deeper, more remote waters due to continued global population declines in traditionally targeted, shallow near-shore stocks, there is an immediate need to direct funding and research effort toward the collection of "baseline" population data for poorly known high latitude species so management strategies can be developed and implemented. This chapter was intended to provide a much-needed summary of what has currently been documented regarding high latitude chondrichthyans, identify pressing research needs, and highlight threats to high latitude chondrichthyan populations that need to be addressed.

Acknowledgments

We would like to thank Joe Bizzarro (Pacific Shark Research Center/Moss Landing Marine Laboratories), Will White and John Stevens (CSIRO), Malcolm Francis (National Institute of Water and Atmospheric Research), Hajime Ishihara (W&I Associates Corporation), Pete Kyne (Charles Darwin University), Dave Kulka (formerly Department of Fisheries and Oceans), Daniel Figueroa (Departamento de Ciencias Marinas, Universidad Nacional de Mar del Plata), and Lisa Natanson (NOAA Fisheries Service, Northeast Fisheries Science Center), who provided invaluable assistance on various portions of this study. Photographs for Figure 3.3 were kindly provided by Monterey Bay Aquarium Research Institute, Joe Bizzarro, and Aaron Carlisle (Stanford University). Funding for this project was provided by NOAA/NMFS to the National Shark Research Consortium and Pacific Shark Research Center and the North Pacific Research Board.

References

Agnew DJ, Nolan CP, Beddington JR, Baranowski R (2000) Approaches to the assessment and management of multispecies skate and ray fisheries using the Falkland Islands fishery as an example. Can J Fish Aquat Sci 57:429–440.

Au DW, Smith SE (1997) A demographic model with population density compensation for estimating productivity and yield per recruit of the leopard shark (*Triakis semifasciata*). Can J Fish Aquat Sci 54:415–420.

Avsar D (2001) Age, growth, reproduction and feeding of spurdog (*Squalus acanthias* Linnaeus, 1758) in the south-eastern Black Sea. Est Coast Shelf Sci 52:269–278.

Barnes RSK, Hughes RN (1982) An introduction to marine ecology. Blackwell Scientific Publications, London.

Barnett LAK, Earley RL, Ebert DA, Cailliet GM (2009) Maturity, fecundity, and reproductive cycle of the spotted ratfish, *Hydrolagus colliei*. Mar Biol 156: 301–316

Beamish RJ, McFarlane GA (1985) Annulus development on the second dorsal spine of the spiny dogfish (*Squalus acanthias*) and its validity for age determination. Can J Fish Aquat Sci 42(11):1799–1805.

Benz GW, Hocking R, Kowunna A, Bullard SA, George JC (2004) A second species of Arctic shark: Pacific sleeper shark *Somniosus pacificus* from Point Hope, Alaska. Polar Biol 27:250–252.

Bigelow HB, Schroeder WC (1948) Sharks. In: Fishes of the Western North Atlantic. Mem Sears Fnd Mar Res 1(1):56–576.

Bigelow HB, Schroeder WC (1953) Sawfishes, guitarfishes, skates and rays. Chimaeroids. In: Fishes of the Western North Atlantic. Mem Sears Fnd Mar Res 1(2):1–562.

Bigelow HB, Schroeder WC (1957) A study of the sharks of the suborder Squaloidea. Bull Mus Comp Zool Harvard 117(1):1–150.

Bovcon ND, Cochia PD (2007) Gula para el reconocimiento de peces capturados por buques pesqueros monitoreados con observadores a bordo. Publicacion especial de la Secretaria de Pesca de la Provincia del Chubut, Rawson.

Brander K (1981) Disappearance of common skate *Raja batis* from the Irish Sea. Nature 5801:48–49.

Briggs JC (1974) Marine zoogeography. McGraw-Hill, New York.

Briggs JC (1995) Global biogeography. Elsevier, Amsterdam.

Cailliet GM (1992) Demography of the central California population of the leopard shark (*Triakis semifasciata*). Aust J Mar Freshw Res 43:183–193.

Cailliet GM, Goldman K (2004) Age determination and validation in chondrichthyan fishes. In: Carrier JC, Musick JA, Heithaus MR (eds) Biology of sharks and their relatives, CRC Press, Boca Raton, FL, pp 399–447.

Cailliet GM, Mollet HF, Pittenger GG, Bedford D, Natanson LJ (1992) Growth and demography of the Pacific angel shark (*Squatina alifornica*), based upon tag returns off California. Aust J Mar Freshw Res 43:1313–1330.

Campana SE, Jones C, McFarlane GA, Myklevoll S (2006) Bomb dating and age validation using the spines of spiny dogfish (*Squalus acanthias*). Environ Biol Fish 77:327–336.

Campana SE, Natanson LJ, Myklevoll S (2002) Bomb dating and age determination of large pelagic sharks. Can J Fish Aquat Sci 59:450–455.

Capape C, Seck AA, Quignard JP (1999) Observations on the reproductive biology of the angular rough shark, *Oxynotus centrina* (Oxynotidae). Cybium 23(3):259–271.

Carrier, JC, Pratt, HL, Castro JI (2004) Reproductive biology of elasmobranchs. In: Carrier JC, Musick JA, and Heithaus MR (eds) Biology of sharks and their relatives, CRC Press, Boca Raton, FL, pp 269–286.

Casey JM, Myers RA (1998) Near extinction of a large, widely distributed fish. Science 28:690–692.

Cavanagh RD, Kyne PM, Fowler SL, Musick JA, Bennett MB (2003) The conservation status of Australasian chondrichthyans: Report of the IUCN Shark Specialist Group Australia and Oceania Regional Red List Workshop, Queensland, Australia, March 7–9, 2003. School of Biomedical Sciences, The University of Queensland, Brisbane.

Chen P, Yuan W (2006) Demographic analysis based on the growth parameter of sharks. Fish Res 78:374–379.

Collette BB, Klein-MacPhee G (2002) Bigelow and Schroeder's fishes of the Gulf of Maine, 3rd edition. Smithsonian Institution Press, Washington, DC.

Colonello JH, Garcia ML, Lasta CA (2007) Reproductive biology of *Rioraja agassizi* from the coastal southwestern Atlantic ecosystem between northern Uruguay (34°S) and northern Argentina (42°S). Environ Biol Fish 80: 277–284.

Compagno LJV (1984) Sharks of the world. An annotated and illustrated catalogue of shark species known to date. Volume 4. Part 1 Hexanchiformes to Lamniformes. Part 2. Carcharhiniformes. FAO Species Catalogue for Fishery Purposes. No. 125, Vol. 4. FAO, Rome.

Compagno LJV (1990) Alternate life history styles of cartilaginous fishes in time and space. Environ Biol Fish 28: 33–75.

Compagno LJV (2001) Sharks of the world. An annotated and illustrated catalogue of shark species known to date. Volume 2. Bullhead, mackerel and carpet sharks (Heterodontiformes, Lamniformes and Orectolobiformes). FAO Species Catalogue for Fishery Purposes. No. 1, Vol. 2. FAO, Rome.

Compagno LJV (2005) Checklist of living chondrichthyans. In: Hamlett WC, Jamieson BGM (eds) Reproductive biology and phylogeny of chondrichthyes: sharks, rays, and chimaeras. Science Publishers, Enfield, New Hampshire, pp 503–538.

Compagno LJV, Ebert DA, Smale MJ (1989) Guide to the sharks and rays of southern Africa. Struik Publishers, Cape Town, South Africa.

Conrath CL, Musick JA (2002) Reproductive biology of the smooth dogfish, *Mustelus canis*, in the northwest Atlantic Ocean. Environ Biol Fish 64:367–377.

Correia JP, Figueiredo IM (1997) A modified decalcification technique for enhancing growth bands in deep-coned vertebrae of elasmobranchs. Environ Biol Fish 50:225–230.

Cortés E (1998) Demographic analysis as an aid in shark stock assessment and management. Fish Res 39:199–208.

Cortés E (1999) Standardized diet compositions and trophic levels of sharks. ICES J. Mar Sci 56:707–719.

Cortés E (2002) Incorporating uncertainty into demographic modeling: application to shark populations and their conservation. Conserv Biol 16:1048–1062.

Cousseau, MB, Figueroa DE, Díaz de Astarloa JM, Mabragaña E, Lucifora LO (2007) Rayas, chuchos y otros batoideos del Atlantico sudoccidental (34°S–55°S). Instituto Nacional de Investigacion y Desarrollo Pesquero, Secretaria de Agriculture, Ganaderia, Pesca y Alimentos, Mar del Plata, Republica, Argentina.

Cox G, Francis M (1997) Sharks and rays of New Zealand. Canterbury University Press, Christchurch, New Zealand.

Cushing DH, Walsh JJ (1976) The ecology of the seas. Blackwell Scientific Publications, London.

Dagit DD, Compagno LJV, Clarke MW (2007) *Hydrolagus mirabilis*. In: 2008 IUCN red list of threatened species. www.iucnredlist.org. Accessed March 7, 2009.

Dagit DD, Hareide N, Clò S (2007) *Chimaera monstrosa*. In: 2008 IUCN red list of threatened species. www.iucnredlist.org. Accessed March 7, 2009.

Dagit DD, Kyne PM (2006) *Hydrolagus ogilbyi*. In: 2008 IUCN red list of threatened species. www.iucnredlist.org. Accessed March 7, 2009.

Davis CD, Cailliet GM, Ebert DA (2007) Age and growth of the roughtail skate *Bathyraja trachura* (Gilbert 1892) from the eastern North Pacific. Environ Biol Fish 80:325–336.

de Carvalho MR (1996) Higher-level elasmobranch phylogeny, basal squaleans, and paraphyly. In: Stiassny MLJ, Parenti LR, Johnson GD (eds) Interrelationships of fishes. Academic Press, San Diego, California.

Dulvy NK, Baum JK, Clarke S, Compagno LJV, Cortés E, Domingo A, Fordham S, Fowler S, Francis MP, Gibson C, Martínez J, Musick JA, Soldo A, Stevens JD, Valenti S (2008) You can swim but you can't hide: the global status and conservation of oceanic pelagic sharks and rays. Aquat Conserv: Mar Freshwat Ecosyst 18:459–482.

Dulvy NK, Metcalfe JD, Glanville J, Pawson MG, Reynolds JD (2000) Fishery stability, local extinctions, and shifts in community structure in skates. Conserv Biol 14(1):283–293.

Dulvy NK, Notobartolo di Sciara G, Serena F, Tinti F, Ungaro N, Mancusi C, Ellis J (2006) *Dipturus batis*. In: 2008 IUCN red list of threatened species. www.iucnredlist.org. Accessed March 9, 2009.

Dulvy NK, Reynolds JD (2002) Predicting extinction vulnerability in skates. Conserv Biol 16(2):440–450.

Ebert DA (1990) The taxonomy, biogeography and biology of cow and frilled sharks (Chondrichthyes: Hexanchiformes). Dissertation, Rhodes University.

Ebert DA (1991) Diet of the sevengill shark *Notorynchus cepedianus* in the temperate coastal waters of southern Africa. S Afr J Mar Sci 11:565–572.

Ebert DA (1994) Diet of the sixgill shark *Hexanchus griseus* off southern Africa. S Afr J Mar Sci 14:213–218.

Ebert DA (2003) Sharks, rays, and chimaeras of California. University of California Press, Berkeley, California.

Ebert DA (2005) Reproductive biology of skates, *Bathyraja* (Ishiyama), along the eastern Bering Sea continental slope. J Fish Biol 66:618–649.

Ebert DA, Bizzarro JJ (2007) Standardized diet composition and trophic levels in skates. Environ Biol Fish 80:221–237.

Ebert DA, Compagno LJV (2007) Biodiversity and systematics of skates (Chondrichthyes: Rajiformes: Rajoidei). Environ Biol Fish 80:111–124.

Ebert DA, Davis CD (2007) Description of skate egg cases (Chondrichthyes: Rajiformes:Rajoidei) from the eastern North Pacific. Zootaxa 1393:1–18.

Ebert DA, Ebert TB (2005) Reproduction, diet and habitat use of leopard sharks, *Triakis semifasciata* (Girard), in Humboldt Bay, California, USA. Mar Freshw Res 56:1089–1098.

Ebert DA, Mollet HF, Baldridge A, Thomas T, Forney K, Ripley WE (2004) Occurrence of the whale shark, *Rhincodon typus* Smith 1828, in California waters. Northwest Nat 85(1):26–28.

Ebert DA, Compagno LJV, Cowley PD (2006) Reproductive biology of catsharks (Chondrichthyes: Scyliorhinidae) off the west coast of southern Africa. ICES J Mar Sci 63:1053–1065.

Ebert DA, Bizzarro JJ, Smith WD, Robinson HJ, Davis CD, Neway AL (2006) Distribution and habitat associations of skates (Rajiformes: Arhychobatidae) along the eastern Bering Sea Continental Slope. Western Groundfish Conference Abstract.

Ebert DA, Smith WD, Haas DL, Ainsley SM, Cailliet GM (2007) Life history and population dynamics of Alaskan skates: providing essential biological information for effective management of bycatch and target species. North Pacific Research Board Project 510 Final Report.

Ebert DA, Smith WD, Cailliet GM (2008) Reproductive biology of two commercially exploited skates, *Raja binoculata* and *R. rhina*, in the western Gulf of Alaska. Fish Res 94:48–57.

Ebert DA. Compagno LJV, Cowley PD (2008) Aspects of the reproductive biology of skates (Chondrichthyes: Rajiformes: Rajoidei) from southern Africa. ICES J Mar Sci 65:81–102.

Ebert DA, White WT, Goldman KJ, Compagno LJV, Daly-Engel TS, Ward RD (In review) Reevaluation and redescription of *Squalus suckleyi* (Girard, 1854) from the North Pacific, with comments on the *Squalus acanthias* subgroup (Squaliformes: Squalidae). Zootaxa.

Ellis J, Ungaro N, Serena F, Dulvy N,Tinti F, Bertozzi M, Pasolini P, Mancusi C, Noarbartolo di Sciara G (2007) *Raja montagui*. In: 2008 IUCN red list of threatened species. www.iucnredlist. org. Accessed March 9, 2009.

Fordham S, Fowler SL, Coelho R, Goldman KJ, Francis M (2006) *Squalus acanthias*. In: 2008 IUCN red list of threatened species. www.iucnredlist.org. Accessed March 7, 2009.

Francis MP (2003) *Mustelus lenticulatus*. In: 2008 IUCN red list of threatened species. www.iucnredlist. org. Accessed March 9, 2009.

Frisk MG (2004) Biology, life history and conservation of elasmobranchs with an emphasis on western Atlantic skates. Dissertation, University of Maryland.

Frisk MG, Miller TJ (2006) Age, growth and latitudinal patterns of two Rajidae species in the northwestern Atlantic: little skate (*Leucoraja erinacea*) and winter skate (*Leucoraja ocellata*). Can J Fish Aquat Sci 63:1078–1091.

Frisk MG, Miller TJ, Fogarty MJ (2002) The population dynamics of the little skate *Leucoraja erinacea*, winter skate *Leucoraja ocellata*, and barndoor skate *Dipturus laevis*: predicting exploitation limits using matrix analyses. ICES J Mar Sci 59:576–586.

Gallagher MJ, Nolan CP (1999) A novel method for the estimation of age and growth in rajids using caudal thorns. Can J Fish Aquat Sci 56(9):1590–1599.

Gallagher MJ, Nolan CP, Jeal F (2005) Age, growth and maturity of the commercial ray species from the Irish Sea. J Northwest Atl Fish Sci 35:47–66.

García VB, Lucifora LO, Myers RA (2008) The importance of habitat and life history to extinction risk in sharks, skates, rays and chimaeras. Proc R Soc B 275:83–89.

Gburski CM, Gaichas SK, Kimura DK (2007) Age and growth of the big skate (*Raja binoculata*) and longnose skate (*R. rhina*) in the Gulf of Alaska. Environ Biol Fish 80:337–349.

Gedamke T, DuPaul WD, Musick JA (2005) Observations on the life history of the barndoor skate, *Dipturus laevis*, on Georges Bank (Western North Atlantic). J Northwest Atl Fish Sci 35:67–78.

Gedamke T, Hoenig JM, Musick JA, DuPaul WD, Gruber SH (2007) Using demographic models to determine intrinsic rate of increase and sustainable fishing for elasmobranchs: pitfalls, advances and applications. N Am J Fish Manage 27:605–618.

Gibson C, Valenti SV, Fowler SL, Fordham SV (2008) The conservation status of northeast Atlantic chondrichthyans: Report of the IUCN Shark Specialist Group, North Atlantic Regional Red List Workshop, Peterborough, UK, February 13–15, 2006.

Goldman KJ, Branstetter S, Musick JA (2006) A re-examination of the age and growth of sand tiger sharks, *Carcharias taurus*, in the western North Atlantic: the importance of aging protocols and use of multiple back-calculation techniques. Environ Biol Fish 77:241–252.

Goldman KJ, Human B (2000) *Lamna ditropis*. In: 2008 IUCN red list of threatened species. www.iucnredlist.org. Accessed March 7, 2009.

Goldman KJ, Musick JA (2008) The biology and ecology of the salmon shark *Lamna ditropis*. In: Camhi MD, Pikitch EK, Babcock EA (eds) Sharks of the open ocean, Blackwell Science, Oxford, UK, pp 95–104.

Gon O, Heemstra PC (1990) Fishes of the southern ocean. J.L.B. Smith Institute of Ichthyology, Grahamstown.

Graham KJ, Andrew NL, Hodgson KE (2001) Changes in relative abundance of sharks and rays on Australian south east fishery trawl grounds after twenty years of fishing. Mar Freshw Res 52:549–561.

Guallart J, Serena F, Mancusi C, Casper BM, Burgess GH, Ebert DA, Clarke M, Stenberg C (2006) *Centrophorus granulosus*. In: 2008 IUCN red list of threatened species. www.iucnredlist.org. Accessed March 7, 2009.

Hartel KE, Kenaley CP, Galbraith JK, Sutton TT (2008) Additional records of deep-sea fishes from off greater New England. Northeast Nat 15(3):317–334.

Hela I, Laevastu T (1962) Fisheries hydrography. Fishing News Books, London.

Hempel G (1990) The Weddell Sea: a high polar ecosystem. In: Sherman K, Alexander LM, Gold BD (eds) Large marine ecosystems: patterns, processes, and yields, American Association for the Advancement of Science, Washington, DC, pp 5–18.

Holden MJ (1974) Problems in the rational exploitation of elasmobranch populations and some suggested solutions. In: Harden FR, Jones C (eds) Sea fisheries research, Logos Press, London, pp 117–138.

Holden MJ, Rout DW, Humphreys CN (1971) The rate of egg laying by three species of ray. J Cons Int Explor Mer 33(3):335–339.

Irvine SB, Stevens JD, Laurenson LJB (2006) Surface bands on deepwater squalid dorsal-fin spines: an alternative method for aging *Centroselachus crepidater*. Can J Fish Aquat Sci 63:617–627.

Ishihara H, Ishiyama R (1985) Two new North Pacific skates (Rajidae) and a revised key to *Bathyraja* in the area. Jpn J Ichthyol 32(2):143–179.

Ishihara H., Ishiyama R (1986) Systematics and distribution of the skates of the North Pacific (Chondrichthyes, Rajoidei). Indo-Pacific Fish Biology: Proc. 2nd Int. Conf. Indo-Pacific Fishes, 269–280.

Ishihara H, Mochizuki T, Homma K, Taniuchi T (2002) Reproductive strategy of the Japanese common skate (Spiny rasp skate) *Okameji kenojei*. In: Fowler SL, Reed TM, Dipper FA (eds) Elasmobranch biodiversity, conservation and management: Proceedings of the International Seminar and Workshop, Sabah, Malaysia, July 1997. IUCN SSC Shark Specialist Group. IUCN, Gland, Switzerland and Cambridge, pp 236–240.

IUCN 2008. 2008 IUCN red list of threatened species. www.iucnredlist.org. Accessed March 7, 2009.

Ivory P, Jeal F, Nolan CP (2005) Age determination, growth and reproduction in the lesser-spotted dogfish, *Scyliorhinus canicula* (L.). J Northwest Atl Fish Sci 35:89–106.

Jones BC, Green GH (1977) Age determination of an elasmobranch (*Squalus acanthias*) by X-ray spectrometry. J Fish Res Board Can 34:44–48.

Joung SJ, Chen CT, Clark E, Uchida S, Huang WYP (1996) The whale shark, *Rhincodon typus*, is a live-bearer: 300 embryos found in one "megamamma supreme." Environ Biol Fish 46:219–223.

Kneebone J, Ferguson DE, Sulikowski JA, Tsang PCW (2007) Endocrinological investigation into the reproductive cycles of two sympatric skate species, *Malacoraja senta* and *Amblyraja radiata*, in the western Gulf of Maine. Environ Biol Fish 80:257–265.

Kusher DI, Smith SE, Cailliet GM (1992) Validated age and growth of the leopard shark, *Triakis semifasciata*, from central California. Environ Biol Fish 35:187–203.

Kyne PM, Simpfendorfer CA (2007) A collation and summarization of available data on deepwater chondrichthyans: biodiversity, life history and fisheries: A report prepared by the IUCN SSC Shark Specialist Group for the Marine Conservation Biology Institute, February 15, 2007.

Lamilla J, Saez S (2003) Clave taxonomica para el reconocimiento de especies de rayas chilenas (Chondrichthyes, Batoidei). Invest Mar Valparaiso 31(2):3–16.

Last PR, Stevens JD (2009) Sharks and rays of Australia, 2nd edition. CSIRO Division of Fisheries, Melbourne, Australia.

Last PR, White WT, Pogonoski JJ (eds) (2007) Descriptions of new dogfishes of the genus *Squalus* (Squaloidea: Squalidae). CSIRO Marine and Atmospheric Research Paper No. 014.

Last PR, White WT, Pogonoski JJ (eds) (2008) Descriptions of new Australian chondrichthyans. CSIRO Marine and Atmospheric Research Paper No. 022.

Last PR, White WT, Pogonoski JJ, Gledhill DC (eds) (2008) Descriptions of new Australian skates (Batoidea: Rajoidei). CSIRO Marine & Atmospheric Research Paper No. 021.

Lindberg GU, Legeza MI (1959) Fishes of the Sea of Japan and the adjacent areas of the Sea of Okhotsk and the Yellow Sea. Part 1. Academy of Sciences of the Union of Soviet Socialist Republics. Translated from Russian by the Israel Program for Scientific Translations, Jerusalem, 1967.

Link LS (2007) Underappreciated species in ecology: "ugly fish" in the northwest Atlantic Ocean. Ecol Appl 17(7):2037–2060.

Link JS, Garrison LP, Almeida FP (2002) Ecological interactions between elasmobranchs and ground-fish species on the northeastern U.S. continental shelf. I. Evaluating predation. North Am J Fish Manage 22:550–562.

Liu KM, Chiang PJ, Chen CT (1998) Age and growth estimates of the bigeye thresher shark, *Alopias superciliosus*, in northeastern Taiwan waters. Fish Bull 96(3):482–491.

Lombardi-Carlson LA, Cortés E, Parsons GR, Manire CA (2003) Latitudinal variation in life-history traits of bonnethead sharks, *Sphyrna tiburo*, (Carcharhiniformes: Sphyrnidae) from the eastern Gulf of Mexico. Mar Freshw Res 54:875–883.

Lucifora LO, Garcia VB (2004) Gastropod predation on egg cases of skates (Chondrichthyes, Rajidae) in the southwestern Atlantic: quantification and life history implications. Mar Biol 145:917–922.

Lucifora LO, Menni RC, Escalante AH (2004) Reproductive biology of the school shark, *Galeorhinus galeus*, off Argentina: support for a single south western Atlantic population with synchronized migratory movements. Environ Biol Fish 71:199–209.

Lucifora LO, Menni RC, Escalante AH (2008) Reproductive ecology and abundance of the sand tiger shark, *Carcharias taurus*, from the southwestern Atlantic. ICES J Mar Sci 59:553–561.

Luning K, Asmus R (1991) Physical characteristics of littoral ecosystems, with special reference to marine plants. In: Mathieson AC, Nienhaus PH (eds) Ecosystems of the world 24: intertidal and littoral ecosystems. Elsevier, Amsterdam.

Matta ME, Gunderson DR (2007) Age, growth, maturity, and mortality of the Alaska skate, *Bathyraja parmifera*, in the eastern Bering Sea. Environ Biol Fish 80:309–323.

McEachran JD, Aschliman N (2004) Phylogeny of batoidea. In: Carrier JC, Musick JA, Heithaus MR (eds) Biology of sharks and their relatives, CRC Press, Boca Raton, FL, pp 79–113.

McFarlane GA, King JR (2006) Age and growth of big skate (*Raja binoculata*) and longnose skate (*Raja rhina*) in British Columbia waters. Fish Res 78:169–178.

McPhie RP, Campana SE (2009) Bomb dating and age determination of skates (family Rajidae) off the eastern coast of Canada. ICES J Mar Sci 66(3):546–560.

Mecklenburg CW, Mecklenburg TA, Thorsteinson LK (2002) Fishes of Alaska. American Fisheries Society, Bethesda, Maryland.

Menni RC, Lucifora LO (2007) Condrictios de la Argentina y Uruguay. Lista de Trabajo. ProBiota, FCNyM, UNLP, Serie Tecnica-Didactica, La Plata, Argentina 11:1–15.

Menni RC, Stehmann MFW (2000) Distribution, environment, and biology of batoid fishes off Argentina, Uruguay, and Brazil. A review. Rev Mus Argentino Cienc Nat 2(1):69–109.

Moore JA, Hartel KE, Craddock JE, Galbraith JK (2003) An annotated list of deepwater fishes from off the New England region, with new area records. Northeast Nat 19(2):159–248.

Musick JA, Ellis JK (2005) Reproductive evolution of chondrichthyans. In: Hamlett WC, Jamieson BGM (eds) Reproductive biology and phylogeny of chondrichthyes: sharks, batoids, and chimaeras. Science Publishers, Enfield, New Hampshire, pp 45–79.

Nakabo T (ed) (2002) Fishes of Japan. Tokai University Press, Tokyo, Japan.

Nakaya K, Shirai S (1992) Fauna and zoogeography of deep-benthic chondrichthyan fishes around the Japanese Archipelago. Jpn J Ichthyol 39(1):37–48.

Natanson LJ, Cailliet GM (1986) Reproduction and development of the Pacific angel shark, *Squatina californica*, off Santa Barbara, California. Copeia 1986(4):987–994.

Natanson LJ, Cailliet GM (1990) Vertebral growth zone deposition in Pacific angel sharks. Copeia 1990(4):1133–1145.

Natanson LJ, Mello JJ, Campana SE (2002) Validated age and growth of the porbeagle shark (*Lamna nasus*) in the western North Atlantic Ocean. Fish Bull 100:266–278.

Natanson LJ, Sulikowski JA, Kneebone JR, Tsang PC (2007) Age and growth estimates for the smooth skate, *Malacoraja senta*, in the Gulf of Maine. Environ Biol Fish 80:293–308.

Natanson LJ, Wintner SP, Johansson F, Piercy A, Campbell P, De Maddalena A, Gulak SJB, Human B, Fulgosi FC, Ebert DA, Hemida F, Mollen FH, Vanni S, Burgess GH, Compagno LJV, Wedderburn-Maxwell A (2008) Ontogenetic vertebral growth patterns in the basking shark *Cetorhinus maximus*. Mar Ecol Prog Ser 361:267–278.

Naylor GJP, Ryburn JA, Fedrigo O, López A (2005) Phylogenetic relationships among the major lineages of sharks and rays. In: Hamlett WC, Jamieson BGM (eds) Reproductive biology and phylogeny of chondrichthyes: sharks, batoids, and chimaeras. Science Publishers, Enfield, New Hampshire, pp 1–25.

Nelson JS (2006) Fishes of the world. John Wiley & Sons, Inc., Hoboken, New Jersey.

Pequeño GR, Lamilla G (1993) Batoideos comunes a las costas de Chile y Argentina-Uruguay (Pisces:Chondrichthyes). Rev Biol Mar, Valparaiso 28(2):203–217.

Peres MB, Vooren CM (1991) Sexual development, reproductive cycle, and fecundity of the school shark *Galeorhinus galeus* off southern Brazil. Fish Bull 89(4):655–667.

Pidwirny M (2006) Introduction to the oceans. In: Fundamentals of physical geography, 2nd edition. http://www.physicalgeography.net/fundamentals/8o.html. Accessed February 18, 2009.

Pogonoski J, Pollard D (2003) *Centrophorus harrissoni*. In: 2008 IUCN red list of threatened species. www.iucnredlist.org. Accessed March 7, 2009.

Quiroz JC, Wiff R (2005) Demographic analysis and exploitation vulnerability of beaked skate (*Dipturus chilensis*) off the Chilean austral zone. ICES CM 2005/N:19.

Roberts CM (2002) Deep impact: the rising toll of fishing in the deep sea. Trends Ecol Evol 17(5):242–245.

Ryland JS, Ajayi TO (1984) Growth and population dynamics of three Raja species (Batoidei) in Carmarthen Bay, British Isles. J Conseil 41:111–120.

Scott WB, Scott MG (1988) Atlantic fishes of Canada. Can Bull Fish Aquat Sci No. 219.

Shirai S (1992) Squalean phylogeny: a new framework of "squaloid" sharks and related taxa. Hokkaido University Press, Sapporo, Japan.

Simpfendorfer CA, Chidlow J, McAuley R, Unsworth P (2000) Age and growth of the whiskery shark, *Furgaleus macki*, from southwestern Australia. Environ Biol Fish 58:335–343.

Simpfendorfer CA, Unsworth P (1998) Reproductive biology of the whiskery shark, *Furgaleus macki*, off south-western Australia. Aust J Mar Freshw Res 49:687–693.

Smith PJ, Steinke D, Mcveagh SM, Stewart AL, Struthers CD, Roberts CD (2008) Molecular analysis of southern Ocean skates (*Bathyraja*) reveals a new species of Antarctic skate. J Fish Biol 73:1170–1182.

Smith SE, Au DW, Show C (1998) Intrinsic rebound potentials of 26 species of Pacific sharks. Mar Freshw Res 49:663–678.

Smith SE, Mitchell RA, Fuller D (2003) Age validation of a leopard shark (*Triakis semifasciata*) recaptured after 20 years. Fish Bull 101:194–198.

Stevens JD, Bonfil R, Dulvy NK, Walker PA (2000) The effects of fishing on sharks, rays, and chimaeras (chondrichthyans), and the implications for marine ecosytems. ICES J Mar Sci 57:476–494.

Stevens J, Fowler SL, Soldo A, McCord M, Baum J, Acuña E, Domingo A, Francis M (2006) *Lamna nasus*. In: 2008 IUCN red list of threatened species. www.iucnredlist.org. Accessed March 7, 2009.

Stevenson DE, Orr JW, Hoff GR, McEachran JD (2007) Field guide to sharks, skates and ratfish of Alaska. Alaska Sea Grant College Program, University of Alaska Fairbanks.

Treloar MA, Laurenson LJB, Stevenson JD (2006) Descriptions of rajid egg cases from southeastern Australian waters. Zootaxa 1231:53–68.

Treloar MA, Laurenson LJB, Stevens JD (2007) Dietary comparisons of six skate species (Rajidae) in southeastern Australian waters. Environ Biol Fish 80:181–196.

Walker PA, Hislop JRG (1998) Sensitive skates or resilient rays? Spatial and temporal shifts in ray species composition in the central and north-western North Sea between 1930 and the present day. ICES J Mar Sci 55:392–402.

Walker TI, Cavanagh RD, Stevens JD (2006) *Galeorhinus galeus*. In: 2008 IUCN red list of threatened species. www.iucnredlist.org. Accessed March 7, 2009.

White WT (2003) *Centrophorus squamosus*. In: 2008 IUCN red list of threatened species. www.iucnredlist.org. Accessed March 2009.

Whitehead PJP, Bauchot ML, Hureau JC, Nielsen J, Tortonese E (eds) (1984) Fishes of the northeastern Atlantic and the Mediterranean. UNESCO, Paris.

Yano K, Sugimoto T, Nomura Y (2003) Capture records of manta ray, *Manta birostris*, and whale shark, *Rhincodon typus*, at Mutsu Bay, Aomori, Japan. Report of Japanese Society for Elasmobranch Studies 39:8–13.

Yudin KG, Cailliet GM (1990) Age and growth of the gray smoothhound *Mustelus californicus* and the brown smoothhound, *M. henlei*, sharks from central California. Copeia 1990(1):191–204.

4

Elasmobranchs of Tropical Marine Ecosystems

William T. White and Emma Sommerville

CONTENTS

4.1 Introduction

The marine ecosystems of the world are vast, with the total volume of water in the worlds' oceans exceeding 1450 km^3 (Briggs 1995), and habitat for marine organisms occurs from the surface to the deepest ocean trench, that is, the Challenger Deep in the Mariana Trench at 10,920 m (Gvirtzman and Stern 2004). This is reflected in the fact that 34 of the 37 recognized animal phyla are found in the marine environment (Nicol 1971). On a global scale, temperature is the primary factor that controls species distributions and the ocean's surface can be divided into four major temperature zones: the tropical, the warm-temperate, the cold-temperate, and the polar zones (see Figure 55, Briggs 1995). These zones then can be subdivided by ocean basins and adjacent landmasses (Parenti 1991, Table 2.1 in Musick, Harbin, and Compagno 2004).

The tropical zone is defined by the 20°C isocryme for the coldest month of the year because it limits the tropical marine fauna (Dana 1853; Briggs 1974). This temperature also seems generally to define the northern- and southernmost limits of coral reef growth (Ekman 1953). The North and South Equatorial currents run from east to west and, as they approach landmasses to the west, the warm currents flow north and south from the equator resulting in a broader tropical zone on the western sides of ocean basins. In contrast, on the eastern sides of the ocean basins, the major currents flow from the higher latitudes bringing in cooler water, resulting in a narrower tropical zone in these regions (Briggs 1995; Longhurst 1998). Thus, using the 20°C isocryme is far more accurate in defining the tropical zone than using straight latitudinal boundaries. The tropical zone can be subdivided into four major biogeographic regions: the Indo–West Pacific, Eastern Pacific, Western Atlantic, and Eastern Atlantic regions.

The Indo–West Pacific bioregion is by far the largest of the four regions, extending more than half of the globe longitudinally and >60° latitudinally, and has by far the greatest species richness, exceeding that of the other three bioregions combined (Briggs 1995). The expansive continental shelf of this region is estimated to cover more than 6.5 million km^2, compared to 2 million km^2 for the other three regions combined (Briggs 1985). It extends from southeastern Africa east to the Central Pacific (including all Pacific island groups), and north to Taiwan. The Eastern Pacific bioregion extends from Magdalena Bay in Mexico south to the Gulf of Guayaquil in Ecuador, and includes five offshore island groups from Clipperton west of Costa Rica to the Galapagos Islands west of Ecuador (Briggs 1995). The Western Atlantic bioregion extends from Bermuda, southern Florida, and the southwestern Gulf of Mexico (United States) to near Rio de Janeiro in Brazil, including the offshore localities Fernando de Noronha and Trindade Islands and St Paul's Rocks. The Eastern Atlantic bioregion extends from the Cape Verde Islands off Cape Verde in Senegal to southern Angola (Briggs 1995).

Based on the distributions of the organisms they contain, the oceans are fundamentally divided into two major Realms, the Pelagic Realm and the Benthic Realm, which are subdivided into a number of depth zones (Briggs 1974, 1995). The Pelagic Realm is treated in Chapter 1. The Benthic Realm can be divided into three main subdivisions: the continental shelf or coastal (0 to 200 m), continental slope or bathyal (upper 200 to 500 m, mid 500 to 1000 m, lower 1000 to 2000 m), and abyssal zones (Briggs 1995; Musick, Harbin, and Compagno 2004).

This chapter examines the biogeography, biodiversity, systematics, ecology, fisheries, management, and conservation of chondrichthyan fishes in the coastal and shelf waters of tropical marine ecosystems. We do not include species that occur entirely below the 200 m depth contour (covered in Chapter 2) or are entirely epipelagic (covered in Chapter 1) or occur entirely in freshwater and/or estuarine habitats.

4.2 Biodiversity and Systematics

4.2.1 Taxonomic Research

The number of new species reported to occur in the tropical marine ecosystems has increased continually since Linnaeus' revolutionary *Systema Naturae* in 1758, with more elasmobranchs described from the tropics (58 species) between 2000 and 2009 than in any previous decade (Figure 4.1). This reflects the recent surge in new chondrichthyan species described from all ecosystems in the last decade (179 species). In previous decades, description of new tropical chondrichthyans has been relatively consistent with the exception of the 1830s when 44 species were identified (Figure 4.1). Almost half of these (20) species were described in Müller and Henle's 1838–1841 *Systematische Beschreibung der Plagiostomen*. Notable surges in tropical chondrichthyan taxonomy are apparent, for example 1758 (Linnaeus 1758), 1801 (mostly Bloch and Schneider 1801), 1839 and 1841 (mostly Müller and Henle 1838–1841), and 2008 (mostly Last, White, and Pogonoski 2008 and Last et al. 2008).

4.2.2 Biodiversity

This chapter focuses on those chondrichthyan species that occur over continental shelf waters of the tropics. Thus, if a species occurs within the tropical zone (see Figure 2.1 in Musick, Harbin, and Compagno 2004) and occurs at a depth of less than 200 m for all or part of its bathymetric range, it is included here (see Table 4.1). A complete checklist of nominal chondrichthyan species occurring in tropical marine ecosystems is provided in Appendix 4.1. Although strict criteria were used to allocate species as occurring in tropical

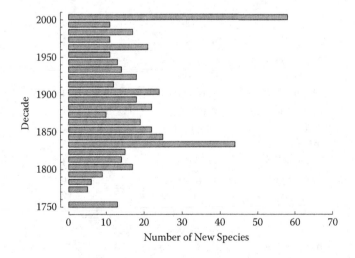

FIGURE 4.1
Number of tropical marine chondrichthyans described per decade since the 1750s.

TABLE 4.1

Diversity of Sharks and Batoids Occurring in Tropical Shelf Waters by Order and Families

Order	Family	Common Name	Described	Undescribed	Total
Sharks					
Hexanchiformes	Hexanchidae	Sixgill and sevengill sharks	2	—	2
Squaliformes	Squalidae	Dogfish sharks	8	—	8
	Centrophoridae	Gulper sharks	2	—	2
	Oxynotidae	Roughsharks	1	—	1
Pristiophoriformes	Pristiophoridae	Sawsharks	2	—	2
Squatiniformes	Squatinidae	Angelsharks	12	—	12
Heterodontiformes	Heterodontidae	Bullhead sharks	7	—	7
Orectolobiformes	Parascylliidae	Collared carpetsharks	2	—	2
	Brachaeluridae	Blind sharks	2	—	2
	Orectolobidae	Wobbegongs	8	—	8
	Hemiscylliidae	Longtailed carpetsharks	14	—	14
	Ginglymostomatidae	Nurse sharks	3	—	3
	Stegostomatidae	Zebra shark	1	—	1
	Rhincodontidae	Whale shark	1	—	1
Lamniformes	Odontaspididae	Sand tiger sharks	2	—	2
	Lamnidae	Mackerel sharks	1	—	1
Carcharhiniformes	Scyliorhinidae	Catsharks	32	—	32
	Proscylliidae	Finback catsharks	4	—	4
	Leptochariidae	Barbeled houndshark	1	—	1
	Triakidae	Houndsharks	29	3	32
	Hemigaleidae	Weasel sharks	8	—	8
	Carcharhinidae	Whaler sharks	49	2	51
	Sphyrnidae	Hammerheads	8	—	8
Batoids					
Rajiformes	Pristidae	Sawfishes	7	—	7
	Rhinidae	Shark ray	1	—	1
	Rhynchobatidae	Wedgefishes	5	2	7
	Rhinobatidae	Guitarfishes	39	—	39
	Platyrhinidae	Thornback rays	1	—	1
	Zanobatidae	Panrays	2	—	2
	Narcinidae	Numbfishes	22	3	25
	Narkidae	Sleeper rays	5	—	5
	Hypnidae	Coffin rays	1	—	1
	Torpedinidae	Torpedo rays	16	—	16
	Arhynchobatidae	Softnose skates	2	—	2
	Rajidae	Hardnose skates	30	—	30
	Urolophidae	Stingarees	7	—	7
	Urotrygonidae	American round stingrays	14	—	14
	Dasyatidae	Stingrays	56	5	61

TABLE 4.1 (*Continued*)

Diversity of Sharks and Batoids Occurring in Tropical Shelf Waters by Order and Families

Order	Family	Common Name	Described	Undescribed	Total
	Gymnuridae	Butterfly rays	11	—	11
	Myliobatidae	Eagle rays	14	—	14
	Rhinopteridae	Cownose rays	9	—	9
	Mobulidae	Devilrays	9	—	9
Total			450	15	465

marine ecosystems based on geographic and depth distribution, in reality the boundary between tropical and warm-temperate zones is not always clear and there are typically areas of significant overlap. For example, where warm currents run along the shelf to higher latitudes, surface waters will cool more rapidly than at depths of >100 m, which can result in tropical outer shelf species extending into higher latitudes than inner shelf species (Briggs 1995). Therefore, many of the species that occur in the tropical zone with a limited distribution in the warm-temperate zone, based on the criteria discussed above, are probably tropical-only species and vice versa.

A total of 450 nominal and 15 undescribed species of chondrichthyans occurs in the tropical marine ecosystems covered in this chapter. This represents ~39% of the global extant chondrichthyan fauna (~1150 species). All orders of sharks (Hexanchiformes, Squaliformes, Squatiniformes, Heterodontiformes, Pristiophoriformes, Lamniformes, Orectolobiformes, and Carcharhiniformes) and skates and rays, that is, batoids (Batoidea), are represented. There are no chimaeroids. The generally deep-water Squaliformes are poorly represented in the shelf ecosystems of the tropics (~9% of global species). Lamniformes are also relatively poorly represented (20% of global species), which is reflected in the predominantly epipelagic and oceanic habits of most lamnoid sharks. In contrast, the Orectolobiformes (77%) and Heterodontiformes (78%) are strongly represented. Of the 450 nominal tropical species in Table 4.1, 199 are sharks and 251 are batoids, which represent ~41% of the global number of shark and batoid species, thus these two major elasmobranch groups are equally represented in this ecosystem. About 49% (226) of these species are found only in the tropics, with an additional 62 species (13%) occurring predominantly in tropical waters but also extending into warm-temperate waters in summer. Since no chimaeroids are found on the tropical shelf regions, the term elasmobranchs is used throughout this chapter.

Of the 34 families of sharks and 23 families of batoids, the shelf ecosystems of tropical marine waters contain representatives from 23 and 19 of these families, respectively. The shark families Scyliorhinidae, Triakidae, and Carcharhinidae are the most speciose, together constituting 58% of the total tropical shark species. All species from eight shark families (Brachaeluridae, Hemiscylliidae, Ginglymostomatidae, Stegostomatidae, Rhincodontidae, Leptochariidae, Hemigaleidae, and Sphyrnidae) have ranges that at least extend into the tropics. The Dasyatidae is the most speciose tropical batoid family by far, followed by the Rhinobatidae, Rajidae, and Narcinidae, which together constitute 61% of the total tropical batoid species. All species from five families (Pristidae, Rhinidae, Rhynchobatidae, Zanobatidae, and Hypnidae) occur at least in part in the tropical marine ecosystems.

The greater diversity of species that occur in the tropics compared to at higher latitudes can be attributed to a number of factors. The high level of species diversity on the continental shelf waters appears to follow the energy-stability-area theory proposed by Wilson

(1992) whereby the more solar energy, the greater the diversity; the more stable the climate, the greater the diversity; the larger the area, the greater the diversity. This theory is applicable both to terrestrial and marine organisms. Another important factor to consider is the age of the tropical marine ecosystems. During the Mesozoic and early Cenozoic Eras (>60 mya), the tropics occupied almost the entire globe and only began to become much reduced in the middle Eocene (~45 mya) (Briggs 1995). Thus, the tropical marine fauna has had a much longer period to evolve and for speciation to occur. The more complex habitats that exist in the tropical regions, for example, diverse coral reef ecosystems, can support a much greater diversity of species due to the complexity of niches available for a species to exploit.

Since tropical marine ecosystems contain such a diversity of habitats and occupy a large geographic area, the species occurring in this ecosystem show a wide diversity of body forms and life histories. Tropical species range from the tiny pygmy ribbontail catshark *Eridacnis radcliffei*, which is demersal on muddy bottoms and attains only 24 cm total length (TL), to the enormous whale shark *Rhincodon typus*, which is pelagic over shelf and oceanic waters and attains more than 1200 cm TL (Compagno, Dando, and Fowler 2005). All of the major body forms are evident in the tropical marine chondrichthyan fauna, with the notable absence of the holocephalan body form, which are a predominantly deepwater group, especially in tropical waters.

4.3 Biogeography

4.3.1 General Biogeography

Of the 465 species occurring in the tropical marine ecosystems, the largest number are distributed in the Indo–West Pacific biogeographic region (301 species) with far less recorded in the Eastern Pacific (81), Western Atlantic (95), and Eastern Atlantic (70) biogeographic regions. In the Indo–West Pacific, the vast majority of sharks (95%) and batoids (84%) occur only within this region, highlighting the geographic barriers present on a global scale. Although the majority of sharks in the other three regions were also restricted to the one biogeographic region, most batoids occurred across more than one region. Of the 42 families found in tropical marine ecosystems, only three are not represented in the Indo–West Pacific. The high diversity of elasmobranchs in the Indo–West Pacific is further highlighted by the breakdown of the number of species occurring in each of the FAO Fishing Areas (Table 4.2), with those areas concentrated in the Indo–West Pacific, that is, 51, 57, 61, and 71, containing the highest number of species.

4.3.2 Zoogeography

In this section, zoogeographical information is provided for each of the major taxonomic groups presented in Table 4.1.

> Hexanchiformes: The Hexanchiformes consists of two families, with only two species belonging to the Hexanchidae occurring in the tropical marine ecosystem. *Hexanchus griseus* and *Heptranchias perlo* are predominantly slope species but are found occasionally on the shelf and are wide ranging but patchy in most temperate and tropical waters.

TABLE 4.2

Diversity of Tropical Marine Elasmobranchs by Major Ocean
Region and FAO Area

Biogeographic Region	FAO Area	Number of Species		
		Sharks	Batoids	Total
Eastern Atlantic	FAO 34	31	42	73
Eastern Atlantic	FAO 47	26	40	66
Eastern Pacific	FAO 77	42	40	82
Eastern Pacific	FAO 87	31	31	62
Indo–West Pacific	FAO 51	69	77	146
Indo–West Pacific	FAO 57	87	89	176
Indo–West Pacific	FAO 61	67	46	113
Indo–West Pacific	FAO 71	103	104	207
Western Atlantic	FAO 31	46	44	90
Western Atlantic	FAO 41	32	24	56

Note: The totals add up to more than the number of known tropical elasmobranchs since some species occur in multiple regions.

Squaliformes: The Squaliformes consists of seven families, three of which are represented in tropical marine ecosystems, that is, Squalidae, Centrophoridae, and
Oxynotidae. These include 11 species all of which occur predominantly on the
upper slope and more than half occur in the Indo–West Pacific.

Pristiophoriformes: The Pristiophoriformes contains a single family and two genera.
Two of the eight species, *Pliotrema warreni* and *Pristiophorus japonicus*, occur in the
tropics. Both of these are restricted to the Indo–West Pacific region.

Squatiniformes: The Squatiniformes consists of a single family and genus. Twelve
of the 20 species occur in tropical marine ecosystems, with between two and four
species in each of the four major biogeographic regions. All are predominantly
shelf species, with several occurring also on the upper slope, and they have relatively restricted ranges.

Heterodontiformes: The Heterodontiformes consists of a single family and a single
genus with seven of the nine species represented in tropical marine ecosystems. Four
heterodontids are confined to the Indo–West Pacific and three to the Eastern Pacific.
Most species occur on the insular shelf in warm-temperate and tropical waters.

Orectolobiformes: The Orectolobiformes consists of seven families all of which are
represented in the tropics. The order is primarily an Indo–West Pacific group, with
only two species of nurse sharks (Ginglymostomatidae) and the cosmopolitan
whale shark (Rhincodontidae) occurring in the other three biogeographic regions.
Thus, of the 31 species belonging to this order, all but one occurs in the Indo–West
Pacific and all but two are restricted to only this region. Tropical shelf members
of the Brachaeluridae (two species), all but one member of the Orectolobidae
(seven species) and all seven species of the genus *Hemiscyllium* (Hemiscylliidae)
are restricted to Australia–New Guinea. Most members of the genus *Chiloscyllium*
are focused around the Indo–Malay Archipelago with only one species reaching
northern Australia and one species restricted to the Western Indian Ocean. The
vast majority of species are demersal on the insular shelf in predominantly tropical waters (Musick, Harbin, and Compagno 2004).

Lamniformes: The Lamniformes are poorly represented in tropical marine ecosystems, with only two of the seven families represented. The three species present (*Carcharodon carcharias*, *Carcharias taurus*, and *Odontaspis ferox*) are wide ranging in all or most temperate and tropical waters.

Carcharhiniformes: The Carcharhiniformes consists of eight families, of which seven are represented in tropical marine ecosystems. The most speciose family, the Carcharhinidae, is well represented in all biogeographic regions with 40 of the 51 species occurring in the Indo–West Pacific. Of these 40 species, 26 are restricted to the Indo–West Pacific, whereas in comparison only one of the 14 species occurring in the Eastern Pacific is restricted to that region. The Western Atlantic has 19 carcharhinids, of which seven are endemic. None of the 13 Eastern Atlantic carcharhinids is restricted to that region. The Scyliorhinidae are also strongly represented in tropical marine ecosystems with a total of 32 species, all of which are restricted to a single biogeographic region. The majority, that is, 22, are restricted to the Indo–West Pacific, with either three or four restricted to each of the other three biogeographic regions. Members of the genera *Asymbolus*, *Atelomycterus*, *Aulohalaelurus*, *Galeus*, *Halaelurus*, *Haploblepharus*, and all but one species of *Cephaloscyllium* are restricted to the Indo–West Pacific. In contrast, tropical members of the *Schroederichthys* and *Scyliorhinus* are restricted to the Atlantic Ocean.

Another diverse family, the Triakidae, has 32 representatives in tropical marine ecosystems, all of which are restricted to a single biogeographic region. The majority (18) are restricted to the Indo–West Pacific, while eight and seven are restricted to the Eastern Pacific and Western Atlantic, respectively. Interestingly, no triakids have been reported from tropical shelf waters of the Eastern Atlantic. Members of the genera *Furgaleus*, *Gogolia*, *Hemitriakis*, *Hypogaleus*, and *Iago* are restricted to the Indo–West Pacific. The only triakids occurring in the Western Atlantic tropical region are members of the genus *Mustelus*, which are also distributed in the Indo–West Pacific and Eastern Pacific. All eight species of the family Hemigaleidae are found predominantly in inshore waters of the tropics, with all but one restricted to the Indo–West Pacific. There are no hemigaleids recorded off either coast of the Americas. All eight species of the family Sphyrnidae are at least occasionally found in tropic marine ecosystems, but they are evenly represented in each of the four regions. Three of the species, *Sphyrna lewini*, *S. mokarran*, and *S. zygaena*, have a circumglobal distribution in tropical waters. The four species of Proscylliidae that are found in the tropics are restricted to the Indo–West Pacific and the single species of Leptochariidae, *Leptocharias smithii*, is restricted to the insular shelves of the Eastern Atlantic coastline.

Batoidea: The Batoidea contains four orders, that is, Pristiformes, Torpediniformes, Rajiformes, and Myliobatiformes, all of which are represented in tropical marine ecosystems (McEachran and Aschliman 2004). In recent years, batoids have been considered to belong to just one order, the Rajiformes. Since a definitive classification of batoids is not available at this stage, the various suborders are used in this section.

The suborder Pristoidei contains a single family, the Pristidae, and seven species all of which occur on insular shelves in tropical waters. Four species are restricted to the Indo–West Pacific and the remaining three species occur in both the Western and Eastern Atlantic. The Rhinoidei contains the monotypic family Rhinidae, which is restricted to tropical insular shelf waters of the Indo–West Pacific. The

Rhynchobatoidei contains a single family with seven species, all occurring on insular shelves of the tropical marine ecosystems. Six of these species are restricted to the Indo–West Pacific, with one species, *Rhynchobatus luebberti*, restricted to predominantly tropical waters in the Eastern Atlantic. The Rhinobatoidei contains a single family with 50 species, with 39 distributed, at least in part, in tropical marine ecosystems. The majority of these, 24, are restricted to the Indo–West Pacific while the remaining species are restricted to either the Eastern Pacific (six species), Western Atlantic (three species) or Eastern Atlantic (five species). The genus *Aptychotrema* (three species) is restricted to waters off Australia, while the genus *Zapteryx* (two species) is restricted to predominantly tropical shelf waters of the Eastern Pacific. The Platyrhinoidei contains a single family with three species, of which only one species occurs in tropical marine ecosystems and is restricted to the Northwest Pacific. The Zanobatoidei contains a single family and genus with two species that are restricted to warm-temperate and tropical waters of the Eastern Atlantic.

The Torpedinoidei consists of four families, all of which are represented in tropical marine ecosystems. The most speciose family, the Narcinidae, is well represented, with 25 of the 33 known species occurring in this zone. The majority of these (17) are restricted to the Indo–West Pacific and they are not found in the tropical Eastern Atlantic. The Torpedinidae are also well represented, with 16 of the 23 species occurring in marine ecosystems of the tropics and are relatively evenly distributed across the four biogeographic regions. Each species, however, is restricted to a single region. Five of the 12 members of the Narkidae occur in tropical marine ecosystems, all of which are restricted to the Indo–West Pacific. The single species belonging to the Hypnidae is restricted to Australia.

The Rajoidei are poorly represented in tropical marine ecosystems, with most members of this group occurring in deeper or temperate waters. Two of the three families occur in tropical marine ecosystems. Only two of the 96 known species of Arhynchobatidae occur in tropical marine ecosystems. One is restricted to the Indo–West Pacific and one to the Eastern Pacific. Of the 168 known species of Rajidae, 30 occur in tropical marine ecosystems. These species are distributed among the four biogeographic regions, with none occurring in more than one region.

The Myliobatoidei are well represented, with 7 of its 10 families occurring in the tropics. The seven tropical species belonging to the Urolophidae are restricted to narrow geographic ranges within the Indo–West Pacific, with all but one species occurring in the Indo–Australian region. The 14 tropical species of the Urotrygonidae are restricted to the Americas, with 10 species found in the Eastern Pacific and 3 in the Western Atlantic. The Dasyatidae are extremely well represented, with 61 of the 81 known species occurring in tropical marine ecosystems. The vast majority (41) occurs in the Indo–West Pacific and all but two of these are restricted to this region. The Eastern Pacific has few dasyatids, with only three species. All members of the genera *Neotrygon, Pastinachus, Urogymnus,* and all but two of the 21 tropical marine species of *Himantura* are restricted to the Indo–West Pacific. Members of the genus *Dasyatis* are more evenly distributed among the four biogeographic regions. The vast majority of the members of the Gymnuridae (11 of 14), are found in tropical marine ecosystems and in all regions except the Eastern Atlantic. Each of the species, however, is restricted to just one biogeographic region. The Myliobatidae are represented in all biogeographic regions with most species (9 of the 14 tropical species) occurring in the Indo–West Pacific. Seven species are restricted to this region. The monogeneric family Rhinopteridae are almost entirely found in the tropics, with 9 of the 10 species occurring in this zone. Six of these are restricted to the Indo–West Pacific and none is found in the Eastern Atlantic. The validity of four of the Indo–West Pacific rhinopterids is uncertain and a detailed revision of this family is required. The vast majority of the members of the Mobulidae occur in the tropics and are evenly distributed in each of the biogeographic regions. Three of the species, *Manta birostris*, *Mobula japanica*, and *Mobula tarapacana*, have circumglobal distributions in predominantly tropical waters and are found from coastal to pelagic habitats.

4.3.3 Centers of Diversity and Endemism

The geography of endemism of marine species is used to divide the biosphere into provinces, which are characterized by a suite of species found only in limited geographical areas, that is, areas of endemism (Briggs 1974; Vermeij 1978). Such areas of endemism are a result of barriers or boundaries to gene flow. The East Indies Triangle—which consists of the islands of Indonesia, the Philippines, the Malay Peninsula, and New Guinea (see Figure 1 in Briggs 2005)—is well documented to be the center of diversity for the marine tropics (Briggs 1995, 2003, 2005). The diversity of the East Indies Triangle is unsurpassed and species diversity decreases notably with distance outside of this region, especially moving eastward. Indeed, the Pacific Islands (with the exception of the extremely isolated islands of the central and Eastern Pacific) exhibit little local endemism (Springer 1982; Briggs 2003). Within the triangle, more than two-thirds of the Indo–Pacific reef fishes are found. In fact, almost a third of the more than 3700 reef fishes occurring in the entire Indo–Pacific occur off the Indonesian island of Flores (Allen and Adrim 2003).

The influence of the East Indies Triangle on tropical marine diversity extends beyond just the Indo–West Pacific region. The two primary barriers separating the Indo–West Pacific region from the other tropical regions are the deep-water East Pacific Barrier which separates Polynesia from the New World to the east and the Africa and Eurasia land masses to the west (Briggs 2003). A number of Indo–West Pacific fish species have managed to successfully migrate and become established in the Eastern Pacific (Leis 1984; Briggs 2003), and similarly to the west into the Atlantic (Briggs 1995). However, these appear to be one-way filters with no successful migrations in the opposite direction, that is, into the Indo–West Pacific.

The center for diversity for the Atlantic Ocean is the Caribbean Sea, with many species evolved in this area having successfully migrated across the Mid-Atlantic Barrier to colonize the tropical Eastern Atlantic (Briggs 2003). The tropical Eastern Atlantic has the poorest biota and there is very little evidence of species from this area successfully colonizing the Western Atlantic (Briggs 1995). Despite the fact that the Caribbean Sea is the Atlantic's center of origin, the species diversity does not compare to that of the East Indies Triangle. For example, the Caribbean contains only ~700 reef fish species (Rocha 2003) compared to >2000 in Indonesian waters alone (Allen and Adrim 2003). Randall (1998) proposed a number of factors that have led to the proliferation of species in the East Indies. The physiography of the region is a major factor whereby the numerous islands, which act as "stepping stones," have formed a buffer discouraging local extinction. The geological history of this region is also complex, having arisen from the merging of different plates due to tectonic movements. The climate of the East Indies region has also been more stable historically compared to other regions with warm sea temperatures consistently over time preventing the mass extinctions that have occurred elsewhere.

In the case of tropical marine elasmobranchs, the East Indies Triangle is clearly the largest center of diversity. A total of 164 elasmobranchs are found in the East Indies Triangle.

This represents 35% of the total number of tropical marine elasmobranchs (465 species) and 54% of the species within the Indo–West Pacific region. Thus, the relatively small area of the East Indies Triangle contains more than double the number of elasmobranch species in the tropical Eastern Pacific. It is also apparent that species diversity decreases with distance away from the East Indies Triangle. For example, the waters off northern Australia contain a further 51 elasmobranch species not found in the East Indies Triangle. Thus, these two regions combined have more than 70% of the Indo–West Pacific elasmobranch species—more than twice that present in the tropical Western Atlantic and three times that in the Eastern Atlantic. Of the 42 families found in tropical marine ecosystems, 31 are represented in the East Indies Triangle and 37 are represented in the East Indies Triangle and northern Australia combined.

As reported for reef fish species (Briggs 2003, 2005), the East Indies Triangle contains a large number of local endemics with restricted ranges. For example, the Bali catshark *Atelomycterus baliensis* is a small reef-dwelling catshark found only around the small Indonesian island of Bali (White, Last, and Dharmadi 2005). The Jimbaran shovelnose ray *Rhinobatos jimbaranensis* is demersal on inshore shelf waters around the Indonesian islands of Bali and Lombok (Last, White, and Fahmi 2006). New Guinea appears to be a center of diversity for the longtail carpetshark genus *Hemiscyllium*. Six of the seven *Hemiscyllium* species occur in waters off New Guinea, five of which are restricted to New Guinea (Compagno, Dando, and Fowler 2005; Allen and Erdmann 2008). Many of these restricted endemics have only been recognized and described in the last decade. Future taxonomic work is required on the elasmobranchs occurring in the East Indies to determine whether more cryptic species exist and a number of species complexes are likely to be encountered.

4.3.4 Habitat Association

A wide variety of habitats exist within the tropical shelf waters of the world, including mangrove forests, seagrass meadows, coral reefs, and unvegetated soft-sediment habitats. Many species of elasmobranchs are relatively mobile and will move among a variety of habitats, but others are specialized for life in a particular habitat and can show strong site fidelity.

Mangrove environments are important nursery areas for many marine species and have a crucial role in nutrient filtering (Valiela, Bowen, and York 2001). Mangrove forests are primarily tropical and are present in all of the biogeographic regions and are found in marine, estuarine, and riverine environments. Many species of elasmobranchs use mangrove ecosystems as either nursery areas or as an important habitat throughout most of their life cycle. In northwestern Australia, the inshore mangrove areas are important nursery areas for juveniles of a number of species, including the giant shovelnose ray *Glaucostegus typus*, the common blacktip shark *Carcharhinus limbatus*, and the sicklefin lemon shark *Negaprion acutidens* (Blaber 1986; White and Potter 2004). Two of these species, *G. typus* and *N. acutidens*, use the mangrove areas as a nursery area for several years but move to different habitats at larger sizes, while *C. limbatus* only utilizes these areas seasonally for a number of months following parturition. Mangrove systems are also important habitats for a number of elasmobranchs throughout most of their life cycle. For example, in northern Australia, the estuary stingray *Dasyatis fluviorum*, the porcupine ray *Urogymnus asperrimus*, the green sawfish *Pristis zijsron*, the nervous shark *Carcharhinus cautus*, and the creek whaler *C. fitzroyensis* use mangrove systems through much of their life cycle (White and Potter 2004; Last and Stevens 2009). There is very limited information on the role of elasmobranchs in mangrove ecosystems and since they are the apex predators within these areas, such studies are required to improve understanding of their role.

Seagrass meadows, which are abundant in tropical and temperate waters, are important nursery areas for a large number of fish species (Walker 1989), and thus support complex and important food webs, of which elasmobranchs are often the predominant apex predators. Many tropical elasmobranchs use these areas throughout all or part of their lifecycle. Shark Bay in northwestern Australia contains some of the world's most extensive and diverse seagrass meadows (Walker 1989) and a number of elasmobranchs make use of these habitats for various reasons including protection from predation and prey resources. White and Potter (2004) found that a number of species use the seagrass meadows in Shark Bay as a nursery area for newborn spinner shark *Carcharhinus brevipinna*, common blacktip shark *C. limbatus*, and sandbar shark *C. plumbeus*. Several species were found to occur over seagrass meadows throughout much of their life cycle, for example, grey carpet shark *Chiloscyllium punctatum*, Australian weasel shark *Hemigaleus australiensis*, whiskery shark *Furgaleus macki*, and milk shark *Rhizoprionodon acutus*. The Shark Bay seagrass meadows are also important for tiger sharks *Galeocerdo cuvier*. Seabirds (pied cormorants), dugongs, and dolphins give up foraging opportunities in shallow seagrass beds to avoid predation by the large number of tiger sharks present (Heithaus and Dill 2002; Heithaus et al. 2002; Heithaus 2005), and therefore tiger sharks may play an important role in the dynamics of these ecosystems (Heithaus et al. 2008).

Coral reefs are extensive throughout the tropical shelf regions of the world and they support an enormous diversity of reef fishes, especially in the Indo–West Pacific region (Allen and Adrim 2003; Briggs 2005). A large number of elasmobranchs utilize the coral reef habitats either as their primary habitat or as a secondary habitat. Many carcharhinid sharks may occur over coral reefs to some extent but several species are considered primarily coral reef dwellers, that is, the whitetip reef shark *Triaenodon obesus*, the grey reef shark *Carcharhinus amblyrhynchos*, the blacktip reef shark *C. melanopterus*, and the Caribbean reef shark *C. perezii*. The whitetip reef shark is well adapted to the coral reef habitat and is rarely seen far off the bottom of the coral reef. It is typically seen resting in caves during the day and at night it forages by using its very slender body to wedge itself into small reef interstices to seek prey (Randall 1977). A large number of benthic shark species are primarily coral reef dwellers. These include many catsharks (Scyliorhinidae), wobbegongs (Orectolobidae), horn sharks (Heterodontidae), zebra shark (*Stegostoma fasciatum*), nurse sharks (Ginglymostomatidae), and many longtailed carpetsharks (Hemiscylliidae). Intense and elaborate color patterns are a feature of many of the benthic coral reef dwelling sharks, such as the wobbegongs, and are used as camouflage against the brightly colored coral reefs. The epaulette shark *Hemiscyllium ocellatum* in waters off northern Australia and New Guinea has adapted to the very shallow reef flats by using their pectoral fins to "walk" around on the reef and can survive long periods of time out of the water if required (Heupel and Bennett 1998; Last and Stevens 2009).

Fewer batoid species use coral reefs as their primary habitat compared to sharks. The best example and most abundant is the blue-spotted fantail ray *Taeniura lymma*, which is commonly seen in shallow coral reef habitats in the Indo–West Pacific (Figure 4.2) (Last and Stevens 2009). Another small species, the blue-spotted maskray *Neotrygon kuhlii*, is also extremely abundant in coral reef habitats of the Indo–West Pacific. The much larger blotched fantail ray *Taeniurops meyeni* is often seen in coral reef habitats, with adults commonly observed at depths greater than 15 m off northern Australia. Large adults (>100 cm disc width) of a number of dasyatid species in the Indo–Australian region, that is, pink whipray *Himantura fai*, leopard whipray *H. leoparda*, reticulate whipray *H. uarnak*, and cowtail stingray *Pastinachus atrus*, are commonly observed in sandy areas within or adjacent

FIGURE 4.2
A color version of this figure follows page 336. The blue-spotted fantail ray *Taeniura lymma* under a coral reef ledge off northeast Efate (Vanuatu) in the Southwest Pacific. (Photo © Will White. All rights reserved.)

to coral reefs. In the Cayman Islands in the Western Atlantic, the Florida torpedo *Torpedo andersoni* is frequently seen on coral reef formations (Carpenter 2002). Members of the pelagic Mobulidae and the semipelagic smalleye stingray *Dasyatis microps* (Pierce, White, and Marshall 2008) are also observed over coral reefs at certain times.

Seasonal changes in productivity on coral reefs can be important for some elasmobranchs. For example, at Ningaloo Reef in northwestern Australia, the whale shark *Rhincodon typus* appears during the Austral autumn when there is a high level of productivity associated with mass synchronous coral spawning events (Taylor 1994). Such massive increases in productivity support a diverse food chain, which includes a number of upper trophic level elasmobranchs, for example, whale sharks and manta rays.

Although focus has been placed on the mangrove, seagrass, and coral reef habitats, a number of other broad habitats are found in tropical shelf waters, each important for a suite of elasmobranch species. For example, soft bottom habitats support a variety of benthic species, for example, angel sharks (Squatinidae), numbfishes (Narcinidae), and shovelnose rays (Rhinobatidae), and benthopelagic species, for example, houndsharks (Triakidae).

4.4 Life History

4.4.1 Reproductive Biology

All of the broad reproductive modes exhibited by elasmobranchs are represented by species occurring in tropical marine ecosystems. The vast majority (371; ~80%) of species, are viviparous (live bearers), with most of the remaining (89; ~19%) species, being oviparous (egg layers) (Table 4.3).

TABLE 4.3

Summary of Reproductive Modes Adopted by the Various Species and Families of Tropical Marine Elasmobranchs

Reproductive Mode	No. of Species	No. of Families	% of Tropical Chondrichthyan Species
Oviparous	89	9	~19
Viviparous, yolk-sac dependent	163	24	~35
Viviparous, with histotrophy	123	7	~27
Viviparous, placental	81	5	~17
Viviparous, oophagy/adelphophagy	4	3	<1
Unknown	5	2	~1

Oviparity is present in at least some species in four of the tropical marine chondrichthyan orders: Heterodontiformes, Orectolobiformes (Parascylliidae, Hemiscylliidae, Stegostomatidae, and probably Ginglymostomatidae), Carcharhiniformes (Scyliorhinidae and Proscylliidae), and Rajiformes (Arhynchobatidae and Rajidae). Oviparous species are typically bottom-dwelling species that deposit eggs externally inside a leathery, structurally complex, and extremely durable case (Musick and Ellis 2005). The developing embryos within these shells obtain all their nourishment from their yolk supply and generally take from a couple of months to a year to hatch (Carrier, Pratt, and Castro 2004). The majority of oviparous species exhibit single (external) oviparity in which one egg is deposited at a time from each oviduct with tens to hundreds of eggs deposited in a spawning season (Musick and Ellis 2005). A small number of scyliorhinid species exhibit multiple (retained) oviparity in which up to about 10 eggs are retained within the oviducts during most of embryonic development and are deposited just prior to hatching, thus reducing the potential predation on deposited egg cases in the external environment (Musick and Ellis 2005). The tropical catshark genus *Halaelurus* exhibits this form of oviparity. For example, the Indonesian speckled catshark *H. maculosus* and the rusty catshark *H. sellus* retain between three and six eggs per uterus during much of the embryonic development (White, Last, and Stevens 2007).

Viviparity is very well represented in the tropical chondrichthyan fauna, with ~80% of species being live bearers compared to only ~57% of species globally. This can be attributed to the large number of carcharhinids and dasyatids as well as the paucity of skates (Arhynchobatidae, Rajidae, and Anacanthobatidae), deepwater catsharks (*Apristurus*), and chimaeras (Holocephali) in the tropical marine ecosystems. Viviparity can be divided into a number of broad categories based on the mode of embryonic nutrition (see Carrier, Pratt, and Castro 2004; Musick and Ellis 2005). The most common reproductive mode among the tropical chondrichthyan fauna is viviparity with yolk-sac dependency, with 163 species representing 14 shark and 10 batoid families. In this mode, the embryos depend solely (or at least primarily) on yolk deposited in the egg during ovulation. In some instances, for example, sawsharks *Pristiophorus*, a small amount of embryonic nourishment comes from mucous secretions from uterine villi, that is, limited histotrophy (Hamlett et al. 2005). Thus, this reproductive mode is primarily lecithotrophic (as is oviparity).

Viviparity with histotrophy is found in ~27% of tropical elasmobranchs, all of which are myliobatoid rays of the families Urolophidae, Urotrygonidae, Dasyatidae, Gymnuridae, Myliobatidae, Rhinopteridae, and Mobulidae. In this mode of reproduction, yolk reserves are depleted by the embryo very early in gestation and subsequently embryos consume a protein- and lipid-rich histotroph (uterine milk) secreted from specialized trophonemata (long uterine villi). Thus, the primary source of embryonic nutrition is matrotrophic via placental analogues (Hamlett et al. 2005). Only three tropical marine Lamniformes and one species of Ginglymostomatidae (*Nebrius ferrugineus*) are viviparous with oophagy/ adelphophagy, in which embryonic nourishment is provided from oocytes ovulated prior to or throughout gestation or from intrauterine cannibalism of siblings. The grey nurse shark *Carcharias taurus* is unique in its mode of oophagy whereby up to a dozen embryos may be produced per oviduct as well as numerous unfertilized eggs. When the largest embryo attains ~100 mm, it kills the other embryos and later in gestation it feeds on its dead siblings and the other eggs (Carrier, Pratt, and Castro 2004). Thus, only one large and well-developed embryo is produced per uterus. The late-term embryos of the single orec- toloboid that displays this mode, *Nebrius ferrugineus*, have been observed to have greatly expanded stomachs full of egg cases and yolky material (Teshima et al. 1995).

Placental viviparity is present in 81 tropical chondrichthyan species, all of which are car- charhinoids (some triakids and all members of the Carcharhinidae, except the tiger shark *Galeocerdo cuvier*). In this mode, the embryos deplete the yolk reserves in the yolk sac in the first few weeks of gestation, after which the exhausted yolk sac elongates and becomes highly vascularized. This modified yolk sac then attaches to the uterine wall to form a yolk sac placental connection through which nutrients are directly provided to the embryo from the maternal bloodstream (Carrier, Pratt, and Castro 2004). One solely tropical species of carcharhinid, the spadenose shark *Scoliodon laticaudus*, has the most highly advanced form of matrotrophy known in fishes and rivals that of mammals. In this species, the eggs are tiny and practically yolk free and implantation of the yolk sac to the uterine wall occurs at very early stages (~3 mm) and gestation is less than six months (Wourms and Demski 1993). Maternal nutrients are attained by hemotrophic placental nutrient transfer across a unique implantation site (a trophonematous cup) allowing maternal blood to bathe the outer epithelium of the yolk sac placenta (Wourms and Demski 1993).

Litter sizes vary greatly within the tropical chondrichthyan fauna, from only one in the manta ray *Manta birostris* (Last and Stevens 2009) to 300 in the whale shark *Rhincodon typus* (Joung et al. 1996). Members of the most speciose shark family in the tropical ecosystem, the Carcharhinidae, generally have a litter size of 2 to 16, with the exception being the tiger shark *Galeocerdo cuvier*, which has litters of up to 82 (mean 33; Compagno, Dando, and Fowler 2005; Whitney and Crow 2007). Members of the most speciose batoid family, the Dasyatidae, have slightly smaller litter sizes of around one to seven (White et al. 2006; Last and Stevens 2009). The litter sizes for many tropical species are largely unknown.

The gestation periods of tropical marine species vary among species and taxonomic groups, with a range of 3 to 22 months but averaging around 12 months (Last and Stevens 2009). In some cases, species occurring in tropical or subtropical ecosystems will have relatively shorter gestation periods than the same or closely related species in temperate waters. For example, the Atlantic stingray *Dasyatis sabina*, which occurs in warm-temperate and tropical waters, can have a reduction in gestation time of up to two weeks with a 1°C increase in temperature (Wallman and Bennett 2006). Furthermore, dasyatid ray gestation periods vary from 3 to 11 months, with those in subtropical waters, for example, *D. sabina*, *D. say*, and *D. guttata* (Thorson 1983; Snelson, Williams-Hooper, and Schmid 1989; Johnson and Snelson 1996), being far shorter than those in temperate waters, for example, *D. cen- troura* (Struhsaker 1969). Information on gestation periods is not available for the majority of the tropical marine elasmobranchs.

Many tropical elasmobranchs display seasonal reproductive cycles, while others give birth throughout the year. Many different reproductive cycles have been recorded in the carcharhinoid sharks in waters off northern Australia. Of the two hemigaleid species, the Australian weasel shark *Hemigaleus australiensis* has a relatively short gestation period of six months with two pregnancies per year, whereas the larger fossil shark *Hemipristis elongata* has only a slightly longer gestation (seven to eight months) but only produces a litter every second year (Stevens and McLoughlin 1991). The reproductive cycles of the carcharhinids vary from species that have an annual cycle, giving birth in the Austral summer (graceful shark *Carcharhinus amblyrhynchoides*, nervous shark *C. cautus*, blacktip reef shark *C. melanopterus*, creek whaler *C. fitzroyensis*, Australian blacktip shark *C. tilstoni*, spot-tail shark *C. sorrah*, and the Australian sharpnose shark *Rhizoprionodon taylori*) to those that give birth throughout the year (whitecheek shark *C. dussumieri*, sliteye shark *Loxodon macrorhinus*, and milk shark *R. acutus*) to those with a biennial cycle and birth during the summer (sandbar shark *C. plumbeus*) or winter (hardnose shark *C. macloti*) (Stevens and McLoughlin 1991).

4.4.2 Age and Growth

Information on longevity, growth characteristics, and age at maturation allows calculation of growth and mortality rates and productivity, information that is critical for estimating the status of a population and assessing the risks of exploitation (Natanson et al. 2007; see Chapter 17). Cailliet and Goldman (2004) provided a review of age determination and validation studies on elasmobranch fishes. Different populations or stocks of a species can often be distinguished by their life history parameters since these parameters are considered to be phenotypic expressions of the interaction between the genotype and the environment (Begg 2005). Thus, age and growth information for a species cannot be assumed to be representative of the entire range for that species, with separate stocks likely to have different age and growth parameters.

A number of studies highlight the differences in age parameters between geographically separated stocks. The blacknose shark *Carcharhinus acronotus* was found to have a significantly lower growth rate (k) and older age at maturity in North Atlantic Ocean populations compared to those in the eastern Gulf of Mexico (Driggers et al. 2004). Likewise, Neer and Thompson (2005) reported that cownose rays *Rhinoptera bonasus* attain maturity earlier in the Gulf of Mexico than in the Chesapeake Bay in the Western Atlantic. Carlson, Sulikowski, and Baremore (2006) reported that blacktip sharks *Carcharhinus limbatus* mature at an earlier age in both sexes in the Gulf of Mexico than in the South Atlantic Bight. The Atlantic sharpnose shark *Rhizoprionodon terranovae* was reported to mature at 2 to 4 years and live to 10 years in the eastern U.S. coast (Loefer and Sedberry 2003), compared to 1.3 to 1.6 years and 5.5 years, respectively, in the Gulf of Mexico (Carlson and Baremore 2003). Although differences could be the result of differences in methodology, low samples sizes, or annual variations, these studies suggest that populations in lower latitudes mature at earlier ages and often grow faster than those in higher latitudes.

Table 4.4 provides a summary of age and growth studies on tropical marine elasmobranchs. Due to latitudinal differences in age parameters, only studies conducted wholly or in part within the tropical zone are included in the summary table. Note that only 35 of the 465 elasmobranchs occurring in tropical marine ecosystems (7.5%) are represented in the summary table. Therefore, studies of age and growth of tropical elasmobranchs are lacking and should be a focus of future research efforts.

The oldest age estimates were for the lemon shark *Negaprion brevirostris* from northeastern Brazil, that is, 45 (Freitas et al. 2006). The oldest age at maturity recorded was 25 years for female nurse sharks *Ginglymostoma cirratum* (Carrier and Luer 1990). The youngest age at maturity recorded was 1.3 years for males of the Atlantic sharpnose sharks *Rhizoprionodon terraenovae*, which only attained three years of age (Carlson and Baremore 2003). Although elasmobranchs are all thought to have relatively slow growth rates (i.e., low k values), there is

TABLE 4.4

Summary of Age and Growth Studies for Elasmobranchs in the Tropical Marine Ecosystem

Family	Species	Location	Method	Sex	von Bertalanffy Growth Parameters			Maximum Age (t_{max})	Age at Maturity (t_{mat})	Reference
					L_∞	k	t_0			
Hemiscylliidae	*Chiloscyllium plagiosum*	Taiwan, NW Pacific	Vertebrae	F	959 mm TL	0.205	−0.95	7	4.5	Chen et al. (2007)
				M	1009 mm TL	0.198	−0.90	7	4.4	
Ginglymostomatidae	*Ginglymostoma cirratum*	Florida, Western Atlantic	Tag/ Recapture	F				35	25	Carrier and Luer (1990)
				M				29	16	
Odontaspididae	*Carcharias taurus*	*Gulf of Mexico and northwest Atlantic	Vertebrae	both	3030 mm TL	0.180	−2.09			Branstetter and Musick (1994)
Carcharhinidae	*Carcharhinus acronotus*	*Gulf of Mexico and northwest Atlantic	Vertebrae	F	2958 mm TL	0.110	−4.20	17	9–10	Goldman (2002)
				M	2495 mm TL	0.160	−3.40	16	4–6	
		NW Florida, Western Atlantic	Vertebrae	F	1137 mm TL	0.352	−1.20	10–16		Carlson, Cortés, and Johnson (1999)
				M	963 mm TL	0.590	−0.75	4.5–9		
		Tampa Bay, Florida, Western Atlantic	Vertebrae	F	1241 mm TL	0.237	−1.54	10–16		Carlson, Cortés, and Johnson (1999)
				M	801 mm TL	0.771	−0.80	4.5–9		
	Carcharhinus amblyrhynchos	Hawaiian Islands, Central Pacific		F				18	6	Radtke and Cailliet (1984); Wetherbee, Crow, and Lowe (1997)
				M				18	8	
	Carcharhinus brevipinna	Taiwan, NW Pacific	Vertebrae	F	2882 mm TL	0.151	−1.99	21	7.8	Joung et al. (2005)
				M	2574 mm TL	0.203	−1.71	17	7.9	
		Gulf of Mexico, United States, Western Atlantic	Vertebrae	F	2790 mm TL	0.070	NA	17.5		Carlson and Baremore (2005)
				M	2027 mm TL	0.110	NA	13.5		
	Carcharhinus cautus	Northwestern Australia, Eastern Indian	Vertebrae	F	1238 mm TL	0.198	−2.52	16	6	White, Hall, and Potter (2002)
				M	1105 mm TL	0.287	−1.75	12	4	
	Carcharhinus falciformis	Yucatan, Gulf of Mexico, Mexico, Western Atlantic	Vertebrae	F				22	12	Bonfil, Mena, and De Anda (1993)
				M				20	10	
		*Gulf of Mexico, United States, Western Atlantic	Vertebrae	F	3110 mm TL	0.101	−2.72	13.8	7–9	Branstetter (1987a)
				M				12.8	6–7	
				both	2905 mm TL	0.153	−2.20			

Continued

TABLE 4.4 (Continued)

Summary of Age and Growth Studies for Elasmobranchs in the Tropical Marine Ecosystem

Family	Species	Location	Method	Sex	von Bertalanffy Growth Parameters			Maximum Age (t_{max})	Age at Maturity (t_{mat})	Reference
					L_∞	k	t_0			
		Pacific Ocean	Vertebrae	F					6–7	Oshitani, Nakano, and Tanaka (2003)
				M					5–6	
				both	2164 mm PCL	0.148	−1.76			
	Carcharhinus isodon	*NE Gulf of Mexico, United States, Western Atlantic	Vertebrae	F	1559 mm TL	0.244	−2.07	8	3.9	Carlson, Cortés, and Bethea (2003)
				M	1338 mm TL	0.412	−1.39	8.1	4.3	
	Carcharhinus leucas	*Northern Gulf of Mexico, United States, Western Atlantic	Vertebrae	F				30		Neer, Thompson, and Carlson (2005)
				M				25		
				both	2289 mm TL	0.089	NA			
		Southern Gulf of Mexico, United States, Western Atlantic	Vertebrae	F	2621 mm TL	0.124	−2.44	28	10	Cruz-Martinez, Chiappa-Carrara, and Arenas-Fuentes (2005)
				M	2484 mm TL	0.169	−1.03	23	9–10	
	Carcharhinus limbatus	Gulf of Mexico, United States, Western Atlantic	Vertebrae	F	1416 mm FL	0.240	−2.18	11.5	5.7	Carlson, Sulikowski, and Baremore (2006)
				M	1260 mm FL	0.270	−2.21	9.5	4.5	
		*NW Gulf of Mexico, United States, Western Atlantic	Vertebrae	B	1760 mm TL	0.274	−1.2	9.3		Branstetter (1987b)
	Carcharhinus obscurus	Western Australia, Eastern Indian	Vertebrae	F	3544 mm FL	0.043	NA	32	17–22	Simpfendorfer et al. (2002)
				M	3365 mm FL	0.045	NA		20–23	
	Carcharhinus plumbeus	Hawaiian Islands, Central Pacific	Vertebrae	F	1528 mm PCL	0.100	NA	23	10	Romine, Grubbs, and Musick (2006)
				M	1385 mm PCL	0.120	NA	19	8	
		Western Australia, Eastern Indian	Vertebrae	F	2458 mm FL	0.039	NA	25	16.2	McAuley et al. (2006)
				M	2263 mm FL	0.044	NA	19	13.8	
		Taiwan, NW Pacific	Vertebrae	F				20.8	7.5–8.2	Joung, Liao, and Chen (2004)
				M				19.8	8.2	
				both	2100 cm TL	0.170	−2.30			

Species	Location	Method	Sex						Reference
	*NW Atlantic, United States, Western Atlantic	Vertebrae	F	1970 mm TL	0.059	−4.80	25	15–16	Sminkey and Musick (1995)
			M	1840 mm TL	0.059	−5.40	18	15–16	Sminkey and Musick (1995)
			F	1650 mm TL	0.086	−3.00	25	15–16	Lessa and Santana (1998)
			M	1660 mm TL	0.087	−3.80	18	15–16	
Carcharhinus porosus	Northern Brazil, Western Atlantic	Vertebrae	both	1251 mm TL	0.101	−2.9	12	6	
Carcharhinus signatus	Northeastern Brazil, Western Atlantic	Vertebrae	F	2654 mm TL	0.114	−2.70			Santana and Lessa (2004)
			M	2565 mm TL	0.124	−2.54		10	
			both	2700 mm TL	0.112	−2.71	17	8	
Carcharhinus sorrah	Northern Australia, Indo-West Pacific	Vertebrae	F	1339 mm TL	0.340	−1.90	7	2–3	Davenport and Stevens (1998)
			M	984 mm TL	1.170	−0.60	7	3	
Carcharhinus tilstoni	Northern Australia, Indo-West Pacific	Vertebrae	F	1942 mm TL	0.140	−2.80	12	3–4	Davenport and Stevens (1998)
			M	1654 mm TL	0.190	−2.60	8	3–4	
Galeocerdo cuvier	United States, Western Atlantic	Tag/Recapture	both	3370 mm FL	0.178	−1.12	27.3	7	Natanson et al. (1999)
Isogomphodon oxyrhynchus	Northern Brazil, Western Atlantic	Vertebrae	F				12	6–7	Lessa et al. (2000)
			M				7	5–6	
Negaprion brevirostris	Northeastern Brazil, Western Atlantic	Tag/Recapture	both	1714 mm TL	0.121	−2.61	45		Freitas et al. (2006)
	Florida and Bahamas, Western Atlantic	Vertebrae	F	3999 mm TL	0.077	−2.16			Brown and Gruber (1988)
			M					12.7	
			both	3175 mm FL	0.057	−2.30	20.2	11.2	
Rhizoprionodon acutus	India, northern Indian	Len. freq.	F	1065 mm TL	0.605	−0.06			Kasim (1991)
			M	1054 mm TL	0.646	−0.05			
Rhizoprionodon taylori	Northeastern Australia, Southwest Pacific	Vertebrae	F	732 mm TL	1.012	0.0455	7	3	Simpfendorfer (1993)
			M	652 mm TL	1.337	0.41	6	3	
Rhizoprionodon terraenovae	*Gulf of Mexico, United States, Western Atlantic	Vertebrae	F	956 mm TL	0.630	−1.03	5.5	1.6	Carlson and Baremore (2003)
			M	919 mm TL	0.850	−0.73	4	1.3	
Scoliodon laticaudus	India, northern Indian	Len. freq.	F	749 mm TL	0.882	0.01			Kasim (1991)
			M	680 mm TL	1.082	−0.01			

Continued

TABLE 4.4 (Continued)

Summary of Age and Growth Studies for Elasmobranchs in the Tropical Marine Ecosystem

Family	Species	Location	Method	Sex	von Bertalanffy Growth Parameters			Maximum Age (t_{max})	Age at Maturity (t_{mat})	Reference
					L_∞	k	t_0			
Sphyrnidae		India, northern Indian	Len. freq.	F	715 mm TL	0.358	0.59			Devadoss (1998)
				M	676 mm TL	0.405	0.59			
		India, northern Indian	Len. freq.	both	740 mm TL	0.682	−0.01	3.1		Mathew and Devaraj (1997)
	Sphyrna lewini	Taiwan, NW Pacific	Vertebrae	F	3197 mm TL	0.249	−0.413	14	4.1	Chen et al. (1990)
				M	3206 mm TL	0.222	−0.746	10.6	3.8	
		Mexico, Eastern Pacific	Vertebrae	F	3533 mm TL	0.153	−0.633	19	8.8	Anislado-Tolentino and Robinson-Mendoza (2001)
				M	3364 mm TL	0.131	−1.09			
		*Gulf of Mexico, United States, Western Atlantic	Vertebrae	F	2331 mm TL	0.090	−2.22	30.5	>15	Piercy et al. (2007)
				M	2148 mm FL	0.130	−1.62	30.5	>15	
		*Gulf of Mexico, United States, Western Atlantic	Vertebrae	F				16.8	15	Branstetter (1987a)
				M					10	
				both	3290 mm TL	0.073	−2.20			
	Sphyrna tiburo	Tampa Bay, Florida, Western Atlantic	Vertebrae	F	1150 mm TL	0.340	−1.1	7	2.2	Parsons (1993)
				M	888 mm TL	0.580	−0.77	6	2	
Pristidae	*Pristis microdon*	Northern Australia, Indo-West Pacific	Vertebrae	both	3625 mm TL	0.070	−4.07	44		Tanaka (1991)
	Pristis perotteti	Florida, Western Atlantic	Vertebrae	both	4560 mm TL	0.089	−1.98	30	10	Simpfendorfer (2000)

Family	Species	Location	Structure	Sex	Size					Reference
Rhinobatidae	*Rhinobatos productus*	Baja California, Eastern Pacific						16	7	Ebert (2003)
								11	8	
Rajidae	"undescribed genus B" *Raja texana*	*Gulf of Mexico, United States, Western Atlantic	Vertebrae	F	722 mm TL	0.187	−0.925	9	5.8	Sulikowski et al. (2007)
				M	512 mm TL	0.334	−0.759	8	5	
Dasyatidae	*Dasyatis Americana*	*Southeastern United States, Western Atlantic	?	F	1500 mm DW	0.539		26	5–6	Henningsen (2002); Cailliet and Goldman (2004)
				M	1125 mm DW	0.206		28	3–4	
	Dasyatis dipterura	Mexico, Eastern Pacific	Vertebrae	F	924 mm DW	0.050	−7.61	28	8–11	Smith, Cailliet, and Melendez (2007)
				M	622 mm DW	0.100	−6.8	19	5–8	
	Neotrygon kuhlii	Jakarta, Indonesia, Indo-West Pacific	Vertebrae	F	313 mm DW	0.311	−1.13	15		White (2003)
				M	257 mm DW	0.831	−0.43	10		
Rhinopteridae	*Rhinoptera bonasus*	*Gulf of Mexico, United States, Western Atlantic	Vertebrae	both	1238 mm DW	0.075	−5.48	18	4–5	Neer and Thompson (2005)
								16	4–5	
Mobulidae	*Mobula japanica*	Mexico, Eastern Pacific	Vertebrae	both	2338 mm DW	0.277	−1.676	14		Cuevas-Zimbrón, Sosa-Nichizaki, and O'Sullivan (2008)

Note: Only age and growth studies conducted whole or in part in tropical waters are included, and an asterisk is used to highlight those that include only tropical waters.

considerable variation among species and even between sexes of the same species. The fastest growth rates were recorded for Australian sharpnose sharks *Rhizoprionodon taylori* with k values of 1.01 to 1.34 (Simpfendorfer 1993) and male spot-tail sharks *Carcharhinus sorrah* with a k of 1.17 (Davenport and Stevens 1998). In general, the fastest growth rates were recorded for small carcharhinid sharks (*C. sorrah*, spadenose shark *Scoliodon laticaudus*, sharpnose sharks *Rhizoprionodon* spp., daggernose shark *Isogomphodon oxyrhynchus*, and blacknose shark *Carcharhinus acronotus*) and the small stingray species *Neotrygon kuhlii* (Table 4.4). The slowest growth rates were recorded for the much larger sandbar shark *Carcharhinus plumbeus* from the eastern Indian Ocean, that is, k values of 0.039 to 0.044 (McAuley et al. 2006). Several large sharks have slow growth rates, the dusky shark *Carcharhinus obscurus* (0.043 to 0.045), scalloped hammerhead *Sphyrna lewini* in the Gulf of Mexico (0.09 to 0.13), as well as the two large sawfish species *Pristis microdon* and *P. perotteti* (0.07 to 0.089).

4.4.3 Feeding Ecology and Behavior

The tropical marine chondrichthyan fauna encompasses 11 ecomorphotypes, as described by Compagno (1990) and includes the only extant archipelagic (superpredator), the white shark *Carcharodon carcharias*, and the filter-feeding giant, the whale shark *Rhincodon typus*. The probenthic, or unspecialized bottom-dwelling sharks, include the horn sharks Heterodontidae and the blind sharks Brachaeluridae, which feed on benthic invertebrates and small teleosts. The Heterodontids have specialized teeth that allow them to crush hard prey such as sea urchins and crabs (Last and Stevens 2009). The leptobenthic (elongated bottom sharks) include tropical species of the families Parascylliidae and Hemiscylliidae. Very little is known about the biology of these species but it is thought that they are opportunistic feeders that prey on benthic invertebrates and small teleosts (Heupel and Bennett 1998; Compagno, Dando, and Fowler 2005). The tropical squatinobenthic sharks have flattened bodies and include 12 species of angel sharks, Squatinidae, and eight species of wobbegongs, Orectolobidae. These species are ambush predators, preying on teleosts, cephalopods, and crustaceans, which swim past as they lie camouflaged on the substrate (Fouts and Nelson 1999; Compagno, Dando, and Fowler 2005).

The littoral species are active coastal sharks and can range from specialist feeders to generalist opportunistic predators. For example, the tiger shark *Galeocerdo cuvier* feeds on a wide variety of prey from invertebrates to marine mammals (Stevens and McLoughlin 1991; Lowe et al. 1996; Simpfendorfer, Goodreid, and McAuley 2001), whereas the hardnose shark *Carcharhinus macloti* and Australian weasel shark *Hemigaleus australiensis*, are highly selective and feed almost exclusively on teleosts and cephalopods, respectively (Stevens and McLoughlin 1991; Salini, Blaber, and Brewer 1992; Taylor and Bennett 2008). The littoral hammerheads, Sphyrnidae, have a laterally expanded head that is thought to increase maneuverability and enhance the sensory capabilities for catching fast prey. Their heads have also been noted to assist in prey handling, as with the great hammerhead *Sphyrna mokarran*, which has been observed to pin stingrays and eagle rays to the bottom using its head (Strong 1990; Chapman and Gruber 2002).

Tropical rhinobenthic rays include all species of the wedgefish family Rhynchobatidae, 39 species of shovelnose rays (Rhinobatidae), and the shark ray *Rhina ancylostoma* (Rhinidae). These rays generally prey on benthic crustaceans but also consume other benthic invertebrates and teleosts, with the rhinobatids using their protrusible jaws to capture and manipulate their prey (Kyne and Bennett 2002; White, Hall, and Potter 2004; Last and Stevens 2009). In some cases they can be very selective in their prey, for example, *Glaucostegus typus* in Shark Bay, Western Australia, feeds almost exclusively on portunid crabs (one species) and penaeid prawns (two species) (White, Hall, and Potter 2004). The tropical pristobenthic species include two species of sawsharks (Pristiophoridae) and all seven species of sawfish (Pristidae). The saw-like snouts of these species are used to stir up invertebrates from the substrate and sometimes act as a weapon to stun fish (Breder 1952; Last and Stevens 2009).

The tropical torpedobenthic rays include the electric ray families Narcinidae, Narkidae, Hypnidae, and Torpedinidae. Species of numbfish (Narcinidae) are known to have highly protrusible jaws to extract benthic invertebrates, while some species of torpedo rays (Torpedinidae) and the coffin ray *Hypnos monopterygius* (Hypnidae) use their electric organs to stun teleosts and invertebrates (Lowe, Bray, and Nelson 1994; Dean and Motta 2004; Last and Stevens 2009). The tropical rajobenthic rays include species of skates (Arhynchobatidae and Rajidae), stingarees (Urolophidae), round rays (Urotrygonidae), stingrays (Dasyatidae), and butterfly rays (Gymnuridae). Most of these species are general opportunistic feeders, with prey ranging from polychaete worms to small fish (Last and Stevens 2009). The Australian butterfly ray *Gymnura australis* feeds mainly on teleosts and is capable of consuming proportionally large prey, occasionally being found to have part of the prey still protruding from the mouth after swallowing (Jacobsen 2007).

The last ecomorph present in tropical marine elasmobranchs are the aquilopelagic or "winged" myliobatoid rays, represented by the eagle rays (Myliobatidae), cownose rays (Rhinopteridae), and manta and devilrays (Mobulidae). The myliobatids and rhinopterids have powerful jaws and plate-like teeth, which are capable of crushing hard-shelled prey such as crustaceans and bivalves, with papillae in the mouth reported to exclude crushed shells from entering the stomach (Summers 2000; Yamaguchi, Kawahara, and Ito 2005; Last and Stevens 2009). In contrast, mobulids are primarily filter feeders, using their large cephalic lobes to guide prey into their mouths where the specialized branchial filter plates on their internal gill arches filter out plankton and small fishes (Compagno 1990; Last and Stevens 2009).

The large diversity of extant elasmobranchs occurring today is probably due partly to resource partitioning. On an interspecific level, dietary partitioning (including spatial and/ or temporal partitioning of food resources) plays an important role in reducing competition for food resources when different elasmobranch species co-occur in the same community (Stevens and McLoughlin 1991; White, Platell, and Potter 2004; Papstamatiou et al. 2006; Navia, Mejia-Falla, and Giraldo 2007). Intraspecific dietary partitioning has been shown in tropical elasmobranch species through ontogenetic dietary shifts (Lowe et al. 1996; Wetherbee, Crow, and Lowe 1997; Heupel and Bennett 1998; Kyne and Bennett 2002; White, Platell, and Potter 2004; McElroy et al. 2006; Bethea et al. 2007) and differences in diet in relation to sex (Klimley 1987; Simpfendorfer, Goodreid, and McAuley 2001). Other studies have reported changes in diet with geographical location (Cortés and Gruber 1990; Simpfendorfer, Goodreid, and McAuley 2001; Bethea, Buckel, and Carlson 2004) and habitat type and/or water depth (Cortés, Manire, and Heuter 1996; McElroy et al. 2006; Bethea et al. 2007). Dietary breadth can also increase or decrease with the size and life stage of individuals (Lowe et al. 1996; Wetherbee, Crow, and Lowe 1996). Factors that may be associated with intraspecific dietary differences include swimming speed, change in habitat occupied, movement patterns, experience with prey, size of jaws, teeth and stomach, predation risks, and energy requirements (Wetherbee and Cortés 2004).

Quantitative dietary analysis of elasmobranchs, including dietary partitioning and competitive interactions, are becoming increasingly popular due to our need to understand the role that these predators play in marine ecosystems (Cortés 1999; Papastamatiou et al. 2006). As some species are apex predators, their removal can initiate trophic cascades throughout a community (Myers et al. 2007; Heithaus et al. 2008). For example, declines in the common blacktip shark *C. limbatus* off the U.S. Atlantic coast led to increases in abundance of cownose rays *Rhinoptera bonasus*, which subsequently led to decreases in commercial catches of scallops due to increased ray predation (Heithaus et al. 2008). A better understanding of the trophic role of elasmobranchs in marine communities is required, but there is little data available for most tropical elasmobranch species.

4.4.4 Population Segregation, Movements, and Migration

Tropical marine elasmobranchs exhibit a diversity of movement patterns across a range of temporal and spatial scales. Short-term processes include diel movements and movements in relation to tidal flow, prey availability, and social factors. Diel movements have been shown for many tropical elasmobranch species, including the Hawaiian stingray *Dasyatis lata* (Cartamil et al. 2003), horn shark *Heterodontus francisci* (Segura-Zarzosa, Abitia-Cárdenas, and Galván-Magaña 1997), swell shark *Cephaloscyllium ventriosum* (Nelson and Johnson 1970), epaulette shark *Hemiscyllium ocellatum* (Heupel and Bennett 1998), scalloped hammerhead *Sphyrna lewini* (Klimley et al. 1988; Bush 2003), common blacktip shark *Carcharhinus limbatus* (Heupel and Simpfendorfer 2005), lemon shark *Negaprion brevirostris* (Sundström et al. 2001), and tiger shark *Galeocerdo cuvier* (Lowe et al. 1996). The majority of these species tend to show increased foraging activity at nighttime, which may be associated with predator avoidance or greater abundance of prey at night. In contrast, tiger sharks in Shark Bay exhibit a more diurnal pattern of foraging (Heithaus 2001).

Tidal flows drive movements of some tropical elasmobranchs, particularly batoids. In Australia, the blue-spotted fantail ray *Taeniura lymma* has been observed to migrate *en masse* onto sandflats and intertidal reefs during high tide to feed on benthic invertebrates (Last and Stevens 2009). The white-spotted eagle ray *Aetobatus narinari* and southern stingray *Dasyatis americana* also move according to the tidal flow in the Bahamas, with high tide providing a greater area for foraging (Gilliam and Sullivan 1993; Silliman and Gruber 1999). Neonate sandbar sharks forage with the tidal flow in their Chesapeake Bay nursery (Grubbs and Musick 2007). The manta ray *Manta birostris* has been observed to forage in three main areas in Komodo Marine Park, Indonesia, where prey is concentrated. These prey concentrations are likely to be driven by the strong tidal currents that produce upwelling and increase primary productivity (Dewar et al. 2008).

Daily social aggregations have been found in the gray reef shark *Carcharhinus amblyrhynchos* and *C. limbatus*. These sharks tend to group during the day and then disperse at night. The aggregations of *C. amblyrhynchos* at Johnston Atoll, Central Pacific Ocean, were correlated with tidal movements, daily fluctuations in water temperature and light levels (Economakis and Lobel 1998).

Tropical elasmobranchs visit cleaning stations to have parasites removed. Juvenile Caribbean reef sharks *Carcharhinus perezii* seek out cleaning stations of yellownose gobies, *Elacatinus randalli*, and then rest on the bottom while being cleaned (Sazima and Moura 2000) and manta rays in the Indo-West Pacific have been observed to return to the same cleaning stations (Homma et al. 1997).

Long-term movement processes can be related to seasonal changes in environmental variables, reproductive behavior, use of nursery areas, and ontogenetic habitat changes. Seasonal temperature differences can influence both the rate of movement of tropical elasmobranchs and their geographic locations. In Kaneohe Bay, Hawaii, Hawaiian stingrays were found to have significantly faster rates of movement during the summer months, when water temperature averages 29°C, than in the winter months when the average water temperature is much lower (19°C) (Cartamil et al. 2003). In Indonesia, manta rays were most abundant in the south during summer and the north during winter and it is specu-lated that these movements are related to water temperatures, with the mantas avoiding temperatures above their upper thermal limit of 30°C (Dewar et al. 2008). Aggregations of common blacktip sharks in Terra Ceia Bay, Florida, were more common in summer and are thought to be due to improved feeding efficiency or predator avoidance (Heupel and Simpfendorfer 2005).

Movements relating to reproductive behavior are important for many tropical elasmo-branchs. The zebra shark *Stegostoma fasciatum* forms aggregations of large adults in south-eastern Queensland over the summer months to mate (Dudgeon, Noad, and Lanyon 2008). Copulation in the nurse shark *Ginglymostoma cirratum* takes place when females enter a shallow lagoon (sometimes being chased in there by males) at Dry Tortugas, Florida Keys (Pratt and Carrier 2001). The scalloped hammerhead *S. lewini* displays sexual segregation, with females moving offshore earlier than males (Klimley 1987).

Inshore waters and estuaries act as nursery areas for many tropical elasmobranchs. A nursery area is an area in which sharks or rays are more common than in other areas, they tend to remain or return there for extended periods, and the area or habitat is used repeat-edly across years (Heupel, Carlson, and Simpfendorfer 2007). Seasonal migration into nurs-ery areas is common for many species of Carcharhinidae and Sphyrnidae (Simpfendorfer and Milward 1993; White and Potter 2004; Heupel and Simpfendorfer 2005; Garla et al. 2006; Yokota and Lessa 2006; Wetherbee, Gruber, and Rosa 2007). Some species such as *Carcharhinus dussumieri*, *C. fitzroyensis*, *C. limbatus*, and *C. tilstoni* use nursery areas for only a few months after birth. Others, such as *C. sorrah*, *Rhizoprionodon acutus*, and *R. taylori* use them until maturity (Simpfendorfer and Milward 1993). Some sharks, which remain in the nursery until maturation, expand their home ranges with time and make increasingly more excursions outside their refuge area (Heupel, Simpfendorfer, and Hueter 2004). The dwarf sawfish *Pristis clavata* uses the estuarine waters of the Fitzroy River in northwestern Australia as a nursery area for at least three years, while the freshwater sawfish *P. micro-don* uses the same river for four to five years, but penetrates much farther into freshwater (Thorburn et al. 2007, 2008).

Large migrations across ocean basins are undertaken by a few tropical elasmobranchs (i.e., the whale shark *Rhincodon typus* and the white shark *Carcharodon carcharias*) and their movements are generally based on oceanographic conditions and seasonal prey abun-dances or prey migrations. However, long-term movement patterns (or lack of patterns) of a few large active species are still not fully understood. For example, five tiger sharks fitted with satellite tags in Shark Bay, Western Australia, showed variable movements, with three remaining in the Shark Bay region, one traveling 500 km out to oceanic waters and then back to the bay and the fifth being tracked 8000 km away to the coastal region of southeast Africa (Heithaus et al. 2007b). It is thought that tiger sharks roam over large areas over a long period of time, but restrict their movements over small temporal scales when they encounter areas of high prey density (Heithaus et al. 2007b).

4.4.5 Demography

Over the past decade, it has become increasingly evident that there have been recent declines in a number of elasmobranch populations due to overexploitation (Simpfendorfer 2000; Baum et al. 2003; Heithaus et al. 2007a). The development of demographic models for elasmobranchs is a useful tool to characterize the vulnerability to exploitation of the populations being stud-ied (Cortés 2004; see Chapter 17). In general, species that have smaller litters, slower growth rates, and later ages at maturity are more vulnerable to extinction due to exploitation than species with faster growth rates, larger litters, etc. (see Chapter 17). Unfortunately, appropriate demographic data are largely lacking for most tropical elasmobranchs.

In an analysis of extinction risks in elasmobranchs, García, Lucifora, and Myers (2008) found that in general, continental shelf elasmobranchs mature earlier, live for fewer years, and have faster growth completion rates than deepwater elasmobranchs. They concluded that conservation priority should not be restricted solely to large species because many small species can also be highly vulnerable to extinction. Many of the fisheries operating in tropical regions involve a large number of species. For example, the Indonesian fishery, which is the largest chondrichthyan fishery in the world, includes catches of at least 105 tropical shelf species (White et al. 2006). Most of these species have very little life history data available, thus demographic models cannot be created for these highly exploited populations. Another consideration is the quality of data available. Although the lack of any data on life history parameters of a species compromises demographic modeling, poor or incorrect data for these parameters can result in incorrect conclusions regarding the potential decline of a shark population.

4.5 Exploitation, Management, and Conservation

4.5.1 Exploitation and Threats

The chondrichthyan fishes occurring in tropical marine ecosystems face the same threats as elasmobranchs on a global scale. These threats can be broken down into two major categories: (1) effects of various fishing activities (direct and indirect), and (2) habitat changes.

4.5.1.1 Fisheries

Due to elasmobranchs' generally slow life history strategies and their high position in food webs, they are more likely to be adversely affected by intense fishing pressure than most teleosts (Stevens et al. 2000). Of the top 20 countries in terms of capture production of elasmobranchs, 10 occur in tropical regions. Eight of these countries, Indonesia, India, Sri Lanka, Pakistan, Taiwan, Malaysia, Thailand, and the Philippines, are in the Indo–West Pacific and in the year 2000 their combined catches constituted more than 40% of the global capture of elasmobranchs (FAO 2002). Although a large proportion of the catches in some of these regions comes from pelagic fisheries (see Chapter 1), there are a wide variety of fisheries operating in the shelf waters throughout the tropical regions of the world. While pelagic fisheries interact with only a small number of chondrichthyan species, many of the shelf fisheries in tropical regions interact with a high diversity of elasmobranchs.

4.5.1.1.1 Case Study: Indonesia

Indonesia has the largest of the world's chondrichthyan fisheries, with more than 122,000 tonnes landed in 2004 (Lack and Sant 2006). Reported chondrichthyan landings in Indonesia were less than 10,000 tonnes from 1950 to 1970, but from the mid-1970s they increased rapidly (FAO 2002). One of the major concerns with the Indonesian fishery is that there is a heavy socioeconomic reliance on elasmobranchs by artisanal fishers (Suzuki 2002). The catches and fishing effort are increasing but the catch per unit effort (CPUE) is decreasing (Monintja and Poernomo 2000), suggesting that the overall abundances of elasmobranchs

FIGURE 4.3
A color version of this figure follows page 336. A typical daily landing of longline caught, large carcharhinid sharks off Lombok in eastern Indonesia. (Photo © Will White. All rights reserved.)

are declining. Of the 465 species of elasmobranchs occurring in tropical marine ecosystems, 105 of these were recorded from numerous fish landing site surveys in southern Indonesia from 2001 to 2006 (White et al. 2006). Prior to these surveys, the only detailed investigation of the species occurring in Indonesian waters was by Pieter Bleeker between 1842 and 1860, before large-scale chondrichthyan fisheries had developed in the region.

Chondrichthyan fisheries in Indonesia can be divided into target, for example, longline, gillnet, and spear, and bycatch fisheries, for example, trawl, trammel net, purse seine, trap, seine, and drop line. Indonesian artisanal fisheries generally use most body parts of elasmobranchs (see Figure 4.3), but in some instances there is considerable discard. For example, although only small numbers of whale sharks are opportunistically taken in Indonesian waters, the few reported to have been landed off Bali in the last decade were simply finned at sea and the carcasses dumped (Figure 4.4; White and Cavanagh 2007).

The increase in chondrichthyan catches in recent decades can be largely attributed to the growing demand and rising prices for fins. Fins of large carcharhinoid sharks and wedgefishes (*Rhynchobatus*) are extremely valuable and are highly sought after, and the increased effort to catch such species has led to localized depletions. The wedgefishes are a particularly good example of such depletions. In the mid-1970s, a large number of vessels from Kalimantan, Sulawesi, and Flores began heavily fishing the waters around the Aru Islands in eastern Indonesia targeting the rhynchobatids for their fins (Amir 1988; Bentley 1996). This fishery reached a peak in 1987 with more than 500 boats (10 to 50 gross tonnes) involved, but by 1996 only ~100 were still operating and these were fishing much farther afield, and occasionally being apprehended in Australian waters (Bentley 1996). Similar trends in rhynchobatid catches were observed during recent surveys of the region, in particular the Muara Angke landing site in Jakarta where large batoids are targeted by fishers using tangle nets (Figure 4.5). Those vessels operating in waters around Java, Sumatra, Kalimantan, and Sulawesi were found to have very low catches of rhynchobatid rays, with

FIGURE 4.4
A color version of this figure follows page 336. Dried caudal fin of a medium-sized whale shark *Rhincodon typus* landed off the Indonesian island of Bali. (Photo © Will White. All rights reserved.)

the catch being dominated by large dasyatid rays (e.g., *Himantura gerrardi*, *H. fai*, *H. jenkinsii*, *Pastinachus sephen*, and *Taeniura meyeni*). In contrast, the few vessels operating much farther afield in the Banda and Arafura Seas had catches dominated by large rhynchobatid rays, and occasionally sawfish (*Pristis microdon*), which are considered extinct in the Java Sea. Furthermore, the vessels were traveling for more than seven days to reach their fishing grounds, indicating large-scale depletion in many areas.

Since there is such a wide variety and number of fishing operations throughout Indonesia, most species are being exploited at every life history stage and there is very little to no chance of recovery. Localized depletions are clearly evident and shifts in species composition at the landing sites are becoming more apparent. A number of species are close to, or have been driven to, local extinction in the waters around the main population centers, for example, Java, Sumatra, Sulawesi, etc. The sicklefin lemon shark *Negaprion acutidens*, for example, was only recorded once during extensive surveys of fish landing sites in the last decade. This species prefers shallow inshore areas around sandy lagoons and turbid mangroves (Last and Stevens 2009), making it highly vulnerable to the fishing pressures in such shallow areas in Indonesia. Sawfish are considered extinct in the Java Sea, and although some dried rostra have been observed recently at various locales in Java, they are very old and the owners simply laugh when asked if they are still caught occasionally. Another example may be a much more dramatic example of local extinction. A species of carcharhinoid shark recently examined at a fish collection in Jakarta is clearly an

FIGURE 4.5
A color version of this figure follows page 336. Large (>300 cm TL) wedgefish, *Rhynchobatus*, and large sting-rays (Dasyatidae) landed at the Muara Angke landing site in Jakarta (Indonesia) from tangle-net fishers. The highly sought after fins were already removed. (Photo © Will White. All rights reserved.)

undescribed species and probably belongs to a new genus. This small but adult specimen was collected from Batavia (Jakarta) more than 100 years ago and is presumably a shallow inshore species given its body form. Given that this species has not been described or observed for over 100 years, despite numerous surveys in the region, it is quite possible that it had already become extinct before it was even recognized as a distinct species.

4.5.1.2 Habitat Changes

Habitat change through removal or degradation is a significant threat facing tropical elasmobranchs throughout the world. This is exacerbated by the fact that those areas with the most intense fishing pressures, for example, Indonesia and India, also have the greatest level of habitat loss and destruction. Those species with important nursery areas in inshore areas close to major population centers are of the most concern. Habitat change can result from both natural and anthropogenic cause. Natural causes, such as severe storms, tsunamis, or massive freshwater discharges, may negatively impact elasmobranchs, but such pulsed disturbances rarely imperil populations over longer time scales. However, elasmobranchs are less resilient in the face of the continuous press of anthropogenic disturbances, or permanent changes (e.g., mangrove removal, dynamiting coral reef systems, seagrass destruction, and pollution).

Climate change is likely to have a strong influence on many marine ecosystems, in particular sensitive habitats such as coral reefs and mangrove ecosystems (Everett 1997; Chin and Kyne 2007; Munday et al. 2008). Coral reefs are threatened worldwide. Current estimates suggest that 19% of the world's coral reefs have been effectively lost, 15% are seriously threatened with loss in the next 10 to 20 years, and 20% in the next 20 to 40 years (Wilkinson 2008). Loss of coral reefs is most significant and increasing around the major population centers, in particular in the "Coral Triangle" region of the Indo–West Pacific biogeographic region where diversity is extremely high. Dynamite and chemical (e.g., cyanide) fishing is a major threat to coral reefs in a number of countries, particularly in South-East Asia (Erdmann 2000; Fox et al. 2003). Increasing sea temperatures are another major concern for coral reefs. Temperature changes of only a few degrees above the average often result in corals becoming stressed, bleaching, and dying, with repeated episodes of mass coral bleaching observed since the 1970s leading to significant declines in coral reefs (Hoegh-Guldberg 1999; Munday et al. 2008). Ocean acidification is also a major threat facing coral reefs as it reduces the abilities of corals to form their carbonate skeletons (Feely et al. 2004; Kleypas et al. 2006; Hoegh-Guldberg et al. 2007). The depletion of coral reefs will have a negative impact on elasmobranch fishes, particularly those that use coral reefs as their primary habitat. It should be noted that some tropical elasmobranchs may actually benefit from climate change, with increasing water temperatures actually extending the available habitat for the species. However, the implications for ecosystem health of such expansions are very unclear.

Mangrove systems have a crucial role in nutrient filtering, thus reducing turbidity, and supporting coral reef fisheries by providing nursery areas for many species (Valiela, Bowen, and York 2001). Those authors reported that ~35% of the world's mangrove forests have been lost and deforestation is still continuing at an alarming rate. Aquaculture industries, primarily shrimp farming, and other agricultural practices are resulting in mass deforestation of mangrove forests (Phillips, Lin, and Beveridge 1993; Primavera 1993; Blasco, Aizpuru, and Gers 2001; Valiela, Bowen, and York 2001). The loss of these mangrove systems is detrimental to those elasmobranch species that utilize these areas, especially those that rely on it as their primary habitat, for example, juvenile porcupine rays *Urogymnus asperrimus*, or as nursery areas, for example, sicklefin lemon sharks *Negaprion acutidens*. Seagrass systems are extremely important in the food webs of the tropical marine ecosystems (Orth et al. 2006). Seagrasses are threatened by direct effects, for example, dredging, fishing, eutrophication, aquaculture, increased turbidity, and coastal construction, and indirect effects, for example, increased ultraviolet irradiance with climate change (Duarte 2002), and seagrass loss has increased almost tenfold over the last 40 years (Orth et al. 2006).

Human activities such as dredging, trawling, and land clearing often result in a significant increase in suspended matter in the water column, thereby increasing turbidity, which can adversely affect seagrass and reefs (Walker 2002). Pollution can also negatively impact habitats, as well as elasmobranchs directly. Heavy metals and hydrocarbons entering the marine environment can impact reefs and seagrass habitats or elasmobranchs directly (see Chapter 12). Mercury is known to reach very high levels in some sharks and although accumulated naturally, human activity results in further elevation of these levels (Walker 1976, 1988).

4.5.2 Management

Management of tropical marine chondrichthyan fisheries is still currently largely insufficient to ensure sustainability of the resource. The Convention on International Trade in Endangered Species of Wild Fauna and Flora (CITES) was established in 1975 to provide a legal framework to provide protection for individual species from overexploitation by restricting international trade (www.cites.org). CITES has become one of the most effective international tools towards protecting such species (Fowler and Cavanagh 2005; Reeve 2002). Currently, Appendix I of CITES lists 892 species that are threatened with extinction and thus international trade is not permitted, Appendix II lists 33,033 species that are subjected to strict trade regulations, and Appendix III lists 161 species that are subjected to regional trade restrictions (www.cites.org). Currently there are only ten elasmobranchs listed by CITES of which all but one occur in tropical marine waters, that is, whale shark *Rhincodon typus*, white shark *Carcharodon carcharias*, and all seven species of sawfish (Fam. Pristidae).

The voluntary International Plan of Action (IPOA) for Sharks was developed by FAO and adopted in 1999. This IPOA for sharks has an overall objective of ensuring the conservation and management of sharks and their long-term sustainable use and it originally called on states to develop and implement a National Plan of Action (NPOA) by early 2001. Despite the poor progress in states developing effective NPOA's and the lack of response of most of the major shark-fishing nations, FAO is continuing its involvement in encouraging implementation of the IPOA for Sharks (Fowler and Cavanagh 2005). Presently, only 7 of the 99 countries with tropical marine ecosystems have developed an NPOA for sharks.

The development of Marine Protected Areas (MPAs) is also an important management tool. Since many species of elasmobranchs are active and mobile species, closed areas will not offer protection for a whole population, but they can be used to protect important segments of the population, for example, inshore pupping grounds and nursery areas (Stevens 2002). For example, strictly protected, large spatial closures in coral reef areas are considered to be the best conservation solution for the declining reef shark populations (Robbins et al. 2006; Dulvy 2006). Such protected areas are likely to be very important for some benthic species, especially rare endemics. In the Dry Tortuga Island group in the Florida Keys, an area known to be a group mating area for nurse sharks *Ginglymostoma cirratum*, and which is part of a heavily used coral reef system, now has seasonal closures to boat traffic to allow for such breeding to go uninterrupted (Carrier and Pratt 1998). Robbins et al. (2006) found that the charismatic grey reef shark *Carcharhinus amblyrhynchos* and whitetip reef shark *Triaenodon obesus* are extremely susceptible to fishing pressure on coral reefs with densities up to 97% and 80%, respectively, lower in fished areas compared to no-take zones. Large, strictly protected, no-entry zones are necessary to protect such species, but such zones are presently few and far between. For example, the Great Barrier Reef of Australia is managed as the largest MPA in the world, but no-entry zones comprise only 1% of this area (Dulvy 2006). Thus, in order to provide a significant management solution for elasmobranchs, protected areas need to be large scale and based on sound scientific studies. Ecotourism has also led to development of protected areas frequented by elasmobranchs in a number of tropical regions. For example, the bull shark *Carcharhinus leucas* is the major attraction at Shark Reef Marine Reserve in Fiji. This reserve was established in 2003 and local villagers cannot fish in the reserve but receive a payment of F$10 per diver per day to compensate (Brunnschweiler and Earle 2006).

Fisheries management in developing countries with high populations and large chondrichthyan landings, for example, India and Indonesia, is extremely difficult due largely to the lack of resources available and the reliance on these fisheries and associated exports by many communities. As previously mentioned, Indonesia has the highest chondrichthyan landings but their fisheries are largely unregulated and underreporting is also a significant problem. In March 2001, the state government of Tamil Nadu in eastern India invoked a ban on fishing for 600 marine species, including 54 elasmobranchs, with large fines and imprisonment for disobeying the ban. However, there was a massive backlash over the ban because more than 50,000 fishermen in the state survive on marine fishing and no compensation or alternatives were provided. At the end of 2001, the ban on shark fishing was drastically reduced to only nine species, including the whale shark *R. typus*.

This emphasizes the importance of a strategic plan toward management of elasmobranchs rather than a reactionary approach that ignores the socioeconomic effects.

4.5.3 Conservation Status

The IUCN *Red List of Threatened Species*™ is widely recognized as the most comprehensive source of information on the conservation status of the world's plant and animal species (IUCN 2001). An analysis of the conservation status of tropical marine elasmobranchs according to the IUCN Red List (IUCN 2008) is provided below.

More than half of the tropical marine elasmobranchs (241 species) have been assessed by the IUCN Shark Specialist Group and published on the Red List (IUCN 2008). At the time of writing, the remaining species are considered Not Evaluated but the process is ongoing and assessments for the rest of these species are scheduled to be published in 2009. About 31% of the evaluated tropical marine elasmobranchs (or 16% of all tropical species) are listed in one of the three threatened categories, that is, Critically Endangered, Endangered, or Vulnerable, ~14% are Near Threatened, 10% are Least Concern, and ~12% are Data Deficient (Figure 4.6). Tropical marine sharks are better represented in the Red List than tropical marine batoids, that is ~61% versus ~45% assessed, respectively. Slightly more batoids are listed in one of the three threatened categories than sharks, ~18% versus ~14%.

The families with the most number of species assessed in one of the three threatened categories are the Rhinobatidae (11 of the 39 tropical species), Carcharhinidae (9 of the 51 tropical species), Pristidae (all 7 species), Narcinidae (6 of the 25 tropical species), and Rhynchobatidae (6 of the 7 tropical species). Of the 16 elasmobranchs globally assessed as Critically Endangered and occurring in tropical marine ecosystems, 7 of these are sawfish (Pristidae). All sawfish species have suffered large population declines throughout the world and in some areas (e.g., Java Sea and South Africa) have suffered local extinctions. Although dried rostra are occasionally observed at some locations in Java, local fishers report that they have not seen any sawfish landings since the 1970s. Two species of river sharks *Glyphis* are also assessed as Critically Endangered due to their rarity and inferred

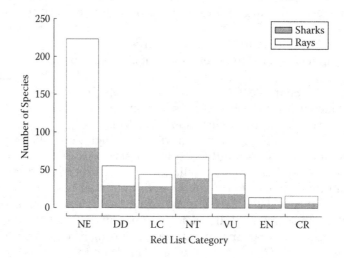

FIGURE 4.6
Number of shark and ray species in each of the Red List categories. NE = Not Evaluated, DD = Data Deficient, LC = Least Concern, NT = Near Threatened, VU = Vulnerable, EN = Endangered, CR = Critically Endangered.

population declines. The river sharks and at least one of the sawfish species are primarily freshwater and estuarine species, but also are found within inshore marine ecosystems. Other species that are listed as Critically Endangered are the sawback angelshark *Squatina aculeata*, the smoothback angelshark *Squatina oculata*, the Pondicherry shark *Carcharhinus hemiodon*, the daggernose shark *Isogomphodon oxyrhynchus*, the Brazilian guitarfish *Rhinobatos horkelii*, the Caribbean numbfish *Narcine bancroftii*, and the Java stingaree *Urolophus javanicus*. Management and recovery plans are urgently required to preserve these species as well as those listed as Endangered and Vulnerable.

4.6 Conclusions

Although the chondrichthyan fauna of the tropical shelf regions of the world is much better known than the deep-water elasmobranchs, there is little or almost nothing known about many of the species occurring in these regions despite increasing research in the past decade. The taxonomy of many taxa is still poorly resolved and a number of species complexes are likely to be found among widespread taxa. For example, the seemingly widespread bluespotted maskray *Neotrygon kuhlii* has many variations across the region and is likely a species complex (Last and Stevens 2009). A large number of elasmobranchs (465 species) occur within the tropical marine ecosystems, the majority of which are found in the Indo–West Pacific biogeographic region. The summary of biological data shows that there is a wide variety of life histories among tropical elasmobranchs with a number of small, fast growing and productive species as well as larger, slower growing and less productive species. Almost all of the major ecomorphotypes are represented by the tropical elasmobranchs. Many threats, mostly anthropogenic, strongly influence tropical marine elasmobranchs. Many of the largest chondrichthyan fisheries in the world occur within the tropical realm, in particular Indonesia and India, which are exclusively tropical. Improved management in many of these regions is required urgently to preserve stocks at a sustainable level. It is our hope that this chapter will allow for a better understanding of the tropical shelf chondrichthyan fauna in terms of their biogeography, biodiversity, life histories, threats, conservation status, and management. Little information is available for a large number of the tropical species and data needs to be improved for the tropical marine elasmobranchs to inform management to ensure sustainability of their stocks into the future.

Acknowledgments

The senior author would like to thank the CSIRO Wealth from Oceans Flagship for providing funding support. Special thanks are extended to David Ebert, John Stevens, Peter Kyne, Colin Simpfendorfer, and Peter Last for their support throughout this project. Thanks also go to Jeff Carrier, Mike Heithaus, and Jack Musick for the opportunity to undertake this project.

References

Allen GR, Adrim M (2003) Coral reef fishes of Indonesia. Zool Stud 42: 1–72.

Allen GR, Erdmann MV (2008) Two new species of bamboo sharks (Orectolobiformes: Hemiscylliidae). aqua. Int J Ichthyol 13: 93–108.

Amir F (1988) Suatu studi tentang pengoperasian shark set bottom gill net di perairan Kepulauan Aru, Maluku. Unpublished thesis, Fisheries Department, Hasanuddin University, Ujung Pandang.

Anislado-Tolentino V, Robinson-Mendoza C (2001) Age and growth for the scalloped hammerhead shark, *Sphyrna lewini* (Griffith and Smith, 1834), along the central Pacific coast of Mexico. Cienc Mar 27: 501–520.

Baum JK, Myers RA, Kehler DG et al. (2003) Collapse and conservation of shark populations in the Northwest Atlantic. Science 299: 389–392.

Bethea DM, Buckel JA, Carlson JK (2004) Foraging ecology of the early life stages of four sympatric shark species. Mar Ecol Prog Ser 268: 245–264.

Bethea DM, Hale L, Carlson JK et al. (2007) Geographic and ontogenetic variation in the diet and daily ration of the bonnethead shark, *Sphyrna tiburo*, from the eastern Gulf of Mexico. Mar Biol 152: 1009–1020.

Begg GA (2005) Life history parameters. In: Cadrin SX, Friedland KD, Waldman JR (eds) Stock identification methods. Elsevier Academic Press, New York, pp 119–150.

Bentley N (1996) Indonesia. In: Chen HK (ed) An overview of shark trade in selected countries of Southeast Asia. TRAFFIC Southeast Asia, Petaling Jaya.

Blaber SJM (1986) Feeding selectivity of a guild of piscivorous fish in mangrove areas of north-west Australia. Aust J Mar Freshwat Res 37: 329–336.

Blasco F, Aizpuru M, Gers C (2001) Depletion of the mangroves of continental Asia. Wetl Ecol Manag 9: 245–256.

Bloch ME, Schneider JG (1801) M.E. Blochii, Systema Ichthyologiae iconibus cx illustratum. Post obitum auctoris opus inchoatum absolvit, correxit, interpolavit Jo. Gottlob Schneider, Saxo. Berolini. Sumtibus Auctoris Impressum et Bibliopolio Sanderiano Commissum. Systema Ichthyologie.

Bonfil R, Mena R, De Anda D (1993) Biological parameters of commercially exploited silky sharks, *Carcharhinus falciformis*, from the Campeche Bank, Mexico. U.S. Department of Commerce, NOAA Technical Report 115: 73–86.

Branstetter S (1987a) Age, growth and reproductive biology of the silky shark, *Carcharhinus falciformis*, and the scalloped hammerhead, *Sphyrna lewini*, from the northwestern Gulf of Mexico. Environ Biol Fish 19: 161–173.

Branstetter S (1987b) Age and growth estimates for blacktip, *Carcharhinus limbatus*, and spinner, *C. brevipinna*, sharks from the northwestern Gulf of Mexico. Copeia 1987: 964–974.

Branstetter S, Musick JA (1994) Age and growth estimates for the sand tiger in the Northwestern Atlantic Ocean. Trans Am Fish Soc 123: 242–254.

Breder CM Jr (1952) On the utility of the saw of the sawfish. Copeia 1952: 90–91.

Briggs JC (1974) Marine zoogeography. McGraw-Hill, New York.

Briggs JC (1985) Species richness among the tropical shelf regions. Biol Morya 6: 3–11.

Briggs JC (1995) Global biogeography. Developments in palaeontology and stratigraphy. Elsevier, New York.

Briggs JC (2003) Marine centres of origin as evolutionary engines. J Biogeogr 30: 1–18.

Briggs JC (2005) The marine East Indies: diversity and speciation. J Biogeogr 32: 1517–1522.

Brown CA, Gruber SH (1988) Age assessment of the lemon shark, *Negaprion brevirostris*, using tetracycline validated vertebral centra. Copeia 1988: 747–753.

Brunnschweiler JM, Earle JL (2006) A contribution to marine life conservation efforts in the South Pacific: the Shark Reef Marine Reserve, Fiji. Cybium 30: 133–139.

Bush A (2003) Diet and diel feeding periodicity of juvenile scalloped hammerhead sharks *Sphyrna lewini*, in Kane'ohe Bay, Oahu, Hawaii. Environ Biol Fish 67: 1–11.

Cailliet GM, Goldman KJ (2004) Age determination and validation in chondrichthyan fishes. In: Carrier JC, Musick JA, Heithaus MR (eds) Biology of sharks and their relatives. CRC Press, Boca Raton, FL, pp 399–447.

Carlson JK, Baremore IE (2003) Changes in the biological parameters of Atlantic sharpnose shark *Rhizoprionodon terraenovae* in the Gulf of Mexico: evidence for density-dependent growth and maturity? Mar Freshwat Res 54: 227–234.

Carlson JK, Baremore IE (2005) Growth dynamics of the spinner shark (*Carcharhinus brevipinna*) off the United States southeast and Gulf of Mexico coasts: a comparison of methods. Fish Bull 103: 280–291.

Carlson JK, Cortés E, Bethea DM (2003) Life history and population dynamics of the finetooth shark (*Carcharhinus isodon*) in the northeastern Gulf of Mexico. Fish Bull 101: 281–292.

Carlson JK, Cortés E, Johnson AG (1999) Age and growth of the blacknose shark, *Carcharhinus acronotus*, in the eastern Gulf of Mexico. Copeia 3: 684–691.

Carlson JK, Sulikowski JR, Baremore IE (2006) Do differences in life history exist for blacktip sharks, *Carcharhinus limbatus*, from the United States South Atlantic Bight and Eastern Gulf of Mexico? Environ Biol Fish 77: 279–292.

Carpenter KE (ed) (2002) The living marine resources of the Western Central Atlantic. Volume 1. Introduction, molluscs, crustaceans, hagfishes, sharks, batoid fishes and chimaeras. FAO Species Identification Guide for Fishery Purposes. FAO, Rome.

Carrier JC, Luer CA (1990) Growth rates in the nurse shark, *Ginglymostoma cirratum*. Copeia 1990: 686–692.

Carrier JC, Pratt HL (1998) Habitat management and closure of a nurse shark breeding and nursery ground. Fish Res 39: 209–213.

Carrier JC, Pratt HL Jr, Castro JI (2004) Reproductive biology of elasmobranchs. In: Carrier JC, Musick JA, Heithaus MR (eds) Biology of sharks and their relatives. CRC Press, Boca Raton, FL, pp 269–286.

Cartamil DP, Vaudo JJ, Lowe CG et al. (2003) Diel movement patterns of the Hawaiian stingray, *Dasyatis lata*: implications for ecological interactions between sympatric elasmobranch species. Mar Biol 142: 841–847.

Chapman DD, Gruber SH (2002) A further observation of the prey-handling behavior of the great hammerhead shark, *Sphyrna mokarran*: predation upon the spotted eagle ray, *Aetobatus narinari*. Bull Mar Sci 70: 947–952.

Chen CT, Leu LC, Joung SJ et al. (1990) Age and growth of the scalloped hammerhead, *Sphyrna lewini*, in northeastern Taiwan waters. Pac Sci 44: 156–170.

Chen WK, Chen PC, Liu KM et al. (2007) Age and growth estimates of the whitespotted bamboo shark, *Chiloscyllium plagiosum*, in the northern waters of Taiwan. Zool Stud 44: 11–27.

Chin A, Kyne PM (2007) Vulnerability of chondrichthyan fishes of the Great Barrier Reef to climate change. In: Johnson JE, Marshall PA (eds) Climate change and the Great Barrier Reef: a vulnerability assessment. Great Barrier Reef Marine Park Authority, Townsville, Australia, pp 393–425.

Compagno LJV (1990) Alternative life-history styles of cartilaginous fishes in time and space. Environ Biol Fish 28: 33–75.

Compagno L, Dando M, Fowler S (2005) A field guide to sharks of the world. Collins, London.

Cortés E (1999) Standardised diet compositions and trophic levels of sharks. ICES J Mar Sci 56: 707–717.

Cortés E (2004) Life history patterns, demography, and population dynamics. In: Carrier JC, Musick JA, Heithaus MR (eds) Biology of sharks and their relatives. CRC Press, Boca Raton, FL, pp 449–469.

Cortés E, Gruber SH (1990) Diet, feeding habits and estimates of daily ration of young lemon sharks, *Negaprion brevirostris* (Poey). Copeia 1990: 204–218.

Cortés E, Manire CA, Heuter RE (1996) Diet, feeding habits and diel feeding chronology of the bonnethead shark, *Sphyrna tiburo*, in southwest Florida. Bull Mar Sci 58: 353–367.

Cruz-Martinez A, Chiappa-Carrara C, Arenas-Fuentes V (2005) Age and growth of the bull shark, *Carcharhinus leucas*, from southern Gulf of Mexico. Northwest Atl Fish Sci 35: 367–374.

Cuevas-Zimbrón E, Sosa-Nichizaki O, O'Sullivan J (2008) Preliminary study on the age and growth of the spinetail mobula, *Mobula japanica* (Müller and Henle, 1841), with comments on its vertebral column structure. Poster Presentation, American Elasmobranch Society, 2008 Annual Meeting, Montreal, Canada.

Dana JD (1853) On an isothermal oceanic chart illustrating the geographical distribution of marine animals. Am J Sci 16: 314–327.

Davenport S, Stevens JD (1998) Age and growth of two commercially important sharks (*Carcharhinus tilstoni* and *C. sorrah*) from northern Australia. Aust J Mar Freshwat Res 39: 417–433.

Dean MN, Motta PJ (2004) Anatomy and functional morphology of the feeding apparatus of the lesser electric ray, *Narcine brasiliensis* (Elasmobranchii: Batoidea). J Morphol 262: 462–483.

Devadoss P (1998) Growth and population parameters of the spade nose shark, *Scoliodon laticaudus* from Calicut coast. Indian J Fish 45: 29–34.

Dewar H, Mous P, Domeier M et al. (2008) Movements and site fidelity of the giant manta ray, *Manta birostris*, in the Komodo Marine Park, Indonesia. Mar Biol 155: 121–133.

Driggers WB III, Carlson JK, Cullum BJ et al. (2004) Age and growth of the blacknose shark, *Carcharhinus acronotus*, in the western North Atlantic Ocean with comments on regional variation in growth rates. Environ Biol Fish 71: 171–178.

Duarte CM (2002) The future of seagrass meadows. Environ Conserv 29: 192–206.

Dudgeon CL, Noad MJ, Lanyon JM (2008) Abundance and demography of a seasonal aggregation of zebra sharks *Stegostoma fasciatum*. Mar Ecol Prog Ser 368: 269–281.

Dulvy NK (2006) Conservation biology: strict marine protected areas prevent reef shark declines. Curr Biol 16: R989–R991.

Ebert DA (2003) Sharks, rays and chimaeras of California. University of California Press, Berkeley.

Economakis AE, Lobel PS (1998) Aggregation behavior of the grey reef shark, *Carcharhinus amblyrhynchos*, at Johnston Atoll, Central Pacific Ocean. Environ Biol Fish 51: 129–139.

Ekman S (1953) Zoogeography of the sea. Sidgwick & Jackson, London.

Erdmann MV (2000) Destructive fishing practices in Indonesian seas. In: Sheppard C (ed) Seas at the millenium: an environmental evaluation, Elsevier Science, New York, pp 392–393.

Everett JT (1997) Fisheries and CLIMATE CHANGE: the IPCC Second Assessment. In: Hancock DA, Smith DC, Grant A et al. (eds) Developing and sustaining world fisheries resources: the state of science and management. Proceedings of the Second World Fisheries Congress, Brisbane.

FAO (2002) FISHSTAT Plus (v. 2.30), Capture Production Database, 1970–2000, and Commodities Trade and Production Database 1976–2000.

Feely RA, Sabine CL, Lee K et al. (2004) Impact of anthropogenic CO_2 on the $CaCO_3$ system in the oceans. Science 305: 362–366.

Fouts WR, Nelson DR (1999) Prey capture by the Pacific angel shark, *Squatina californica*: visually mediated strikes and ambush-site characteristics. Copeia 1999: 304–312.

Fowler SL, Cavanagh RD (2005) International conservation and management initiatives for chondrichthyan fishes. In: Fowler SL, Cavanagh RD, Camhi M et al. (eds) Sharks, rays and chimaeras: the status of the chondrichthyan fishes. IUCN/SSC Shark Specialist Group, IUCN, Gland, Switzerland and Cambridge, UK, pp 58–69.

Fox HE, Pet JS, Dahuri R et al. (2003) Recovery in rubble fields: long-term impacts of blast fishing. Mar Pollut Bull 46: 1024–1031.

Freitas RHA, Rosa RS, Gruber SH, Wetherbee BM (2006) Early growth and juvenile population structure of lemon sharks *Negaprion brevirostris* in the Atol das Rocas Biological Reserve, off northeast Brazil. J Fish Biol 68: 1319–1332.

García VB, Lucifora LO, Myers RA (2008) The importance of habitat and life history to extinction risk in sharks, skates, rays and chimaeras. Proc R Soc B 275: 83–89.

Garla RC, Chapman DD, Wetherbee BM et al. (2006) Movement patterns of young Caribbean reef sharks, *Carcharhinus perezi*, at Fernando de Noronha Archipelago, Brazil: the potential of marine protected areas for conservation of a nursery ground. Mar Biol 149: 189–199.

Gilliam D, Sullivan K (1993) Diet and feeding habits of the southern stingray, *Dasyatis americana*, in the central Bahamas. Bull Mar Sci 52: 1007–1013.

Goldman KJ (2002) Aspects of age, growth, demographics and thermal biology of two lamniform shark species. PhD Dissertation, College of William and Mary, School of Marine Science, Virginia Institute of Marine Science, Williamsburg, VA.

Grubbs RD, Musick JA (2007) Spatial delineation of summer nursery areas for juvenile sandbar sharks in Chesapeake Bay, Virginia. In: McCandless CT, Kohler NE, Pratt HL Jr (eds) Shark nursery grounds of the Gulf of Mexico and the East Coast waters of the United States. American Fisheries Society Symposium 50: 63–86.

Gvirtzman Z, Stern RJ (2004) Bathymetry of Mariana trench-arc system and formation of the Challenger Deep as a consequence of weak plate coupling. Tectonics 23: TC2011, doi:10.1029/2003TC001581.

Hamlett WC, Kormanik G, Storrie M et al. (2005) Chondrichthyan parity, lecithotrophy and matrotrophy. In: Hamlett WC (ed) Reproductive biology and phylogeny of chondrichthyes: sharks, batoids and chimaeras. Science Publisher, Enfield, New Hampshire, pp 395–434.

Heithaus MR (2001) The biology of tiger sharks, *Galeocerdo cuvier*, in Shark Bay, Western Australia: sex ratio, size distribution, diet, and seasonal changes in catch rates. Environ Biol Fish 61: 25–36.

Heithaus MR (2005) Habitat use and group size of pied cormorants (*Phalacrocorax varius*) in a seagrass ecosystem: possible effects of food abundance and predation risk. Mar Biol 147: 27–35.

Heithaus MR, Burkholder D, Hueter RE et al. (2007a) Spatial and temporal variation in shark communities of the lower Florida Keys and evidence for historical population declines. Can J Fish Aquat Sci 64: 1302–1313.

Heithaus MR, Dill LM (2002) Food availability and tiger shark predation risk in Xuence bottlenose dolphin habitat use. Ecology 83: 480–491.

Heithaus MR, Dill LM, Marshall GJ, Buhleier B (2002) Habitat use and foraging behavior of tiger sharks (*Galeocerdo cuvier*) in a seagrass ecosystem. Mar Biol 140: 237–248.

Heithaus MR, Frid A, Wirsing AJ, Worm B (2008) Predicting ecological consequences of marine top predator declines. Trend Ecol Evol 23: 202–210.

Heithaus MR, Wirsing AJ, Dill LM et al. (2007b) Long-term movements of tiger sharks satellite-tagged in Shark Bay, Western Australia. Mar Biol 151: 1455–1461.

Henningsen AD (2002) Age and growth in captive southern stingrays, *Dasyatis americana*. Abstract, American Elasmobranch Society, 2002 Annual Meeting, Kansas City, MO.

Heupel MR, Bennett MB (1998) Observations on the diet and feeding habits of the epaulette shark, *Hemiscyllium ocellatum* (Bonnaterre) on Heron Island reef, Great Barrier Reef, Australia. Mar Freshwat Res 49: 753–756.

Heupel MR, Carlson JK, Simpfendorfer CA (2007) Shark nursery areas: concepts, definition, characterization and assumptions. Mar Ecol Prog Ser 337: 287–297.

Heupel MR, Simpfendorfer CA (2005) Quantitative analysis of aggregation behaviour in juvenile blacktip sharks. Mar Biol 147: 1239–1249.

Heupel MR, Simpfendorfer CA, Hueter RE (2004) Estimation of shark home ranges using passive monitoring techniques. Environ Biol Fish 71: 135–142.

Hoegh-Guldberg O (1999) Climate change, coral bleaching and the future of the world's coral reefs. Mar Freshwat Res 50: 839–866.

Hoegh-Guldberg O, Mumby PJ, Hooten AJ et al. (2007) Coral reefs under rapid climate change and ocean acidification. Science 318: 1737–1742.

Homma K, Maruyama T, Itoh T et al. (1997) Biology of the manta ray, *Manta birostris*, Walbaum, in the Indo–Pacific. In: Séret B, Sire JY (eds) Proceedings of the 5th Indo–Pacific Fish Conference, Nouméa. Soc Fr Ichthyol, Paris, pp 209–216.

IUCN (2001) IUCN Red List Categories and Criteria: Version 3.1. IUCN Species Survival Commission. IUCN, Gland, Switzerland and Cambridge, UK.

IUCN (2008) 2008 IUCN Red List of Threatened Species. http://www.redlist.org. Accessed December 20, 2008.

Jacobsen IP (2007) The biology of five benthic elasmobranch species from northern Australia and north-east Australia, incorporating a taxonomic review of the Indo–West Pacific Gymnuridae. PhD Thesis, University of Queensland.

Johnson MR, Snelson FF Jr (1996) Reproductive life history of the Atlantic stingray, *Dasyatis sabina* (Pisces, Dasyatidae), in the freshwater St. Johns River, Florida. Bull Mar Sci 59: 74–88.

Joung SJ, Chen CT, Clark E et al. (1996) The whale shark, *Rhincodon typus*, is a livebearer: 300 embryos found in one "megamamma" supreme. Environ Biol Fish 46: 219–223.

Joung SJ, Liao YY, Chen CT (2004) Age and growth of sandbar shark, *Carcharhinus plumbeus*, in north-eastern Taiwan waters. Fish Res 70: 83–96.

Joung SJ, Liao YY, Liu KM et al. (2005) Age, growth, and reproduction of the spinner shark, *Carcharhinus brevipinna*, in the northeastern waters of Taiwan. Zool Stud 44: 102–110.

Kasim HM (1991) Shark fishery of Veraval coast with special reference to population dynamics of *Scoliodon laticaudus* (Muller and Henle) and *Rhizoprionodon acutus* (Ruppell). J Mar Biol Assoc India 33: 213–228.

Kleypas JA, Feely RA, Fabry VJ et al. (2006) Impacts of ocean acidification on coral reefs and other marine calcifiers: a guide for future research. Proceedings Workshop April 18–20, 2005, St Petersburg, FL, sponsored by NSF, NOAA, and the US Geological Survey.

Klimley AP (1987) The determinants of sexual segregation in the scalloped hammerhead shark, *Sphyrna lewini*. Environ Biol Fish 18: 27–40.

Klimley AP, Butler SB, Nelson NR et al. (1988) Diel movements of scalloped hammerhead shark, *Sphyrna lewini* Griffith and Smith, to and from a seamouth in the Gulf of California. J Fish Biol 33: 751–761.

Kyne PM, Bennett MB (2002) Diet of the eastern shovelnose ray, *Aptychotrema rostrata* (Shaw and Nodder; 1794), from Moreton Bay, Queensland, Australia. Mar Freshwat Res 53: 679–686.

Lack M, Sant G (2006) Confronting shark conservation head on! TRAFFIC International, Cambridge, UK.

Last PR, Stevens JD (2009) Sharks and rays of Australia, 2nd edition. CSIRO Publishing, Australia.

Last PR, White WT, Fahmi (2006) *Rhinobatos jimbaranensis* sp. nov. and *R. penggali* sp. nov., two new shovelnose rays (Batoidea: Rhinobatidae) from eastern Indonesia. Cybium 30: 235–246.

Last PR, White WT, Pogonoski JJ (eds) (2008a) Descriptions of new Australian chondrichthyans. CSIRO Marine and Atmospheric Research Paper 022.

Last PR, White WT, Pogonoski JJ et al. (eds) (2008b) Descriptions of new Australian skates (Batoidea: Rajoidei). CSIRO Marine and Atmospheric Research Paper 021.

Leis JM (1984) Larval fish dispersal and the East Pacific Barrier. Oceanogr Trop 19: 181–192.

Lessa R, Santana FM (1998) Age determination and growth of the smalltail shark, *Carcharhinus porosus*, from northern Brazil. Mar Freshwat Res 49: 705–111.

Lessa R, Santana FM, Batista V et al. (2000) Age and growth of the daggernose shark, *Isogomphodon oxyrhynchus*, from northern Brazil. Mar Freshwat Res 51: 339–347.

Linnaeus C (1758) Systema Naturae, Ed. X. (Systema naturae per regna tria naturae, secundum classes, ordines, genera, species, cum characteribus, differentiis, synonymis, locis. Tomus I. Editio decima, reformata.) Holmiae. Systema Naturae ed. 10, 1.

Loefer JK, Sedberry GR (2003) Life history of the Atlantic sharpnose shark (*Rhizoprionodon terraeno-vae*) (Richardson, 1836) off the southeastern United States. Fish Bull 101: 75–88.

Longhurst A (1998) Ecological geography of the sea. Academic Press, San Diego.

Lowe CG, Bray RN, Nelson DR (1994) Feeding and associated electrical behavior of the Pacific electric ray *Torpedo californica* in the field. Mar Biol 120: 161–169.

Lowe CG, Wetherbee BM, Crow GL et al. (1996) Ontogenetic dietary shift and feeding behaviour of the tiger shark, *Galeocerdo cuvier*, in Hawaiian waters. Environ Biol Fish 47: 203–211.

Mathew CJ, Devaraj M (1997) The biology and population dynamics of the spadenose shark *Scoliodon laticaudus* in the coastal waters of Maharashtra State, India. Indian J Fish 44: 11–27.

McAuley RB, Simpfendorfer CA, Hyndes GA et al. (2006) Validated age and growth of the sandbar shark, *Carcharhinus plumbeus* (Nardo 1827) in the waters off Western Australia. Environ Biol Fish 77: 385–400.

McEachran JD, Aschliman N (2004) Phylogeny of Batoidea. In: Carrier JC, Musick JA, Heithaus MR (eds) Biology of sharks and their relatives. CRC Press, Boca Raton, FL, pp 79–113.

McElroy WD, Wetherbee BM, Mostello CS et al. (2006) Food habits and ontogenetic changes in the diet of the sandbar shark, *Carcharhinus plumbeus*, in Hawaii. Environ Biol Fish 76: 81–92.

Monintja DR, Poernomo RP (2000) Proposed concept for catch policy on shark and tuna including southern bluefin tuna in Indonesia. Paper presented at Indonesian-Australian Workshop on Shark and Tuna, Denpasar, March 2000.

Müller J, Henle FGJ (1838–1841) Systematische Beschreibung der Plagiostomen. Berlin. Plagiostomen.

Munday PL, Jones GP, Pratchett MS et al. (2008) Climate change and the future for coral reef fishes. Fish Fish 9: 261–285.

Musick JA, Ellis JK (2005) Reproductive evolution of chondrichthyans. In: Hamlett WC (ed) Reproductive biology and phylogeny of chondrichthyes: sharks, batoids and chimaeras. Science Publisher, Enfield, New Hampshire, pp. 45–71.

Musick JA, Harbin MM, Compagno LJVC (2004) Historical zoogeography of the Selachii. In: Carrier JC, Musick JA, Heithaus MR (eds) Biology of sharks and their relatives. CRC Press, Boca Raton, FL, pp 33–78.

Myers RA, Baum JK, Shepherd TD et al. (2007) Cascading effects of the loss of apex predatory sharks from a coastal ocean. Science 315: 1846–1850.

Natanson LJ, Casey JG, Kohler NE et al. (1999) Growth of the tiger shark, *Galeocerdo cuvier*, in the western North Atlantic based on tag returns and length frequencies; and a note on the effects of tagging. Fish Bull 97: 944–953.

Natanson LJ, Sulikowski JA, Kneebone JR et al. (2007) Age and growth estimates for the smooth skate, *Malacoraja senta*, in the Gulf of Maine. Environ Biol Fish 80: 293–308.

Navia AF, Mejia-Falla PA, Giraldo A (2007) Feeding ecology of elasmobranch fishes in coastal waters of the Columbian Eastern Tropical Pacific. BMC Ecol 7: 8 doi:10.1186/1472–6785–7–8. Accessed March 20, 2009.

Neer JA, Thompson BA (2005) Life history of the cownose ray, *Rhinoptera bonasus*, in the northern Gulf of Mexico, with comments on geographic variability in life history traits. Environ Biol Fish 73: 321–331.

Neer JA, Thompson BA, Carlson JK (2005) Age and growth of *Carcharhinus leucas* in the northern Gulf of Mexico: incorporating variability in size at birth. J Fish Biol 67: 370–383.

Nelson D, Johnson R (1970) Diel activity rhythms in the nocturnal bottom-dwelling sharks, *Heterodontus francisci* and *Cephaloscyllium ventriosum*. Copeia 1970: 732–739.

Nicol D (1971) Species, class and phylum diversity of animals. Q J Florida Acad Sci 34: 191–194.

Orth RJ, Carruthers TJB, Dennison WC et al. (2006) A global crisis for seagrass ecosystems. Bioscience 56: 987–996.

Oshitani S, Nakano H, Tanaka S (2003) Age and growth of the silky shark *Carcharhinus falciformis* from the Pacific Ocean. Fish Sci 69: 456–464.

Papastamatiou YP, Wetherbee BM, Lowe CG et al. (2006) Distribution and diet of four species of carcharhinid shark in the Hawaiian Islands: evidence for resource partitioning and competitive exclusion. Mar Ecol Prog Ser 320: 239–251.

Parenti LR (1991) Ocean basins and the biogeography of freshwater fishes. Aust Syst Bot 4: 137–149.

Parsons GR (1993) Age determination and growth of the bonnethead shark *Sphyrna tiburo*: a comparison of two populations. Mar Biol 117: 23–31.

Phillips MJ, Lin CK, Beveridge MCM (1993) Shrimp culture and the environment: lessons from the world's most rapidly expanding warmwater aquaculture sector. In: Pullin RSV, Rosenthal H, Maclean JL (eds) Environment and aquaculture in developing countries. ICLARM Conference Proceedings 31.

Pierce SJ, White WT, Marshall AD (2008) New record of the smalleye stingray, *Dasyatis microps* (Myliobatiformes: Dasyatidae) from the western Indian Ocean. Zootaxa 1734: 65–68.

Piercy AN, Carlson JK, Sulikowski JA, Burgess GA (2007) Age and growth of the scalloped hammerhead shark, *Sphyrna lewini*, in the northwest Atlantic Ocean and Gulf of Mexico. Mar Freshwat Res 58: 34–40.

Pratt HL Jr, Carrier JC (2001) A review of elasmobranch reproductive behavior with a case study on the nurse shark, *Ginglymostoma cirratum*. Environ Biol Fish 60: 157–188.

Primavera JH (1993) A critical review of shrimp pond culture in the Philippines. Rev Fish Sci 1: 151–201.

Radtke RL, Cailliet GM (1984) Age estimation and growth of the gray reef shark Carcharhinus ambly-rhynchos from the northwestern Hawaiian Islands. In: Proceedings of the Second Symposium on Resource Investigations of the NW Hawaiian Islands. Volume 2. University of Hawaii Sea Grant: Honolulu.

Randall JE (1977) Contribution to the biology of the whitetip reef shark (*Triaenodon obesus*). Pac Sci 31: 143–164.

Randall JE (1998) Zoogeography of shore fishes of the Indo-Pacific region. Zool Stud 37: 227–268.

Reeve R (2002) Policing international trade in endangered species: The CITES Treaty and compliance. The Royal Institute of International Affairs and Earthscan, London.

Robbins WD, Hisano M, Connolly SR et al. (2006) Ongoing collapse of coral reef shark populations. Curr Biol 16: 2314–2319.

Rocha LA (2003) Patterns of distribution and processes of speciation in Brazilian reef fishes. J Biogeogr 30: 1161–1171.

Romine JG, Grubbs RD, Musick JA (2006) Age and growth of the sandbar shark, *Carcharhinus plumbeus*, in Hawaiian waters through vertebral analysis. Environ Biol Fish 77: 229–239.

Salini JP, Blaber SJM, Brewer DT (1992) Diets of sharks from estuaries and adjacent waters of the North-eastern Gulf of Carpentaria, Australia. Aust J Mar Freshwat Res 43: 87–96.

Santana FM, Lessa R (2004) Age determination and growth of the night shark (*Carcharhinus signatus*) off the northeastern Brazilian coast. Fish Bull 102: 156–167.

Sazima I, Moura RL (2000) Shark (*Carcharhinus perezi*), cleaned by the goby (*Elacatinus randalli*), at Fernando de Noronha Archipelago, western South Atlantic. Copeia 2000: 297–299.

Segura-Zarzosa JC, Abitia-Cárdenas LA, Galván-Magaña F (1997) Observations on the feeding hab-its of the shark *Heterodontus francisci* Girard 1854 (Chondrichthyes: Heterodontidae), in San Ignacio Lagoon, Baja California Sur, Mexico. Cienc Mar 23: 111–128.

Silliman W, Gruber S (1999) Behavioral biology of the spotted eagle ray, *Aetobatus narinari*. Bahamas J Sci 7: 13–20.

Simpfendorfer CA (1993) Age and growth of the Australian sharpnose shark, *Rhizoprionodon taylori*, from north Queensland, Australia. Environ Biol Fish 36: 233–241.

Simpfendorfer CA (2000) Predicting population recovery rates for endangered western Atlantic saw-fishes using demographic analysis. Environ Biol Fish 58: 371–377.

Simpfendorfer CA, Goodreid AB, McAuley RB (2001) Size, sex and geographic variation in the diet of the tiger shark, *Galeocerdo cuvier*, from Western Australian waters. Environ Biol Fish 61: 37–46.

Simpfendorfer CA, McAuley RB, Chidlow J et al. (2002) Validated age and growth of the dusky shark, *Carcharhinus obscurus*, from Western Australia waters. Mar Freshwat Res 53: 567–573.

Simpfendorfer CA, Milward NE (1993) Utilisation of a tropical bay as a nursery area by sharks of the families Carcharhinidae and Sphyrnidae. Environ Biol Fishes 37: 337–345.

Sminkey TR, Musick JA (1995) Age and growth of the sandbar shark, *Carcharhinus plumbeus*, before and after population depletion. Copeia 1995: 871–883.

Smith WD, Cailliet GM, Melendez EM (2007) Maturity and growth characteristics of a commercially exploited stingray, *Dasyatis dipterura*. Mar Freshwat Res 58: 54–66.

Snelson FF Jr, Williams-Hooper SE, Schmid TH (1989) Biology of the bluntnose stingray, *Dasyatis sayi*, in Florida coastal lagoons. Bull Mar Sci 45: 15–25.

Springer VG (1982) Pacific plate biogeography with special reference to shorefishes. Smithsonian Contrib Zool 465: 1–82.

Stevens JD (2002) The role of protected areas in elasmobranch fisheries management and conserva-tion. Occasional Paper of the IUCN Species Survival Commission 25: 241–242.

Stevens JD, Bonfil R, Dulvy NK et al. (2000) The effects of fishing on sharks, rays, and chimaeras (chondrichthyans), and the implications for marine ecosystems. ICES J Mar Sci 57: 476–494.

Stevens JD, McLoughlin KJ (1991) Distribution, size and sex composition, reproductive biology, and diet of sharks from northern Australia. Aust J Mar Freshwat Res 42: 151–199.

Strong WR Jr (1990) Hammerhead shark predation on stingrays: an observation of prey handling by *Sphyrna mokarran*. Copeia 1990: 836–840.

Struhsaker P (1969) Observations on the biology and distribution of the thorny stingray, *Dasyatis centroura* (Pisces: Dasyatidae). Bull Mar Sci 19: 456–481.

Sulikowski JA, Irvine SB, DeValerio KC et al. (2007) Age, growth and maturity of the rounded skate, *Raja texana*, from the Gulf of Mexico, USA. Mar Freshwat Res 58: 41–53.

Summers AP (2000) Stiffening the stingray skeleton: an investigation of durophagy in myliobatid stingrays (Chondrichthyes, Batoidea, Myliobatidae). J Morphol 243: 113–126.

Sundström LF, Gruber SH, Clermont SM et al. (2001) Review of elasmobranch behavioral studies using ultrasonic telemetry with special reference to the lemon shark, *Negaprion brevirostris*, around Bimini Islands, Bahamas. Environ Biol Fish 60: 225–250.

Suzuki T (2002) Development of shark fisheries and shark fin export in Indonesia. Case study of Karangsong Village, Indramayu, West Java. Elasmobranch biodiversity, conservation and management. Occasional Paper of the IUCN Species Survival Commission 25:149–157.

Tanaka S (1991) Age estimation of freshwater sawfish and sharks in northern Australia and Papua New Guinea. Univ Mus Univ Tokyo Nat Cult 3: 71–82.

Taylor G (1994) Whale sharks, the giants of Ningaloo Reef. Angus & Robertson, Sydney.

Taylor SM, Bennett MB (2008) Cephalopod dietary specialization and ontogenetic partitioning of the Australian weasel shark *Hemigaleus australiensis* White, Last & Compagno. J Fish Biol 72: 917–936.

Teshima K, Kamei Y, Toda M et al. (1995) Reproductive mode of the tawny nurse shark taken from the Yaeyama Islands, Okinawa, Japan with comments on individuals lacking the second dorsal fin. Bull Seikai Natl Fish Res Inst 73: 1–12.

Thorburn DC, Morgan DL, Rowland AJ et al. (2007) Freshwater sawfish *Pristis microdon* Latham, 1794 (Chondrichthyes: Pristidae) in the Kimberley region of Western Australia. Zootaxa 1471: 27–41.

Thorburn, DC, Morgan DL, Rowland AJ et al. (2008) Life history notes of the critically endangered dwarf sawfish, *Pristis clavata*, Garman, 1906 from the Kimberley region of Western Australia. Environ Biol Fish 83: 139–145.

Thorson TB (1983) Observations on the morphology, ecology and life history of the euryhaline stingray, *Dasyatis guttata* (Bloch and Schneider) 1801. Acta Biol Venez 11: 95–125.

Valiela I, Bowen JL, York JK (2001) Mangrove forests: one of the world's threatened major tropical environments. Bioscience 51: 807–815.

Vermeij GJ (1978) Biogeography and adaptation patterns of marine life. Harvard University Press, Cambridge, MA.

Walker TI (1976) Effects of species, sex, length and locality on the mercury content of school shark *Galeorhinus australis* (Macleay) and gummy shark *Mustelus antarcticus* Guenther from south-eastern Australian waters. Aust J Mar Freshwat Res 27: 603–616.

Walker TI (1988) Mercury concentrations in edible tissues of elasmobranchs, teleosts, crustaceans and molluscs from south-eastern Australian waters. Aust J Mar Freshwat Res 39: 39–49.

Walker TI (1989) Regional studies: seagrass in Shark Bay, the foundation of an ecosystem. In: Larkum AWD, McComb AJ, Shepherd SA (eds) Biology of seagrasses. Elsevier, New York.

Walker, TI (2002) Review of fisheries and processes impacting shark populations of the world. Elasmobranch biodiversity, conservation and management. Occasional Paper of the IUCN Species Survival Commission 25: 149–157.

Wallman HL, Bennett WA (2006) Effects of parturition and feeding on thermal preference of Atlantic stingray, *Dasyatis sabina* (Lesueur). Environ Biol Fish 75: 259–267.

Wetherbee BM, Cortés E (2004) Food consumption and feeding habits. In: Carrier JC, Musick JA, Heithaus MR (eds) Biology of sharks and their relatives. CRC Press, Boca Raton, FL, pp 225–246.

Wetherbee BM, Crow GL, Lowe CG (1996) Biology of the Galapagos shark, *Carcharhinus galapagensis*, in Hawai'i. Environ Biol Fish 45: 299–310.

Wetherbee BM, Crow GI, Lowe CG (1997) Distribution, reproduction and diet of the gray reef shark *Carcharhinus amblyrhynchos* in Hawaii. Mar Ecol Prog Ser 151: 181–189.

Wetherbee BM, Gruber SH, Rosa RS (2007) Movement patterns of juvenile lemon sharks *Negaprion brevirostris* within Atol das Rocas, Brazil: a nursery characterized by tidal extremes. Mar Ecol Prog Ser 343: 283–293.

White WT (2003) Aspects of the biology of elasmobranchs in a subtropical embayment in Western Australia and of chondrichthyan fisheries in Indonesia. PhD thesis, Murdoch University.

White WT, Cavanagh R (2007) Whale shark landings in Indonesian artisanal shark and ray fisheries. In: Irvine TR, Keesing JK (eds) Whale sharks: science, conservation and management. Proceedings of the First International Whale Shark Conference, May 9–12, 2005 Australia. Fish Res 84: 128–131.

White WT, Hall NG, Potter IC (2002) Size and age compositions and reproductive biology of the nervous shark, *Carcharhinus cautus*, in a large subtropical embayment, including an analysis of growth during pre- and postnatal life. Mar Biol 141: 1153–1166.

White WT, Last PR, Dharmadi (2005) Description of a new species of catshark, *Atelomycterus baliensis* (Carcharhiniformes: Scyliorhinidae) from eastern Indonesia. Cybium 29: 33–40.

White WT, Last PR, Stevens JD (2007) *Halaelurus maculosus* n. sp. and *H. sellus* n. sp., two new species of catshark (Carcharhiniformes: Scyliorhinidae) from the Indo–West Pacific. Zootaxa 1639: 1–21.

White WT, Last PR, Stevens JD et al. (2006) Economically important sharks and rays of Indonesia. ACIAR Publishing, Canberra.

White WT, Platell ME, Potter IC (2004) Comparisons between the diets of four abundant species of elasmobranchs in a subtropical embayment: implications for resource partitioning. Mar Biol 144: 439–448.

White WT, Potter IC (2004) Habitat partitioning among four species of elasmobranch in nearshore, shallow waters of a sub-tropical embayment. Mar Biol 145: 1023–1032.

Whitney NM, Crow GL (2007) Reproductive biology of the tiger shark (*Galeocerdo cuvier*) in Hawaii. Mar Biol 151: 63–70.

Wilkinson C (2008) Status of coral reefs of the world: 2008. Global Coral Reef Monitoring Network and Rainforest Research Centre, Townsville, Australia.

Wilson EO (1992) The diversity of life. Belknap Press, Harvard University, Cambridge, MA.

Wourms JP, Demski LS (1993) The reproduction and development of sharks, skates, rays and ratfishes: introduction, history, overview, and future prospects. Environ Biol Fish 38: 7–21.

Yamaguchi A, Kawahara I, Ito S (2005) Occurrence, growth and food of longheaded eagle ray, *Aetobatus flagellum*, in Ariake Sound, Kyushu, Japan. Environ Biol Fish 74: 229–238.

Yokota L, Lessa RP (2006) A nursery area for sharks and rays in Northeastern Brazil. Environ Biol Fish 75: 349–360.

Appendix 4.1

Checklist of Chondrichthyans of the Tropical Marine Ecosystem

This checklist of chondrichthyans occurring within the tropical marine ecosystem includes all nominal (described) valid species, but not undescribed taxa or synonyms. The number of undescribed taxa in each family, if any, is also provided. The only undescribed species taken into account are those which have been used in the primary literature. For each species, a summary of its geographic range, the FAO Fishing Areas (tropical ones only) in which it occurs, habitat zones, and depth range is provided. Reference sources utilized to acquire this information is included at the end of the checklist. It includes species described up to December 31, 2008. Doubtful species are preceded by a question mark and species that have only a very limited distribution in tropical waters, that is, "edge" species, are preceded by an asterisk.

FAO Fisheries Areas (the United Nation's Food and Agriculture Organization's Major Fishing Areas for Statistical Purposes) that overlap the tropical marine ecosystem are outlined below:

31 Western Central Atlantic
34 Eastern Central Atlantic
41 Southwest Atlantic
47 Southeast Atlantic
51 Western Indian Ocean
57 Eastern Indian Ocean
61 Northwest Pacific
71 Western Central Pacific
77 Eastern Central Pacific
87 Southeast Pacific

Maps of these areas can be located at ftp://ftp.fao.org/fi/maps/Default.htm#CURRENT. The habitat zones are described in Section 4.3.4.

Class Chondrichthyes: Cartilaginous Fishes

Subclass Elasmobranchii: Sharks and Rays

Order Hexanchiformes: Cow and Frilled Sharks

Family Hexanchidae: Sixgill and Sevengill Sharks
Heptranchias perlo (Bonnaterre, 1788) – Sharpnose Sevengill Shark
Wide-ranging but patchy in temperate and tropical waters of the Atlantic and Indo–Pacific. Tropical FAO Areas: 31, 34, 41, 47, 51, 57, 61, 71. Shelf (occasional) and upper to mid slope. 27–1000 m.
Hexanchus nakamurai Teng, 1962 – Bigeye Sixgill Shark

Wide-ranging but patchy in warm-temperate and tropical waters of the Atlantic and Indo–Pacific. Tropical FAO Areas: 31, 34(?), 47, 51, 57, 61, 71. Shelf (occasional) and upper slope. 90–621 m.

Order Squaliformes: Dogfish Sharks

Family Squalidae: Dogfish Sharks

Cirrhigaleus asper (Merrett, 1973) – Roughskin Spurdog

Wide-ranging but patchy in warm-temperate and tropical waters of the Atlantic and Indo–Pacific. Tropical FAO Areas: 31, 41, 47, 51, 57(?), 77. Shelf and upper slope. 73–600 m.

*Squalus albifrons Last, White & Stevens, 2007 – Eastern Highfin Spurdog

Warm-temperate and tropical Southwest Pacific: Endemic to eastern Australia. Tropical FAO Areas: 71. Outer shelf and upper slope. 131–450 m.

*Squalus blainville (Risso, 1827) – Longnose Spurdog

Nominally from temperate and tropical Eastern Atlantic (and the Mediterranean), but not well defined due to confusion with other species and taxonomic issues. Tropical FAO Areas: 34, 47. Shelf and upper slope. 16–440 m.

*Squalus crassispinus Last, Edmunds & Yearsley, 2007 – Fatspine Spurdog

Tropical Eastern Indian: Endemic to northwestern Australia. Tropical FAO Areas: 57. Outermost shelf and upper slope. 187–262 m.

*Squalus cubensis Howell-Rivero, 1936 – Cuban Dogfish

Warm-temperate and tropical Western Atlantic. Tropical FAO Areas: 31, 41. Shelf and upper slope. 60–380 m.

*Squalus japonicus Ishikawa, 1908 – Japanese Spurdog

Warm-temperate and tropical Northwest Pacific: East Asia. Tropical FAO Areas: 61, 71. Outer shelf and upper slope. 120–340 m.

*Squalus melanurus Fourmanoir, 1979 – Blacktail Spurdog

Western Central Pacific: Endemic to New Caledonia. Tropical FAO Areas: 71. Shelf and upper slope. 34–480 m.

*Squalus mitsukurii Jordan & Snyder, *in* Jordan & Fowler, 1903. Shortspine Spurdog

Wide-ranging, but patchy in temperate and tropical waters of the Northwest Pacific, but not well defined due to taxonomic issues. Populations in Atlantic, Indian and eastern Pacific likely to be separate species. Tropical FAO Areas: 61. Shelf and upper to mid slope. 100–500 m. Resolution of considerable taxonomic issues is ongoing.

Family Centrophoridae: Gulper Sharks

*Centrophorus atromarginatus Garman, 1913 – Dwarf Gulper Shark

Patchy in warm-temperate and tropical waters of the Indo–West Pacific, but not well defined. Tropical FAO Areas: 51, 57, 61, 71. Outer shelf and upper slope. 150–450 m. Possibly represents a species complex.

*Centrophorus moluccensis Bleeker, 1860 – Smallfin Gulper Shark

Patchy in warm-temperate and tropical waters of the Indo–West Pacific. Tropical FAO Areas: 51, 57, 61, 71. Outer shelf and upper to mid slope. 125–820 m. Possibly represents a species complex.

Family Oxynotidae: Roughsharks

Oxynotus centrina (Linnaeus, 1758) – Angular Roughshark

Wide-ranging temperate and tropical waters in the Eastern Atlantic (and the Mediterranean). Tropical FAO Areas: 34, 47. Shelf and upper slope. 50–660 m (mostly >100 m).

Order Pristiphoriformes: Sawsharks

Family Pristiophoridae: Sawsharks

Pliotrema warreni Regan, 1906 – Sixgill Sawshark

Warm-temperate and tropical waters in the Southeast Atlantic and Western Indian: narrow ranging off southern Africa. Tropical FAO Areas: 51. Shelf and upper slope. 37–500 m.

Pristiophorus japonicus Günther, 1870 – Japanese Sawshark

Warm-temperate and tropical (limited) waters in the Northwest Pacific. Tropical FAO Areas: 61. Habitat and depth unknown, presumably shelf and/or upper slope.

Order Squatiniformes: Angelsharks

Family Squatinidae: Angelsharks

Squatina aculeata Cuvier, 1829 – Sawback Angelshark

Warm temperate and tropical waters in the Eastern Atlantic (and the Western Mediterranean). Tropical FAO Areas: 34, 47. Shelf and upper slope. 30–500 m.

Squatina africana Regan, 1908 – African Angelshark

Western Indian: warm temperate and tropical waters off East Africa, including southern Madagascar. Tropical FAO Areas: 51. Shelf and upper slope. 0–494 m.

Squatina armata (Philippi, 1887) – Chilean Angelshark

Eastern Pacific: Temperate, and probably tropical waters, off South America. Tropical FAO Areas: 87. Insular shelf. 30–75 m.

Squatina californica Ayres, 1859 – Pacific Angelshark

Temperate and tropical waters in the Eastern Pacific. Tropical FAO Areas: 77, 87. Shelf. 1–200 m.

Squatina dumeril Lesueur, 1818 – Sand Devil

Temperate and tropical waters in the Northwest Atlantic. Tropical FAO Areas: 31. Shelf and upper slope. 40–250 m.

Squatina formosa Shen & Ting, 1972 – Taiwan Angelshark

Northwest Pacific: restricted to Taiwan and the Philippines. Tropical FAO Areas: 61, 71. Outer shelf and upper slope. 183–385 m.

Squatina heteroptera Castro-Aguirre, Peréz & Campos, 2007

Western Atlantic: known only from Gulf of Mexico. Tropical FAO Areas: 31. Shelf. 101 m. Validity of species not yet confirmed.

Squatina japonica Bleeker, 1858 – Japanese Angelshark

Warm-temperate and tropical (limited) waters in the Northwest Pacific. Tropical FAO Areas: 61. Shelf (and possibly upper slope).

Squatina mexicana Castro-Aguirre, Peréz & Campos, 2007

Western Atlantic: known only from Gulf of Mexico. Tropical FAO Areas: 31. Shelf. 96–101 m. Validity of species not yet confirmed.

Squatina oculata Bonaparte, 1840 – Smoothback Angelshark

Warm-temperate and tropical waters in the Western Atlantic (and the Mediterranean). Tropical FAO Areas: 34, 47. Shelf and upper slope. 20–500 m.

Squatina pseudocellata Last & White, 2008 – Western Angelshark

Eastern Indian: Endemic to tropical waters of northwestern Australia. Tropical FAO Areas: 57. Outer shelf and upper slope. 150–312 m.

Squatina punctata Marini, 1936 – Angular Angelshark

Warm-temperate and tropical waters in the Southwest Atlantic. Tropical FAO Areas: 41. Insular shelf. 10–80 m.

Order Heterodontiformes: Bullhead Sharks

Family Heterodontidae: Bullhead Sharks

Heterodontus francisci (Girard, 1855) – Horn Shark

Warm-temperate and tropical waters in the Northeast Pacific. Tropical FAO Areas: 77. Shelf. 0–150 m.

Heterodontus japonicus Maclay & Macleay, 1884 – Japanese Bullhead Shark

Temperate and tropical (limited) waters in the Northwest Pacific. Tropical FAO Areas: 61. Insular shelf. 6–37 m.

Heterodontus mexicanus Taylor & Castro-Aguirre, 1972 – Mexican Hornshark

Warm-temperate and tropical waters in the Eastern Pacific. Tropical FAO Areas: 77, 87. Insular shelf. 0–50 m.

Heterodontus omanensis Baldwin, 2005 – Oman Bullhead Shark

Tropical waters in the Western Indian: Endemic to waters off Oman. Tropical FAO Areas: 51. Insular shelf. 0–80 m.

Heterodontus quoyi (Fréminville, 1840) – Galapagos Bullhead Shark

Warm-temperate and tropical waters in the Eastern Pacific. Tropical FAO Areas: 51. Insular shelf. 16–30 m.

Heterodontus ramalheira (Smith, 1949) – Whitespotted Bullhead Shark

Warm-temperate and tropical waters in the Western Indian: East Africa and Arabian Sea. Tropical FAO Areas: 51. Outer shelf and uppermost slope. 40–275 m.

Heterodontus zebra (Gray, 1831) – Zebra Horn Shark

Warm-temperate and tropical waters in the Indo–West Pacific. Tropical FAO Areas: 57, 61, 71. Shelf. 0–200 m.

Order Orectolobiformes: Carpetsharks

Family Parascylliidae: Collared Carpetsharks

Cirrhoscyllium expolitum Smith & Radcliffe, *in* Smith, 1913 – Barbelthroat Carpet Shark

Tropical waters in the Northwest Pacific and South China Sea. Tropical FAO Areas: 61, 71. Outer shelf. 183–190 m.

Cirrhoscyllium formosanum Teng, 1959 – Taiwan Saddled Carpet Shark

Northwest Pacific: Endemic to Taiwan. Tropical FAO Areas: 61. Outer shelf. 110 m.

Family Brachaeluridae: Blind Sharks

Brachaelurus colcloughi Ogilby, 1908 – Colcloughs Shark

Warm-temperate and tropical waters in the Southwest Pacific: Endemic to eastern Australia. Tropical FAO Areas: 71. Shelf. 0–100 m.

*Brachaelurus waddi (Bloch & Schneider, 1801) – Blind Shark

Warm-temperate and tropical (limited) waters in the Southwest Pacific: Endemic to eastern Australia (records from northern and northwestern Australia require confirmation). Tropical FAO Areas: 71. Shelf. 0–140 m.

Family Orectolobidae: Wobbegongs

Eucrossorhinus dasypogon (Bleeker, 1867) – Tasselled Wobbegong

Tropical waters in the Indo–West Pacific: Endemic to northern Australia, eastern Indonesia and New Guinea. Tropical FAO Areas: 57, 71. Insular shelf. 0–50 m.

*Orectolobus halei Whitley, 1940 – Gulf Wobbegong

Temperate and tropical (limited) waters in the Eastern Indian and Southeast Pacific: Endemic to southern Australia. Tropical FAO Areas: 57, 71. Shelf. 0–100 m.

Orectolobus hutchinsi Last & Chidlow, 2006 – Western Wobbegong

Warm-temperate and tropical waters in the Eastern Indian: Endemic to Western Australia. Tropical FAO Areas: 57. Shelf. 5–105 m.

Orectolobus japonicus Regan, 1906 – Japanese Wobbegong

Warm-temperate and tropical waters in the Northwest Pacific. Tropical FAO Areas: 61, 71. Shelf.

*Orectolobus maculatus (Bonnaterre, 1788) – Spotted Wobbegong

Temperate and tropical (limited) waters in the Eastern Indian and Southeast Pacific: Endemic to southern Australia. Tropical FAO Areas: 57, 71. Shelf and upper slope. 0–218 m.

Orectolobus ornatus (de Vis, 1883) – Ornate Wobbegong

Warm-temperate and tropical waters in the Southeast Pacific: Endemic to eastern Australia. Tropical FAO Areas: 71. Shelf. 0–100 m.

Orectolobus reticulatus Last, Pogonoski & White, 2008 – Network Wobbegong

Tropical waters in the Indo–West Pacific: Endemic to northern Australia. Tropical FAO Areas: 57, 71. Insular shelf. 0–20 m.

Orectolobus wardi Whitley, 1939 – Northern Wobbegong

Tropical waters in the Indo–West Pacific: Endemic to northern Australia. Tropical FAO Areas: 57, 71. Insular shelf. 0–20 m.

Family Hemiscylliidae: Longtailed Carpetsharks

Chiloscyllium arabicum Gubanov, *in* Gubanov & Schleib, 1980 – Arabian Carpet Shark

Tropical waters in the Northwest Indian. Tropical FAO Areas: 51. Shelf. 3–100 m.

Chiloscyllium burmensis Dingerkus & DeFino, 1983 – Burmese Bamboo Shark

Tropical waters in the Northeast Indian. Tropical FAO Areas: 57. Insular shelf?

Chiloscyllium griseum Müller & Henle, 1838 – Grey Bamboo Shark

Tropical waters in the Indo–West Pacific. Tropical FAO Areas: 57, 71. Insular shelf. 5–80 m.

Chiloscyllium hasselti Bleeker, 1852 – Indonesian Bamboo Shark

Tropical waters in the Indo–West Pacific. Tropical FAO Areas: 57, 61, 71. Insular shelf. 0–12 m.

Chiloscyllium indicum (Gmelin, 1788) – Slender Bamboo Shark

Warm-temperate (limited) and tropical waters in the Indo–West Pacific. Tropical FAO Areas: 57, 61, 71. Insular shelf. 0–20 m.

Chiloscyllium plagiosum (Anonymous [Bennett], 1830) – Whitespotted Bamboo Shark

Warm-temperate (limited) and tropical waters in the Indo–West Pacific. Tropical FAO Areas: 51, 57, 61, 71. Insular shelf.

Chiloscyllium punctatum Müller & Henle, 1838 – Grey Carpet Shark

Warm-temperate (limited) and tropical waters in the Indo–West Pacific. Tropical FAO Areas: 57, 61, 71. Insular shelf. 0–85 m.

Hemiscyllium freycineti Quoy & Gaimard, 1824 – Indonesian Speckled Carpet Shark

Indo–West Pacific: Endemic to New Guinea. Tropical FAO Areas: 71. Insular shelf. 0–40 m.

Hemiscyllium galei Allen & Erdmann, 2008

Indo–West Pacific: Endemic to New Guinea. Tropical FAO Areas: 71. Insular shelf. 0–20 m.

Hemiscyllium hallstromi Whitley, 1967 – Papuan Epaluette Shark

Indo–West Pacific: Endemic to New Guinea. Tropical FAO Areas: 71. Insular shelf. 0–20 m.

Hemiscyllium henryi Allen & Erdmann, 2008

Indo–West Pacific: Endemic to New Guinea. Tropical FAO Areas: 71. Insular shelf. 0–30 m.

Hemiscyllium ocellatum (Bonnaterre, 1788) – Epaluette Shark

Warm-temperate (limited) and tropical waters in the Indo–West Pacific: Northern Australia, New Guinea and eastern Indonesia. Tropical FAO Areas: 57, 71. Insular shelf. 0–10 m.

Hemiscyllium strahani Whitley, 1967 – Hooded Carpet Shark

Indo–West Pacific: New Guinea. Tropical FAO Areas: 71. Insular shelf. 3–13 m.

Hemiscyllium trispeculare Richardson, 1843 – Speckled Carpet Shark

Indo–West Pacific: Endemic to northern Australia. Tropical FAO Areas: 57, 71. Insular shelf. 0–20 m.

Family Ginglymostomatidae: Nurse Sharks

Ginglymostoma cirratum (Bonnaterre, 1788) – Nurse Shark

Warm-temperate and tropical waters in the Eastern Pacific and Atlantic. Tropical FAO Areas: 31, 34, 41, 47, 77, 87. Shelf. 0–140 m.

Nebrius ferrugineus (Lesson, 1831) – Tawny Shark

Warm-temperate (limited) and tropical waters in the Indo–West Pacific. Tropical FAO Areas: 51, 57, 61, 71. Insular shelf. 0–70 m.

Pseudoginglymostoma brevicaudatum Günther, *in* Playfair & Günther, 1867 – Short-tail Nurse Shark

Tropical waters in the Western Indian. Tropical FAO Areas: 51. Presumably shelf.

Family Stegostomatidae: Zebra Shark

Stegostoma fasciatum (Hermann, 1783) – Zebra Shark

Warm-temperate (limited) and tropical waters in the Indo–West Pacific. Tropical FAO Areas: 51, 57, 61, 71. Insular shelf. 0–62 m.

Family Rhincodontidae: Whale Shark

Rhincodon typus Smith, 1828 – Whale Shark

Circumglobal in all warm-temperate and tropical waters (except Mediterranean). Tropical FAO Areas: 31, 34, 41, 47, 51, 57, 61, 71, 77, 87. Pelagic in open ocean and close inshore. 0–700 m.

Order Lamniformes: Mackerel Sharks

Family Odontaspididae: Sand Tiger Sharks

Carcharias taurus Rafinesque, 1810 – Grey Nurse Shark

Temperate and tropical waters in the Atlantic and Indo–West Pacific. Tropical FAO Areas: 31, 34, 47, 51, 57, 61, 71. Shelf. 1–191 m.

Odontaspis ferox (Risso, 1810) – Sand Tiger Shark

Wide-ranging but patchy in the Atlantic, including the Mediterranean, and the Indo–Pacific. Possibly circumglobal in warm-temperate and tropical waters. Tropical FAO Areas: 31, 34, 51, 57, 71. Shelf and upper slope. 13–850 m.

Family Lamnidae: Mackerel Sharks

Carcharodon carcharias (Linnaeus, 1758) – White Shark

Circumglobal in temperate and tropical waters. Tropical FAO Areas: 31, 34, 41, 47, 51, 57, 61, 71, 77, 87. Shelf, slope and oceanic. 0–1300 m.

Order Carcharhiniformes: Ground Sharks

Family Scyliorhinidae: Catsharks

Asymbolus parvus Compagno, Stevens & Last, 1999 – Dwarf Catshark

Indo–West Pacific: Endemic to northwestern Australia. Tropical FAO Areas: 57. Outer shelf and upper slope. 160–260 m.

Atelomycterus baliensis White, Last & Dharmadi, 2005 – Bali Catshark

Indo–West Pacific: Endemic to Bali in Indonesia. Tropical FAO Areas: 57. Habitat and depth unknown. Presumably on shelf.

Atelomycterus fasciatus Compagno & Stevens, 1993 – Banded Catshark

Indo–West Pacific: Endemic to northwestern Australia. Tropical FAO Areas: 57. Shelf. 30–120 m.

Atelomycterus macleayi Whitley, 1939 – Marbled Catshark

Indo–West Pacific: Endemic to northern Australia. Tropical FAO Areas: 57, 71. Insular shelf. 0–4 m.

Atelomycterus marmoratus (Anonymous [Bennett], 1830) – Coral Catshark

Tropical waters in the Indo–West Pacific. Tropical FAO Areas: 51, 57, 61, 71. Insular shelf.

Atelomycterus marnkalha Jacobsen & Bennett, 2007 – Eastern Banded Catshark

Indo–West Pacific: Endemic to northeastern Australia. Tropical FAO Areas: 71. Insular shelf. 10–74 m.

Aulohalaelurus kanakorum Séret, 1990 – New Caledonia Catshark

Western Pacific: Endemic to New Caledonia. Tropical FAO Areas: 71. Insular shelf. 0–50 m.

Cephaloscyllium circulopullum Yano, Ahmad & Gambang, 2005 – Circle-blotch Pygmy Swellshark

Indo–West Pacific: Only known from northern Borneo. Uncertain as to whether this species is clearly separable from *C. sarawakensis*. Tropical FAO Areas: 71. Outer shelf. 118–165 m.

Cephaloscyllium maculatum Schaaf-Da Silva & Ebert, 2008 – Spotted Swellshark

Indo–West Pacific: Only known from Taiwan. Tropical FAO Areas: 61. Habitat and depth unknown. Probably on shelf.

Cephaloscyllium pardelotum Schaaf-Da Silva & Ebert, 2008 – Leopard-spotted Swellshark

Indo–West Pacific: Only known from Taiwan. Tropical FAO Areas: 61. Habitat and depth unknown. Probably on shelf.

Cephaloscyllium sarawakensis Yano, Ahmad & Gambang, 2005 – Sarawak Pygmy Swellshark

Indo–West Pacific: Only known from northern Borneo. Uncertain as to whether this species is clearly separable from *C. circulopullum*. Tropical FAO Areas: 71. Outer shelf. 118–165 m.

Cephaloscyllium speccum Last, Séret & White, 2008 – Speckled Swellshark

Western Pacific: Endemic to northeastern Australia, possibly also Vanuatu. Tropical FAO Areas: 71. Outer shelf and upper slope. 150–455 m.

Cephaloscyllium sufflans (Regan, 1921) – Balloon Shark

Western Indian: Endemic to South Africa and Mozambique. Tropical FAO Areas: 51. Shelf and upper slope. 40–440 m.

*Cephaloscyllium umbratile Jordan & Fowler, 1903 – Japanese Swellshark

Temperate and tropical (limited) waters in the Northwest Pacific. Tropical FAO Areas: 61. Shelf. 20–200 m.

Cephaloscyllium ventriosum (Garman, 1880) – Swellshark

Warm-temperate and tropical waters of the Eastern Pacific. Tropical FAO Areas: 77, 87. Shelf and upper slope. 5–457 m.

*Cephalurus cephalus (Gilbert, 1892) – Lollipop Catshark

Warm-temperate and tropical waters of the Eastern Pacific: Mexico. Tropical FAO Areas: 77. Outer shelf and upper and mid slope. 155–927 m.

Galeus sauteri (Jordan & Richardson, 1909) – Blacktip Sawtail Catshark

Warm-temperate and tropical waters of the Northwest Pacific. Tropical FAO Areas: 61, 71. Insular shelf. 60–90 m.

Halaelurus boesemani Springer & D'Aubrey, 1972 – Speckled Catshark

Western Indian: Somalia and Gulf of Aden. Tropical FAO Areas: 51. Insular shelf. 37–91 m.

Halaelurus buergeri (Müller & Henle, 1838) – Darkspot Catshark

Warm-temperate and tropical waters in the Northwest Pacific. Tropical FAO Areas: 61, 71. Shelf. 80–100 m.

Halaelurus lineatus Bass, D'Aubrey & Kistnasamy, 1975 – Lined Catshark

Western Indian: South Africa and Mozambique. Tropical FAO Areas: 51. Shelf and upper slope. 0–290 m.

Halaelurus maculosus White, Last & Stevens, 2007 – Indonesian Speckled Catshark

Indo–West Pacific: Endemic to Indonesia. Tropical FAO Areas: 57, 71(?). Habitat and depth unknown, presumably shelf.

Halaelurus quagga (Alcock, 1899) – Quagga Catshark

Western Indian: Known from Somalia and India. Tropical FAO Areas: 51. Shelf. 54–186 m.

Halaelurus sellus White, Last & Stevens, 2007 – Australian Speckled Catshark

Eastern Indian: Endemic to northwestern Australia. Tropical FAO Areas: 57. Shelf. 60–265 m.

*Haploblepharus edwardsii (Schinz, 1822) – Puffadder Shyshark

Eastern Indian: Endemic to South Africa. Tropical FAO Areas: 51. Shelf. 0–133 m.

Parmaturus xaniurus (Gilbert, 1892) – Filetail Catshark

Warm-temperate and tropical waters in the Eastern Pacific: United States and Mexico. Tropical FAO Areas: 77. Shelf and slope. 91–1251 m.

*Schroederichthys maculatus Springer, 1966 – Narrowtail Catshark

Tropical waters in the Western Atlantic: Central and northern South America. Tropical FAO Areas: 31. Outer shelf and upper slope. 190–410 m.

Schroederichthys tenuis Springer, 1966 – Slender Catshark

Tropical waters in the Western Atlantic: northern South America. Tropical FAO Areas: 31. Shelf and upper slope. 72–450 m.

*Scyliorhinus canicula (Linnaeus, 1758) – Smallspotted Catshark

Temperate and tropical (limited) waters in the Northeast and central Eastern Atlantic. Tropical FAO Areas: 34. Shelf. 0–110 m.

Scyliorhinus cervigoni Maurin & Bonnet, 1970 – West African Catshark

Tropical waters in the Eastern Atlantic. Tropical FAO Areas: 34, 47. Shelf and slope. 45–500 m.

Scyliorhinus haeckelii Miranda Ribeiro, 1907 – Freckled Catshark

Warm-temperate and tropical waters in the Western Atlantic: east coast of South America. Tropical FAO Areas: 31, 41. Shelf and slope. 37–439 m.

*Scyliorhinus retifer (Garman, 1881) – Chain Catshark

Warm-temperate and tropical waters in the Northwest Atlantic. Tropical FAO Areas: 31. Shelf and slope. 73–754 m.

Scyliorhinus stellaris (Linnaeus, 1758) – Nursehound

Temperate and tropical waters in the Northeast Atlantic. Tropical FAO Areas: 34. Shelf. 1–125 m.

Family Proscylliidae: Finback Catsharks

Ctenacis fehlmanni (Springer, 1968) – Harlequin Catshark

Western Indian: Known only off Somalia. Tropical FAO Areas: 51. Outer shelf.

Eridacnis radcliffei Smith, 1913 – Pygmy Ribbontail Catshark

Tropical waters in the Indo–West Pacific. Tropical FAO Areas: 51, 57, 71. Shelf and slope. 71–766 m.

*Eridacnis sinuans (Smith, 1957) – African Ribbontail Catshark

Warm-temperate and tropical waters in the Western Indian: South Africa, Mozambique and Tanzania. Tropical FAO Areas: 51. Outer shelf and upper slope. 180–480 m.

Proscyllium habereri Hilgendorf, 1904 – Graceful Catshark

Warm-temperate and tropical waters in the Indo–West Pacific: Java north to Japan. Tropical FAO Areas: 61, 71. Shelf. 50–100 m.

Family Leptochariidae: Barbeled Houndshark

Leptocharias smithii (Müller & Henle, 1839) – Barbeled Houndshark

Warm-temperate and tropical waters in the Eastern Atlantic. Tropical FAO Areas: 61, 71. Shelf. 10–75 m.

Family Triakidae: Houndsharks

Furgaleus macki (Whitley, 1943) – Whiskery Shark

Temperate and tropical waters in the Eastern Indian: southern and western Australia. Tropical FAO Areas: 57. Shelf and upper slope. 0–220 m.

Gogolia filewoodi Compagno, 1973 – Sailback Houndshark

Central Western Pacific: Known only from northern New Guinea. Tropical FAO Areas: 71. Shelf. 73 m.

Hemitriakis complicofasciata Takahashi & Nakaya, 2004 – Ocellate Topeshark

Warm-temperate and tropical waters in the Northwest Pacific: Known only from northern New Guinea. Tropical FAO Areas: 71. Shelf. ~100 m.

Hemitriakis falcata Compagno & Stevens, 1993 – Sicklefin Houndshark

Tropical waters in the Eastern Indian: Known only from northwestern Australia. Tropical FAO Areas: 57. Outer shelf. 146–197 m.

Hemitriakis japanica (Müller & Henle, 1839) – Japanese Topeshark

Warm-temperate and tropical waters in the Northwest Pacific. Tropical FAO Areas: 61. Insular shelf. 0–48 m.

Hypogaleus hyugaensis (Miyosi, 1939) – Pencil Shark

Warm-temperate and tropical waters in the Indo–West Pacific. Tropical FAO Areas: 51, 61. Shelf and upper slope. 40–230 m.

Mustelus albipinnis Castro-Aguirre, Atuna-Mendiola, Gonzáz-Acosta & De la Cruz-Agüero, 2005 – White Margin Hound Shark

Warm-temperate and tropical waters in the Northeast Pacific: Mexico. Tropical FAO Areas: 77. Outer shelf and upper slope. 103–281 m.

Mustelus californicus Gill, 1864 – Grey Smoothhound

Warm-temperate and tropical waters in the Northeast Pacific: northern California to Mexico. Tropical FAO Areas: 77. Shelf. 0–46 m.

Mustelus canis (Mitchill, 1815) – Dusky Smoothhound

Warm-temperate and tropical waters in the Western Atlantic. Tropical FAO Areas: 31, 41(?). Shelf and upper slope. 0–360 m.

Mustelus dorsalis Gill, 1864 – Sharpnose Smoothhound

Tropical waters in the Central Eastern Pacific: Mexico to Ecuador. Tropical FAO Areas: 77, 87. Shelf.

Mustelus griseus Pietschmann, 1908 – Spotless Smoothhound

Warm-temperate and tropical waters in the Northwest Pacific. Tropical FAO Areas: 61, 71. Insular shelf. 0–51 m.

Mustelus henlei (Gill, 1863) – Brown Smoothhound

Warm-temperate and tropical waters in the Eastern Pacific. Tropical FAO Areas: 77, 87. Shelf. 0–200 m.

Mustelus higmani Springer & Lowe, 1963 – Smalleye Smoothhound

Warm-temperate and tropical waters in the Western Atlantic: northern Gulf of Mexico to southern Brazil. Tropical FAO Areas: 31, 41. Shelf and slope. 0–1281 m.

Mustelus lunulatus Jordan & Gilbert, 1882 – Sicklefin Smoothhound

Warm-temperate and tropical waters in the Eastern Pacific: southern California to Panama. Tropical FAO Areas: 77. Shelf.

Mustelus manazo Bleeker, 1854 – Starspotted Smoothhound

Warm-temperate and tropical waters in the Northwest Pacific. Tropical FAO Areas: 61, 71. Shelf.

*Mustelus mento Cope, 1877 – Speckled Smoothhound

Temperate and tropical (limited) waters off Pacific and Atlantic coasts of South America. Tropical FAO Areas: 87. Insular shelf. 16–50 m.

Mustelus minicanis Heemstra, 1997 – Venezuelan Dwarf Smoothhound

Tropical waters in the Western Atlantic: Colombia and Venezuela. Tropical FAO Areas: 31. Shelf. 71–183 m.

Mustelus mosis Hemprich & Ehrenberg, 1899 – Arabian Smoothhound

Warm-temperate (limited) and tropical waters in the Indo–West Pacific. Tropical FAO Areas: 51, 57. Shelf.

Mustelus norrisi Springer, 1939 – Narrowfin Smoothhound

Warm-temperate and tropical waters in the Western Atlantic. Tropical FAO Areas: 31. Shelf. 0–84 m.

Mustelus ravidus White & Last, 2006 – Grey Gummy Shark

Indo–West Pacific: Endemic to northwestern Australia. Tropical FAO Areas: 57. Outer shelf and upper slope. 100–300 m.

Mustelus sinusmexicanus Heemstra, 1997 – Gulf Smoothhound

Warm-temperate and tropical waters in the Western Atlantic: Endemic to the Gulf of Mexico. Tropical FAO Areas: 31. Shelf and upper slope. 36–229 m.

Mustelus stevensi White & Last, 2008 – Western Spotted Gummy Shark

Tropical waters in the Eastern Indian: Endemic to northwestern Australia. Tropical FAO Areas: 57. Outer shelf and upper slope. 120–400 m.

Mustelus walkeri White & Last, 2008 – Eastern Spotted Gummy Shark

Tropical waters in the Southwest Pacific: Endemic to northeastern Australia. Tropical FAO Areas: 71. Shelf and upper slope. 52–403 m.

Mustelus widodoi White & Last, 2006 – White-fin Smoothhound

Tropical waters in the Indo–West Pacific: Endemic to Indonesia. Tropical FAO Areas: 57, 71. Habitat and depth unknown, presumably shelf and possibly upper slope.

Triakis acutipinna Kato, 1968 – Sharpfin Houndshark

Tropical waters in the Southeast Pacific: Known only from Ecuador. Tropical FAO Areas: 87. Habitat and depth unknown.

*Triakis maculata Kner & Steindachner, 1867 – Spotted Houndshark

Warm-temperate and tropical (limited) waters in the Southeast Pacific: Galapagos Islands to Chile. Tropical FAO Areas: 87. Shelf. Depth unknown.

Triakis scyllium Müller & Henle, 1839 – Banded Houndshark

Warm-temperate and tropical waters in the Northwest Pacific. Tropical FAO Areas: 61, 71(?). Shelf. Depth unknown.

Triakis semifasciata Girard, 1855 – Leopard Shark

Warm-temperate and tropical waters in the Northeast Pacific. Tropical FAO Areas: 77. Shelf. 0–91 m.

** plus an additional three undescribed species.

Family Hemigaleidae: Weasel Sharks

Chaenogaleus macrostoma (Bleeker, 1852) – Hooktooth Shark

Tropical waters in the Indo–West Pacific. Tropical FAO Areas: 51, 57, 61, 71. Insular shelf. 0–59 m.

Hemigaleus australiensis White, Last & Compagno, 2005 – Australian Weasel Shark

Warm-temperate (limited) and tropical waters in the Eastern Indian and Southwest Pacific: Endemic to northern Australia and possibly New Guinea. Tropical FAO Areas: 57, 71. Shelf. 0–170 m.

Hemigaleus microstoma Bleeker, 1852 – Sicklefin Weasel Shark

Tropical waters in the Indo–West Pacific. Tropical FAO Areas: 51, 57, 61, 71. Shelf. 0–100 m.

Hemipristis elongata (Klunzinger, 1871) – Fossil Shark

Warm-temperate and tropical waters in the Indo–West Pacific. Tropical FAO Areas: 51, 57, 61, 71. Shelf. 1–132 m.

Paragaleus leucolomatus Compagno & Smale, 1985 – Whitetip Weasel Shark

Tropical waters in the Western Indian: Known only from southeastern Australia. Tropical FAO Areas: 51. Insular shelf. 0–20 m.

Paragaleus pectoralis (Garman, 1906) – Atlantic Weasel Shark

Warm-temperate (limited) and tropical waters in the Eastern Atlantic. Tropical FAO Areas: 34, 47. Shelf. 0–100 m.

Paragaleus randalli Compagno, Krupp & Carpenter, 1996 – Slender Weasel Shark

Tropical waters in the Indian: Arabian Gulf to India. Tropical FAO Areas: 51, 57. Insular shelf. 0–18 m.

Paragaleus tengi (Chen, 1963) – Straighttooth Weasel Shark

Warm-temperate and tropical waters in the Northeast Pacific. Tropical FAO Areas: 61. Insular shelf. Depth unknown.

Family Carcharhinidae: Requiem Sharks

Carcharhinus acronotus (Poey, 1860) – Blacknose Shark

Warm-temperate and tropical waters in the Western Atlantic. Tropical FAO Areas: 31, 41. Insular shelf. 18–64 m.

Carcharhinus albimarginatus (Rüppell, 1837) – Silvertip Shark

Tropical waters in the Indo–West Pacific and Eastern Pacific (unconfirmed in Western Atlantic). Tropical FAO Areas: 51, 57, 61, 71, 77, 87. Inshore and offshore pelagic (not oceanic). Surface to 800 m.

Carcharhinus altimus (Springer, 1950) – Bignose Shark

Wide-ranging but patchy in warm-temperate and tropical waters in the Indo–Pacific and Atlantic. Tropical FAO Areas: 31, 34, 51, 57, 61, 71, 77, 87. Shelf and upper slope. 25–430 m (mostly >90 m, young shallower, to 25 m).

Carcharhinus amblyrhynchoides (Whitley, 1934) – Graceful Shark

Tropical waters in the Indo–West Pacific. Tropical FAO Areas: 51, 57, 71. Insular shelf. 0–50 m.

Carcharhinus amblyrhynchos (Bleeker, 1856) – Grey Reef Shark

Tropical waters in the Indo–West to Central Pacific. Tropical FAO Areas: 51, 57, 71, 77. Shelf. 0–140 m.

Carcharhinus amboinensis (Müller & Henle, 1839) – Pigeye Shark

Warm-temperate (limited) and tropical waters in the Indo–West Pacific and Eastern Atlantic (Nigeria). Tropical FAO Areas: 34, 51, 57, 71, 77. Insular shelf. 0–60 m.

Carcharhinus borneensis (Bleeker, 1859) – Borneo Shark

Tropical waters in the Indo–West Pacific: Indonesia (Borneo and possibly Java) to southern China. Tropical FAO Areas: 61, 71. Insular shelf. 0–40 m.

*Carcharhinus brachyurus (Günther, 1870) – Bronze Whaler

Warm-temperate and tropical (limited) waters in most waters in the Indo–Pacific and Atlantic (including the Mediterranean). Tropical FAO Areas: 34, 57, 61, 71, 77. Shelf. 0–100 m.

Carcharhinus brevipinna (Müller & Henle, 1839) – Spinner Shark

Warm-temperate and tropical waters in the Indo–West Pacific and Atlantic. Tropical FAO Areas: 31, 34, 41, 47, 51, 57, 61, 71. Insular shelf. 0–75 m.

Carcharhinus cautus (Whitley, 1945) – Nervous Shark

Tropical waters in the Eastern Indian and Southwest Pacific. Tropical FAO Areas: 57, 71. Insular shelf. 0–20 m.

Carcharhinus dussumieri (Müller & Henle, 1839) – Whitecheek Shark

Warm-temperate (limited) and tropical waters in the Indo–West Pacific. Tropical FAO Areas: 51, 57, 61, 71. Shelf. 0–170 m.

Carcharhinus falciformis (Müller & Henle, 1839) – Silky Shark

Circumglobal in all tropical waters and occasionally in warm-temperate waters. Tropical FAO Areas: 31, 34, 41, 47, 51, 57, 61, 71, 77, 87. Mostly oceanic and epipelagic, occasionally occurring inshore. Surface to at least 500 m.

Carcharhinus fitzroyensis (Whitley, 1943) – Creek Shark

Tropical waters in the Indo–West Pacific: Endemic to northern Australia. Tropical FAO Areas: 57, 71. Insular shelf. 0–40 m.

Carcharhinus galapagensis (Snodgrass & Heller, 1905) – Galapagos Shark

Circumglobal but patchy in tropical waters, mainly around oceanic islands. Tropical FAO Areas: 31, 34, 47, 51, 71, 77, 87. Coastal pelagic, found well offshore but not oceanic. 0–180 m.

Carcharhinus hemiodon (Valenciennes, 1839) – Pondicherry Shark

Tropical waters in the Indo–West Pacific. Tropical FAO Areas: 51, 57, 61, 71. Insular shelf. 0–40 m.

Carcharhinus isodon (Valenciennes, 1839) – Finetooth Shark

Warm-temperate and tropical waters in the Western Atlantic. Tropical FAO Areas: 31. Insular shelf. 0–20 m.

Carcharhinus leiodon Garrick, 1985 – Smalltooth Blacktip

Western Indian: Known only from the Gulf of Aden. Tropical FAO Areas: 51. Habitat and depth unknown, probably inshore.

Carcharhinus leucas (Müller & Henle, 1839) – Bull Shark

Warm-temperate and tropical waters in the Indo–Pacific and Western Atlantic. Tropical FAO Areas: 31, 41, 51, 57, 71, 77, 87. Shelf, and in freshwater. 0–152 m.

Carcharhinus limbatus (Müller & Henle, 1839) – Common Blacktip Shark

Circumglobal in warm-temperate and tropical waters. Tropical FAO Areas: 31, 34, 41, 47, 51, 57, 61, 71, 77, 87. Shelf. 0–100 m.

Carcharhinus macloti (Müller & Henle, 1839) – Hardnose Shark

Tropical waters in the Indo–West Pacific. Tropical FAO Areas: 51, 57, 61, 71. Shelf. 0–170 m.

Carcharhinus melanopterus (Quoy & Gaimard, 1824) – Blacktip Reef Shark

Warm-temperate (limited) and tropical waters in the Indo–West Pacific. Tropical FAO Areas: 51, 57, 61, 71, 77. Insular shelf. 0–20 m.

Carcharhinus obscurus (Lesueur, 1818) – Dusky Shark

Probably circumglobal in warm-temperate and tropical waters. Tropical FAO Areas: 31, 34, 41, 47, 51, 57, 61, 71, 77. Shelf and pelagic, well offshore to oceanic. 0–400 m.

Carcharhinus perezii (Poey, 1876) – Caribbean Reef Shark

Warm-temperate and tropical waters in the Western Atlantic: North Carolina (United States) to Brazil. Tropical FAO Areas: 31, 41. Insular shelf. 0–30 m.

Carcharhinus plumbeus (Nardo, 1827) – Sandbar Shark

Circumglobal in warm-temperate and tropical waters. Tropical FAO Areas: 31, 34, 41, 47, 51, 57, 61, 71, 77, 87. Shelf and upper slope. 1–280 m.

Carcharhinus porosus (Ranzani, 1839) – Smalltail Shark

Warm-temperate and tropical waters in the Eastern Pacific and Western Atlantic. Tropical FAO Areas: 31, 41, 77, 87. Insular shelf. 0–36 m.

Carcharhinus sealei (Pietschmann, 1916) – Blackspot Shark

Tropical waters in the Indo–West Pacific. Tropical FAO Areas: 51, 57, 61, 71. Insular shelf. 0–40 m.

Carcharhinus signatus (Poey, 1868) – Night Shark

Warm-temperate and tropical waters in the Western and Eastern Atlantic. Tropical FAO Areas: 31, 34, 41(?), 47. Shelf. 50–100 m.

Carcharhinus sorrah (Müller & Henle, 1839) – Spot-tail Shark

Warm-temperate (limited) and tropical waters in the Indo–West Pacific. Tropical FAO Areas: 51, 57, 61, 71. Shelf. 0–140 m.

Carcharhinus tilstoni (Whitley, 1950) – Australian Blacktip Shark

Tropical waters in the Indo–West Pacific: Endemic to northern Australia. Tropical FAO Areas: 57, 71. Shelf. 0–150 m.

Galeocerdo cuvier (Péron & Lesueur, 1822) – Tiger Shark

Circumglobal in warm-temperate and tropical waters. Tropical FAO Areas: 31, 34, 41, 47, 51, 57, 61, 71, 77, 87. Shelf. 0–140 m.

Glyphis gangeticus (Müller & Henle, 1839) – Ganges River Shark

Tropical waters in the Indo–West Pacific: Known from river and estuarine systems and adjacent coastal areas of India and possibly Pakistan. Tropical FAO Areas: 51, 57. Freshwater rivers, estuaries and coastal marine.

Glyphis garricki Compagno, White & Last, 2008 – Northern River Shark

Tropical waters in the Indo–West Pacific: Known from river and estuarine systems and adjacent coastal areas of northern Australia and New Guinea. Tropical FAO Areas: 57, 71. Freshwater rivers, estuaries, and coastal marine.

Glyphis glyphis (Müller & Henle, 1839) – Speartooth Shark

Tropical waters in the Indo–West Pacific: Known from river and estuarine systems and adjacent coastal areas of northern Australia and New Guinea. Tropical FAO Areas: 57, 71. Freshwater rivers, estuaries, and coastal marine.

*Glyphis siamensis (Steindachner, 1896) – Irrawaddy River Shark

Tropical waters in the Indo–West Pacific: Known only from single specimen off Myanmar. Tropical FAO Areas: 57. Estuarine and riverine, and possibly coastal marine.

Isogomphodon oxyrhynchus (Müller & Henle, 1839) – Daggernose Shark

Tropical waters in the Western Atlantic: northern South America. Tropical FAO Areas: 31, 41. Insular shelf. 4–40 m.

Lamiopsis temmincki (Müller & Henle, 1839) – Broadfin Shark

Tropical waters in the Indo–West Pacific. Tropical FAO Areas: 51, 57, 61, 71. Insular shelf. Possibly a species complex.

Loxodon macrorhinus Müller & Henle, 1839 – Sliteye Shark

Warm-temperate (limited) and tropical waters in the Indo–West Pacific. Tropical FAO Areas: 51, 57, 61, 71. Insular shelf. 7–80 m.

Nasolamia velox (Gilbert, 1898) – Whitenose Shark

Warm-temperate and tropical waters in the Eastern Pacific: Mexico to Peru. Tropical FAO Areas: 77, 87. Shelf. 15–192 m.

Negaprion acutidens (Rüppell, 1837) – Sicklefin Lemon Shark

Tropical waters in the Indo–West and Central Pacific. Tropical FAO Areas: 51, 57, 71, 77. Insular shelf. 0–30 m.

Negaprion brevirostris (Poey, 1868) – Lemon Shark

Warm-temperate (limited) and tropical waters in the Western and Eastern Atlantic, and Eastern Pacific. Tropical FAO Areas: 31, 34, 41, 47, 77, 87. Shelf. 0–92 m.

Rhizoprionodon acutus (Rüppell, 1837) – Milk Shark

Warm-temperate (limited) and tropical waters in the Indo–West Pacific and Eastern Atlantic. Tropical FAO Areas: 34, 47, 51, 57, 61, 71. Shelf. 1–200 m.

Rhizoprionodon lalandii (Valenciennes, 1839) – Brazilian Sharpnose Shark

Tropical waters in the Western Atlantic: Panama to Southern Brazil. Tropical FAO Areas: 31, 41. Insular shelf. 3–70 m.

Rhizoprionodon longurio (Jordan & Gilbert, 1882) – Pacific Sharpnose Shark

Warm-temperate (limited) and tropical waters in the Eastern Pacific: California to Peru. Tropical FAO Areas: 77, 87. Insular shelf. 0–27 m.

Rhizoprionodon oligolinx Springer, 1964 – Grey Sharpnose Shark

Tropical waters in the Indo–West Pacific. Tropical FAO Areas: 51, 57, 61, 71. Shelf.

Rhizoprionodon porosus (Poey, 1861) – Caribbean Sharpnose Shark

Warm-temperate (limited) and tropical waters in the Western Atlantic: Caribbean and South America. Tropical FAO Areas: 31, 41. Shelf and upper slope. 0–500 m.

Rhizoprionodon taylori (Ogilby, 1915) – Australian Sharpnose Shark

Tropical waters in the Eastern Indian and Southwest Pacific: Endemic to northern Australia and southern New Guinea. Tropical FAO Areas: 57, 71. Shelf. 0–110 m.

Rhizoprionodon terraenovae (Richardson, 1836) – Atlantic Sharpnose Shark

Warm-temperate and tropical waters in the Northwest Atlantic. Tropical FAO Areas: 31. Shelf and upper slope. 0–280 m.

Scoliodon laticaudus Müller & Henle, 1838 – Spadenose Shark

Warm-temperate (limited) and tropical waters in the Indo–West Pacific. Tropical FAO Areas: 51, 57, 61, 71. Coastal inshore.

Triaenodon obesus (Rüppell, 1837) – Whitetip Reef Shark

Tropical waters in the Indo–Pacific. Tropical FAO Areas: 51, 57, 61, 71, 77, 87. Shelf and upper slope. 1–330 m.

** plus an additional two undescribed species.

Family Sphyrnidae: Hammerheads

Eusphyra blochii (Cuvier, 1816) – Winghead Shark

Tropical waters in the Indo–West Pacific. Tropical FAO Areas: 51, 57, 61, 71. Shallow insular shelf.

Sphyrna corona Springer, 1940 – Scalloped Bonnethead

Tropical waters in the Eastern Pacific: Mexico to Peru. Tropical FAO Areas: 77, 87. Shelf, presumably inshore.

Sphyrna lewini (Griffith & Smith, 1834) – Scalloped Hammerhead

Circumglobal in warm-temperate and tropical waters. Tropical FAO Areas: 31, 34, 41, 47, 51, 57, 61, 71, 77, 87. Shelf and upper slope, semi-oceanic. 0–275 m.

Sphyrna media Springer, 1940 – Scoophead

Warm-temperate (limited) and tropical waters in the Eastern Pacific and Western Atlantic. Tropical FAO Areas: 31, 41, 77, 87. Shelf.

Sphyrna mokarran (Rüppell, 1837) – Great Hammerhead

Circumglobal in warm-temperate and tropical waters. Tropical FAO Areas: 31, 34, 41, 47, 51, 57, 61, 71, 77, 87. Coastal pelagic and semi-oceanic, close inshore to well offshore. 0–80 m.

Sphyrna tiburo (Linnaeus, 1758) – Bonnethead

Warm-temperate and tropical waters in the Eastern Pacific and Western Atlantic. Tropical FAO Areas: 31, 41, 77, 87. Insular shelf. 0–80 m.

Sphyrna tudes (Valenciennes, 1822) – Smalleye Hammerhead

Warm-temperate (limited) and tropical waters in the Western Atlantic: Venezuela to Uruguay. Tropical FAO Areas: 31, 41. Insular shelf. 0–20 m.

Sphyrna zygaena (Linnaeus, 1837) – Smooth Hammerhead

Circumglobal in temperate and tropical waters. Tropical FAO Areas: 31, 34, 41(?), 47(?), 51, 57, 61, 71, 77, 87. Close inshore to well offshore. 0–20+ m.

Order Rajiformes: Rays (or batoids)
  Suborder Pristoidei: Sawfishes

  Family Pristidae: Sawfishes

Anoxypristis cuspidata (Latham, 1794) – Narrow Sawfish

Warm-temperate (limited) and tropical waters in the Indo–West Pacific. Tropical FAO Areas: 51, 57, 61, 71. Insular shelf. 0–40 m.

Pristis clavata Garman, 1906 – Dwarf Sawfish

Tropical waters in the Indo–West Pacific. Tropical FAO Areas: 57, 71. Insular shelf, including estuaries and rivers. 0–20 m.

Pristis microdon Latham, 1794 – Freshwater Sawfish

Tropical waters in the Indo–West Pacific. Tropical FAO Areas: 57, 71. Predominantly in riverine and estuarine areas, also insular shelf.

Pristis pectinata Latham, 1794 – Smalltooth Sawfish

Warm-temperate and tropical waters in the Western and Eastern Atlantic. Tropical FAO Areas: 31, 34, 41, 47. Shelf, including estuaries and brackish rivers. 0–100 m.

Pristis perotteti Valenciennes, in Müller & Henle, 1841 – Largetooth Sawfish

Warm-temperate (limited) and tropical waters in the Western and Eastern Atlantic. Tropical FAO Areas: 31, 34, 41, 47. Insular shelf, including estuaries and brackish rivers.

Pristis pristis (Linnaeus, 1758) – Common Sawfish

Warm-temperate and tropical waters in the Western and Eastern Atlantic. Tropical FAO Areas: 31, 34, 41, 47. Predominantly in riverine and estuarine areas, also insular shelf. Nomenclature currently under review and a name change is pending.

Pristis zijsron Bleeker, 1851 – Green Sawfish

Warm-temperate (limited) and tropical waters in the Indo–West Pacific. Tropical FAO Areas: 51, 57, 71. Insular shelf.

Suborder Rhinoidei: Sharkrays

Family Rhinidae: Sharkrays

Rhina ancylostoma Bloch & Schneider, 1801 – Shark Ray

Warm-temperate (limited) and tropical waters in the Indo–West Pacific. Tropical FAO Areas: 51, 57, 61, 71. Insular shelf. 0–70 m.

Suborder Rhynchobatoidei: Wedgefishes

Family Rhynchobatidae: Wedgefishes

Rhynchobatus australiae Whitley, 1939 – White-spotted Guitarfish

Warm-temperate (limited) and tropical waters in the Indo–West Pacific. Tropical FAO Areas: 51, 57, 61, 71. Insular shelf. 0–60 m.

Rhynchobatus djiddensis (Forsskål, 1775) – Whitespotted Wedgefish

Warm-temperate (limited) and tropical waters in the Western Indian: South Africa to Oman. Tropical FAO Areas: 51. Insular shelf. 0–30 m.

Rhynchobatus laevis (Bloch & Schneider, 1801) – Smoothnose Wedgefish

Warm-temperate (limited) and tropical waters in the Indo–West Pacific. Tropical FAO Areas: 51, 57, 61, 71. Insular shelf. 0–60 m.

Rhynchobatus luebberti Ehrenbaum, 1914 – African Wedgefish

Warm-temperate (limited) and tropical waters in the Eastern Atlantic: Mauritania to Angola. Tropical FAO Areas: 34, 47. Insular shelf. 0–70 m.

Rhynchobatus palpebratus Compagno & Last, 2008 – Eyebrow Wedgefish

Tropical waters in the Indo–West Pacific: northern Australia through to at least Thailand. Tropical FAO Areas: 57, 71. Insular shelf. 5–60 m.

** plus an additional two undescribed species.
Suborder Rhinobatoidei: Guitarfishes

Family Rhinobatidae: Guitarfishes

Aptychotrema rostrata (Shaw & Nodder, 1794) – Eastern Shovelnose Ray

Warm-temperate and tropical waters in the Southwest Pacific: Endemic to eastern Australia. Tropical FAO Areas: 71. Shelf and upper slope. 0–220 m.

Aptychotrema timorensis Last, 2004 – Spotted Shovelnose Ray

Tropical waters in the Indo–West Pacific: Known only from the Timor Sea, northern Australia. Tropical FAO Areas: 71. Outer shelf. 120 m.

Aptychotrema vincentiana (Haacke, 1885) – Western Shovelnose Ray

Warm-temperate and tropical waters in the Eastern Indian and Southern Ocean: Endemic to southern and western Australia. Tropical FAO Areas: 57. Shelf. 0–125 m.

Glaucostegus cemiculus (St. Hilaire, 1817) – Blackchin Guitarfish

Warm-temperate and tropical waters in the Eastern Atlantic (including the Mediterranean). Tropical FAO Areas: 34, 47. Shelf. 0–100 m.

Glaucostegus formosensis (Norman, 1926) – Taiwan Guitarfish

Tropical waters in the Northwest Pacific: Taiwan, and possibly the Philippines. Tropical FAO Areas: 61, 71. Shelf. 0–119 m.

Glaucostegus granulatus (Cuvier, 1829) – Sharpnose Guitarfish

Tropical waters in the Indo–West Pacific, but true range poorly defined. Tropical FAO Areas: 57, 71. Shelf. 0–119 m.

Glaucostegus halavi (Forsskål, 1775) – Halavi Guitarfish

Tropical waters in the Western Indian: Red Sea to Gulf of Oman. Tropical FAO Areas: 51. Insular shelf. 0–40 m.

Glaucostegus obtusus (Müller & Henle, 1841) – Widenose Guitarfish

Tropical waters in the Indo–West Pacific: Pakistan to Thailand. Tropical FAO Areas: 51, 57, 71. Shelf.

Glaucostegus petiti (Chabanaud, 1929) – Madagascar Guitarfish

Tropical waters in the Western Indian: Known only from Madagascar. Tropical FAO Areas: 51. Presumably shelf.

Glaucostegus thouin (Anonymous, 1798) – Clubnose Guitarfish

Warm-temperate (limited) and tropical waters in the Indo–West Pacific. Tropical FAO Areas: 51, 57, 71. Insular shelf. 0–60 m.

Glaucostegus typus (Bennett, 1830) – Giant Shovelnose Ray

Warm-temperate (limited) and tropical waters in the Indo–West Pacific. Tropical FAO Areas: 51, 57, 71. Shelf. 0–100 m.

Rhinobatos albomaculatus Norman, 1930 – Whitespotted Guitarfish

Tropical waters in the Eastern Atlantic: Gulf of Guinea to Angola. Tropical FAO Areas: 34, 47. Insular shelf. 0–35 m.

Rhinobatos annandalei Norman, 1926 – Bengal Guitarfish

Tropical waters in the Northern Indian: India and Sri Lanka. Tropical FAO Areas: 51, 57. Insular shelf.

*Rhinobatos (Acroteriobatus) annulatus Smith, *in* Müller & Henle, 1841 – Lesser Guitarfish

Warm-temperate and tropical (limited) waters off southern Africa: Angola to South Africa (Natal). Tropical FAO Areas: 47. Shelf. 0–100 m.

Rhinobatos glaucostigmus Jordan & Gilbert, 1884 – Slatyspotted Guitarfish

Warm-temperate (limited) and tropical waters in the Eastern Pacific: Gulf of California to Ecuador. Tropical FAO Areas: 77, 87. Insular shelf.

Rhinobatos holcorhynchus Norman, 1922 – Slender Guitarfish

Warm-temperate and tropical waters in the Western Indian: Kenya to South Africa. Tropical FAO Areas: 51. Shelf and upper slope. 75–253 m.

Rhinobatos horkelii Müller & Henle, 1841 – Brazilian Guitarfish

Warm-temperate and tropical waters in the Southwest Atlantic: Brazil and Argentina. Tropical FAO Areas: 41. Shelf. 0–150 m.

Rhinobatos hynnicephalus Richardson, 1846 – Ringstraked Guitarfish

Warm-temperate and tropical waters in the Northwest Pacific. Tropical FAO Areas: 61, 71. Shelf.

Rhinobatos irvinei Norman, 1931 – Spineback Guitarfish

Warm-temperate and tropical waters in the Eastern Atlantic: Morocco to Namibia. Tropical FAO Areas: 34, 47. Insular shelf.

Rhinobatos jimbaranensis Last, White & Fahmi, 2006 – Jimbaran Shovelnose Ray

Tropical waters in the Indo–West Pacific: Known only from Indonesia. Tropical FAO Areas: 57, 71(?). Presumably shelf.

Rhinobatos lentiginosus Garman, 1880 – Freckled Guitarfish

Warm-temperate and tropical waters in the Western Central Atlantic. Tropical FAO Areas: 31. Insular shelf. 0–18 m.

Rhinobatos leucorhynchus Günther, 1867 – Whitenose Guitarfish

Tropical waters in the Eastern Central Atlantic: Mexico to Ecuador. Tropical FAO Areas: 77, 87. Insular shelf. 0–50 m.

Rhinobatos (Acroteriobatus) leucospilus Norman, 1926 – Greyspot Guitarfish

Warm-temperate and tropical waters in the Western Indian: South Africa and Mozambique. Tropical FAO Areas: 51. Shelf. 0–100 m.

Rhinobatos lionotus Norman, 1926 – Smoothback Guitarfish

Tropical waters in the Northern Indian: Known only from India and Sri Lanka. Tropical FAO Areas: 51(?), 57. Insular shelf. 0–20 m.

Rhinobatos nudidorsalis Last, Compagno & Nakaya, 2004 – Bareback Shovelnose Ray

Tropical waters in the Central Indian: Known only from the Mascarine Ridge. Tropical FAO Areas: 51. Outer shelf. 87–125 m.

Rhinobatos (Acroteriobatus) ocellatus Norman, 1926 – Speckled Guitarfish

Warm-temperate and tropical waters in the Western Indian: South Africa and Mozambique. Tropical FAO Areas: 51. Shelf. 60–185 m.

Rhinobatos penggali Last, White & Fahmi, 2006 – Indonesian Shovelnose Ray

Tropical waters in the Indo–West Pacific: Known only from Indonesia. Tropical FAO Areas: 57, 71(?). Presumably shelf.

Rhinobatos percellens (Walbaum, 1792) – Southern Guitarfish

Warm-temperate and tropical waters in the Western Atlantic: Jamaica to northern Argentina. Tropical FAO Areas: 31, 41. Shelf. 0–110 m.

Rhinobatos planiceps Garman, 1880 – Flathead Guitarfish

Warm-temperate and tropical (limited) waters in the Eastern Pacific: Ecuador to Chile. Tropical FAO Areas: 87. Insular shelf.

Rhinobatos prahli Acero & Franke, 1995 – Gorgona Guitarfish

Warm-temperate (limited) and tropical waters in the Eastern Pacific: Costa Rica to Peru. Tropical FAO Areas: 77, 87. Insular shelf. 0–70 m.

Rhinobatos productus Girard, 1854 – Shovelnose Guitarfish

Warm-temperate and tropical (limited) waters in the Northeast Pacific: United States and Mexico. Tropical FAO Areas: 77. Shelf. 0–92 m.

Rhinobatos punctifer Compagno & Randall, 1987 – Spotted Guitarfish

Tropical waters in the Western Indian: Red Sea and Oman. Tropical FAO Areas: 51. Shelf.

Rhinobatos rhinobatos (Linnaeus, 1758) – Common Guitarfish

Temperate and tropical waters in the Eastern Atlantic (including the Mediterranean). Tropical FAO Areas: 34, 47. Shelf. 0–180 m.

Rhinobatos sainsburyi Last, 2004 – Goldeneye Shovelnose Ray

Tropical waters in the Indo–West Pacific: Northwestern Australia. Tropical FAO Areas: 57, 71. Shelf. 70–200 m.

Rhinobatos (Acroteriobatus) salalah Randall & Compagno, 1995 – Salalah Guitarfish

Tropical waters in the Western Indian: Known only from off Oman. Tropical FAO Areas: 51. Shelf.

Rhinobatos schlegelii Müller & Henle, 1841 – Brown Guitarfish

Warm-temperate and tropical waters in the Northwest Pacific. Tropical FAO Areas: 61, 71. Shelf.

Rhinobatos (Acroteriobatus) zanzibarensis Norman, 1926 – Zanzibar Guitarfish

Tropical waters in the Western Indian: Known only from Zanzibar. Tropical FAO Areas: 51. Shelf.

Zapteryx exasperata (Jordan & Gilbert, 1880) – Banded Guitarfish

Warm-temperate (limited) and tropical waters in the Eastern Pacific: California to Peru. Tropical FAO Areas: 77, 87. Shelf. 0–200 m.

Zapteryx xyster Jordan & Evermann, 1896 – Southern Banded Guitarfish

Tropical waters in the Eastern Pacific: Known only from Colombia and Panama. Tropical FAO Areas: 77, 87. Shelf.

Suborder Platyrhinoidei: Thornback Rays

Family Platyrhinidae: Thornback Rays

Platyrhina sinensis (Bloch & Schneider, 1801) – Fanray

Warm-temperate and tropical waters in the Northwest Pacific: Vietnam to Japan. Tropical FAO Areas: 61, 71. Insular shelf. 0–60 m.

Suborder Zanobatoidei: Panrays

Family Zanobatidae: Panrays

Zanobatus atlantica? (Chabanaud, 1928) – Atlantic Panray

Warm-temperate and tropical waters in the Eastern Atlantic. Tropical FAO Areas: 34. Shelf. 40–100 m. Questionably valid.

Zanobatus schoenleinii (Müller & Henle, 1841) – Striped Panray

Warm-temperate and tropical waters in the Eastern Atlantic. Tropical FAO Areas: 34, 47. Shelf. 40–100 m.

Suborder Torpedinoidei: Electric Rays

Family Narcinidae: Numbfishes

Diplobatis colombiensis Fechhelm & McEachran, 1984 – Colombian Painted Electric Ray

Tropical waters in the Western Central Atlantic: Known only from off northern Colombia in the Caribbean Sea. Tropical FAO Areas: 31. Shelf. 30–100 m.

Diplobatis guamachensis Martin Salazar, 1957 – Venezuelan Painted Electric Ray

Tropical waters in the Western Central Atlantic: Known only from off Colombia, Trinidad and Venezuela. Tropical FAO Areas: 31. Shelf. 30–183 m.

Diplobatis ommata (Jordan & Gilbert *in* Jordan & Bollman, 1889) – Target Ray

Warm temperate (limited) and tropical waters in the Eastern Pacific: California to Ecuador. Tropical FAO Areas: 77, 87. Insular shelf. 0–64 m.

Diplobatis pictus Palmer, 1950 – Painted Electric Ray

Tropical waters in the Western Central Atlantic: Venezuela to northeastern Brazil. Tropical FAO Areas: 31, 41. Shelf. 2–130 m.

Narcine atzi Carvalho & Randall, 2003 – Oman Numbfish

Tropical waters in the Western and Eastern Indian: Oman and Bay of Bengal. Tropical FAO Areas: 51. Insular shelf. 27 m.

Narcine bancroftii (Griffith, 1834) – Caribbean Numbfish

Warm-temperate (limited) and tropical waters in the Western Atlantic: North Carolina (United States) to northeastern Brazil. Tropical FAO Areas: 31, 41. Insular shelf. 0–35 m.

Narcine brevilabiata Bessednov, 1966 – Shortlip Numbfish

Tropical waters in the Northwest Pacific: Malaysia to China. Tropical FAO Areas: 61(?), 71. Insular shelf. 41–70 m.

Narcine brunnea Annandale, 1909 – Brown Numbfish

Tropical waters in the Indo–West Pacific. Tropical FAO Areas: 51, 57, 71. Presumably shelf, possibly also upper slope.

Narcine entemedor Jordan & Starks, 1895 – Cortez Numbfish

Warm-temperate (limited) and tropical waters in the Eastern Pacific: Gulf of California to Peru. Tropical FAO Areas: 77, 87. Insular shelf.

Narcine insolita Carvalho, Seret & Compagno, 2002 – Madagascar Numbfish

Tropical waters in the Western Indian: Known only from off Madagascar. Tropical FAO Areas: 51. Outer shelf. 150–175 m.

Narcine lasti Carvalho & Séret, 2002 – Western Numbfish

Eastern Indian and Western Central Pacific: Eastern Indonesia and western Australia. Tropical FAO Areas: 57, 71. Outer shelf and upper slope. 178–333 m.

Narcine leoparda Carvalho, 2001 – Colombian Numbfish

Tropical waters in the Southeast Pacific: Known only from off Colombia. Tropical FAO Areas: 87. Insular shelf. 1–33 m.

Narcine lingula Richardson, 1840 – Chinese Numbfish

Tropical waters in the Indo–West Pacific. Tropical FAO Areas: 51(?), 57, 61(?), 71. Shelf.

Narcine maculata (Shaw, 1804) – Darkfinned Numbfish

Tropical waters in the Northwest Pacific: Philippines to Taiwan. Tropical FAO Areas: 61, 71. Insular shelf. 30–80 m.

Narcine nelsoni Carvalho, 2008 – Eastern Numbfish

Tropical waters in the Western Central Pacific: Endemic to northeastern Australia. Tropical FAO Areas: 71. Outer shelf and upper slope. 140–540 m.

Narcine oculifera Carvalho, Compagno & Mee, 2002 – Bigeye Numbfish

Tropical waters in the Western Indian: Gulf of Aden and Gulf of Oman. Tropical FAO Areas: 51. Shelf. 20–152 m.

Narcine ornata Carvalho, 2008 – Ornate Numbfish

Tropical waters in the Indo–West Pacific: Endemic to northern Australia. Tropical FAO Areas: 57, 71. Shelf. 48–132 m.

Narcine prodorsalis Bessednov, 1966 – Tonkin Numbfish

Tropical waters in the Western Pacific: Indonesia to China, possibly wider ranging. Tropical FAO Areas: 57(?), 61(?), 71. Insular shelf. 50 m.

Narcine rierai (Lloris & Rucabado, 1991) – Mozambique Numbfish

Tropical waters in the Western Indian: Somalia to Mozambique. Tropical FAO Areas: 51. Outer shelf and upper slope. 169–214 m.

Narcine timlei (Bloch & Schneider, 1801) – Blackspotted Numbfish

Warm-temperate (limited) and tropical waters in the Indo–West Pacific: Pakistan to Japan. Tropical FAO Areas: 51, 57, 61, 71. Shelf.

Narcine vermiculatus Breder, 1926 – Vermiculated Numbfish

Warm-temperate (limited) and tropical waters in the Eastern Pacific: Gulf of California to Panama. Tropical FAO Areas: 77. Insular shelf.

Narcine westraliensis McKay, 1966 – Banded Numbfish

Tropical waters in the Eastern Indian: Endemic to northwestern Australia. Tropical FAO Areas: 57. Insular shelf. 10–70 m.

** plus an additional three undescribed species.

Family Narkidae: Sleeper Rays

Heteronarce bentuvai (Baranes & Randall, 1989) – Elat Electric Ray

Tropical waters in the Western Indian: Gulf of Aqaba. Tropical FAO Areas: 51. Outer shelf. 80–200 m.

Heteronarce mollis (Lloyd, 1907) – Soft Sleeper Ray

Tropical waters in the Western Indian: Gulf of Aden. Tropical FAO Areas: 51. Shelf and upper slope. 73–346 m.

Narke dipterygia (Bloch & Schneider, 1801) – Spottail Electric Ray

Warm-temperate (limited) and tropical waters in the Indo–West Pacific. Tropical FAO Areas: 51, 57, 61, 71. Shelf, and possibly upper slope.

Narke japonica (Temminck & Schlegel, 1850) – Japanese Spotted Torpedo

Warm-temperate and tropical waters in the Northwest Pacific. Tropical FAO Areas: 61, 71. Insular shelf. 12–23 m.

Temera hardwickii Gray, 1831 – Finless Electric Ray

Tropical waters in the Indo–West Pacific: Andaman Sea to Viet Nam. Tropical FAO Areas: 57, 71. Shelf, and possibly upper slope.

Family Hypnidae: Coffin Rays

Hypnos monopterygius (Shaw & Nodder, 1795) – Coffin Ray

Warm-temperate and tropical (limited) waters in the Indo–West Pacific: Endemic to Australia. Tropical FAO Areas: 57, 71. Shelf and upper slope. 0–220 m.

Family Torpedinidae: Torpedo Rays

Torpedo adenensis Carvalho, Stehmann & Manilo, 2002 – Aden Gulf Torpedo

Tropical waters in the Western Indian: Endemic to the Gulf of Aden. Tropical FAO Areas: 51. Shelf. 6–140 m.

Torpedo andersoni Bullis, 1962 – Florida Torpedo

Tropical waters in the Western Central Atlantic: Cayman Islands. Tropical FAO Areas: 31. Shelf and upper slope. 11–229 m.

Torpedo bauchotae Cadenat, Capape & Desoutter, 1978 – Rosette Torpedo

Tropical waters in the Eastern Atlantic. Tropical FAO Areas: 34, 47(?). Shelf and upper slope. 10–300 m.

Torpedo fuscomaculata Peters, 1855 – Blackspotted Torpedo

Warm-temperate and tropical (limited) waters in the Southwest Indian: South Africa and Mozambique. Tropical FAO Areas: 51. Shelf and upper slope. 0–439 m.

Torpedo mackayana Metzelaar, 1919 – Ringed Torpedo

Tropical waters in the Eastern Atlantic: Senegal to Angola. Tropical FAO Areas: 34, 47. Insular shelf. 30–50 m.

Torpedo macneilli (Whitley, 1932) – Short-tailed Torpedo

Temperate and tropical (limited) waters in the Western Indian and Eastern Pacific: Endemic to southern Australia. Tropical FAO Areas: 57, 71. Outer shelf and slope. 90–825 m.

Torpedo marmorata Risso, 1810 – Spotted Torpedo

Temperate and tropical waters in the Eastern Atlantic. Tropical FAO Areas: 34, 47. Shelf. 2–200 m.

Torpedo nobiliana Bonaparte, 1835 – Great Torpedo

Wide-ranging in the Eastern and Western Atlantic. Tropical FAO Areas: 31, 34, 41, 47. Shelf and upper slope, adults semipelagic to pelagic. 2–530 m.

Torpedo panthera Olfers, 1831 – Leopard Torpedo

Tropical waters in the Western Indian: Somalia to India. Tropical FAO Areas: 51. Shelf. 0–110 m.

Torpedo peruana Chirichigno, 1963 – Peruvian Torpedo

Warm-temperate and tropical waters in the Southeast Pacific: Peru to Colombia (including Galapagos Islands). Tropical FAO Areas: 87. Shelf. 24–168 m.

Torpedo puelcha Lahille, 1928 – Argentine Torpedo

Warm-temperate and tropical (limited) waters in the Southwest Atlantic: Brazil to Argentina. Tropical FAO Areas: 41. Shelf and upper slope. 0–600 m.

Torpedo sinuspersici Olfers, 1831 – Gulf Torpedo

Warm-temperate (limited) and tropical waters in the Western Indian: South Africa to India. Tropical FAO Areas: 51. Shelf. 0–200 m.

Torpedo suessii Steindachner, 1898 – Red Sea Torpedo

Tropical waters in the Western Indian: Known only from the Red Sea. Tropical FAO Areas: 51. Presumably shelf.

Torpedo torpedo (Linnaeus, 1758) – Ocellate Torpedo

Warm-temperate and tropical waters in the Eastern Atlantic: Bay of Biscay to Angola (most common in tropical waters). Tropical FAO Areas: 34, 47. Insular shelf. 2–70 m.

Torpedo tremens de Buen, 1959 – Chilean Torpedo

Warm-temperate and tropical waters in the Eastern Pacific: Chile to Costa Rica. Tropical FAO Areas: 77, 87. Shelf and upper slope. 20–700 m.

Torpedo zugmayeri Engelhardt, 1912 – Baluchistan Torpedo

Tropical waters in the Western Indian: Described from southwestern Pakistan. Tropical FAO Areas: 77, 87. Validity of this species needs to be ascertained.

Suborder Rajoidei: Skates

Family Arhynchobatidae: Softnose Skates

Irolita westraliensis Last & Gledhill, 2008 – Western Round Skate

Tropical waters in the Eastern Indian: Endemic to northwestern Australia. Tropical FAO Areas: 57. Outer shelf and upper slope. 140–210 m.

Sympterygia brevicaudata Cope, 1877 – Western Round Skate

Warm-temperate and tropical waters in the Eastern Pacific: Chile to Ecuador. Tropical FAO Areas: 87. Insular shelf. 18–38 m.

Family Rajidae: Hardnose Skates

Dipturus campbelli (Wallace, 1967) – Blackspot Skate

Warm-temperate and tropical waters in the Western Indian: South Africa and Mozambique. Tropical FAO Areas: 51. Outer shelf and upper slope. 137–403 m.

Dipturus endeavouri Last, 2008 – Endeavour Skate

Warm-temperate and tropical waters in the Southwest Pacific: Endemic to eastern Australia. Tropical FAO Areas: 71. Outer shelf and upper slope. 125–290 m.

Dipturus falloargus Last, 2008 – False Argus Skate

Tropical waters in the Eastern Indian: Endemic to northwestern Australia. Tropical FAO Areas: 57, 71. Outer shelf and upper slope. 120–255 m.

Dipturus kwangtungensis (Chu, 1960) – Kwangtung Skate

Warm-temperate and tropical waters in the Northwest Pacific. Tropical FAO Areas: 61, 71. Shelf and upper slope. 30–240 m.

Dipturus olseni (Bigelow & Schroeder, 1951) – Spreadfin Skate

Warm-temperate and tropical waters in the Western Central Atlantic: northern Gulf of Mexico. Tropical FAO Areas: 31. Shelf and upper slope. 55–384 m.

Dipturus polyommata (Ogilby, 1911) – Argus Skate

Tropical waters in the Western Central Pacific: Endemic to northeastern Australia. Tropical FAO Areas: 71. Outer shelf and upper slope. 135–320 m.

*Dipturus tengu (Jordan & Fowler, 1903) – Acutenose Skate

Warm-temperate and tropical (limited) waters in the Northwest Pacific: Japan to Philippines. Tropical FAO Areas: 61, 71. Shelf and upper slope. 45–300 m.

Fenestraja sinusmexicanus (Bigelow & Schroeder, 1950) – Gulf Pygmy Skate

Patchy in the Western Central Atlantic. Tropical FAO Areas: 31. Shelf and upper to mid slope. 56–1096 m.

Leucoraja garmani (Whitley, 1939) – Rosette Skate

Warm-temperate (limited) and tropical waters in the Northwest and Western Central Atlantic: Cape Cod to Florida (United States). Tropical FAO Areas: 31. Shelf and upper slope. 37–530 m.

Leucoraja lentiginosa (Bigelow & Schroeder, 1951) – Freckled Skate

Warm-temperate (limited) and tropical waters in the Western Central Atlantic: northern Gulf of Mexico. Tropical FAO Areas: 31. Shelf and upper slope. 53–588 m.

Leucoraja leucosticta Stehmann, 1971 – Whitedappled Skate

Tropical waters in the Eastern Central and Southeast Atlantic: West Africa. Tropical FAO Areas: 34, 47. Shelf and upper slope. 70–600 m.

*Okamejei arafurensis Last & Gledhill, 2008 – Arafura Skate

Tropical waters in the Western Central Pacific and Eastern Indian: Endemic to northern Australia. Tropical FAO Areas: 57, 71. Outermost shelf and upper slope. 179–298 m.

Okamejei boesemani (Ishihara, 1987) – Black Sand Skate

Warm-temperate (limited) and tropical waters in the Western Pacific. Tropical FAO Areas: 61, 71. Insular shelf. 20–90 m.

Okamejei hollandi (Jordan & Richardson, 1909) – Yellow-spotted Skate

Warm-temperate and tropical waters in the Northwest Pacific. Tropical FAO Areas: 61, 71(?). Insular shelf. 67–87 m.

Okamejei pita (Fricke & Al-Hussar, 1995) – Pita Skate

Tropical waters in the Western Indian: western Arabian Sea. Tropical FAO Areas: 51. Insular shelf. 0–15 m.

Okamejei powelli Alcock, 1898 – Indian Ringed Skate

Tropical waters in the northern Indian. Tropical FAO Areas: 51, 57. Outer shelf and upper slope. 122–244 m.

Raja clavata Linnaeus, 1758 – Thornback Skate

Temperate and tropical waters in the Eastern Atlantic. Tropical FAO Areas: 34, 47. Shelf and upper slope. 20–577 m.

Raja herwigi Krefft, 1965 – Cape Verde Skate

Tropical waters in the Eastern Central Atlantic: Endemic to around Cape Verde. Tropical FAO Areas: 34. Shelf. 55–102 m.

Raja miraletus Linnaeus, 1758 – Brown Skate

Warm-temperate and tropical waters in the Eastern Atlantic. Tropical FAO Areas: 34, 47. Shelf and upper slope. 17–462 m.

Raja straeleni Poll, 1951 – Biscuit Skate

Warm-temperate (limited) and tropical waters in the Eastern Atlantic: Western Sahara to South Africa. Tropical FAO Areas: 34, 47. Outer shelf and upper slope. 100–300 m.

*Rajella barnardi (Norman, 1935) – Bigthorn Skate

Warm-temperate (limited) and tropical waters in the Eastern Atlantic: Morocco to South Africa. Tropical FAO Areas: 34, 47. Outer shelf and slope. 170–1700 m.

*Rajella leopardus (von Bonde & Swart, 1923) – Leopard Skate

Patchily recorded in warm-temperate and tropical (limited) waters in the Eastern Atlantic: West Africa. Tropical FAO Areas: 34(?), 47. Outer shelf and slope. 170–1920 m.

Rostroraja alba (Lacepede, 1803) – White Skate

Wide ranging in temperate and tropical waters in the Eastern Atlantic. Tropical FAO Areas: 34, 47. Shelf and upper slope. 50–500 m.

* "undescribed genus A" *Raja rhina* Jordan & Gilbert, 1880 – Longnose Skate

Temperate and tropical (limited) waters in the Northeast Pacific: Bering Sea to Mexico. Tropical FAO Areas: 77. Shelf and upper slope. 25–675 m.

"undescribed genus B" *Raja ackleyi* Garman, 1881 – Ocellate Skate

Tropical waters in the Western Central Atlantic: Florida to Yucatan (Mexico). Tropical FAO Areas: 31. Shelf and upper slope. 32–384 m.

"undescribed genus B" *Raja cervigoni* Bigelow & Schroeder, 1964 – Venezuela Skate

Tropical waters in the Western Central Atlantic: northern South America. Tropical FAO Areas: 31. Shelf. 37–174 m.

"undescribed genus B" *Raja eglanteria* Bosc, 1800 – Clearnose Skate

Temperate and tropical waters in the Northwest Atlantic. Tropical FAO Areas: 31. Shelf. 0–111 m.

"undescribed genus B" *Raja equatorialis* Jordan & Bollman, 1890 – Equatorial Skate

Warm-temperate (limited) and tropical waters in the Eastern Pacific. Tropical FAO Areas: 77, 87. Insular shelf. 60 m.

"undescribed genus B" *Raja texana* Chandler, 1921 – Roundel Skate

Warm-temperate and tropical waters in the Western Central Atlantic. Tropical FAO Areas: 31. Insular shelf. 15–110 m.

"undescribed genus B" *Raja velezi* Chirichigno, 1973 – Rasptail Skate

Warm-temperate and tropical waters in the Eastern Central Pacific. Tropical FAO Areas: 77, 87. Shelf and upper slope. 35–300 m.

Suborder Myliobatoidei: Stingrays

Family Urolophidae: Stingarees

Trygonoptera testacea (Müller & Henle, 1841) – Common Stingaree

Warm-temperate and tropical (limited) waters in the Southwest Pacific: Endemic to eastern Australia. Tropical FAO Areas: 71. Insular shelf. 0–60 m.

Urolophus aurantiacus Müller & Henle, 1841 – Sepia Stingaree

Warm-temperate and tropical waters in the Northwest Pacific. Tropical FAO Areas: 61. Shelf. 10–200 m.

Urolophus flavomosaicus Last & Gomon, 1987 – Patchwork Stingaree

Tropical waters in the Eastern Indian and Southwest Pacific: Disjunct in northeastern and northwestern Australia. Tropical FAO Areas: 57, 71. Shelf and upper slope. 60–320 m.

Urolophus javanicus (Martens, 1864) – Java Stingaree

Tropical waters in the Indo–West Pacific: Known only from off Jakarta in Indonesia. Tropical FAO Areas: 57(?), 71(?). Presumably shelf, possibly upper slope.

Urolophus kapalensis Yearsley & Last, 2006 – Kapala Stingaree

Warm-temperate and tropical (limited) waters in the Southwest Pacific: Endemic to eastern Australia. Tropical FAO Areas: 71. Shelf. 10–130 m.

Urolophus piperatus Séret & Last, 2003 – Coral Sea Stingaree

Tropical waters in the Southwest Pacific: Endemic to northeastern Australia. Tropical FAO Areas: 71. Outer shelf and upper slope. 170–370 m.

Urolophus westraliensis Last & Gomon, 1987 – Brown Stingaree

Tropical waters in the Eastern Indian: Endemic to northwestern Australia. Tropical FAO Areas: 57. Shelf and upper slope. 60–220 m.

Family Urotrygonidae: American Round Stingrays

Urobatis concentricus Osburn & Nichols, 1916 – Spot-on-spot Round Ray

Warm-temperate and tropical waters in the Eastern Central Pacific: Endemic to Mexico. Tropical FAO Areas: 77. Estuarine and coastal inshore waters.

Urobatis halleri (Cooper, 1863) – Haller's Round Ray

Warm-temperate and tropical waters in the Eastern Central Pacific: California to Panama. Tropical FAO Areas: 77. Insular shelf. 0–91 m.

Urobatis jamaicensis (Cuvier, 1816) – Yellow Shovelnose Stingaree

Warm-temperate (limited) and tropical waters in the Western Central Atlantic. Tropical FAO Areas: 31. Estuarine and coastal inshore waters.

Urobatis maculatus Garman, 1913 – Spotted Round Ray

Tropical waters in the Eastern Central Pacific: Endemic to Mexico. Tropical FAO Areas: 77. Estuarine and coastal inshore waters.

Urotrygon aspidura (Jordan & Gilbert, 1882) – Spiny-tail Round Ray

Tropical waters in the Eastern Central Pacific: Costa Rica and Panama. Tropical FAO Areas: 77. Insular shelf.

Urotrygon chilensis (Günther, 1872) – Chilean Round Ray

Warm-temperate and tropical waters in the Eastern Pacific: Mexico to Chile. Tropical FAO Areas: 77, 87. Insular shelf. 0–60 m.

Urotrygon microphthalmum Delsman, 1941 – Smalleyed Round Stingray

Tropical waters in the Western Central Atlantic: Venezuela to Brazil. Tropical FAO Areas: 31. Insular shelf. 16–54 m.

Urotrygon munda Gill, 1863 – Munda Round Ray

Tropical waters in the Eastern Central Pacific: Endemic to Central America. Tropical FAO Areas: 77. Coastal inshore.

Urotrygon nana Miyake & McEachran, 1988 – Dwarf Round Ray

Tropical waters in the Eastern Central Pacific: Mexico to Costa Rica. Tropical FAO Areas: 77. Presumably insular shelf.

Urotrygon reticulata Miyake & McEachran, 1988 – Reticulate Round Ray

Tropical waters in the Eastern Central Pacific: Known only from Panama. Tropical FAO Areas: 77. Presumably insular shelf.

Urotrygon rogersi (Jordan & Starks *in* Jordan, 1895) – Roger's Round Ray

Tropical waters in the Eastern Central Pacific: Mexico to Ecuador. Tropical FAO Areas: 77, 87. Coastal inshore.

Urotrygon simulatrix Miyake & McEachran, 1988 – Fake Round Ray

Tropical waters in the Eastern Central Pacific: Known only from Gulf of Panama. Tropical FAO Areas: 77. Presumably insular shelf.

Urotrygon venezuelae Schultz, 1949 – Venezuela Round Stingray

Tropical waters in the Western Central Atlantic: Known only from Venezuela. Tropical FAO Areas: 31. Coastal inshore.

Family Dasyatidae: Stingrays

Dasyatis akajei (Müller & Henle, 1841) – Red Stingray

Warm-temperate and tropical waters in the Northwest Pacific. Tropical FAO Areas: 61, 71. Shelf.

Dasyatis americana Hildebrand & Schroeder, 1928 – Southern Stingray

Warm-temperate and tropical waters in the Western Atlantic. Tropical FAO Areas: 31. Insular shelf. 0–53 m.

Dasyatis bennetti (Müller & Henle, 1841) – Bennett's Cowtail Stingray

Warm-temperate and tropical waters in the Indo–West Pacific. Tropical FAO Areas: 57, 61, 71. Shelf.

Dasyatis centroura (Mitchill, 1815) – Roughtail Stingray

Warm-temperate and tropical waters in the Western and Eastern Atlantic. Tropical FAO Areas: 31, 34, 41, 47. Shelf and upper slope. 0–274 m.

Dasyatis chrysonota (Smith, 1828) – Blue Stingray

Warm-temperate and tropical waters in the Eastern Atlantic and Western Indian. Tropical FAO Areas: 47, 51. Shelf. 0–100 m.

Dasyatis colarensis Santos, Gomes & Charvet-Almeida, 2004 – Colares Stingray

Tropical waters in the Western Atlantic: Brazil and Venezuela. Tropical FAO Areas: 31. Insular shelf. 0–20 m.

Dasyatis dipterura Jordan & Gilbert, 1880 – Diamond Stingray

Warm-temperate and tropical waters in the Eastern Pacific. Tropical FAO Areas: 77, 87. Insular shelf. 0–50 m.

Dasyatis fluviorum Ogilby, 1908 – Estuary Stingray

Warm-temperate (limited) and tropical waters in the Southwest Pacific: Probably endemic to northeastern Australia. Tropical FAO Areas: 71. Insular shelf, including estuaries. 0–20 m.

Dasyatis geijskesi Boeseman, 1948 – Wingfin Stingray

Tropical waters in the Western Central Atlantic: Venezuela to Brazil. Tropical FAO Areas: 31. Insular shelf. 8–20 m.

Dasyatis guttata (Bloch & Schneider, 1801) – Longnose Stingray

Tropical waters in the Western Central Atlantic: Mexico to Brazil. Tropical FAO Areas: 31. Insular shelf. 0–36 m.

Dasyatis hypostigma Santos & Carvalho, 2004 – Groovebelly Stingray

Warm-temperate and tropical waters in the Southwest Atlantic: mainly Brazil. Tropical FAO Areas: 31(?), 41. Insular shelf. 5–80 m.

Dasyatis laevigata Chu, 1960 – Yantai Stingray

Warm-temperate and tropical waters in the Northwest Pacific: China and Taiwan. Tropical FAO Areas: 61. Shelf.

Dasyatis lata (Garman, 1880) – Brown Stingray

Tropical waters in the Central Pacific: Known from the Hawaiian Islands. Tropical FAO Areas: 77. Shelf and upper slope. 40–357 m.

Dasyatis longa (Garman, 1880) – Longtail Stingray

Tropical waters in the Eastern Pacific: Mexico to Colombia. Tropical FAO Areas: 77, 87. Insular shelf. 0–90 m.

Dasyatis margarita (Günther, 1870) – Daisy Stingray

Tropical waters in the Eastern Atlantic. Tropical FAO Areas: 34, 47. Insular shelf. 0–60 m.

Dasyatis margaritella Compagno & Roberts, 1984 – Pearl Stingray

Tropical waters in the Eastern Atlantic. Tropical FAO Areas: 34, 47. Shelf.

Dasyatis marianae Gomes, Rosa & Gadig, 2000 – Brazilian Large-eyed Stingray

Tropical waters in the Southwest Atlantic: Endemic to Brazil. Tropical FAO Areas: 31, 41. Insular shelf. 8–15 m.

Dasyatis marmorata Steindachner, 1892 – Marbled Stingray

Warm-temperate and tropical waters in the Eastern Atlantic. Tropical FAO Areas: 34, 47. Insular shelf. 12–65 m.

Dasyatis microps (Annandale, 1908) – Smalleye Stingray

Warm-temperate and tropical waters in the Eastern Atlantic. Tropical FAO Areas: 34, 47. Shelf. 0–200 m.

Dasyatis pastinaca (Linnaeus, 1758) – Common Stingray

Temperate and tropical waters in the Eastern Atlantic. Tropical FAO Areas: 34, 47. Shelf. 5–200 m.

Dasyatis parvonigra Last & White, 2008 – Dwarf Black Stingray

Tropical waters in the Eastern Indian: Known only from northwestern Australia and possibly southern Indonesia. Tropical FAO Areas: 57. Outer shelf. 125–185 m.

Dasyatis rudis (Günther, 1870) – Smalltooth Stingray

Tropical waters in the Eastern Central Atlantic. Tropical FAO Areas: 34. Insular shelf.

Dasyatis sabina (Lesueur, 1824) – Atlantic Stingray

Warm-temperate and tropical waters in the Western Central Atlantic: Chesapeake Bay to Gulf of Mexico. Tropical FAO Areas: 31. Insular shelf, also in freshwater and estuaries. 0–25 m.

Dasyatis say (Lesueur, 1817) – Bluntnose Stingray

Warm-temperate and tropical waters in the Western Central and Northwest Atlantic. Tropical FAO Areas: 31. Insular shelf, including estuaries. 0–20 m.

*Dasyatis thetidis Ogilby, 1899 – Black Stingray

Warm-temperate and tropical (limited) waters in the Indo–Pacific: southeastern Africa, southern Australia and New Zealand. Tropical FAO Areas: 51, 57, 71. Shelf and upper slope. 0–360 m.

Dasyatis ushiei Jordan & Hubbs, 1925 – Cow Stingray

Warm-temperate and tropical waters in the Indo–West Pacific: Indonesia to Japan. Tropical FAO Areas: 61, 71. Shelf and upper slope.

Dasyatis zugei (Müller & Henle, 1841) – Pale-edged Stingray

Warm-temperate and tropical waters in the Indo–West Pacific: Sri Lanka to Japan. Tropical FAO Areas: 57, 61, 71. Shelf. 0–100 m.

Himantura astra Last, Manjaji-Matsumoto & Pogonoski, 2008 – Black-spotted Whipray

Tropical waters in the Eastern Indian and Western Pacific: Endemic to northern Australia and New Guinea. Tropical FAO Areas: 57, 71. Shelf. 0–140 m.

Himantura fai Jordan & Seale, 1906 – Pink Whipray

Widespread in tropical waters in the Indo–West Pacific. Tropical FAO Areas: 51, 57, 71. Shelf. 0–200 m.

Himantura gerrardi (Gray, 1851) – Sharpnose Whipray

Widespread in tropical waters in the Indo–West Pacific. Possibly a species complex. Tropical FAO Areas: 51(?), 57, 61, 71. Insular shelf. 0–60 m.

Himantura granulata (Macleay, 1883) – Mangrove Whipray

Possibly widespread in tropical waters in the Indo–West Pacific. Tropical FAO Areas: 51, 57, 71. Insular shelf. 0–85 m.

Himantura hortlei Last, Manjaji-Matsumoto & Kailola, 2006 – Hortles Whipray

Tropical waters in the Indo–West Pacific: Endemic to southern New Guinea. Tropical FAO Areas: 71. Coastal inshore and estuarine. 0–10 m.

Himantura imbricata (Bloch & Schneider, 1801) – Scaly Whipray

Tropical waters in the Indo–West Pacific: Red Sea to Java, but poorly defined. Tropical FAO Areas: 51, 57, 71. Insular shelf.

Himantura jenkinsii (Annandale, 1909) – Jenkins Whipray

Possibly widespread in tropical waters in the Indo–West Pacific, but poorly defined. Tropical FAO Areas: 51, 57, 71. Insular shelf. 25–90 m.

Himantura leoparda Manjaji-Matsumoto & Last, 2008 – Leopard Whipray

Possibly widespread in tropical waters in the Indo–West Pacific, but poorly defined due to confusion with *H. undulata* and *H. uarnak*. Tropical FAO Areas: 51, 57, 61, 71. Insular shelf. 0–70 m.

Himantura lobistoma Manjaji-Matsumoto & Last, 2006 – Tubemouth Whipray

Tropical waters in the Indo–West Pacific: Known only from Borneo, and possibly Sumatra. Tropical FAO Areas: 71. Coastal inshore and estuarine.

Himantura marginata (Blyth, 1860) – Blackedge Whipray

Tropical waters in the Indo–West Pacific. Poorly known species. Tropical FAO Areas: 51, 57. Insular shelf.

Himantura pacifica? (Beebe & Tee-Van, 1941) – Pacific Chupare

Tropical waters in the Eastern Central Pacific: Costa Rica and Galapagos Islands. Tropical FAO Areas: 77, 87. Insular shelf. Questionably valid.

Himantura pastinacoides (Bleeker, 1852) – Round Whipray

Tropical waters in the Indo–West Pacific: Indo–Malay Archipelago. Tropical FAO Areas: 57(?), 71. Insular shelf, including estuaries.

Himantura schmardae (Werner, 1904) – Chupare Stingray

Tropical waters in the Western Central Atlantic. Tropical FAO Areas: 31. Insular shelf.

Himantura toshi Whitley, 1939 – Brown Whipray

Tropical waters in the Eastern Indian and Southwest Pacific: Endemic to northern Australia. Tropical FAO Areas: 57, 71. Insular shelf.

Himantura uarnacoides (Bleeker, 1852) – Whitenose Whipray

Tropical waters in the Indo–West Pacific: India to Java (Indonesia). Tropical FAO Areas: 51(?), 57, 71. Insular shelf. 0–30 m.

Himantura uarnak (Forsskål, 1775) – Reticulate Whipray

Widespread in tropical waters in the Indo–West Pacific. Often confused with *H. leoparda* and *H. undulata*. Tropical FAO Areas: 51, 57, 61, 71. Insular shelf. 0–45 m.

Himantura undulata (Bleeker, 1852) – Leopard Whipray

Possibly widespread in tropical waters in the Indo–West Pacific, but poorly defined due to confusion with *H. leoparda* and *H. uarnak*. Tropical FAO Areas: 51, 57, 61, 71. Insular shelf.

Himantura walga (Müller & Henle, 1841) – Dwarf Whipray

Tropical waters in the Indo–West Pacific: Thailand, Indonesia and Malaysia, possibly to India. Tropical FAO Areas: 57, 71. Insular shelf.

Neotrygon annotata (Last, 1987) – Plain Maskray

Tropical waters in the Eastern Indian and Southwest Pacific: endemic to northern Australia. Tropical FAO Areas: 57, 71. Insular shelf. 12–62 m.

Neotrygon kuhlii (Müller & Henle, 1841) – Blue-spotted Maskray

Widespread in tropical waters in the Indo–West Pacific, but likely to be a species complex. Tropical FAO Areas: 51, 57, 61, 71. Insular shelf. 0–90 m.

Neotrygon leylandi (Last, 1987) – Painted Maskray

Tropical waters in the Eastern Indian: Endemic to northwestern Australia. Tropical FAO Areas: 57. Insular shelf. 15–90 m.

Neotrygon picta Last & White, 2008 – Speckled Maskray

Tropical waters in the Southwest Pacific: Endemic to northern and northeastern Australia. Tropical FAO Areas: 71. Shelf. 5–100 m.

Pastinachus atrus (Macleay, 1883) – Cowtail Stingray

Found in tropical and warm temperate (limited) waters in the Indo–West Pacific. Not well defined and taxonomic issues within this genus need resolving. Tropical FAO Areas: 57, 61, 71. Insular shelf. 0–60 m.

Pastinachus sephen (Forsskål, 1775) – Cowtail Stingray

Probably restricted to the Western Indian Ocean tropical waters. Taxonomic issues within this genus need resolving. Tropical FAO Areas: 51. Insular shelf.

Pastinachus solocirostris Last, Manjaji-Matsumoto & Yearsley, 2005 – Roughnose Stingray

Tropical waters in the Indo–West Pacific: Known only from Indonesia and Malaysian Borneo. Tropical FAO Areas: 57(?), 71. Presumably insular shelf.

Taeniura grabata Geoffroy Saint-Hilaire, 1817 – Round Stingray

Warm-temperate and tropical waters in the Eastern Atlantic. Tropical FAO Areas: 34, 47. Shelf and upper slope. 10–300 m.

Taeniura lymma (Forsskål, 1775) – Blue-spotted Fantail Ray

Widespread in tropical waters in the Indo–West Pacific. Tropical FAO Areas: 51, 57, 71. Insular shelf. 0–20 m.

Taeniurops meyeni (Müller & Henle, 1841) – Blotched Fantail Ray

Widespread in warm-temperate (limited) and tropical waters in the Indo–West Pacific. Tropical FAO Areas: 51, 57, 61, 71. Shelf. 5–100 m.

Urogymnus asperrimus (Bloch & Schneider, 1801) – Porcupine Ray

Widespread in warm-temperate (limited) and tropical waters in the Indo–West Pacific and Eastern Atlantic. Tropical FAO Areas: 34, 47(?), 51, 57, 71. Insular shelf. 0–20 m.

** plus an additional three undescribed species.

Family Gymnuridae: Butterfly Rays

Gymnura afuerae (Hildebrand, 1946) – Peruvian Butterfly Ray

Warm-temperate and tropical waters in the Southeast Pacific: Ecuador and Peru. Tropical FAO Areas: 87. Presumably shelf.

Gymnura altavela (Linnaeus, 1758) – Spiny Butterfly Ray

Warm-temperate and tropical waters in the Western and Eastern Atlantic. Tropical FAO Areas: 31, 34, 41, 47. Insular shelf. 0–55 m.

Gymnura australis (Ramsay & Ogilby, 1886) – Australian Butterfly Ray

Warm-temperate (limited) and tropical waters in the Eastern Indian and Western Pacific: Endemic to northern Australia and New Guinea. Tropical FAO Areas: 57, 71. Shelf and upper slope. 0–250 m.

*Gymnura bimaculata (Norman, 1925) – Twinspot Butterfly Ray

Warm-temperate and tropical (?) waters in the Northwest Pacific. Tropical FAO Areas: 61, 71(?). Presumably shelf.

Gymnura crebripunctata (Peters, 1869) – Longsnout Butterfly Ray

Tropical waters in the Eastern Central Pacific: Gulf of California to Panama. Tropical FAO Areas: 77. Shelf, including estuaries.

Gymnura japonica (Schlegal, 1850) – Japanese Butterfly Ray

Warm-temperate and tropical waters in the Northwest Pacific. Tropical FAO Areas: 61, 71. Shelf.

Gymnura marmorata (Cooper, 1863) – California Butterfly Ray

Warm-temperate (limited) and tropical waters in the Eastern Central Pacific: California to Peru. Tropical FAO Areas: 77, 87. Insular shelf. 0–22 m.

Gymnura micrura (Bloch & Schneider, 1801) – Smooth Butterfly Ray

Warm-temperate and tropical waters in the Western Atlantic: Chesapeake Bay to Brazil. Tropical FAO Areas: 31, 41. Insular shelf. 0–40 m.

Gymnura poecilura (Shaw, 1804) – Longtail Butterfly Ray

Warm-temperate (limited) and tropical waters in the Indo–West Pacific: Red Sea to Japan. Tropical FAO Areas: 51, 57, 61, 71. Insular shelf. 0–30 m.

Gymnura tentaculata (Valenciennes *in* Müller & Henle, 1841) – Tentacled Butterfly Ray

Tropical waters in the Western Central Pacific: Known only from Papua New Guinea. Tropical FAO Areas: 71. Presumably insular shelf.

Gymnura zonura (Bleeker, 1852) – Zonetail Butterfly Ray

Tropical waters in the Indo–West Pacific: India to Indonesia. Tropical FAO Areas: 51(?), 57, 71. Insular shelf. 0–37 m.

Family Myliobatidae: Eagle Rays

Aetobatus flagellum (Bloch & Schneider, 1801) – Longheaded Eagle Ray

Tropical waters in the Indo–West Pacific, but poorly defined due to taxonomic issues: Arabian Sea to Indonesia. Tropical FAO Areas: 51, 57, 71. Insular shelf.

Aetobatus narinari (Euphrasen, 1790) – White-spotted Eagle Ray

Widespread in all warm-temperate (limited) and tropical seas. Possibly a species complex. Tropical FAO Areas: 31, 34, 41, 47, 51, 57, 61, 71, 77, 87. Inshore to well offshore, semipelagic.

Aetobatus ocellatus? (Kuhl *in* van Hasselt, 1823) – Ocellate Eagle Ray

Tropical waters in the Indo–West Pacific. Tropical FAO Areas: 57, 71. Presumably insular shelf. Questionably valid.

Aetomylaeus maculatus (Gray, 1832) – Mottled Eagle Ray

Tropical waters in the Indo–West Pacific: India to Taiwan. Tropical FAO Areas: 51(?), 57, 61, 71. Presumably insular shelf.

Aetomylaeus nichofii (Schneider, 1801) – Banded Eagle Ray

Widespread in warm-temperate (limited) and tropical waters in the Indo–West Pacific: Persian Gulf to Japan and northern Australia. Tropical FAO Areas: 51, 57, 61, 71. Shelf. 0–115 m.

Aetomylaeus vespertilio (Bleeker, 1852) – Ornate Eagle Ray

Widespread, but patchily reported, in tropical waters in the Indo–West Pacific: Mozambique to Taiwan and northern Australia. Tropical FAO Areas: 51, 57, 61, 71. Shelf. 0–110 m.

Myliobatis aquila (Linnaeus, 1758) – Common Eagle Ray

Warm-temperate and tropical waters in the Eastern Atlantic. Tropical FAO Areas: 34, 47. Shelf. 1–95 m.

Myliobatis freminvillii Lesueur, 1824 – Bullnose Ray

Warm-temperate and tropical waters in the Western Atlantic. Tropical FAO Areas: 31, 41. Insular shelf and estuaries. 0–10 m.

Myliobatis goodei Garman, 1885 – Southern Eagle Ray

Warm-temperate and tropical waters in the Western Atlantic. Tropical FAO Areas: 31, 41. Insular shelf.

Myliobatis hamlyni Ogilby, 1911 – Purple Eagle Ray

Warm-temperate (limited) and tropical waters in the Eastern Indian and Southwest Pacific: Known only from northern Australia but poorly known and possibly wider ranging. Tropical FAO Areas: 57, 71. Outer shelf and upper slope. 120–350 m.

Myliobatis longirostris Applegate & Fitch, 1964 – Longnose Eagle Ray

Warm-temperate and tropical waters in the Eastern Central Pacific: Mexico to Peru. Tropical FAO Areas: 77, 87. Insular shelf. 0–64 m.

Myliobatis tobijei Bleeker, 1854 – Kite Ray

Warm-temperate and tropical waters in the Northwest and Western Central Pacific: Japan to Indonesia. Tropical FAO Areas: 57(?), 61, 71. Shelf and upper slope. 0–220 m.

Pteromylaeus asperrimus (Jordan & Evermann, 1898) – Roughskin Bullray

Tropical waters in the Eastern Central Pacific: Panama and Galapagos Islands. Tropical FAO Areas: 77, 87. Insular shelf.

Pteromylaeus bovinus (Geoffroy St. Hilaire, 1817) – Bullray

Warm-temperate and tropical waters in the Eastern Atlantic and Western Indian. Tropical FAO Areas: 34, 47, 51. Insular shelf, including estuaries. 0–65 m.

Family Rhinopteridae: Cownose Rays

Rhinoptera adspersa? Valenciennes *in* Müller & Henle, 1841 – Rough Cownose Ray

Tropical waters in the Indo–West Pacific, but poorly defined. Tropical FAO Areas: 57, 71. Presumably insular shelf. Questionably valid.

Rhinoptera bonasus (Mitchill, 1815) – Cownose Ray

Warm-temperate and tropical waters in the Western Atlantic. Tropical FAO Areas: 31, 41. Insular shelf. 0–22 m.

Rhinoptera brasiliensis Müller & Henle, 1841 – Brazilian Cownose Ray

Tropical waters in the Western Atlantic: Gulf of Mexico to Brazil. Tropical FAO Areas: 31, 41. Insular shelf. 0–20 m.

Rhinoptera javanica (Müller & Henle, 1841) – Javanese Cownose Ray

Widespread, but not well defined, in warm-temperate (limited) and tropical waters in the Indo–West Pacific. Tropical FAO Areas: 51, 57, 61, 71. Insular shelf. 0–20 m.

Rhinoptera jayakari? Boulenger, 1895 – Oman Cownose Ray

Tropical waters in the Western Indian: Oman. Tropical FAO Areas: 51. Presumably insular shelf. Questionably valid.

Rhinoptera neglecta Ogilby, 1912 – Australian Cownose Ray

Tropical waters in the Eastern Indian and Western Pacific: northern Australia and Indonesia, probably more widespread. Tropical FAO Areas: 57(?), 71. Insular shelf.

Rhinoptera peli? Bleeker, 1863 – African Cownose Ray

Tropical waters in the Eastern Central Atlantic: Described off Guinea. Tropical FAO Areas: 34. Questionably valid.

Rhinoptera sewelli? Misra, 1947 – Indian Cownose Ray

Tropical waters in the Western Indian: Arabian Sea coast of India. Tropical FAO Areas: 51. Questionably valid.

Rhinoptera steindachneri Evermann & Jenkins, 1891 – Hawkray Cownose Ray

Tropical waters in the Eastern Central Pacific. Tropical FAO Areas: 77, 87. Insular Shelf. 0–65 m.

Family Mobulidae: Devilrays

Manta birostris (Donndorff, 1798) – Manta Ray

Circumglobal in all warm-temperate (limited) and tropical waters. Recent studies have shown it to be a species complex of at least two species. Tropical FAO Areas: 31, 34, 41, 47, 51, 57, 61, 71, 77, 87. Pelagic, mainly over continental shelf.

Mobula eregoodootenkee (Cuvier, 1829) – Pygmy Devilray

Tropical waters in the northern Indian and Western Central Pacific: Red Sea to Australia. Tropical FAO Areas: 51, 57, 71. Pelagic in inshore waters. 0–50 m.

Mobula hypostoma (Bancroft, 1831) – Atlantic Devilray

Warm-temperate (limited) and tropical waters in the Western Atlantic. Tropical FAO Areas: 31, 41. Pelagic in coastal and oceanic waters.

Mobula japanica (Muller & Henle, 1841) – Japanese Devilray

Circumglobal in all warm-temperate (limited) and tropical waters. Tropical FAO Areas: 31, 34, 41, 47(?), 51, 57, 61, 71, 77, 87. Pelagic in inshore, offshore and oceanic waters.

Mobula kuhlii (Müller & Henle, 1841) – Shortfin Devilray

Tropical waters in the northern Indian and Western Central Pacific: South Africa to Indonesia. Tropical FAO Areas: 51, 57, 71. Pelagic in inshore waters.

Mobula munkiana Notarbartolo-di-Sciara, 1987 – Pygmy Devilray

Warm-temperate and tropical waters in the Eastern Pacific: Mexico to Peru. Tropical FAO Areas: 77, 87. Pelagic in inshore waters.

Mobula rochebrunei (Vaillant, 1879) – Lesser Guinean Devilray

Tropical waters in the Eastern Atlantic: Mauritania to Angola. Tropical FAO Areas: 34, 47. Pelagic in inshore waters.

Mobula tarapacana (Philippi, 1892) – Chilean Devilray

Possibly circumglobal in warm-temperate (limited) and tropical waters. Tropical FAO Areas: 31, 34, 41(?), 47, 51, 57, 61, 71, 77, 87. Pelagic in inshore, offshore, and oceanic waters.

Mobula thurstoni (Lloyd, 1908) – Bentfin Devilray

Circumglobal in all warm-temperate (limited) and tropical waters. Tropical FAO Areas: 31, 34, 41, 47(?), 51, 57, 61, 71, 77, 87. Pelagic in inshore waters. 0–100 m.

References

Allen GR, Erdmann MV (2008) Two new species of bamboo sharks (Orectolobiformes: Hemiscylliidae). Aqua. Int J Ichthyol 13: 93–108.

Aguirre H, Madrid VJ, Virgen JA (2002) Presence of *Echinorhinus cookei* off central Pacific Mexico. J Fish Biol 61: 1403–1409.

Bonfil R, Abdallah M (2004) Field identification guide to the sharks and rays of the Red Sea and Gulf of Aden. FAO Species Identification Guide for Fishery Purposes. Food and Agricultural Organization of the United Nations. Rome.

Carpenter KE (ed) (2002) The living marine resources of the Western Central Atlantic. Volume 1. Introduction, molluscs, crustaceans, hagfishes, sharks, batoid fishes and chimaeras. FAO Species Identification Guide for Fishery Purposes. FAO, Rome.

Carpenter KE, Niem VH (eds) (1998) The living marine resources of the Western Central Pacific. Volume 2. Cephalopods, crustaceans, holothurians and sharks. FAO Species Identification Guide for Fishery Purposes. FAO, Rome.

Carpenter KE, Niem VH (eds) (1999) The living marine resources of the Western Central Pacific. Volume 3. Batoid fishes, chimaeras and bony fishes, part 1 (Elopidae to Linophrynidae). FAO Species Identification Guide for Fishery Purposes. FAO, Rome.

Castro Aguirre JL, Mendiola AA, Acosta AFG et al. (2005) *Mustelus albipinnis* sp. nov. (Chondrichthyes: Carcharhiniformes: Triakidae) de la cost suroccidental de Baja California sur, Mexico. Hidrobiologica 15: 123–130.

Castro Aguirre JL, Pérez HE, Campos LH (2006) Two new species of the genus *Squatina* (Chondrichthyes: Squatinidae) from the Gulf of Mexico. Rev Biol Trop 54(3): 1031–1040.

Compagno LJV (2001) Sharks of the world. An annotated and illustrated catalogue of shark species known to date. Volume 2. Bullhead, mackerel and carpet sharks (Heterodontiformes, Lamniformes and Orectolobiformes). FAO Species Catalogue for Fishery Purposes. No. 1, Vol. 2. FAO, Rome.

Compagno LJV (2005) Global checklist of living chondrichthyan fishes. In: Fowler SL, Cavanagh RD, Camhi M, Burgess GH, Cailliet GM, Fordham SV, Simpfendorfer CA, Musick CA (eds) Sharks, rays and chimaeras: the status of chondrichthyan fishes. IUCN/SSC Shark Specialist Group. IUCN, Gland, Switzerland and Cambridge, UK, pp. 401–423.

Compagno L, Dando M, Fowler S (2005) A field guide to sharks of the world. Collins, London.

de Carvalho MR (1999) A systematic revision of the electric ray genus Narcine Henle, 1834 (Chondrichthyes: Torpediniformes: Narcinidae), and the higher-level relationships of the elasmobranch fishes (Chondrichthyes). PhD thesis, City University of New York.

de Carvalho MR, Compagno LJV, Mee JKL (2002) *Narcine oculifera*, a new species of electric ray from the gulfs of Oman and Aden (Chondrichthyes: Torpediniformes: Narcinidae). Copeia 2002: 137–145.

de Carvalho MR, Randall JE (2003) Numbfishes from the Arabian Sea and surrounding gulfs, with the description of a new species from Oman (Chondrichthyes: Torpediniformes: Narcinidae). Ichthyol Res 50: 59–66.

de Carvalho MR, Séret B, Compagno LJV (2002) A new species of electric ray of the genus *Narcine* Henle, 1834 from the southwestern Indian Ocean (Chondrichthyes: Torpediniformes: Narcinidae). S Afr J Mar Sci 24: 135–149.

Ebert DA (2003) Sharks, rays and chimaeras of California. University of California Press, Berkeley.

Eschmeyer WN (2008) The catalogue of fishes online. California Academy of Sciences: San Francisco. Available from: http://www.calacademy.org/research/ichthyology/catalog/fishcatmain.asp (accessed December 2008).

Froese R, Pauly D (eds) (2008) FishBase. Electronic publication. www.fishbase.org, version (11/2008).

IUCN (2008) The IUCN Red List of Threatened Species. www.iucnredlist.org (accessed December 2008).

Jacobsen IP, Bennett MB (2007) Description of a new species of catshark, *Atelomycterus marnkalha* n. sp. (Carcharhiniformes: Scyliorhinidae) from north-east Australia. Zootaxa 1520: 19–36.

Jimenez JCP, Nishizaki OS, Castillo-Geniz JL (2005) A new eastern North Pacific smoothhound shark (genus *Mustelus*, family Triakidae) from the Gulf of California. Copeia 2005(4): 834–845.

Kyne PM, Simpfendorfer CA (2007) A collation and summarization of available data on deepwater chondrichthyans: biodiversity, life history and fisheries. IUCN Shark Specialist Group report.

Last PR, Stevens JD (1994) Sharks and rays of Australia. CSIRO Publishing, Melbourne.

Last PR, Stevens JD (2009) Sharks and rays of Australia, 2nd edition. CSIRO Publishing, Australia.

Manjaji BM (2004) Taxonomy and phylogenetic systematics of the Indo–Pacific Whip-Tailed Stingray genus *Himantura* Müller & Henle 1837 (Chondrichthyes: Myliobatiformes: Dasyatidae). PhD Thesis, University of Tasmania, Volumes 1 and 2.

Michael SW (1993) Reef sharks and rays of the world: a guide to their identification, behavior, and ecology. Sea Challengers, Monterey, CA.

Takahashi M, Nakaya K (2004) *Hemitriakis complicofasciata*, a new whitefin topeshark (Carcharhiniformes: Triakidae) from Japan. Ichthyol Res 51: 248–255.

White WT (2007) Aspects of the biology of carcharhiniform sharks in Indonesian waters. J Mar Biol Assoc UK 87: 1269–1275.

White WT, Last PR (2006) Description of two new species of smooth-hounds, *Mustelus widodoi* and *M. ravidus* (Carcharhiniformes: Triakidae) from the Western Central Pacific. Cybium 30: 235–246.

White WT, Last PR, Dharmadi (2005) Description of a new species of catshark, *Atelomycterus baliensis* (Carcharhiniformes: Scyliorhinidae) from eastern Indonesia. Cybium 29: 33–40.

White WT, Last PR, Stevens JD (2007) *Halaelurus maculosus* n. sp. and *H. sellus* n. sp., two new species of catshark (Carcharhiniformes: Scyliorhinidae) from the Indo–West Pacific. Zootaxa 1639: 1–21.

White WT, Last PR, Stevens JD et al. (2006) Economically important sharks and rays of Indonesia. Australian Centre for International Agricultural Research, Canberra.

Yano K, Ahmad A, Gambang AC et al. (2005) Sharks and rays of Malaysia and Brunei Darussalam. SEAFDEC, MFRDMD.

5

Biology of the South American Potamotrygonid Stingrays

Ricardo S. Rosa, Patricia Charvet-Almeida, and Carla Christie Diban Quijada

CONTENTS

5.1 Introduction

The family Potamotrygonidae was described by Garman (1877) to include freshwater stingrays restricted to the South American continent. This is the most speciose extant group of elasmobranchs living in fresh waters, with approximately 20 valid species (Charvet-Almeida et al. 2002; de Carvalho, Lovejoy, and Rosa 2003; Rosa, de Carvalho, and Wanderley 2008). Potamotrygonid stingrays occur in most of the major river systems of South America draining to the Atlantic Ocean and the Caribbean Sea, with the exception of the São Francisco River and all coastal drainages south of the Parnaíba River in Brazil (Rosa 1985a).

In spite of the general strictly freshwater habit (a few species may enter estuarine waters), recent authors have proposed, based on cladistic analyses, that two marine species previously assigned to the genus *Himantura* (*H. schmardae* and *H. pacifica*) should be included in the Potamotrygonidae (McEachran and Aschliman 2004), while other authors prefer to treat these species as *incertae sedis* (de Carvalho, Maisey, and Grande 2004). According to these latter authors, there is no strong support for the sister-group relationship between the amphi-American *Himantura* and the potamotrygonids, as originally proposed by Lovejoy (1996). This chapter follows them by treating the freshwater potamotrygonids exclusively.

The potamotrygonids can be distinguished from marine and freshwater stingrays of other families by the well-developed anteromedian projection on their pelvic girdle, called the prepelvic process (Garman 1877, 1913; Bigelow and Schroeder 1953; Rosa, Castello, and Thorson 1987). Other synapomorphies include their loss of urea retention and the reduction of the rectal gland (Rosa 1985a; Rosa, Castello, and Thorson 1987; de Carvalho, Maisey, and Grande 2004).

After the initial period of taxonomic and morphological descriptions of potamotrygonid stingrays, which started in the early nineteenth century and culminated with Garman's (1913) monograph, this group has received little attention in the literature until recent decades, when finally it became the object of biological and medical research, as well as of the aquarium fish trade. Several of their aspects have been investigated so far, such as parasitology (e.g., Mayes, Brooks, and Thorson 1978), reproduction and development (e.g., Thorson, Langhammer, and Oetinger 1983; Lasso, Rial, and Lasso-Alcalá 1996), physiology (e.g., Thorson, Cowan, and Watson 1967), and phylogeny (e.g., Lovejoy 1996; de Carvalho, Maisey, and Grande 2004).

Biologists have often encountered difficulties in identifying potamotrygonid stingrays, and the same happened with researchers working on medical aspects of the group, not to mention fish traders and breeders, usually less concerned with taxonomy. This situation resulted from the lack of thorough taxonomic investigation that could provide sound specific and even generic diagnoses. The two partial reviews in the literature (Garman 1913; Castex 1964a) failed to include all known species and type specimens. The taxonomic revision of the family by Rosa (1985a) remained unpublished for the most part and is outdated.

The objective of this chapter is to present a historical perspective of the study of the Potamotrygonidae, including a review of their taxonomic diversity and phylogenetic relationships, as well as a synthesis of the knowledge on various biological aspects of the group, including feeding, reproduction, and conservation.

5.2 South American Paleogeography and the Occurrence of Potamotrygonid Stingrays

The paleogeographic development of the South American hydrographic basins has been a primary determinant both of the evolutionary history of potamotrygonid stingrays and the major ecological features of their current habitats. Indeed, successive drowning of several South American hydrographic basins by continental seaways from the Cretaceous to the Miocene (Harrington 1962; Räsänen et al. 1995) was crucial to the establishment of the putative marine ancestor of potamotrygonids on the South American continent, as well as to the evolutionary and biogeographic differentiation of the group in fresh water (Brooks, Thorson, and Mayes 1981; Rosa 1985a; Lovejoy 1997; Lovejoy, Bermingham, and Andrew 1998; de Carvalho, Maisey, and Grande 2004).

The paleogeography and development of the major South American hydrographic systems were reviewed by Lundberg et al. (1998). One major paleogeographic feature of the Amazon region is Pebas Sea or Pebas Lake, a lacustrine environment that covered most of the western Amazon region, extending north to Colombia and Venezuela. Pebas Lake was subjected to marine ingressions, which possibly lasted until the mid or late Miocene (Hoorn 1996; Lundberg et al. 1998; Räsänen et al. 1998; Wesselingh 2006) and likely is the pathway for the establishment, and subsequent evolution, of marine biota into the Amazonian freshwater biotopes (Wesselingh 2006). During the Plio-Pleistocene the Pebas system became entirely fluvial, which might explain subandine biogeographic patterns, especially those of the aquatic biota (Räsänen et al. 1995).

Whether the Pebas Sea was joined with the Paraná-Plata basin during the Miocene is uncertain, and the evidence for such connection is lacking both on paleontological and geological grounds (Lundberg et al. 1998; Roddaz et al. 2006; Wesselingh 2006). Nonetheless, the presence of fossils assigned to the Potamotrygonidae in Acre and in Rio Paraná (Larrazet 1886 as *Dynatobatis* and *Raja*; Deynat and Brito 1994; Räsänen et al. 1995; Brito and Deynat 2004) suggests a previous connection and an upper Miocene minimal age for the group. Major transgressions drowned the Paraná-Plata basin and southern Argentina from 5 to 4.2 Ma reestablishing the Paranean Sea (Hubert and Henno 2006), and led to local extinctions of freshwater biota and subsequent invasion by Amazonian species.

The timing of establishment of the putative potamotrygonid ancestor in South America is still debated. The marine ancestor could have entered South America as early as the Cretaceous (Lovejoy 1997), but DNA sequences and a molecular clock place the origin of the potamotrygonids in the Early Miocene (Lovejoy, Bermingham, and Andrew 1998). Fossil evidence and dating of the separation between the Amazon and the Paraná-Plata basins, however, suggest that the origin of the potamotrygonids must have occurred before the Miocene (Brito and Deynat 2004). De Carvalho, Maisey, and Grande (2004) also argued for an earlier establishment of potamotrygonids and dismissed the role of the marine ingressions into the Pebas Sea for the establishment of the marine ancestor, due to its recency. Instead, they claimed that the Late Cretaceous transgressions were more consistent with the minimum age derived from their phylogeny. Because there are no clearcut biogeographic scenarios, hypotheses about the origin of the potamotrygonids should only be proposed with a sound phylogeny and precise dating of fossil evidence (De Carvalho, Maisey, and Grande 2004).

Historical variation in water levels and water flow also are important for current biogeographic patterns within the potamotrygonids. Indeed, operative connections between the Amazon and the Paraná-Plata basin may have existed within the last 10 Ma through headwater capture (Lundberg et al. 1998; Hubert and Henno 2006), such as those occurring at moderately elevated terrains (ca. 1000 m) of the Guaporé and Paraguay upper reaches (Innocêncio 1977). The Orinoco-Negro connection, through the Rio Casiquiare (Lundberg et al. 1998), for example, may explain distribution patterns of shared species within the Potamotrygonidae (at least five taxa). Other river basin connections that were operative between the Amazon and the northern South American drainages are mentioned in Sioli (1967) and Soares (1977).

5.3 History of the Investigation of Potamotrygonid Stingrays

The first mention of potamotrygonid stingrays in the literature possibly is that of the Spanish explorer Álvar Nuñez, known as "Cabeza de Vaca," who walked from southern Brazil to Paraguay in the early sixteenth century, and referred to the presence of stingrays in the rivers he crossed on his way (Cabeza de Vaca 1999).

The earliest known descriptive account and illustration of a South American freshwater stingray appeared in the early seventeenth century, under the native name *yabeburapeni*, in a manuscript by the Jesuit Cristóvão de Lisboa (Lisboa 1967). Other manuscript accounts on several species of stingrays from Rio Paraguay were written by the Jesuit Sanchez Labrador in the mid-seventeenth century. Gumilia (1791) mentioned the presence of stingrays in Rio Orinoco. The manuscripts of Alexandre Rodrigues Ferreira, from the late eighteenth century, recorded a freshwater stingray from Pará, Brazil (Ferreira 1972). Other early vernacular accounts of freshwater stingrays, especially from Rio Paraná and Rio Paraguay drainages, were treated in detail by Castex (1963a, 1963b, 1963c).

The first published species description of a freshwater stingray, Pastenague de Humboldt (Roulin 1829), still did not follow strictly binominal nomenclature. Alfred Russel Wallace illustrated four species of freshwater stingrays during his exploration of Rio Negro and Rio Uaupés during 1850 to 1852, but these species were not formally named or described, and the illustrations remained unpublished until recently (Wallace 2002). The binominal taxonomic history of potamotrygonids started in 1834, with the publication of the plate of *Trygon histrix* by d'Orbigny. Müller and Henle (1841) provided descriptions for *T. histrix* and for two new species from Brazil, *Trygon motoro* and *Trygon aiereba*, which were not illustrated. Schomburgk (1843) described and illustrated the new species *Trygon garrapa*, *Trygon strogylopterus*, and the new genus and species *Elipesurus spinicauda*, all from Rio Branco, Brazil. Castelnau (1855) described and illustrated four new species from Brazil: *Trygon dumerilii* from Rio Araguaia, *T. mulleri* from Rio Crixas and Rio Araguaia, *T. henlei* from Rio Tocantins, and *T. d'orbignyi*, also from Rio Tocantins. Duméril (1865) established the new subgenus *Paratrygon* for *T. aiereba* Müller & Henle, and described *Taeniura magdalenae*, from Rio Magdalena in Colombia. Garman (1877) established two new genera, *Potamotrygon* and *Disceus*, to include all South American freshwater stingrays, and presented a key to the species of *Potamotrygon*. Günther (1880) described two new species, *Trygon brachyurus* from Buenos Aires and *T. reticulatus* from Surinam. Vaillant (1880) described the new species *Taeniura constellata* from Rio Amazonas, Brazil.

Garman (1913) reviewed the previous descriptions of potamotrygonids, but overlooked *T. constellata* Vaillant and *T. histrix ocellata*, a subspecies described by Engelhardt (1912) from the mouth of Rio Amazonas. Garman (1913) also described five new species of *Potamotrygon* (*P. circularis*, *P. humerosus*, *P. laticeps*, *P. scobina*, and *P. signatus*) based on specimens collected by the Thayer Expedition in Brazil.

Devicenzi and Teague (1942) described *Potamotrygon brumi* from Rio Uruguay and Arambourg (1947) erroneously assigned a new fossil species from Africa, *Potamotrygon africana*, to the Potamotrygonidae.

The taxonomic work on potamotrygonids was undertaken mostly by South American authors from the late 1950s to the early 1970s, but was restricted to descriptions of new species of *Potamotrygon* (Fernández-Yépez 1957; Castex 1963c, 1963d, 1964b; Achenbach 1967; Castello and Yagolkowski 1969; Castex and Castello 1970a, 1970b), and another partial review (Castex 1964a). Stauch and Blanc (1962) described a new species, *Potamotrygon garouaensis*, from the Bénoué River in Nigeria, which was later assigned to the genus *Dasyatis* by Castello (1973) and by Thorson and Watson (1975).

Rosa (1985a) presented an unpublished dissertation on the taxonomy of the group. Three genera were recognized in the family, one of which subsequently was described as new (Rosa, Castello, and Thorson 1987). Other papers regarding nomenclatural aspects of potamotrygonids (Rosa 1985b, 1991) originated from that dissertation, but the taxonomic revision itself was never published. Nonetheless, several subsequent authors followed most of Rosa's unpublished taxonomic decisions, with respect to valid genera and species, synonymies and type designations (Séret and McEachran 1986; Compagno and Cook 1995; Compagno 1999; de Carvalho, Lovejoy, and Rosa 2003).

Interest in potamotrygonid stingrays beyond their taxonomy began with investigations of their venomous properties (Vellard 1931), and continued with more general studies of their biology in Argentina in the 1960s, by Mariano Castex and collaborators (e.g., Castex 1963e; Achenbach 1969, 1972). They were followed shortly by Thomas Thorson's pioneering work on the physiology of osmoregulation, demonstrating that potamotrygonids possessed low urea content when compared with marine elasmobranchs, and were unable to raise urea levels when exposed to salt water (Thorson, Cowan, and Watson 1967; Thorson 1970; Gerst and Thorson 1977). Another prominent investigation that followed and still is an ongoing research program is that of parasitology of potamotrygonids (e.g., Brooks and Thorson 1976; Mayes, Brooks, and Thorson 1978; Brooks, Mayes, and Thorson 1981; Domingues and Marques 2007). This research has led to more comprehensive studies on the comparative biology of potamotrygonids, such as host-parasite coevolution and historical biogeography (e.g., Brooks, Thorson, and Mayes 1981; Marques 2000).

5.4 Systematics of Potamotrygonid Stingrays

Potamotrygonids were viewed as a taxonomically problematic group, mainly due to the poor original descriptions and illustrations of most species (Rosa 1985a; Zorzi 1995). Rosa (1985a) recognized three genera (*Potamotrygon*, *Paratrygon*, and *Plesiotrygon*) and 20 species. Nonetheless, nomenclatural problems and taxonomic decisions are still pending, particularly at the specific level. Moreover, there are several undescribed taxa under study (Ross and Schäfer 2000; Charvet-Almeida et al. 2002; de Carvalho, Lovejoy, and Rosa 2003). Two new species of *Potamotrygon* recently were described by Deynat (2006) and Rosa, de Carvalho, and Wanderley (2008). No additional taxonomic clarification is attempted in this chapter, but an overview of all nominal taxa in the form of an annotated list may provide elements for a future revision.

5.4.1 Taxonomic Account

5.4.1.1 *Potamotrygonidae* Garman, 1877

Trygonidae [in part]. Günther, 1870: 471; Garman, 1877: 208; Steindachner, 1878: 72.

Potamotrygones. Garman, 1877: 208.

Paratrygoninae. Gill, 1893: 130.

Dasyatidae [in part]. Jordan, 1887: 557; Ribeiro, 1907: 39; Bertin, 1939a: 21.

Potamotrygonidae. Garman, 1913: 415.

Elipesuridae. Jordan, 1923: 104.

Paratrygonidae. Fowler, 1948: 4; Fowler, 1970: 42.

Type genus: *Potamotrygon*. Garman, 1877.

Remarks: Garman (1877) established the name Potamotrygones to denote a suprageneric taxon for the South American freshwater stingrays, giving its original date and authorship to the family name Potamotrygonidae (Article 11.7.1.2. of the ICZN). Independent familial status for the South American freshwater stingrays was recognized in most of the recent fish classifications (Eschmeyer 1998; Compagno 1999; Nelson 2006).

5.4.1.2 *The Genera of Potamotrygonidae*

Four nominal genera (*Elipesurus*, *Paratrygon*, *Potamotrygon*, and *Disceus*) have been used exclusively for South American freshwater stingrays, and four other generic names (*Pastinaca*, *Trygon*, *Taeniura*, and *Himantura*) both for freshwater and marine stingrays (Rosa 1985a). Garman's genera *Potamotrygon* and *Disceus* have been widely accepted, while the validity of *Elipesurus* and *Paratrygon* has been debated.

Steindachner (1878) suggested that *Taeniura* should have priority over *Potamotrygon*, but the former genus had been established for marine stingrays. Jordan (1887), Eigenmann and Eigenmann (1891), Eigenmann (1910), and Rosa (1985a, 1991) correctly treated *Paratrygon* as the senior synonym of *Disceus* [see Rosa (1991) for a discussion]. Ribeiro (1907) used *Elipesurus* (amended to *Ellipesurus*) as a senior synonym of *Potamotrygon*.

Garman (1913) used the genera *Disceus* and *Potamotrygon* in his revision, and cited *Elipesurus* as doubtful. Bertin (1939a) and Fowler (1948, 1970) mistakenly synonymized *Paratrygon* and *Potamotrygon*. Fowler (1948, 1970) also maintained both *Elipesurus* and *Disceus* as valid genera in his classification.

Castex (1968, 1969) considered *Potamotrygon* and *Disceus* as the valid genera, and recommended the rejection of *Elipesurus* as a doubtful name. Bailey (1969) considered that *Elipesurus* should be used as the senior synonym of *Disceus*. The genus *Elipesurus* and its type species *E. spinicauda* were based on a single specimen, with a very short tail, lacking the caudal sting usually found in other species of the family. Several authors (Garman 1877, 1913; Vaillant 1880; Eigenmann and Eigenmann 1891; Castex 1964a, 1968, 1969; Castex and Castello 1969; Bailey 1969) regarded these morphological conditions as the probable result of a mutilation of the tail. The original specimen of *E. spinicauda* apparently was not preserved, and no similar specimens have been collected since. The poor original description and illustration of *E. spinicauda* do not contain diagnostic characters; therefore, *Elipesurus* and its type species are preferably treated as doubtful names (Rosa 1985b).

One additional potamotrygonid genus and species, *Plesiotrygon iwamae*, was included in Rosa (1985a) and formally described in Rosa, Castello, and Thorson (1987). Ishihara and Taniuchi (1995) presented evidence and descriptive data of a specimen from Venezuela, possibly belonging to a distinct potamotrygonid genus, which they did not formally name. de Carvalho (1996) commented on that specimen and indicated that it probably could be identified as *Paratrygon*. Similar specimens have been collected in the Amazon drainage and are currently under study. Although these specimens share several synapomorphies with *Paratrygon*, they also have unique features [Figures 5.A.6 to 5.A.8 (see Appendix 5.1)].

5.4.1.3 Key to the Genera of Potamotrygonidae

1A. Distance from mouth to anterior margin of disc relatively long, 2.2 to 3.3 times in disc width; pelvic fins dorsally covered by the disc 2

1B. Distance from mouth to anterior margin of disc relatively short, 3.6 to 5.6 times in disc width; pelvic fins exposed behind posterior margin of the disc.......................................3

2A. A knob-shaped process on external margin of spiracles; anterior margin of disc concave, lacking anteromedian prominence; caudal sting present ....*Paratrygon* Duméril

2B. External spiracular process absent; anterior margin of disc convex, with a small anteromedian prominence; caudal sting absent or minute........................ undescribed genus

3A. Tail relatively short, less than two times disc width; tail with dorsal and ventral finfolds; eyes relatively large and pedunculate, eye diameter usually less than 4.0 in interorbital distance...*Potamotrygon* Garman

3B. Tail relatively long and filiform, more than two times disc width; only a ventral finfold on tail; eyes minute, non-pedunculate, eye diameter at least 4.4 in interorbital distance ....*Plesiotrygon* Rosa, Castello, and Thorson

5.4.1.3.1 Paratrygon Duméril

(Figures 5.A.1 to 5.A.3)

> *Trygon* [in part]. Müller and Henle, 1841: 196.
>
> *Paratrygon* Duméril, 1865: 594.
>
> *Disceus* Garman, 1877: 208.
>
> Type species: *Trygon aiereba* Müller and Henle, 1841 by monotypy.

Remarks: Duméril (1865) established *Paratrygon* as a subgenus of *Trygon*, for the single species *Trygon aiereba* Mülller and Henle. Günther (1870) used *Paratrygon* at the generic level. Rosa (1985a, 1991) treated *Paratrygon* as the senior synonym of *Disceus*. *Paratrygon* is currently monotypic although there is evidence of additional undescribed species (M.R. de Carvalho, personal communication).

Geographic distribution: Northern Bolivia, eastern Peru and Ecuador, northern Brazil (Amazonas and Pará) in Rio Amazonas and major tributaries including the lower Rio Tocantins, and Venezuela, in the Rio Orinoco drainage (Figure 5.1).

5.4.1.3.2 Plesiotrygon Rosa, Castello and Thorson

(Figures 5.A.4 to 5.A.5)

> *Plesiotrygon* Rosa, Castello, and Thorson, 1987.
>
> Type species: *Plesiotrygon iwamae* Rosa, Castello, and Thorson by monotypy.

Remarks: *Plesiotrygon* is currently monotypic although there is evidence of one additional undescribed species (M.R. de Carvalho, personal communication; Ross and Schäfer 2000; Toffoli 2006).

FIGURE 5.1
Geographic distribution of the potamotrygonid genera based on collection and literature records, modified from Rosa (1985a). Squares = *Paratrygon;* triangles = *Plesiotrygon;* circles = *Potamotrygon;* asterisk = undescribed genus.

Geographic distribution: Endemic to the Amazon drainage, from Rio Napo in Ecuador to the mouth of Rio Amazonas, in Pará, Brazil, and including the lower Rio Tocantins (Figure 5.1).

5.4.1.3.3 *Potamotrygon* Garman
(Figures 5.A.9 to 5.A.30)

> *Trygon* [in part]. Müller and Henle, 1841: 167, 197; Castelnau, 1855: 100-103; Duméril, 1865: 582; Günther, 1870: 472-483.
>
> *Taeniura* [in part]. Müller and Henle, 1841: 197; Duméril, 1865: 619-625.
>
> *Potamotrygon* Garman, 1877: 210.
>
> Type species: *Potamotrygon histrix* (Müller and Henle, in d'Orbigny, 1834) by subsequent selection of Jordan (1919).

Remarks: A polytypic genus with 16 to 19 species recognized in recent accounts (Rosa 1985a; Compagno 1999; de Carvalho, Lovejoy, and Rosa 2003; Rosa, de Carvalho, and Wanderley 2008) and at least four undescribed species (de Carvalho, Lovejoy, and Rosa 2003). These numbers are presently viewed as conservative, and the genus may be more diverse.

Geographic distribution: Species of *Potamotrygon* occur in all South American countries with the exception of Chile, and are found in the drainages of Rio Atrato and Magdalena (Colombia), Rio Orinoco and Maracaibo (Venezuela), coastal rivers of the Guianas, Surinam, and Amapá State in Brazil, Rio Amazonas and tributaries (Bolivia, Brazil, Colombia,

Ecuador, Peru, Venezuela), including Rio Tocantins, coastal rivers of Maranhão and Piauí States in Brazil, including Rio Parnaíba, and the Paraná-Paraguay basin (Argentina, Brazil, Paraguay, and Uruguay), including Rio Uruguay (Figure 5.1).

The species of *Potamotrygon*—Thirty-one nominal species can be assigned to the genus *Potamotrygon*, 29 of which are available names and two of which are unavailable names (*Potamotrygon labratoris* Castex, 1963 *nomen nudum* and *Potamotrygon pauckei* Castex, 1963 *nomen nudum*) first published as figure legends, without a formal description and type designation. Of these, 23 possibly are valid species. The following annotated list of available names in alphabetical order may provide elements for further taxonomic revision of the genus. Institutional abbreviations follow Eschmeyer (1998).

Potamotrygon alba Castex 1963. Described from three specimens collected in Asunción, Paraguay (Castex 1963c). The original description is very poor, lacking an illustration, morphometric data, and type designation. Castex (1964a) placed *P. alba* in the synonymy of *Potamotrygon motoro*, stating that the former was an albino form of the latter. No similar specimens have been collected since and the present location of the original specimens is unknown; therefore, *P. alba* preferably should be treated as a doubtful species.

Potamotrygon boesemani Rosa, de Carvalho, and Wanderley 2008. Described from and possibly endemic to the Corantijn river drainage in Surinam.

Potamotrygon brachyura (Günther 1880). Originally described as *Trygon brachyurus* from Buenos Aires, Argentina. No holotype designation was given in the original description. A male specimen (BMNH 1879.2.12.4) is cited as type on the label and museum catalog. It is possible that this specimen was part of the original material examined by Günther, although it was not cited in the original publication. This is a valid species, endemic to the Paraná-Paraguay basin, occurring in Argentina, Brazil, Paraguay, and Uruguay.

Potamotrygon brumi Devicenzi and Teague 1942. Described from Isla Queguay Grande, Rio Uruguay, this species was correctly placed in the synonymy of *P. brachyura* by Castex (1966). The type of *P. brumi* was reported lost from the MNM in Montevideo (Luengo 1966; Olazarri et al. 1970).

Potamotrygon castexi Castello and Yagolkowski 1969. Described from Rosario, Argentina. Holotype: MACN 5777, by original designation, presently lost. Paratypes also possibly lost. One additional specimen from Bolivia (UMMZ 204551) was identified from a photograph and included in the original description without type status. Rosa (1985a) suggested that *P. castexi* might represent a color morph of *Potamotrygon falkneri*, but the taxonomic status of the Amazonian form of *P. castexi*, reported from Bolivia, Brazil, and Peru, remains unclear.

Potamotrygon circularis Garman 1913. Described from Teffé, Amazonas, Brazil. Syntypes: MCZ 291-S, MCZ 295-S (one male and one female) and MCZ 296-S, possibly representing more than one species. Rosa (1985a) synonymized *Potamotrygon circularis* and *Potamotrygon constellata* (Vaillant 1880) and was followed by de Carvalho, Lovejoy, and Rosa (2003). This synonymy was based on the presence of tubercles dorsally on the disc, but later it was found that such tubercles occur in other species, including *Potamotrygon humerosa*, *Potamotrygon motoro*, and *Potamotrygon orbignyi*. Further comparisons are necessary to confirm the validity of this species.

Potamotrygon constellata (Vaillant 1880). Originally described as *Taeniura constellata* from Calderão, Amazonas, Brazil. Lectotype: MNHN A.1010, female (Bertin 1939b; Rosa 1985a; Séret and McEachran 1986). This species resembles *P. circularis* in the possession of tubercles on the disc and might be the junior synonym of the latter.

Potamotrygon dumerilii (Castelnau 1855). Originally described as *Trygon* (*Taenura*) *dumerilii* from Rio Araguaia, Brazil. Holotype: MNHN 2367, female, by original indication (Bertin 1939b; Rosa 1985a; Séret and McEachran 1986). This species was placed in the synonymy of

Potamotrygon motoro by several authors (Günther 1870; Ribeiro 1907; Garman 1913; Bertin 1939a; Fowler 1948, 1970), but it is clearly distinct from the latter. Rosa (1985a) cited the species for the Rio Paraná drainage, but de Carvalho, Lovejoy, and Rosa (2003) treated the species as doubtful or as a possible synonym of *Potamotrygon orbignyi*. Further studies are necessary to confirm the validity of this species.

Potamotrygon falkneri Castex and Maciel in Castex 1963c. Described from Rio Paraná, Argentina. Holotype: MFA 236, female, by original indication. This is a valid species, known from the Paraná-Paraguay basin in Argentina, Brazil, and Paraguay.

Potamotrygon garrapa (Schomburgk 1843). Originally described as *Trygon garrapa* from Rio Branco, Brazil. Type: ZMB 4624, female. It has been placed in the synonymy of *Potamotrygon motoro* by several authors (Eigenmann and Eigenmann 1891; Garman 1877, 1913; Berg 1895; Eigenmann 1910; Fowler 1948; Castex 1964a), but its original illustration differs from the latter species in showing the yellow ocelli continuing distally on the tail. In this character it resembles *Potamotrygon henlei* and *P. leopoldi*. The brief description of the teeth also resembles *P. henlei* but the latter is presently unknown from Rio Branco. Further collection in Rio Branco and comparative study of the type specimen might eventually prove the validity of this species.

Potamotrygon henlei (Castelnau 1855). Originally described as *Trygon (Taenura) henlei* from Rio Tocantins, Brazil. Holotype: MNHN 2353, by original indication (Bertin 1939b; Rosa 1985a; Séret and McEachran 1986). It was placed in the synonymy of *Potamotrygon motoro* by Günther (1870), Garman (1877, 1913), Berg (1895), Fowler (1948), and Castex (1964a), but is clearly distinct from the latter species in having ocelli on the tail, multiple rows of middorsal tail spines, and relatively larger teeth. This is a valid species, endemic to the Rio Tocantins drainage. Another species with similar color pattern possibly occurs in Rio Araguaia.

Potamotrygon histrix (Müller and Henle 1834). This specific name was established as *Trygon histrix* by indication only, as the legend of a stingray illustration from Buenos Aires, issued by d'Orbigny in 1834 (Sheborn and Griffin 1934; Rosa 1985a). Dating of the plate was interpreted as 1839 by Séret and McEachran (1986). Holotype: MNHN 2449, female, by original indication (Rosa 1985a; Séret and McEachran 1986). The specific name should retain its original spelling as *T. histrix* instead of *T. hystrix* used later by Müller and Henle (1841) and subsequent authors. Müller and Henle (1841) provided a written description for *T. histrix* (with the new spelling *T. hystrix*), and included five additional specimens from Surinam, Venezuela, and Rio de Janeiro (questionable locality data). This heterogeneous series of specimens later has been interpreted as the type material (syntypes) of *T. histrix*, and Surinam and Venezuela mistakenly added to the range of the species. Only the specimen from Buenos Aires should retain type status (Rosa, de Carvalho, and Wanderley 2008). This is a valid species endemic to the Paraná-Paraguay basin, occurring in Argentina, Brazil, and Paraguay. A morphological and color pattern redescription is pending.

Potamotrygon humboldtii (Roulin in Duméril 1865). Originally published as *Pastinachus humboldtii* based on the vernacular account by Roulin (1829) from Rio Meta in Colombia, under the vernacular name Pastenague de Humboldt. The brief repetition and reference to Roulin's description given by Duméril provided an indication to the name and authorship. No type is known. This species was treated as doubtful by Rosa (1985a) and de Carvalho, Lovejoy, and Rosa (2003). Further studies are necessary to confirm its validity.

Potamotrygon humerosa Garman 1913. Described from Monte Alegre, Rio Amazonas, Pará, Brazil. Data on the holotype (MCZ 299-S) were given by Rosa (1985a). Castex (1967a) mistakenly reported that the type of *P. humerosa* was identical to *Potamotrygon motoro*. This species was treated as doubtful by de Carvalho, Lovejoy, and Rosa (2003) but valid by Rosa (1985a) and Rosa and de Carvalho (2007).

Potamotrygon laticeps (Garman 1913). Described from Tefé and Óbidos, Brazil. Syntypes: MCZ 290-S (two specimens from Tefé), by original indication. Castex (1967a) and Rosa (1985a) placed this species in the synonymy of *Potamotrygon motoro*. Pending a taxonomic revision, this specific name is available for the Amazonian form of *P. motoro*.

Potamotrygon leopoldi Castex and Castello 1970. Described from the upper Rio Xingu, below the confluence of Rio Suiá-Missú, Mato Grosso, Brazil (Castex and Castello 1970b). Holotype: IRSNB 23936, male, by original designation. This is a valid species endemic to the Rio Xingu drainage.

Potamotrygon magdalenae (Valenciennes in Duméril 1865). Originally described as *Taeniura magdalenae* from Rio Magdalena, Colombia. Holotype: MNHN 2368, by original indication (Bertin 1939b; Rosa 1985a; Séret and McEachran 1986). A valid species possibly endemic to the Magdalena river drainage, but also cited for the Atrato and Catatumbo rivers (Galvis, Mojica, and Camargo 1997). The presently isolated forms in the Atrato and Catatumbo rivers may represent vicariant species.

Potamotrygon marinae Deynat 2006. Described from the Maroni River in French Guiana. Holotype: MNHN 1998-0119, male. Further study is necessary to evaluate the reported absence of the prepelvic process in this species.

Potamotrygon menchacai Achenbach 1967. Described from Rio Colastiné, Santa Fé, Argentina. Holotype: MFA 289, male, by original designation. Rosa (1985a) placed this species in the synonymy of *P. falkneri*. De Carvalho, Lovejoy, and Rosa (2003) followed this synonymy but indicated that the species might be valid. Examination of the holotype conformed to *P. falkneri* in all examined characters (dentition, tail spines, morphometrics, and meristics) but coloration, which is interpreted as an intraspecific variation of the latter species.

Potamotrygon motoro (Müller and Henle 1841). Originally described as *Raja motoro* from Cuiabá, Mato Grosso, Brazil. Lectotype: ZMB 4662, male, selected by Rosa (1985a), based on indications of type status on the label and catalog. *P. motoro* is currently considered to be a wide-ranging species, but might represent a species complex requiring subdivision (de Carvalho, Lovejoy, and Rosa 2003). Further studies are necessary to confirm the identity of morphs inhabiting the Amazon, Orinoco, lower Paraná, and Uruguay river drainages, as well as the isolated form found in Rio Mearim, Maranhão, Brazil.

Potamotrygon mulleri (Castelnau 1855). Originally described as *Trygon (Taenura) mulleri* [sic] from Rio Crixas and Rio Araguaia, Brazil. Holotype: MNHN 2354, female, by original indication (Bertin 1939b; Rosa 1985a; Séret and McEachran 1986). Castelnau did not use the umlaut in the specific name and although this was an incorrect latinization, the original spelling should be retained (Art. 32.5.1.1. of the ICZN). This species was placed in the synonymy of *P. motoro* by several authors (Günther 1870; Castex 1964a; Rosa 1985a; Séret and McEachran 1986; de Carvalho, Lovejoy, and Rosa 2003) but the latter species is unknown from the type locality of *P. mulleri* (Rio Araguaia). The name is available for the species similar to *P. henlei* that occurs in Rio Araguaia.

Potamotrygon ocellata (Engelhardt 1912). Originally described as *Trygon hystrix ocellata* from the southern coast of Ilha Mexiana, at the mouth of Rio Amazonas, Brazil. The type specimen is considered lost in World War II (Neumann 2006). Rosa (1985a) provisionally treated this species as valid, based on the examination of a specimen (MNRJ 10620) from Amapá, Brazil, with a color pattern similar to the original description. Additional specimens are necessary to confirm its taxonomic status.

Potamotrygon orbignyi (Castelnau 1855). Originally described as *Trygon (Taenura) d'orbignyi* [sic] from Rio Tocantins, Brazil. Holotype: MNHN 2333, female, by original indication (Bertin 1939b; Rosa 1985a; Séret and McEachran 1986). Castelnau (1855) used two different spellings of the specific name (*d'orbignyi* and *orbignyi*) in the original description. The spelling *orbignyi* has been correctly adopted by subsequent authors (Duméril 1865; Günther 1870; Ribeiro 1907).

Potamotrygon reticulata (Günther 1880). Originally described as *Trygon reticulatus* from Surinam. Holotype: BMNH 1870.3.10.1, male, by original indication. Rosa (1985a) placed *T. reticulatus* in the synonymy of *P. orbignyi* based on color pattern, and was followed by de Carvalho, Lovejoy, and Rosa (2003) and Rosa and de Carvalho (2007). As pointed out above, *P. orbignyi* recognized by Rosa (1985a) possibly represents a species complex with reticulate color pattern, and further studies are necessary to confirm the validity of *P. reticulata.*

Potamotrygon schroederi Fernández-Yépez 1957. Described from Boca Apurito, Rio Apure, Venezuela. Holotype: collection of Augustín Fernandez-Yépez 51289, not examined. This is a valid species, endemic to the Orinoco and Negro river drainages in Venezuela and Amazonas, Brazil, recorded to the mouth of Rio Negro in Manaus. Rosa (1985a) mistakenly identified specimens of an undescribed species from the Rio Negro as the young of *P. schroederi.*

Potamotrygon schuehmacheri Castex 1964. Described from Rio Colastiné Sur, Santa Fe, Argentina (Castex 1964b). Holotype: MFA 269, male, by original designation. Two different spellings (*schuhmacheri* and *schühmacheri*) were used in the original publication. Taniuchi (1982) changed the specific name to *schuemacheri*, which was treated as an incorrect subsequent spelling by de Carvalho, Lovejoy, and Rosa (2003) (Art. 33.5 of the ICZN). *Potamotrygon schuehmacheri* is known only from the holotype. One additional specimen (USNM 181767) from Paraguay resembles the holotype in coloration. Further study is necessary to confirm the validity of this species or its synonymy with *P. histrix.*

Potamotrygon scobina Garman 1913. Described from Cametá, Rio Tocantins, Pará, Brazil. Holotype: MCZ 602-S, male, by original indication. This is a valid species, found in Rio Tocantins and in the Rio Amazonas and its major tributaries.

Potamotrygon signata Garman 1913. Described from São Gonçalo, Rio Parnaíba, Piauí, Brazil. Lectotype: MCZ 600-S, male, selected by Rosa (1985a). This is a valid species, endemic to the Rio Parnaíba drainage in Brazil.

Potamotrygon yepezi Castex and Castello 1970. Described from Rio Palmar, Maracaibo drainage, Venezuela (Castex and Castello 1970a). Holotype USNM 121662, male, by original designation. This is a valid species, endemic to the Maracaibo drainage in Venezuela.

5.4.2 Phylogeny

In spite of many studies of potamotrygonid relationships (e.g., Brooks, Thorson, and Mayes 1981; Thorson, Brooks, and Mayes 1983; Rosa 1985a; Rosa, Castello, and Thorson 1987; Nishida 1990; Dingerkus 1995; McEachran, Dunn, and Miyake 1996; Lovejoy 1996; Hoberg et al. 1998; Marques 2000; Quijada 2003; Dunn, McEachran, and Honeycutt 2003; de Carvalho, Maisey, and Grande 2004; Toffoli 2006), there is no consensus on the sister group of potamotrygonids or relationships within the clade (with both *Plesiotrygon* and *Paratrygon* considered as the basal genus of the family). The Urotrygonidae or Dasyatidae generally are postulated to be the sister group of potamotrygonids with consensus only that potamotrygonids evolved from a marine ancestor that invaded South America. Whether it was via the Atlantic or Pacific Ocean, and dating of this invasion (see above) remain controversial.

The Potamotrygonidae were treated as close relatives of Dasyatidae by Garman (1913) and Bigelow and Schroeder (1953), and, based on the presence of a cartilage between the hyomandibular and mandibular cartilages (possibly the angular cartilage), Miyake (1988) also suggested that potamotrygonids were related to dasyatids. Parasitological data, however, suggests that the Potamotrygonidae represent a monophyletic group closely related to the Urolophidae (*sensu* Compagno 1999) (Brooks, Thorson, and Mayes 1981; Thorson, Brooks, and Mayes 1983). Specifically, all species of potamotrygonids studied by Thorson, Brooks, and Mayes (1983) and some urolophids harbor parasites derived from a common ancestral helminth fauna that may have occurred in a common ancestor host. These data suggest a vicariance hypothesis for the origin of potamotrygonids consisting of an urolophid ancestor trapped in South America by Andean orogeny.

Rosa (1985a) proposed a hypothesis of internal phylogenetic relationships of potamotrygonids at the generic level, as well of the relationships between potamotrygonids and other myliobatiforms. The clade [*Hexatrygon* (*Urolophus* + *Urotrygon*)] was treated as the sister group of Potamotrygonidae, based on presence of cartilaginous rays supporting the caudal fin or caudal elements (ontogenetically lost in some taxa).

According to Rosa (1985a) and Rosa, Castello, and Thorson (1987), the monophyly of the family is widely corroborated by presence of the prepelvic (or prepubic) process on the ischiopubic bar, ontogenetic loss of the rostral appendix cartilage, reduction of the caudal fin skeletal support, adaptation to freshwater as evidenced by the reduction of the rectal gland and suppression of urea retention, reduction in the number of diplospondilous vertebrae, and reduction of the ampullae of Lorenzini and associated canals to a microampullary system. Additionally, these analyses suggested that *Potamotrygon* and *Paratrygon* form a derived sister group to *Plesiotrygon*. The listed synapomorphies of *Potamotrygon* and *Paratrygon* included a high modal number of pectoral fin radials, the reduction of the tail length, a high number of branchial rays on the ventral pseudohyoid, and the reduction of the posteroventral fenestra of the scapulocoracoid. According to Rosa (1985a), *Paratrygon* shows the highest degree of anagenetic changes among potamotrygonids, reflected by its great number of autapomorphies (enlargement of the synarcual plates, postspiracular and scapulocoracoid cartilages), and by its morphological divergence from the basic oval and anterior prominent disc form, found both in other potamotrygonid genera and out-groups.

Nishida (1990) investigated the internal relationships of the Myliobatidoidei. One of the basal subclades in his phylogeny (B node) included Urolophidae (*Urolophus* + *Urotrygon*) as sister to an unresolved polytomy with *Potamotrygon*, *Taeniura*, and (*Dasyatis* + *Himantura*). This substem is characterized by a single synapomorphy, the presence of a ligament between the mandibular and hyomandibular cartilages (character 71 of Nishida), and it is separated from *Urotrygon* and *Urolophus* by the absence of a posterodorsal fenestra in the scapulocoracoid. Additionally, Nishida (1990) observed that some species of *Urotrygon* (*U. asterias*, *U. mundus*, and *U. testaceus*) have inner margins of ceratobranchials 1 and 2 fused proximally, similar to some species of *Potamotrygon* and *Paratrygon*. Nishida (1990) corroborated the monophyly of Potamotrygonidae, based on five characters, but *Potamotrygon* was the only potamotrgonid genus included in his cladistic analysis.

Dingerkus (1995) analyzed myliobatiform relationships based on the morphology of the visceral arches. His phylogeny suggested that *Taeniura* is the sister group of *Potamotrygon*, the only potamotrygonid genus analyzed. That group formed a polytomy with basal *Trygonoptera*, *Urotrygon*, and *Urolophus* (= *Urobatis*) *jamaicensis*.

McEachran, Dunn, and Miyake (1996) considered four families within the Myliobatoidei and included *Taeniura*, amphi-American *Himantura* species (*H. schmardae* and *H. pacifica*), and potamotrygonids in the Potamotrygonidae. The resulting phylogeny was very similar to that of Lovejoy (1996), with *Potamotrygon* and *Himantura* considered sister groups based on the presence of angular cartilages between the hiomandibular and Meckel's cartilages.

Lovejoy (1996) analyzing myliobatiforms, proposed a clade composed of *Taeniura* + [amphi-American *Himantura* + (*Paratrygon* + {*Potamotrygon* and *Plesiotrygon*})]. *Taeniura* was the sister group of the remaining clade based on the extension of the medial component of spiracularis muscle. The clade [amphi-American *Himantura* + (*Paratrygon* + {*Potamotrygon* and *Plesiotrygon*})] was supported by the extension of the spiracularis muscle (his character 36) and the presence of angular cartilages (his character 12).

According to Lovejoy (1996), *Paratrygon* is the basal genus of Potamotrygonidae because the anterior base of the lateral stay joins the synarcual ventral to the spinal nerve foramina, and thus the lateral stay is pierced along its entire length by large foramina, while *Potamotrygon*, *Plesiotrygon*, and *Urobatis jamaicensis* have the lateral stay joined above or dorsal to these foramina, which are smaller than in *Paratrygon*.

Hoberg et al. (1998) provided a reexamination of the historical biogeographic and coevolutionary relationships among gnathostomatids and their hosts. The phylogenetic tree for *Echinocephalus* may provide support for Lovejoy's hypothesis of stingray relationships, but it does not provide evidence for an Atlantic basin origin of potamotrygonids; rather, it strongly supports a Pacific origin.

Marques (2000), using molecular and parasitological data to study the origin and diversification patterns of Potamotrygonidae, corroborated the monophyly of the family and suggested amphi-American *Himantura* as the closest marine relative. His work also suggested that *Potamotrygon* is a nonmonophyletic group, and that two species of this genus (*Potamotrygon yepezi* and an undescribed species of *Potamotryon* from Rio Negro) shared with *Plesiotrygon* a more recent ancestry.

Quijada (2003) provided a phylogenetic analysis of Potamotrygonidae, including all genera of the family and additional information for an undescribed genus cited by Ishihara and Taniuchi (1995). The analysis resulted in a clade formed by [*Taeniura* (*Urobatis* {*Urolophus*})] as the sister group of potamotrygonids (Figure 5.2). Quijada (2003) indicated that the undescribed genus is more closely related to *Paratrygon* than to other genera of the family, as previously noted by Ishihara and Taniuchi (1995) and Charvet-Almeida (2001).

In a molecular analysis, Dunn, McEachran, and Honeycutt (2003) used DNA sequences to investigate the relationships of myliobatoid genera and the sister group of Potamotrygonidae. The cladogram obtained by maximum parsimony analysis suggested that a clade composed of (*Dasyatis* + *Taeniura*) is sister to a clade composed of *Urobatis* + (*Potamotrygon* + *Gymnura*), with bootstrap support less than 50% to the latter clade, but with bootstrap support greater than 50% in maximum likelihood analysis, with *Urobatis* as sister group of *Potamotrygon*. In another dataset, the maximum likelihood indicated a sister relationship between *Himantura* and *Potamotrygon*.

In McEachran and Aschliman (2004), amphi-American *Himantura* + *Potamotrygon* clade comprised a polytomy with other dasyatids and pelagic myliobatiforms. *Taeniura* and amphi-American *Himantura* + *Potamotrygon* were united in the same clade, but the relationships differed and were less resolved than in McEachran, Dunn, and Miyake (1996). The relationships between *Himantura schmardae* and *Potamotrygon* in McEachran and Aschliman (2004) were based on presence of two small angular cartilages, following Nishida (1990), Lovejoy (1996), and McEachran, Dunn, and Miyake (1996). However, the angular cartilages of *H. schmardae* were not illustrated.

De Carvalho, Maisey, and Grande (2004) used both extant and fossil genera to assess the relationships among Myliobatiforms. The strict consensus tree showed the fossil *Heliobatis* as the most basal genus of Myliobatoidea, sister to the clade Potamotrygonidae + [Dasyatidae + (Gymnuridae + Myliobatidae)]. *Potamotrygon* and *Plesiotrygon* were considered derived genera within the Potamotrygonidae clade by the presence of angular cartilages, by the lateral stay of synarcual originated dorsal to spinal nerve foramina, by the contact between propterygium and mesopterygium, and by the spiracularis muscle condition.

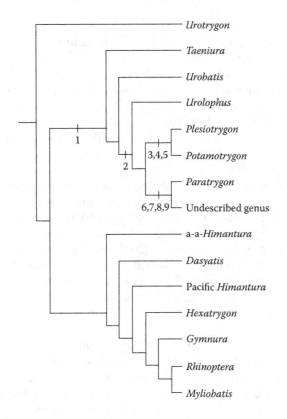

FIGURE 5.2
Strict consensus of the three equally parsimonious trees obtained in Quijada (2003) representing phylogenetic relationships of potamotrygonids. Length = 227; consistency index = 80; retention index = 90; a-a =
amphi-American. Synapomorphies: 1, posterior projection of mesopterygium; 2, absence of dorsal foramen in
scapulocoracoid cartilage; 3, ontogenetic loss of rostral appendix cartilage; 4, cartilaginous transverse bar at
symphysis of palatoquadrate; 5, reduction of lateral process of Meckel's cartilage; 6, enlargement of scapulocoracoid cartilages; 7, rostral appendix cartilage absent in embryos and adults; 8, reduction or loss of preorbital
process; 9, reduction or loss of anterior postorbital process. A full discussion of synapomorphies is found in
Quijada (2003).

Toffoli (2006) proposed a phylogenetic hypothesis of the Potamotrygonidae based on
mithochondrial genes of nine species of the family. His analysis indicated that the family is monophyletic and that *Paratrygon* is the basal genus and *Potamotrygon* is the more
derived genus. The position of *Plesiotrygon* varied according to the reconstruction method
and the database, and some data suggested that this genus may be part of *Potamotrygon*, a
result corroborated by Toffoli et al. (2008).

The affinities of Potamotrygonidae with amphi-American *Himantura* species were proposed based on the insertion of the spiracularis muscle in the otic region of the neurocra-
nium and the presence of angular cartilages in the Meckelian-hyomandibular connection
(Lovejoy 1996; Mc Eachran, Dunn, and Miyake 1996).

Lovejoy (1996) used an ordered character, the extension of the medial component of
the spiracularis muscle, to support *Taeniura* and amphi-American *Himantura* relationships with the clade [*Paratrygon* + (*Potamotrygon* and *Plesiotrygon*)]. De Carvalho, Maisey,
and Grande (2004) suggested that the spiracularis muscle is more complex than previously interpreted by Lovejoy (1996) and by McEachran, Dunn, and Miyake (1996) and

rejected *Taeniura*, amphi-American *Himantura* and potamotrygonids as a monophyletic group. De Carvalho, Maisey, and Grande (2004) also pointed out that in some species of *Potamotrygon* and in *Himantura schmardae*, the spicularis and the depressor hyomandibularis muscles are difficult to separate from each other. Additionally, the width of the spiracular muscle is not easy to quantify, and in some taxa it is not continuous with depressor hyomandibularis (e.g., *Paratrygon*). In *Plesiotrygon* this muscle is more slender than in *Potamotrygon* (but more robust than in *Paratrygon*) and is more difficult to delimit due to an abundance of connective tissue (de Carvalho, Maisey, and Grande 2004). Due to the complexity of the spiracular muscle, more comparative descriptions are needed to corroborate its use in phylogeny.

Lovejoy (1996) suggested that the ligamentous connection between the hyomandibular and Meckel's cartilages is the plesiomorphic state and the presence of robust angular cartilages is the derived state. Additionally, Lovejoy (1996) suggested the strengthening of the connection with multiple small cartilages observed in amphi-American *Himantura* as the intermediate state. According to Quijada (2003) and de Carvalho, Maisey, and Grande (2004), this character shows intrageneric and intraspecific variation. De Carvalho, Maisey, and Grande (2004) observed two angular cartilages of dissimilar sizes in *P. signata* and one angular cartilage in *Plesiotrygon*. Quijada (2003) observed usually two, but also only one or three angular cartilages, even in the same species of *Potamotrygon* and up to two cartilages in *Plesiotrygon*.

Lovejoy (1996) observed a small angular cartilage in *Paratrygon*, but mistakenly pictured it as two angular cartilages (his Figure 6D). None of these findings was corroborated by de Carvalho, Maisey, and Grande (2004). Quijada (2003) also did not find angular cartilages in *Paratygon* and in the undescribed genus. According to de Carvalho, Maisey, and Grande (2004), the angular cartilages occurring in *Potamotrygon* and *Plesiotrygon* are much larger and discretely shaped, very distinct from those of *H. schmardae* and *H. pacifica*. They pointed that *H. schmardae* has minute angular cartilages, difficult if not impossible to discern in radiographs, and this may vary intraspecifically. Also, some specimens of *H. schmardae* possess scattered nonprismatic calcification within the ligament, in addition to or replacing the small angular cartilages. Lovejoy (1996, citing de Carvalho's unpublished data) pointed out an area of slight chondrification in the hyomandibular/mandibular connective ligament in *T. lymma*. Angular elements prismatically calcified also were observed in the fossil genus *Asterotrygon* (de Carvalho, Maisey, and Grande 2004).

The homology between the discrete and robust angular cartilages of *Plesiotrygon* and *Potamotrygon* and the "angular cartilage collection" of amphi-American *Himantura* is not clear, and the two conditions should not be polarized *a priori* as the derived condition, such as interpreted in Lovejoy's (1996) transformation series. To Quijada (2003) the lack of angular cartilages in *Paratrygon* is a derived condition probably associated with the enlargement of mandibular cartilages.

According to Hoberg et al. (1998), there is no support for the relationship between Potamotrygonidae and amphi-american *Himantura* species, because this relation was based on a single homoplastic character. According to their phylogeny of myliobatiform parasites, Potamotrygonidae is more closely related to *Urolophus* and *Taeniura* that occur in the Pacific Ocean.

Finally, among the various characters to be further investigated in a phylogenetic perspective, the presence and homology of the accessory hyomandibular cartilage (*sensu* Lovejoy 1996) remain to be clarified in several stingray taxa, because Garman (1913) illustrated such cartilage and referred to it as the angular in *Myliobatis*, *Aetomyleus*, *Aetobatus*, *Rhinoptera*, and *Mobula*.

5.5 Habitat and Ecology

In spite of being considered completely adapted to fresh water, the ecology of the potamotrygonids, however, remains poorly known. In many cases they were assumed to be similar to their potential marine and estuarine relatives, including the Urolophidae, Urotrygonidae, and Dasyatidae. However, this group exhibits many unique characteristics.

The first brief comments on freshwater stingray ecology were sparsely included in species descriptions (e.g., Schomburgk 1843). Later, general observations regarding the natural history of potamotrygonids were published (Castex 1963a, 1963b, 1964a; Castex and Maciel 1965; Achenbach and Achenbach 1976). However, only recently have detailed studies been carried out on the biology of some freshwater stingray species (Teshima and Takeshita 1992; Lasso, Rial, and Lasso-Alcalá 1996; Araújo 1998; Almeida 2003, 2008; Charvet-Almeida 2001, 2006; Charvet-Almeida, Araújo, and Almeida 2005; Rincon 2006). Most information comes from the Brazilian Amazon region but more studies are being pursued in other South American countries.

Potamotrygonids vary considerably in habitat specialization from widely distributed habitat generalists (e.g., *Potamotrygon motoro* and *P. orbignyi*), to species with narrow ranges (e.g., *Potamotrygon humerosa*) or that are endemic (e.g., *Potamotrygon leopoldi* and *P. henlei*). The degree of endemism is high, which likely is due to the great variation and subdivision in hydrological systems in South America. In fact, many tributary basins of the Amazon River have one or two endemic species each. Freshwater stingrays are found in all river types of the Amazon region (Sioli 1967) including white water (murky, sediment rich), clear (bluish or greenish, sediment poor), and black water (dark tea-like color, humic acid rich) rivers. In other basins, potamotrygonids also are found in diverse habitat types. Water temperature, pH, and conductivity seem to be the main physical factors that vary among habitats of these basins (mainly in the Paraná and Prata River systems) and the Amazonian rivers. Even among rivers within the Amazon region, these factors likely are the principal abiotic conditions that influence species distributions (Charvet-Almeida and Almeida 2008).

Potamotrygonids may be limited by high salinities, but exact tolerances are still unresolved. Thorson, Brooks, and Mayes (1983) reported that wild *Potamotrygon* spp. tolerate salinities up to 3.4, but in the Amazon estuary potamotrygonids have been observed to tolerate 4.5 ppt (Charvet-Almeida 2001) and to occur at salinites up to 12.4 ppt (Almeida 2008; Almeida et al. 2009). *Plesiotrygon iwamae* seems to be one of the most salt-tolerant species (Charvet-Almeida 2001). The rectal gland, a salt excretory organ, is found in all species of potamotrygonids. The size, however, is reduced and it may not retain an excretory function (Thorson, Wotton, and Georgi 1978) although some species tolerate brackish conditions much better than expected for a freshwater organism (Charvet-Almeida 2001; Almeida 2008; Almeida et al. 2009). Physiological studies on the role and function of the rectal gland in potamotrygonids are still required as are investigations of salinity tolerances.

Freshwater stingrays are found over a great diversity of substrates, including sandy, muddy, and rocky bottoms and all possible combinations (Figure 5.3; Charvet-Almeida 2006). Some species, such as an undescribed species of *Potamotrygon* from the Rio Negro basin, live among leafs and plant fragments (Araújo 1998).

Potamotrygonids are found in a wide variety of habitats, including both lentic and lotic habitats and even in strong currents (*P. motoro*) (Charvet-Almeida 2001; Almeida 2003, 2008; Almeida et al. 2009) or near waterfalls and rapids (*P. leopoldi*) (Zuanon 1999; Charvet-Almeida 2006). Some species may have narrow habitat affinities while others occur in a variety of

FIGURE 5.3
A color version of this figure follows page 336. Potamotrygon leopoldi in its natural habitat at the Rio Xingu, Brazil. (Photo © P. Charvet-Almeida. All rights reserved.)

habitats. In rivers, potamotrygonids can be found from headwaters to mouth, but barriers, such as waterfalls and dams, are very important in determining species distributions and often limit access to upstream areas (Charvet-Almeida 2006; Garrone Neto et al. 2007).

Once natural barriers are erected due to man-made interference (e.g., damming), some species have shown the ability to invade new habitats (Garrone Neto et al. 2007). This requires attention when considering future impacts of environmental changes and when releasing captive specimens in nature.

Recent studies in the Amazon region show that potamotrygonids exhibit sexual and ontogenetic segregation (Charvet-Almeida 2006; Almeida 2008). Such behavior probably also occurs in other river basins. Neonate and juvenile specimens tend to stay in shallow areas close to the shore, in nursery areas where they probably remain protected from larger channel predators. Sometimes adults and larger individuals enter shallow areas to feed, but generally remain in deeper waters. The mechanism for sexual segregation is not yet clearly understood but predominantly males or females are caught at specific locations along the reproductive cycle steps (Charvet-Almeida 2006).

Unquestionably, further studies are needed to better understand the ecology of this unique group of stingrays. Murky waters, strong currents, and risk of entanglement or accidents in many river basins make these stingrays a particularly difficult group to study in most areas of their distribution.

5.6 Life History

5.6.1 Diet and Feeding

Potamotrygonids feed mainly on small invertebrates and vertebrates, including insects, annelids, molluscs, crustaceans, and bony fishes and even take catfish without being harmed by their spines (Achenbach and Achenbach 1976; Rosa, Castello, and Thorson 1987;

Lasso, Rial, and Lasso-Alcalá 1996; Zuanon 1999; Charvet-Almeida 2001, 2006; Pântano-Neto 2001; Pântano-Neto and Souza 2002; Bragança, Charvet-Almeida, and Barthem 2004; Charvet-Almeida 2006; Rincon 2006; Lonardoni et al. 2006; Silva and Uieda 2007; Almeida 2008). Most species of potamotrygonids have a diversified diet. When several species coexist, there may be a high degree of dietary overlap but in other cases there appears to be trophic partitioning (Charvet-Almeida 2006; Lonardoni et al. 2006; Silva and Uieda 2007). In general, Rincon (2006) considered Potamotrygonidae to feed at an intermediate trophic level, but some species (e.g., *P. aiereba*) are upper trophic level predators in riverine systems (Charvet-Almeida 2006).

The diet of potamotrygonids varies among species and geographically within species. Most regional variation in diets can be attributed to variation in the abundance of particular prey types within a general prey class (Almeida 2008). For example, an insectivorous species could have Ephemeroptera as its main food item in one location and Trichoptera in another. Other species are much more specialized and may feed almost exclusively on a single prey; for example, *P. leopoldi* feeding on a single species of gastropod in areas where there are two to three other snail species (Charvet-Almeida 2006).

Empty stomachs or stomachs with little food are often observed in feeding studies. Quite often plant fragments (parts of leaves and stems) are found among the stomach contents of freshwater stingrays (Charvet-Almeida 2001; Rincon 2006; Almeida 2008). It is unlikely that these fragments provide nourishment and they are probably accidentally ingested during feeding (Bragança, Charvet-Almeida, and Barthem 2004; Charvet-Almeida 2006). This may be the case for sand, soil, and stone fragments also observed in stomachs of several species.

A pioneering study of oro-branchial anatomy (Pântano-Neto 2001; Pântano-Neto and Souza 2002) indicated that cephalic muscles and skeletal characteristics were related to the diet of *P. motoro* and *P. henlei*. Pântano-Neto and Souza (2002) showed that anatomical differences (cephalic muscles and muscle mass) could correspond to functional adaptations to the diet and feeding behavior in these stingrays. Further and detailed anatomical studies are still needed for most species.

Ontogenetic shifts in diets, which are common in elasmobranchs, occur in *P. motoro* and possibly other potamotrygonids (Achenbach and Achenbach 1976; Yossa 2004; Charvet-Almeida 2006; Rincon 2006; Almeida 2008). Seasonal shifts in diets (Lonardoni et al. 2006) and sex differences in feeding (Lasso, Rial, and Lasso-Alcalá 1996; Charvet-Almeida 2006) also are apparent.

Interestingly, the time of day at which particular species feed varies within the potamotrygonids. Some species, like *P. leopoldi* and *P. aiereba*, feed mainly at night while others feed throughout the day (e.g., *P. orbignyi*) (Charvet-Almeida 2006). In other situations, feeding behavior may be tied to other environmental fluctuations. For example, feeding activity of *P. iwamae* in the lower Amazon seems to be concentrated during slack tides when water velocities are low (Charvet-Almeida 2001). Finally, some species possibly ingest small quantities of prey with no noticeable patterns in feeding (Rincon 2006; Almeida 2008).

In captivity, perhaps one of the greatest challenges is to provide adequate nutrition and to induce stingrays to feed. Nevertheless, some breeders indicate that even species with highly specialized diets in the wild (e.g., *Potamotrygon leopoldi*) (Charvet-Almeida 2006) can adjust well to artificial food, even if completely different from its natural diet.

5.6.2 Mating, Reproduction, and Reproductive Cycle

Freshwater stingray mating is rarely observed in the wild. However, mating is reported to occur in a ventral-to-ventral position with the female being constrained by the male against the bottom substrate (Castex 1963b) or aligned in a side-by-side position (Thorson, Langhammer, and Oetinger 1983). Castex (1963b) also indicated that during copulation, tails overlapped and the male's pectoral fins partially embraced the female. Males bite and hold on to the female's pectoral fins leaving small clear scars and abrasion marks (Thorson, Langhammer, and Oetinger 1983; Lasso 1985; Charvet-Almeida 2006). In potamotrygonids the teeth of males are more pointed than those of females (Rosa, Castello, and Thorson 1987; Charvet-Almeida 2006), which helps them grip a female's pectoral fin during court-ship and copulation (Thorson, Langhammer, and Oetinger 1983; Lasso 1985). Based on captive observations, Thorson, Langhammer, and Oetinger (1983) suggested that *P. motoro* copulation occurs at night. However, they pointed out that behavior could be different in the wild.

As with all Myliobatiformes viviparity with matrotrophy has been identified as the repro-ductive mode of all potamotrygonid species studied so far (Thorson, Langhammer, and Oetinger 1983; Teshima and Takeshita 1992; Araújo 1998; Charvet-Almeida 2001, 2006; Charvet-Almeida, Araújo, and Almeida 2005; Rincon 2006; Almeida 2008). Also, females tend to be larger (disc width) than males (Lasso, Rial, and Lasso-Alcalá 1996; Araújo 1998; Almeida 2003, 2008; Charvet-Almeida 2001, 2006; Charvet-Almeida, Araújo, and Almeida 2005). Population sex ratios tend to be very close to 1:1. However, for some species (e.g., *P. aiereba* and *P. motoro*) and regions, females tend to outnumber males (Charvet-Almeida 2006; Almeida 2008).

Potamotrygonid clasper morphology was studied by Rosa (1985a), Taniuchi and Ishihara (1990), and McEachran and Aschliman (2004). Thorson, Langhammer, and Oetinger (1983) and Lasso, Rial, and Lasso-Alcalá (1996) suggested that proportional clasper length is one of the best external indicators of sexual maturity for males. In adult males clasper length as percent disc width ranges from 20% to 28% (Lasso, Rial, and Lasso-Alcalá 1996; Rosa, Castello, and Thorson 1987; Charvet-Almeida 2001, 2006; Almeida 2008). This characteris-tic, when associated to clasper calcification, provides helpful evidence to determine sexual maturity (Charvet-Almeida 2001, 2006; Almeida 2008). Testes in adult males have been reported to be well lobulated and both functional (Figure 5.4). Folds of the seminal vesicle are easily observed in adults, while in juvenile specimens this structure is filiform.

Females have asymmetric reproductive organs (Figure 5.5). In many cases there is atro-phy of the right ovary with follicle production restricted to the left ovary (Castex 1963e; Castex and Maciel 1965; Achenbach and Achenbach 1976; Thorson, Langhammer, and Oetinger 1983; Lasso, Rial, and Lasso-Alcalá 1996; Araújo 1998; Charvet-Almeida 2001, 2006; Charvet-Almeida, Araújo, and Almeida 2005; Rincon 2006; Almeida 2008). However, there are reported exceptions for *P. magdalenae* (both ovaries functional; Teshima and Takeshita 1992) and few individuals of other species have been reported that exceptionally produced follicles in the right ovary (Charvet-Almeida 2006; Almeida 2008).

Both uteri have been observed to be functional in the species studied and the average fecundity differs among species. Low fecundity has been indicated for *P. orbignyi* [two embryos, Winemiller and Taphorn (1989); and one embryo, Lasso, Rial, and Lasso-Alcalá 1996; Rincon 2006). The species with the lowest average uterine fecundity (two embryos) were reported to be *P. iwamae*, *P. aiereba*, *P. orbignyi*, *P. schroederi*, and *Potamotrygon* sp. (Araújo 1998; Charvet-Almeida 2001; Charvet-Almeida, Araújo, and Almeida 2005). Some species have been reported to bear up to 15 (*P. motoro*; Achenbach and Achenbach 1976), 16 (*P. scobina*; Charvet-Almeida, Araújo, and Almeida 2005), 19 (*P. brachyura*; Achenbach and Achenbach 1976), and even 21 (*P. motoro*; Almeida 2008) pups per litter.

Gestation periods are little known and vary among species (Charvet-Almeida, Araújo, and Almeida 2005). Available estimates range from 3 months (*Potamotrygon* sp.; Araújo 1998) to 8 (*P. iwamae*; Charvet-Almeida 2001) or even 11 months (*P. orbignyi*; Winemiller 1989).

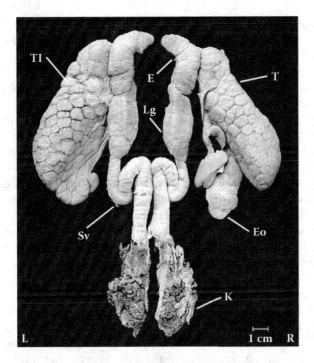

FIGURE 5.4
A color version of this figure follows page 336. Reproductive system of an adult male specimen of *Potamotrygon leopoldi*. T = testicle, Tl = testicular lobe, E = epididymis, Lg = Leydig gland, Sv = seminal vesicle, K = kidney, and Eo = epigonal organ. (Photo © P. Charvet-Almeida. All rights reserved.)

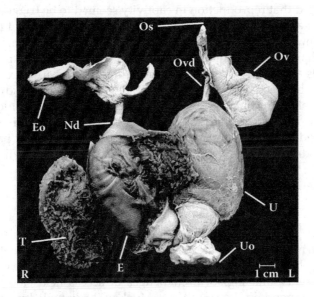

FIGURE 5.5
A color version of this figure follows page 336. Reproductive system of an adult pregnant female specimen of *Potamotrygon orbignyi*. Os = ostium, Ov = ovary, Ovd = oviduct, Nd = nidamentle gland, U = uterus, T = trophonema, E = embryo, Uo = urogenital opening, and Eo = epigonal organ. (Photo © P. Charvet-Almeida. All rights reserved.)

Shallow beach areas are considered as nursery grounds (Charvet-Almeida 2001, 2006), but birthing seasons vary. Potamotrygonids seem to have two strategies regarding birth season. Some potamotrygonid species give birth during the rainy season, and others during the dry season (Araújo 1998; Charvet-Almeida 2006; Rincon 2006; Almeida 2008). Pupping seasons last from two to four months (Castex 1963b; Achenbach and Achenbach 1976; Charvet-Almeida, Araújo, and Almeida 2005). The timing of birth may vary intraspecifically across a species' range. A recent study (Charvet-Almeida 2006) suggested that during the rainy season pups probably could take shelter in the flooded forest (*igapó*), but possibly at a cost of more dispersed food. In contrast, with lower water levels, pups born in the dry season probably would be subject to higher risk of predation in narrower river channels but would benefit from increased access to, or ease of finding, food (Charvet-Almeida 2006). Thus, the two pupping alternatives offer differing trade-offs and it is not surprising that sympatric species may have different pupping times (Charvet-Almeida 2006).

The age of sexual maturity has only been estimated for few species and varies from two up to 35 years (Castex 1963b; Achenbach and Achenbach 1976; Charvet-Almeida, Araújo, and Almeida 2005; Charvet-Almeida 2006; Rincon 2006), but most species seem to reach sexual maturity between three and eight years of age.

Most potamotrygonid species have a yearly reproductive cycle that is often associated with the hydrologic cycle of each region or river basin (Achenbach and Achenbach 1976; Thorson, Langhammer, and Oetinger 1983; Lasso, Rial, and Lasso-Alcalá 1996; Araújo 1998; Charvet-Almeida 2001, 2006; Charvet-Almeida, Araújo, and Almeida 2005; Rincon 2006; Almeida 2008). *P. aiereba* may only reproduce every other year (Charvet-Almeida 2006) while *Potamotrygon magdalenae* does not show a well-defined reproductive period (Teshima and Takeshita 1992). Changes in water characteristics associated with the hydrologic cycle probably trigger the beginning and subsequent steps of most species' reproductive cycles (Charvet-Almeida, Araújo, and Almeida 2005). Thorson, Langhammer, and Oetinger (1983) noted that reproduction in captivity seemed to be triggered by water temperature changes. Lasso, Rial, and Lasso-Alcalá (1996) pointed out that physico-chemical water changes followed the beginning of the rainy season. Climatic variations, such as El Niño, also cause changes in the potamotrygonid reproductive cycle (Araújo 1998).

Potamotrygonids sometimes seem to hybridize in the wild (Castex and Maciel 1965; Achenbach and Achenbach 1976). Lasso (1985) suggested that hybridization may help explain polychromatism in potamotrygonids. Hybrid individuals could also contribute to the difficulties in species identification (Almeida et al. 2008). In captivity, fertile hybrids are being both accidentally and intentionally produced; valued color patterns are being obtained by hybridization mainly in Asian breeding farms. Amazingly, descendents turned out to be fertile hybrids that are rebred with their parents to achieve even more valuable color patterns.

5.6.3 Movements and Behavior

Very little is known about freshwater stingray habitat use, movements, and behavior. Direct underwater observations of behavior patterns are difficult in the wild due to murky waters, strong currents, and logistical challenges of access field sites. Few species are found in waters with good diving conditions. Much of the knowledge about the behavior of freshwater stingrays, therefore, relies on captive observations.

Freshwater stingrays have long been feared among fishermen, Indians, riparian inhabitants, and tourists in South American riverine regions. Castex and Loza (1964) described the stinging mechanism as a direct action and reaction process where the stingray would direct its sting toward the stimulated region. This reaction causes many injuries to people as well as to cattle and horses that, unaware of freshwater stingrays, step on their disc (Schomburgk 1843; Castex 1963b; Halstead 1970; Achenbach and Achenbach 1976; Lasso 1985; Pierini et al. 1996; Pardal and Rezende 1997; Haddad 2000; Charvet-Almeida 2001). Stings are made of vitrodentine covered by enamel (Halstead 1970) and are periodically shed (Thorson, Langhammer, and Oetinger 1988). Embryos are born with functional stings (Castex 1963a) and female uteri are occasionally punctured by the stings of embryos (Charvet-Almeida 2001).

Freshwater stingray's venom was studied by Castex and Loza (1964) and recently was indicated as more powerful and produced in higher volumes than their marine counterparts (Barbaro et al. 2007). Sting morphology studies showed that this is due to differences in the distribution of the venom secretory cells (Pedroso et al. 2007). Biological activities of the venom of marine and freshwater stingrays also differ (Barbaro et al. 2007).

Probably Schomburgk (1843) first reported the feeding strategy of freshwater stingrays. He indicated that freshwater stingrays remained hidden in the river's sandy substrate to surprise and capture their prey. Achenbach and Achenbach (1976) described some food searching strategies and mentioned that some species from the Paraná River basin attacked small fish schools (*Astyanax* and *Aphyocharax*). These authors also reported that potamotrygonids remained indifferent to the presence of strong spines on a small catfish (*Pimelodella gracilis*) they preyed upon. A similar behavior pattern of attacking fish schools was also observed in the shallow areas of the Xingu River (Charvet-Almeida 2006). Potamotrygonids have been observed often searching for food and disturbing the substrate in the rapids of the Xingu River (Zuanon 1999).

In general, freshwater stingrays have two main foraging tactics that are related to mouth anatomy (Pântano-Neto 2001; Pântano-Neto and Souza 2002; Charvet-Almeida 2006; Rincon 2006), visual conditions and swimming characteristics (Charvet-Almeida 2006). In one group are the stingrays that search carefully and slowly for food. These "searcher" species prey mainly on food items unlikely to move rapidly and flee (e.g., insect larvae, gastropods, etc.). They move vertically against the substrate (submerged logs, branches, and rocks) and horizontally, turning over the bottom substrate in search of prey. Adults of some species (e.g., *P. leopoldi*) can even turn over large rocks (over 1 kg) while searching for food (Charvet-Almeida 2006). Although species belonging to this group tend to have delicate but extremely agile mouth muscles (e.g., *P. orbignyi*), some of them have very strong musculature used to crush hard gastropod shells (e.g., *P. leopoldi*). Suction plays an important role in prey capture and ingestion in this group (Charvet-Almeida 2006). The second group, "active hunters," usually are larger species that can swim fast when charging at a fish school or shrimp. The species in this group tend to move fast when searching for food to the point of creating small waves in shallow waters (Charvet-Almeida 2006; Almeida 2008). Their tooth morphology and muscles are adapted to hold and immobilize prey prior to swallowing it (Charvet-Almeida 2006). In some of these hunting species, vision plays an important role in prey capture (Charvet-Almeida 2006).

Potamotrygonids are fascinating because they may be the only elasmobranchs that provide maternal care. Achenbach and Achenbach (1976) described newborns staying around and on top of their mother's dorsal surface after birth. This same pattern has been observed in other species and regions (Araújo 1998; Charvet-Almeida 2006) but not everywhere and for all species (Charvet-Almeida 2001; Almeida 2008). Unfortunately, this behavior is very hard to document in the wild and does not seem to have been observed in captive specimens.

One common behavior among potamotrygonids is to swim close to the surface whenever they are exposed to hypoxic conditions. In some areas, when the water level drops after the flood season, stingrays get trapped in marginal lakes and pools. There, the water level further decreases and sometimes the habitat dries out. In this process it is possible to observe potamotrygonids swimming at the surface, where the dissolved oxygen levels tend to be higher.

Movement patterns of freshwater stingrays have not been studied in detail. In the Amazon estuary it is clear that when salinity levels begin to rise during the dry season, some species (e.g., *Plesiotrygon iwamae*) migrate upstream (Charvet-Almeida 2001). In this same region, some species also move toward the center of the Marajó Island through rivers and streams (Almeida 2008). These movements are linked to environment changes and probably trigger reproduction for some species (Charvet-Almeida 2001; Almeida 2003; Charvet-Almeida, Araújo, and Almeida 2005). Seasonal movements associated with reproduction were also reported by Achenbach and Achenbach (1976) for *P. motoro*, *P. brachyura*, and *P. histrix*. In more restricted areas (rivers), these movements are much more difficult to be identified and seem to be related to water temperature changes. Rincon (2006) suggested that species from the Paraná River, however, did not perform long distance movements within the study area.

Captive studies of spatial memory and orientation strategies of juvenile *P. motoro* feeding within a maze (Schluessel and Bleckmann 2005) showed that this species constructs a visual cognitive map of its environment and employs a variety of orientation strategies based on the complexity of the task. Such a spatial memory could aid in recognizing terrain, locating food, and quickly moving to specific places (as when hiding from predators) (Schluessel and Bleckmann 2005).

5.7 Fisheries and Uses

Undoubtedly the commercialization of freshwater stingrays for ornamental purposes has had important socioeconomic impacts on many riparian communities. The capture of rays for the ornamental trade is an artisanal activity that usually involves family members (Araújo 1998; Araújo et al. 2004; Glémet 2004; Charvet-Almeida 2006). The ornamental trade of these stingrays has been studied in more detail in the Rio Negro (Araújo 1998) and Xingu River (Glémet 2004; Charvet-Almeida 2006) regions. Unlike many other ornamental freshwater fishes, potamotrygonids are traded individually and have a high unit value for fishermen, exporters, and importers. In this sense, handling and transportation care differ from other fish that are usually taken by the thousands (Glémet 2004; Charvet-Almeida 2006).

The main countries involved in the Potamotrygonidae trade were reported to be the United States, Japan, Taiwan, and Germany (Charvet-Almeida et al. 2002) but nowadays the growing Asian market has changed this scenario. Potamotrygonids are usually commercialized according to dorsal color codes, called the P-color system (Ross and Schäffer 2000). Because polychromatism is very common within this group (Charvet-Almeida 2001; Almeida 2003) often different codes correspond to the same species and vice versa (Charvet-Almeida et al. 2002). Ornamental fisheries are directed mainly at neonate and juvenile specimens, which require specific management measures. Trade demands individuals with perfect disc condition, resulting in the return of captured healthy specimens with minor disc damage (cuts and bites) to the population (Charvet-Almeida 2006).

Until now only Brazil has had specific regulations regarding the exportation of these stingrays. The regulations established a species-specific export quota and limited maximum disc width size. Despite this, some adult specimens were illegally exported and captive breeding of potamotrygonids has been estalished mainly in Asia during the present decade. Other South American countries do not have specific export regulations for potamotrygonids. There are species that cannot be taken from Brazil (e.g., *P. aiereba*) that are exported from neighboring countries, and some South American border areas are extremely difficult or even unsafe to monitor (Araújo et al. 2004). In this sense, an international effort is recommended to help regulate this ornamental trade.

Destructive negative fisheries for elasmobranchs involve purposeful mutilations or death of specimens (Castex 1963b; Compagno 1990; Compagno and Cook 1995). Freshwater stingrays are much feared for the painful wounds they cause. Consequently, in many regions they are killed and discarded, or have their tails cut off or stings removed. Right after mutilation takes place, bleeding makes them vulnerable to the attack of piranhas (mainly *Serrasalmus* spp.) and other predators (Charvet-Almeida 2006). Araújo (1998) reported the killing of freshwater stingrays in the Rio Negro region as a practice of "beach cleaning" prior to tourists' arrival. Mutilated potamotrygonids have been observed in many other regions (Castex 1963b; Charvet-Almeida 2001; Almeida 2003, 2008; Rincon 2006).

Schomburgk (1843) observed that freshwater stingrays could be used as a food source but historically were not valued by fishermen in the Amazon region as a food fish (Ferreira, Zuanon, and dos Santos 1998). Potamotrygonids were recorded in the diet of the Suyá Indians (Ribeiro 1979) in the Xingu River basin, while Lasso (1985) noted that potamotrygonids are abundant in some regions but only occasionally consumed. In some areas they are taken as a food source only when a better option is not available (Castex 1963b; Ferreira, Zuanon, and dos Santos 1998; Charvet-Almeida 2001; Charvet-Almeida and Almeida 2003), while in other regions they do not seem to be taken as a food source at all (Araújo 1998; Charvet-Almeida 2006; Rincon 2006). Nevertheless, there are regions where freshwater stingrays are caught and consumed regularly as a subsistence food source (Santos, Jegu, and Merona 1984; Charvet-Almeida 2001). Around 2001–2002 a directed industrial fleet started catching freshwater stingrays in the Solimões/Amazonas River channel to be processed as minced fish and fish fillet. Landings have increased significantly (Pró-Várzea/IBAMA 2002) and these catches (reaching over 1000 tons/year) have been a matter of concern since they started (Charvet-Almeida and Almeida 2008). Potamotrygonid sport fishing has not been a common practice but seems to be slowly gaining popularity in some regions of South America. Many sport fishermen who intentionally or accidentally catch a freshwater stingray end up killing the specimen to free it from the hook and avoid handling a live ray with a dangerous sting. In some areas sport fishing may end up negativly impacting stingrays.

Apart from the more common uses described above, freshwater stingrays are taken for some bizarre purposes that are associated with folklore and popular beliefs. Their effectiveness is debatable, but probably associated with psychosomatic effects (Charvet-Almeida 2001, 2006). Perhaps in the future, scientific studies might help evaluate the effectiveness of stingray folk remedies. Popular medicine in the Amazon region cites potamotrygonid liver oil as efficient in the treatment of inflammatory processes, especially those associated with the respiratory system, and asthma (Charvet-Almeida 2001, 2006; Charvet-Almeida and Almeida 2003). This substance in some regions is used to help heal cuts and bruises (Charvet-Almeida 2001). Anecdotal reports also suggest that liver oil aids in child delivery to reduce labor pain. It is possible that the belief that liver oil could be used for medicinal purposes originated in the use of squalene.

Lasso (1985) mentioned that freshwater stingray skin had been used as shagreen in the past. Stings were reported to be used as ornaments, arrowheads, and as tattoo and body perforation instruments (Schomburgk 1843; Castex 1963b; Charvet-Almeida 2001, 2006). Since there is a negative image (fear) associated with freshwater stingrays, in some religious rituals or ceremonies, parts (mainly tail and sting) have been reported to be used as a negative or harmful element (Charvet-Almeida and Almeida 2003). The serrated edges of dried stings are used in the removal of lice and lice eggs (Charvet-Almeida 2001, 2006) among riparian inhabitants. On some islands, sting powder (obtained through drying and grinding) when mixed with gunpowder has been suggested as efficient in killing prey while hunting (Charvet-Almeida 2001).

5.8 Conservation and Management Implications

In spite of the various uses of potamotrygonids in fisheries and the ornamental trade, population estimates and monitoring are lacking for most species. Only five species are currently listed in the 2008 IUCN Red List of Threatened Species, four of them as Data Deficient (*P. leopoldi, P. motoro, P. schuemacheri,* and *P. scobina*) and one as Least Concern (*P. henlei*). Rincon (2004) reported that the population of *P. henlei* showed an increase after the construction of the Tucuruí dam in the Tocantins River, but there had been no population assessments prior to damming.

Potential threats to potamotrygonids come from the various forms of removal of individuals by fisheries and culls and from environmental impacts on their habitats such as damming, mining, and water pollution (Araújo et al. 2004). Presently there are no thorough studies monitoring the effects of these threats on stingray populations. As is the case with most elasmobranchs, intrinsic biological features of potamotrygnids, such as low fecundity, may exacerbate population declines and slow their recoveries. Additional threats are the restricted freshwater habitat and endemic distributions (Araújo et al. 2004).

Among the management requirements to ensure the conservation of potamotrygonids, monitoring the fisheries is the most urgent. This should include the landings both from directed fisheries and by-catch, and the number of individuals captured for the ornamental trade, including the losses by mortality in handling, transportation, and commercialization. Such data would permit the regulation of fisheries through the establishment of quotas. So far, only Brazil has adopted regulations on the ornamental trade, by a system of export quotas of authorized species (Charvet-Almeida et al. 2002).

Population monitoring should be undertaken in areas prone to be affected by major development plans, such as the construction of dams and waterways. Obviously, the acquisition of data on the population biology, ecology, and reproduction of all species is essential for the establishment of conservation plans and actions.

5.9 Conclusions

Despite the presence of freshwater stingrays in fossil and extant biotas other than South American, potamotrygonids comparatively have an intriguingly high diversity that matches the geographic and ecological diversification of South American river drainages. The assessment of their internal and external relationships and the reconstruction of their biogeographic history are still constrained by the lack of a clear taxonomy, and of the need for further studies on anatomy, genetics, paleontology, and paleogeography.

Potamotrygonids present life history characteristics that are somewhat similar to marine stingrays, but at the same time are unique for their complete adaptation to freshwater environments. The fact that they are of medical and economic importance, taken both as an ornamental and food fish, and at the same time, potentially affected by the drastic man-made changes taking place in South America, poses urgent questions regarding their exploitation potential and conservation. Further studies are required to better understand the evolutionary and natural history of these elasmobranchs, and for their adequate management and conservation.

Acknowledgments

We thank the fish curators and technicians of the mentioned institutions for the access to collections, photographs, and specimen loans. The late Thomas B. Thorson forwarded specimens and photographs; Hugo P. Castello forwarded original manuscript notes and photographs by M.N. Castex. Seiichi Watanabe of the Museum of Tokyo University of Fisheries loaned the specimen of the undescribed genus reported from Venezuela. Fernando P. Marques and M.R. de Carvalho permitted acess to the specimens collected by them. Camm Swift, G.M. Santos, H.-J. Paepke, L.N. Chao, and M.L.G. Araújo provided photos of specimens. We thank Mauricio Pinto de Almeida for his constant support in field work and contribution to the biological section of the chapter. Finally, we thank José Lima de Figueiredo for his long-term encouragement of our elasmobranch studies. CNPq and CAPES supported RSR with research grants for elasmobranch studies.

References

Achenbach GM (1967) Nota sobre una nueva especie de raya fluvial (Batoidei, Potamotrygonidae) pescada en el río Colastiné (Paraná medio, Departamento La Capital, Provincia de Santa Fe, Republica Argentina). J Com Mus prov Cienc nat F Ameghino 1: 1–7.

Achenbach GM (1969) Algunos aspectos en la respiración de la raya fluvial (Chondrichthyes, Potamotrygonidae). Com Mus prov Cienc nat F Ameghino 3: 1–12.

Achenbach GM (1972) Algunos aspectos en la respiración de la raya fluvial (Chondrichthyes, *Potamotrygon*). Acta Zool Liloana 29: 107–119.

Achenbach GM, Achenbach SVM (1976) Notas acerca de algunas especies de raya fluvial (Batoidei, Potamotrygonidae), que frecuentan el sistema hidrográfico del rio Parana medio en el Departamento La Capital (Santa Fe-Argentina). Com Mus prov Cienc nat F Ameghino 8: 1–34.

Almeida MP (2003) Pesca, policromatismo e aspectos sistemáticos de *Potamotrygon scobina* Garman, 1913 (Chondrichthyes: Potamotrygonidae) da região da ilha de Colares—baía de Marajó—Pará. Dissertation, Museu Paraense Emílio Goeldi, Universidade Federal do Pará.

Almeida MP (2008) História natural das raias de água doce (Chondrichthyes: Potamotrygonidae) na ilha de Marajó (Pará-Brasil). Dissertation, Museu Paraense Emílio Goeldi, Universidade Federal do Pará.

Almeida MP, Barthem RB, Viana AS, Charvet-Almeida P (2008) Diversidade de raias de água doce—
    (Chondrichthyes: Potamotrygonidae) no estuário amazônico. Arq Ciên Mar 41: 82–89.
Almeida MP, Barthem RB, Viana AS, Charvet-Almeida P (2009) Factors affecting the distribution
    and abundance of freshwater stingrays (Chondrichthyes: Potamotrygonidae) at Marajó island,
    mouth of the Amazon River. Pan J Aquatic Sci 4: 1–11.
Arambourg C (1947) Contribution a l'étude géologique et paléontologique du bassin du Lac Rodolphe
    et de la basse vallée de l'Omo. In: Mission scientifique de l'Omo (1932–1933), Vol. 1, Geologie—
    Anthropologie. Muséum National d'Histoire Naturelle, Paris, pp 469–471.
Araújo MLG (1998) Biologia reprodutiva e pesca de *Potamotrygon* sp. C (Chondrichthyes—
    Potamotrygonidae), no médio rio Negro, Amazonas. Dissertation, Instituto Nacional de
    Pesquisas da Amazônia, Universidade do Amazonas.
Araújo MLG, Charvet-Almeida P, Almeida MP, Pereira H (2004) Conservation perspectives and
    management challenges for freshwater stingrays. Shark News 16: 12–13.
Bailey RM (1969) Comment on the proposed suppression of *Elipesurus spinicauda* Schomburgk
    (Pisces) Z. N. (S.) 1825. Bull Zool Nomencl 25: 133–134.
Barbaro KC, Lira MS, Malta MB, Soares SL, Garrone-Neto D, Cardoso JLC, Santoro ML, Haddad
    ML Jr (2007) Comparative study on extracts from the tissue covering the stinger of freshwater
    (*Potamotrygon falkneri*) and marine (*Dasyatis guttata*) stingrays. Toxicon 50: 676–687.
Berg C (1895) Enumeración sistemática y sinonímica de los peces de las costas argentina y uruguaya.
    Ann Mus Nac Buenos Aires 4: 1–120.
Bertin L (1939a) Essai de classification et de nomenclature des poissons de la sous-classe des séla-
    ciens. Bull Inst Oceanogr Monaco 775: 1–24.
Bertin L (1939b) Catalogue des types de poissons du Muséum National d'Histoire Naturelle. 1re.
    Partie. Cyclostomes et Sélaciens. Bull Mus Nat Hist Nat, Ser. 2, 11: 51–98.
Bigelow HB, Schroeder WC (1953) Sawfishes, guitarfishes, skates and rays; chimaeroids. In Fishes
    of the Western North Atlantic. Vol. 1, Part 2, Sears Foundation for Marine Research, Yale
    University, New Haven, CT.
Bragança AJM, Charvet-Almeida P, Barthem RB (2004) Preliminary observations on the feeding of
    the freshwater stingrays *Potamotrygon orbignyi*, *Potamotrygon scobina* and *Plesiotrygon iwamae*
    (Chondrichthyes: Potamotrygonidae) in the Cotijuba Island region, Pará, Brazil. In: Martin
    RA, MacKinlay D (eds) Biology and conservation of elasmobranchs: symposium proceed-
    ings. VI International Congress on the Biology of Fish, AFS – Physiology Section, Manaus, pp
    49–59.
Brito PM, Deynat PP (2004) Freshwater stingrays from the Miocene of South America with com-
    ments on the rise of potamotrygonids (Batoidea, Myliobatiformes). In: Arratia G, Wilson MVH,
    Cloutier R (eds) Recent advances in the origin and early radiation of vertebrates. Verlag Dr.
    Friedrich Pfeil, München, pp 575–582.
Brooks DR, Mayes MA, Thorson TB (1981) Systematic review of cestodes infecting freshwater sting-
    rays (Chondrichthyes: Potamotrygonidae) including four new species from Venezuela. Proc
    Helminthol Soc Wash 48: 43–64.
Brooks DR, Thorson TB (1976) Two tetraphyllidean cestodes from the freshwater stingray *Potamotrygon
    magdalenae* Duméril 1852 (Chondrichthyes: Potamotrygonidae) from Colombia. J Parasitol 62:
    943–947.
Brooks DR, Thorson TB, Mayes MA (1981) Freshwater stingrays (Potamotrygonidae) and their helm-
    inth parasites: testing hypotheses of evolution and coevolution. In: Funk VA, Brooks DR (eds)
    Advances in cladistics, Proceedings of the First Meeting of the Willi Hennig Society. New York
    Botanical Garden, New York, pp 147–175.
Cabeza de Vaca (pseud.) (1999) Naufrágios e comentários Álvar Nunes. L&PM, Porto Alegre.
Castello HP (1973) Sobre la correcta posición sistemática de la raya de agua dulce africana
    (Condrichthyes, Dasyatidae) (Republica Federal del Camerun). Trab V Congr Latinoamer Zool,
    Montevideo, 1: 67–71.
Castello HP, Yagolkowski DR (1969) *Potamotrygon castexi* n. sp., una nueva especie de raya de agua
    dulce del Rio Parana. Acta Sci Inst Latinoamer Fisiol Reprod 6: 1–21.

Castelnau FL (1855) Animaux nouveaux ou rares recueillis pendant l'expédition dans les parties centrales de l'Amerique du Sud, de Rio de Janeiro a Lima, et de Lima au Para. Vol. 2. P. Bertrand, Paris.

Castex MN (1963a) Notas heurísticas sobre el género *Potamotrygon*. Museo Argentino de Ciencias Naturales Bernardino Rivadavia, Buenos Aires.

Castex MN (1963b) La raya fluvial. Notas histórico-geográficas. Librería y Editorial Castellví, Santa Fe.

Castex MN (1963c) El gênero *Potamotrygon* en el Parana medio. An Mus Prov Cienc Nat F Ameghino 2: 1–86.

Castex MN (1963d) Una nueva especie de raya fluvial: *Potamotrygon pauckei*. Notas distintivas. Bol Acad Nac Cienc 43: 289–294.

Castex MN (1963e) Observaciones sobre la raya de río *Potamotrygon motoro* (Müller y Henle). Com Mus Arg Cienc Nat B Rivadavia, Hidrobiología 1: 7–14.

Castex MN (1964a) Estado actual de los estudios sobre la raya fluvial neotropical. Rev Mus Prov Cienc Nat F Ameghino, número extraordinario del cincuentenario: 9–49.

Castex MN (1964b) Una nueva especie de raya fluvial americana: *Potamotrygon schuhmacheri* sp. n. Neotropica 10: 92–94.

Castex MN (1966) Observaciones en torno al género *Elipesurus* Schomburgk 1843 y nueva sinonimia de *Potamotrygon brachyurus* (Günther, 1880) (Chondrichthyes, Potamotrygonidae). Physis 26: 33–38.

Castex MN (1967a) Bases para el estudio de las rayas de água dulce del sistema amazonico. Nuevas sinonimias de "*P. motoro*" (M. H., 1841). In: Atas do Simpósio sobre a Biota Amazônica (Limnologia), Rio de Janeiro, 3: 89–92.

Castex MN (1967b) Freshwater venomous rays. In: Animal toxins, First International Symposium on Animal Toxins, Atlantic City, New Jersey. Pergamon Press, New York, pp 167–176.

Castex MN (1968) *Elipesurus* Schomburgk 1843 (Pisces): proposed suppression under the plenary powers. Z. N. (S.) 1825. Bull Zool Nomencl 24: 353–355.

Castex MN (1969) Comment on the objections forwarded by R. M. Bailey to the proposed suppression of *Elipesurus spinicauda* Schomburgk (Pisces) Z. N. (S.) 1825. Bull Zool Nomencl 26: 68–69.

Castex MN, Castello HP (1969) Nuevas sinonimias para el género monotipico *Disceus* Garman 1877 (Potamotrygonidae) y observaciones sistematicas a la familia Paratrygonidae Fowler 1948 (dubit.). Acta Sci Inst Latinoamer Fisiol Reprod 7: 1–43.

Castex MN, Castello HP (1970a) *Potamotrygon yepezi*, n sp. (Condrichthyes), una nueva especie de raya de agua dulce para los rios venezolanos. Acta Sci Inst Latinoamer Fisiol Reprod 8: 15–39.

Castex MN, Castello HP (1970b) *Potamotrygon leopoldi*, una nueva especie de raya de agua dulce para el Rio Xingú, Brasil (Chondrichthyes, Potamotrygonidae). Acta Sci Inst Latinoamer Fisiol Reprod 10: 1–16.

Castex MN, Loza F (1964) Etiologia de la enfermedad paratrygonica: estudio anatomico, histologico y funcional del aparato agressor de la raya fluvial americana (gén. *Potamotrygon*). Rev Asoc Med Argent 50: 551–554.

Castex MN, Maciel IM (1965) Notas sobre la familia Potamotrygonidae Garman 1913. Dirección General de Recursos Naturales, Publicacion Tecnica, Santa Fé, 14: 1–23.

Charvet-Almeida P (2001). Ocorrência, biologia e uso das raias de água doce na baía de Marajó (Pará, Brasil), com ênfase na biologia de *Plesiotrygon iwamae* (Chondrichthyes: Potamotrygonidae). Dissertation, Museu Paraense Emílio Goeldi, Universidade Federal do Pará.

Charvet-Almeida P (2006) História natural e conservação das raias de água doce (Chondrichthyes: Potamotrygonidae) no médio rio Xingu, área de influência do projeto hidrelétrico de Belo Monte (Pará, Brasil). Dissertation, Universidade Federal da Paraíba.

Charvet-Almeida P, Almeida MP (2003) Fishery, uses and conservation of freshwater stingrays (Chondrichthyes: Potamotrygonidae) in the Marajó Bay (Brazil). Abstract of the Joint Meeting of Ichthyologists and Herpetologists, Manaus.

Charvet-Almeida P, Almeida MP (2008) Contribuição ao conhecimento, distribuição e aos desafios para a conservação dos elasmobrânquios (raias e tubarões) no sistema Solimões-Amazonas. In: Albernaz, ALKM (org) Conservação da várzea: identificação e caracterização de regiões biogeográficas. IBAMA/ProVárzea, Manaus, pp 199–235.

Charvet-Almeida P, Araújo MLG, Almeida MP (2005) Reproductive aspects of freshwater stingrays (Chondrichthyes: Potamotrygonidae) in the Brazilian Amazon Basin. J Northw Atlan Fish Sci 35: 165–171.

Charvet-Almeida P, Araújo MLG, Rosa RS, Rincón G (2002) Neotropical freshwater stingrays: diversity and conservation status. Shark News 14: 1–2.

Compagno LJV (1990) Shark exploration and conservation. In: Pratt HL Jr, Gruber SH, Taniuchi T (eds) Elasmobranch as living resources: advances in biology, ecology and systematics, and the status of fisheries. NOAA Tech Rep NMFS 90: 391–414.

Compagno LJV (1999) Checklist of living elasmobranchs. In: Hamlett WC (ed) Sharks, skates and rays, the biology of elasmobranch fishes. John Hopkins University Press, Baltimore, pp 471–498.

Compagno LJV, Cook SF (1995) The exploitation and conservation of freshwater elasmobranchs: status of taxa and prospects for the future. In: Oetinger MI, Zorzi GD (eds) The biology of freshwater elasmobranchs, a symposium to honor Thomas B. Thorson. J Aquaricult Aqua Sci 7: 62–90.

de Carvalho MR (1996) Biology of freshwater elasmobranchs. In: Oetinger MI, Zorzi GD (eds) The biology of freshwater elasmobranchs, a symposium to honor Thomas B. Thorson. (Review). Copeia 1996: 1047–1050.

de Carvalho MR , Lovejoy NR, Rosa RS (2003) Potamotrygonidae. In: Reis RE, Ferraris CJ Jr, Kullander SO (eds) Checklist of freshwater fishes of South and Central America. Editora da Pontifícia Universidade Católica, Porto Alegre, pp 22–29.

de Carvalho MR, Maisey JG, Grande L (2004) Freshwater stingrays of the Green River Formation of Wyoming (Early Eocene), with the description of a new genus and species and an analysis of its phylogenetic relationships (Chondrichthyes: Myliobatiformes). Bull Am Mus Nat Hist 284: 1–136.

d'Orbigny A (1844) Voyage dans l'Amerique meridionale, 4 vols. P. Bertrand, Paris.

Devicenzi GJ, Teague GW (1942) Ictiofauna del Uruguay Medio. An Mus Hist Nat Montevideo 5: 1–103.

Deynat P (2006) *Potamotrygon marinae* n. sp., une nouvelle espèce de raies d'eau douce de Guyane (Myliobatiformes, Potamotrygonidae). CR Biol 329: 483–493.

Deynat P, Brito PM (1994) Révision des tubercules cutanés des raies (Chondrichthyes, Batoidea) du basin du Paraná, Tertiaire d'Amerique du Sud. Ann Paleontol 80: 237–251.

Dingerkus G (1995) Relationships of potamotrygonin stingrays (Chondrichthyes: Batiformes: Myliobatidae). In: Oetinger MI, Zorzi GD (eds) The biology of freshwater elasmobranchs, a symposium to honor Thomas B. Thorson. J Aquaricult Aqua Sci 7: 32–37.

Domingues MV, Marques FPL (2007) Revision of *Potamotrygonocotyle* Mayes, Brooks &Thorson, 1981 (Platyhelminthes: Monogenoidea: Monocotylidae), with descriptions of four new species from the gills of the freshwater stingrays *Potamotrygon* spp. (Rajiformes: Potamotrygonidae) from the La Plata river Basin. Syst Parasitol 67: 157–174.

Duméril A (1865) Histoire naturelle des poissons ou ichthyologie générale. Vol. 1, Elasmobranches plagiostomes et holocéphales ou chiméres. Librairie Encyclopédique de Roret, Paris.

Dunn KA, McEachran JD, Honeycutt RL (2003) Molecular phylogenies of myliobatiform fishes (Chondrichthyes: Myliobatiformes), with comments on the effects of missing data on parsimony and likelihood. Mol Phylog Evol 27: 259–270.

Eigenmann CH (1910) Catalogue of the fresh-water fishes of tropical and south temperate America. Rep Princeton Univ Exped Patagonia 1896–1899 (Zool.) 3: 375–511.

Eigenmann CH, Eigenmann RS (1891) A catalogue of the freshwater fishes of South America. Proc U S Natl Mus 14: 1–81.

Engelhardt R (1912) Uber einige neue Selachier-Formen. Zool Anz 39: 643–648.

Eschmeyer WN (1998) Catalog of fishes, 3 vols. California Academy of Sciences, San Francisco.

Fernández-Yépez A (1957) Nueva raya para la ciencia *Potamotrygon schroederi*, n. sp. Bol Mus Cienc Nat 1/2: 7–11.

Ferreira AR (1972) Viagem filosófica pelas capitanias do Grão Pará, Rio Negro, Mato Grosso e Cuiabá. Memórias, Zoologia, Botânica. Conselho Federal de Cultura, Rio de Janeiro.

Ferreira EJG, Zuanon JAS, dos Santos GM (1998) Peixes Comerciais do Médio Amazonas: região de Santarém, Pará. IBAMA, Brazil.

Fowler HW (1948) Os peixes de água doce do Brasil. Arq Zool São Paulo 6: 1–204.

Fowler HW (1970) A catalog of world fishes (XII). Quart J Taiwan Mus 23: 39–126.

Galvis G, Mojica JI, Camargo M (1997) Peces del Catatumbo. D'Vinni Editorial, Bogota.

Garman S (1877) On the pelvis and external sexual organs of selachiens, with special reference to the new genera *Potamotrygon* and *Disceus*. Proc Bost Soc Nat Hist 19: 197–215.

Garman S (1913) The Plagiostomia (sharks, skates and rays). Mem Mus Comp Zool 36: i–xiii + 1–515, 75 pls.

Garrone Neto D, Haddad V Jr, Vilela MJA, Uieda VS (2007) Registro de ocorrência de duas espécies de potamotrigonídeos na região do alto rio Paraná e algumas considerações sobre sua biologia. Biota Neotrop 7: 1–4.

Gerst JW, Thorson TB (1977) Effects of saline acclimation on plasma electrolytes, urea excretion and hepatic urea biosynthesis in a freshwater stingray, *Potamotrygon* sp. Garman, 1877. Comp Biochem Physiol 56A: 87–93.

Gill TN (1893) Families and subfamilies of fishes. Mem Natl Acad Sci 6: 127–138.

Glémet R (2004) La pêcherie ornamentale du rio Xingu (Pará, Brésil): approches gestionnaire et de conservation avec accentuation sur la pêche dês Potamotrygonidae. Rapport de stage de DESS Gestion des Zones Humides, Université D'Angers, Angers.

Gumilia J (1791) Historia natural, civil y geográfica de las naciones situadas en las riveras del Rio Orinoco. Barcelona.

Günther A (1870) Catalogue of fishes in the British Museum. Vol. 8. Taylor and Francis, London.

Günther A (1880) A contribution to the knowledge of the fish-fauna of the Rio de la Plata. Ann Mag Nat Hist 5: 7–15.

Haddad V Jr (2000) Atlas de Animais Aquáticos Perigosos do Brasil: guia médico de diagnóstico e tratamento de acidentes. Roca Ltda., São Paulo.

Halstead BW (1970) Poisonous and venomous marine animals of the world. Vol. 3: Vertebrates. United States Government Printing Office, Washington, DC.

Harrington HJ (1962) Paleogeographic development of South America. Bull Amer Assoc Petrol Geol 46: 1773–1814.

Hoberg EP, Brooks DR, Molina-Ureña H, Erbe E (1998) *Echinocephalus janzeni* n. sp. (Nematoda: Gnathostomatidae) in *Himantura pacifica* (Chondrichthyes: Myliobatiformes) from the Pacific coast of Costa Rica and Mexico, with historical biogeographical analysis of the genus. J Parasitol 84: 571–581.

Hoorn C (1996) Miocene deposits in the Amazonian foreland basin. Science 273: 122–123.

Hubert N, Renno JF (2006) Historical biogeography of South American freshwater fishes. J Biogeogr 33: 1414–1436.

Innocêncio NR (1977) Hidrografia. In: Geografia do Brasil. Vol. 4, Região Centro-Oeste. Instituto Brasileiro de Geografia e Estatística, Rio de Janeiro, pp 85–112.

Ishihara H, Taniuchi T (1995) A strange potamotrygonid ray (Chondrichthyes: Potamotrygonidae) from the Orinoco river system. In: Oetinger MI, Zorzi GD (eds) The biology of freshwater elasmobranchs, a symposium to honor Thomas B. Thorson. J Aquaricult Aqua Sci 7: 91–97.

Jordan DS (1887) A preliminary list of the fishes of the West Indies. Proc US Natl Mus 9: 554–608.

Jordan DS (1919) The genera of fishes, Part 3. Leland Stanford Junior Univ. Publ. Univ. Ser., i–xv + 285–410.

Jordan DS (1923) A classification of fishes including families and genera as far as known. Stanford Univ Publ Biol Sci 3: 77–243.

Larrazet J (1886) Des pièces de la peau de quelques Sélaciens fossiles. Bull Soc Geol France, Ser 3 14: 255–277.

Lasso CA (1985) Las rayas de agua dulce. Natura 77: 6–9.

Lasso CA, Rial AB, Lasso-Alcalá O (1996) Notes on the biology of the freshwater stingrays *Paratrygon aiereba* (Müller & Henle, 1841) and *Potamotrygon orbignyi* (Castelnau, 1855) (Chondrichthyes: Potamotrygonidae) in the Venezuelan llanos. Aqua 2: 39–52.

Lisboa FC (1967) História dos animais e árvores do Maranhão. Estudo e notas do Dr. Jaime Walter. Arquivo Histórico Ultramarino, Lisboa.

Lonardoni AP, Goulart E, Oliveira EF, Abelha M (2006) Hábitos alimentares e sobreposição trófica das raias *Potamotrygon falkneri* e *Potamotrygon motoro* (Chondrichthyes: Potamotrygonidae) na planície alagável do alto rio Paraná, Brasil. Acta Sci Biol Sci 28: 195–202.

Lovejoy NR (1996) Systematics of myliobatoid elasmobranches: with emphasis on the phylogeny and historical biogeography of neotropical freshwater stingrays (Potamotrygonidae: Rajiformes). Zool J Linn Soc 117: 207–257.

Lovejoy NR (1997) Stingrays, parasites, and neotropical biogeography: a closer look at Brooks et al.'s hypotheses concerning the origin of neotropical freshwater rays (Potamotrygonidae). Syst Biol 46: 218–230.

Lovejoy NR, Bermingham E, Andrew AP (1998) Marine incursion into South America. Nature 396: 421–422.

Luengo JA (1966) Relación de los géneros y especies de peces descritos por Garibaldi J. Devicenzi y de los tipos depositados en el Museo Nacional de Historia Natural de Montevideo. Atas Soc Biol Rio de Janeiro 10: 19–21.

Lundberg JG, Marshall JG, Guerrero J, Horton B, Malabarba MCSL, Wesselingh F (1998) The stage for neotropical fish diversification: A history of tropical South American Rivers. In: Malabarba LR, Reis RE, Vari RP, Lucena ZM, Lucena CAS (eds) Phylogeny and classification of neotropical fishes. Edipucrs, Porto Alegre, pp 13–48.

Marques F (2000) Evolution of neotropical freshwater stingrays and their parasites: taking into account space and time. Dissertation, University of Toronto.

Mayes MA, Brooks DR, Thorson TB (1978) Two new species of *Acanthobothrium* van Beneden 1849 (Cestoidea: Tetraphyllidea) from freshwater stingrays in South America. J Parasitol 64: 838–841.

McEachran JD, Aschliman N (2004) Phylogeny of Batoidea. In: Carrier JC, Musick JA, Heithaus MR (eds) Biology of sharks and their relatives. CRC Press, Boca Raton, FL, pp 79–113.

McEachran JD, Dunn KA, Miyake T (1996) Interrelationships of the batoid fishes (Chondrichthyes: Batoidea). In: Stiassny MLJ, Parenti LR, Johnson GD (eds) Interrelationships of fishes. Academic Press, London, pp 63–82.

Miyake T (1988) The systematics of the stingray genus *Urotrygon* with comments on the interrelationships within Urolophidae (Chondrichthyes, Myliobatiformes). Dissertation, Texas A&M University.

Müller J, Henle FGJ (1841) Systematische beschreibung der Plagiostomen. Verlag von Veit, Berlin.

Nelson JS (2006) Fishes of the world, 4th edition. John Wiley & Sons, Hoboken, NJ.

Neumann D (2006) Type Catalogue of the Ichthyological Collection of the Zoologische Staatssammlung München. Part I: Historic type material from the "Old Collection," destroyed in the night 24/25 April 1944. Spixiana 29: 259–285.

Nishida K (1990) Phylogeny of the suborder Myliobatidoidei. Mem Fac Fish Hokkaido Univ 37:1–108.

Olazarri J, Mones A, Ximénez A, Philippi ME (1970) Lista de los exemplares-tipo depositados en el Museo Nacional de Historia Natural de Montevideo, Uruguay. Com Zool Mus Hist Nat Montevideo 10: 1–12.

Pântano-Neto J (2001) Estudo preliminar da anatomia descritiva e funcional associada à alimentação em raias de água-doce (Potamotrygonidae, Myliobatiformes, Elasmobranchii). Dissertation, Universidade de São Paulo.

Pântano-Neto J, Souza AM (2002) Anatomia da musculatura oro-braquial associada à alimentação de duas espécies de raias de água doce (Potamotrygonidae; Elasmobranchii). Publ Avulsas Inst Pau Bras Hist Nat 5: 53–65.

Pardal PPO, Rezende MB (1997) Acidentes por peixes. In: Leão RNQ (ed) Doenças Infecciosas e Parasitárias: enfoque amazônico. Cejup Ltda., Belém, pp 813–817.

Pedroso CM, Jared C, Charvet-Almeida P, Almeida MP, Garrone Neto D, Lira MS, Haddad V Jr, Bárbaro KC, Antoniazzi MM (2007) Morphological characterization of the venom secretory epidermal cells in the stinger of marine and freshwater stingrays. Toxicon 50: 688–697.

Pierini SV, Warrell DA, de Paulo A, Theakston RDG (1996) High incidence of bites and stings by snakes and other animals among rubber tappers and Amazonian Indians of the Juruá valley, Acre State, Brazil. Toxicon 34(2): 225–236.

Pró-Várzea/IBAMA (2002) Estatística Pesqueira do Amazonas e Pará—2001. IBAMA/Pró-Várzea, Manaus.

Quijada CCD (2003) Relações filogenéticas intergenéricas de raias neotropicais de água doce (Chondrichthyes: Potamotrygonidae). Dissertation, Universidade Federal da Paraíba.

Räsänen ME, Linna AM, Santos JCR, Negri FR (1995) Late Miocene tidal deposits in the Amazonian foreland basin. Science 269: 386–390.

Räsänen M, Linna A, Irion G, Hernani LR, Huaman RV, Wesselingh F (1998) Geología y geoformas de la zona de Iquitos. In: Kalliola R, Paitán SF (eds) Geoecología y Desarrollo Amazónico: Estudio Integrado en la Zona de Iquitos, Perú. Turun Yliopisto, Turku, pp 59–137.

Ribeiro AM (1907) Fauna brasiliense. Peixes II (Desmobranchios). Archos Mus Nac Rio de Janeiro 14: 129–217.

Ribeiro BG (1979) Diário do Xingu. Paz e Terra, Rio de Janeiro.

Rincon G (2004) *Potamotrygon henlei*. In: 2009 IUCN Red List of Threatened Species. Available on http://www.iucnredlist.org/details/39402/0. Acessed January 27, 2009.

Rincon G (2006) Aspectos taxonômicos, alimentação e reprodução da raia de água doce *Potamotrygon orbignyi* (Castelnau) (Elasmobranchii: Potamotrygonidae) no rio Paraná—Tocantins. Dissertation, Universidade Estadual Paulista.

Roddaz M, Brusset S, Baby P, Hérail G (2006) Miocene tidal-influenced sedimentation to continental Pliocene sedimentation in the forebulge–backbulge depozones of the Beni–Mamore foreland Basin (northern Bolivia). J S Amer Earth Sci 20: 351–368.

Rosa RS (1985a) A systematic revision of the South American freshwater stingrays (Chondrichthyes: Potamotrygonidae). Dissertation, College of William and Mary.

Rosa RS (1985b) Further comment on the nomenclature of the freshwater stingray *Elipesurus spinicauda* Schomburgk, 1843 (Chondrichthyes: Potamotrygonidae). Revta Bras Zool 3: 27–31.

Rosa RS (1991) *Paratrygon aiereba* (Muller and Henle, 1841) the senior synonym of the freshwater stingray *Disceus thayeri* Garman, 1913 (Chondrichthyes: Potamotrygonidae). Revta bras Zool 7: 425–437.

Rosa RS, Castello HP, Thorson TB (1987) *Plesiotrygon iwamae*, a new genus and species of Neotropical freshwater stingray (Chondrichthyes: Potamotrygonidae). Copeia 1987: 447–458.

Rosa RS, de Carvalho MR (2007) Família Potamotrygonidae. In: Buckup PA, Menezes NA, Ghazzi MS (orgs) Catálogo das espécies de peixes de água doce do Brasil. Museu Nacional, Rio de Janeiro, pp 17–18.

Rosa RS, de Carvalho MR, Wanderley CA (2008) *Potamotrygon boesemani* (Chondrichthyes: Myliobatiformes: Potamotrygonidae), a new species of neotropical freshwater stingray from Surinam. Neotr Ichthyol 6: 1–8.

Ross RA, Schäfer F (2000) Aqualog Süsswasser Rochen: Freshwater Rays. Verlag ACS, Mörfelden-Walldorf.

Roulin M (1829) Description d'une pastenague fluviatile du Meta (Pastenague de Humboldt). Ann Sci Nat 6: 104–107.

Santos GM, Jegu M, Merona B (1984) Catálogo de Peixes Comerciais do baixo Rio Tocantins: projeto Tucuruí. Eletronorte, CNPq and INPA, Manaus, pp 15–16.

Schluessel V, Bleckmann H (2005) Spatial memory and orientation strategies in the elasmobranch *Potamotrygon motoro*. J Comp Physiol A 191: 695–706.

Schomburgk RH (1843) Fishes of British Guiana, Part 2. In: Jardine W (ed) Naturalist's Library, Vol. 40. W.H. Lizars, Edinburgh.

Séret B, McEachran JD (1986) Catalogue critique des types de Poissons du Muséum national d'Histoire naturelle (Suite). Poissons Batoides (Chondricthyes, Elasmobranchii, Batoidea). Bull Mus Nat Hist Nat 8: 3–50.

Sheborn CD, Griffin JF (1934) On the dates of publication of the natural history portions of Alcide d'Orbigny's "Voyage Amerique Méridionale." Ann Mag Nat Hist 10: 130–134.

Silva TB, Uieda VS (2007) Preliminary data on the feeding habits of the freshwater stingrays *Potamotrygon falkneri* and *Potamotrygon motoro* (Potamotrygonidae) from the Upper Paraná River basin, Brazil. Biota Neotr 7: 221–226.

Sioli H (1967) Studies in Amazonian waters. In: Atas do Simpósio sobre a biota amazônica, Vol. 3 (Limnologia). CNPq, Museu Paraense Emílio Goeldi, Belém, pp 9–50.

Soares LC (1977) Hidrografia. In: Geografia do Brasil. Vol. 1, Região Norte. Instituto Brasileiro de Geografia e Estatística, Rio de Janeiro, pp 95–166.

Stauch A, Blanc M (1962) Description d'un selacien rajiforme des eaux douces du Nord Cameroun: *Potamotrygon garouaensis* n. sp. Bull Mus Nat Hist Nat 34: 166–171.

Steindachner F (1878) Zur Fisch-fauna des Magdalenen-Stromes. Denk K akad Wiss Wien 39: 19–78.

Taniuchi T (1982) (in Japanese) Investigational report of freshwater stingrays in South America. In: Studies on the adaptability and phylogenetic evolution of freshwater elasmobranchs. Scientific Research Team on Freshwater Elasmobranchs. University of Tokyo, Tokyo, pp 21–58.

Taniuchi T, Ishihara H (1990) Anatomical comparison of claspers of freshwater stingrays (Dasyatidae and Potamotrygonidae). Jap J Ichthyol 37–: 10–16.

Teshima K, Takeshita K (1992) Reproduction of the freshwater stingray, *Potamotrygon magdalenae* taken from the Magdalenae River system in Colombia, South America. Bull Seikai Natl Fish Res Inst 70: 11–27.

Thorson TB (1970) Freshwater stingrays, *Potamotrygon* spp.: failure to concentrate urea when exposed to saline medium. Life Sci 9: 893–900.

Thorson TB, Brooks DR, Mayes MA (1983) The evolution of freshwater adaptation in stingrays. Nat Geog Res Rep 15: 663–694.

Thorson TB, Cowan CM, Watson DE (1967) *Potamotrygon* spp.: elasmobranchs with low urea content. Science 158: 375–377.

Thorson TB, Langhammer JK, Oetinger MI (1983) Reproduction and development of the South American freshwater stingrays, *Potamotrygon circularis* and *P. motoro*. Env Biol Fish 9: 3–24

Thorson TB, Langhammer JK, Oetinger MI (1988) Periodic shedding and replacement of venomous caudal spines, with special reference to South American freshwater stingrays, *Potamotrygon* spp. Environ Biol Fish 23: 299–314.

Thorson TB, Watson DE (1975) Reassignment of the African freshwater stingray, *Potamotrygon garouaensis*, to the genus *Dasyatis*, on physiologic and morphologic grounds. Copeia 1975: 701–712.

Thorson TB, Wotton RM, Georgi TA (1978) Rectal gland of freshwater stingrays, *Potamotrygon* spp. (Chondrichthyes: Potamotrygonidae). Biol Bull 154: 508–516.

Toffoli D (2006) História evolutiva de espécies do gênero *Potamotrygon* Garman, 1877 (Potamotrygonidae) na Bacia Amazônica. Dissertation, Instituto Nacional de Pesquisas da Amazônia, Universidade Federal do Amazonas.

Toffoli D, Hrbek T, Araújo MLG, Almeida MP, Charvet-Almeida P, Farias IP (2008) A test of the utility of DNA barcoding in the radiation of the freshwater stingray genus *Potamotrygon* (Potamotrygonidae, Myliobatiformes). Genet Mol Biol 31: 324–336.

Vaillant L (1880) Sur les raies recueillies dans l'Amazone par M. le Dr. Jobert. Bull Soc Philom Paris 7: 251–252.

Vellard J (1931) Venim des raies (*Taeniura*) du Rio Araguaya (Brésil). CR Acad Sci Paris 192: 1279–1281.

Wallace AR (2002) Fishes of the Rio Negro. Monica Toledo-Piza (org). Museu de Zoologia Universidade de São Paulo, São Paulo.

Wesselingh FP (2006) Miocene long-lived Lake Pebas as a stage of mollusc radiations, with implications for landscape evolution in western Amazonia. Scripta Geol 133: 1–17.

Winemiller K (1989) Patterns of variation in life history among South American fishes in seasonal environments. Oecologia 81: 225–241.

Winemiller K, Taphorn D (1989) La evolución de las estrategias de vida en los peces de los llanos ocidentales de Venezuela. Biollania 6: 77–122.

Yossa MI (2004) Nutritional value of *Potamotrygon motoro* diet (Chondrichthyes - Potamotrygonidae). In: Martin RA, MacKinlay D (eds) Biology and conservation of elasmobranchs: symposium proceedings. VI International Congress on the Biology of Fish, AFS – Physiology Section, Manaus, p 76.

Zorzi GD (1995) The biology of freshwater elasmobranchs: an historical perspective. In: Oetinger MI, Zorzi GD (eds) The biology of freshwater elasmobranchs, a symposium to honor Thomas B. Thorson. J Aquaricult Aqua Sci 7: 10–31.

Zuanon JAS (1999) História natural da ictiofauna de corredeiras do rio Xingu, na região de Altamira, Pará. Dissertation, Universidade Estadual de Campinas.

Appendix 5.1

FIGURE 5.A.1–5.A.10 (See facing page.)
A color version of this figure follows page 336. Specimens of Potamotrygonidae. **5.A.1** *Paratrygon aiereba*, MCZ 297-S, syntype of *Disceus thayeri*, dorsal view. (Photo by R.S. Rosa.) **5.A.2** *Paratrygon aiereba*, ZMB 4632, type of *Trygon strogylopterus*, dorsal view. (Photo courtesy of H.-J. Paepke.) **5.A.3** *Paratrygon aiereba*, detail of dorsal disc, showing eyes and spiracular projection. (Photo by R.S. Rosa.) **5.A.4** *Plesiotrygon iwamae*, MZUSP 10153, holotype, dorsal view. (Photo by R.S. Rosa.) **5.A.5** *Plesiotrygon iwamae*, dorsal view of freshly collected specimen. (Photo courtesy of L.N. Chao.) **5.A.6** Undescribed genus and species, dorsal view of freshly collected specimen, approximately 718 mm disc length. (Photo by P.C. Almeida.) **5.A.7** Undescribed genus and species, dorsal view of disc with detail of eyes and spiracles. (Photo by R.S. Rosa.) **5.A.8** Comparative view of anterior portion of disc between *Paratrygon aiereba* (left) and the undescribed genus and species (right). (Photo by P.C. Almeida.) **5.A.9** *Potamotrygon boesemani*, USNM225574, holotype, dorsal view. (Photo by R.S. Rosa.) **5.A.10** *Potamotrygon brachyura*, MFA uncatalogued specimen, dorsal view. (Photo by R.S. Rosa. All rights reserved.)

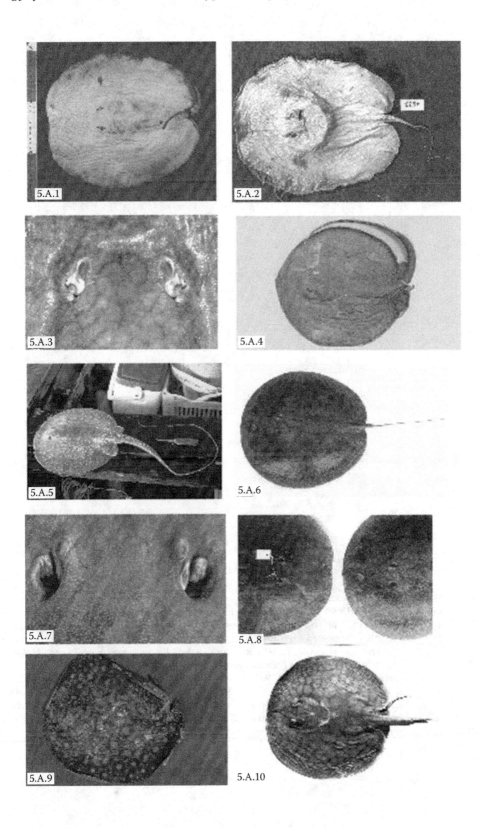

5.A.1

5.A.2

5.A.3

5.A.4

5.A.5

5.A.6

5.A.7

5.A.8

5.A.9

5.A.10

FIGURE 5.A.11–5.A.20 (See facing page.)
A color version of this figure follows page 336. Specimens of Potamotrygonidae. **5.A.11** *Potamotrygon castexi*, MACN 5777, holotype, dorsal view of freshly collected specimen. (Photo by H.P. Castello.) **5.A.12** *Potamotrygon* cf. *castexi*, dorsal view of freshly collected specimen in Peru. (Photo courtesy of C. Swift.) **5.A.13** *Potamotrygon circularis*, MCZ 291-S, syntype, dorsal view. (Photo by R.S. Rosa.) **5.A.14** *Potamotrygon constellata* MNHN A.1010. type, dorsal view. (Photo by R.S. Rosa.) **5.A.15** *Potamotrygon dumerilii*, MNHN 2367, type, dorsal view. (Photo by R.S. Rosa.) **5.A.16** *Potamotrygon* cf. *dumerilii*, MFA 268, specimen mistakenly selected as lectotype of *P. histrix* by M.N. Castex, dorsal view. (Photo by M.N. Castex.) **5.A.17** *Potamotrygon falkneri*, MFA 235, paratype, dorsal view. (Photo by R.S. Rosa.) **5.A.18** *Potamotrygon henlei*, dorsal view of freshly collected female specimen and aborted embryo, in Rio Tocantins, Brazil. (Photo by G.M. Santos.) **5.A.19** *Potamotrygon histrix*, MNHN 2449, type, dorsal view. (Photo by R.S. Rosa.) **5.A.20** *Potamotrygon humerosa*, MCZ 299-S, holotype, dorsal view. (Photo by R.S. Rosa. All rights reserved.)

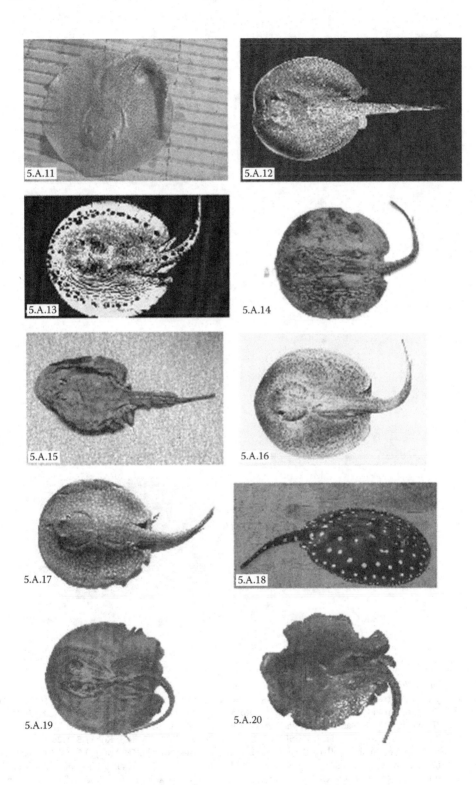

5.A.11

5.A.12

5.A.13

5.A.14

5.A.15

5.A.16

5.A.17

5.A.18

5.A.19

5.A.20

FIGURE 5.A.21–5.A.30 (See facing page.)
A color version of this figure follows page 336. Specimens of Potamotrygonidae. **5.A.21** *Potamotrygon laticeps*, MCZ 605-S, syntype, dorsal view. (Photo by R.S. Rosa.) **5.A.22** *Potamotrygon magdalenae*, dorsal view of specimen collected by T.B. Thorson in Colombia. (Photo by R.S. Rosa.) **5.A.23** *Potamotrygon menchacai*, MFA 289, holotype, dorsal view. (Photo by R.S. Rosa.) **5.A.24** *Potamotrygon motoro*, ZMB 4662, type, dorsal view. (Photo by M.N. Castex.) **5.A.25** *Potamotrygon orbignyi*, MZUSP 14790, dorsal view. (Photo by R.S. Rosa.) **5.A.26** *Potamotrygon schroederi*, dorsal view of freshly collected specimen. (Photo courtesy of M.L.G. Araújo.) **5.A.27** *Potamotrygon schuehmacheri*, MFA 269, holotype, dorsal view. (Photo by R.S. Rosa.) **5.A.28** *Potamotrygon scobina*, LACM 39925, dorsal view. (Photo by R.S. Rosa.) **5.A.29** *Potamotrygon signata* dorsal view of freshly collected specimen in Rio Parnaíba drainage, Brazil. (Photo by C.C.D. Quijada.) **5.A.30** *Potamotrygon yepezi*, USNM 121668, paratype, dorsal view. (Photo by H.P. Castello. All rights reserved.)

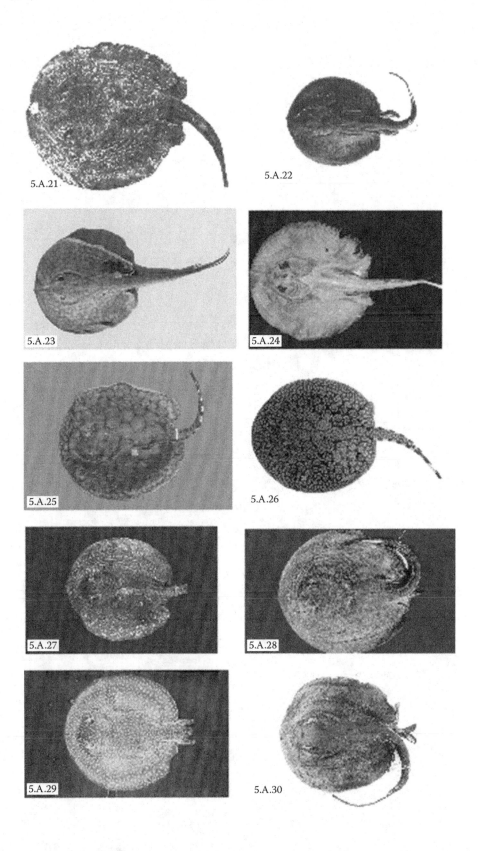

5.A.21

5.A.22

5.A.23

5.A.24

5.A.25

5.A.26

5.A.27

5.A.28

5.A.29

5.A.30

6

Life History Strategies of Batoids

Michael G. Frisk

CONTENTS

6.1 Introduction

Extant batoid species are grouped into 20 families and six orders that contain at least 513 species (McEachran and Dunn 1998; McEachran and Fechhelm 1998). The Rajidae is the most speciose family, with at least 232 species. Most batoids are benthic, feeding on crustaceans, mollusks, polychaetes, and fishes (McEachran and Dunn 1998; Ebert and Bizzarro 2007; Mabragana and Giberto 2007). Batoids have adapted to a wide range of habitats, occurring in all the oceans, and they are commonly found in shallow estuarine, coastal, and shelf regions and in depths up to 3000 m (McEachran and Fechhelm 1998; McEachran and Aschliman 2004). Although most batoids are marine, the family Potamotrygonidae and some species in the family Pristidae occur in fresh water (McEachran and Fechhelm 1998). Interest in batoid fishes has grown and is exemplified by developing fisheries (NEFSC 1999), population collapses (Brander 1981; McPhie and Campana 2009), and interactions with commercially important species (Brander 1981; Link, Garrison, and Almeida 2002; Myers et al. 2007). More broadly, the importance of batoids has been highlighted by their possible role in structuring ecosystem dynamics (Murawski 1991; Link, Garrison, and Almeida 2002; Heithaus 2004; Frisk et al. 2008).

The comparative life histories of batoids have received less attention than teleost and other elasmobranch species. Here I will review life histories of batoid orders, and when possible provide life history analyses to draw comparisons with other phylogenetic groups. The aim of the chapter is to review the following subjects: (1) the current understanding of batoid life histories in relation to other phylogenetic groups; (2) the potential for compensatory population dynamics; and (3) migration patterns. Several batoid orders suffer from a paucity of published data and for those groups only a brief biological description is provided. One consequence of the taxonomic distribution of published data is that most of our understanding of the biology, ecology, and population dynamics of the cohort is derived from a single family, the Rajidae.

6.2 Demographics

Data on age at maturity (T_{mat}), longevity (T_{max}), and von Bertalanffy growth coefficients (k) were obtained from existing comparative analyses and databases including the following: (1) Frisk, Miller, and Fogarty (2001); (2) Ebert, Compagno, and Cowley (2008); (3) Cailliet and Goldman (2004); and (4) the Pacific shark database (web page: psrc.mlml.calstate.edu/current-research/shark-fisheries-database/). Additionally, searches were conducted in ISI Web of Knowledge for the years 2000 to 2008 with the following search terms: Batoid(s); Rajiform(es); skate(s); ray(s), sawfish(es), guitarfish(es) and all family names. The grey literature was not specifically searched, but was included if it was used in previous peer-reviewed research.

6.2.1 Definition of Vital Rates

I followed standard definitions of vital rates, including the following: (1) maximum size (L_{max}) is based on observations of maximum total length for a species; (2) length at maturity (L_{mat}) is generally estimated at the point when 50% of individuals in a species or stock reach maturity; (3) longevity (T_{max}) is the maximum age observed for a species or stock within a species; (4) age at maturity (T_{mat}) is estimated as the age when 50% of individuals in a species or stock reach maturity; and (5) growth rate (k) is the von Bertalanffy growth coefficient. Theoretical estimates of maximum size or longevity based on analysis of growth parameters were not used in life history comparisons unless specifically identified. For all analyses, appropriate units will be reported.

6.2.2 Rajiformes: Rajidae (Skates), Rhinobatidae (Guitarfish)

6.2.2.1 Rajidae: Skates

Skates are bottom-dwelling marine species that feed on various demersal prey, including polychaetes, amphipods, decapods, and fishes (Richards, Merriman, and Calhoun 1963; McHugh 2001; Gedamke 2006; Ebert and Bizzarro 2007). For marine taxa, they are one of the most vulnerable to population declines and extirpations as a result of exploitation (Dulvy et al. 2000; Dulvy and Reynolds 2002).

Longevity and maximum size: Direct observations of longevities for Rajidae species overall were as high as 20 to 28 years (Table 6.1). Where sex-specific estimates were provided females live an average of 17.5 (SD = 5.7) years, while males live on average 17.3 (SD = 4.94) years (Table 6.2). Individual estimates for the family range from 8 years for male Roundel skate, *Raja texana* (Sulikowski et al. 2007b), to 50 years for the common skate, *Dipturus batis*, in the Celtic Sea (DuBuit 1972). However, the latter estimate was based on extrapolation of von Bertalanffy growth parameters from the oldest aged specimen (23 years). Maximum total lengths in Rajidae range from 31 cm for the Argentinean zipper sand skate, *Psammobatis extent*, to the Chilean rough skin skate, *Dipturus trachyderma*, which reaches a theoretical maximum size of 265 cm (Licandeo, Cerna, and Cespedes 2007).

Age and size of maturation: Based on a handful of estimates, females mature at an average age of 10.1 (SD = 3.26) years, about 64% of maximum age. Males mature at an average of 9.1 (SD = 3.30) years, approximately 60% of maximum age (Table 6.2). Length at maturation in skates ranges between 60% and 90% of maximum length (Holden 1973, 1974; Frisk, Miller, and Fogarty 2001; Ebert, Compagno, and Cowley 2008; Ebert, Smith, and Cailliet 2008).

Growth: Rajidae are slow-growing fishes. Estimates of the von Bertalanffy growth coefficient k, which expresses the rate at which individuals achieve maximum size, ranges from very slow in the cuckoo ray, *Leucoraja naevus*, with $k = 0.019$ yr$^{-(1)}$ to relatively fast $k = 0.50$ yr^{-1} in the brown ray, *Raja miraletus* (Zeiner and Wolf 1993). Average growth rates for Rajidae were ($k_{male} = 0.14$, SD = 0.10; k_{female} 0.12, SD = 0.07).

Reproduction: All skate species are oviparous and produce egg cases. On average, they have higher fecundity than other elasmobranchs (Dulvy et al. 2000), with annual fecundity ranging from 17 in thorny skate, *Amblyraja radiate*, to 153 in thornback ray, *Raja clavata* (Holden 1972; Ryland and Ajayi 1984). Generally, fecundity is thought to range from 40 to 150 eggs per year (Hoenig and Gruber 1990; Ellis and Shackley 1995; Musick and Ellis 2005). However, estimating fecundity in skates is difficult because egg-laying rates in the wild are difficult to measure and fecundity is often reported as ovarian fecundity. The Dolfinarium Harderwijk's aquarium "Ray Reef" exhibit contained five species of skate from 1993 to 2001 and provided an opportunity to observe egg production in captivity (Koop 2005). Over this period, 41,772 eggs were produced, of which 39% contained yolk sacs, and only 3% hatched. All species combined resulted in an annual average fecundity of 117 eggs per female. Field-based and captive studies appear to predict similar fecundities (Koop 2005). However, a better understanding of hatching success and the proportion of eggs with yolk sacs is necessary for some population models (Frisk, Miller, and Fogarty 2002). Previous estimates of hatching success in oviparous elasmobranchs ranged from 20% to 60% (Grover 1972; Smith and Griffiths 1997).

6.2.2.2 Rhinobatidae: Guitarfish

The Rhinobatidae contains about 40 small- to medium-bodied, benthic species (Last and Stevens 1994; McEachran and Fechhelm 1998). Very little is known regarding the family's life history strategies. All species in the family are yolksac viviparous, with reported litter sizes of 1 to 28 (mean: 6.7) (Musick and Ellis 2005). Timmons and Bray (1997) reported that the shovelnose guitarfish, *Rhinobatos productus*, lives 11 years, female first reproduction occurs at eight years, and the species has a slow growth coefficient ($k_{female} = 0.016$; $k_{male} = 0.095$). Females of the species produce 4 to 18 pups in a litter (Villavicencio-Garayzar 1995; Dowton-Hoffman 1996).

TABLE 6.1

Data for Comparative Analyses

Family	Name	Species	Region	T_{max} Both	Female	Male	T_{mat} Both	Female	Male	k Both	Female	Male	Source
Pristidae	Largetooth sawfish	Pristis microdon	Australia and Papua New Guinea	44.0						0.070			Tanaka 1991
Pristidae	Smalltooth sawfish	Pristis pectinata	Southeastern U.S.	30.0			10.0			0.080			Simpfendorfer 2000
Pristidae	Large-tooth sawfish	Pristis perotteti	Southeastern U.S.	30.0			10.0			0.080			Simpfendorfer 2000
Dasyatidae	Southern stingray	Dasyatis americana	Southeastern U.S.		26.0	28.0		5.5	3.5		0.539	0.206	Henningsen 2002
Dasyatidae	Blue stingray	Dasyatis chrysonota	South Africa		14.0	9.0		4.0	4.0		0.070	0.170	Cowley 1997
Dasyatidae	Common stingray	Dasyatis pastinaca	Eastern Mediterranean	10.0			4.5			0.089			Ismen 2003
Dasyatidae	pelagic stingray	Dasyatis violacea	Monterey Bay Aquarium	10.0							0.200	0.350	Mollet et al. 2002; Mollet and Cailliet 2002
Dasyatidae	pelagic stingray	Dasyatis violacea	Pacific	12.0			3.0						Neer 2008
Dasyatidae	pelagic stingray	Dasyatis violacea	Central America					3.0	2.0				Mollet 2002
Dasyatidae	Freshwater whipray	Himantura chaophraya	Rivers in South Asia				7.0						White et al. 2001
Dasyatidae	Sharpnose stingray	Himantura gerrardi	Rivers in South Asia	7.0			5.0						White et al. 2001
Dasyatidae	Scaly whipray	Himantura imbricata	Rivers in South Asia	3.0			2.0						Tanaka and Ohnishi 1998
Dasyatidae		Himantura laosensis	Rivers in South Asia				2.5						Tanaka and Ohnishi 1998
Dasyatidae	White-edge freshwater whip ray	Himantura signifer	Rivers in South Asia	4.0	5.0	5.0		1.5	2.5				Tanaka and Ohnishi 1998

Family	Common name	Species	Location								Reference
Dasyatidae	Cowtail stingray	Pastinachus sephen	Rivers in South Asia	19.5		7.0	5.0				Tanaka and Ohnishi 1998
Mobulidae	Manta ray	Manta birostris	Eastern North Pacific								Homma et al. 1997
Myliobatidae	Bat ray	Myliobatis californica	Elkhorn Slough, California	24.0		6.0	2.5				Martin and Cailliet 1988a; Martin and Cailliet 1988b
Narcinidae	Cortez electric ray	Narcine entemedor	Baja California Sur, Mexico	15.0		11.0			0.302	0.104	Villavicencio-Garayzar 2000
Rajidae	Thorny skate	Amblyraja radiata	Eastern Canadian coast	28.0		10.7	14.7	0.070	0.060	0.090	McPhie and Campana 2009
Rajidae	Thorny skate	Amblyraja radiata	Gulf of Maine	16.0		11.0	10.9	0.130	0.130	0.110	Sulikowski et al. 2006; Sulikowski et al. 2005
Rajidae	Alaska skate	Bathyraja parmifera	Bering Sea	17.0	15.0	10.0	9.0	0.087	0.087	0.120	Matta and Gunderson 2007
Rajidae	Blue skate	Dipturus batis	Celtic Sea	23.0		11.0		0.057			DuBuit 1972
Rajidae	New Zealand smooth skate	Dipturus innominatus	New Zealand	24.0	24.0	13.0	8.0	0.095			Francis et al. 2001
Rajidae	barndoor skate	Dipturus laevis	Georges Bank			6.5	5.8	0.140			Gedamke et al. 2005
Rajidae	New Zealand rough skate	Dipturus nasutus	New Zealand	9.0	9.0	6.0	4.0	0.160			Francis et al. 2001.
Rajidae	Slime skate	Dipturus pullopunctata	South Africa	14.0		5.0		0.050			Walmsley-Hart et al. 1999
Rajidae	Roughskin skate	Dipturus trachyderma	Chile	26.0	25.0	17.0	15.0		0.079	0.087	Licandeo et al. 2007
Rajidae	Little skate	Leucoraja erinacea	NW Atlantic	12.5		7.3	7.1	0.190	0.200	0.170	Frisk and Miller 2006
Rajidae	Little skate	Leucoraja erinacea	Eastern Canadian coast	12.0		6.9		0.190	0.200	0.170	McPhie and Campana 2009
Rajidae	Cuckoo ray	Leucoraja naevus	Celtic Sea	13.5	9.0			0.019			DuBuit 1972
Rajidae	Winter skate	Leucoraja ocellata	Gulf of Maine	18.0		11.5	11.0		0.059	0.074	Sulikowski et al. 2003; Sulikowski et al. 2005

Continued

TABLE 6.1 (Continued)

Data for Comparative Analyses

Family	Name	Species	Region	T_{max} Both	T_{max} Female	T_{max} Male	T_{mat} Both	T_{mat} Female	T_{mat} Male	k Both	k Female	k Male	Source
Rajidae	Winter skate	Leucoraja ocellata	Georges Bank		19.5	20.5		12.5		0.070	0.070	0.080	Frisk and Miller 2006
Rajidae	Winter skate	Leucoraja ocellata	Eastern Canadian coast	19.0				13.4	11.3	0.180	0.240	0.180	McPhie and Campana 2009
Rajidae	Yellow spotted skate	Leucoraja wallacei	South Africa		12.0			7.0			0.260		Walmsley-Hart et al. 1999
Rajidae	Smooth skate	Malacoraja senta	Eastern Canadian coast	15.0				10.1	11.7	0.120	0.100	0.150	McPhie and Campana 2009
Rajidae	Smooth skate	Malacoraja senta	Gulf of Maine		14.0	15.0					0.120	0.120	Natanson et al. 2007
Rajidae	Big skate	Raja binoculata	Monterey Bay, CA		12.0	11.0		11.0	7.5				Zeiner and Wolf 1993
Rajidae	Big skate	Raja binoculata	Gulf of Alaska		14.0	15.0					0.080	0.152	Gburski et al. 2007
Rajidae	Big skate	Raja binoculata	British Columbia		26.0	25.0					0.040	0.050	Gburski et al. 2007
Rajidae	Brown ray	Raja miraletus	Off Egypt	18.0	17.2	15.0	5.0				0.250	0.502	Abdel-Aziz 1992
Rajidae	Longnose skate	Raja rhina	Gulf of Alaska		24.0	25.0					0.037	0.056	Gburski et al. 2007
Rajidae	Longnose skate	Raja rhina	Monterey Bay, CA	17.0	12.0	13.0		10.5	7.0	0.260	0.160	0.250	Zeiner and Wolf 1993
Rajidae	Longnose skate	Raja rhina	West coast		22.0	20.0					0.051	0.042	Gburski et al. 2007
Rajidae	Longnose skate	Raja rhina	British Columbia		26.0	23.0					0.060	0.070	Gburski et al. 2007
Rajidae	Roundel skate	Raja texana	Gulf of Mexico, U.S.		9.0	8.0		5.8	5.0		0.229	0.286	Sulikowski et al. 2007
Rajidae	Yellownose skate	Dipturus chilensis	Golfo de Arauco, Chile		20.0	14.0					0.127	0.123	Licandeo and Cerna 2007

Family	Common name	Species	Location	T_{max}			T_{mat}			k			Reference
				female	male	both	female	male	both	female	male	both	
Rajidae	Yellownose skate	*Dipturus chilensis*	Valdivia, Chile	21.0	18.0		14.4	11.2		0.112	0.134		Licandeo and Cerna 2007
Rajidae	Yellownose skate	*Dipturus chilensis*	Southern fjords, Chile	21.0	17.0		13.5	10.7		0.104	0.116		Licandeo and Cerna 2007
Rajidae	Yellownose skate	*Dipturus chilensis*	Patagonian fjords, Chile	22.0	19.0		12.8	10.3		0.087	0.110		Licandeo and Cerna 2007
Rajidae	New Zealand rough skate	*Zearaja nasutus*	New Zealand			9.0	6.0	4.0				0.160	Francis et al. 2001
Urolophidae	Western shovelnose stingaree	*Trygonoptera mucosa*	Western Australia	17.0	12.0		5.0	2.0		0.241	0.493		White et al. 2002
Urolophidae	Masked stingaree	*Trygonoptera personalis*	Western Australia	16.0	10.0		4.0	4.0		0.143	0.203		White et al. 2002
Urolophidae	Round Stingray	*Urobatis halleri*	Eastern North Pacific				2.6	2.6					Babel 1967
Urolophidae	Lobed stingaree	*Urolophus lobatus*	New Zealand	14.0	13.0		4.0	2.0		0.369	0.514		White et al. 2001
Torpedinidae	Pacific electric ray	*Torpedo californica*	California, NE Pacific	16.0	14.0		9.0	6.0		0.078	0.137		Neer and Cailliet 2001
Torpedinidae	Spotted torpedo	*Torpedo marmorata*	Bay of Biscay, France	20.0	12.5		12.0	5.0					Mellinger 1971
Rhinobatidae	shovelnose guitarfish	*Rhinobatos productos*	Southern California	11.0	11.0	11.0	8.0	8.0	8.0	0.016	0.095	0.047	Timmons and Bray 1997

Note: T_{max} is longevity measured as the maximum age observed for a species or stock within a species; T_{mat} is the age of maturity estimated as the age when 50% of individuals in a species or stock reach maturity; k is the von Bertalanffy growth coefficient; both, indicates the estimate is not sex specific.

TABLE 6.2

Summary of Demographic Data*

Sex	Variable	N	Mean	SD	Minimum	Maximum
Batoids						
Both	T_{max}	23	16.85	9.62	3	44
Female	T_{max}	35	17.12	5.68	5	26
Male	T_{max}	31	15.10	5.97	5	28
Both	T_{mat}	11	6.00	2.96	2	10
Female	T_{mat}	34	8.57	3.90	2	17
Male	T_{mat}	31	6.88	3.83	2	15
Both	k	18	0.11	0.06	0.02	0.26
Female	k	34	0.15	0.11	0.02	0.54
Male	k	32	0.17	0.13	0.04	0.51
Rajidae						
Both	T_{max}	12	17.25	5.53	9	28
Female	T_{max}	24	17.55	5.70	9	26
Male	T_{max}	19	17.34	4.94	8	25
Female	T_{mat}	23	10.12	3.26	5	17
Male	T_{mat}	18	9.12	3.30	4	15
Female	k	25	0.12	0.07	0.04	0.26
Male	k	23	0.14	0.10	0.04	0.50

* Where sex represents sex-specific values, variable represents vital rates as defined in Section 6.2. N = sample size; mean = numeric average; SD = standard deviation; minimum = the smallest value; maximum = the largest value.

6.2.3 Myliobatiformes

The Myliobatiformes comprise Dasyatidae (stingrays), Gymnuridae (butterfly rays), Hexatrygonidae, Mobulidae (Manta rays), Myliobatidae (eagle rays), Platyrhinidae, Plesiobatidae, Potamotrygonidae (river stingrays), Urolophidae, Urotrygonidae, and Zanobatidae.

The majority of data for this order are for species from the families Dasyatidae (stingrays), and Myliobatidae (eagle rays). The stingrays represent approximately 60 medium- to large-bodied fish that occur in fresh, estuarine, and marine waters (Last and Stevens 1994). Stingrays commonly feed on worms, mollusks, crustaceans, and small fish (Bigelow and Schroeder 1953). Generally, members of the family live in areas such as bottom habitat near river mouths and in bays (Last 1994). There is a paucity of demographic information for the family as a whole. All dasyatids have histotrophic viviparity, and litter sizes have been reported to range from 1 to 13 (mean = 4.4) (Musick and Ellis 2005). Mollet and Cailliet (2002) reported that the pelagic stingray, *Pteroplatytrygon violacea*, reproduces first at three years and has a reported longevity of 10 to 12 years (Mollet, Ezcurra, and O'Sullivan 2002; Neer 2008). The southern stingray, *Dasyatis americana*, has a reported longevity of 28 years for males (Henningsen 2002).

The eagle rays are medium- to large-bodied batoids comprising approximately 22 species (Last and Stevens 1994). Last and Stevens (1994) report that some species reach a disc width of 3.0 m. They occur in temperate and tropical seas (Last and Stevens 1994; McEachran and Fechhelm 1998). Reproduction in the family is histotrophic viviparity, with litter sizes in the range of 2 to 12 (Baxter 1980; Martin and Cailliet 1988; Last and Stevens 1994; Ebert 2003). The bat eagle ray, *Myliobatis californica*, is reported to live as long as 24 years, with first reproduction at 2.5 years (Martin and Cailliet 1988).

6.2.4 Torpediniformes: Hypnidae, Narcinidae (Electric Rays); Narkidae (Sleeper Rays); Torpedinidae (Electric Rays)

The order consists of four families ranging from the large-bodied torpedo rays to the smaller-bodied electric rays (Last and Stevens 1994). Bigelow and Schroeder (1953) report that some species reach total lengths of 1.5 to 1.8 m. Very little is known regarding the life history strategies of this order. Most species feed on crustaceans, mollusks, worms, and small fish (Bigelow and Schroeder 1953). All species in the order are yolksac viviparous (Musick and Ellis 2005). The Pacific electric ray, *Torpedo californica*, is one of the best-studied. It lives at least 16 years, with males maturing at six years, and females at nine years (Neer and Cailliet 2001). Males reach a smaller total length (92 cm) and have a faster growth rate (k_{male} = 0.14), while females grow at a slower rate (k_{female} = 0.08) and reach a larger total length (137 cm) (Ebert 2003).

6.2.5 Pristiformes: Pristidae (Sawfishes)

The order consists of one family containing four to seven species of sawfish (Last and Stevens 1994). Sawfish are large-bodied, long-lived, and slow-growing demersal species. Reproduction is yolksac viviparity (Musick and Ellis 2005). All species in the order are vulnerable to population declines from overfishing (Simpfendorfer 2000). They live in estuaries and sheltered bays (Bigelow and Schroeder 1953). Last and Stevens (1994) report that some species can reach a total length of 7.0 m. Data from three species suggest that longevity ranges from 30 to 44 years, growth is slow (k = 0.07 to 0.08), and maturation is late at 10 years (Table 6.1) (Tanaka 1991; Simpfendorfer 2000).

6.3 Comparative Life History Analyses

Trade-offs are a central feature of life history traits. For example, along the "fast-slow continuum," "slow" species tend to be larger bodied, slower growing, and often have reduced population productivity at low abundances in contrast with "fast" species, which tend to be smaller-bodied, faster growing, and have a high capacity to recover from low population sizes (Charnov 1993; Frisk, Miller, and Fogarty 2001; Dulvy and Reynolds 2002; Roff 2002; Frisk, Miller, and Dulvy 2005). These theoretical generalizations are often observed in relationships between longevity and growth, and age/size of maturity and reproductive output (Charnov 1993). Furthermore, trade-offs are often constrained by phylogeny; such that a fast-slow continuum exists within many groups (Charnov and Berrigan 1991). A fast-slow continuum has been recognized in elasmobranchs based primarily on data from sharks (Cortes 2000, 2002; Frisk, Miller, and Dulvy 2005). A similar pattern may also characterize the batoids. For example, larger batoid species appear to be longer-lived, slower growing, and have lower reproductive output (Frisk, Miller, and Fogarty 2001; Dulvy and Reynolds 2002; Frisk Miller, and Dulvy 2005). Over the last decade, estimates of vital rates in batoids have increased, further refining our understanding of growth, longevity, and age at maturity.

As a group, batoids are longer-lived and have delayed maturity compared to teleost fishes (Figure 6.1). Data for teleost fishes was obtained from the meta-analysis of Winemiller and Rose (1992). On average, female batoids mature at an age of 8.6 (SD = 3.9) years and males slightly younger at 6.9 (SD = 3.83) years (Table 6.2). Average batoid longevity is 17.1 (SD = 5.68) and 15.1 (SD = 5.97) years for females and males, respectively. The ratio of age at maturity to longevity in teleost fishes ranges between 0.2 and 0.4, well below that of sharks (Cortes 2000) and batoids (Figure 6.1). This suggests that elasmobranchs reach maturity at a later relative age and have fewer reproductive years, relative to the length of their juvenile

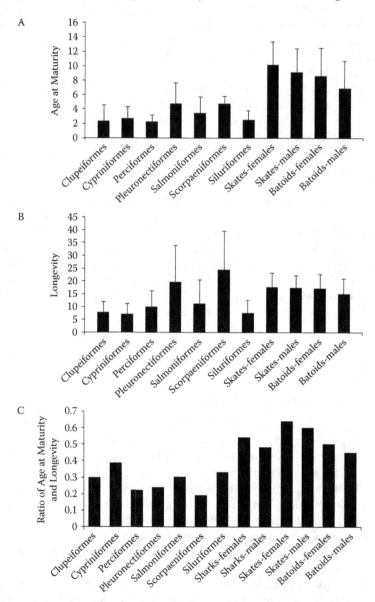

FIGURE 6.1
Comparison of batoid life histories with teleosts and sharks. Panel A displays the average age at maturity, panel B displays the average longevity, and panel C displays the ratio of age at maturity and longevity.

stage, than most teleosts. The apparent delay in maturity in batoids, however, is dependent to an extent on the approach used to estimate ages. For example, Frisk, Miller, and Fogarty (2001) reported that elasmobranchs have a smaller ratio (0.38) of age at maturity and longevity when they included theoretical estimates of maximum age extrapolated from von Bertalanffy growth parameters compared to Cortés (2002), who used age estimates based on direct analysis of hard parts. Hence, the choice of methodology may lead to different results regarding the "true" maximum age. Further, both approaches introduce sources of uncertainty in the derived estimates that are specific to each approach. For example, estimating maximum longevity for exploited species is problematic because few individuals likely attain the maximum age of a species and thus the probability that a fish near maximum age is captured, and aged, is low. Even after many years of sampling an abundant skate species, the low probability of capturing and aging a fish near the maximum age can be profound. Templeman (1984a) tagged 722 thorny skate in the Newfoundland region, with some individuals at liberty for 20 years. The age of individuals when they were tagged and released was unclear, so an estimate of longevity beyond 23+ years cannot be determined. Yet, hard part aging of thorny skate, based on specimens captured following a large population decline, provided lower estimates of longevity of 16 years for males and females (Sulikowski et al. 2005b). McPhie and Campana (2009) estimated longevity of thorny skates to be 19 years based on hard parts and 47 years using model-based estimates. McPhie and Campana also used radiocarbon bomb dating techniques to estimate the age of an individual thorny skate captured in 1988 to be at least 28 years (McPhie and Campana 2009). Assuming an age at maturity of 10.7 (McPhie and Campana 2009), the ratio of age at maturity to longevity would be 0.67 for hard part aging, 0.46 based on ages from tagging data, and 0.38 based on carbon bomb dating. Unfortunately, there is no ideal approach and the method selected for a particular application must weigh the potential biases of each.

Only species in the family Rajidae provided sufficient data to permit comparative life history analyses. A significant positive relationship was found between longevity and age at maturation for females and males (females: $T_{mat} = 0.57 \cdot T_{max} + 1.02$, $r^2 = 0.78$, $p > 0.000$; males: $T_{mat} = 0.57 \cdot T_{max} + 0.47$, $r^2 = 0.95$, $p > 0.000$; Figure 6.2; Table 6.3). The von Bertalanffy growth coefficient was significantly and negatively related to longevity for both sexes (females: $k = -0.009 \cdot T_{max} + 0.276$, $r^2 = 0.431$, $p > 0.002$; males: $k = -0.014 \cdot T_{max} + 0.385$, $r^2 = 0.362$, $p > 0.008$; Figure 6.2; Table 6.3). Together these figures suggest that Rajidae follow the pattern exhibited by other taxa, such that longer-lived species mature later and grow slower compared to shorter-lived species. Skates appear to have life spans and reach maturity at ages similar to other elasmobranch groups (Figure 6.1). Based on analysis of ages derived from hard parts, the ratio of age at maturity and longevity is generally higher for skates than for other elasmobranch groups and teleosts generally.

An important characteristic of elasmobranch life histories is that females mature at a larger size than males in order to carry large young (Cortes 2000). However, in oviparous species this trend is not as clear, with the sexes maturing at similar sizes in several species (Lucifora and Garcia 2004; Ebert 2005; Ebert, Compagno, and Cowley 2008). This may occur in egg laying species because females do not need to attain relatively larger body sizes to carry young (Ebert, Compagno, and Cowley 2008). A review of the literature on batoids suggests it is also common for males or females to reach maturity at differing sizes (Braccini and Chiaramonte 2002a, 2002b; Mabragana and Cousseau 2004; Frisk and Miller 2006, 2009; Ruocco et al. 2006; Colonello, Garcia, and Lasta 2007). Selection for a larger female size in oviparous species may result in increased internal volume and a greater number of offspring (Frisk, Miller, and Dulvy 2005).

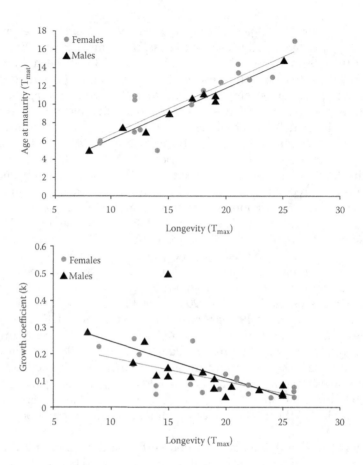

FIGURE 6.2

The relationships of longevity, maturation, and growth for the family Rajidae. Panel A displays age at maturity versus longevity and panel B displays the von Bertalanffy growth coefficient versus longevity.

TABLE 6.3

Regression Analysis Results for the Family Rajidae*

Sex	Model	n	a	b	F value	r^2	$p > F$
Female	T_{max} vs. k	20	−0.009	0.276	13.643	0.431	0.002
Male	T_{max} vs. k	18	−0.014	0.385	9.10	0.362	0.008
Female	T_{mat} vs. T_{max}	16	0.569	1.017	51.260	0.785	0.000
Male	T_{mat} vs. T_{max}	9	0.569	0.475	149.960	0.955	0.000

* Where sex represents gender-specific values. n = sample size; a = the slope of the regression; b = the intercept of the regression; F value = the F statistic; r^2 = the proportion of the error explained by the explanatory variable; p = the significance level.

Comparative analyses have many potential sources of uncertainty that may give rise to spurious results. Two important sources of error are methodological variation and the assumption that a group's vital rates are accurately reflected in existing published research. Larger species tend to be studied more frequently for a variety of reasons, including the following: (1) vulnerability to exploitation; (2) ease of capture; and (3) appeal to the researcher. Because size influences many life history traits, there could be bias in comparative analyses if larger species are overly represented in datasets. Methodological differences between laboratories and researchers may result in different interpretations of age, growth, etc., adding uncertainty to geographic comparisons and comparative analyses (Cailliet et al. 1990; Tanaka, Cailliet, and Yudin 1990; Lombardi-Carlson et al. 2003; Cailliet and Goldman 2004; McPhie and Campana 2009).

6.4 Geographic Variation in Life History Traits

Local races, or subpopulations, within a species have been recognized in fish populations for more than a century (Heincke 1898). Often patterns of life history variation develop along latitudinal gradients (Shine and Charnov 1992), which are a result of temperature-induced changes in growth and metabolism (Conover, Brown, and Ehtisham 1997; Brown, Ehtisham, and Conover 1998; Conover 1998; Billerbeck, Schultz, and Conover 2000). In elasmobranchs, several studies have shown that populations in higher latitudes, and colder environments, display slower growth, reach a larger size, and have a longer life span than populations in lower latitudes (Tanaka and Mizue 1979; Branstetter et al. 1987; Parsons 1993; Carlson and Parsons 1997; Carlson, Cortes, and Johnson 1999; Yamaguchi, Taniuchi, and Shimizu 2000; Driggers et al. 2004; Frisk and Miller 2006, 2009), but whether these differences are genetic or the result of phenotypic plasticity remains unclear.

Geographic trends in vital rates are relatively unstudied in batoids; however, several regional variations in life history patterns have been investigated in skates and rays. Licandeo and Cerna (2007) compared vital rates of kite skate, *Dipturus chilensis*, in the channels and fjords in southern Chile. Their results suggested a latitudinal gradient, with individuals near Patagonian fjords (higher latitudes) attaining a larger size, longer life span, and slower growth rate compared to individuals captured in southern fjords at lower latitudes (Licandeo and Cerna 2007). Similarly, in the western Atlantic, size and longevity in little skate, *Leucoraja erinacea*, increases with latitude (Frisk and Miller 2006, 2009). McPhie and Campana (2009) found that thorny skate, little skate, and smooth skate, *Malacoraja senta*, on the Scotian Shelf appeared to grow slower, reach a larger size, and live longer compared to southern populations. Another western Atlantic species, winter skate, *Leucoraja ocellata*, captured between the mid-Atlantic and Georges Bank, showed no patterns in vital rates (Frisk and Miller 2006) while growth of individuals captured on the Scotian Shelf were faster compared to lower latitudes (McPhie and Campana 2009). The higher growth rate on the Scotian Shelf may be an artifact of sampling from a depleted population or differing methods used in each study (McPhie and Campana 2009).

In a series of studies conducted between 1947 and 1982 in the Newfoundland, Labrador and Nova Scotia regions, Templeman (1982a, 1982b, 1984a, 1984b, 1987a, 1987b) provided evidence that thorny skate showed a high degree of geographic variation in vital rates and morphometric characters, including the following: (1) size of maturation, (2) number of tooth rows, (3) thorniness, (4) length–weight relationships, (5) egg capsule size, (6) weight of egg capsule, (7) albumen volume, and (8) egg volume. Generally, these factors were related to geographic differences in maximum length of thorny skate. In addition, skate size at sexual maturity did not increase with latitude; instead, thorny skate displayed local adaptation, with individuals off of Labrador reaching maturity at smaller sizes than on the Grand Banks (Templeman 1987a). Templeman (1984b, 1987b) argued that meristic differences provided evidence that thorny skate have separate stocks in the western Atlantic. In contrast, genetic analysis found that thorny skates between Newfoundland, Iceland, Kattegat, Norway, and the central North Sea were not genetically differentiated, indicating mixing between the regions (Chevolot et al. 2007).

In a study on the biology of skates in southern Africa, Ebert, Compagno, and Cowley (2008) documented regional variation in vital rates for several species. Their study compared vital rates between the western and southern coasts of southern Africa. Specifically, they found a difference in total length and weight of male *Leurcoraja wallacei* and *R. straeleni* and for female *R. straeleni*, with individuals attaining a larger size on the west coast. Similarly, size at maturation was larger for male and female *L. wallacei* and *D. pullopunctata*. However, not all of the 22 species examined exhibited differences in vital rates between the west and south coasts of southern Africa. Mechanisms underlying observed patterns were unknown and the authors proposed that discrete population structure or latitudinal variation could be possible explanations for differences in vital rates.

Outside Rajidae, little research has been conducted on geographic variation of vital rates in batoids, leaving a critical gap in knowledge needed for management. However, cownose rays, *Rhinoptera bonasus*, in the Gulf of Mexico reached maturity at a smaller size and younger age, grew slower, and reached a smaller maximum size compared to individuals in the Atlantic coast of the United States (Neer and Thompson 2005). These regions may represent separate populations with unique migratory behaviors (Collins, Heupel, and Motta 2007). Regional differences in vital rates can also result from fishery-induced changes through compensatory processes and selective harvest (Walker and Hislop 1998; Frisk and Miller 2006). Whether genetic or phenotypic, the presence of geographic differences in vital rates must be taken into account when developing management policies for batoids (Dulvy and Reynolds 2002; Frisk and Miller 2006; Ebert, Compagno, and Cowley 2008).

6.5 Compensatory Dynamics

Individuals have evolved the capacity to overcompensate reproductive output in the face of variability in annual mortality (Pitcher and Hart 1982). Compensation can result from changes in individual growth rates, age and size of maturation, reproductive output, and survival. For example, following decades of exploitation in the northwestern Atlantic, spiny dogfish, *Squalus acanthias*, have experienced reductions in neonate size, fecundity, age and size at maturity (Rago et al. 1998; Sosebee 2005). However, there is a paucity of information on compensatory behavior in batoids and what is known has largely been from life history analyses and not from direct field observation.

Batoids generally have delayed age of maturation, low fecundity, and slow growth rates, making them a vulnerable group to overexploitation (e.g., Musick 1999; Dulvy et al. 2008). Indeed, batoids appear to be among the most susceptible marine taxa to fisheries exploitation (Dulvy and Reynolds 2002; Dulvy et al. 2008). Within batoids, large body size, associated with increased age at maturity, longevity, and size at maturation, appears to correlate with a higher risk of overexploitation, extirpation, and in some cases extinction (Dulvy and Reynolds 2002). Juvenile and adult survival and age at maturity appear to be the demographic rates that are most important in determining population growth rates, and thus the probability of persisting under exploitation (Walker and Hislop 1998; Heppell et al. 1999; Musick 1999; Bewster-Geisz and Miller 2000; Frisk, Miller, and Fogarty 2002; Frisk, Miller, and Dulvy 2005). These also are expected to be parameters that are under strong selection for exploited batoid populations.

The net reproductive rate (R_0), defined as the lifetime production of offspring, is a common measure of evolutionary fitness (Roff 2002). This measure of fitness is calculated as the sum of the product of age-specific survival and fecundity rates. In addition, the net reproductive rate can be used to estimate the optimal timing of first reproduction given survival and fecundity estimates. Thus, optimal age at maturity can be estimated from age-specific schedules of survival and reproduction given the following function:

$$R_0 = \int_{x=T_{mat}}^{\infty} (m_t) \cdot \exp^{-M \cdot t} = \frac{1}{M} \cdot (m_t) \exp^{-M \cdot T_{mat}} \qquad (6.1)$$

where R_0 is the net reproductive rate, T_{mat} is age of maturity, M is natural mortality, and m_t is age-specific fecundity (Roff 2002). The theoretical age of maturity is estimated as the T_{mat} that satisfies:

$$\frac{\partial R_0}{\partial T_{mat}} = 0 \qquad (6.2)$$

This condition can be estimated by plotting R_0 against increasing age at maturation (T_{mat}) and finding the point of the function where R_0 is maximized.

Ages at maturity that optimize evolutionary fitness (R_0) and those that arise under fishing pressure can be quite different. As an illustration of the potential fishery impacts on the compensatory dynamics of skates, the optimal age of maturation was estimated for little skate and winter skate assuming no fishing and an instantaneous fishing mortality rate of 0.35. Little skate and winter skate were chosen because they are sympatric species with contrasting life histories along the fast-slow continuum in elasmobranchs (Frisk, Miller, and Dulvy 2005). Little skate is a smaller bodied (L_{max} = 57 cm), short-lived (12.5 years), relatively fast growing (k = 0.19 years⁻¹) species with lower annual fecundity (21 to 57 year⁻¹), compared to winter skate, which is a larger bodied (L_{max} = 111 cm), longer-lived (20.5+ years), slower growing (k = 0.07 year⁻¹) species with higher annual fecundity (26 to 100 year⁻¹) (Frisk 2004; Frisk and Miller 2006, 2009). In the following analyses natural mortality was assumed to be 0.35 for little skate and 0.22 for winter skate (Frisk 2004). Here I examine patterns of selection for age of maturity in the virgin and exploited states:

Net reproductive rate (R_0) with no exploitation: Optimal ages at maturation in little skate and winter skate were 7.0 and 14.5 years, respectively, corresponding to the peak in egg production (Figure 6.3). In little skate, the predicted age at maturation was nearly identical to observed values (6.5 to 7.0 years; Figure 6.3A). In winter skate, the predicted optimal age of maturation was 1.5 years older than observed (Figure 6.3B).

Net reproductive rate (R_0) with exploitation: Optimal ages at maturation in little skate and winter skate were 6.5 and 10.5 years, respectively (Figure 6.3). Under exploitation,

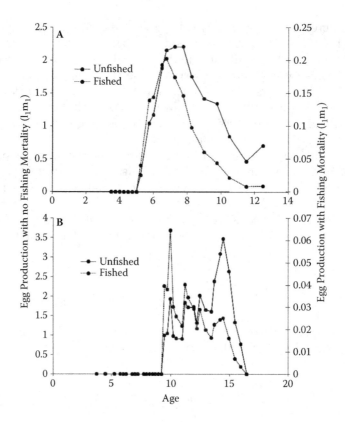

FIGURE 6.3
Age-specific egg production with fishing and no fishing ($F = 0.35$) for little skate (panel A) and winter skate (panel B). Note that peak egg production occurs several years earlier when fishing mortality is assumed for winter skate.

optimal age at maturation remained relatively constant for little skate (Figure 6.3A), while exploitation led to a four-year decrease in the optimal age at maturation for winter skate (Figure 6.3B). Lifetime egg production decreased in both species.

On an evolutionary scale, these simple analyses demonstrate trade-offs between survival, age at maturation, and reproductive output. Demographic analyses of elasmobranch populations have shown that shorter-lived species have strong selection pressure on age at maturation, while longer-lived species have comparatively greater selection toward adult survival (Frisk, Miller, and Dulvy 2005). Species that have adapted to more opportunistic strategies with faster life histories are well suited to respond to environmental and anthropogenic increases in mortality (Winemiller and Rose 1992). On the other hand, species with slower life histories perform well in stable environments and are vulnerable to sudden, in the biological sense, increases in mortality such as fishing (Winemiller and Rose 1992). Little skate's life history was already "faster" than that of winter skate. Thus, under the new exploitation level the theoretical optimal age at maturation in little skate only decreased by 0.5 years, while winter skate would have to decrease age at maturation by four years. The different magnitudes of compensation between a small bodied and large bodied species demonstrate the constraints longer-lived species face when exploited. While only one life history trait was allowed to vary, the example is illustrative of the contrasting challenges species along the fast-slow continuum face when exploited. Finally, harvest is often selective for large individuals and here we assumed no size selectivity that may impact the response of the population to harvest.

Exploitation can lead to changes in demographic rates either through compensatory responses or through the action of direct selection of specific life history traits. Recent attention has been given to the long-term consequences of selective harvest (Conover and Munch 2002). Although fisheries-induced evolutionary changes in vital rates of batoids have not been observed, this may be due to a lack of data. Indeed, there have been documented changes in vital rates of batoids induced by exploitation altering age structure and length frequencies of populations (Walker and Heessen 1996; Walker and Hislop 1998; Oddone, Paesch, and Norbis 2005; Frisk et al. 2008). In a size selective fishery the actual impacts of exploitation may counter the force of selection by disproportionately removing faster-growing individuals leaving slower-growing individuals to reproduce, potentially leading to long-term shifts in the genetic structure of the population (Conover and Munch 2002). Because batoid life history is not suited for long-term laboratory experiments and because it is difficult to use field data to decipher between compensatory or genetic changes or changes to vital rates that result from alteration of the age/length structure of a population, it is likely that the evolutionary impacts of fishing on batoid life histories may remain uncertain for some time to come.

6.6 Spatial Strategies

The extent of individual movements and species ranges have been related to a species' life history (Miller 1979; Rohde 1992; Brown 1995; Gaston 1996; Roy et al. 1998; Goodwin, Dulvy, and Reynolds 2005). In a meta-analysis, Goodwin, Dulvy, and Reynolds (2005) found that egg-laying teleosts had larger distributional ranges compared to live-bearing species and argued that this resulted from post-spawning dispersal of eggs leading to range expansion. However, in elasmobranchs live bearers had greater geographic distribution compared to egg layers (Goodwin, Dulvy, and Reynolds 2005). An explanation for this pattern is that egg-laying elasmobranchs are smaller than live-bearing species (Musick and Ellis 2005) and vagility and range size are positively related to a species total length (Musick, Harbin, and Compagno 2004). In addition, elasmobranch egg layers may have to confine their ranges to suitable spawning habitat because egg cases are spatially anchored and hatchlings do not disperse as widely as larval teleosts. In contrast, live-bearing pregnant females may move over larger ranges (and *in utero* young are protected from predators and buffered from resource shortages) allowing for larger distributional ranges compared to their egg-laying relatives (Shine 1978; Shine and Bull 1979; Clutton-Brock 1991; Pope, Shepherd, and Webb 1994; Qualls and Shine 1998; Goodwin, Dulvy, and Reynolds 2005).

Migration to spawning grounds where environmental conditions are favorable for young and maximize offspring survival has been used to explain patterns across a broad range of animal taxa (Roff 2002; Jorgensen et al. 2008). These migrations often involve movements between foraging and breeding grounds to optimize energy resources of adult and offspring survival (Roff 2002). Often adult migration benefits offspring directly while adults endure the energy costs of migration and lost time for foraging (Jorgensen et al. 2008). Evidence is growing that skates undergo seasonal migrations often linked to reproduction (Hunter et al. 2005a, 2005b). For example, Hunter et al. (2005b) tagged thornback rays with electronic data storage and conventional tags in the Thames Estuary, UK. Of the tagged rays, 51% were captured and returned. Reconstruction of thornback ray movements indicated that the species migrates out of the estuary to the North Sea in the winter and returns during the spawning season in the spring and summer. The authors speculated that the movements are linked to reproduction. Many skates, however, are serial spawners and produce eggs over the entire year. Indeed, a survey of the literature suggests that a majority of Rajidae produce eggs throughout the year, with many displaying one or two peak seasons (Fitz and Daiber 1963; Richards, Tsang, and Howell 1963; Johnson 1979; Braccini and Chiaramonte 2002b; Sulikowski, Tsang, and Howell 2004; San Martin, Perez, and Chiaramonte 2005; Sulikowski et al. 2005a; Ruocco et al. 2006; Colonello, Garcia, and Lasta 2007; Kneebone et al. 2007; Oddone et al. 2007; Sulikowski et al. 2007a; Frisk and Miller 2009). In such species, migrations are tied to seasonal peaks in spawning with at least a minimal level of reproduction throughout the rest of the year. This strategy appears to be unique to some egg-laying batoids and could be an example of "bet hedging" outside optimal reproductive conditions.

In contrast to Rajidae, the families Dasyatidae, Myliobatidae, Rhinobatidae, Urolophidae, and Torpedinidae produce litters during one or two seasons (Capape and Zaouali 1994; Fahy, Spieler, and Hamlett 2007). The migration patterns of these species are poorly understood but many species show seasonal and reproductively linked migrations (Smith and Merriner 1985; Blaylock 1993; Murdy, Birdsong, and Musick 1997; Grusha 2005; Collins, Garcia, and Lasta 2007). In addition, other environmental cues for migrations have been observed. The shovelnose guitarfish enters the waters near Bahia Almejas, Baja California, en masse, related to changes in season, lunar cycles, sea temperature, and perhaps prey availability. Understanding the trade-offs that determine whether a species evolves to migrate short or long distances, or not at all, requires detailed knowledge of the entire life cycle. Unfortunately, in batoids this level of knowledge does not exist.

6.6.1 Migration Patterns of Western Atlantic Skates

In the western Atlantic, five Rajidae species show a range of migratory patterns from very little movement in smooth skate and thorny skate to long distance migration in winter skate (McEachran and Musick 1975; Templeman 1984a). In this section, I review the distribution patterns and reproductive seasonality of clearnose skate, *Raja eglanteria*, little skate, thorny skate, smooth skate, and winter skate based on the annual spring, fall, and winter National Marine Fisheries Service's (NMFS) bottom trawl surveys. The NMFS spring and fall survey have been conducted annually since 1968 in the western Atlantic covering the mid-Atlantic, southern New England, the Gulf of Maine, and Georges Bank regions (see Sosebee and Cadrin 2006). The winter survey was limited to 1992–1999. For this analysis, the mid-Atlantic was defined as the area between Cape Hatteras and the Hudson Canyon, southern New England as the area from the Hudson Canyon to Georges Bank, and the Gulf of Maine as the area east of Boston and north of Georges Bank (Figure 6.4). The spring and fall surveys cover all four regions each year while the winter survey does not sample the Gulf of Maine.

Clearnose skate: The species is a relatively large skate and the largest specimen captured during the survey had a total length of 78 cm (NEFSC 1999). Of the five species, clearnose skate has the most southern distribution with a proportion of the population south of the sampling region. In the spring, the species is abundant along the coast of North Carolina, extending north to Delaware Bay, and is abundant at the edge of the mid-Atlantic shelf off Virginia and North Carolina (Figure 6.5). During the winter, the population congregates

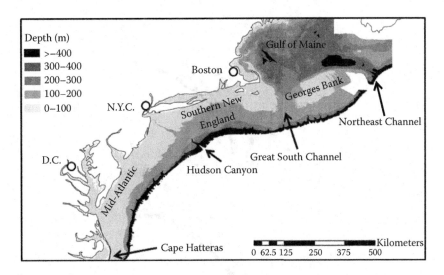

FIGURE 6.4
The western Atlantic showing the region covered by the NMFS bottom trawl survey conducted during the fall, winter, and spring, where the "Mid-Atlantic" region is the area between Cape Hatteras and the Hudson Canyon; the "Southern New England" region is the area between the Hudson Canyon and the Great South Channel; "Georges Bank" is the area between the Great South Channel and the Northeast Channel; the Gulf of Maine is the area north of the Northeast Channel southwest to Cape Cod.

at the edge of the shelf and inshore off of Virginia and North Carolina. In autumn, the species distribution is concentrated inshore and extends north to Long Island, New York. No direct estimates exist of the distance clearnose skate travel during their seasonal migrations. Assuming a straight line of travel, it appears some clearnose skate seasonally move 120 km on and off shore and 580 km north and south along the coast. Fitz and Daiber (1963) reported that in the Delaware region egg production exhibited a strong peak in the late spring. It seems likely that early spring inshore movement observed in the survey corresponds to the species' spawning season.

Little skate: The species is a relatively small skate and the largest specimen captured during the survey had a total length of 62 cm (NEFSC 1999). Little skate is widely distributed, with the center of abundance occurring between the southern New England and Georges Bank regions and a range that extends north of the survey area (Figure 6.6). Little skate's wide distribution and apparently weak seasonal distributional patterns make deciphering migratory behavior difficult. Generally, little skate moves inshore in the spring and to deeper shelf waters during the autumn and winter (NEFSC 1999; McEachran 2002). It is possible the species makes partial migrations; specimens are routinely captured over the entire survey area in all seasons. The species is reported to release egg capsules year round, with peaks in the winter and mid-summer (Fitz and Daiber 1963; Richards, Tsang, and Howell 1963; Scott and Scott 1988; Frisk 2004). As a result of the species' weak seasonal patterns it is difficult to estimate movement distances. Additionally, this species may have multiple stocks with separate demographics (Frisk and Miller 2006, 2009), possibly with differing migration patterns.

Thorny skate: The species is a relatively large skate and the largest specimen captured during the survey was a total length of 111 cm (NEFSC 1999). Thorny skate are only abundant within the survey area in the Gulf Maine and sections of Georges Bank. The species shows no seasonal distributional changes and appears not to make any seasonal migration

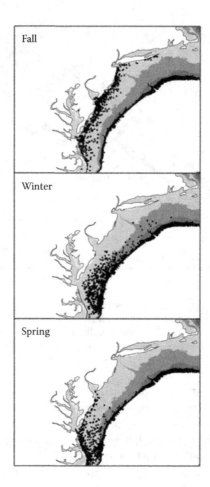

FIGURE 6.5
The seasonal distribution trends of adult clearnose skate, where the size of the dots indicates increasing numbers of skates captured per station.

(NEFSC 1999). These results are supported by Templeman (1984) who tagged 722 thorny skate off Newfoundland, Canada, and found most moved less than 111 km and caught some near tagging sites 15 and 16 years later. McEachran (2002) reported that reproduction occurs throughout the year, with a possible peak in the summer. In contrast, endocrinological analyses do not indicate seasonal trends (Sulikowski et al. 2005a, 2005b; Kneebone et al. 2007). Thorny skate may have regionally unique populations with differing demographics (Templeman 1984a, 1984b, 1987a, 1987b).

Smooth skate: The species is a medium-sized skate and the largest specimen captured during the survey had a total length of 71 cm (NEFSC 1999). Smooth skate is only abundant within the survey area in the Gulf of Maine and to a lesser extent the Georges Bank region. Smooth skate does not appear to make seasonal migrations. McEachran (2002) reported females carrying egg cases in the summer and fall. However, endocrinological analyses do not indicate seasonal trends (Sulikowski et al. 2007a).

Winter skate: The species is a relatively large skate and the largest specimen captured during the survey had a total length of 113 cm (NEFSC 1999). The distribution of winter skate extends north of the survey area and connectivity between regions has been hypothesized (Frisk et al. 2008; Frisk et al., in press). During the autumn, winter skate are

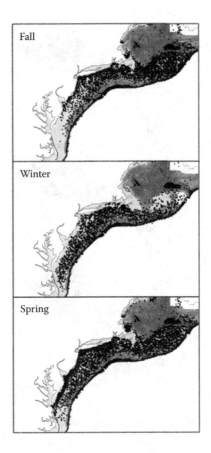

FIGURE 6.6
The seasonal distribution trends of adult little skate, where the size of the dots indicates increasing numbers of skates captured per station.

aggregated on Georges Bank and near Stellwagen Bank in the Gulf of Maine (Figure 6.7). In the winter and spring, winter skate make an extensive southern migration along the coast to North Carolina. However, a large portion of individuals in the population do not make the migration on an annual basis. The winter survey indicates winter skate move offshore and are distributed throughout the mid-Atlantic, southern New England, and southern Georges Bank. Assuming a straight line of movement, it appears some winter skate migrate 430 km between Cape Hatteras and the Hudson Canyon and some move 940 km between Cape Hatteras and Georges Bank. Winter skate appear to exhibit maximum egg production in the autumn and winter when the species is distributed offshore and on Georges Bank, a potential spawning ground (Frisk 2004).

Of the five Rajidae species examined here, clearnose skate, little skate, and winter skate have been observed to make longer seasonal migrations compared to species without clear seasonal trends in reproduction. Little skate and winter skate appear to make partial migrations, but further work is needed to determine if partial migration can explain the observed distribution patterns or if they result from stock-specific migratory behavior. The deep watered thorny skate and smooth skate did not exhibit strong seasonal migration patterns and reproduction does not appear strongly related to season. It should be emphasized that our understating of skate migration is largely limited to data collected during scientific surveys, which may not be adequate for deciphering seasonal movements

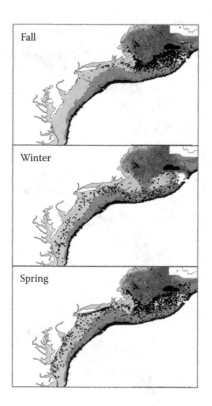

FIGURE 6.7

The seasonal distribution trends of adult winter skate, where the size of the dots indicates increasing numbers of skates captured per station.

(e.g., Dunton et al., in review) and further studies are needed to resolve migrations of the above-mentioned species.

6.6.2 Migration and Ecosystem Management

Ecosystems that have supported important fisheries for centuries are increasingly under pressure from high exploitation levels (Pauly et al. 1998; Pauly and Palomares 2005) and dramatic changes in abundances of batoids have occurred in some ecosystems. In some cases, batoid species have declined as a result of direct targeting by individual fisheries (e.g., common skate in the Irish Sea; Brander 1981). However a lack of directed fisheries for batoids does not preclude fisheries-induced changes in their abundances. In particular, two observed abundance increases of batoids (winter skate and cownose ray) in the western Atlantic have been at the center of debate over declines, or continued poor condition, of major fisheries (Murawski 1991; Mayo, Fogarty, and Serchuk 1992; Murawski and Idoine 1992; Safina 1995; Fogarty and Murawski 1998; Myers et al. 2007).

6.6.3 Winter Skate and Georges Bank

Winter skate dramatically increased in the Georges Bank region in the 1980s following declines of important commercial species (Frisk et al. 2008). The increase in medium-sized elasmobranchs led to the hypothesis that winter skate benefited from competitive release of resources after the decline of groundfish (Murawski 1991; Mayo, Fogarty, and Serchuk

1992; Murawski and Idoine 1992; Safina 1995; Fogarty and Murawski 1998; Myers et al. 2007). A recent analysis of length frequency data indicated the increase in abundance resulted from a sudden emergence of all length classes of winter skate (Frisk et al. 2008). Further, population modeling and observed length frequency data suggested that the increase was incompatible with the life history of winter skate assuming a closed system (Frisk et al., in press). Collectively, these two recent papers hypothesized that the increase in abundance was not the result of internal processes within the Georges Bank ecosystem; but instead, a more parsimonious explanation was a migration from surrounding regions.

The increase of winter skate on Georges Bank corresponded to their decline on the Scotian Shelf (Frisk et al. 2008). The Scotian Shelf, which is separated from Georges Bank by the Northeast Channel, was hypothesized as the source of winter skate on Georges Bank (Frisk et al. 2008). Supporting evidence consisted of a decline in physiological condition indices for winter skate on the Scotian Shelf beginning in the late 1970s and lasting through the 1980s. The decline in physiological condition could have resulted from warming temperatures and changing trophic structure on the Scotian Shelf during the late 1970s and 1980s (Choi et al. 2005; Frisk et al. 2008). Did winter skate migrate to Georges Bank as a result of more favorable competitive/environmental conditions compared to the Scotian Shelf?

The cues for large-scale distributional shifts of animal populations are poorly understood. However, several mechanisms have been hypothesized, including the following: climate (Frisk et al. 2008), density-dependent migratory waves (Fauchald, Mauritzen, and Gjosaeter 2006), and food availability (Olsson et al. 2006). Understanding the mechanism behind the increase in winter skate abundance has implications for single-species and ecosystem-based management. If the increase resulted from a competitive release internal to the Georges Bank system, increasing removals of winter skate could rectify the perceived imbalance. If migration explains the increase, additional removals could prove devastating for a population where mature females have drastically declined throughout the species' range (Frisk et al. 2008; McPhie and Campana 2009).

What other potential source could have supplied enough adults to explain the increase of winter skate in the 1980s on Georges Bank? In 2007, 277 barndoor skate and 146 large winter skate were captured during the spring NMFS survey on the southeastern flank of Georges Bank at a depth of 296 m. The tow represents 277 of the 325 barndoor skate and a significant portion of total winter skate captured during the survey. This single tow raises the possibility that portions of the populations of barndoor skate, which has rapidly increased on Georges Bank in recent years (Gedamke et al. 2007), and winter skate may move to deep water outside the survey range. If portions of the populations of these skate species are moving in or out of the Georges Bank system, it would fundamentally change views of the population dynamics of winter skate and barndoor skate and the potential mechanisms underlying ecosystem processes. Clearly, additional research is needed to better understand the cues for skate migration and the implications for ecosystem structure. As fisheries move toward ecosystem management, difficult decisions will include increasing harvest of select species to shift resources for a more desired ecosystem state (Walters and Martell 2004). However, until we understand the extent of species range size, migration behavior, population connectivity, and boundaries of ecosystems, management is bound to make poor decisions (Jordaan et al. 2007).

6.6.4 Cownose Rays

Cownose rays are a highly migratory species with a general northern spring migration along the eastern United States. The rays are common visitors to bays and estuaries along

FIGURE 6.8
A color version of this figure follows page 336. Cownose rays photographed on their annual migration in the Gulf of Mexico. The rays are thought to make an annual migration between the Yucatan Peninsula to Florida. (Photos © Sandra Critelli. All rights reserved.)

the eastern seaboard including the Chesapeake Bay (Murdy, Birdsong, and Musick 1997). Migration appears tied to seasonal changes and reproductive activity (Murdy, Birdsong, and Musick 1997; Collins, Heupel, and Motta 2007). In the Chesapeake Bay, cownose rays release pups in the early summer and mate toward the end of the season (Murdy, Birdsong, and Musick 1997). In the autumn, the species completes its annual migration returning to warmer southern waters. A fascinating aspect of the species' migratory behavior is that they often form large schools of millions of individuals, which may segregate by sex and age (Figure 6.8; Clarke 1963; Smith and Merriner 1985; Blaylock 1989, 1993; Rogers et al. 1990; Murdy, Birdsong, and Musick 1997). While this represents the general migratory behavior of the species, movement patterns are poorly known and several migratory patterns have been proposed. Acoustic tagging studies reveal that some individuals remain in Pine Island Sound, Florida, during winter months, which is inconsistent with the long distance migration hypothesis (Collins, Heupel, and Motta 2007). The authors argued that temperatures in southern regions may not reach low enough levels to cue seasonal migration (Hopkins and Cech 2003). It has also been proposed that a separate population in the Gulf of Mexico makes unique seasonal migrations from the Yucatan peninsula to Florida (Collins, Heupel, and Motta 2007).

The need for understanding cownose ray migrations has been highlighted by the recent suggestion that the decline of large sharks in the northwest Atlantic has led to an increase in small elasmobranchs, including cownose rays, through relaxation of predation pressure (Myers et al. 2007). Further, the cownose ray increase has been tied to the decline of a century-old bay scallop fishery in the southeastern United States that provides a food source for humans and cownose rays alike (Myers et al. 2007). Here again, a species with a "slow" life history has been hypothesized to have increased several orders of magnitude since the mid-1970s (Myers et al. 2007). The observation of an increase in population size has not been adequately tested against all possible hypotheses. Just as in the case of winter skate on Georges Bank, the actual mechanism explaining the observed increase has a fundamental impact on management decisions and conservation of cownose rays. Clearly, whether or not the increases in winter skate and cownose ray were real or simply a distributional shift, it highlights the role batoids play in structuring ecosystems.

6.7 Conclusions

Batoids are long-lived, slow-growing species with a diverse range of life histories and spatial behaviors. Still, within the general long-lived strategy, the group has a continuum of "slow" to "fast" life histories with differing responses to exploitation. Batoids appear to reach maturation and have similar longevities as observed for other elasmobranch groups. Generally, faster-growing and shorter-lived species fare better under exploitation. Evidence suggests batoids display regional trends in vital rates often observed over latitudinal gradients. Understanding the stock structure and whether geographic differences in vital rates result from phenotypic or genetic differences is critical to developing management strategies. Several aspects of batoid life history have not been adequately addressed including the following: (1) the density-dependence of life history traits and (2) the relationship between adult abundance and recruitment (see Gedamke 2006).

Batoids have evolved life history strategies that provide a great deal of control over spawning locations and population connectivity mediated through adult placement of young or eggs (Compagno 1990; Goodwin, Dulvy, and Reynolds 2005; Frisk et al. 2008). Even with only sparse data, evidence suggests that not only are live-bearing batoids highly migratory; but many egg-laying species have complex and, in some cases, moderately long distance migratory behavior. A great deal of work is needed to better understand spawning behavior in batoids and it should be cautioned that these results are based on only a handful of studies. Critical information needed to interpret abundance estimates and set appropriate management policies are missing for most species, including the following: (1) stock structure, (2) causes of partial migration, (3) triggers for timing of migration, (4) mechanisms behind mass migrations, and (5) energetic trade-offs associated with long distance movement.

References

Abdel-Aziz SH (1992) The use of vertebral ring of the brown ray *Raja miraletus* off Egyptian Mediterranean coast for estimation of age and growth. Cybium 16: 121–132.

Babel JS (1967) Reproduction, life history and ecology of the round stingray *Urolophus halleri* Cooper. Fish-B NOAA 137: 104.

Baxter JL (1980) Inshore fishes of California. Resources Agency, Dept. of Fish and Game. Sacramento, CA.

Beverton RJH (1992) Patterns of reproductive strategy parameters in some marine teleost fishes. J Fish Biol 41(Supplement B): 137–160.

Beverton RJH, Holt SJ (1959) A review of the life-spans and mortality rates of fish in nature, and their relationship to growth and other physiological characteristics. CIBA Foundation Colloquia on Ageing 54: 142–180.

Bewster-Geisz KK, Miller TJ (2000) Management of the sandbar shark: implications of a stage-based model. Fish-B NOAA 98: 236–249.

Bigelow HB, Schroeder WC (1953) Fishes of the western north Atlantic, Part 2. Mem. Sears Found. Marine Research, No. 2, Chap. 1, Batoidae; Chap. 2, Holocephali, xv, 562 pp.

Billerbeck JM, Schultz ET, Conover DO (2000) Adaptive variation in energy acquisition and allocation among latitudinal populations of the Atlantic silverside. Oecologia 122: 210–219.

Blaylock RA (1989) A massive school of cownose rays, *Rhinoptera-bonasus* (Rhinopteridae), in lower Chesapeake Bay, Virginia. Copeia: 744–748.

Blaylock RA (1993) Distribution and abundance of the cownose ray, *Rhinoptera-bonasus*, in lower Chesapeake Bay. Estuaries 16: 255–263.

Braccini JM, Chiaramonte GE (2002) Biology of the skate *Psammobatis extenta* (Garman, 1913) (Batoidea: Rajidae). Rev Chil Hist Nat 75: 179–188.

Braccini JM, Chiaramonte GE (2002) Reproductive biology of *Psammobatis extenta*. J Fish Biol 61: 272–288.

Brander K (1981) Disappearance of common skate *Raia batis* from the Irish Sea. Nature 290: 48–49.

Branstetter S, Musick JA, Colvocoresses JA (1987) A comparison of age and growth of the tiger shark, *Galeocerdo cuvieri*, from off Virginia and from the northwestern Gulf of Mexico. Fish-B NOAA. 85: 269–279.

Bremer JRA, Frisk MG, Miller TJ, Turner J, Vinas J, Kwil K (2005) Genetic identification of cryptic juveniles of little skate and winter skate. J Fish Biol 66: 1177–1182.

Brown JH (1995) Macroecology. University of Chicago Press, Chicago.

Brown JJ, Ehtisham A, Conover DO (1998) Variation in larval growth rate among striped bass stocks from different latitudes. J Am Fish Soc 127: 598–610.

Cailliet GM, Goldman KJ (2004) Age determination and validation in Chondrichthyan fishes. In: Carrier JC, Musick JA, Heithaus MR (eds) Biology of sharks and their relatives, CRC Press, Boca Raton, FL.

Cailliet GM, Yudin KG, Tanaka S, Taniuchi T (1990) Growth characteristics of two populations of *Mustelus manazo* from Japan based upon cross-readings of vertebral growth bands. In: Pratt HL, Gruber SH, Taniuchi T (eds) Elasmobranchs as living resources: advances in the biology, ecology, systematics, and status of the fisheries, NOAA Technical Report 90, National Marine Fisheries Service, Washington, DC, pp 167–176.

Capape C, Zaouali J (1994) Distribution and reproductive biology of the blackchin guitarfish, *Rhinobatos-cemiculus* (Pisces, Rhinobatidae), in Tunisian waters (Central Mediterranean). Aust J Mar Freshwater Res 45: 551–561.

Carlson JK, Cortes E, Johnson AG (1999) Age and growth of the blacknose shark, *Carcharhinus acronotus*, in the eastern Gulf of Mexico. Copeia: 684–691.

Carlson JK, Parsons GR (1997) Age and growth of the bonnethead shark, *Sphyrna tiburo*, from northwest Florida, with comments on clinal variation. Environ Biol Fish 50: 331–341.

Charnov EL (1993) Life history invariants: some explorations of symmetry in evolutionary ecology. Oxford University Press, Oxford.

Charnov EL, Berrigan D (1990) Dimensionless numbers and life history evolution: age of maturity versus the adult lifespan. Evol Ecol 4: 273–275.

Charnov EL, Berrigan D (1991) Dimensionless numbers and the assembly rules for life histories. Philos Trans R Soc London 332: 41–48.

Chevolot M, Hoarau G, Rijnsdorp AD, Stam WT, Olsen JL (2006) Phylogeography and population structure of thornback rays (*Raja clavata* L., Rajidae). Mol Ecol 15: 3693–3705.

Chevolot M, Wolfs PHJ, Palsson J, Rijnsdorp AD, Stam WT, Olsen JL (2007) Population structure and historical demography of the thorny skate (*Amblyraja radiata*, Rajidae) in the north Atlantic. Mar Biol 151: 1275–1286.

Choi JS, Frank KT, Petrie BD, Leggett WC (2005) Integrated assessment of a large marine ecosystem: a case study of the devolution of the Eastern Scotian Shelf, Canada. Oceanogr Mar Biol Annu Rev 43: 47–67.

Clarke E (1963) Massive aggregations of large rays and sharks in and near Sarasota, Florida. Zoologica 48: 61–64.

Clutton-Brock TH (1991) The evolution of parental care. Princeton University Press, Princeton, NJ.

Collins AB, Heupel MR, Motta PJ (2007) Residence and movement patterns of cownose rays *Rhinoptera bonasus* within a south-west Florida estuary. J Fish Biol 71: 1159–1178.

Colonello JH, Garcia ML, Lasta CA (2007) Reproductive biology of *Rioraja agassizi* from the coastal southwestern Atlantic ecosystem between northern Uruguay (34 degrees S) and northern Argentina (42 degrees S). Environ Biol Fish 80: 277–284.

Compagno LJV (1990) Alternative life-history styles of cartilaginous fishes in time and space. Environ Biol Fish 28: 33–75.

Conover DO (1998) Local adaptation in marine fishes: evidence and implications for stock enhancement. Bull Mar Sci 62: 477–493.

Conover DO (2003) Countergradient variation and the evolution of growth rate: lessons from silverside fishes. Integr Comp Biol 43: 922.

Conover DO, Brown JJ, Ehtisham A (1997) Countergradient variation in growth of young striped bass (*Morone saxatilis*) from different latitudes. Can J Fish Aquat Sci 54: 2401–2409.

Conover DO, Munch SB (2002) Sustaining fisheries yields over evolutionary time scales. Science 297: 94–96.

Cortés E (2000) Life history patterns and correlations in sharks. Rev Fish Sci 8: 299–344.

Cortés E (2002) Incorporating uncertainty into demographic modeling: application to shark populations and their conservation. Conserv Biol 16: 1048–1062.

Cowley PD (1997) Age and growth of the blue stingray *Dasyatis chrysonota* from the south-eastern Cape coast of South Africa. S Afr J Marine Sci 18: 31–38.

de Kroon H, Groenendael J, Ehrlen J (2000) Elasticities: a review of methods and model limitation. Ecology 81: 607–618.

Dowton-Hoffman CA (1996) Estrategia reproductive de la guitarra *Rhinobatos productus* (Ayres 1856) en la costa occidental de Baja California Sur, Mexico, Tesis de Licenciatura.

Driggers WB, Oakley DA, Ulrich G, Carlson JK, Cullum BJ, Dean JM (2004) Reproductive biology of *Carcharhinus acronotus* in the coastal waters of South Carolina. J Fish Biol 64: 1540–1551.

DuBuit MH (1972) Age et croissance de *Raja batis* et de *Raja naevus* en Mer Celtique. J Conseil 37: 261–265.

Dulvy NK, Baum JK, Clarke S, Compagno LJV, Cortes E, Domingo A, Fordham S, Fowler S, Francis MP, Gibson C, Martinez J, Musick JA, Soldo A, Stevens JD, Valenti S (2008) You can swim but you can't hide: the global status and conservation of oceanic pelagic sharks and rays. Aquatic Conserv 18: 459–482.

Dulvy NK, Metcalfe JD, Glanville J, Pawson MG, Reynolds JD (2000) Fishery stability, local extinctions, and shifts in community structure in skates. Conserv Biol 14: 283–293.

Dulvy NK, Reynolds JD (2002) Predicting extinction vulnerability in skates. Conserv Biol 16: 440–450.

Dunton KJ, Jordaan A, Conover DO, Frisk MG (in review) Use of fishery independent surveys to determine abundance, habitat, and distribution of Atlantic Sturgeon (*Acipenser oxyrinchus*) within the Atlantic Ocean.

Ebert DA (2003) Sharks, rays and chimaeras of California. University of California Press, Berkeley, California.

Ebert DA (2005) Reproductive biology of skates, Bathyraja (Ishiyama), along the eastern Bering Sea continental slope. J Fish Biol 66: 618–649.

Ebert DA, Bizzarro JJ (2007) Standardized diet compositions and trophic levels of skates (Chondrichthyes : Rajiformes : Rajoidei). Environ Biol Fish 80: 221–237.

Ebert DA, Compagno LJV, Cowley PD (2008) Aspects of the reproductive biology of skates (Chondrichthyes : Rajiformes : Rajoidei) from southern Africa. ICES J Mar Sci 65: 81–102.

Ebert DA, Smith WD, Cailliet GM (2008) Reproductive biology of two commercially exploited skates, *Raja binoculata* and *R. rhina*, in the western Gulf of Alaska. Fish Res 94: 48–57.

Ellis JR, Shackley SE (1995) Observations on egg-laying in the thornback ray. J Fish Biol 46: 903–904.

Fahy DP, Spieler RE, Hamlett WC (2007) Preliminary observations on the reproductive cycle and uterine fecundity of the yellow stingray, *Urobatis jamaicensis* (Elasmobranchii: Myliobatiformes: Urolophidae) in Southeast Florida, USA. Raffles B Zool: 131–139.

Fauchald P, Mauritzen M, Gjosaeter H (2006) Density-dependent migratory waves in the marine pelagic ecosystem. Ecology 87: 2915–2924.

Fitz ES, Daiber FC (1963) An introduction to the biology of *Raja eglanteria* Bosc 1802 and *Raja erinacea* Mitchill 1825 as occur in Delaware Bay. Bulletin of the Bingham Oceanographic Collection, no. 18.

Fogarty MJ, Murawski SA (1998) Large-scale disturbance and the structure of marine system: fishery impacts on Georges Bank. Ecol Appl 8: S6–S22.

Francis MP, Maolagain CO, Stevens D (2001) Age, growth, and sexual maturity of two New Zealand endemic skates, *Dipturus nasutus* and *D. innominatus*. NZ J Mar Freshwater Res 35: 831–842.

Frisk MG (2004) Biology, life history and conservation of elasmobranchs with an emphasis on western Atlantic skates. Dissertation, University of Maryland.

Frisk MG, Martell SJD, Miller TJ, Sosebee K (in press) Exploring the population dynamics of winter skate (*Leucoraja ocellata*) in the Georges Bank region using a statistical catch-at-age model incorporating length, migration, and recruitment process errors. CJFAS.

Frisk MG, Miller TJ (2006) Age, growth, and latitudinal patterns of two Rajidae species in the northwestern Atlantic: little skate (*Leucoraja erinacea*) and winter skate (*Leucoraja ocellata*). Can J Fish Aquat Sci 63: 1078–1091.

Frisk MG, Miller TJ (2009) Maturation of little skate and winter skate in the western Atlantic from Cape Hatteras to Georges Bank. Mar Coast Fish Dynam Manage Ecosystem Sci 1: 1–11.

Frisk MG, Miller TJ, Dulvy NK (2005) Life histories and vulnerability to exploitation of elasmobranchs: inferences from elasticity, perturbation and phylogenetic analyses. J Northwest Atl Fish Soc 35: 27–45.

Frisk MG, Miller TJ, Fogarty MJ (2001) Estimation and analysis of biological parameters in elasmobranch fishes: a comparative life history study. Can J Fish Aquat Sci 58: 969–981.

Frisk MG, Miller TJ, Fogarty MJ (2002) The population dynamics of little skate *Leucoraja erinacea*, winter skate *Leucoraja ocellata*, and barndoor skate *Dipturus laevis*: predicting exploitation limits using matrix analyses. ICES J Mar Sci 59: 576–586.

Frisk MG, Miller TJ, Martell SJD, Sosebee K (2008) New hypothesis helps explain elasmobranch "outburst" on Georges Bank in the 1980s. Ecol Appl 18: 234–245.

Gaston KJ (1996) Species-range-size distributions: patterns, mechanisms and implications. Trends Ecol Evol 11: 197–201.

Gburski CM, Gaichas SK, Kimura DK (2007) Age and growth of big skate (*Raja binoculata*) and longnose skate (*R. rhina*) in the Gulf of Alaska. Environ Biol Fish 80: 337–349.

Gedamke T (2006) Developing a stock assessment for the barndoor skate, *Dipturus laevis*, in the northeast United States. College of William and Mary. Virginia Institute of Marine Science, Gloucester Point, VA.

Gedamke T, DuPaul WD, Musick JA (2005) Observations on the life history of the barndoor skate, *Dipturus laevis*, on Georges Bank (Western North Atlantic). J Northwest Atl Fish Soc 35: 67–78.

Gedamke T, Hoenig JM, Musick JA, DuPaul WD, Gruber SH (2007) Using demographic models to determine intrinsic rate of increase and sustainable fishing for elasmobranchs: pitfalls, advances, and applications. North Am J Fish Manage 27: 605–618.

Goodwin NB, Dulvy NK, Reynolds JD (2005) Macroecology of live-bearing in fishes: latitudinal and depth range comparisons with egg-laying relatives. Oikos 110: 209–218.

Grover CA (1972) Predation on egg-cases of swell shark, *Cephaloscyllium ventriosum*. Copeia: 871–872.

Grusha DS (2005) Investigation of the life history of the cownose ray, *Rhinoptera bonasus* (Mitchill 1815). College of William and Mary. Virginia Institute of Marine Science, Gloucester Point, VA.

Heincke F (1898) Naturgeschichte des Herings. Die Lokalformen und die Wanderungen des Herings in den europäischen Meeren. Teil 1. Otto Salle, Berlin. (Abhandlungen des Deutschen Seefischerei-Vereins, Bd. 2).

Heithaus MR (2004) Predator-prey interactions. In: Carrier JC, Musick JA, Heithaus MR (eds) Biology of sharks and their relatives, CRC Press, Boca Raton, FL.

Henningsen AD (2002) Age and growth in captive southern stingrays, *Dasyafis americana*. Abstract, American Elasmobranch Society, Annual Meeting, Kansas City.

Heppell SS, Crowder LB, Menzel TR, Musick JA (1999) Life table analysis of long-lived marine species with implications for conservation and management. In: Musick JA (ed) Life in the slow lane: ecology and conservation of long-lived marine animals. J Am Fish Soc 23: 137–146.

Hoenig JM, Gruber SH (1990) Life history patterns in the elasmobranchs: implications for fisheries management. In: Pratt HL, Gruber SH, Taniuchi T (eds) Elasmobranchs as living resources: advances in the biology, ecology, systematics, and the status of the fisheries, NOAA Tech. Rep. NMFS 90, U.S. Department of Commerce, Washington, DC, pp 1–16.

Holden MJ (1972) The growth rates of *raja brachyura*, *R. Clavata* and *R. Montagui* as determined from tagging data. J Conseil 34: 161–168.

Holden MJ (1973) Are long-term sustainable fisheries for elasmobranchs possiple? Rap Proces 164: 360–367.

Holden MJ (1974) Problems in the rational exploitation of elasmobranch populations and some suggested solutions. In: Harden-Jones FR (ed) Sea fisheries research. Halstead Press/John Wiley, New York, pp 117–137.

Holden MJ (1975) The fecundity of *Raja clavata* in British waters. J Conseil 36(2): 110–118.

Holden MJ, Vince MR (1973) Age validation studies on the centra of *Raja clavata* using tetracycline. J Conseil 35: 13–17.

Homma K, Maruyama T, Itoh T, Ishihara H, Uchida S (1997) Biology of the manta ray, *Manta birostris* Walbaum, in the Indo-Pacific. Proceedings of the 5th Indo-Pacific Fisheries Conference, pp 209–216.

Hopkins TE, Cech JJ (2003) The influence of environmental variables on the distribution and abundance of three elasmobranchs in Tomales Bay, California. Environ Biol Fish 66: 279–291.

Hunter E, Blickley AA, Stewart C, Metcalf JD (2005a) Repeated seasonal migration by a thornback ray in the southern North Sea. J Mar Biol Assoc UK 85: 1199–1200.

Hunter E, Buckley AA, Stewart C, Metcalfe JD (2005b) Migratory behaviour of the thornback ray *Raja clavata*, in the southern North Sea. J Mar Biol Assoc UK 85: 1095–1105.

Ismen A (2003) Age, growth, reproduction and food of common stingray (*Dasyatis pastinaca* L., 1758) in Iskenderun Bay, the eastern Mediterranean. Fish Res 60: PII S0165-7836(02)00058-9.

Johnson GF (1979) The biology of the little skate, *Raja Erinacea* Mitchill 1825, in Block Island Sound, Rhode Island. University of Rhode Island, Narragansett.

Jordaan A, Frisk MG, Wolff NH, Incze LS, Hamlin L, Chen Y (2007) Structure of fish assemblages along the northeastern United States based on trawl survey data: indicators of biodiversity and a basis for ecosystem and area-based management. ICES CM 2007 A:05.

Jordaan A, Kling LJ (2003) Determining the optimal temperature range for Atlantic cod (*Gadus morhua*) during early life. In: Browman HI, Skiftesvik AB (eds) The big fish bang. Proceedings of the 26th Annual Larval Fish Conference, Institute of Marine Research, Bergen, Norway, pp 45–62.

Jorgensen C, Dunlop ES, Opdal AF, Fiksen O (2008) The evolution of spawning migrations: state dependence and fishing-induced changes. Ecology 89: 3436–3448.

Kinnison MT, Unwin MJ, Hendry AP, Quinn TP (2001) Migratory costs and the evolution of egg size and number in introduced and indigenous salmon populations. Evolution 55: 1656–1667.

Kneebone J, Ferguson DE, Sulikowski JA, Tsang PCW (2007) Endocrinological investigation into the reproductive cycles of two sympatric skate species, *Malacoraja senta* and *Amblyraja radiata*, in the western Gulf of Maine. Environ Biol Fish 80: 257–265.

Koop JH (2005) Reproduction of captive *Raja* spp. in the Dolfinarium Harderwijk. J Mar Biol Assoc UK 85: 1201–1202.

Last PR, Stevens JD (1994) Sharks and Rays of Australia. CSIRO, Australia.

Licandeo R, Cerna FT (2007) Geographic variation in life-history traits of the endemic kite skate *Dipturus chilensis* (Batoidea : Rajidae), along its distribution in the fjords and channels of southern Chile. J Fish Biol 71: 421–440.

Licandeo R, Cerna F, Cespedes R (2007) Age, growth, and reproduction of the roughskin skate, *Dipturus trachyderma*, from the southeastern Pacific. ICES J Mar Sci 64: 141–148.

Licandeo RR, Lamilla JG, Rubilar PG, Vega RM (2006) Age, growth, and sexual maturity of the yellownose skate *Dipturus chilensis* in the south-eastern Pacific. J Fish Biol 68: 488–506.

Link JS, Garrison LP, Almeida FP (2002) Ecological interactions between elasmobranchs and groundfish species on the northeastern U.S. continental shelf. I. Evaluating predation. North Am J Fish Manage 22: 550–562.

Little W (1995) Common skate and tope: first results of Glasgow Museum's tagging study. Glasg Nat 22: 455–466.

Lombardi-Carlson LA, Cortes E, Parsons GR, Manire CA (2003) Latitudinal variation in life-history traits of bonnethead sharks, *Sphyrna tiburo* (Carcharhiniformes: Sphyrnidae) from the eastern Gulf of Mexico. Mar Freshwater Res 54: 875–883.

Lucifora LO, Garcia VB (2004) Gastropod predation on egg cases of skates (Chondrichthyes, Rajidae) in the southwestern Atlantic: quantification and life history implications. Mar Biol 145: 917–922.

Mabragana E, Cousseau MB (2004) Reproductive biology of two sympatric skates in the south-west Atlantic: *Psammobatis rudis* and *Psammobatis normani*. J Fish Biol 65: 559–573.

Mabragana E, Giberto DA (2007) Feeding ecology and abundance of two sympatric skates, the shortfin sand skate *Psammobatis normani* McEachran, and the smallthorn sand skate *P. rudis* Gunther (Chondrichthyes, Rajidae), in the southwest Atlantic. ICES J Mar Sci 64: 1017–1027.

Martin LK, Cailliet GM (1988a) Age and growth determination of the bat ray, *Myliobatis californica* Gill, in Central California. Copeia 3: 762–773.

Martin LK, Cailliet GM (1988b) Aspects of the reproduction of the bat ray, *Myliobatis californica*, in Central California. Copeia 3: 754–762.

Matta ME, Gunderson DR (2007) Age, growth, maturity, and mortality of the Alaska skate, *Bathyraja parmifera*, in the eastern Bering Sea. Environ Biol Fish 80: 309–323.

Mayo RM, Fogarty MJ, Serchuk F (1992) Changes in biomass, production and species composition of the fish populations in the northwest Atlantic over the last 30 years and their possible causes. J Northwest Atl Fish Soc 14: 59–78.

McEachran JD (2002) Skates. Family Rajidae. In: Collete BB, Klein-MacPhee G (eds) Fishes of the Gulf of Maine. Smithsonian Institution Press, Washington, DC.

McEachran JD, Aschliman N (2004) Phylogeny of Batoidea. In: Carrier JC, Musick JA, Heithaus MR (eds) Biology of sharks and their relatives. CRC Press, Boca Raton, FL, pp 79–113.

McEachran JD, Boesch DF, Musick JA (1976) Food division within two sympatric species-pairs of skates (Pisces-Rajidae). Mar Biol 35: 301–317.

McEachran JD, Dunn KA (1998) Phylogenetic analysis of skates, a morphologically conservative clade of elasmobranchs (Chondrichthyes: Rajidae). Copeia: 271–290.

McEachran JD, Dunn KA, Miyake T (1996) Interrelationships of the batoid fishes (Chrondrichthyes: Batoidea). In: Stiassny MLJ, Parenti LR, Johnson GD (eds) Interrelationships of fishes. Academic Press, San Diego, pp 63–84.

McEachran JD, Fechhelm JD (1998) Fishes of the Gulf of Mexico: Myxiniformes to Gasterosteiformes. University of Texas Press, Austin.

McEachran JD, Musick JA (1975) Distribution and relative abundance of seven species of skates (Pisces: Rajdae) which occur between Nova Scotia and Cape Hatteras. Fish B-NOAA no. 73.

McHugh NJ (2001) A thesis in marine biology: diet overlap of little skate (*Leucoraja erinacea*), winter skate (*Leucoraja ocellata*) and haddock (*Melanogrammus aeglefinus*) at two sites on Georges Bank. University of Massachusetts.

McPhie RP, Campana SE (2009) Bomb dating and age determination of skates (family Rajidae) off the eastern coast of Canada. ICES J Mar Sci 66: 546–560.

McPhie RP, Campana SE (2009) Reproductive characteristics and population decline of four species of skate (Rajidae) off the eastern coast of Canada. J Fish Biol 75: 223–246.

Mellinger J (1971) Croissance et reproduction de la torpille (*Torpedo marmorata*). I. Introduction, ecologie, croissance generale et dimorphisme sexual, cycle, fecondite [Growth and reproduction of the electric ray (*Topedo marmorata*) I. Introduction. ecology, body growth and sexual dimorphism, cycle, fecundity]. Bull Biol France Belgique 105: 167–215.

Miller PJ (1979) Adaptiveness and implications of small size in teleosts. Sym Zool S 44: 263–306.

Mollet HF (2002) Distribution of the pelagic stingray, *Dasyatis violacea* (Bonaparte, 1832), off California, Central America, and worldwide. Mar Freshwater Res 53: 525–530.

Mollet HF, Cailliet GM (2002) Comparative population demography of elasmobranchs using life history tables, Leslie matrices and stage-based matrix models. Mar Freshwater Res 53: 503–516.

Mollet HF, Ezcurra JM, O'Sullivan JB (2002) Captive biology of the pelagic stingray, Dasyatis violacea (Bonaparte, 1832). Mar Freshwater Res 53: 531–541.

Murawski SA (1991) Can we manage our multispecies fisheries? Fisheries 16: 5–13.

Murawski SA, Idoine J (1992) Multi species size composition: a conservative property of exploited fishery systems? J Northwest Atl Fish Soc 14: 79–85.

Murdy EO, Birdsong RS, Musick JA (1997) Fishes of Chesapeake Bay. Smithsonian Institution Press, Washington, DC.

Musick JA (1999) Ecology and conservation of long-lived marine animals. Am Fish Soc S 23: 1–7.

Musick JA, Ellis JK (2005) Reproductive evolution of chondrichthyes. In: Hamlett WC (ed) Reproductive biology and phylogeny of chondrichthyes: sharks, batoids and chimaeras. Science Publishers, Plymouth, UK, pp 45–79.

Musick JA, Harbin MM, Compagno LJV (2004) Historical zoogeography of the Selachii. In: Carrier JC, Musick JA, Heithaus MR (eds) Biology of sharks and their relatives. CRC Press, Boca Raton, FL.

Myers RA, Baum JK, Shepherd TD, Powers SP, Peterson CH (2007) Cascading effects of the loss of apex predatory sharks from a coastal ocean. Science 315: 1846–1850.

Natanson LJ, Sulikowski JA, Kneebone JR, Tsang PC (2007) Age and growth estimates for the smooth skate, *Malacoraja senta*, in the Gulf of Maine. Environ Biol Fish 80: 293–308.

Neer JA (2008) The Biology and Ecology of the Pelagic Stingray, *Pteroplatytrygon violacea* (Bonaparte, 1832). In: Camhi M, Pikitch EK, Babcock E (eds) Sharks of the open ocean. Blackwell Scientific,

Neer JA, Cailliet GM (2001) Aspects of the life history of the Pacific electric ray, *Torpedo californica* (Ayers). Copeia 2001: 842–847.

Neer JA, Thompson BA (2005) Life history of the cownose ray, *Rhinoptera bonasus*, in the northern Gulf of Mexico, with comments on geographic variability in life history traits. Environ Biol Fish 73: 321–331.

NEFSC (1999) Assessment of the northeast region skate complex for 1999. Stock Assessment Review Committee Southern Demersal Working Group.

Oddone MC, Amorim AF, Mancini PL, Norbis W, Velasco G (2007) The reproductive biology and cycle of *Rioraja agassizi* (Muller and Henle, 1841) (Chondrichthyes: Rajidae) in southeastern Brazil, SW Atlantic Ocean. Sci Mar 71: 593–604.

Oddone MC, Paesch L, Norbis W (2005) Size at first sexual maturity of two species of rajoid skates, genera Atlantoraja and Dipturus (Pisces, Elasmobranchii, Rajidae), from the south-western Atlantic Ocean. J Appl Ichthyol 21: 70–72.

Olsson, IC, Greenberg, LA, Bergman, E, and Wysujack, K. 2006. Environmentally induced migration: the importance of food. Ecology Letters 9: 645–651.

Ottersen G, Planque B, Belgrano A, Post E, Reid PC, Stenseth NC (2001) Ecological effects of the North Atlantic Oscillation. Oecologia 128: 1–14.

Parsons G (1993) Age determination and growth of the bonnethead shark: a comparison of two populations. Mar Biol 117: 23–31.

Pauly D, Christensen V, Dalsgaard J, Froese R, Torres F (1998) Fishing down marine food webs. Science 279: 860–863.

Pauly D, Palomares ML (2005) Fishing down marine food web: it is far more pervasive than we thought. J Mar Sci 76: 197–211.

Pitcher TJ, Hart PJB (1982) Fisheries ecology. AVI Publishing Company, West Port, CT.

Pope JG, Shepherd JG, Webb J (1994) Successful surf-riding on size spectra: the secret of survival in the sea. Philos Trans Roy Soc London Ser B 343: 41–49.

Qualls CP, Shine R (1998) Costs of reproduction in conspecific oviparous and viviparous lizards, *Lerista bougainvillii*. Oikos 82: 539–551.

Rago PJ, Sosebee KA, Brodziak JKT, Murawski SA, Anderson ED (1998) Implications of recent increases in catches on the dynamics of northwest Atlantic spiny dogfish (*Squalus acanthias*). Fish Res 39: 165–181.

Richards SW, Merriman D, Calhoun LH (1963) The biology of the little skate, *raja erinacea* Mitchill. Bull Bingham Oceanogr Collect 18, 5–67.

Roff DA (1988) The evolution of migration and some life history parameters in marine fishes. Environ Biol Fish 22: 133–146.

Roff D (2002) Life history evolution. Sinauer Associates, Sunderland, MA.

Rogers C, Roden C, Lohoefener R, Mullin K, Hoggard W (1990) Behavior, distribution, and relative abundance of cownose ray schools *Rhinoptera bonasus* in the northern Gulf of Mexico. Northeast Gulf Science 11: 69–76.

Rohde K (1992) Latitudinal gradients in species-diversity: The search for the primary cause. Oikos 65: 514–527.

Roy K, Jablonski D, Valentine JW, Rosenberg G (1998) Marine latitudinal diversity gradients: tests of causal hypotheses. Proc Natl Acad Sci USA 95: 3699–3702.

Ruocco NL, Lucifora LO, de Astarloa JMD, Wohler O (2006) Reproductive biology and abundance of the white-dotted skate, *Bathyraja albomaculata*, in the southwest Atlantic. ICES J Mar Sci 63: 105–116.

Ryland JS, Ajayi TO (1984) Growth and population-dynamics of three Raja species (Batoidei) in Carmarthen Bay, British-Isles. J Conseil 41: 111–120.

Safina C (1995) The world's imperiled fish. Sci Am 273: 46–53.

Salazar-Hermoso F, Villavicencio-Garayzar C (1999) Relative abundance of the shovelnose guitarfish *Rhinobatos productus* (Ayres, 1856) (Pisces: Rhinobatidae) in Bahla Almejas, Baja California Sur, from 1991 to 1995. Cienc Mar 25: 401–422.

San Martin MJ, Perez JE, Chiaramonte GE (2005) Reproductive biology of the south west Atlantic marbled sand skate *Psammobatis bergi* Marini, 1932 (Elasmobranchii, Rajidae). J Appl Ichthyol 21: 504–510.

Schwartz FJ (1965) Inter-American migrations and systematics of the western Atlantic cownose ray, *Rhinoptera bonasus*. 6th Meeting, Association of Island Marine Laboratories of the Caribbean, Isla Margarita, Venezuela, pp 20–22.

Schwartz FJ (1990) Mass migratory congregations and movements of several species of cownose rays, genus Rhinoptera: a worldwide review. J Eliza Mitchell Sci Soc 106: 10–13.

Scott WB, Scott MG (1988) Atlantic fishes of Canada. Can J Fish Aquat Sci 219:

Shine R (1978) Propagule size and parental care: safe-harbor hypothesis. J Theor Biol 75: 417–424.

Shine R (1989) Alternative models for the evolution of offspring size. Am Nat 134: 311–317.

Shine R, Bull JJ (1979) Evolution of live-bearing in lizards and snakes. Am Nat 113: 905–923.

Shine R, Charnov EL (1992) Patterns of survival, growth, and maturation in snakes and lizards. Am Nat 139: 1257–1269.

Simpfendorfer CA (2000) Predicting population recovery rates for endangered western Atlantic sawfishes using demographic analysis. Environ Biol Fish 58: 371–377.

Smith C, Griffiths C (1997) Shark and skate egg-cases cast up on two South African beaches and their rates of hatching success, or causes of death. S Afr J Zool 32: 112–117.

Smith JW, Merriner JV (1985) Food-habits and feeding-behavior of the cownose ray, *Rhinoptera bonasus*, in Lower Chesapeake Bay. Estuaries 8: 305–310.

Sosebee KA (2002) Maturity of skates in northeast United States waters (Elasmobranch Fisheries: Poster). Sci Counc Res Doc NAFO 02: 17.

Sosebee KA (2005) Are density-dependent effects on elasmobranch maturity possible? J Northwest Atl Fish Soc 35: 115–124.

Sosebee KA, Cadrin SX (2006) Historical perspectives on the abundance and biomass of some northeast demersal finfish stocks from NMFS and Massachusetts inshore bottom trawl surveys, 1963–2002. NOAA/NMFS/NEFSC, Woods Hole, MA.

Sulikowski JA, Elzey S, Kneebone J, Jurek J, Howell WH, Tsang PCW (2007a) The reproductive cycle of the smooth skate, *Malacoraja senta*, in the Gulf of Maine. Mar Freshwater Res 58: 98–103.

Sulikowski JA, Irvine SB, DeValerio KC, Carlson JK (2007b) Age, growth and maturity of the roundel skate, *Raja texana*, from the Gulf of Mexico, USA. Mar Freshwater Res 58: 41–53.

Sulikowski JA, Kneebone J, Elzey S, Danley PD, Howell WH, Jurek J, Tsang PCW (2005a) The reproductive cycle of the thorny skate (*Amblyraid radiata*) in the western Gulf of Maine. Fish B-NOAA 103: 536–543.

Sulikowski JA, Kneebone J, Elzey S, Jurek J, Danley PD, Howell WH, Tsang PCW (2005b) Age and growth estimates of the thorny skate (*Amblyraja radiata*) in the western Gulf of Maine. Fish B-NOAA 103: 161–168.

Sulikowski JA, Kneebone J, Elzey S, Jurek J, Howell WH, Tsang PCW (2006) Using the composite variables of reproductive morphology, histology and steroid hormones to determine age and size at sexual maturity for the thorny skate *Amblyraja radiata* in the western Gulf of Maine. J Fish Biol 69: 1449–1465.

Sulikowski JA, Morin MD, Suk SH, Howell WH (2003) Age and growth estimates of the winter skate (*Leucoraja ocellata*) in the western Gulf of Maine. Fish B-NOAA 101: 405–413.

Sulikowski JA, Tsang PCW, Howell WH (2004) An annual cycle of steroid hormone concentrations and gonad development in the winter skate, *Leucoraja ocellata*, from the western Gulf of Maine. Mar Biol 144: 845–853.

Sulikowski JA, Tsang PCW, Howell WH (2005) Age and size at sexual maturity for the winter skate, *Leucoraja ocellata*, in the western Gulf of Maine based on morphological, histological and steroid hormone analyses. Environ Biol Fish 72: 429–441.

Tanaka S (1991) Age estimation of freshwater sawfish and sharks in northern Australia and Papua New Guinea. Univ Mus Univ Tokyo, Nat Cult 3: 71–82.

Tanaka S, Cailliet GM, Yudin KG (1990) Differences in growth of the blue shark, Prionace glauca: technique or population? In: Pratt HL, Gruber SH, Taniuchi T (eds) Elasmobranchs as living resources: advances in the biology, ecology, systematics, and the status of the fisheries. NOAA Technical Report 90. National Marine Fisheries Service, Washington, DC, pp 177–187.

Tanaka S, Mizue K (1979) Studies on sharks 15. Age and growth of Japanese dogfish *Mustelus manazo* Bleeker in the East China Sea. J Jpn Soc Sci Fish 45: 43–50.

Tanaka S, Ohnishi (1998) Some biological aspects of freshwater stingrays collected from Chao Phraya, Mekong, and Ganges River systems. Adaptibility and conservation of freshwater elasmobranchs, report of research project, grant-in-aid for international scientific research (field research) in the financial year of 1996 and 1997, pp 102–119.

Templeman W (1982a) Development, occurrence and characteristics of egg capsules of the thorny skate, *Raja radiata*, in the northwest Atlantic. J Northwest Atl Fish Soc 3: 47–56.

Templeman W (1982b) Stomach contents of the thorny skate, *Raja radiata*, from the northwest Atlantic. J Northwest Atl Fish Soc 3: 123–126.

Templeman W (1984a) Migrations of thorny skate, *Raja radiata*, tagged in the Newfoundland area. J Northwest Atl Fish Soc 5: 55–63.

Templeman W (1984b) Variations in numbers of median dorsal thorns and rows of teeth in thorny skate (*Raja radiata*) of the northwest Atlantic. J Northwest Atl Fish Soc 5: 171–179.

Templeman W (1987a) Differences in sexual maturity and related characteristics between populations of thorny skate (*Raja radiata*) in the northwest Atlantic. J Northwest Atl Fish Soc 7: 155–167.

Templeman W (1987b) Length-weight relationships, morphometric characteristics and thorniness of thorny skate (*Raja radiata*) from the northwest Atlantic. J Northwest Atl Fish Soc 7: 89–98.

Timmons M, Bray RN (1997) Age, growth, and sexual maturity of shovelnose guitarfish, *Rhinobatos productus* (Ayres). Fish B-NOAA 95: 349–359.

Villavicencio-Garayzar CJ (1995) Reproductive-biology of the banded guitarfish, *Zapterix exasperata* (Pisces, Rhinobatidae), in Bahia-Almejas, Baja-California-Sur, Mexico. Cienc Mar 21: 141–153.

Villavicencio-Garayzar CJ (2000) Taxonomia, abundancia estacional, edad y crecimiento y biologia reproductive de *Narcine entemedor* Jordan y Starks (Chondrichthyes: Narcinidae), en Bahia Almejas, B.C.S., Mexico. PhD dissertation, Universidad Autonoma de Nuevo Leon.

Walker PA, Heessen HJL (1996) Long-term changes in ray populations in the North Sea. ICES 53: 1085–1093.

Walker PA, Hislop JRG (1998) Sensitive skates or resilient rays? Spatial and temporal shifts in ray species composition in the central and north-western North Sea between 1930 and the present day. ICES 55: 392–402.

Walker P, Howlett G, Millner R (1997) Distribution, movement and stock structure of three ray species in the North Sea and eastern English Channel. ICES 54: 797–808.

Walmsley-Hart SA, Sauer WHH, Buxton CD (1999) The biology of the skates *Raja wallacei* and *R. pullopunctata* (Batoidea: Rajidae) on the Agulhas Bank, South Africa. S Afr J Marine Sci 21: 165–179.

Walters CJ, Martell SJD (2004) Fisheries ecology and management. Princeton University Press, Princeton.

White WT, Hall NG, Potter IC (2002) Reproductive biology and growth during pre- and postnatal life of *Trygonoptera personata* and *T. mucosa* (Batoidea: Urolophidae). Mar Biol 140: 699–712.

White WT, Platell ME, Potter IC (2001) Relationship between reproductive biology and age composition and growth in *Urolophus lobatus* (Batoidea: Urolophidae). Mar Biol 138: 135–147.

White WT, Potter IC (2005) Reproductive biology, size and age compositions and growth of the batoid *Urolophus paucimaculatus*, including comparisons with other species of the Urolophidae. Mar Freshwater Res 56: 101–110.

Winemiller K, Rose K (1992) Patterns of life-history diversification in North American fishes: implications for population regulation. Can J Fish Aquat Sci 49: 2196–2218.

Yamaguchi A, Taniuchi T, Shimizu M (2000) Geographic variations in reproductive parameters of the starspotted dogfish, *Mustelus manazo*, from five localities in Japan and in Taiwan. Environ Biol Fish 57: 221–233.

Zeiner SJ, Wolf P (1993) Growth characteristics and esimates of age at maturity of two species of skates (*Raja binoculata* and *Raja rhina*) from Monterey Bay, California. NOAA Technical Report NMFS 115: Conservation Biology of Elasmobranchs: 87–99.

Section II

Adaptive Physiology

7

Ontogenetic Shifts in Movements and Habitat Use

R. Dean Grubbs

CONTENTS

7.1 Introduction

7.1.1 Background and Concepts

Habitat use and selection influence population dynamics, interspecific and intraspecific interactions, ecosystem structure, and biodiversity (Morris 2003). Ontogenetic shifts in movements and habitat use are common if not ubiquitous among animal taxa (e.g., Wilbur 1980; Hart 1983; Vuorinen, Rajasilta, and Salo 1983; Werner and Hall 1988; Dickman 1992), especially fishes (Werner and Gilliam 1984; Werner 1986; Werner and Hall 1988; M. Jones et al. 2003). In teleosts with indirect development, individuals increase in mass by three or four orders of magnitude over their lifetimes and undergo complete changes in feeding modes and patterns of habitat use. Ontogenetic shifts in diet are common among many vertebrate taxa, including elasmobranchs (Jackson et al. 2004; Wetherbee and Cortés 2004). The diets of fishes are more easily studied than spatial and temporal patterns of micro-habitat use, and therefore dietary shifts are the most commonly cited ontogenetic changes in fish ecology. Among fishes, these shifts are most distinct in taxa with dramatic increases in body size through ontogeny (e.g., Marks and Conover 1993). The ontogenetic shifts of many teleosts fishes from planktivorous larvae to piscivorous adults are the most commonly studied (Werner and Gilliam 1984; Mittelbach and Persson 1998). Such shifts may also occur early in ontogeny, especially in pelagic fishes. For example, Llopiz and Cowen (2008) found that larval sailfish, *Istiophorus platypterus*, and blue marlin, *Makaira nigricans*, shift from planktivorous feeding on crustaceans (e.g., copepods) to feeding on fish lar-vae following flexion of the notochord (~5 mm body length). Shifts in feeding mode also occur in elasmobranchs but are less dramatic than in teleost fishes due in part to direct development and the comparatively limited overall growth and ontogenetic changes to morphology.

Ontogenetic shifts in diet are often linked proximately to changes in habitat and may also be related to changes in prey selectivity, energetics (Juanes and Conover 1994), metab-olism (Jackson et al. 2004), or changes in foraging ability due to increased gape and swim-ming speed. Ontogenetic changes in feeding mode and prey capture due to changes in jaw anatomy (Stoner and Livingston 1984) and biomechanics facilitate shifts in diet and, there-fore, habitat use (Lowe et al. 1996; Lucifora et al. 2006; Lowry and Motta 2007). Lowry and Motta (2008), for example, demonstrated that interspecific variation in ontogenetic shifts in diet between the strictly suction feeding whitespotted bamboo shark (*Chiloscyllium pla-giosum*) and the leopard shark (*Triakis semifasciata*), which uses a combination of ram and suction feeding, are predictable based on ontogenetic changes in suction feeding ability.

Prey and habitat selection are ultimately functions of the availability of resources (includ-ing food) and mortality risk (Lima and Dill 1990). Often, substantial energetic costs as well as risks of starvation and reproductive failure are associated with changing habitats. It is also well established that predators affect the movements of prey, as well as habitat use by prey (Werner et al. 1983; Olson, Mittelbach, and Osenberg 1995; Lima 1998). In the pres-ence of predators, fishes often use areas of increased forage only when structure or habitat complexity allows concealment (Fraser and Cerri 1982; Gotceitas and Colgan 1990) and in many situations, organisms are predicted to seek habitats that minimize the ratio of mor-tality risk to growth (Werner and Gilliam 1984; Werner and Hall 1988). Therefore, organ-isms may select habitats with lower forage availability if protection from predators is high and in many cases ontogenetic niche shifts (ONS) are life-stage specific adaptive strate-gies to minimize the ratio between mortality risk and net energetic gain or growth (see Heithaus 2007 for a review of foraging-safety currencies as they relate to elasmobranch use of nursery habitats).

Growth and mortality rates are often inversely correlated with body size and are also functions of the availability of food, density of predators, refuges from predation, and the environmental conditions that affect physiological performance. According to ONS theory, animals are expected to shift habitats with growth in response to changes in the above parameters. Theoretically, a threshold body size exists above which predation risk decreases sufficiently to allow expansion into habitats that were previously high risk, but have greater forage or reproductive rewards. In other words, the increased fitness achieved by exploiting additional resources (e.g., mates, food, or habitat) offsets the costs and risks associated with changing habitats as well as the costs and risks associated with remaining in the original habitat (Gilliam and Fraser 1987). Limited experimental field work on coral reef teleosts has supported this hypothesis (Dahlgren and Eggleston 2000).

Significant plasticity exists in the timing of niche shifts by fishes based on predation risk (Holbrook and Schmitt 1988; He and Kitchell 1990; L'Abe'e-Lund et al. 1993; M. Jones et al. 2003) as well as interspecific competition (Werner and Hall 1977; Bergman and Greenberg 1994), intraspecific competition (Holbrooks and Schmitt 1992; Persson and Bronmark 2002), prey dynamics, or abiotic factors (Langeland et al. 1991). Ontogenetic shifts in space and resource use may transform the structure of fish guilds and have profound effects on a species' role in an ecosystem (Muñoz and Ojeda 1998). Adaptive plasticity in these shifts (Takimoto 2003) likely has a stabilizing effect on community dynamics through a density-dependent negative feedback involving resource availability.

All elasmobranch fishes have direct development; therefore, lifetime growth rarely includes increases in mass greater than two orders of magnitude. This is likely the reason that ontogenetic shifts in aspects such as foraging mode are less dramatic than in the teleosts. Shifts from herbivory to carnivory or from planktivory to piscivory rarely if ever occur in the elasmobranchs. Nevertheless, size-based and sex-based segregation by habitat is one of the more notable patterns in the distributions of many species of sharks and rays (Bass 1978; Sims, Nash, and Morritt 2001; Sims 2005). While the study of ontogenetic niche shifts in the elasmobranchs is in its infancy, significant work has been done on a few coastal species. I will draw on the literature involving many species to illustrate patterns, but I will also draw heavily from my work and the work of others on the sandbar shark, *Carcharhinus plumbeus,* one of the most studied species of elasmobranch fishes. In this chapter I will show that ontogenetic changes in habitat use by elasmobranch fishes can be quite dramatic and likely have important implications for population dynamics and community structure.

7.1.2 Habitat Use and Selection

The terms *habitat use, habitat utilization, habitat selection,* and *habitat preference* are often used interchangeably in the elasmobranch ecology literature. *Utilization* is invariably a polysyllabic synonym of *use* and should be avoided in scientific writing (Eschmeyer 1990). In this chapter I will employ *habitat use* when relative availability of habitats is unknown or not considered. Morris (2003) defined habitat selection as "the process whereby individuals preferentially use, or occupy, a non-random set of available habitats." By this definition *selection* and *preference* are synonymous. However, an organism may *prefer* a resource without *selecting* its use. Therefore, I will use *habitat selection* when habitats are reported to be used disproportionately to their availability.

7.1.3 Home Range and Activity Space

The concept of a *home range* was developed in reference to terrestrial mammals (Burt 1943) and birds (Odum and Kuenzler 1955). Burt (1943) defined a home range as "the area, usually around a home site, over which the animal normally travels in search of food." An animal's home range may be stationary or may shift with seasons or resource distribution but should be persistent. Many animals make occasional forays outside of their normal range; movements that Burt (1943) suggested are not part of the animal's home range. For this reason, home ranges are often estimated using a portion of the location data (usually 95%) that encompasses the regular movements of the organism. See Simpfendorfer and Heupel (2004) for an additional discussion.

In elasmobranch fishes, home ranges are most often estimated using active (Lagrangian) and more recently passive (Eulerian) telemetry. As an animal is tracked using either method, the cumulative area used by an animal increases until an asymptote is reached. If this asymptote is reached over an ecologically meaningful time frame (days to weeks) then the area defined may be interpreted as the home range. Using active telemetry, tracking durations are often limited to hours or a few days and this asymptote is never reached; therefore, the defined activity spaces do not qualify as home ranges. Conversely, passive telemetry allows much longer monitoring durations because it is much less labor intensive. However, activity space can only be estimated using these methods while the tagged animal is within an array of acoustic receivers with overlapping zones of detection. These estimates are only home ranges if the subject remained entirely within the array during the majority (preferably >95%) of the monitoring period. Often, the core area defined using telemetry represent small parts of the home range (Klimley and Nelson 1984), though these activity spaces are commonly referred to as home ranges. In this chapter, I will primarily use *activity space* and reserve *home range* for the very few studies to which it applies.

7.1.4 Taxonomic Breadth of Published Studies

Investigation of ontogenetic niche shifts inherently requires the study of habitat use and movement patterns across life stages of a given species. Such patterns have been explicitly investigated in few species of elasmobranch fishes due to difficulties in conducting such studies of large wide-ranging animals in a concealing environment. The subjects of a majority of the detailed studies of habitat use, primarily using sonic telemetry, have been coastal carcharhiniform sharks. Far less is known about the ecology of pelagic and deep sea elasmobranchs. However, using information such as fishery-dependent and fishery-independent catch rates, trophic ecology, and the results of mark-recapture and telemetry studies we can infer patterns of ontogenetic niche shifts for a diverse array of batoid, squaloid, and galeoid taxa.

7.2 Ontogenetic Shifts in Habitat Use

7.2.1 Patterns in Juvenile Activity Space

It is likely that ontogenetic changes in habitat use occur nearly universally across elasmobranch taxa, though this prediction has been poorly investigated beyond juvenile life stages due in large part to difficulties in quantifying habitat use for adults of many taxa. Many elasmobranchs are seasonally migratory. Alimentary and climatic migrations between summer and winter areas are common and the ontogenetic development of seasonal long-distance migrations characteristic of many elasmobranch species are dramatic examples of ontogenetic niche switching.

Patterns of habitat use by sharks and rays can be coarsely assessed using catch rate data and conventional mark-recapture methods; however, biotelemetry methods are needed to assess fine-scale patterns. Unfortunately, most telemetry methods cannot be used to investigate habitat use across multiple life stages. Biotelemetry methods are of two major forms. Active telemetry employs a Lagrangian approach by following individual animals while passive telemetry uses the Eulerian model of monitoring animals as they pass fixed locations. Using active telemetry, investigators are able to gain insights into fine-scale habitat use, and potentially estimate activity spaces and seasonal home ranges (Morrissey and Gruber 1993a); however, tracking durations are often limited. Passive telemetry has been heavily used with elasmobranchs in recent years in part because it is less labor intensive than active telemetry and provides much longer monitoring durations. These methods provide great insight into diel, seasonal, and interannual patterns of movement and residency as well as broader-scale habitat use (e.g., Garla et al. 2006). Passive techniques have also been used to estimate activity spaces; however, these estimates often do not truly represent home range estimates (e.g., Carlson et al. 2008) unless the subject remains within the array during the majority of the monitoring period (e.g., Heupel, Simpfendorfer, and Heuter 2004; Heupel et al. 2006).

Most studies of habitat use in elasmobranchs have investigated nearshore nurseries of carcharhinid and sphyrnid sharks (Simpfendorfer and Heupel 2004), though a number of recent studies have involved other taxa and life stages (e.g., Sims, Nash, and Morritt 2001; Heithaus et al. 2002; Carrarro and Gladstone 2006; Collins, Heupel, and Motta 2007; Powter and Gladstone 2009). The movements, activity spaces, and patterns of habitat use have been studied in more detail for lemon sharks (*Negaprion brevirostris*) and sandbar sharks (*Carcharhinus plumbeus*) than any other species. Much of the work on lemon sharks has been conducted in a single locality (Bimini, Bahamas). Morrissey and Gruber (1993a) were the first to estimate the activity space of an elasmobranch and found that juvenile lemon sharks establish very small, confined, and persistent home ranges in a subtropical lagoon, the North Sound of Bimini. Morrissey and Gruber (1993b) suggested that very young lemon sharks use shallow warm waters over sandy/rocky substrates disproportionately to their availability to avoid predators. Franks (2007) demonstrated that presence of juvenile lemon sharks was negatively correlated to distance from shore and depth and positively correlated with prey abundance and seagrass density. Whereas juveniles are primarily restricted to mangrove-fringed shorelines and adjacent seagrass flats, adults use a much wider array of habitats, including coral reefs and sandy bottom habitats to depths of more than 50 meters (Cortés and Gruber 1990). Similarly, in a tropical atoll in the equatorial Atlantic, Wetherbee, Gruber, and Rosa (2007) reported young-of-year lemon sharks were restricted to movements between tidal pools in the reef flat while large lemon sharks were more common in larger pools and tidal creeks. The reef tidal pools in this atoll serve the same refuge function as red mangrove (*Rhizophora mangle*) prop roots in other regions. Cannibalism is common in many shark species, including lemon sharks (Vorenberg 1962; Wetherbee, Gruber, and Cortés 1990), thus the restricted range and habitats of young-of-year sharks may minimize interaction with large conspecifics. Similarly, Chapman et al. (2007) found evidence of ontogenetic changes in temporal use of habitats in Caribbean reef sharks (*Carcharhinus perezi*) at an atoll off Belize. Smaller sharks were common on the outside fore-reef slope during the day but would retreat to the relatively protected lagoon during the night. Larger sharks, in contrast, were most common at night on the fore-reef. Papastamatiou et al. (2009) reported a similar, but spatially constrained, pattern of ontogenetic variation in habitat use among blacktip reef sharks (*Carcharhinus melanopterus*) associated with a small equatorial atoll.

In most vertebrates, activity space increases with body size (McNab 1963), a trend that is most evident in carnivores (Harestad and Bunnell 1979). Habitat selection for very young sharks is a balance between maximizing access to available resources (e.g., minimizing competition) and avoiding predators, most often larger sharks (Springer 1967; Branstetter 1990; Heithaus 2007; Heupel, Carlson, and Simpfendorfer 2007). Predation risk is often lower in shallower habitats typically used as nurseries but may vary considerably. These shallower waters are also warmer and often more productive (Beck et al. 2001) than adjacent deeper regions, thereby imparting an energetic advantage to juvenile elasmobranchs in facilitating faster growth rates. Predictably, species that have large birth sizes [e.g., *Carcharhinus obscurus*, maximum birth size = 90 cm total length (TL)] and/or fast juvenile growth rates (e.g., *Rhizoprionodon terraenovae*, K = 0.35 to 0.50) often inhabit unprotected coastal areas as juveniles while those that have small birth sizes and slow growth rates (e.g., *C. plumbeus*, maximum birth size = 65 cm TL, K = 0.05) typically use more protected areas such as estuaries as nurseries (Branstetter 1990). With growth, predation risk declines while movement rates and forage capabilities often increase. These factors combined with likely increases in energetic demands and intraspecific competition drive older individuals to exploit additional habitats should result in increased dietary breadth and activity space.

Ontogenetic increases in activity spaces have been documented in several shark species (Morrissey and Gruber 1993a; Wetherbee, Gruber, and Rosa 2007). Morrissey and Gruber (1993a) and Franks (2007) demonstrate significant ontogenetic increases in the size of home ranges of lemon sharks in two nurseries in Bimini, Bahamas (Figure 7.1). Home range is expanded over the first three years of life as the sharks grow and begin exploiting habitats farther away from the protection of the mangrove fringe. Franks (2007) reported that home ranges of juvenile lemon sharks increase substantially after four years of age because they are no longer bound to the nursery areas and move into seagrass, bank, and reef habitats. The home ranges of subadult lemon sharks in Bimini are reportedly an order of magnitude larger than those of juveniles (deMarignac 2000). Similarly, the activity space estimates for large juvenile bull sharks in a Florida estuary (Yeiser, Heupel, and Simfendorfer 2008) were approximately an order of magnitude greater than

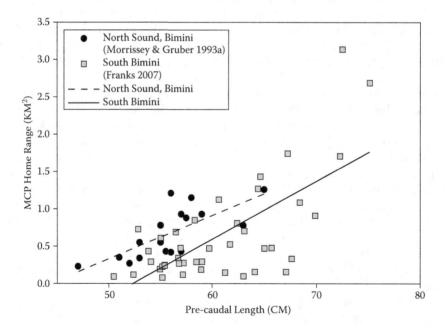

FIGURE 7.1
Ontogenetic increase in home range of juvenile lemon sharks (home range versus precaudal length) tracked using active telemetry in Bimini, Bahamas. North Sound data are from Morrissey and Gruber (1993) and South Bimini data are from Franks (2007).

those estimated for small juvenile bull sharks upriver in the same estuarine system. A few studies of juvenile sharks reported no correlation between activity spaces and shark size. Often, this lack of correlation is a function of active tracking durations that were too short to appropriately estimate home range (e.g., Grubbs 2001; Rechisky and Wetherbee 2003; Papastamatiou et al. 2009) or studies that had relatively little variability in the size of sharks monitored (Holland et al. 1993; Heupel et al. 2006; Carlson et al. 2008). There are few data concerning changes in batoid activity spaces with growth. Collins, Heupel, and Motta (2007) used passive telemetry to study short-term patterns of residency and activity space in cownose rays (*Rhinoptera bonasus*) in a Florida estuary. They reported a positive correlation between disc width and activity space. One of the limitations of passive telemetry is that the spatial resolution is often insufficient to discern whether an activity space expansion such as this translates into use of new habitats or simply a greater area of the same habitats occupied at smaller sizes.

Latitudinal variation in productivity and environmental variability have important consequences for ontogenetic changes in habitat use. Inshore juvenile habitats in tropical and subtropical regions may be occupied year-round (Yokota and Lessa 2006) or seasonally (e.g., Simpfendorfer and Milward 1993) while year-round occupancy is not an option in temperate regions due to the wide fluctuation in temperature (Musick and Colvocoresses 1986). Activity spaces of juvenile carcharhinid and sphyrnid sharks (Table 7.1) are substantially larger at higher latitudes (Figure 7.2). This is a similar but more dramatic trend than demonstrated in the home range of mammals (Harestad and Bunnell 1979). This simple meta-analysis (Table 7.1, Figure 7.2) for sharks suffers because many of the estimates do not represent home ranges, variable telemetry methods were used (active and passive), and varying analysis techniques were employed (polygon and kernel methods). Nevertheless, the trend may be conservative since the studies with sufficient sample sizes and tracking durations to accurately estimate home range were conducted in the subtropics (e.g., Morrissey and Gruber 1993a; Heupel, Simpfendorfer, and Heuter 2004) where home ranges are typically small.

There are three major ecosystems represented in Figure 7.2. Sharks in very confined tropical or subtropical atolls or lagoons have very small (~1 km²) activity spaces (Franks 2007; Garla et al. 2006; Holland et al. 1993; Morrissey and Gruber 1993a; Papastamatiou et al. 2009). Those associated with moderate-sized subtropical estuaries have small to moderate (1 to 30 km²) activity spaces (Heupel, Simpfendorfer, and Heuter 2004; Heupel et al. 2006; Carlson et al. 2008; Yeiser, Heupel, and Simfendorfer 2008; Ortega et al. 2009). Unfortunately, only two studies, both on sandbar sharks, have been conducted in large temperate estuaries (Grubbs 2001; Rechisky and Wetherbee 2003). Short-term activity spaces in these estuaries were very large (~100 km²) compared to those estimated in tropical and subtropical regions. The latitudinal differences in activity space, therefore, may be due to behavioral differences among the species (i.e., sandbar sharks behave differently), a hypothesis that could be easily tested (see Section 7.8, Future Research). Likely, the correlation observed in sharks reflects latitudinal variation in the distribution of resources (see below).

7.2.2 Changes between Juvenile and Adult Habitats

Habitats used by juvenile sharks and rays are generally inshore of adult habitats (Smale 1991; Yokota and Lessa 2006) though there are exceptions (e.g., Ebert and Cowley 2003; Carraro and Gladstone 2006). Parturition usually takes place in or adjacent to inshore primary nurseries and with growth, juveniles move progressively away from nurseries and

TABLE 7.1

Summary of Published Activity Space Estimates for Juvenile Carcharhinid and Sphyrnid Sharks Using Active and Passive Acoustic Telemetry

Species	Code[a]	MCP (km²) Mean	MCP (km²) Range	KUD (km²) Mean	KUD (km²) Range	N >24 h (total)	Monitoring Duration	Latitude (Degrees)	Telemetry Method	Reference	
Carcharhinus perezi	Cper	0.79 max.				14	0–400+ days	4.0°S	Passive	Garla et al. 2006	Tropical/Subtropical Lagoons
Carcharhinus melanopterus	Cmel	0.48 (0.55)	0.19–0.91	0.34 (0.33)	0.20–0.94	6 (7)	17–72 hours	5.00°N	Active	Papastomatiou et al. 2009	Tropical/Subtropical Lagoons
Sphyrna lewini	Slew	1.52 (1.26)	0.46–3.52			4 (6)	9–72 hours	21.5°N	Active	Holland et al. 1993	Tropical/Subtropical Lagoons
Negaprion brevirostris	Nbre	0.68	0.23–1.26			17	1–153 days	25.75°N	Active	Morrissey and Gruber 1993a	Tropical/Subtropical Lagoons
Negaprion brevirostris	Nbre2	0.80	0.14–2.6			28	90–700+ days	25.75°N	Active	Franks 2007	Tropical/Subtropical Lagoons
Negaprion brevirostris	Nbre3	0.65	0.08–3.14			41	90-700+ days	25.68°N	Active	Franks 2007	Tropical/Subtropical Lagoons
Sphyrna tiburo	Stib			18.62	1.22–60.31	36	1–173 days	26.50°N	Passive	Heupel et al. 2006	Subtropical Estuaries
Carcharhinus leucas	Cleu	(2.5)	1.2–4.3			5 (8)	6–24 hours	26.60°N	Active	Ortega et al. 2009	Subtropical Estuaries
Carcharhinus leucas	Cleu2			31.51 22.66	9.54–68.82 1.86–55.43	19	8–89 days	26.50°N	Passive	Yeiser, Heupel, and Simfendorfer 2008	Subtropical Estuaries
Negaprion brevirostris	Nbre4			31.32	6.07–63.57	5	12–83 days	26.50°N	Passive	Yeiser, Heupel, and Simfendorfer 2008	Subtropical Estuaries
Carcharhinus limbatus	Clim		0.02–15.68	0.7 1.2	0.02–10.36	74	1–167 days	27.50°N	Passive	Heupel, Simpfendorfer, and Heuter 2004	Subtropical Estuaries
Rhizoprionodon terraenovae	Rter			8.41 7.38 5.06	0.004–17.50	56	1–37 days	30.00°N	Passive	Carlson et al. 2008	Subtropical Estuaries
Carcharhinus plumbeus	Cplu	110.26 (87.30)	39.6–275.8	140.1 (119.65)	62.28–382.7	7 (10)	10–64 hours	37.00°N	Active	Grubbs 2001	Temperate Estuaries
Carcharhinus plumbeus	Cplu2	78.47 (55.7)	6.7–339.9	85.45 (62.7)	12.5–315.4	15 (25)	3–70 hours	39.00°N	Active	Rechisky and Wetherbee 2003	Temperate Estuaries

[a] The codes correspond to those used in Figure 7.2. MCP = Minimum Convex Polygon, KUD = 95% Kernal Utilization Distribution.

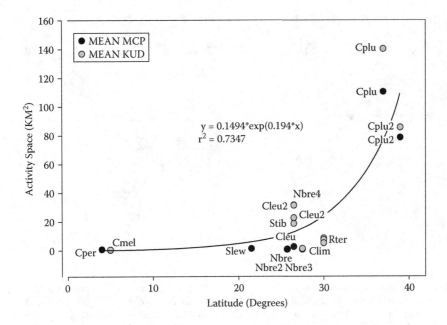

FIGURE 7.2
Activity spaces reported in the literature versus latitude for juveniles of ten species of carcharhinid and sphyrnid sharks. MCP = Minimum Convex Polygon, KUD = 95% Kernal Utilization Distribution. Species abbreviations: Cleu = *Carcharhinus leucas*, Clim = *C. limbatus*, Cmel = *C. melanopterus*, Cper = *C. perezi*, Cplu = *C. plumbeus*, Nbre = *Negaprion brevirostris*, Rter = *Rhizoprionodon terraenovae*, Slew = *Sphyrna lewini*, Stib = *S. tiburo*. Abbreviations for species with multiple estimates are followed by a number corresponding to details and data sources listed in Table 7.1.

closer to adult habitats. Bass (1978) reported that bull sharks (*C. leucas*) in the St. Lucia estuarine system of South Africa exhibit this pattern. The upper estuary was populated primarily by juveniles, with increasing prevalence of large juveniles and adults near the mouth. Adjacent ocean waters were populated by subadult and adult bull sharks. Similarly, Ebert (1989) reported that broadnose sevengill sharks, *Notorynchus cepedianus*, use near-shore bays and estuaries as nurseries and move offshore as adults.

As discussed above for lemon sharks, activity space in many elasmobranchs increases by more than an order of magnitude over the juvenile life stages. Adult activity spaces are very difficult to assess but often encompass thousands if not tens of thousands of square kilometers (Sundstrom et al. 2001). The age-specific distribution and differential migration patterns of sandbar sharks (*C. plumbeus*) in the Northwest Atlantic (Figure 7.3) illustrate this general pattern. Parturition takes place in warm-temperate estuaries (e.g., Chesapeake Bay and Delaware Bay) that serve as primary and secondary nurseries (Grubbs and Musick 2007; McCandless et al. 2007), providing forage and presumably refuge from pre-dation during the first summer. In fall, decreasing photoperiod (likely signaling impend-ing temperature declines) triggers a climatic migration south to wintering areas off North Carolina (Grubbs et al. 2007). A return migration to the natal estuary occurs the follow-ing spring. This migration pattern is repeated annually until at least seven years of age but shifts offshore and increases spatially with growth (Figure 7.3a). As subadults and adults, this migration is greatly expanded to include overwintering nearshore off the east coast of Florida and along the edge of the continental shelf in the eastern Gulf of Mexico (Figure 7.3b). Some adult sandbar sharks remain in southern waters year-round while

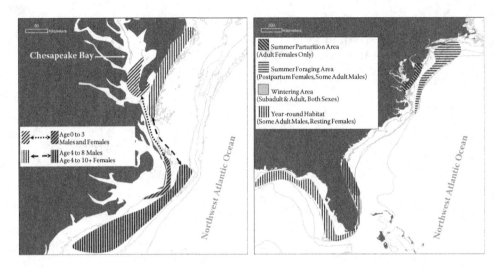

FIGURE 7.3
Ontogenetic and sexually differential shifts in movements and migration in sandbar sharks, *Carcharhinus plumbeus*, in the Northwest Atlantic Ocean (based on Grubbs et al. 2007; Kohler, Casey, and Turner 1998; Grubbs and Musick, unpublished data). (a) Comparison of the summer and winter distributions of young (0–3 years old) and older (4+ years old) juvenile sandbar sharks showing an ontogenetic shift offshore. (b) Distribution of adult sandbar shark habitats including areas of summer foraging, overwintering, and parturition.

others migrate to foraging areas along the edge of the continental shelf off the northeast coast of the United States during summer. This pattern varies between males and females. I will return to this example later in this chapter.

The general pattern observed in coastal sharks likely applies to coastal batoids as well, though there has been limited research on these taxa. Round stingrays (*Urobatis halleri*) are primarily restricted to shallow coastal waters (<20 m depth) with soft sediments and water temperatures above 10°C, and juveniles are confined to shallower, warmer waters than adults (Babel 1967). While adults migrate offshore in winter, juveniles remain inshore, but move to warmer areas near inlets (Babel 1967; Hoisington and Lowe 2005). Yokota and Lessa (2006) reported that juvenile *Gymnura micrura*, *Dasyatis guttata*, *Rhinobatos percellens*, and *Narcine brasiliensis* occur primarily in shallow turbid waters very close to shore in Brazil while adults use slightly deeper waters offshore. While the data are limited, this study included three orders of batoids, suggesting the general pattern observed in most coastal sharks also applies to coastal batoids in the tropics. The endangered smalltooth sawfish (*Pristis pectinata*) exhibits an ontogenetic niche shift comparable to some large coastal sharks. In the United States, juveniles are restricted to shallow mud and sand flats adjacent to mangrove-fringed shorelines in southern Florida (Simpfendorfer 2005). As adults, however, smalltooth sawfish historically ranged from New York to Brazil, including waters deeper than 100 m (National Marine Fisheries Service 2009).

Many coastal pelagic and oceanic sharks are wide ranging and adult habitats may include entire ocean basins, making studies of ontogenetic changes in movements and patterns of habitat use especially difficult. Data sources are usually limited to catch-composition data and conventional tag-recapture data. The available data suggest these taxa also show ontogenetic changes in habitats and activity space. Holland et al. (1993) reported that juvenile scalloped hammerhead sharks (*Sphyrna lewini*) had very small activity spaces (mean = 1.26 km²) while using a protected subtropical bay as a nursery, and Duncan and Holland (2006) reported that these sharks remain in the bay for at least one year with no evidence of an ontogenetic shift in habitat during that period. As adults, however, scalloped hammerheads are wide-ranging oceanic sharks with likely activity spaces in the thousands of square kilometers. Similarly, tagging and tracking data show that tiger sharks are very wide ranging (Kohler, Casey, and Turner 1998), whereas nursery areas have been documented in near-coastal waters in very restricted portions of the overall range (Natanson et al. 1999; Driggers et al. 2008). The ecological characteristics of these nursery habitats for tiger sharks have not been documented. The blue shark (*Prionace glauca*) is likely the widest ranging of all elasmobranch species and is also one of the most abundant. Kohler et al. (2002) reported on the results of more than 90,000 blue sharks tagged in the Atlantic Ocean by the Cooperative Shark Tagging Program. Recapture data suggest that adult habitat includes all Atlantic waters, but as in coastal species, habitats for juvenile blue sharks are much more restricted than adult habitats. Similarly, silky sharks (*C. falciformis*) are widespread throughout the tropical Pacific, but juveniles occur in dense aggregations on equatorial reefs and atolls.

7.3 Vertical Shifts in Habitat Use

Ontogenetic changes in depth ranges (i.e., use of vertical habitats) are also common and may be more physiologically demanding than the changes associated with large-scale seasonal migrations. Ontogenetic niche shifts in the third dimension take two primary forms, changes in short-term patterns of vertical movement (e.g., daily vertical migrations) and long-term changes in the depths occupied.

7.3.1 Changes in Vertical Movement Patterns

Many pelagic fishes exhibit diel vertical migrations (or commutes) generally related to the diel movements of mesopelagic boundary communities (MBC), also called the deep-scattering layer (DSL), on which they feed. These communities, which include many fish and invertebrate taxa, spend daylight hours 500 to 1000 m deep, but migrate en masse to depths between 100 and 300 m at sunset and then return to deeper habitats at sunrise. Mesopelagic fishes that migrate vertically tend to have higher metabolic rates and higher daily rations than those that are nonmigratory (Pearre 2003). The sunset migration to shallow waters is related to forage, whereas the return migration (deeper at sunrise) is linked to decreased predation risk and lower metabolic rates in darker, colder waters (Pearre 2003). Many large pelagic fishes, including sharks, that are predators of MBC taxa also undergo diel migrations coincident or opposite (i.e., from above) the vertical migrations of the MBC to intercept these organisms when they are concentrated. Many species of pelagic sharks, from the smallest, *Euprotomicrus bispinatus* (Hubbs, Iwai, and Matsubara 1967), to among the largest, *Cetorhinus maximus* (Sims et al. 2005), exhibit diel vertical migrations or at least periodic bounce dives to exploit mesopelagic prey (Carey and Scharold 1990).

    Vertical movements often require tolerance of changes in the physical environment (temperature, light, pressure) that far exceed those encountered during seasonal "horizontal" migrations. Ontogenetic changes in vertical movements are common among pelagic fishes and are related to ontogenetic variation in physiological tolerances. For example, with increased thermal inertia and endothermic capabilities, juvenile yellowfin tuna (*Thunnus albacares*) increase diving depth with ontogeny, which facilitates a significant niche expansion (Graham et al. 2007). As diets shift from epipelagic prey to a mix of epipe-lagic and mesopelagic taxa, stable isotopes suggest these tunas elevate two trophic levels over a very short period of ontogeny (Graham et al. 2007). Similar niche expansions likely occur in pelagic sharks, though this area is relatively unstudied. While active and archi-val telemetry methods have been used to investigate the vertical movements of a number of pelagic shark species, sample sizes and size distributions are generally insufficient to assess ontogenetic patterns. Most of the studies that exist are restricted to endothermic lamniforms. Sepulveda et al. (2004) reported a positive correlation between fork length and maximum diving depth in juvenile shortfin mako sharks (*Isurus oxyrhynchus*). In addi-tion to increased thermal inertia with ontogeny, Bernal, Sepulveda, and Graham (2001) suggested there are ontogenetic changes in endothermy and thermoregulatory abilities that may be related to this shift in vertical behavior. However, this pattern is far from con-sistent. Telemetry data for juvenile and adult bigeye thresher sharks, *Alopias superciliosus* (Weng and Block 2004; Nakano et al. 2003) suggest adults and juveniles display similar diel patterns of vertical migration, remaining in the upper mixed layer during night but diving to 300–600 m deep during the day where temperatures were 6°C to 8°C. Sample sizes were very low and the tracks occurred in three regions so inferences to ontogenetic changes in habitat use are tenuous.

White sharks (*Carcharodon carcharias*) are among the most studied of all large sharks. Data from Weng et al. (2007) and Boustany et al. (2002) suggest an ontogenetic shift toward deeper and cooler water occurs in white sharks in the eastern Pacific, consistent with an increase in thermal inertia and endothermic capacity. Young-of-year and three-year-old white sharks were tracked deeper during daylight and crepuscular periods than dur-ing the night (Weng et al. 2007). Nighttime depths were in the upper mixed layer while the average daytime dives for three-year-olds were to 226 m deep and 9.2°C but those of young-of-year sharks were only 100 m deep and 11.2°C. The average maximum depth and minimum temperature experienced by these white sharks was 394 m and 8.4°C for three-year-olds and only 241 m and 9.4°C for young-of-year. The vertical ranges of six adult white sharks while near shore were more limited than for juveniles (0 to 75 m deep and 10°C to 14°C; Boustany et al. 2002). However, most of these sharks made long migrations across a substantial portion of the Pacific basin before settling for several months in subtropical waters as far west as Hawaii. During this migration, sea-surface temperatures increased from 12°C to 25°C and swimming depths were bimodal between the surface swimming and 300+ m deep. During the deeper phase, these adult sharks experienced temperatures as low as 6°C. Such long-distance migrations are not known for juvenile white sharks, although subadults as small as 3.3 m have been caught in Hawaii (Ikehara 1961).

Shepard et al. (2006) used fast Fourier transformation to seek dominant peaks in peri-odicity of vertical movements in basking sharks (*Cetorhinus maximus*) tagged with archi-val transmitters. Five of six sharks tracked showed a dominant spectral peak in diving behavior near 24 hours, suggesting a diel pattern in vertical movements. All five of these sharks were 450 to 650 cm in total length. One basking shark showed no dominant period of diving behavior and this was the only definite juvenile in the study (250 cm). Though sample sizes were quite low, this is preliminary evidence of an ontogenetic shift toward diel movements of vertical migration in this species as well.

7.3.2 Changes in Habitat Depth

Heinke's law, often called the "bigger-deeper" phenomenon, refers to an ontogenetic shift toward deeper habitats by bigger, older fishes. Although it is often applied to deep-sea fishes associated with continental slopes, the concept, if not the mechanisms, can be expanded to other environments. Ontogenetic shifts from shallow to deeper habitats appear to be the norm among elasmobranch taxa associated with the continental shelf and upper continental slope, but there is variation in this pattern.

Coastal elasmobranchs often show a distinct shift to deeper waters through ontogeneny. This is illustrated by the offshore expansion of sandbar sharks in the northwest Atlantic during summer (Figure 7.4). In this species there is an ontogenetic and sexual shift in habitat use. North of Cape Hatteras, North Carolina, estuaries and near-shore waters less than 10 m deep are populated with neonate to three-year-old sharks of both sexes. Coastal waters 10 to 20 m deep are summer habitat for juvenile males between three and six years of age but for females from three to at least ten years of age as well as pregnant and post-partum adults. Offshore habitats >20 meters deep are predominantly inhabited by large juvenile and adult females. Juvenile sandbar sharks overwinter off the coast of North Carolina (Figure 7.3a) and exhibit a similar pattern. The ontogenetic separation continues during migration. Juveniles from age zero to three overwinter close to shore while those between four and seven years old are in deeper waters offshore (Grubbs et al. 2007).

A majority of all shark species occur only in waters deeper than 200 m along continental and insular slopes (Kyne and Simpfendorfer 2007; Chapter 2), yet very little is known about habitat use in most of these taxa. From fisheries data we know more about the distribution of species associated with the upper slope (200 to 750 m) than we do for mid-slope (750 to 1500 m) and deep slope (1500 to 2250 m) taxa though no consistent patterns emerge. Many rajids are known to shift from juvenile habitats on continental shelves to adult habitats along the upper slope (Brickle et al. 2003). Often, these taxa undergo seasonal inshore-offshore migrations (alimentary or gametic) resulting in great overlap in the annual depth distributions of juveniles and adults but the ranges of these migrations often shift ontogenetically deeper. Although Macpherson and Duarte (1991) found no correlation between depth and size in *Raja clavata* between 68 and 357 m deep and *Raja confundus* between 203 and 805 m, there were positive correlations in the eastern Atlantic between depth and size in two species of squaloid sharks, *Deania calcea* between 375 and 836 m and *Squalus blainvillei* between 68 and 437 m. They also found a strong positive correlation between size and depth for the scyliorhinid shark *Galeus melastomus* between 260 and 650 m deep in the northwest Mediterranean but no such relationship in *Galeus polli* in the southeast Atlantic between 167 and 622 m. Richardson et al. (2000) reported that male *Holohalaelurus regani* exhibited a positive ontogenetic shift to deeper habitats in the southeast Atlantic. Juveniles and adult females occurred primarily in slope waters shallower than 300 m while adult males were deeper. In the same region, Macpherson and Duarte (1991) found no significant trend in depth and size in *H. regani*, but their samples included few juveniles, suggesting a difference in size-based gear recruitment between the studies.

The bigger-deeper prediction often fails and sometimes reverses (i.e., smaller-deeper) at mid-slope and deep-slope depths for teleosts (e.g., Stefanescu, Rucabado, and Lloris 1992). Similarly, adult and juveniles of many of the mid-slope skates co-occur (e.g., Orlov, Cotton, and Byrkjedal 2006) and most studies have found little evidence of ontogenetic shifts in habitat depths for deepwater squaloids found below 1000 m (e.g., Jakobsdottir 2001). Macpherson and Duarte (1991) reported a significant positive correlation between size and depth for the squaloid shark *Etmopterus spinax*, on the upper slope (between 334 and 648 m) in the northwest Mediterranean, but E.G. Jones et al. (2003), using bait cameras in the eastern Mediterranean, reported no significant change in size of *E. spinax* on the deep slope (between 1500 and 2490 m). Close examination of their data suggests a smaller-deeper trend may occur on the deep slope. Sharks in the largest size classes (>400 mm

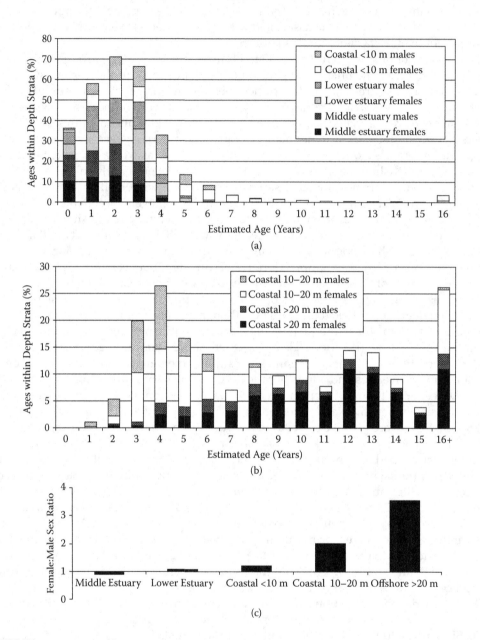

FIGURE 7.4
Age frequencies of male and female sandbar sharks (*Carcharhinus plumbeus*) captured in a fishery-independent longline survey conducted by the Virginia Institute of Marine Science. Ages are based on length-at-age data from Sminkey (1994). (a) Sandbar sharks captured in the Chesapeake Bay (middle and lower estuary) and coastal waters shallower than 10 m. (b) Sandbar sharks captured in coastal waters between 10 and 20 m deep and offshore waters greater than 20 m deep. (c) Sex ratio (females:males) for each of the strata presented inshore to offshore.

TL) were only seen in the shallowest depth strata (1500 and 1800 m) while those in the smallest size class (<260 mm TL) were only seen on cameras deployed deeper than 2200 m. Jakobsdottir (2001) found a similar smaller-deeper trend in this species on the upper slope southeast of Iceland (436 to 1653 m).

7.4 Shifts in Patterns of Residency

We are only beginning to study ontogenetic shifts in duration of residency in particular habitats (e.g., estuarine nurseries). Size-sorted schools are common in elasmobranchs and these may be maintained by ontogenetic differences in swim speed. However, some evidence suggests there are age-based differences in temporal aspects of migration in addition to the spatial differences discussed earlier. For example, Smith and Merriner (1987) reported that cownose rays (*Rhinoptera bonasus*) emigrate from the Chesapeake Bay in late September, but the juveniles are the last to leave and may linger until late October. Catch composition data from a fishery-independent survey suggest the same trend applies to sandbar sharks using the Chesapeake Bay as a nursery (Grubbs and Musick, unpublished data). The youngest sandbar sharks recruit to the estuary first (through birth or return migration). During June, sandbar sharks two years of age or younger make up ~80% of the population in the lower Chesapeake Bay. Between dispersal of smaller sharks up the estuary and return of older age classes, these younger sharks only make up 40% of the lower estuary population from July through early September. The older juveniles also emigrate first and by late September 80% to 90% of the remaining sandbar sharks are two years or younger. If we interpret age class presence in catch at length data to represent seasonal residency, these data suggest that during the summer, age-1 sandbar sharks are present in the Chesapeake Bay for about 120 days while age-3 sandbar sharks are only present for about 60 days.

7.5 Ontogenetic Shifts in Diet: Niche Expansion or Contraction

Changes in diet are the most common ontogenetic shifts reported in fishes and are very common in elasmobranchs as well. I will not cover the great diversity of dietary shifts here, but refer the reader to Wetherbee, Gruber, and Cortés (1990) and Wetherbee and Cortés (2004) for thorough reviews. In teleost fishes, ontogenetic shifts in diet may include complete changes in feeding mode such as shifts from planktivory to piscivory (Werner and Gilliam 1984; Mittelbach and Persson 1998) or carnivory to herbivory (Tibbetts and Carseldine 2005). Shifts in feeding mode also occur in elasmobranchs but are less dramatic due in part to direct development resulting in comparatively limited overall growth and ontogenetic changes in morphology. For example, no elasmobranchs shift from herbivory to carnivory (but see Bethea et al. 2007) or from true planktivory to piscivory. The closest to these extremes are species that shift from feeding on pelagic organisms to benthic prey. For example, some *Squalus acanthias* populations shift from feeding in part on planktonic ctenophores as juveniles to feeding on demersal crustaceans and fishes as adults (Alonso et al. 2002).

Trophic position is positively correlated with body size in many taxa. Therefore, the trophic position of most predators, including elasmobranchs, usually increases ontogenetically as larger individuals include larger prey in their diets (Cortés 1999; Ebert and Bizarro 2007). Niche breadth, as estimated from diet data, often increases ontogenetically (e.g., Laptikhovksy, Arkhipkin, and Henderson 2001) due to the addition of prey taxa while comparatively few taxa are eliminated. However, there is great variability in the patterns observed. For example, Ellis, Pawson, and Shackley (1996) reported increases in dietary niche breadth in *Scyliorhinus canicula* and *Raja clavata* but highest dietary niche breadth in the smallest size class of *Squalus acanthias*. In some cases, ontogenetic shift towards trophic specialization even results in niche contraction (Mahe et al. 2007)

Shifts in the diets of most sharks result from size-based expansions of available habitats but not all ontogenetic shifts in diet reflect shifts in habitat and conversely, shifts in habitat use may not be reflected in the diet. Ontogenetic shifts in the diets of demersal elasmobranchs, especially batoids, are often functions of the ontogeny of durophagy. For example, Babel (1967) reported a dramatic shift in the diet of the round stingray, *Urobatis halleri*, from a diet dominated by crustaceans (primarily decapods) and polychaetes to primarily feeding on pelycepod (bivalve) mollusks. Dietary shifts are also common in the sharks associated with deep slopes that show little evidence of habitat shifts (e.g., Carrassón, Stefanescu, and Cartes 1992). These shifts may be related to increased gape, foraging modes or processes such as the entrainment of mesopelagic taxa by geomorphological features (e.g., seamounts, steep slopes) during diel migrations (Holland and Grubbs 2007). Interpretation of dietary shifts can also be complex. Ebert and Cowley (2003) described the diet of the blue stingray, *Dasyatis chrysonata*, across four size classes and between surf zone, near-shore, and offshore habitats and reported that ontogenetic shifts in diet were zone dependent.

Elasmobranchs are commonly referred to as opportunistic predators and prey selection is often highly plastic, varying on multiple temporal and spatial scales. Most species have diverse diets that allow them to switch prey taxa in response to changes in their relative or absolute abundance (Braccini and Perez 2005). Size-dependent predation has been reported in elasmobranchs (Cortés, Manire, and Hueter 1996; Scharf, Juanes, and Rountree 2000; Bethea, Buckel, and Carlson 2004); however, in most cases the mean size of prey consumed remains quite small through ontogeny (Braccini, Gillanders, and Walker 2005). With growth, the upper size limit of prey consumed increases but smaller prey consumed at smaller sizes often remain in the diet. As such, the correlation between predator size and prey size is usually weak with notable exceptions (e.g., white sharks and tiger sharks; see below).

Many studies have supported the hypothesis that elasmobranchs consume the most abundant prey (Wetherbee, Gruber, and Cortés 1990; Hanchet 1991; Ellis, Pawson, and Shackley 1996; Alonso et al. 2002; Braccini, Gillanders, and Walker 2005). However, some evidence suggests elasmobranchs are selective for large or energy-rich prey. Sims and Quayle (1998) reported that basking sharks selectively chose areas of high densities of large copepod species and low densities of smaller species. Stillwell and Kohler (1982) reported that shortfin mako sharks (*Isurus oxyrhynchus*) fed primarily on cephalopods offshore but fed heavily on bluefish (*Pomatomus saltatrix*) when inshore in spite of high abundances of cephalopod prey inshore. The ontogenetic diet shifts that occur in tiger sharks and white sharks are among the most dramatic of the sharks and might reflect selection for energy-rich prey. Both species feed predominately on demersal teleosts, cephalopods, and other invertebrates as juveniles. As adults, common prey of tiger sharks include elasmobranchs, sea turtles, birds, and mammals (Rancurel and Intes 1982; Witzell 1987; Simpfendorfer 1992; Lowe et al. 1996; Simpfendorfer, Goodreid, and McAuley 2001). Similarly, marine mammals become a major component of white shark diets as adults (see Long and Jones 1996 for review). The teeth of white sharks even undergo an ontogenetic transformation to accommodate this dietary shift (Hubbell 1996). Ebert (2002) reported a similar ontogenetic pattern in the diet of the broadnose sevengill shark (*Notorynchus cepedianus*), with juveniles feeding primarily on teleost fishes and adults feeding on elasmobranch fishes and marine mammals. A switch from ectothermic prey (fishes and invertebrates) to energy-rich homeotherms (mammals and bird) could be due to selection for more profitable (i.e., higher energy content) prey in these species (Wirsing, Heithaus, and Dill 2006).

It is likely that prey density and availability play major roles in habitat use decisions by elasmobranchs (Sims 2003). Ellis and Musick (2006) documented a distinct ontogenetic shift in the diet of juvenile sandbar sharks from crustaceans (primarily portunid crabs and squillid stomatopods) to teleosts and elasmobranchs (mostly rajids). This diet shift took place during the first six years of life and Ellis and Musick (2006) hypothesized that the diet reflected a shift to deeper habitats. Over the same life stages, Dowd et al. (2006) estimated the daily ration decreased from 2.17% to 1.30% body weight per day but the daily energy ration increased from 233 to 784 kJ/day. The caloric content of these primary prey increased ontogenetically as well; 4810 J/g (wet weight) for the crustaceans, 5010 J/g for the teleosts, 5400 J/g for the elasmobranchs (Dowd et al. 2006), indicating a trend in consumption of, but not necessarily selection for, larger and more energetic prey with ontogeny. Coupled with manual telemetry data (Grubbs 2001) that suggest nighttime activity spaces for juvenile sandbar sharks were larger, shallower, and warmer than daytime activity spaces, these results support Sims' (2003) hypothesis that sharks should actively select habitats that maximize net energy gain and that daytime and nighttime activity may differ.

7.6 Adaptive Functions and Drivers of Ontogenetic Shifts

Ontogenetic shifts in habitats, usually involving niche expansion, often result from intense competition for limited resources or from changes in habitat-specific mortality risk (Knudesen et al. 2006). The use of nurseries by elasmobranchs has been hypothesized to date back at least to the Paleozoic Era (Lund 1990), suggesting it is an evolutionarily stable strategy of size-based habitat selection that may have evolved to minimize predation risk or because smaller sharks cannot compete with larger conspecifics (Springer 1967; Branstetter 1990; Heithaus 2007; Heupel, Carlson, and Simpfendorfer 2007). With growth, older conspecifics move to adjacent, more expansive habitats due to trophic requirements and size-based reductions in mortality risk. Alternatively, competition in a habitat with low mortality risk may result in state-dependent differential migration of inferior competitors (Werner et al. 1983; Clark 1994). Patterns of habitat use and selection in elasmobranch fishes are proximately affected by a range of abiotic and biotic factors. Abiotic factors play major roles in bioenergetics and habitat suitability, whereas biotic factors such as prey density and availability, predation risk, and competitor densities also affect the quality of a habitat.

7.6.1 Physiological (Abiotic) Drivers

Temperature, salinity, and dissolved oxygen are likely the most important abiotic factors affecting habitat selection in elasmobranchs, though factors such as currents (Ackerman et al. 2000; Carraro and Gladstone 2006), photoperiod (Grubbs et al. 2007), turbidity (Blaber and Blaber 1980; Cyrus and Blaber 1992), and light may also be influential. Temperature controls rates of physiological processes, including digestion and growth, and most fishes live within relatively narrow thermal regimes. As such, water temperature is likely the most influential variable affecting elasmobranch movements on migratory scales (e.g., Fitz and Daiber 1963; Grubbs et al. 2007) and also affects habitat selection on much shorter temporal scales (Morrissey and Gruber 1993b; Wallman and Bennett 2006). Behavioral thermoregulation whereby animals use thermal gradients to their energetic advantage (e.g., forage in warm waters but rest in cooler waters) has been suggested for a number of elasmobranch taxa. Selection for temperature can occur both through horizontal movement (e.g., Matern, Cech, and Hopkins 2000; Wallman and Bennett 2006) and vertical commuting (Carey and Scharold 1990). Sims (2003) estimated that *Scyliorhinus canicula* realized significant metabolic savings by feeding in shallow waters only at night when temperatures have cooled slightly and were closer to those in the deeper waters occupied during the day, and suggested that these sharks select temporal use of habitats and foraging to maximize net energy gain.

The habitats of juveniles of many fish taxa, including elasmobranchs, are often warmer than adult habitats (McCauley and Huggins 1979). Tolerance of temperature fluctuations often increases ontogenetically due to changes in physiological function and increases in thermal inertia. Ontogenetic downshifts occur in the optimal temperature for growth in some teleost taxa (Pedersen and Jobling 1989; Fonds et al. 1992; Imsland et al. 1996; Jonassen, Imsland, and Stefansson 1999) although this has not been explicitly tested in elasmobranchs. The thermal environment also may result in variation in age-based and sex-based segregation between populations of a given species. For example, in eastern Atlantic waters, Muñoz-Chápuli (1984) found that juvenile and subadult *Scyliorhinus canicula* mostly occurred at depths shallower than 100 m, whereas adults were most common below 400 m. However, D'Onghia et al. (1995) found no age-based segregation in the Aegean Sea population of *S. canicula*. D'Onghia et al. (1995) hypothesized that the thermal stratification of Atlantic waters prevented juveniles from occurring deeper while waters in the north Aegean Sea are well mixed below 100 m and quite warm (13.5°C to 14.0°C).

Salinity varies little throughout most of the world's oceans and likely has little influence on habitat selection and movements of oceanic elasmobranch species. However, salinity directly regulates distribution of many fishes in estuaries (Gunter 1961; Loneragan et al. 1987). Temperature and salinity have been reported as the factors that limit the distribution of elasmobranchs in some estuaries (Recksiek and McCleave 1973; Snelson and Williams 1981; Cyrus and Blaber 1992; Hopkins and Cech 2003). Whereas temperature likely controls seasonal use of estuaries, Grubbs and Musick (2007) reported that salinity was the dominant factor determining habitat suitability for juvenile sandbar sharks during summer in the Chesapeake Bay. Predictably, sandbar sharks were more widely distributed in the Chesapeake Bay during drought years than during wet years. Catch composition data from the Chesapeake Bay suggests young-of-year sandbar sharks occur farther up the estuary in lower salinity waters than older conspecifics. It is unknown if osmoregulatory abilities vary ontogenetically or between sympatric species, but variation in halal scope may be an adaptation to decrease intraspecific and interspecific competition and/or predation.

Dissolved oxygen often influences the distribution of fishes, especially in estuaries and regions of the continental shelf prone to hypoxia (Eby and Crowder 2004). Seasonal and episodic hypoxia that could be stressful to elasmobranchs (Carlson and Parsons 2001) is common in summer months in many estuaries. Elasmobranch fishes react in

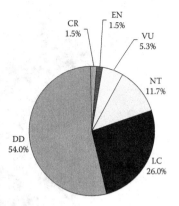

FIGURE 2.1

The conservation status of deepwater chondrichthyans according to the IUCN *Red List of Threatened Species*™. Percentages of species within each Red List category are shown: CR, Critically Endangered; EN, Endangered; VU, Vulnerable; NT, Near Threatened; LC, Least Concern; DD, Data Deficient.

FIGURE 3.3

Representatives of the five major groups inhabiting high latitude seas going clockwise starting in the upper left-hand corner. Rajiformes: Rajidae (*Amblyraja badia*); Carcharhiniformes: Scyliorhinidae (*Apristurus kampae*); Chimaeriformes: Chimaeridae (*Hydrolagus colliei*); Squaliformes: Somniosidae (*Somniosus pacificus*); and Carcharhiniformes: Triakidae (*Triakis semifasciata*). (Photos © Monterey Bay Aquarium Research Institute (cat shark, skate, sleeper shark); Joseph J. Bizzarro (chimaera); Aaron B. Carlisle (leopard shark). All rights reserved.)

FIGURE 4.2

The blue-spotted fantail ray *Taeniura lymma* under a coral reef ledge off northeast Efate (Vanuatu) in the Southwest Pacific. (Photo © Will White. All rights reserved.)

FIGURE 4.3
A typical daily landing of longline caught, large carcharhinid sharks off Lombok in eastern Indonesia. (Photo
© Will White. All rights reserved.)

FIGURE 4.4
Dried caudal fin of a medium-sized whale shark *Rhincodon typus* landed off the Indonesian island of Bali. (Photo
© Will White. All rights reserved.)

FIGURE 4.5
Large (>300 cm TL) wedgefish, *Rhynchobatus*, and large stingrays (Dasyatidae) landed at the Muara Angke landing site in Jakarta (Indonesia) from tangle-net fishers. The highly sought after fins were already removed. (Photo © Will White. All rights reserved.)

FIGURE 5.3
Potamotrygon leopoldi in its natural habitat at the Rio Xingu, Brazil. (Photo © P. Charvet-Almeida. All rights reserved.)

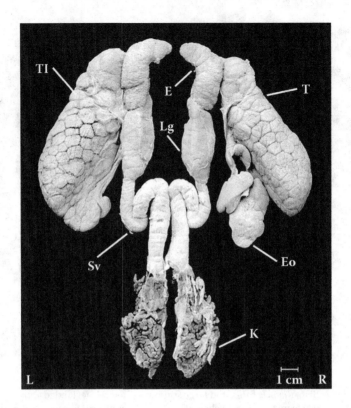

FIGURE 5.4
Reproductive system of an adult male specimen of *Potamotrygon leopoldi*. T = testicle, Tl = testicular lobe, E = epididymis, Lg = Leydig gland, Sv = seminal vesicle, K = kidney, and Eo = epigonal organ. (Photo © P. Charvet-Almeida. All rights reserved.)

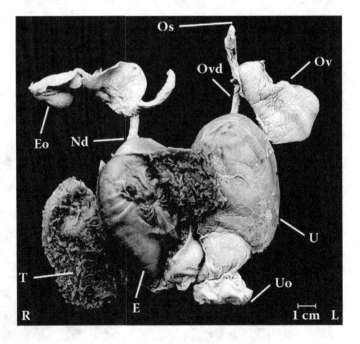

FIGURE 5.5
Reproductive system of an adult pregnant female specimen of *Potamotrygon orbignyi*. Os = ostium, Ov = ovary, Ovd = oviduct, Nd = nidamentle gland, U = uterus, T = trophonema, E = embryo, Uo = urogenital opening, and Eo = epigonal organ. (Photo © P. Charvet-Almeida. All rights reserved.)

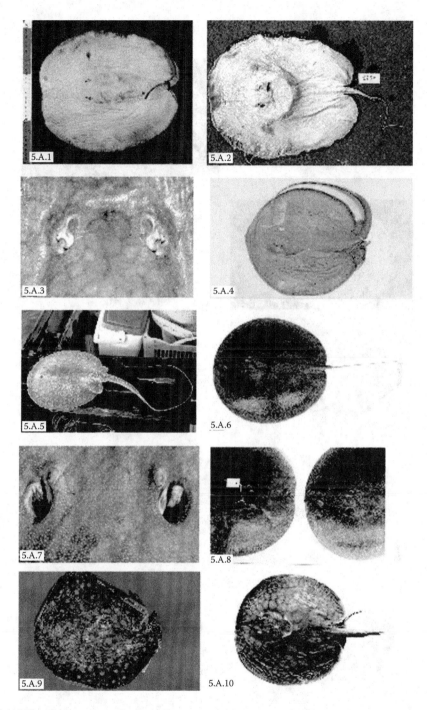

FIGURE 5.A.1–5.A.10

Specimens of Potamotrygonidae. **5.A.1** *Paratrygon aiereba*, MCZ 297-S, syntype of *Disceus thayeri*, dorsal view. (Photo by R.S. Rosa.) **5.A.2** *Paratrygon aiereba*, ZMB 4632, type of *Trygon strogylopterus*, dorsal view. (Photo courtesy of H.-J. Paepke.) **5.A.3** *Paratrygon aiereba*, detail of dorsal disc, showing eyes and spiracular projection. (Photo by R.S. Rosa.) **5.A.4** *Plesiotrygon iwamae*, MZUSP 10153, holotype, dorsal view. (Photo by R.S. Rosa.) **5.A.5** *Plesiotrygon iwamae*, dorsal view of freshly collected specimen. (Photo courtesy of L.N. Chao.) **5.A.6** Undescribed genus and species, dorsal view of freshly collected specimen, approximately 718 mm disc length. (Photo by P.C. Almeida.) **5.A.7** Undescribed genus and species, dorsal view of disc with detail of eyes and spiracles. (Photo by R.S. Rosa.) **5.A.8** Comparative view of anterior portion of disc between *Paratrygon aiereba* (left) and the undescribed genus and species (right). (Photo by P.C. Almeida.) **5.A.9** *Potamotrygon boesemani*, USNM225574, holotype, dorsal view. (Photo by R.S. Rosa.) **5.A.10** *Potamotrygon brachyura*, MFA uncatalogued specimen, dorsal view. (Photo by R.S. Rosa. All rights reserved.)

FIGURE 5.A.11–5.A.20

Specimens of Potamotrygonidae. **5.A.11** *Potamotrygon castexi*, MACN 5777, holotype, dorsal view of freshly collected specimen. (Photo by H.P. Castello.) **5.A.12** *Potamotrygon* cf. *castexi*, dorsal view of freshly collected specimen in Peru. (Photo courtesy of C. Swift.) **5.A.13** *Potamotrygon circularis*, MCZ 291-S, syntype, dorsal view. (Photo by R.S. Rosa.) **5.A.14** *Potamotrygon constellata* MNHN A.1010. type, dorsal view. (Photo by R.S. Rosa.) **5.A.15** *Potamotrygon dumerilii*, MNHN 2367, type, dorsal view. (Photo by R.S. Rosa.) **5.A.16** *Potamotrygon* cf. *dumerilii*, MFA 268, specimen mistakenly selected as lectotype of *P. histrix* by M.N. Castex, dorsal view. (Photo by M.N. Castex.) **5.A.17** *Potamotrygon falkneri*, MFA 235, paratype, dorsal view. (Photo by R.S. Rosa.) **5.A.18** *Potamotrygon henlei*, dorsal view of freshly collected female specimen and aborted embryo, in Rio Tocantins, Brazil. (Photo by G.M. Santos.) **5.A.19** *Potamotrygon histrix*, MNHN 2449, type, dorsal view. (Photo by R.S. Rosa.) **5.A.20** *Potamotrygon humerosa*, MCZ 299-S, holotype, dorsal view. (Photo by R.S. Rosa.)

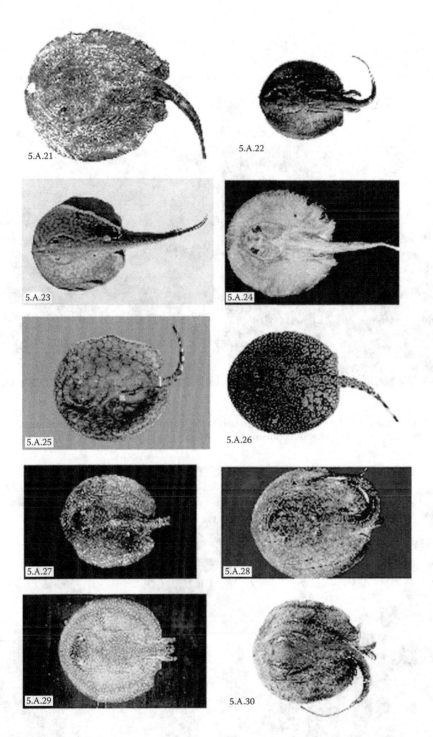

FIGURE 5.A.21–5.A.30

Specimens of Potamotrygonidae. **5.A.21** *Potamotrygon laticeps*, MCZ 605-S, syntype, dorsal view. (Photo by R.S. Rosa.) **5.A.22** *Potamotrygon magdalenae*, dorsal view of specimen collected by T.B. Thorson in Colombia. (Photo by R.S. Rosa.) **5.A.23** *Potamotrygon menchacai*, MFA 289, holotype, dorsal view. (Photo by R.S. Rosa.) **5.A.24** *Potamotrygon motoro*, ZMB 4662, type, dorsal view. (Photo by M.N. Castex.) **5.A.25** *Potamotrygon orbignyi*, MZUSP 14790, dorsal view. (Photo by R.S. Rosa.) **5.A.26** *Potamotrygon schroederi*, dorsal view of freshly collected specimen. (Photo courtesy of M.L.G. Araújo.) **5.A.27** *Potamotrygon schuehmacheri*, MFA 269, holotype, dorsal view. (Photo by R.S. Rosa.) **5.A.28** *Potamotrygon scobina*, LACM 39925, dorsal view. (Photo by R.S. Rosa.) **5.A.29** *Potamotrygon signata* dorsal view of freshly collected specimen in Rio Parnaíba drainage, Brazil. (Photo by C.C.D. Quijada.) **5.A.30** *Potamotrygon yepezi*, USNM 121668, paratype, dorsal view. (Photo by H.P. Castello. All rights reserved.)

FIGURE 6.8
Cownose rays photographed on their annual migration in the Gulf of Mexico. The rays are thought to make an annual migration between the Yucatan Peninsula to Florida. (Photos © Sandra Critelli. All rights reserved.)

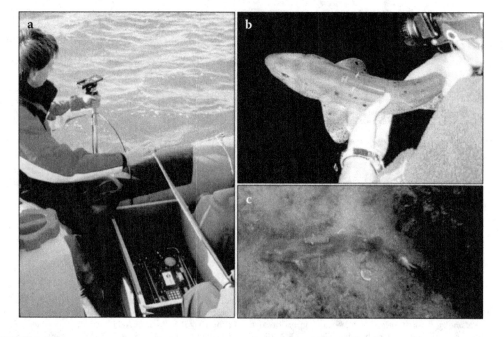

FIGURE 8.1
Acoustic telemetry of sharks and other fish. (a) A directional hydrophone and receiver is used to locate the sound source (b, c) of acoustic transmitters fitted to free-ranging fish. The transmitters shown here are microprocessor controlled, weigh a few grams in water and can remain transmitting at one pulse per second for over a year. (Photos courtesy of MBA, Plymouth, UK.)

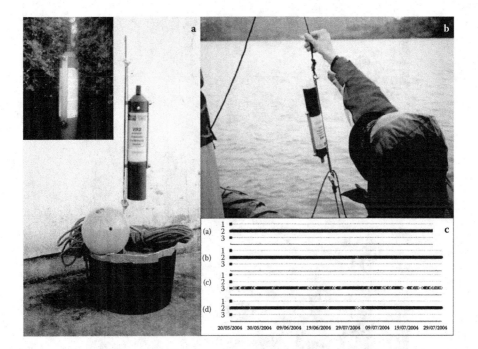

FIGURE 8.2
Data-logging acoustic receivers for long-term monitoring of fish presence or absence at a particular location. (a) Receivers can be moored on the seabed with the receiver uppermost, or (b) suspended a few meters below a surface buoy with the receiver pointing down. (c) Example of recorded data from four small spotted catshark *Scyliorhinus canicula* (inset a–d) at each of three moored receivers (inset 1–3 for each fish). Circles denote presence of a fish at a receiver. Note the almost continuous presence of catshark a, b, and d at receiver 2 for at least two months, whereas fish c was present intermittently at receiver 3. (Modified from Sims DW et al. (2006a) J Anim Ecol 75:176–190. Photos courtesy of MBA, Plymouth, UK.)

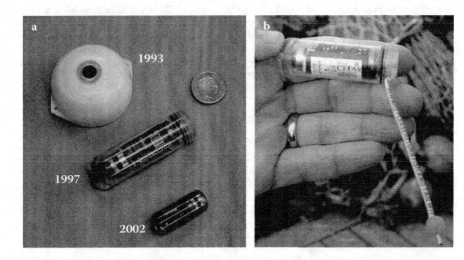

FIGURE 8.3
Examples of data-logging tags used to track the movements and behavior of marine fish including sharks. (a) Data storage tags (DST) show the increased miniaturization over time. The Mk4 DST of 2002 has about four times the memory storage capacity of the Mk1 tag of 1993. All tags were designed and built by the Cefas team in Lowestoft, UK. The two later versions have both been used to study shark behavior. (b) A DST for internal placement with an external marker to alert fishers. This can also incorporate a light sensor at the end of a "stalk" for recording of light level for geolocation purposes. The DST is placed inside the body cavity of the fish and the stalk remains on the outside to record ambient light levels in the sea. (Photos courtesy of J. Metcalfe, Cefas Laboratory, Lowestoft, UK.)

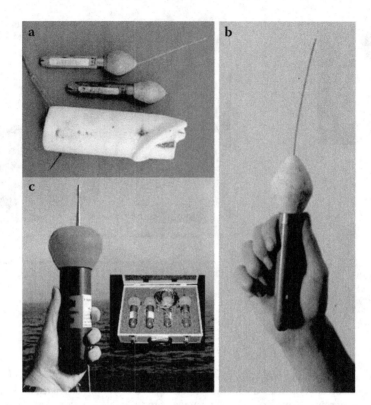

FIGURE 8.4
Examples of satellite-linked and mobile telephone-linked archival tags used for tracking highly migratory shark species. (a) The large white tag (lower) is a satellite-linked depth recorder modified to incorporate a custom-made, timed-release mechanism. It was deployed on a basking shark by the author in July 1999 and was physically returned after washing ashore in France five years later; it was one of the first PSATs attached to a shark. The smaller tags [pop-up archival transmitting (PAT) tags, PAT Mk3, middle, and Mk4, top] transmitted data via satellite and were found and returned to the MBA for full archival data download after one year floating at sea. (b) A closer view of a PAT Mk3 tag comprising a mini-data-logger (gray body of tag) interfaced with an Argos transmitter with protruding antenna, and white syntactic foam for flotation. On the gray body of the tag the sensors can be seen, from top to bottom: pressure (depth), light, and temperature. (c) A GSM mobile telephone-linked tag deployed between 2004 and 2006 on basking and blue sharks in the English Channel. (Photographs courtesy of MBA, Plymouth, UK.)

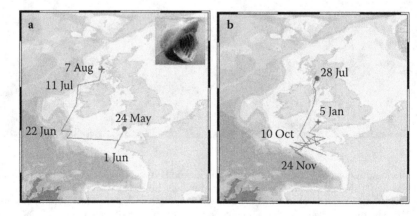

FIGURE 8.5
Example trackings of PSAT-tagged basking sharks (a) between southwest England and northwest Scotland during summer, and (b) vice versa movement between summer and winter. (Modified from Sims DW et al. 2003. Mar Ecol Prog Ser 248:187–196.)

FIGURE 8.6
Different-sized fin-mounted Argos satellite transmitters (PTT, platform terminal transmitter) used to (left photograph) track subadult blue sharks released in the English Channel, and (right) fitted to juvenile blues off southern Portugal. (Photos © (top) J. Stafford-Dietsch, (left) MBA, and (right) N. Queiroz. All rights reserved.)

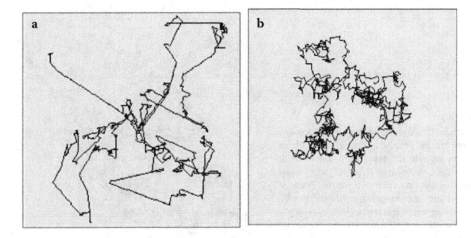

FIGURE 8.11
(a) A computer-simulated Lévy flight search pattern compared to (b) simple Brownian motion. Note the long reorientation jumps in (a) between clusters of smaller steps, with this pattern occurring across all scales, characterizes Lévy motion.

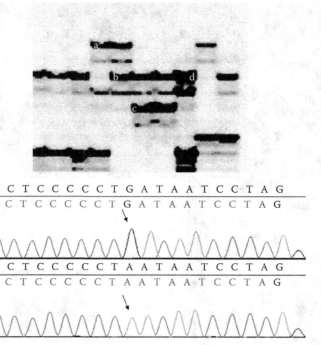

T C T C C C C T G A T A A T C C T A G

1   T C T C C C C C T G A T A A T C C T A G

T C T C C C C T A A T A A T C C T A G

2   T C T C C C C C T A A T A A T C C T A G

FIGURE 10.1

A comparison of microsatellite data and SNP data. Top panel is a gel image generated on a LiCor 4200 Global IR² system. Each band is a different allele detected in an array of juvenile *Carcharhinus plumbeus* and each column is one individual. The relationship between alleles marked a, b, and c is assumed to be the result of either the loss or gain of repetitive motif units but is only scored as a size polymorphism so one cannot be sure. Allele d shows a strong stutter band, which is a PCR artifact that may make accurate scoring difficult. Bottom panel is an image of a 20 base pair read that differs by one base pair substitution. In this case the relationship between alleles is clear; 1 has a G at site 10 and 2 has an A at site 10. When scored as an SNP all individuals sampled will show one state or the other for each allele at this locus. This provides less information than a polymorphic microsatellite locus.

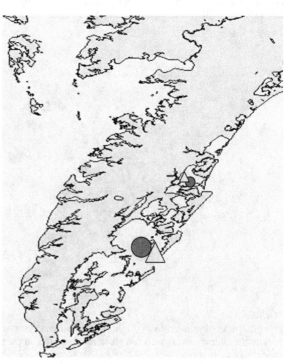

FIGURE 10.3

The reproductive output of females of known genotype can be detected through the use of multilocus genetic fingerprinting. Here adult and juvenile *Carcharhinus plumbeus* were sampled in the lagoons of the Eastern Shore of Virginia. A mother–offspring relationship was detected using eight microsatellites between a sampled juvenile (small triangle) and a postpartum female (large triangle) sampled two years later. A postpartum female (large circle) and two juveniles (small circles) identified as the female's progeny were caught several months apart in the same year.

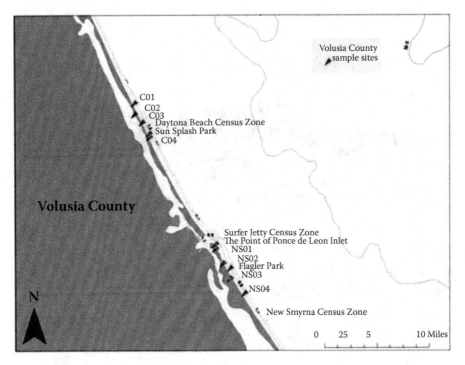

FIGURE 13.1
Volusia County, Florida, beaches and aquatic user visual census stations used in study.

FIGURE 13.11
Blacktip shark (*Carcharhinus limbatus*) jumping out of the water in a spinning motion near surfer in New Smyrna Beach, Volusia County, Florida, 2008. (Photo © KemMcNair.com. All rights reserved.)

FIGURE 14.2
Shark-proof bathing enclosure at Manly, Sydney Harbor, 2008.

FIGURE 14.3
Shark spotter overlooking Fish Hoek Beach, Cape Town, 2009. (Photo by Darryl Colenbrander.)

FIGURE 14.4
Shark net off Durban.

FIGURE 14.5
Drumline with a shark net in the background off Cairns, Queensland.

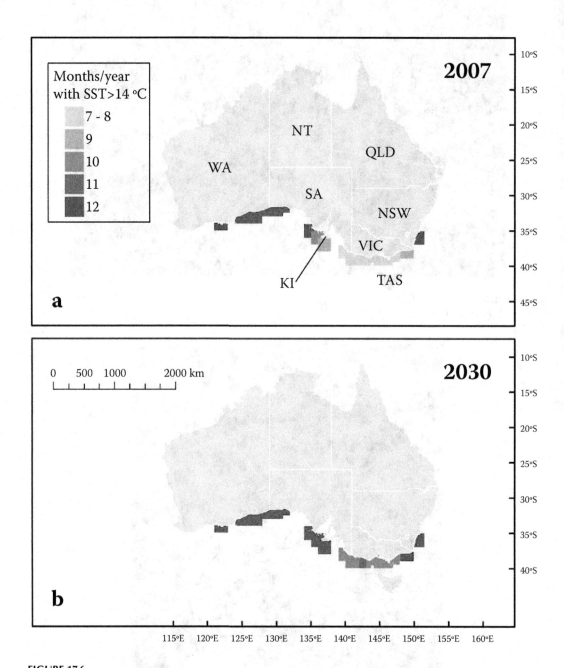

FIGURE 17.6
Climate change and the increasing distribution of suitable thermal habitat for the grey nurse shark. Present day (A) and predicted to 2030 (B) estimates of the number of months each year where annual minimum monthly sea surface temperature averages are greater than 14°C in 1-degree blocks are along the south Australian coast. Predictions for 2030 are derived from the CSIRO Mk3 model. (Redrawn from Bradshaw CJA, Peddemors VM, Mcauley RB, Harcourt RG (2008) Final Report to the Commonwealth of Australia, Department of the Environment, Water, Heritage and the Arts.)

varied ways to depressed dissolved oxygen (Carlson et al. 2004) and some species may be more tolerant of hypoxia than others (Carlson and Parsons 2003). Some species that are not obligate ram ventilators decrease metabolic and activity rates while increasing buccal pumping, whereas some obligate ram-ventilating species actually increase metabolism and swimming speeds as potential flight responses (Carlson and Parsons 2001). Dissolved oxygen appears to be a primary abiotic factor influencing the distribution of juvenile bull sharks (*C. leucas*) in a subtropical estuary (Heithaus et al. 2009), and low oxygen likely limits the northern distribution of juvenile sandbar sharks in the Chesapeake Bay (Grubbs and Musick 2007). In contrast, the distributions of some elasmobranchs appear to be unaffected by hypoxia in the Gulf of Mexico and these species may exploit hypoxic regions for trophic benefit (Kevin Craig, personal communication). For example, Atlantic sharpnose sharks feed on species that concentrate along the edges of hypoxic zones while cownose rays forage on infaunal invertebrates that may be forced closer to the water–seabed interface by hypoxia (Kevin Craig, personal communication). Whether elasmobranchs show ontogenetic changes in hypoxia tolerance has not been investigated.

7.6.2 Competition and Distribution of Resources

The most important biotic factors influencing habitat selection by juvenile animals are prey availability and predation risk (Lima and Dill 1990). The overall availability of prey to an individual will be influenced by prey density, the probability of capturing prey, and the presence of conspecific and heterospecific competitors. Thus, high densities of elasmobranchs might be facilitated by resource partitioning (interspecific specialization) and ontogenetic niche shifts (intraspecific specialization). Taylor and Bennett (2008) reported that *Hemigaleus australiensis* is a highly specialized cephalopod predator that co-occurs with numerous species of piscivorous sharks. A significant ontogenetic shift in habitat from sand and seagrass to coral reefs facilitates a shift in the dominant families of cephalopod prey consumed in this species. Trophic specialization combined with ontogenetic shifts in habitat and diet may thereby decrease competition for available resources.

Ontogenetic shifts in the diets of skates (Rajidae) have been widely documented (e.g., Braccini and Perez 2005; Alonso et al. 2002). Brickle et al. (2003) compared the diets of three sympatric species of *Bathyraja* occurring on the edge of the Falkland Island shelf. *Bathyraja albomaculata* shifted from gammarid amphipod prey as juveniles to polychaetes as adults, *B. brachyurops* fed on amphipods, euphausids, and polychaetes as juveniles and shifted to teleost fishes and cephalopods as adults, and *B. griseocola* fed predominantly on isopods and amphipods as juveniles but were largely piscivorous as adults. *Bathyraja albomaculata* is a more specialized predator than *B. brachyurops* and *B. griseocola*, and while all three species prey heavily on gammarid amphipods as juveniles, differences in diet, dietary shifts, and depth distribution are likely sufficient to minimize interspecific competition and allow coexistence. Similar results have been reported for sympatric species of skates in other regions (e.g., McEachran, Boesch, and Musick 1976).

Activity spaces and latitude are positively correlated in terrestrial mammals as well as sharks (see Section 7.2.2). In mammals, this correlation has been attributed to lower productivity in higher latitudes, which results in increased ranging to locate suitable resources (Harestad and Bunnell 1979). In marine and estuarine ecosystems, however, seasonal productivity is higher at temperate than tropical latitudes. The latitudinal variations in activity space reported for sharks (Figure 7.2) are likely functions of prey density, prey patchiness, and competition for these resources rather than productivity. A comparison of lemon sharks in Bimini, Bahamas, and sandbar sharks in the Chesapeake Bay illustrate this hypothesis. Productivity in tropical ecosystems such as Bimini, Bahamas, is relatively low compared to temperate regions, but is elevated in communities such as mangrove-fringed shorelines and seagrass meadows where most production is through the detrital pathway rather than from phytoplankton. Abundance of prey organisms around Bimini is elevated in mangrove communities relative to adjacent seagrasses, but they are distributed relatively uniformly along the mangrove fringe and vary little seasonally (Newman, Handy, and Gruber 2007). Morrissey and Gruber (1993a) found that juvenile lemon sharks in this area established small, overlapping home ranges that stretched along the mangrove fringe, an efficient strategy for taking advantage of linearly but uniformly distributed resources while minimizing intraspecific competition. As the sharks grew and predation risk decreased, activity spaces increased and expanded into deeper water, providing additional mechanisms for minimizing competition. By contrast, production in temperate estuaries such as the Chesapeake Bay is very high, driven by spring and fall phytoplankton blooms, and supports a tremendous biomass of consumers. Most inhabitants, including all elasmobranchs, are only seasonally resident due to the variable productivity and climate. All of the primary prey for juvenile sandbar sharks (see Medved, Stillwell, and Casey 1985 and Ellis and Musick 2006) are absent in winter. During summer, prey are extremely abundant but their distributions are patchy and these patches move frequently; therefore, juvenile sandbar sharks do not establish persistent home ranges. Juvenile sandbar sharks must locate and exploit profitable patches of prey (e.g., sciaenid fish aggregations). Once profitability declines as functions of prey density and competition, the sharks must make a decision to seek the next prey patch (perhaps an aggregation of portunid crabs). A more nomadic pattern of movement allows these small sharks to exploit and compete for spatially and temporally variable resources; therefore, even short-term activity spaces are very large and may encompass a large portion of the estuary (Grubbs 2001; Rechisky and Wetherbee 2003).

7.6.3 Predation

Predation risk is a primary driver in habitat selection by fishes and may directly influence ontogenetic shifts in patterns of habitat use. Reduction in predation risk is often cited as a major factor influencing habitat preferences of juvenile sharks (Morrissey and Gruber 1993a; Heupel and Hueter 2002; Rechisky and Wetherbee 2003; Grubbs and Musick 2007). Refuge or nursery habitats may be inaccessible to predators due to physical limitations (e.g., depth, structural complexity) or physiological tolerances (e.g., for temperature or salinity). Heupel and Hueter (2002) reported that habitat use by juvenile blacktip sharks (*C. limbatus*) in a known nursery area was not related to prey abundance, and suggested that minimizing predation risk may be a primary driver of habitat selection. Such selection in elasmobranchs is not necessarily restricted to juveniles. Carraro and Gladstone (2006) reported that the same structurally complex habitats were selected by small and large ornate wobbegong sharks, *Orectolobus ornatus*. These habitats were relatively devoid of prey, leading the authors to hypothesize that these habitats serve as structural refuges.

Often, size-dependent predation is hypothesized as the primary factor influencing the correlation between growth and survival, though this relationship is difficult to demonstrate empirically (Anderson 1988) and is confounded by density-dependent effects (e.g., population thinning; Craig et al. 2006). With growth, it is assumed that predation risk and the need to use refuge habitats declines, though this assumption has rarely been explicitly investigated. DiBattista et al. (2007) suggested there was no ontogenetic change in mortality and there was selection against faster growth (larger size) in juvenile lemon sharks over the first three years of life. However, most large species of sharks that use nurseries remain in those nurseries until at least age four and the authors recognized that the apparent selection that is counter to conventional theory may be balanced by opposite selection at a later life stage.

Behavioral avoidance of or submission to larger conspecifics has been noted in multiple species of carcharhiniform sharks (Allee and Dickinson 1954; Springer 1967; Myrberg and Gruber 1974; Weihs, Keyes, and Stalls 1981; Chapman et al. 2007). Franks (2007) is one of few studies to explicitly test whether habitat selection by an elasmobranch is influenced by both prey abundance and predation risk. His results support the hypothesis that juvenile lemon sharks selected areas of abundant prey and avoided areas inhabited by common predators. Similarly, Conrath and Musick (2007) examined the use of coastal lagoons in Virginia by juvenile sandbar sharks. Unlike the Chesapeake Bay, salinity in these bays is comparable to adjacent coastal waters and therefore was not a factor in shark distribution. Relative abundance of neonates and small juveniles (less than three years of age) was correlated with distance inshore from the inlets (i.e., small sharks avoided the inlets). The inlets are known habitats for potential predators, including larger sharks of multiple species. Relative abundance of large juveniles (three plus years of age) was not correlated with any physical factor, including distance from the inlets. As in the lemon shark example, these data suggest activity spaces increase substantially after about three years of age and is likely related to size-specific reductions in predation risk.

7.6.4 Sex-Mediated Differences in Ontogenetic Shifts

There are often sex-mediated components to ontogenetic patterns of habitat use in elasmobranchs. These differences could be functions of sexual dimorphism. Size dimorphism may result in higher mobility of the larger sex or dimorphism in tooth morphology may affect prey selectivity. However, it is likely that many sex-mediated differences in habitat use are ultimately related to reproduction. Movements and migration patterns that vary as a function of sex (differential migration) can be observed in both immature and reproductive life stages. In juvenile fishes including elasmobranchs, migration behaviors serve to balance trophic advantages and physical protection (refuge). At mature stages, sexually differential migration is usually a function of differing energetic commitments to reproduction. The galeoid sharks exhibit striking examples of sexually differential migration because reproductive modes employ internal fertilization and direct development. Mating necessarily takes place in locations common to the migration paths of both sexes. Since the gametic contribution of the males ends at mating, all other phases of their migration are likely related to foraging. Females have an additional component of gametic migration to specific regions for parturition.

For example, sandbar sharks, *Carcharhinus plumbeus*, in the western North Atlantic exhibit ontogenetically and sexually differential migrations. The sex ratios of sandbar sharks off Virginia (Figure 7.4c) are close to 1:1 in the estuaries and coastal waters less than 10 m deep but are heavily skewed toward females offshore (2:1 at 10 to 20 m, 3.5:1 at >20 m). Juvenile male sandbar sharks born in the Chesapeake Bay return to Virginia waters during summer months for the first six years of life (Figures 7.4a and 7.3b) but remain in warmer waters to the south thereafter. As adults, most males remain in waters from the Atlantic coast of Florida through the Gulf of Mexico, though some make northern foraging migrations during summer along the edge of the continental shelf. By contrast, juvenile females continue to return to Virginia waters for at least 10 years in an apparent "dummy" migration in preparation for adult migrations when they will return on a two-year cycle for parturition (Figure 7.3b).

7.7 Implications for Management

An understanding of ontogenetic shifts in habitat use and movement patterns is critical to effective management of elasmobranch populations from at least three perspectives. First, juvenile habitats for many species occur in near-shore waters. Some species are obligately tied to specific habitats that are among the most vulnerable to anthropogenic degradation (e.g., smalltooth sawfish, *Pristis pectinata*). An understanding of ontogenetic patterns of habitat use is requisite to identifying and protecting vulnerable habitats. Second, the functional role of elasmobranch species in marine ecosystems changes ontogenetically with changes in habitat use, movement patterns, and trophic ecology. Different life stages of a given species may be functionally different species in terms of their influences on the trophic dynamics of an ecosystem (see Lucifora et al. 2009). As managers move toward implementing the approaches of ecosystem-based management and the use of mass balance models (e.g., Ecopath) becomes more commonplace, these changing roles must be considered. Third, as a result of spatial and temporal changes in habitat use, the agencies responsible for managing stocks often change with the ontogeny of the organism. Thus, fisheries managers must consider that an individual agency may be managing only a single life stage. Management of the sandbar shark stock in the northwest Atlantic (NWA) illustrates this third issue and the cost of failing to account for a species' ontogenetic shifts in its management. The majority of sandbar sharks born to this stock are restricted to state waters (<3 miles from shore) from Delaware to Virginia (during summer months) and North Carolina (during winter months) and fall under the jurisdiction of state agencies and the Atlantic States Marine Fisheries Commission (ASMFC). By four years of age, these juvenile sharks are using state waters along the East Coast of the United States from Delaware to South Carolina, and by age six or seven they are using state and federal waters from New Jersey to Florida. By 10 years of age, some sandbar sharks born in estuaries of the Mid-Atlantic Bight are overwintering offshore as far as the edge of the continental shelf from South Carolina in the Atlantic to at least Texas in the Gulf of Mexico (Grubbs et al. 2007). Thus, sandbar sharks at this stage occur in the waters of up to 14 states, multiple fishery commissions, as well as federal waters. As adults, the distribution of sandbar sharks includes federal and potentially international waters from off New York to Mexico and Cuba (Springer 1967).

The NWA sandbar shark population declined dramatically through the 1980s and likely reached a minimum in the early 1990s. In spite of progressively more aggressive federal management since 1993, the population has not shown significant signs of recovery. Until recently all life stages of sandbar sharks were being targeted by regionally directed fisheries. Adults and subadults were taken in directed commercial longline fisheries off the Gulf and Atlantic coasts of Florida; large juveniles were targeted by multiple fisheries off North Carolina, and small juveniles were targeted by gillnet fisheries in state waters, especially in Virginia. While federal quotas were continually reduced for the directed longline fishery, a largely unregulated gillnet fishery in Virginia waters continued to harvest juvenile sandbar sharks, harvests that did not count against the federal quota because much of it was landed as "unclassified shark." From 2004 through 2007, the federal quota for large coastal sharks (predominantly sandbar sharks) harvested in U.S. Atlantic waters was ~41 MT for waters from Virginia to Maine and ~549 MT from North Carolina to Florida. During that period an average of ~100 MT of large coastal sharks (~80% were sandbar sharks) were landed from state waters of Virginia alone, more than twice the total federal quota for the region. Fortunately, by 2009, coordination between federal and state agencies and the regional fishery commissions resulted in regulations that prohibited the harvest of sandbar sharks in state and federal waters. Under these regulations, the only U.S. harvest of this species is through a small, federally regulated research fishery with an annual quota comparable to the biomass landed in state waters of Virginia a few years prior (see National Marine Fisheries Service 2007 for regulations).

7.8 Future Research

Additional studies are needed that explicitly investigate movements and habitat use by sharks and batoids across life stages. Studies of habitat use and activity spaces using acoustic telemetry have predominantly been limited to juvenile sharks, largely due to the difficulty in actively tracking large sharks for significant periods, and the expense of deploying arrays of acoustic receivers that are spatially comparable to the movement patterns of larger sharks. Satellite and archival telemetry have greatly enhanced our ability to investigate long-term movements of larger sharks. Unfortunately, the spatial resolution of these techniques often is inadequate to assess habitat use and even activity space. SPOT tags that transmit location of the subject *in situ* offer great promise but are only applicable to species that inhabit very shallow waters or regularly swim near the surface. Most studies to date have been in relatively few coastal regions and have only considered the behavior of the animal relative to its abiotic environment. Most of the elasmobranch fishes occur only in waters deeper than 200 m (Musick, Harbin, and Compagno 2004) yet we know very little about movements and habitat use of any of their life stages. Studies are needed across a broad array of ecological landscapes (i.e., tropical to boreal climates, littoral to abyssal depths). Future research into habitat selection also needs to be conducted in the context of the community and ecosystem (Heithaus 2001). Simultaneously collected data concerning prey and predator abundances and behaviors are needed to investigate the relative balance between predation risk, energetic benefits, competition, and physiological processes and their effects on ontogenetic patterns of habitat selection.

Acknowledgments

I thank Bryan Franks and John Morrissey for graciously allowing me to use their data in Figure 7.1. I thank Chris Stallings for assistance and critical review of the figures. I also thank Kelly Preston for helping with bibliographic searches and aiding in assembly of the References section.

References

Ackerman JT, Kondratieff MC, Matern SA, Cech JJ (2000) Tidal influence on spatial dynamics of leopard sharks, *Triakis semifasciata*, in Tomales Bay, California. Environ Biol Fish 58:33–43.

Allee WC, Dickinson JC (1954) Dominance and subordination in the smooth dogfish *Mustelus canis* (Mitchill). Physiol Zool 27(4):356–364.

Alonso MK, Crespo EA, García NA, Pedraza SN, Mariotti PA, Mora NJ (2002) Fishery and ontogenetic driven changes in the diet of the spiny dogfish, *Squalus acanthias*, in Patagonian waters, Argentina. Environ Biol Fish 63:193–202.

Anderson JT (1988) A review of size dependent survival during pre-recruit stages of fishes in relation to recruitment. J Northw Atl Fish Sci 8:55–66.

Babel JS (1967) Reproduction, life history, and ecology of the round stingray *Urolophus halleri*. Cooper. Calif Fish Game 137:1–104.

Bass AJ (1978) Problems in studies of sharks in the southwest Indian Ocean. In: Hodgson ES, Mathewson RF (eds) Sensory biology of sharks, skates, and rays. Arlington, VA: Office of Naval Research, pp 545–594.

Beck MW, Heck KL, Able KW, Childers DL, Eggleston DB, Gillanders BM, Halpern B, Hays CG, Hoshino K, Minello TJ, Orth RJ, Sheridan PF, Weinstein MR (2001) The identification, conservation, and management of estuarine and marine nurseries for fish and invertebrates. Bioscience 51:633–641.

Bergman E, Greenberg LA (1994) Competition between a planktivore, a benthivore, and a species with ontogenic diet shifts. Ecology 75:1233–1245.

Bernal D, Sepulveda C, Graham JB (2001) Water-tunnel studies of heat balance in swimming mako sharks. J Exp Biol 204:4043–4054.

Bethea DM, Buckel JA, Carlson JK (2004) Foraging ecology of the early life stages of four sympatric shark species. Mar Ecol Prog Ser 268:245–264.

Bethea DM, Hale L, Carlson JK, Contés E, Manire CA, Gelsleichter J (2007) Geographic and ontogenetic variation in the diet and daily ration of the bonnethead shark, *Sphyrna tiburo*, from the eastern Gulf of Mexico. Mar Biol 152:1009–1020.

Blaber SJM, Blaber TG (1980) Factors affecting the distribution of juvenile estuarine and inshore fish. J Fish Biol 17(2):143–162.

Boustany AM, Davis SF, Pyle P, Anderson SD, LeBoeuf BJ, Block BA (2002) Satellite tagging — expanded niche for white sharks. Nature 415(6867):35–36.

Braccini JM, Gillanders BM, Walker TI (2005) Sources of variation in the feeding ecology of the piked spurdog (*Squalus megalops*): implications for inferring predator-prey interactions from overall dietary composition. ICES J Mar Sci 62:1076–1094.

Braccini JM, Perez JE (2005) Feeding habits of the sandskate *Psammobatis extenta* (Garman, 1913): sources of variation in dietary composition. Mar Freshw Res 56:395–403.

Branstetter, S (1990) Early life-history strategies of carcharhinoid and lamnoid sharks of the northwest Atlantic. In: Pratt HL Jr., Gruber SH, Taniuchi Y (eds) Elasmobranchs as living resources: advances in the biology, ecology, systematics and the status of the fisheries. Washington, DC: U.S. Department of Commerce, NOAA Technical Report- NMFS 90, pp 17–28.

Brickle P, Laptikohovsky V, Pompert J, Bishop A (2003) Ontogenetic changes in the feeding habits and dietary overlap between three abundant rajid species on the Falkland Islands' shelf. J Mar Biol Assoc UK 83:1119–1125.

Burt WH (1943) Territoriality and home range concepts as applied to mammals. J Mammal 24:346–352.

Carey FG, Scharold JV (1990) Movements of blue sharks (*Prionace glauca*) in depth and course. Mar Biol 106:329–342.

Carlson JK, Goldman KJ, Lowe, CG (2004) Metabolism, energetic demand, and endothermy. In: Carrier JC, Musick JA, Heithaus MR (eds.) Biology of sharks and their relatives. Boca Raton, FL: CRC Press, pp 203–224.

Carlson JK, Heupel MR, Bethea DM, Hollensead LD (2008) Coastal habitat use and residency of juvenile Atlantic sharpnose sharks (*Rhizoprionodon terraenovae*). Estuaries Coasts 31:931–940.

Carlson JK, Parsons GR (2001) The effects of hypoxia on three sympatric shark species: physiological and behavioral responses. Environ Biol Fish 61(4):427–433.

Carlson, JK, Parsons GR (2003) Respiratory and hematological responses of the bonnethead shark, *Sphyrna tiburo*, to acute changes in dissolved oxygen. J Exp Mar Biol Ecol 294:15–26.

Carraro R, Gladstone W (2006) Habitat preferences and site fidelity of the ornate wobbegong shark (*Orectolobus ornatus*) on rocky reefs of New South Wales. Pac Sci 60(2):207–223.

Carrassón M, Stefanescu C, Cartes JE (1992) Diets and bathymetric distributions of two bathyal sharks of the Catalan deep-sea (Western Mediterranean). Mar Ecol Prog Ser 82:21–30.

Chapman DD, Pikitch EK, Babcock EA, Shivji MS (2007) Deep-diving and diel changes in vertical habitat use by Caribbean reef sharks *Carcharhinus perezi*. Mar Ecol Prog Ser 344:271–275.

Clark CW (1994) Antipredator behavior and the asset-protection principle. Behav Ecol 5:159–170.

Collins AB, Heupel MR, Motta PJ (2007) Residence and movement patterns of cownose rays *Rhinoptera bonasus* within a south-west Florida estuary. J Fish Biol 71:1159–1178.

Conrath CL, Musick JA (2007) The sandbar shark summer nursery within the bays and lagoons of the Eastern Shore of Virginia. Trans Am Fish Soc 136:999–1007.

Cortés E (1999) Standardized diet compositions and trophic levels of sharks. ICES J Mar Sci 56:707–717.

Cortés E, Gruber SH (1990) Feeding habits and estimates of daily ration of young lemon sharks, *Negaprion brevirostris* (Poey). Copeia 1990(1):204–218.

Cortés E, Manire CA, Hueter RE (1996) Diet, feeding habits, and diel feeding chronology of the bonnethead shark, *Sphyrna tiburo*, in southwest Florida. Bull Mar Sci 58:353–367.

Craig JK, Burke BJ, Crowder LB, Rice JA (2006) Prey growth and size-dependent predation in juvenile estuarine fishes: experimental and model analyses. Ecology 87(9):2366–2377.

Cyrus DP, Blaber SJM (1992) Turbidity and salinity in a tropical northern Australian estuary and their influence on fish distribution. Estuar Coast Shelf Sci 35:545–563.

Dahlgren CP, Eggleston DB (2000) Ecological processes underlying ontogenetic habitat shifts in a coral reef fish. Ecology 81:2227–2240.

D'Onghia G, Matarrese A, Tursi A, Sion L (1995) Observations on depth distribution pattern of the small-spotted catshark in the North Aegean Sea. J Fish Biol 47:421–426.

de Marignac JR (2000) Home range and diel movement patterns of subadult lemon sharks, *Negaprion brevirostris*, in a shallow tropical lagoon, Bimini, Bahamas. Master's Thesis, San Jose State University, San Jose, CA.

DiBattista JD, Feldheim KA, Gruber SH, Hendry AP (2007) When bigger is not better: selection against large size, high condition and fast growth in juvenile lemon sharks. J Evol Biol 20:201–212.

Dickman CR (1992) Predation and habitat shift in the house mouse, *Mus domesticus*. Ecology 73(1):313–322.

Dowd WW, Brill RW, Bushnell PG, Musick JA (2006) Estimating consumption rates of juvenile sandbar sharks (*Carcharhinus plumbeus*) in Chesapeake Bay, Virginia, using a bioenergetics model. Fish Bull 104(3):332–342.

Driggers WB III, Ingram GW Jr, Grace MA, Gledhill CT, Henwood TA, Horton CN, Jones CM (2008) Pupping areas and mortality rates of young tiger sharks *Galeocerdo cuvier* in the western North Atlantic Ocean. Aquat Biol 2:161–170.

Duncan KM, Holland KN (2006) Habitat use, growth rates and dispersal patterns of juvenile scalloped hammerhead sharks *Sphyrna lewini* in a nursery habitat. Mar Ecol Prog Ser 312:211–221.

Ebert DA (1989) Life-history of the sevengill shark — *Notorynchus cepedianus* Peron, in two northern California bays. Calif Fish Game 75(2):102–112.

Ebert DA (2002) Ontogenetic changes in the diet of the sevengill shark (*Notorynchus cepedianus*). Mar Freshw Res 53:517–523.

Ebert DA, Bizzarro JJ (2007) Standardized diet compositions and trophic levels of skates (Chondrichthyes: Rajiformes: Rajoidei). Environ Biol Fish 80:221–237.

Ebert DA, Cowley PD (2003) Diet, feeding behaviour and habitat utilisation of the blue stingray *Dasyatis chrysonota* (Smith, 1828) in South African waters. Mar Freshw Res 54:957–965.

Eby LA, Crowder LB (2004) Effects of hypoxic disturbances on an estuarine nekton assemblage across multiple scales. Estuaries 27:342–351.

Ellis JK, Musick JA (2006) Ontogenetic changes in the diet of the sandbar shark, *Carcharhinus plumbeus*, in lower Chesapeake Bay and Virginia (USA) coastal waters Environ Biol Fish 80:51–67.

Ellis JR, Pawson MG, Shackley SE (1996) The comparative feeding ecology of six species of shark and four species of ray (Elasmobranchii) in the northeast Atlantic. J Mar Biol Assoc UK 76(1):89–106.

Eschmeyer PH (1990) Usage and style in writing fishery manuscripts. In: Hunter J (ed) Writing for fishery journals. Bethesda, MD: American Fisheries Society.

Fitz ES Jr, Daiber FC (1963) An introduction to the biology of *Raja eglanteria* Bosc 1802 and *Raja erinacea* Mitchill 1825 as they occur in Delaware Bay. Bull Bingham Ocean Coll 18:69–97.

Fonds M, Cronie R, Vethaak AD, Van der Puyl P (1992) Metabolism, food-consumption and growth of plaice (*Pleuronectes platessa*) and flounder (*Platichthys flesus*) in relation to fish size and temperature. Neth J Sea Res 29:127–143.

Franks BR (2007) The spatial ecology and resource selection of juvenile lemon sharks (*Negaprion brevirostris*) in their primary nursery areas. Ph.D. Dissertation. Drexel University, Philadelphia, PA.

Fraser DR, Cerri RD (1982) Experimental evaluation of predator-prey relationships in a patchy environment: consequences for habitat use patterns in minnow. Ecology 63:307–313.

Garla RC, Chapman DD, Wetherbee BM, Shivji M (2006) Movement patterns of young Caribbean reef sharks, *Carcharhinus perezi*, at Fernando de Noronha Archipelago, Brazil: the potential of marine protected areas for conservation of a nursery ground. Mar Biol 149:189–199.

Gilliam JF, Fraser DF (1987) Habitat selection under predation hazard: test of a model with foraging minnows. Ecology 68:1856–1862.

Gotceitas V, Colgan P (1990) The effects of prey availability and predation risk on habitat selection by juvenile bluegill sunfish. Copeia 2:409–417.

Graham B, Grubbs D, Holland K, Popp B (2007) A rapid ontogenetic shift in the diet of juvenile yellowfin tuna from Hawai'i. Mar Biol 150:647–658.

Grubbs RD (2001) Nursery delineation, habitat utilization, movements, and migration of juvenile *Carcharhinus plumbeus* in Chesapeake Bay, Virginia, USA. Ph.D. Dissertation, College of William and Mary, Gloucester Point, VA.

Grubbs RD, Musick JA (2007) Spatial delineation of summer nursery areas for juvenile sandbar sharks in Chesapeake Bay, Virginia. Am Fish Soc Symp 50:63–86.

Grubbs RD, Musick JA, Conrath CL, Romine JG (2007) Long-term movements, migration, and temporal delineation of summer nurseries for juvenile sandbar sharks in the Chesapeake Bay region. Am Fish Soc Symp 50:87–108.

Gunter G (1961) Some relations of estuarine organisms to salinity. Limnol. Oceanogr. 6:182–190.

Hanchet S (1991) Diet of spiny dogfish, *Squalus acanthias* Linnaeus, on the east-coast, south-island, New-Zealand. J Fish Biol 39:313–323.

Harestad AS, Bunnell FL (1979) Home range and body-weight: re-evaluation. Ecology 60(2):389–402.

Hart DR (1983) Dietary and habitat shift with size of red-eared turtles (*Pseudemys scripta*) in a southern Louisiana population. Herpetologica 39(3):285–290.

He X, Kitchell JF (1990) Direct and indirect effects of predation on a fish community: a whole-lake experiment. Trans Am Fish Soc 119:825–835.

Heithaus MR (2001) Habitat selection by predators and prey in communities with asymmetrical intraguild predation. Oikos 92(3):542–554.

Heithaus MR (2007) Nursery areas as essential shark habitats: a theoretical perspective. Am Fish Soc Symp 50:3–13.

Heithaus MR, Delius BK, Wirsing AJ, Dunphy-Daly MM (2009) Physical factors influencing the distribution of a top predator in a subtropical oligotrophic estuary. Limnol Oceanogr 54:472–482.

Heithaus MR, Dill LM, Marshall GJ, Buhleier B (2002) Habitat use and foraging behavior of tiger sharks (*Galeocerdo cuvier*) in a seagrass ecosystem. Mar Biol 140:237–248.

Heupel MR, Carlson JK, Simpfendorfer CA (2007) Shark nursery areas: concepts, definition, characterization and assumptions. Mar Ecol Prog Ser 337:287–297.

Heupel MR, Hueter RE (2002) Importance of prey density in relation to the movement patterns of juvenile blacktip sharks (*Carcharhinus limbatus*) within a coastal nursery area. Mar Freshw Res 53:543–550.

Heupel MR, Simpfendorfer CA, Collins AB, Tyminski JP (2006) Residency and movement patterns of bonnethead sharks, *Sphyrna tiburo*, in a large Florida estuary. Environ Biol Fish 76:47–67.

Heupel MR, Simpfendorfer CA, Heuter RE (2004) Estimation of shark home ranges using passive monitoring techniques. Environ Biol Fish 71:135–142.

Hoisington G, Lowe CG (2005) Abundance and distribution of the round stingray, *Urobatis halleri*, near a heated effluent outfall. Mar Environ Res 60:437–453.

Holbrook SJ, Schmitt RJ (1988) The combined effects of predation risk and food reward on patch selection. Ecology 69:125–134.

Holbrook SJ, Schmitt RJ (1992) Causes and consequences of dietary specialization in surfperches: patch choice and intraspecific competition. Ecology 73:402–412.

Holland KN, Grubbs RD (2007) Fish visitors to seamounts, Section A: Tunas and billfish. In: Pitcher T, Morato T, Hart P, Clark M, Haggan N, Santos R (eds) Seamounts: ecology, fisheries and conservation. Fish and Aquatic Resources Series 12. Oxford, UK: Blackwell Publishing, pp 189–207.

Holland KN, Wetherbee BM, Peterson JD, Lowe CG (1993) Movements and distribution of hammerhead shark pups on their natal grounds. Copeia 1993:495–502.

Hopkins TE, Cech JJ Jr (2003) The influence of environmental variables on the distribution and abundance of three elasmobranchs in Tomales Bay, California. Environ Biol Fish 66:279–291.

Hubbell G (1996) Using tooth structure to determine the evolutionary history of the white shark. In: Klimley AP, Ainley DG (eds) Great white sharks: the biology of *Carcharodon carcharias*. San Diego, CA: Academic Press, pp 9–18.

Hubbs CL, Iwai T, Matsubara K (1967) External and internal characters, horizontal and vertical distribution, luminescence, and food of the dwarf pelagic shark *Euprotomicrus bispinatus*. Bull Scripps Institution Oceanogr 10:1–64.

Ikehara II (1961) Billy Weaver shark research and control program final report. Division of Fish and Game, Department of Agriculture and Conservation, State of Hawai'i, Honolulu.

Imsland AK, Sunde LM, Flokvord A, Stefansson SO (1996) The interaction of temperature and fish size on growth of juvenile turbot. J Fish Biol 49:926–940.

Jackson AC, Rundle SD, Attrill MJ, Cotton PA (2004) Ontogenetic changes in metabolism may determine diet shifts for a sit-and-wait predator. J Anim Ecol 73:536–545.

Jakobsdottir KB (2001) Biological aspects of two deep-water squalid sharks: *Centroscyllium fabricii* (Reinhardt, 1825) and *Etmopterus princeps* (Collett, 1904) in Icelandic waters. Fish Res 51:247–265.

Jonassen TM, Imsland AK, Stefansson SO (1999) The interaction of temperature and fish size on growth of juvenile halibut. J Fish Biol 54:556–572.

Jones EG, Tselepides A, Bahley PM, Collins MA, Priede IG (2003) Bathymetric distribution of some benthic and benthopelagic species attracted to bated cameras and traps in the deep eastern Mediterranean. Mar Ecol Prog Ser 251:75–86.

Jones M, Laurila A, Peuhkuri N, Püronen J, Seppä T (2003a) Timing an ontogenetic niche shift: responses of emerging salmon alevins to chemical cues from predators and competitors. Oikos 102:155–163.

Juanes F, Conover DO (1994) Piscivory and prey size selection in young-of-the-year bluefish: predator preference or size-dependent capture success. Mar Ecol Prog Ser 114:59–69.

Klimley AP, Nelson DR (1984) Diel movement patterns of the scalloped hammerhead shark (*Sphyrna lewini*) in relation to El Bajo Espiritu Santo: a refuging central-position social system. Behav Ecol Sociobiol 15:45–54.

Knudesen R, Klemetsen A, Amundsen P-A, Hermansen B (2006) Incipient speciation through niche expansion: an example from the Arctic charr in a subarctic lake. Proc R Soc London Ser B 273:2291–2298.

Kohler NE, Casey JG, Turner PA (1998) NMFS Cooperative Shark Tagging Program, 1962–93: an atlas of shark tag and recapture data. Mar Fish Rev 60:1–87.

Kohler NE, Turner PA, Hoey JJ, Natanson LJ, Briggs R (2002) Tag and recapture data for three pelagic shark species: blue shark (*Prionace glauca*), shortfin mako (*Isurus oxyrinchus*) and porbeagle (*Lamna nasus*) in the North Atlantic Ocean. Coll Vol Sci Pap ICCAT 54:1231–1260.

Kyne PM, Simpfendorfer CA (2007) A collation and summarization of available data on deepwater Chondrichthyans: biodiversity, life history and fisheries. IUCN SSC Shark Specialist Group for the Marine Conservation Biology Institute.

L'Abe´e-Lund JH, Langeland A, Jonsson B, Ugedal O (1993) Spatial segregation by age and size in Arctic charr: a trade-off between feeding possibility and risk of predation. J Anim Ecol 62:160–168.

Langeland A, L'Abe´e-Lund JH, Jonsson B, Jonsson N (1991). Resource partitioning and niche shift in Arctic charr, *Salvelinus alpinus* and brown trout, *Salmo trutta*. J Anim Ecol 60:895–912.

Laptikhovsky VV, Arkhipkin AI, Henderson AC (2001) Feeding habits and dietary overlap in spiny dogfish *Squalus acanthias* (Squalidae) and narrowmouth catshark *Schroederichthys bivius* (Scyliorhinidae). J Mar Biol Assoc UK 81:1015–1018.

Lima, SL (1998) Nonlethal effects in the ecology of predator–prey interactions. BioScience 48:25–34.

Lima SL, Dill LM (1990) Behavioral decisions made under risk of predation: a review and prospectus. Can J Zool 68:619–640.

Llopiz L, Cowen RK (2008) Precocious, selective and successful feeding of larval billfishes in the oceanic Straits of Florida. Mar Ecol Prog Ser 358:231–244.

Loneragan NR, Potter IC, Lenanton RCJ, Caputi N (1987) Influence of environmental variables on the fish fauna of the deeper waters of a large Australian estuary. Mar Biol 94:631–641.

Long DJ, Jones RE (1996) White shark predation and scavenging on cetaceans in the eastern North Pacific Ocean. In: Klimley AP, Ainley DG (eds) Great white sharks: the biology of *Carcharodon carcharias*. San Diego, CA: Academic Press, pp 293–307.

Lowe CG, Wetherbee BM, Crow GL, Tester AL (1996) Ontogenetic dietary shifts and feeding behavior of the tiger shark, *Galeocerdo cuvier*, in Hawaiian waters. Environ Biol Fish 47:203–211.

Lowry D, Motta PJ (2007) Ontogeny of feeding behavior and cranial morphology in the whitespotted bambooshark *Chiloscyllium plagiosum*. Mar Biol 151:2013–2023.

Lowry D, Motta PJ (2008) Relative importance of growth and behaviour to elasmobranch suction-feeding performance over early ontogeny. J R Soc London Interface 5:641–652.

Lucifora LO, García VB, Menni RC, Escalante AH (2006) Food habits, selectivity, and foraging modes of the school shark *Galeorhinus galeus*. Mar Ecol Prog Ser 315:259–270.

Lucifora LO, Garcia VB, Menni RC, Escalante AH, Hozbor NM (2009) Effects of body size, age and maturity stage on diet in a large shark: ecological and applied implications. Ecol Res 24:109–118.

Lund R (1990) Chondrichthyan life history styles as revealed by the 320 million years old Mississippian of Montana. Environ Biol Fish 27:1–19.

Macpherson E, Duarte CM (1991) Bathymetric trends in demersal fish size: is there a general relationship? Mar Ecol Prog Ser 71:103–112.

Mahe K, Amara R, Bryckaert T, Kacher M, Brylinksi JM (2007) Ontogenetic and spatial variation in the diet of hake (*Merluccius merluccius*) in the Bay of Biscay and the Celtic Sea. ICES J Mar Sci 64:1210–1219.

Marks RE, Conover DO (1993) Ontogenetic shift in the diet of young-of-year bluefish *Pomatomus saltatrix* during the oceanic phase of the early life history. Fish Bull 91:97–106.

Matern SA, Cech JJ, Hopkins TE (2000) Diel movements of bat rays, *Myliobatis californica*, in Tomales Bay, California: evidence for behavioral thermoregulation. Environ Biol Fish 58:173–182.

McCandless CT, Pratt HL, Kohler NE, Merson RR, Recksiek CW (2007) Distribution, localized abundance, movements, and migrations of juvenile sandbar sharks tagged in Delaware Bay. Am Fish Soc Symp 50:45–62.

McCauley RW, Huggins NW (1979) Ontogenetic and non-thermal seasonal effects on thermal preferenda of fish. Amer Zool 19:267–271.

McEachran JD, Boesch DF, Musick JA (1976) Food division within two sympatric species-pairs of skates (Pisces: Rajidae). Mar Biol 35:301–317.

McNab BK (1963) Bioenergetics and determination of home range size. Amer Nat 97:133–140.

Medved RJ, Stillwell CE, Casey JJ (1985) Stomach contents of young sandbar sharks, *Carcharhinus plumbeus*, in Chincoteague Bay, Virginia. Fish Bull 83:395–402.

Mittelbach GG (1981) Foraging efficiency and body size: a study of optimal diet and habitat use by bluegills. Ecology 62:1370–1386.

Mittelbach GG, Persson L (1998) The ontogeny of piscivory and its ecological consequences. Can J Fish Aquat Sci 55:1454–1465.

Morris DW (2003) Toward an ecological synthesis: a case for habitat selection. Oecologia 136:1–13.

Morrissey JF, Gruber SH (1993a) Home range of juvenile lemon sharks, *Negaprion brevirostris*. Copeia 1993:425–434.

Morrissey JF, Gruber SH (1993b) Habitat selection by the juvenile lemon shark, *Negaprion brevirostris*. Environ Biol Fish 38:311–319.

Muñoz AA, Ojeda FP (1998) Guild structure of carnivorous intertidal fishes of the Chilean coast: implications of ontogenetic dietary shifts. Oecologia 114:563–573.

Muñoz-Chápuli R (1984). Ethologie de la reproduction chez quelques requins del'Atlantique Nord-Est. Cybium 8:1–14.

Musick JA, Colvocoresses JA (1986) Seasonal recruitment of subtropical sharks in Chesapeake Bight, U.S.A. In: Yanez y Arancibia A, Pauley D (eds) Workshop on recruitment in tropical coastal demersal communities, FAO/UNESCO, Campeche, Mexico, April 21–25, 1986. I.O.C. Workshop Rep. 44, pp 301–311.

Musick JA, Harbin MM, Compagno LJV (2004) Historical zoogeography of the Selachii. In: Carrier JF, Musick JA, Heithaus MR (eds) Biology of sharks and their relatives. Boca Raton, FL: CRC Press, pp 33–78.

Myrberg AA, Gruber SH (1974) Behavior of bonnethead shark, *Sphyrna tiburo*. Copeia 2:358–374.

Nakano H, Matsunaga H, Okamoto H, Okazaki M (2003) Acoustic tracking of bigeye thresher shark *Alopias superciliosus* in the eastern Pacific Ocean. Mar Ecol Prog Ser 265:255–261.

Natanson LJ, Casey JG, Kohler NE, Colket T (1999) Growth of the tiger shark, *Galeocerdo cuvier*, in the western North Atlantic based on tag returns and length frequencies; and a note on the effects of tagging. Fish Bull 97:944–953.

National Marine Fisheries Service (2007) Final Amendment 2 to the Consolidated Atlantic Highly Migratory Species, Fishery Management Plan. National Oceanic and Atmospheric Administration, National Marine Fisheries Service, Office of Sustainable Fisheries, Highly Migratory Species Management Division, Silver Spring, MD.

National Marine Fisheries Service (2009) Recovery Plan for Smalltooth Sawfish (*Pristis pectinata*). Prepared by the Smalltooth Sawfish Recovery Team for the National Marine Fisheries Service, Silver Spring, MD.

Newman SP, Handy RD, Gruber SH (2007) Spatial and temporal variations in mangrove and seagrass faunal communities at Bimini, Bahamas. Bull Mar Sci 80:529–553.

Odum EP, Kuenzler EJ (1955) Measurement of territory and home range size in birds. Auk 72:128–137.

Olson MH, Mittelbach GG, Osenberg CW (1995) Competition between predator and prey: resource-based mechanisms and implications for stage-structured dynamics. Ecology 76:1758–1771.

Orlov A, Cotton C, Byrkjedal I (2006) Deepwater skates (Rajidae) collected during the 2004 cruises of R.V. G.O. Sars and M.S. Loran in the Mid-Atlantic Ridge area. Cybium 30(Suppl 4):35–48.

Ortega LA, Heupel MR, Beynen PV, Motta PJ (2009) Movement patterns and water quality preferences of juvenile bull sharks (*Carcharhinus leucas*) in a Florida estuary. Environ Biol Fish 84:361–373.

Papastamatiou YP, Lowe CG, Caselle JE, Friedlander AM (2009) Scale-dependent effects of habitat on movements and path structure of reef sharks at a predator-dominated atoll. Ecology 90:996–1008.

Pearre S Jr (2003) Eat and run? The hunger-satiation hypothesis in vertical migration: history, evidence and consequences. Biol Rev 78:1–79.

Pedersen T, Jobling M (1989) Growth-rates of large, sexually mature cod, *Gadus morhua*, in relation to condition and temperature during an annual cycle. Aquaculture 81(2):161–168.

Persson A, Brönmark C (2002) Foraging capacities and effects of competitive release on ontogenetic diet shift in bream, *Abramis brama*. Oikos 97:271–281.

Powter DM, Gladstone W (2009) Habitat-mediated use of space by juvenile and mating adult Port Jackson sharks, *Heterodontus portusjacksoni*, in Eastern Australia. Pac Sci 63:1–14.

Rancurel P, Intes A (1982) Le requin tigre, *Galeocerdo cuvier* Lacepède, des eaux Neo-caledoniennes examen des contenus stomacaux. Tethys 10:195–199.

Rechisky EL, Wetherbee BM (2003) Short-term movements of juvenile and neonate sandbar sharks, *Carcharhinus plumbeus*, on their nursery grounds in Delaware Bay. Environ Biol Fish 68:113–128.

Recksiek CW, McCleave JD (1973) Distribution of pelagic fishes in the Sheepscot River-Back River estuary, Wiscasset, Maine. Trans Am Fish Soc 102:541–551.

Richardson AJ, Maharah G, Compagno LJV, Leslie RW, Eberts DA, Gibbons MJ (2000) Abundance, distribution, morphometrics, reproduction and diet of the *Izak* catshark. J Fish Biol 56:552–576.

Scharf FS, Juanes F, Rountree RA (2000) Predator size: prey size relationships of marine fish predators: interspecific variation and effects of ontogeny and body size on trophic-niche breadth. Mar Ecol Prog Ser 208:229–248.

Sepulveda CA, Kohin S, Chan C, Vetter R, Graham JB (2004) Movement patterns, depth preferences, and stomach temperatures of free-swimming juvenile mako sharks, *Isurus oxyrinchus*, in the Southern California Bight. Mar Biol 145:191–199.

Shepard ELC, Mohammed ZA, Southall EJ, Witt MJ, Metcalfe JD, Sims DW (2006) Diel and tidal rhythms in diving behaviour of pelagic sharks identified by signal processing of archival tagging data. Mar Ecol Prog Ser 228:205–213.

Simpfendorfer CA (1992) Biology of tiger sharks (*Galeocerdo cuvier*) caught by the Queensland shark meshing program off Townsville, Australia. Aust J Mar Freshw Res 43:33–43.

Simpfendorfer CA (2005) Threatened fishes of the world: *Pristis pectinata* Latham, 1794 (Pristidae). Environ Biol Fish 73:20.

Simpfendorfer CA, Goodreid AB, McAuley RB (2001) Size, sex and geographic variation in the diet of the tiger shark, *Galeocerdo cuvier*, from Western Australian waters. Environ Biol Fish 61:37–46.

Simpfendorfer CA, Heupel MR (2004) Assessing habitat use and movement. In: Carrier JC, Musick JA, Heithaus MR (eds) Biology of sharks and their relatives. Boca Raton, FL: CRC Press, pp 553–572.

Simpfendorfer CA, Milward NE (1993) Utilisation of a tropical bay as a nursery area by sharks of the families Carcharhinidae and Sphyrnidae. Environ Biol Fish 37:337–345.

Sims DW (2003) Tractable models for testing theories about natural strategies: foraging behaviour and habitat selection of free-ranging sharks. J Fish Biol 63(Suppl A):53–73.

Sims DW (2005) Differences in habitat selection and reproductive strategies of male and female sharks. In: Ruckstuhl KE, Neuhaus P (eds) Sexual segregation in vertebrates. Cambridge University Press, Cambridge, UK, pp 127–147.

Sims DW, Merrett DA (1997) Determination of zooplankton characteristics in the presence of surface feeding basking sharks (*Cetorhinus maximus*). Mar Ecol Prog Ser 158:297–302.

Sims DW, Nash JP, Morritt D (2001) Movements and activity of male and female dogfish in a tidal sea lough: alternative behavioural strategies and apparent sexual segregation. Mar Biol 139:1165–1175.

Sims DW, Quayle VA (1998) Selective foraging behaviour of basking sharks on zooplankton in a small-scale front. Nature 393:460–464.

Sims DW, Southall EJ, Tarling GA, Metcalfe JD (2005) Habitat specific normal and reverse diel vertical migration in the plankton-feeding basking shark. J Anim Ecol 74:755–761.

Smale MJ (1991) Occurrence and feeding of three shark species, *Carcharhinus brachyurus*, *C. obscurus* and *Sphyrna zygaena*, on the Eastern Cape Coast of South-Africa. South African J Mar Sci 11:31–42.

Sminkey TR (1994) Age growth, and population dynamics of the sandbar shark, *Carcharhinus plumbeus*, at different population levels. Ph.D. Dissertation, Virginia Institute of Marine Science, College of William and Mary, Gloucester Point, VA.

Smith JW, Merriner JV (1987) Age and growth, movements and distribution of the cownose ray, *Rhinoptera bonasus*, in Chesapeake Bay. Estuaries 10:153–164.

Snelson FF Jr, Williams SE (1981) Notes on the occurrence, distribution, and biology of elasmobranch fishes in the Indian River lagoon system, Florida. Estuaries 4:110–120.

Springer S (1967) Social organization of shark populations. In: Gilbert PW, Mathewson RF, Rall DP (eds) Sharks, skates and rays, Baltimore: Johns Hopkins University Press, pp 149–174.

Stefanescu C, Rucabado J, Lloris D (1992) Depth-size trends in western Mediterranean demersal deep-sea fishes. Mar Ecol Prog Ser 81:205–213.

Stillwell CE, Kohler NE (1982) Food, feeding-habits, and estimates of daily ration of the shortfin mako (*Isurus oxyrinchus*) in the northwest Atlantic. Can J Fish Aquat Sci 39:407–414.

Stoner AW, Livingston RJ (1984) Ontogenetic patterns in diet and feeding morphology in sympatric sparid fishes from seagrass meadows. Copeia 1984:174–187.

Sundstrom LF, Gruber SH, Clermont SM, Correia JPS, de Marignac JRC, Morrissey JF, Lowrance CR, Thomassen L, Oliveira MT (2001) Review of elasmobranch behavioral studies using ultrasonic telemetry with special reference to the lemon shark, *Negaprion brevirostris*, around Bimini Islands, Bahamas. Environ Biol Fish 60:225–250.

Takimoto G (2003) Adaptive plasticity in ontogenetic niche shifts stabilizes consumer-resource dynamics. Am Nat 162(1):93–109.

Taylor SM, Bennett MB (2008) Cephalopod dietary specialization and ontogenetic partitioning of the Australian weasel shark *Hemigaleus australiensis* Whit, Last & Compagno. J Fish Biol 72:917–936.

Tibbetts IR, Carseldine L (2005) Trophic shifts in three subtropical Australian halfbeak (Teleostei: Hemiramphidae). Mar Freshw Res 56:925–932.

Vorenberg MM (1962) Cannibalistic tendencies of lemon and bull sharks. Copeia 1962:455–456.

Vuorinen I, Rajasilta M, Salo J (1983) Selective predation and habitat shift in a copepod species: support for the predation hypothesis. Oecologia 59:62–64.

Wallman HL, Bennett WA (2006) Effects of parturition and feeding on thermal preference of Atlantic stingray *Dasyatis Sabina* (Lesueur). Environ Biol Fish 75:259–267.

Walters CJ, Juanes F (1993) Recruitment limitation as a consequence of natural-selection for use of restricted feeding habits and predation risk-taking by juvenile fishes. Can J Fish Aquat Sci 50:2058–2070.

Weihs D, Keyes RS, Stalls DM (1981) Voluntary swimming speeds of two species of large carcharhinid sharks. Copeia 1981:219–222.

Weng KC, Block BA (2004) Diel vertical migration of the bigeye thresher shark (*Alopias superciliosus*), a species possessing orbital retia mirabilia. Fish Bull 102:221–229.

Weng KC, O'Sullivan JB, Lowe CG, Winkler CE, Dewar H, Block BA (2007) Movements, behavior and habitat preferences of juvenile white sharks *Carcharodon carcharias* in the eastern Pacific. Mar Ecol Prog Ser 338:211–224.

Werner EE (1986) Species interactions in freshwater fish communities. In: Diamond J and Case TJ (eds) Community ecology. New York: Harper and Row, pp 344–358.

Werner EE, Gilliam JF (1984) The ontogenetic niche and species interactions in size-structured populations. Annu Rev Ecol Syst 15:393–425.

Werner EE, Gilliam JF, Hall DJ, Mittlebach GG (1983) An experimental test of the effects of predation risk on habitat use in fish. Ecology 64:1540–1548.

Werner EE, Hall DJ (1977) Competition and habitat shift in two sunfishes (Centrarchidae). Ecology 58:869–876.

Werner EE, Hall DJ (1988) Ontogenetic habitat shifts in bluegill: the foraging rate predation risk trade-off. Ecology 69:1352–1366.

Wetherbee BM, Cortés E (2004) Food consumption and feeding habits. In: Carrier JF, Musick JA, Heithaus MR (eds) Biology of sharks and their relatives. Boca Raton, FL: CRC Press, pp 225–246.

Wetherbee BM, Gruber SH, Cortés E (1990) Diet, feeding habits, digestion, and consumption in sharks, with special reference to the lemon shark, *Negaprion brevirostris*. In: Pratt HL, Gruber SH, Taniuchi T (eds) Elasmobranchs as living resources: advances in the biology, ecology, systematics, and the status of the fisheries. NOAA Tech. Rep. NMFS 90. Seattle, WA: U.S. Department of Commerce, pp 29–47.

Wetherbee BM, Gruber SH, Rosa RS (2007) Movement patterns of juvenile lemon sharks *Negaprion brevirostris* within Atol das Rocas, Brazil: a nursery characterized by tidal extremes. Mar Ecol Prog Ser 343:283–293.

Wilbur HM (1980) Complex life-cycles. Annu Rev Ecol Syst 11:67–93.

Wirsing AJ, Heithaus MR, Dill LM (2006) Tiger shark (*Galeocerdo cuvier*) abundance and growth in a subtropical embayment: evidence from seven years of standardized fishing effort. Mar Biol 149:961–968.

Witzell WN (1987) Selective predation on large cheloniid sea turtles by tiger sharks (*Galeocerdo cuvier*). Jpn J Herpetol 12:22–29.

Yeiser BG, Heupel MR, Simfendorfer CA (2008) Occurrence, home range and movement patterns of juvenile bull (*Carcharhinus leucas*) and lemon (*Negaprion brevirostris*) sharks within a Florida estuary. Mar Freshw Res 59:489–501.

Yokota L, Lessa RP (2006) A nursery area for sharks and rays in Northeastern Brazil. Environ Biol Fish 75:349–360.

8

Tracking and Analysis Techniques for Understanding Free-Ranging Shark Movements and Behavior

David W. Sims

CONTENTS

8.1 Introduction

Sharks are important living resources to human societies globally in cultural, economic, health, biodiversity, and conservation contexts. They can play a key role in the functioning of marine ecosystems and may provide higher trophic-level indication of ocean ecosystem state (Sims and Quayle 1998; Myers et al. 2007). However, rapid declines in some large pelagic sharks are occurring on a worldwide scale due to overfishing (Baum et al. 2003; Myers et al. 2007). There is particular concern that target and by-catch fisheries that have been well developed for at least the past century in the Atlantic Ocean, for example, are depleting shark populations below sustainable levels where recovery may not be possible, or at best may be very slow, even if fishing pressure is removed (Pauly et al. 2002; Clarke et al. 2006). Although some fisheries assessments indicate less pronounced declines for large pelagic (Sibert et al. 2006) and coastal sharks (Burgess et al. 2005), undoubtedly they are particularly susceptible to overharvesting on account of slow growth rates, late age at sexual maturity, and relatively low fecundity. During the twentieth century some large skate species were eliminated from areas where they were once very common and have not returned (Brander 1981; Casey and Myers 1998), suggesting large sharks, having at least some similarities in life history to skates, are also likely to be at risk of regional extinction (Chapter 17). Many pelagic sharks are now red-listed by the International Union for Conservation of Nature (IUCN), with some now at a fraction of their historical biomass (Dulvy et al. 2008). In the face of these apparently dramatic declines and the need for prompt action aimed at securing the future of species and populations, biologists have developed over the last 30 years or so numerous techniques for tracking and analyzing the movements of individual sharks in the natural environment. But how is this linked to broader issues in understanding animal ecology, and how can this help in applied settings, such as in shark fisheries management and species conservation?

Ideally the sustainable management of shark populations requires a detailed knowledge of their spatial ecology. How do spatial movements of individuals within a species affect a population's distribution pattern and temporal dynamics across a wide range of scales in relation to their physical and biotic environment, including species interactions? The spatial distribution of a shark species will be influenced by how activity and behavior (linked to motivational and energy requirements) affects rates of key movements, such as search strategies and encounter rates with prey, the location of mates and timing of courtship, and occupation times in preferred habitats. Therefore, day-to-day changes in movement and behavior will influence broader-scale patterns in distribution and population structure, but will also signal a shark's response to environmental fluctuations (Sims and Quayle 1998). An ability first to understand and then to predict shark space-use patterns and responses to important variables should help identify the extent and dynamics of population distributions and essential habitats, ultimately providing spatial and temporal foci for management in relation to fishing activity and distribution (Mucientes et al. 2009). It may also serve to identify predator hotspots in the ocean requiring special conservation initiatives (Worm, Lotze, and Myers 2003).

A key problem in determining the usefulness of management or conservation measures for highly mobile and wide-ranging species such as sharks is that their movements, behavior, and distribution patterns remain enigmatic for the majority of species. Without knowing where and what habitats sharks occupy and over what time scales, it is extremely difficult to determine not only how many sharks there are in a population, but to predict how they will respond to future environmental changes. How shark species will respond to climate change that is predicted to increase sea surface temperatures by about 4°C over the next century remains unclear. Equally, how sharks are distributed in relation to fishing locations and effort is important to determine, not least because if remaining aggregations occur in areas where fishing activity is most intense, this will further exacerbate declines (Wearmouth and Sims 2008; Mucientes et al. 2009). Understanding these processes relies on knowledge about how movements and behavior in relation to environment influence population structuring and abundance changes of fish (Metcalfe 2006).

Central to this problem is that sharks spend most, if not all, of their time below the sea surface where they cannot be observed directly or followed to elucidate movements. Until recently only very coarse and simplistic data on fish movements and activities were available for identifying putative fish stocks and possible migrations (Harden Jones 1968; Guénette, Pitcher, and Walters 2000). What we are learning from the application of new technologies is that fish such as sharks have complicated spatial and temporal behavioral dynamics, characterized by daily and seasonal migrations, regional differences in behavior, distinct habitat preferences, and age and sexual segregation (Wearmouth and Sims 2008). The picture is much more behaviorally complex than previously thought (Harden Jones 1968). Linked to the deficiency in knowing little about shark movements and behavior in relation to environment is the difficulty in identifying why such behaviors occur when and where they do.

In behavioral ecology much has been learnt from optimality models seeking to reveal the decision-making processes of animals (Krebs and Davies 1997), that is, why animals behave a certain way when faced with a set of (necessarily) changing conditions. In formulating these models it is usual to define the likely decision being made by the animal (e.g., time spent in a prey patch), then to define particular currencies (e.g., maximize net rate of gain) and constraints (e.g., travel or handling time, environmental factors) for testing hypotheses about how these different constraints influence the costs and benefits of decision making. The problem with applying this framework to the behavioral study of sharks is that such decisions, currencies, and constraints may be relatively straightforward to formalize theoretically, but in practice they are very difficult to measure accurately (if at all) for long enough time periods over which natural changes in behavior and energetics can be recorded. Therefore, empirical tests of optimality models for shark behavior are not in widespread use (Sims 2003). So how can we begin to understand what behavioral mechanisms underlie shark spatial movements given these logistical difficulties?

This chapter will describe how new technologies have provided the means to track shark movements and behavior in the natural environment. It will consider what data these remote sensing technologies provide, and how these data can be used to understand how sharks use their environment. The review of the technologies provided here is not exhaustive, but rather is focused on the principal techniques currently being used to understand shark movements and behavior in the wild. In addition to electronic tracking devices, the focus here is also on how new analysis procedures can yield insights into shark habitat selection, patterns and processes of behavior, how these may alter with environmental changes, and whether certain patterns can be considered adaptive programs or are emergent properties. How these results are already being used to aid shark conservation will be highlighted, and future perspectives in this fast-developing field will be explored.

8.2 Tracking Techniques

8.2.1 Historical Perspective: The Advent of Electronic Tagging

Among the most well documented of historical fisheries are those that have targeted catches of seasonal runs or migrations of fishes such as the bluefin tuna *Thunnus thynnus*, Atlantic herring *Clupea harengus*, or the Atlantic salmon *Salmo salar*. These population-level spatial movements of fish often act to concentrate vast shoals into areas of sea or estuary that are readily accessible to large groups of fishers and therefore yield high capture rates. Perhaps in response to perceived declines in key commercial stocks (Sims and Southward 2006), it was recognized in the nineteenth century that understanding these large-scale distribution changes necessitates knowing where individuals have come from. Although the tagging of fish has a long history, dating back to at least the 1600s (McFarlane, Wydoski, and Prince 1990), extensive tagging of a broader range of species using numbered or colored markers, or with barbed hooks or fin clips only commenced in the 1800s, and expanded quite slowly until about 1930 (Kohler and Turner 2001). The concept of the technique is for large numbers of fish to be tagged or marked in some way so they are individually identifiable, released back to the natural environment, then during normal fishing operations recaptures are made and information is returned to the tagger. This simple technique, known as capture-mark-recapture tagging, has been used extensively up to the present time for providing data on the dispersal characteristics of regional populations, large-scale migrations, and estimates of population size.

The expansion of sea fisheries in the decades after World War II and the increased need for the monitoring and regulation of stocks saw number tagging studies gain considerable momentum, particularly in the Atlantic and Pacific oceans. Commercially important fish captured by government survey vessels or in scientific studies were individually marked with numbered tags and in the 1940s tagging studies of sharks (e.g., spurdog *Squalus acanthias*) and skates were already established (Kohler and Turner 2001). A review of these studies is beyond the scope of this chapter; however, the results for shark species are thoroughly described elsewhere (e.g., Kohler and Turner 2001). Nevertheless, although mark-recapture tagging gave fascinating new insights into the spatiotemporal movements and distribution of important fish groups such as sharks (e.g., M.J. Holden 1967), these data were, of course, rudimentary in helping to understand spatial ecology. Using the mark-recapture tagging approach, scientists knew where geographically the number-tagged fish had been originally marked, and from the fishery-dependent returns they knew whereabouts it had been caught; nothing was known, however, about where a fish had been between release (hypothetical position A) and capture (hypothetical position Z). Theoretically, individual fish could have undertaken complicated movement patterns or long-distance migrations in this time to areas not known or exploited by fishers. Therefore, not knowing a fish's journey from positions B through to Y masked important components of its life history and, potentially, could hold the key to some reasons why fish populations were already declining in some regions by the start of the twentieth century (Sims and Southward 2006; Wearmouth and Sims 2008).

Before the advent of electronic devices capable of tracking shark spatial movements, the study of wild shark behavior was limited to brief descriptions from divers or from observers aboard vessels. It was only when animal-attached electronic tags were developed that it became possible to monitor in detail a greater range of movements and behavior exhibited by sharks in the natural environment. Over the last 20 years, these data have contributed fundamental insights about shark behavior and have helped provide basic data to inform management and conservation initiatives.

8.2.2 Acoustic Telemetry

8.2.2.1 Focal Tracking

Acoustic telemetry being used as a tool to study the individual behavior of fish in the natural environment dates back to the 1950s (Arnold and Dewar 2001). It was in the 1960s, however, when electronic devices capable of emitting or transponding sound energy first came to be used to track individual sharks (Carey 1992). Sound is the only practical means of transmitting a signal through seawater over distances greater than a few meters because radio waves do not propagate sufficiently in such a conducting medium, at least at the frequencies used in biotelemetry (Nelson 1990; Priede 1992). Radio transmitters can be used to track fish in shallow fresh water but through-the-water telemetry of marine fish must be made by acoustic transmission (Nelson 1990). Ultrasonic frequencies are used (34 to 84 kHz) because frequency is proportional to transducer diameter, and is an important factor in reducing the size of the transmitter (thus the "package" to be carried by a fish). Lower frequency transmitters of greater size have been used to track large fish (e.g., Carey and Scharold 1990), whereas higher frequency tags are reserved for the tracking of smaller species (Sims, Nash, and Morritt 2001). A wide array of sensors was made available for acoustic tags to transmit data on water temperature, swimming depth, fish muscle temperature, cranial temperature, swim speed, tail beat frequency, and heart rate. Multiplexed tags were capable of transmitting sensor data for up to three channels (Carey and Scharold 1990).

In practical terms, sound pulses at a known frequency emitted from a transmitter attached to a fish are received using a directional hydrophone and portable receiver such that a vessel can follow the moves made by the transmitter, and hence the individual fish, including any encoded signals from sensors (Figure 8.1). The main problem with this technique for tracking large, highly mobile species that traverse tens of kilometers per day is that tracking individuals continuously in the open sea far from land becomes prohibitively expensive, especially since only one fish at a time can be tracked by a ship. Because of this, even by the 1990s there were relatively few horizontal and vertical tracks of sharks available for analysis, and these were understandably of short duration (hours to a few days; e.g., Carey et al. 1982; Klimley 1993).

During the 1970s and 1980s, the blue shark *Prionace glauca* was one pelagic shark species that was tracked regularly in the open sea by acoustic telemetry, yielding insights into the pattern and range of its vertical movements (for overview, see Carey and Scharold 1990). In general it was found that blue sharks tracked on the northeast U.S. continental shelf and shelf edge exhibited depth oscillations and that this diving pattern was seasonal. During late summer, autumn, and winter, sharks showed remarkably regular vertical oscillations, particularly during daylight hours, from the surface or near surface to about 400 m depth and back, whereas at night oscillations were usually confined to the top 100 m. Interestingly, sharks tracked in summer further inshore (water depth <80 m) did not show these large, regular changes in swimming depth. In early and mid summer, blue sharks remained in the upper 10 m, seldom diving deeper into colder water below the thermocline at 15 m (Carey and Scharold 1990). More recent work on other large sharks have demonstrated marked species-level differences that would have been impossible to detect without focal acoustic tracking (sometimes called "active tracking") of individual sharks equipped with depth-sensing acoustic transmitters. For example, transmitter-

FIGURE 8.1
A color version of this figure follows page 336. Acoustic telemetry of sharks and other fish. (a) A directional hydrophone and receiver is used to locate the sound source (b, c) of acoustic transmitters fitted to free-ranging fish. The transmitters shown here are microprocessor controlled, weigh a few grams in water, and can remain transmitting at one pulse per second for over a year. (Photos courtesy of MBA, Plymouth, UK.)

tagged tiger sharks (*Galeocerdo cuvier*) in the Pacific Ocean tracked off the south coast of Oahu, Hawaii, across deep water to the Penguin Banks did not show night-day differences in depth oscillations or overall swimming depth (Holland et al. 1999). In water >100 m depth they swam predominantly close to the seabed, whereas in deeper water (>300 m) they remained within the top 100 m in the warm, mixed layer above the thermocline (at 60 to 80 m depth) irrespective of time of day. Nevertheless, tiger sharks did undertake oscillations in swimming depth during their horizontal movements, albeit of relatively small amplitude (~50 m) compared with blue sharks swimming over deep water.

Despite these successes, the short duration of focal acoustic tracking of large pelagic sharks was a distinct limitation since tracking time periods greater than hours to a few days are generally needed to capture broadly representative behavior. Perhaps even more of a problem in such studies was tracking individuals that were captured on lines prior to tagging. Klimley (1993) attached tags underwater to free-swimming sharks without the need for capture, but other studies requiring fish to be hooked on lines are naturally dealing with more distressed individuals. Horizontal acoustic trackings of blue shark off the U.S. northeast coast showed that in late summer and autumn they generally moved south and southeast offshore following tagging and release (Carey and Scharold 1990). It was suggested this consistent pattern of movement may have been related to seasonal migration, but it was also recognized that in part these movements may be a general reaction of pelagic fish to move into deep water offshore after being captured for tagging (Carey and Scharold 1990). For longer trackings where the shark was tagged without the need for capture, horizontal movements are more likely to provide insights into natural foraging or ranging movements. For example, a white shark (*Carcharodon carcharias*) that was tagged feeding in the vicinity of a whale carcass off Long Island, New York, continued feeding and swimming within a 3 km radius of the whale for 1.5 days (Carey et al. 1982). The shark then moved southwest following quite closely the 25 m isobath for a further two days.

Although acoustic telemetry of focal individuals provided some considerable detail over the short term, it is not a method for continuous recording of long-term details about shark movements and behavior. Some smaller shark species (e.g., *Scyliorhinus canicula*, total length ~0.7 m) have been tracked continuously for up to 2 weeks, but even so, horizontal and vertical movement data over longer periods at the level of spatial resolution possible with this technique is not feasible.

8.2.2.2 Static Array Monitoring

Acoustic tracking using stationary, recording receiver arrays, sometimes called passive tracking, provided the first data on the longer-term movements and behavior of free-ranging sharks (for overview, see Voegeli et al. 2001). First used in the 1980s, modern versions of such devices are widely used today for long-term monitoring of shark movements, from studies of small catsharks in isolated lagoons to large sharks in estuarine and in reef-fringing open sea habitats. In such studies, data-logging acoustic receivers are usually moored at a specific site, on a buoyed line, or attached to ballast on the seabed (Figure 8.2). These stationary receivers detect the *pings* from shark-attached acoustic transmitters within a circular area around the receiver, the radius of which can fluctuate due to physical changes

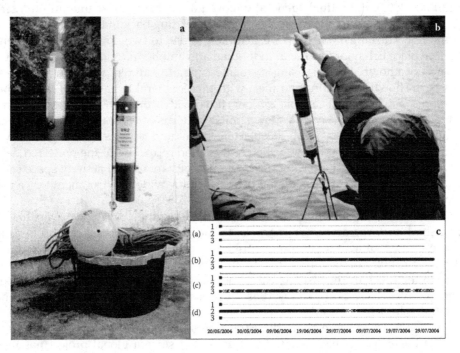

FIGURE 8.2
A color version of this figure follows page 336. Data-logging acoustic receivers for long-term monitoring of fish presence or absence at a particular location. (a) Receivers can be moored on the seabed with the receiver uppermost, or (b) suspended a few meters below a surface buoy with the receiver pointing down. (c) Example of recorded data from four small spotted catshark *Scyliorhinus canicula* (inset a–d) at each of three moored receivers (inset 1–3 for each fish). Circles denote presence of a fish at a receiver. Note the almost continuous presence of catshark a, b, and d at receiver 2 for at least two months, whereas fish c was present intermittently at receiver 3. (Modified from Sims DW et al. (2006a) J Anim Ecol 75: 176–190. Photos courtesy of MBA, Plymouth, UK.)

in the water column that affect the acoustic signal-to-noise ratio, for example, turbidity, and the background noise. The detection distance, hence radius, for the small, high frequency (69 kHz) transmitters generally is 250 to 500 m, but much shorter or longer ranges are possible, depending on underwater conditions. Hence, although these receivers record only presence or absence of a tag within its detection area, a distinct advantage of this system is that large numbers of transmitters can be monitored simultaneously because the transmitters have unique identification numbers encoded in the sound pulse train. Thus, since lithium battery life in the receivers can be anything from 6 months to several years, depending on number of detections received per unit time, this technique is well suited to long-term monitoring of sites that form part of the core activity space of individual sharks, or of specific locations to which sharks return after much longer-range movements, either on a short-term or longer-term basis (Simpfendorfer, Heupel, and Hueter 2002; Bonfil et al. 2005; Sims et al. 2006b).

In pioneering studies in the eastern Pacific, the movements of transmitter-tagged scalloped hammerhead sharks *Sphyrna lewini* were monitored using data-logging acoustic receivers moored along the plateau of a seamount (Klimley et al. 1988). The departures, arrivals, and occupancy times of individual sharks were recorded. These data showed that individuals grouped at the seamount during the day, but departed before dusk, moving separately into the pelagic environment at night and returned near dawn the next day. Acoustic tracking of individual *S. lewini* using a vessel resolved those nocturnal trajectories (Klimley 1993). Individuals generally moved away from the seamount on a straight line or meandering course into deep water before heading back in the general direction of the seamount. Trackings, however, were limited to one to two days so the persistence of this central place behavior within an individual and among individuals across the annual cycle was not known. Studies using the same type of technology but on smaller shark species have shown interesting parallels with these general behaviors of larger species. Small spotted catsharks (*Scyliorhinus canicula*) fitted with acoustic transmitters and tracked continuously for up to two weeks show similar patterns of central place refuging (Sims, Nash, and Morritt 2001; Sims et al. 2006b). Here, males and females remain generally inactive during the day, leaving to forage before dusk and returning to the preferred daytime locations near dawn. The pattern of daytime inactivity in a small activity space centered on rock refuges and nocturnal foraging behavior in the wider environment was a pattern sustained over many months (Sims 2003; Sims et al. 2006b).

Spatial arrays of larger numbers of stationary data-logging acoustic receivers have been used in shark nursery areas, for example, to monitor long-term space use, to estimate short-term centers of activity (Simpfendorfer, Heupel, and Hueter 2002), and as a means to estimate mortality of juveniles prior to migration away from the nursery (Heupel and Simpfendorfer 2002). In the last 10 years, linear arrays of large numbers of data-logging acoustic receivers have been deployed as "curtains" at key positions along coastlines or across straits, to detect the migrations of fish such as Pacific salmon (Welch et al. 2008). This methodology has yet to be used for tracking large numbers of sharks during migration, but it has considerable potential in this regard since a global project has recently invested in the deployment of many long-term acoustic receiver networks on continental shelves and in the open ocean worldwide (see http://oceantrackingnetwork.org).

8.2.2.3 Combined Radio-Acoustic Systems

Radio-acoustic positioning systems were developed in the 1980s for the detailed study of transmitter-tagged marine animal movements in an area of a few square kilometers to which individuals show some fidelity (O'Dor et al. 1998). One of the most commonly used commercial systems comprises three buoys containing radio-controlled ultrasonic receivers that are positioned in a triangular array (Voegeli et al. 2001). A fish fitted with an acoustic transmitter (frequency ranges, 25 to 80 kHz) is then released into the natural environment monitored within the array. Tracking individual fish in real time is achieved by the receivers on each of the buoys selecting the same transmitter frequency and detecting the arrival times of each sound pulse received over a specified period, whereupon the arrival times from all three buoys are sent by radio signal to a base station on shore that can be several kilometers away (Voegeli et al. 2001). At the base station radio receiver, a linked PC running Windows-based software determines when the same pulses arrive at the three buoys and calculates each tag's position with respect to the three buoys (for overview, see Klimley et al. 2001a). Tracking four fish simultaneously, for example, enables a position for each to be calculated approximately every 1.25 min, as the buoy receivers cycle through the transmitter frequencies in turn. During deployment, the geographical positions of the buoys are recorded accurately using a Global Positioning System (GPS) receiver such that real-time fish tracks can be displayed on a geo-referenced map of bathymetry, for example. Although this system limits tracking of fish to the area within and immediately around the triangle of the array, the principal advantage is the high accuracy of the positions calculated (to within 1 m inside the triangle) and, for a low number of tags monitored, the high repetition rate of positions determined.

This system has been used to study the movements and behavior of large sharks such as the great white, as well as small catsharks, but is not as widely used as focal acoustic tracking or archival tagging techniques (see the next section). However, where they have been used, new insights into shark behavior have been obtained. Using a radio-acoustic positioning system deployed off Ano Nuevo Island, California, six great white sharks were tracked intermittently as they undertook movements near the island, which is a haul-out site for their seal prey (Klimley et al. 2001a, 2001b). The three buoys were positioned about 400 m apart and some 500 m off the island. Even though the 4.5 to 5.2-m long sharks were capable of making extensive ranging movements away from the island, about 40% of their time was spent within the small area monitored by the buoy array, and they remained nearby the rest of the time (Klimley et al. 2001b). Sharks patrolled back and forth parallel to the shoreline with certain areas patrolled preferentially by certain individuals. Interestingly, during the total 78-day tracking time for all individuals, only two likely feeding bouts occurred. This suggested that although white sharks remain close to their prey and continually undertake patrols, their rate of feeding, or at least their rate of launching a full attack that could be measured by the tracking system, is probably quite low.

8.2.3 Archival Tagging

8.2.3.1 Data-Logging Tags

The widespread availability of miniaturized data-logging computers in the 1980s revolutionized the study of wild fish behavior. During the early 1990s, data-loggers were developed that were small enough not to impede the natural swimming behavior of the fish but with powerful batteries and memory sizes capable of recording and storing large amounts of high-quality data (termed archival data). They were also relatively cheap so large numbers could be deployed, and for commercially important species at least, tags could be returned to scientists through the fishery. Early tags were programmed to record pressure (depth) and later models also incorporated temperature and ambient light level sensors

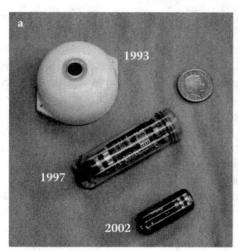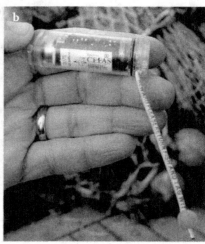

FIGURE 8.3

A color version of this figure follows page 336. Examples of data-logging tags used to track the movements and behavior of marine fish including sharks. (a) Data storage tags (DST) show the increased miniaturization over time. The Mk4 DST of 2002 has about 30 times the memory storage capacity of the Mk1 tag of 1993. All tags were designed and built by the Cefas team in Lowestoft, UK. The two later versions have both been used to study shark behavior. (b) A DST for internal placement with an external marker to alert fishers. This can also incorporate a light sensor at the end of a "stalk" for recording light level for geolocation purposes. The DST is placed inside the body cavity of the fish and the stalk remains on the outside to record ambient light levels in the sea. (Photos courtesy of J. Metcalfe, Cefas Laboratory, Lowestoft, UK.)

(Arnold and Dewar 2001; Figure 8.3). Large-scale deployments of electronic tags on North Sea plaice (*Pleuronectes platessa*) (Metcalfe and Arnold 1997) and on Atlantic bluefin tuna (*Thunnus thynnus*) (Block et al. 2001) were among the first studies to successfully use this technology for tracking movements and behavior of hundreds of individual fish. Studies followed that fitted depth/temperature/light-level-logging tags, also known as data storage tags (DSTs), to sharks and other large pelagic fish (e.g., Schaefer and Fuller 2002), revealing new patterns of behavior. However, one drawback was that data retrieval was unpredictable since it relied on tags being returned through developed fisheries where return rates can vary widely depending on fishing activity in a given region, and may even be quite low (~5% to 10% of tagged individuals released; Metcalfe and Arnold 1997).

The use of data storage tags on sharks has provided more detailed temporal information on their depth use over longer time periods than was possible with focal acoustic telemetry (see Section 8.2.2.1). Viewed at this high resolution over days to months, longer-term and sometimes quite different patterns in shark swimming behavior are evident. For example, DST tracking of school shark (*Galeorhinus galeus*) showed vertical rhythms in swimming depth over the diel cycle. In continental shelf waters off southern Australia, *G. galeus* spent the day at depth before ascending at night into shallower water often for several hours (West and Stevens 2001). When *G. galeus* was in deeper water off the continental shelf this pattern was maintained, with descent at dawn to depths of up to 600 m before ascent to shallower waters at dusk. In maintaining this pattern it appears the school shark exhibits what is termed a "normal" diel vertical migration (DVM; dusk ascent, dawn descent) for much of the time when ranging and foraging off southern Australia. This pattern may represent prey tracking of vertically migrating populations of squid, for example, that also undertake similar daily vertical movements as those observed in school sharks (West and Stevens 2001).

By contrast, great white sharks (*Carcharodon carcharias*) tagged and tracked off southern Australia appear to exhibit at least three different vertical movement patterns. A pattern of regular vertical oscillations from about 50 m depth to the surface irrespective of day or night was apparent during directed westerly movements of white sharks (Bruce, Stevens, and Malcolm 2006). These swimming movements were reminiscent of blue shark oscillations during straight-line movements. However, this pattern was not maintained when ranging within the Spencer Gulf inlet of South Australia, where bottom-oriented behavior with few ascents to the surface was the dominant pattern. Furthermore, a third pattern comprised diel vertical movements with shallow depths between the surface and 25 m depth occupied principally during the day, and depths around 50 m selected at night (Bruce, Stevens, and Malcolm 2006). This diel vertical movement pattern occurred when the white shark was present at offshore islands. Therefore, it appears that large sharks display different vertical movements as a function of habitat and also in response to differing prey distribution, abundance, and availability.

Clearly, DSTs are able to provide detailed records of shark vertical movements; however, determining the horizontal (geographic) locations during diving behavior is rather inaccurate. DSTs have no in-built transmitters for in-air or underwater position finding; rather, light or pressure (depth) sensor data recorded by these tags may permit geolocation using light-level (Wilson et al. 1992; Bradshaw, Sims, and Hays 2007), sea-bed depth (West and Stevens 2001), or tidal (Hunter et al. 2004) methodologies. However, none of these are particularly accurate and vary in accuracy with latitude (e.g., high latitude light-level geolocation accuracy, ±75 km; Pade et al. 2009). So, this tracking technique is relevant only to larger species that regularly traverse distances greater than the error field of such estimations. Furthermore, DSTs are generally recovered and returned to researchers through a developed fishery for the species in question, and so may be less useful for sharks that move into areas not exploited by fishers (species with little or no commercial value, or protected by conservation legislation, such that they are not retained even if caught, like basking shark *Cetorhinus maximus*). Thus, by the late 1990s, there was a need for an electronic tag that was capable of recording sensor data at fine temporal resolution over long time periods, but that could then transmit such data remotely back to the scientists.

8.2.3.2 Satellite-Linked Archival Tracking

Animal tracking methods using satellites were developed in the late 1970s to provide a means of collecting data at precise times, over large spatiotemporal scales, and, for fish, that was independent of fisheries returns. The Argos system of satellites made remote animal tracking at the large scale possible. The system comprises Argos receivers aboard U.S. National Oceanic and Atmospheric Administration (NOAA) satellites that orbit the poles 850 km above the earth. An Argos platform transmitter terminal (PTT) attached to an animal transmits at ultra-high frequency (401.65 MHz) and at these wavelengths antennas are conveniently short (150 mm for an Argos quarter wavelength whip; Priede 1992). An Argos satellite pass over a given position lasts about 10 to 12 min on average and for geolocation of the tag at least two 360-ms messages must be received. The geolocation principle of Argos PTTs by overpassing satellites is that fields of equal Doppler shift correspond to "cones" symmetrical about the orbit track. The two cones intersect at two points on the earth's surface, giving two possible PTT locations (Taillade 1992), one of which can be ruled out on the basis that the tagged animal could not possibly cover such distances in the time between locations.

A study in the early 1980s showed that direct satellite tracking of a shark was feasible (Priede 1984). In a pioneering study in the Clyde Sea off Scotland, Monty Priede of Aberdeen University attached an Argos PTT molded into a large buoyant float to a 7-m long basking shark (*Cetorhinus maximus*) via a 10-m long tether. Whenever the shark swam near the surface the tag was able to break the surface and transmit to overpassing Argos receivers. The shark was tracked moving in an approximately circular course for 17 days along a biologically productive thermal front (Priede 1984; Priede and Miller 2009). Despite this early success, there were no further studies to satellite track individual sharks for about another 10 years. Tags at this time were large; therefore, only the largest planktivorous species such as whale shark, *Rhincodon typus* (Eckert and Stewart 2001) and basking shark were capable of towing these tags through water. Despite a successful attempt by Francis Carey of the Woods Hole Oceanographic Institution to satellite track predatory sharks (a blue shark, *Prionace glauca*) using a satellite transmitter mounted on a dorsal fin "saddle" (Kingman 1996), by the late 1990s the problem still largely remained: how could open-ocean shark movements and behavior be tracked at reasonable spatial resolution over longer time periods?

A solution to this problem was developed in the form of a hydrid electronic tag that combined sensor data-logging with satellite transmission (Block et al. 1998; for review, see Arnold and Dewar 2001) (Figure 8.4). This new tag, termed a pop-off satellite archival transmitter (PSAT), essentially comprised an Argos PTT linked to a data-storage tag. This first tag was attached to a fish "host" like an external parasite, during which time it recorded 61 hourly or daily water temperature measurements (Arnold and Dewar 2001). After a preprogrammed time the tag would release from the fish, that is "pop off," float to the surface, and begin transmitting messages approximately every 60 s to Argos receivers that geolocated the tag's position and received the temperature data. By the late 1990s the technique of animal tracking by geolocating Argos PTTs was not new, having become a central technique for tracking the movements of flying (e.g., albatross, swan) and air-breathing aquatic animals (cetacean, turtle, seal; Tanaka 1987; Jouventin and Weimerskirch 1990; for review, see Priede and French 1991; Watkins et al. 1999). Nevertheless, its application to geolocating a single position of moderate to large-sized fish remotely, and tracking its habitat changes (depth and temperature) was a real innovation, and precipitated a flurry of discoveries. For example, bluefin tuna fitted with PSATs off the North Carolina coast were found to travel over 3000 km east across the North Atlantic Ocean in two to three months (Block et al. 1998; Lutcavage et al. 1999). In addition, movements of the great white shark were generally unknown, but those aggregating around small islands in the coastal waters of California were generally thought to remain in shelf waters, moving up and down the coast according to season. However, PSAT tagging revealed some incredible movements of these apex predators; tagged sharks moved into the open ocean away from the coast and one shark made the journey from California to Hawaii (some 3800 km across the North Pacific Ocean) in less than six months, making dives down to 700 m and experiencing temperatures down to 5°C (Boustany et al. 2002).

As impressive as these findings were, the details they provided about movements were still quite rudimentary. In terms of horizontal movements, PSATs at this stage were capable of providing essentially the same locational information as that obtained from mark-recapture studies; that is, where the fish was at the time it was tagged and released, and where it was after a known time period. Clearly, the transmission of hourly temperature data was useful in summarizing the gross thermal habitat occupied by the fish during its time at liberty, but elucidating the actual trajectories of large fish in space and

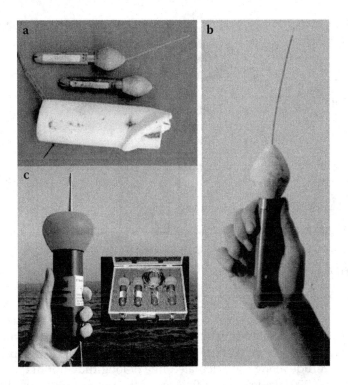

FIGURE 8.4
A color version of this figure follows page 336. Examples of satellite-linked and mobile telephone-linked archival tags used for tracking highly migratory shark species. (a) The large white tag (lower) is a satellite-linked depth recorder modified to incorporate a custom-made, timed-release mechanism. It was deployed on a basking shark by the author in July 1999 and was physically returned after washing ashore in France five years later; it was one of the first PSATs attached to a shark. The smaller tags [pop-up archival transmitting (PAT) tags, PAT Mk3, middle, and Mk4, top] transmitted data via satellite and were found and returned to the MBA for full archival data download after one year floating at sea. (b) A closer view of a PAT Mk3 tag comprising a mini-data-logger (gray body of tag) interfaced with an Argos transmitter with protruding antenna, and white syntactic foam for flotation. On the gray body of the tag the sensors can be seen, from top to bottom: pressure (depth), light, and temperature. (c) A GSM mobile telephone-linked tag deployed between 2004 and 2006 on basking and blue sharks in the English Channel. (Photographs courtesy of MBA, Plymouth, UK.)

time was still the most desired objective. The next advancement made in tracking fish was fundamental and has transformed our understanding of large fish behavior.

Later generations of PSAT tags were increasingly capable of recording and storing larger amounts of data and from more sensors; pressure (depth) and light-level sensors were added in subsequent models. It was around 1999 to 2000 that PSATs with depth, temperature, and ambient light-level sensors became available. The latter was an important addition because longitude can be estimated by comparing the time of local midnight or midday with that at Greenwich, and latitude from estimates of day length (Wilson et al. 1992). Therefore, electronic tags capable of recording light level could provide data amenable to calculations of geolocation anywhere on the earth's surface, and thereby allow reconstruction of a fish's movement track from data retrieved from remote locations by Argos satellite. Furthermore, because latitude estimates are particularly sensitive to small differences in estimation of day length, water temperature data recorded by the tag could be used to bias-reduce latitude estimations. By comparing the tag-recorded water temperature when the fish was near the surface (identified from tag depth data) with that

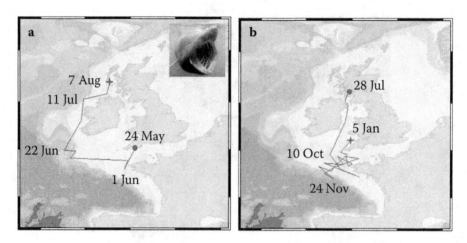

FIGURE 8.5

A color version of this figure follows page 336. Example trackings of PSAT-tagged basking sharks (a) between southwest England and northwest Scotland during summer, and (b) vice versa movement between summer and winter. (Modified from Sims DW et al. (2003) Mar Ecol Prog Ser 248:187–196.)

determined from satellite remote-sensing images of sea surface temperature (e.g., advanced very high resolution radiometer, AVHRR, images), it was possible to determine more accurate estimates of latitude along the calculated line of longitude (for details, see Sims et al. 2003, 2006a; Teo et al. 2004; Pade et al. 2009). Among the first studies to use light-level intensity changes to calculate large-scale horizontal movements of pelagic fish were those with bluefin tuna (Block et al. 2001) and basking shark (Sims et al. 2003). Reconstructed tracks of bluefin tuna tagged off North Carolina showed movements between foraging grounds in the western Atlantic prior to transatlantic migrations into the Mediterranean Sea for spawning. Bluefins were also shown to dive to over 1000 m depth during these excursions (Block et al. 2001). The question of diving activity and what large fish in temperate waters might do when, with changing season, food levels decline, is particularly relevant to the plankton-feeding basking shark (for review, see Sims 2008). This species, the world's second largest fish, is most frequently seen in coastal waters feeding at the waters' surface during summer months (Sims and Quayle 1998). In the northwest and northeast Atlantic, surface foraging occurs from around April to October, usually with a peak in sightings from May until August. The seasonal increase in the surface sightings of C. *maximus* coincides with increased zooplankton abundance at this time (Sims, Fox, and Merrett 1997). Basking sharks tagged with PSATs were tracked moving between waters off southwest England to Scotland, and vice versa, sometimes over periods of only a few weeks (Sims et al. 2003; Figure 8.5). Sharks travelled long distances (390 to 460 km) to locate temporally discrete productivity hotspots along tidal fronts and on shelf-break fronts. It was also shown from basking shark trackings over seasonal scales (up to 7.5 months) that they were active during winter and do not hibernate, as was once supposed (Matthews 1962). Instead, they conduct extensive horizontal (up to 3400 km) and vertical (>750 m depth) movements to use productive continental-shelf and shelf-edge habitats during summer, autumn, and winter (Sims et al. 2003).

Despite being the most readily available and widely used geolocation system for animal tracking studies at the global scale, there are some persistent disadvantages of the Argos system. The narrow Argos receiver bandwidth means transmitted message lengths from PSAT tags that have been attached to sharks are necessarily short (360 ms), with relatively small amounts of data capable of being transmitted (32 bytes) per message. The transmission times of these tags after pop-off is limited due to rapid battery exhaustion, usually ranging from 0.5 to 1 month, so different methods have been employed by the tag manufacturers to increase rates of data recovery. Hence, for one tag type (see www.wildlifecomputers.com) comprehensive summary data in the form of histogram messages (swimming depth, water temperature, profiles of water temperature with depth, times of sunrise/sunset for use in geolocation) derived from archival sensor records are transmitted remotely via satellites, whereas if a tag is physically recovered the entire archived dataset can be downloaded. For another tag type (see www.microwavetelemetry.com), times of sunrise and sunset together with hourly temperature and pressure readings can be recorded for over a year then transmitted after pop-off. In this tag, a special duty cycle timer extends transmission to one month for uploading of this archival dataset by activating the PTT only when Argos satellites are most likely to be in view of the tag.

The availability of PSAT tags, with the ability to geolocate a single position of a fish remotely and track its behavior and habitat changes (depth and temperature) between tagging and pop-off has been a landmark development. The findings from PSATs have been impressive, particularly for sharks; for example, with an individual white shark return migration being tracked over some 10,000 km distance (Bonfil et al. 2005) and dives of a whale shark (*Rhincodon typus*) reaching to nearly 1.3 km depth (Brunnschweiler et al. 2009). Nonetheless, much recent research effort has been aimed at improving accuracy of the estimates of light-level-derived locations of fish during their free-ranging movements (e.g., Teo et al. 2004; Nielsen et al. 2006). It appears, however, that there are distinct limits to the spatial accuracy of these estimates (~60 to 180 km), even in higher latitudes where errors are reduced, such that a spatial error of only 10% of a shark's daily movement distance renders detection of specific behaviors within the track prone to error (Bradshaw, Sims, and Hays 2007). Hence, light-level geolocation appears appropriate for tracking large-scale movements, but is less able to resolve the specific pattern of smaller-scale movements typical of most sharks during much of their annual cycle. As a consequence, researchers have rediscovered Priede's idea of direct satellite tracking of sharks, and as Francis Carey pioneered in the 1990s, have begun attaching Argos PTTs directly to the first dorsal fin of large sharks; for example, the blue shark (Figure 8.6), salmon shark *Lamna ditropis* (Weng et al. 2005), white shark (Bruce, Stevens, and Malcolm 2006), and tiger shark (Heithaus et al. 2007).

The long-range movements of the salmon shark as determined from fin-mounted Argos PTTs appear, like those of the basking shark, for example, to be linked to seasonal changes in environmental conditions. Salmon sharks were tracked from summer feeding locations in Alaskan coastal waters southward to overwintering areas encompassing a wide range of habitats from Hawaii to the North American Pacific coast (Weng et al. 2005). Some individuals, however, overwintered in Alaskan waters and in so doing experienced water temperatures between 2°C and 8°C and dived no deeper than 400 m. Those sharks migrating south by contrast, occupied depths where seawater temperatures were up to 24°C and regularly dived into cooler waters to depths over 500 m (Weng et al. 2005). Because many large species of shark regularly come to the surface, as do many large bony fishes such as swordfish (*Xiphias gladius*; Carey and Robison 1981) and ocean sunfish (*Mola mola*; Sims et al. 2009), the use of fin-mounted Argos PTTs presents a new opportunity for resolving more accurately the horizontal movements of sharks across a broader range of spatial and temporal scales.

FIGURE 8.6
A color version of this figure follows page 336. Different sized fin-mounted Argos satellite transmitters (PTT, platform terminal transmitter) used to (left photograph) track subadult blue sharks released in the English Channel, and (right) fitted to juvenile blues off southern Portugal. (Photos © (top) J. Stafford-Dietsch, (left) MBA, and (right) N. Queiroz. All rights reserved.)

8.2.4 Recent Tag Developments

One of the principal issues driving the development of new generations of electronic tags for marine fish tracking is the need for improved spatial accuracy of horizontal trajectories with few temporal gaps. Because a crucial set of questions about shark behavior, habitat use, and conservation rely on knowing where individuals are located, good spatial data is a key requirement. Present methods provided by tags, however, are deficient in several respects. For instance, shark horizontal movements can be tracked indirectly with data-logging tags using light-level geolocation. The spatial accuracy of this method is only very coarse, which may prevent its reliable use with some analysis techniques (see Section 8.3, Movement Analysis). Furthermore, there will be temporal gaps in the time series of light-level geolocated positions because sharks often dive at dawn and dusk, just at the time when tags need to record increasing/decreasing light levels without changes in shark depth interfering with baseline values. As mentioned above, shark horizontal movements can be tracked directly in real time using fin-mounted Argos PTTs. The accuracy of this method is better than that of light-level geolocation, but at best it is still only accurate to within 150 m of a tag's actual location, and at worst this error can be greater than many kilometers. As described previously, relatively small errors in location accuracy can result in data being unreliable to use in some spatial analyses, and could result in observed movements or behavior patterns being interpreted incorrectly (Hays et al. 2001; Bradshaw, Sims, and Hays 2007; Royer and Lutcavage 2008). So, improving spatial accuracy and obtaining time series of locations without gaps is important.

The GPS constellation of satellites has been used successfully to track, with greater accuracy than the Argos system, the movements of turtles, pinnipeds, and seabirds (e.g.,

Weimerskirch et al. 2002; Schofield et al. 2007). Radiowaves do not penetrate seawater significantly so GPS receivers carried by fully aquatic marine animals must operate in air to obtain positional information. Clearly, marine vertebrates with lungs such as turtles and seals surface regularly to breathe air, allowing the device (if fitted on or near the head) to become dry and start acquiring satellites in the GPS constellation for calculation of latitude and longitude. Successful demonstration of a GPS receiver for long-term tracking of large fish (>90 days) has already taken place on ocean sunfish off Portugal (Sims et al. 2009). This prototype device links a Fastloc GPS receiver with an Argos PTT, the idea being that the Fastloc GPS enables fast acquisition of the satellite constellation for location fixing (in less than 80 ms), with the satellite ephemeris data transmitted via Argos for remote recovery of data. Sensor data on some devices can also be transmitted. The spatial accuracy of Fastloc GPS receivers is high, with 95% of positions occurring within 18 m of the tag's actual location if up to 10 satellites are acquired for the calculation of tag position (see www.wildtracker.com). Given that many species of shark come to the surface more often than was previously thought (Sims et al. 2005b; Weng et al. 2005), albeit for sometimes short periods of time, the advent of this new tag will undoubtedly enable more accurate tracks of shark movements to be obtained in due course.

Despite the improved location accuracy of GPS tags, the problem of temporal gaps in the track still remain because fish may not surface on a daily basis, perhaps remaining at depth for days, weeks, or months. Clearly, while GPS or Argos tags that provide tracks with gaps are quite sufficient to identify general habitats occupied, broader-scale movements, including migrations, and changes in distributions, they may be less useful in supporting deeper analysis aimed at understanding the behavioral mechanisms underlying observed patterns. Some movement analysis procedures require continuous-time tracks (tracks with fixed time intervals between locations), rather than discontinuous tracks with "jumps" due to nonsurfacing behavior, during which no spatial locations are possible. This gap problem cannot be resolved using GPS or Argos tags; however, there is another method that may work well for some species of sharks. Dead reckoning is a technique used for estimating your position through time based on measures of your movement rate and turning angles, that is, independent of knowing your actual location with respect to your surrounding environment. For example, the straight-line distance covered by an animal that starts from position x and proceeds at a constant known speed for t time to position y, can be calculated easily by multiplying the velocity by the time taken to move from x to y. Incorporating a turn into this trajectory presents no problem for determining relative position so long as the turn angle, speed of movement, and time between all subsequent turns are known. In practical terms, turn angle is determined from change in compass heading. Changes in swimming depth or flying altitude can also be incorporated into the vector calculations to provide three-dimensional trajectories. Marine animals are subject to drift in water currents that cannot be measured using dead reckoning without another point of reference (Wilson, Shepard, and Liebsch 2008). Therefore, it also becomes important to know the end point of the individual's track, that is, the time and position when the device is removed. Corrections for drift can then be applied and the estimated trajectory reconstructed.

Measuring animal speed, compass heading, and time between compass heading changes enables an animal's trajectory to be reconstructed with good accuracy. This technique has been used to track the fine-scale foraging movements of basking sharks responding to zooplankton density gradients (e.g., Sims and Quayle 1998). Importantly, new data-logging tags have been developed that incorporate the necessary sensors for deriving trajectories by dead reckoning (Wilson, Shepard, and Liebsch 2008). By logging 14 parameters at infra-second frequencies, including channels for a tri-axial accelerometer, these so-called "daily diary" tags can record movement, behavior, and energy expenditure (when appropriately validated/calibrated), and physical characteristics of the animal's environment. They have been trialed successfully on captive lemon sharks *Negaprion brevirostris* and free-living whale sharks, which suggests these tags may be at least one practical means of obtaining continuous time tracks with high spatial accuracy. One disadvantage of this tag will be that most large fish such as sharks do not return regularly to precise locations where daily diary tags can be retrieved reliably, unlike the situation for seals (e.g., at haul-out sites, breeding beaches) or seabirds (e.g., at nests). Although this tag holds particular promise for revealing fine-scale data on movement, feeding events, energy expenditure, and so on over long time scales, its impact on wild shark behavior studies may be limited only by this difficulty.

Archival tagging of sharks presents itself as an opportunity for retrieval of large data-sets of swimming depth, water temperature, and light level, among other parameters, with no gaps in the time series. Some archival tags can record at high resolution for over a year and store 65 Mbytes of data in nonvolatile Flash memory for many years. While retrieval of these tags at sea is unlikely, in a study of basking shark in the northeast Atlantic, about 30% of drifting PSATs, which may or may not have transmitted summary data to Argos satellites, were found on beaches in the United Kingdom, France, and the Netherlands and were returned to researchers (Shepard et al. 2006). Given that this level of tag return is similar to studies on elasmobranchs where tags are returned through well-developed fisheries, the concept of "pop-off and drift" (Sims et al. 2005a) may represent a cost effective method of data retrieval in populated regions, especially as the unit price of data storage tags of specific size decreases over time and with further advances in technology. It may also be a particularly appropriate method for shark species that are not fished commercially or are protected by legislation, where fishery-dependent tag returns would not be a viable mechanism for data retrieval.

To further ensure archival data retrieval, pop-off and drift tags that were attached to blue and basking sharks have recently been fitted with a GSM (Global System for Mobile communications) mobile telephone communicator (Sims et al. 2005a). Instead of transmitting data to Argos satellites, the GSM unit attempts to call a base station each week until such time that it drifts within range of GSM receiver masts that are part of the publicly available mobile phone network. Once communication is established and maintained the full archival dataset can be downloaded as long as the line remains connected. The increased bandwidth of the mobile phone network enables this high volume data transfer, rates of transmission that could not be supported by the limited bandwidth of the Argos system.

Up to now we have discussed how electronic tags can be used to track the movements, behavior, and in some cases the physiology of sharks. Clearly, knowing the movements of sharks is necessary to understand habitat selection, but would not the picture become so much clearer if it was known from remote telemetry where, when, and how much prey was consumed by free-ranging sharks? An advance of this type would allow scientists to ask more in-depth questions about shark behavior; for example, how often do large sharks feed, what is the prey capture success rate, and are particular foraging movements optimal? A temperature-recording archival tag placed in the stomach of an endothermic shark (e.g., porbeagle *Lamna nasus*), which maintains a higher body temperature compared to external seawater, enables estimates of individual feeding events on ectothermic (cold) prey to be made. Stomach temperature data can also be transferred to an externally attached PSAT tag for remote transmission of such feeding events via Argos satellites. However, it was

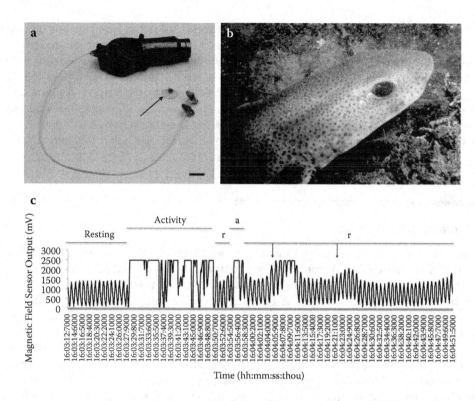

FIGURE 8.7
A jaw-mounted respiratory, feeding, and activity sensor for sharks. (a) A modified inter-mandibular sensor (IMASEN) comprising a data-logger with Hall-effect sensor and boron neodinimum magnet (arrowed) was used in 2004 to record body movements of catshark (b). (c) Example high-resolution trace of catshark jaw movements during resting, and showing breathing rhythms, and swimming activity. Arrows denote changes in baseline due to subtle shifts in body posture during resting. (Photos courtesy of MBA, Plymouth, UK.)

only very recently that an archival tag was developed to be capable of direct monitoring of key behaviors such as feeding.

The feeding sensor tag now developed for fish is based on the intermandibular angle sensor (IMASEN) used previously in penguin feeding studies (Wilson et al. 2002). The data-logging IMASEN works by measuring (and recording) the change in voltage that occurs as a consequence of relative movement between a Hall-effect magnetic field sensor and a small, rare earth (neodinium boron) magnet, which is used to produce a magnetic field. For obtaining information on jaw movements, including ventilation and feeding, the concept is to fit a magnet and a sensor to the upper and lower jaws, respectively, of a fish such that when the mouth opens the sensor and magnet become farther apart, and the recorded voltage drops, whereas when the mouth closes the jaws meet and the voltage increases. This technique was first tried out on a fish in 2004, when a small spotted catshark (*Scyliorhinus canicula*) was fitted with an IMASEN and jaw movements were recorded at 10 Hz temporal resolution over a 24-h period (D.W. Sims and E.J Southall, unpublished data) (Figure 8.7). This showed ventilation rhythms could be recorded in fine detail, in addition to periods of activity and post-activity ventilation frequency and amplitude (Figure 8.7). However, it was not until recently that the technique was improved with a smaller, more robust and reliable archival tag, that actual feeding events were recorded in a captive fish, an Atlantic cod *Gadus morhua* (Metcalfe et al. 2009). This technique has great potential for studying in detail the foraging behavior of free-ranging sharks. Linking feeding sensor tags to tri-axial accelerometer tags that measure fine-scales changes in body position and speed at high resolution could provide a means to separate feeding capture events from feeding attempts (failed captures). The addition of a still or video camera to such data loggers is also a distinct possibility as cameras become increasingly miniaturized. As long as tags can be recovered from sharks, or a remote means of downloading data is developed for the tag, it seems it will likely lead to many intriguing insights about shark foraging, how sharks structure search patterns, and why they select particular habitats at certain times.

8.3 Movement Analysis

The quantitative analysis of animal movement as a formal subject is relatively new (e.g., Turchin 1998). It has emerged from the need to understand population redistribution patterns, that is, how the spatial and temporal abundance of organisms change with respect to environment. Individual movement is a key driver of distribution changes in animal populations and the interplay of movement with environmental heterogeneity is an extremely important aspect of ecological dynamics (Turchin 1998). Unlike the well-developed concepts and methodologies supporting studies in population ecology, animal movement, or movement ecology as it is becoming known (C. Holden 2006), has been largely neglected as a quantitative subject where explicit models can be tested (Turchin 1998). Only very recently has it emerged as a coherent paradigm providing a conceptual framework for unifying animal movements, behavior, and physiological states, with the framing of behavioral questions and the testing of hypotheses (Nathan 2008; Nathan et al. 2008). One reason for this late development is the complex nature of animal movement; it involves at least two scales—spatial and temporal—which makes it intrinsically more difficult to study than the largely one-dimensional problems associated with aspects of studying population density, such as birth and death processes, for example (Turchin 1998).

Movement ecology proposes a paradigm that integrates conceptual, theoretical, methodological, and empirical frameworks for studying movement in all organisms (Nathan et al. 2008). It is based on a depiction of the interplay between internal states, motion itself, orientation and navigation, and the role of external factors affecting movement. Prior to the emergence of a movement ecology framework there were, in essence, four existing paradigms representing different scientific disciplines in which the movements of organisms have been studied, namely biomechanical, optimality, random, and cognitive (e.g., Stephens and Krebs 1986; Berg 1993; Dukas 1998). Movement ecology attempts to unify these disciplines by proposing a practical framework for generating hypotheses and to facilitate an understanding of the causes, mechanisms, and spatiotemporal patterns of movement and their role in various ecological and evolutionary processes (Nathan et al. 2008). While an exhaustive discussion of the overarching use or value of the movement ecology paradigm is beyond the scope of this chapter, it is pertinent to emphasize here that formulating hypotheses and designing robust tests is key to revealing the causes and mechanisms of animal movement. This is especially relevant to the study of shark movements because there are particular difficulties associated with testing hypotheses about the free-ranging behavior of sharks (for overview, see Sims 2003). However, recent advances in electronic tagging will facilitate the analysis of shark movements and testing of explicit hypotheses of shark behavior. Moreover, and importantly, the quantitative analysis of animal movement allows the same approach and set of methods to be used on different individuals and species, allowing comparisons between individuals and across species (Rutz and Hays 2009). In so doing, this presents an opportunity for revealing general features or "rules" of movement across diverse taxa, and elucidating the factors driving the evolution of behavior.

In this part of the chapter, I will describe some of the types of questions about shark movements and the underlying behavior that have been tested, and the techniques that have been used. The techniques described do not represent a complete list but instead consider some of the more useful procedures that have been used to aid understanding of shark behavior, and to highlight those techniques that show some particular promise. For detailed review and discussion of quantitative movement analysis techniques, especially for those procedures not explicitly considered here (e.g., autocorrelation functions, net squared displacement), the treatment by Turchin (1998) is recommended. It is also worth noting that perhaps one ultimate purpose of tracking studies of sharks is to identify potential mechanisms or general features of movement that will allow their realistic prediction across a broad range of scales and in the face of environmental change. The capability to predict when and where sharks move will provide information crucial to improved shark fisheries management and species conservation, which is urgently needed for a range of large shark species.

8.3.1 Types of Data

The development of electronic tags has proceeded largely independently of considerations about the data type and quality required to support robust tests of hypotheses about spatial movements, behavior, and habitat selection. This is natural given that the pace of technical development in biotelemetry as a field, or biologging science as it is now generically known, has advanced rapidly (Ropert-Coudert and Wilson 2005; Rutz and Hays 2009). As the descriptive component of biologging science becomes mature (e.g., where do animals go? When do they go there? How long do they stay?), so new frontiers of process-based inquiry have opened up in just the last few years (e.g., what habitat types are they occupying? What are the animals doing? How much energy do they expend? Why are they doing those behaviors and in those habitats at that time? Are general features apparent between individuals and across species? Are the patterns adaptive?). While the descriptive components are well supported by data collected by the tags described earlier in this chapter, process-based questions appear less so, although several new technical achievements should bridge some key gaps (see Section 8.2.5).

At present, the main types of field data available for analysis of shark behavior fall into three broad categories: (1) time-series of spatial geolocations of animal positions—some may be short and spatially accurate (to a few meters) within a limited area, while longer tracks may have gaps of varying time period with relatively large error fields; (2) time-series of pressure (depth) readings—often at very high temporal resolution (e.g., 1 s) with no gaps (archival tags), and sometimes at lower resolution (minutes to hours) with temporal gaps (PSAT tags); (3) intermittent monitoring of presence/absence data within a known area(s) across longer time periods (up to multiple years). Quantitative analysis of animal movement for the first two categories is developing, but formal analysis procedures for the third data type have not developed so rapidly.

Analysis of data types such as these essentially requires movement and activity patterns (broadly termed behavioral signals) to be identified from collected data. Field deployments of electronic tags that can also monitor environmental variables (i.e., water temperature, salinity, and light level) can be complemented by laboratory manipulations to determine important factors underlying the behavioral process. However, an experimental approach is largely feasible for only the smaller-bodied shark species. Nevertheless, even with what appear to be imperfect and perhaps limited datasets in terms of the actual behavioral data recorded, movement analysis techniques allow these data to be dissected for salient patterns and features, by descriptive analysis, and by process-based hypothesis testing. The following sections provide some background to the analysis techniques and then give an example(s) of where they have been used to study sharks, and what this approach has delivered in terms of an understanding of an aspect of shark behavior.

8.3.2 Descriptive Techniques

These techniques generally describe quantitatively where different patterns of movement (activity) occur within a data time-series, and therefore where, temporally and spatially, they are located within a tracking dataset in relation to environmental variables. The manner in which a pattern is identified differs in each case, but each technique essentially detects differences in the quantity of some specified movement parameter. From an underlying knowledge of how movement relates to this parameter, a type of behavior can be inferred, for example, whether an animal is foraging, commuting, or station keeping.

8.3.2.1 Tortuosity

Trajectories of free-ranging animals such as sharks comprise a complex of movement types (straight line, looping, convoluted paths, etc.) that may be related to a suite of different behaviors (foraging, resting, courting, fleeing, etc.). To understand what remotely tracked, free-ranging animals are doing requires inferring these behaviors from identification of different movement types. Because animals generally respond to resources by spending more time in the vicinity of those resources until they move on, one of the most ready methods for interpreting movement is to determine how much of the trajectory occurs within a specified spatial unit, or, equivalently, if travel speed is constant, how much time is spent per unit area. Area-restricted searching behavior, a type of intensive movement, is a good example in this context. It has been observed directly in many organisms (e.g., amoeba, insects, birds, fish) that where food density is high, foragers remain within the area as they tend to turn more often and more sharply compared to movements in low density areas (e.g., Smith 1974). Thus, in food-rich areas the trajectory of the forager is more tortuous (highly convoluted), comprising more frequent, large angle turns, than in areas with lower food densities typified by few turns with small deviations in path heading. A simple index is often used to quantify the degree of tortuosity (IOT, index of tortuosity) such that the "bee-line" distance from the track's start to finish point is divided by the distance moved. Here then, straight-line (ballistic) trajectories will have tortuosity scores approaching 1.0, where the bee line is nearly equal to the distance moved, whereas highly tortuous, area-restricted movement will have values closer to zero, since the bee-line distance will be an underestimate of the distance moved. Clearly, by calculating IOT for successive portions of an entire track it is possible to determine how tortuosity changes as the animal moves along a path (see Section 8.3.2.3). Another related method is to determine the time taken for an animal to pass across a circle of given radius, the time taken being proportional to the tortuosity of the movement within that area.

Measures of tortuosity describe the degree of area restrictedness of an animal's movement in relation to some resource, but these movements have been validated in relatively

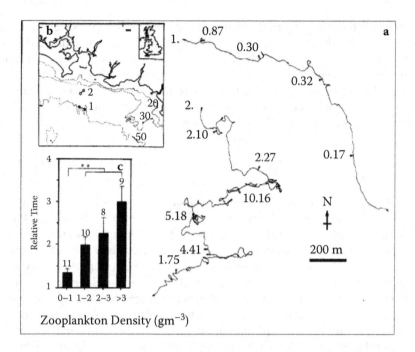

FIGURE 8.8
Fine-scale foraging tracks of basking shark showing (a) intensive and extensive movement patterns linked to changes in zooplankton prey density (g m^{-3}) along the tracks off Plymouth, UK (b). Note high prey densities characterize area-restricted movements (c). (From Sims DW, Quayle VA (1998) Nature 393:460–464. With permission.)

few animal systems to date. Validations require that the area-restricted searching (ARS) movements are related quantitatively to the amount of resource. One such study has been undertaken on a shark species to explore its foraging behavior. The surface foraging movements of basking shark (*Cetorhinus maximus*) have been demonstrated to display ARS patterns. Near small-scale thermal fronts in the western English Channel basking sharks were tracked while foraging on zooplankton, where densities of zooplankton were measured along the sharks' feeding paths (Sims and Quayle 1998). Swimming speeds of feeding sharks were measured and found to be constant at around 1 m s^{-1} (Sims 1999) so the time taken to traverse a circle of known radius was calculated (a circle's center being located at the point on the track where each zooplankton sample was taken). The results showed that shark tracks were more tortuous where zooplankton densities were higher, with the time spent in an area increasing threefold as zooplankton increased from 0-1 g m^{-3} to >3 g m^{-3} (Sims and Quayle 1998; Figure 8.8). This highlighted the importance of oceanographic features such as fronts to basking sharks, and provided a clear numerical relationship between zooplankton density and shark behavior (the likelihood of remaining in an area) that can be used to map key habitats of this vulnerable species (Sims 2008).

Movement analysis proved useful in the basking shark–zooplankton system because the tracks were spatially two-dimensional and zooplankton can be effectively measured with simple nets. However, for the relatively few studies where the horizontal movements of sharks have been determined with good accuracy (see Section 8.2.4), prey fields or gradients have not been possible to measure for logistical reasons. Hence, very little is known about how the tortuosity of macropredatory shark paths relates to the locations and densities of schooling fish prey, for example. It is clear, therefore, that in measuring tortuosity of shark movement paths, we are for the majority of species going to be limited to inferring what the movement pattern signifies, and to assume the relationship of it with some resource. Where it is likely that sharks are located on their seasonal foraging grounds, or where movements occur during a time of day when a shark species is known to forage, it will be reasonable to infer that where the trajectory becomes more tortuous, could represent a time when prey were encountered (and perhaps captured). This simple analysis may allow an examination of how track tortuosity varies according to different habitat types.

8.3.2.2 Home Range

The degree of area restrictedness (or intensiveness) of movements is also relevant to estimates of a shark's home range. There have been a number of individual-based telemetry studies that have examined the home range of sharks but generally these have been aimed at smaller-bodied species (e.g., *Scyliorhinus canicula*; Sims et al. 2006b) or juveniles of larger species (e.g., *Negaprion brevirostris*; Morrissey and Gruber 1993). Only a few telemetry studies have estimated core areas occupied by coastal-pelagic species (e.g., *Lamna nasus*; Pade et al. 2009), perhaps because it is often unclear what represents a "home" range in highly mobile sharks, as opposed to what might be seen as simply a transient activity space. Although the definition of home range varies in its species-specific details, there is a general understanding that it comprises the relatively circumscribed area over which an organism travels to acquire the resources it needs for survival and reproduction (Dingle 1996). Seen in this context, the long-term tracking of movements of small-bodied shark species or juveniles of larger species, perhaps occupying a nursery area for some years, may well encompass a definable home range. In contrast, the home range of a pelagic species such as the blue shark could well encompass an entire ocean basin, given that it may feed in certain locations spatially well separated from where it conducts mating, for example.

In telemetry studies, the locations of where individual animals occur over time are used to build a picture of the home range of the species in question. It is generally considered that geolocations taken over a long time period are required to estimate accurately the home range. The home range will comprise areas that are used with varying intensity, with activity often concentrated around feeding or resting sites (Wray, Cresswell, and Rogers 1992). If tracking studies are short, therefore, it seems likely that what can be estimated is more reasonably described as an activity space (Sims, Nash, and Morritt 2001) that may or may not constitute a core space. Nevertheless, if sufficient geolocations are obtained over time it is possible to calculate what is understood to represent a home range, either of an individual, or of a group, using a range of different approaches (for review, see Worton 1987).

One of the earliest nonparametric techniques for estimating home range was the minimum convex polygon (MCP) approach, which is essentially an outline method. MCPs are constructed around the outermost locations within the total cluster of animal locations. Clearly, this provides a quantifiable areal range for an animal, but it provides no internal structure of the home range, such as the location of core areas. The MCP method may also overestimate the home range because it takes no account of the usable space, which for coastal sharks could include land masses, for instance (Morrissey and Gruber 1993). Methods for determining home range and its internal structure have been developed, and include the kernal density approach, Dirichlet tesselations, and Delauney triangulations (Wray, Cresswell, and Rogers 1992; Wray et al. 1992; Burgman and Fox 2003).

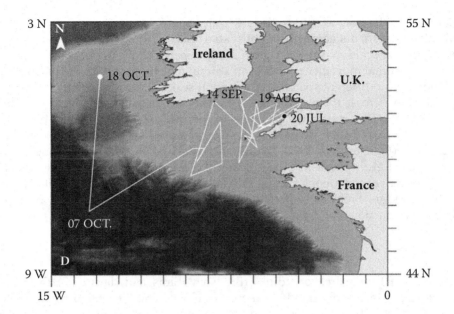

FIGURE 8.9

Satellite-tracked movement of a male, 1.8 m long (fork length) porbeagle shark (*Lamna nasus*). Delauney tessela-tion analysis to derive an α-hull bounding polygon (see text) indicates a core activity space over 90 days of 90,153 km². (Adapted from Pade NG et al. (2009) J Exp Mar Biol Ecol 370:64–74.)

Some of these techniques have been used to study shark home range and core areas. MCPs were used to examine the home range of individual juvenile lemon sharks at Bimini, Bahamas. It was found that area used varied among tracked individuals, ranging between 0.23 and 1.26 km² (mean, 0.68 km²), but that the variability in activity space area was accounted for by a shark's size, with larger individuals occupying a larger activity space (Morrissey and Gruber 1993). At the larger spatial scale, the seasonal core-habitat space use of porbeagle shark (*L. nasus*) was estimated from α-hull bounding polygons (Pade et al. 2009). Briefly, with this technique, a Delauney tesselation from each shark's geolocation pattern is constructed to form a network of triangular polygons. All connec-tions in the triangulation are then measured and connections with a length greater than the mean plus 1.5 standard deviations are excluded. The areas of remaining triangles are then calculated and summed. The core-habitat space usages by a male and a female por-beagle shark tracked during a key foraging period in late summer in coastal waters off southwest England were 8602 and 9961 km², respectively. The space use of another male tracked for longer, between July and October, was much higher at 90,153 km², typifying wider-ranging movements toward the continental shelf edge in autumn (Pade et al. 2009; Figure 8.9).

8.3.2.3 The Question of Scale: First-Passage Time Analysis

In Section 8.3.2.1 we considered a simple measure of tortuosity or the time spent per unit area as two ways to examine the changing structure of an animal's movement path in rela-tion to environment. Such measures can be used, for example, to determine where along an animal's track more time was spent conducting ARS behavior, which may signify that an animal was foraging, resting, or had located mates, for instance. Now we turn our attention to consider how to determine how time spent in an area of known size changes not only along a path, but across different spatial scales.

The question of scale in ecology (Levin 1992) is particularly important to consider in the context of animal movement because an animal may interact with the patch structure of the landscape at different spatial and temporal scales (With 1994). Paths that are more tortuous tend to increase the time along the path at various scales of assessment (Fauchald and Tveraa 2003). This reasoning gave rise to the method known as first-passage time (FPT) analysis, where FPT is defined as the time required for an animal to cross a circle of given radius (Johnson et al. 1992). To determine the first passage times along a path, the circle of smallest given radius is moved along the path at equidistant points (usually by creating intermittent steps along the tracked path), with this procedure repeated for circles of increasing radius. From these iterations, the estimated relative variance, $\hat{S}_r$, in FPT is calculated as a function of r:

$$\hat{S}_r = Var \log(t_r)$$

where t_r is the FPT for a circle of radius r (Bradshaw, Sims, and Hays 2007). Expressing $\hat{S}_r$ as a function of r therefore provides a means of identifying the spatial scales associated with an ARS, that is, the increased rate of turning and large turn angles where resources are plentiful (Fauchald and Tveraa 2003; Pinaud and Weimerskirch 2005).

The value of this technique for analyzing shark tracks is only now beginning to be recognized. Studies using FPT have been undertaken principally on terrestrial and flying species. Analyses of satellite-tracked yellow-nosed albatross, for example, showed that 22 birds adopted an ARS at the scale of about 130 km, with half of these adopting a second, nested ARS around the 30 km scale (Pinaud and Weimerskirch 2005). Interestingly, at the larger scale, ARS was associated with pelagic, subtropical waters, whereas at the smaller scale it was evident that ARS was linked to productive cyclonic eddies. Albatross movement paths analyzed using FPT demonstrated scale-dependent adjustments in searching behavior in relation to different features in the oceanic environment. It seems likely that using FPT with shark trajectories will allow a quantitative description of important behaviors associated with key habitats. This method may become particularly useful in exploring shark behavior as the accuracy and length of shark tracks progress with the development of new tag technologies (see Section 8.2.5), as shown for GPS-tracked ocean sunfish (Sims et al. 2009b).

8.3.2.4 Pattern Identification: Signal Processing

Pattern as well as scale is an important problem in ecology (Levin 1992). A central tenet of animal movement studies is the identification and understanding of observed patterns. Signal processing of animal movement data is a technique that enables patterns to be quantified directly and then related to where they occur in the time-series of move displacements. Shark diving behavior has been the recent subject of such analyses (Graham, Roberts, and Smart 2006; Shepard et al. 2006).

Signal processing techniques such as fast Fourier transform (FFT) analysis accurately summarize the relative importance of periodic components within time series, and are well suited to the analysis of archival tagging records as they achieve rapid throughput of high-resolution data (Graham, Roberts, and Smart 2006). They also provide advantages over techniques such as autocorrelation functions (ACF), which are likely to require preliminary de-trending of the depth record (Neat et al. 2006). The FFT operates by approximating a function with a sum of different sine and cosine terms (Chatfield 1996). The influence of each periodic component is indicated by the magnitude of the corresponding spectral peak in the periodogram. Archival tagging datasets of depth (pressure) time series are particularly appropriate because the resolution and range of frequencies are directly related to the sampling frequency and duration (Graham, Roberts, and Smart 2006). Specifically, FFTs can identify periodicities up to the Nyquist frequency, which is half the sampling rate. Thus, for tracking studies recording pressure at 1 s intervals this represents detection of periodicities of 1 cycle per 30 s^{-1} (Shepard et al. 2006).

FFT analysis has been used to investigate whale shark *Rhincodon typus* (Graham, Roberts, and Smart 2006) and basking shark (Shepard et al. 2006) diving behavior. Diel and lunar periodicity in the vertical movements were identified for a single whale shark in Belize waters, in relation to a predictable food pulse (spawning snappers; Graham, Roberts, and Smart 2006). In contrast to local foraging movements, during the long-distance, directed movement of a 7-m long female whale shark from Mozambique to Madagascar, high frequency oscillations of many cycles per day were detected using FFT (Brunnschweiler and Sims, unpublished data). This individual also conducted many deep dives, some to nearly 1.3 km depth (Brunnschweiler et al. 2009), and the sub-diel oscillations reflected short-term, lower displacement vertical movements occurring within those larger-scale displacements.

The vertical movements of basking sharks recorded with archival tags in the Northeast Atlantic were found to be consistent with those associated with foraging on diel vertically migrating zooplankton prey (Sims et al. 2005b). In deep, thermally stratified waters sharks exhibited normal diel vertical migration (nDVM), comprising a dusk ascent into surface waters followed by a dawn descent to deeper depths (Figure 8.10). This corresponded closely with migrating sound-scattering layers made up of *Calanus* and euphausiids that moved into surface waters at dusk, returning to depths of 50 to 80 m at dawn where

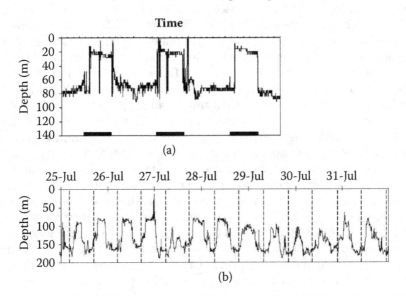

(a)

(b)

FIGURE 8.10
Diving behavior of tagged basking shark showing (a) the "normal" pattern of diel vertical migration that is linked to zooplankton prey movements in thermally stratified waters. (b) A tidal pattern of two vertical movement cycles per day exhibited by a basking shark in tidally mixed waters above a large-scale topographical feature in the English Channel. (Adapted from Sims et al. (2005) J Anim Ecol 74: 755–761, and Shepard et al. (2006) Mar Ecol Prog Ser 328:205–213.)

they remained during the day until dusk. Basking sharks occupying thermally stratified waters were recorded undertaking an nDVM pattern tracking zooplankton movements for up to a month before moving to new areas or changing their vertical movement pattern. Interestingly, the vertical pattern of movement was found to vary between different oceanographic habitat types. In contrast to basking sharks in stratified waters, individuals occupying shallow, inner-shelf areas near thermal fronts conducted reverse DVM, comprising a dusk descent to depths between 20 and 80 m before ascending at dawn into surface waters where they remained during the day (Sims et al. 2005b). This difference in shark swimming movements in fronts compared with thermally stratified waters was due to induction of reverse DVM in *Calanus* by the presence of high concentrations of chaetognaths (important predators of calanoid copepods) at the surface during the night, followed by downward migrations of chaetgnaths during the day (i.e., the normal DVM of the predators produced a behavioral switch to reverse DVM in the copepod prey).

FFT analysis of basking shark vertical movements for multiple individuals across seasons showed that normal and reverse DVM represented the main periodic dive behavior, occurring for 11% to 72% of individual track times (Shepard et al. 2006). However, a tidal pattern (two cycles per day) of vertical movement was also identified, appearing related to the basking shark exploiting tidally mediated migrations of zooplankton. The tidal pattern was seen to alternate with a diel pattern as the shark moved from vertically mixed (tidal) waters to frontal and fully stratified waters, respectively, showing behavioral flexibility of basking shark to environmental changes in the availability of prey at depth (Shepard et al. 2006; Figure 8.10).

8.3.3 Process-Based Techniques

Central to an understanding of shark space use is the identification of the role of environment and habitat characteristics in shaping movement and behavior patterns. For such an understanding it becomes necessary to investigate closely the links between different behavior types and concomitant changes in the environment, and to try to separate habitat *selection* or preferences from habitat *correlations*. Detecting selection of sharks and other animals for particular habitats is difficult since movements generally convey little about why those habitats have been selected (Kramer, Rangeley, and Chapman 1997). Simple mapping of shark movements onto environmental fields, for example a horizontal trajectory overlaid on a satellite remote-sensing map of sea surface temperature, provides a view of where and in what type of habitat the individual was located, but actually tells us nothing in itself about the habitat selection processes underlying the movement; that is, why the habitat was selected from those habitats available. To understand why sharks go where they do, it becomes important to be able to move beyond simple shark–habitat correlations, and to monitor where the sharks are in comparison with where they are not. Comparing the habitat types where sharks are located to the types of other (presumably) equally available habitats where they are not located at a given time, allows us to delve into the dynamics of habitat *selection*. As mentioned previously (see Introduction), an understanding of habitat choices moves us closer to being able to predict the movements and behavior of sharks when faced with a particular set of environmental conditions.

8.3.3.1 Testing for Habitat Selection

There is a growing literature documenting the locations of sharks through time in relation to different habitat types encountered. Generally, the aim has been to describe the habitat correlations of the various species studied to try to gain a mechanistic insight into what may control movement patterns. Less attention, however, has been paid to comparing occupied habitat types with other habitats available to detect potential differences regarding biotic and abiotic factors, such as prey presence or water temperature. In addressing the former, horizontal trajectories have been plotted on maps of sea surface temperature (e.g., Priede 1984; Skomal, Wood, and Caloyianis 2004; Weng et al. 2005; Bonfil et al. 2005), bathymetry (e.g., Holland et al. 1999; Sims et al. 2003), geomagnetic anomalies (Carey and Scharold 1990), and primary (Sims et al. 2003) and secondary productivity (Sims and Quayle 1998; Sims et al. 2006a). Vertical movements of sharks have been mapped onto variations in vertical thermal structure (e.g., Carey and Scharold 1990), in relation to seabed depth (e.g., Gunn et al. 1999) and with respect to concentrations of prey species (Sims et al. 2005b; Figure 8.10). Assessing movements in relation to prey is perhaps one of the most useful environmental fields to use, at least over the short term, because the distribution of prey is a key factor influencing predator movements. It is expected that clear and persistent changes in movements should be related to specific changes in prey abundance and availability.

Relating shark movements to environmental fields such as those in the studies mentioned above has revealed much about the diversity of behaviors present in the wild. It has provided insights into what environmental factors may influence observed patterns. However, the findings of such studies often represent simple habitat correlations that may or may not be the product of habitat selection. In such studies, random correlation may be an equally likely explanation of the observed patterns because other areas not occupied by the shark are not included in the analysis, only those where the shark occurs. The question is, what is it about the location where the shark occurs that renders it different from the surrounding habitats that are not apparently selected? Explicit tests of habitat preferences of sharks are much less common in the literature, although some useful recent examples suggest what is possible to gain from such an approach.

The general approach used in preference-testing studies is to compare the amount of time spent or prey encountered in each habitat as a function of the movement track observed, compared with predicted values for that individual based on random walks through the same environment. The habitat preferences of tiger sharks in a shallow, seagrass ecosystem in Australia were examined by acoustic tracking and animal-borne cameras as they moved through different habitats (Heithaus et al. 2001, 2002, 2006). Tracks of actual sharks were used to estimate the proportion of time spent in two different habitats, shallow and deep. These were then compared with habitats visited during random walks. To produce a random walk, the distances moved between five-minute position fixes in an actual tiger shark track, termed the move step lengths, were randomized to form a new track of the same length, but with a new direction of travel angle assigned to each move step taken from the distribution of commonly observed travel angles of tiger shark tracks. A particular strength of this approach is that actual shark tracks were each compared with 500 random tracks so significance levels of actual versus random could be calculated. The analysis showed that although there was individual variation in habitat use, tiger sharks preferred shallow habitat where prey was more abundant (Heithaus et al. 2002, 2006).

Assessing the habitat selection of sharks in relation to dynamic prey "landscapes" has been a more difficult goal to achieve. The principal reason for this is that temporally changing prey-density fields are not available for the vast majority of marine predators. One of the few cases where this is possible, however, is for filter-feeding sharks. As mentioned previously, basking sharks feed selectively on large calanoid copepods in specific assemblages of zooplankton (Sims and Quayle 1998). In the north Atlantic the Continuous

Plankton Recorder (CPR) survey has for over 50 years undertaken broad-scale measurements of zooplankton abundance to species level (Richardson et al. 2006). The large-scale spatial coverage of individual plankton samples (minimum spatial and temporal resolutions of 56 × 36 km and 14 days, respectively) means it was possible to relate the broad-scale foraging movements of satellite-tracked basking sharks to the spatiotemporal abundance of their copepod prey. Basking shark tracks were mapped onto time-referenced copepod abundance fields in a recent study, with the amount of zooplankton "encountered" estimated for each movement track (Sims et al. 2006a). The total prey encountered along each track was compared with that encountered by 1000 random walks of model sharks, where the move steps taken by each random walker were drawn from the distribution of move steps observed for real sharks. The study showed that movements of adult and subadult basking sharks yielded consistently higher prey encounter rates than 90% of random-walk simulations. This suggested that the structure of movements undertaken by basking sharks were aimed at exploiting preferred habitats with the richest zooplankton densities available across their range.

8.3.3.2 Exploring Behavioral "Rules"

It is relatively straightforward to see how measurements of energy expenditure, manipulative laboratory experiments, and field tracking of smaller-bodied sharks can be combined to test explicit hypotheses in behavioral ecology (e.g., Sims 2003; Sims et al. 2006b). Doing the same for large sharks, however, is logistically prohibitive. As we have seen, large sharks often undertake complicated horizontal and vertical movements but as has been mentioned already, what is less well known is what they are doing during such movements and why they undertake the structure of movement pattern they do? Generally speaking it has proved very difficult to identify what behaviors sharks are exhibiting, whether they are searching, feeding, commuting, resting, or migrating, and so on. Until an electronic device capable of providing a "daily diary" of shark activities and energy expenditure becomes a practical reality (see Section 8.2.4), we are reliant upon using inferential approaches to help tease out particular behaviors and, in some cases, the strategies potentially used by sharks to find resources (food, mates, refuge, etc.). In this section I will describe some analysis techniques taken from the field of statistical physics and indicate where these can provide a new perspective for interpreting and understanding shark movement patterns in the wild, in particular those associated with searching.

8.3.3.3 Searching Natural Environments

A central issue in behavioral ecology is understanding how organisms search for resources within heterogeneous natural environments (MacArthur and Pianka 1966; Stephens and Krebs 1986). Organisms are often assumed to move through an environment in a manner that optimizes their chances of encountering resource targets, such as food, potential mates, or preferred refuging locations. For a forager searching for prey in a stable, unchanging environment, prior expectation of when and where to find items will inform a deterministic search pattern (Stephens and Krebs 1986; Houston and McNamara 1999). However, foragers in environments that couple complex prey distributions with stochastic dynamics will not be able to attain a universal knowledge of prey availability. This raises the question of how should a forager best search across complex landscapes to optimize the probability of encountering suitable prey densities? Because nearly all motile animals face this same problem it suggests the possibility that a general foraging rule for optimizing search patterns has emerged in animals by natural selection. Pelagic sharks such as the blue shark (*Prionace glauca*) appear well designed for sustained cruising at low swimming speeds and energy cost, presumably in part as a consequence of their need to search large areas and depths to locate sparse resources in sufficient quantities. Hence, pelagic sharks may be ideal candidates for testing ideas about optimal searching by applying Lévy statistics to movement patterns, and in so doing this may reveal insights about what governs aspects of their behavior.

8.3.3.4 Optimal Lévy Flights

Recent progress in optimal foraging theory has focused on probabilistic searches described by a category of random-walk models known as Lévy flights (Viswanathan et al. 2000; Bartumeus et al. 2005). Lévy flights are specialized random walks that comprise "walk clusters" of relatively short step lengths, or flight intervals (distances between turns), connected by longer movement "jumps" between them, with this pattern repeated across all scales (Bartumeus et al. 2005; Figure 8.11). In a Lévy flight the move step lengths are chosen from a probability distribution with a power-law tail, resulting in step lengths with no characteristic scale:

$$P(l_j) \sim l_j^{-\mu}, \text{ with } 1 < \mu \le 3$$

where l_j is the flight length and μ is the exponent (or Lévy exponent) of the power law. Theoretical studies indicate Lévy flights represent an optimal solution to the biological search problem in complex landscapes where prey are sparsely and randomly distributed outside an organism's sensory detection range (Viswanathan et al. 1999, 2000; Bartumeus et al. 2005). Simulation studies indicate an optimal search has a Lévy exponent of $\mu \cong 2$ (Viswanathan et al. 1999). The advantage to predators of selecting step lengths with a Lévy distribution compared with simple Brownian motion for example, is that Lévy flight increases the probability of encountering new patches compared with other types of searches (Viswanathan et al. 2000; Bartumeus et al. 2002; Figure 8.11). Recent studies (Sims, Righton, and Pitchford 2007; Benhamou 2007; Edwards et al. 2007) contend Lévy flight

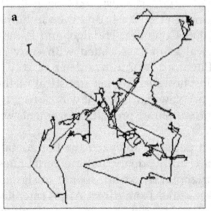
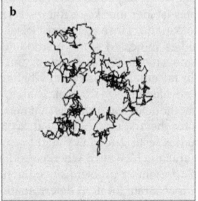

FIGURE 8.11
A color version of this figure follows page 336. (a) A computer-simulated Lévy flight search pattern compared to (b) simple Brownian motion. Note the long reorientation jumps in (a) between clusters of smaller steps, with this pattern occurring across all scales, characterizes Lévy motion.

patterns have been wrongly ascribed to some species through use of incorrect methods, while others indicate Lévy-like behavior with optimal power-law exponents (Bartumeus et al. 2005), supporting the hypothesis that $\mu \cong 2$ may represent an evolutionary optimal value of the Lévy exponent (Bartumeus 2007).

Here then, is a theoretical prediction about searching movements that can be tested with empirical data. The approach is to ask whether individual animals exhibit movement patterns that are consistent with Lévy behavior, that is, whether the move step length frequency distribution is well described by a power law form with a heavy tail. If there is sufficient support for observed movements to be approximated by a Lévy distribution of move step lengths, where $1 < \mu \leq 3$, then we can infer that the animal might be adopting a probabilistic, Lévy-like searching pattern. Furthermore, if the calculated exponent lies close to $\mu \cong 2$, it is possible the structure of movements undertaken by an animal may be optimal. Similarly, fluctuations or changes in the Lévy exponent of movements could signal ecologically important shifts in behavior, or expose different search strategies used by different shark species.

8.3.3.5 *Testing Empirical Data*

Marine vertebrates such as large sharks that feed on ephemeral resources like zooplankton and smaller pelagic fish typify the type of predator that might undertake such probabilistic searches described with Lévy distributions. This is principally because they have sensory detection ranges limited by the seawater medium and experience extreme variability in food supply over a broad range of spatiotemporal scales (Mackus and Boyd 1979; Makris et al. 2006). At near-distance scales sharks use sensory information on resource abundance and distribution, and at very broad scales may have awareness of seasonal and geographical prey distributions, but across the broad range of mesoscale boundaries (one to hundreds of kilometers) pelagic sharks, are more like probabilistic or "blind" hunters than deterministic foragers. Across such scales, the necessary spatial knowledge required for successful foraging will depend largely on the search strategy employed.

Lévy-like movement behavior has apparently been detected among diverse organisms, including amoeba (Schuster and Levandowsky 1996), zooplankton (Bartumeus et al. 2003), insects (honeybees; Reynolds et al. 2007), social canids (jackals; Atkinson et al. 2002), arboreal primates (spider monkeys; Ramos-Fernandez et al. 2004); and even in human movements (Brockmann, Hufnagel, and Geisel 2006; Gonzalez, Hidalgo, and Barabasi 2008). Recent studies indicated some methodological errors associated with early studies of organism movements (Sims, Righton, and Pitchford 2007; Edwards et al. 2007), perhaps resulting in false detections of Lévy behavior; however, robust statistical methods have now been developed (e.g., Clauset, Shalizi, and Newman 2007). It was during this recent period of methodological progress that marine vertebrate predator movements were tested rigorously for the first time (Sims et al. 2008).

Most studies testing for Lévy flight search patterns analyzed the horizontal trajectories of organisms. However, when considering fully aquatic marine vertebrates such as sharks, this presents a problem since the horizontal tracks are subject to significant spatial errors. Inaccurate location determinations, either from direct Argos satellite tracking or from reconstructed tracks using light-level geolocation, are not necessarily important when considering large-scale movements such as migration. This is because the gross movement displacement is greater than the quantified error field. However, in testing for the presence of Lévy flights the move step lengths whether they be small or large are important to measure accurately as they form the frequency distribution, and it is from this that the Lévy exponent is determined. Simulation studies show that a Lévy exponent of a move step frequency distribution cannot be recovered from the original Lévy flight movement when locations are subjected to spatial errors of about 10% of the maximum daily movement distance (Bradshaw, Sims, and Hays 2007). This means that for a shark moving at 50 km per day, for example, the spatial location error during tracking can be no greater, and ideally much less, than 5 km, otherwise a Lévy behavior pattern that is present is unlikely to be reliably detected. This is clearly a problem if light-level geolocations of shark trajectories are used because error fields are large for this method (e.g., ~60 to 180 km). Although Argos satellite geolocations (class 1 to 3) of fin-mounted transmitters are much more accurate (Weng et al. 2005), the gaps in transmissions caused by a shark not necessarily surfacing regularly enough are a problem since move step lengths cannot be determined accurately if locations in the trajectory are missing.

The limitations location errors or gaps put on using horizontal movement data from pelagic sharks has precluded, so far, their use in testing for Lévy flight phenomena. Recent progress, however, has been made using vertical movements to test for the presence of Lévy behavior (Sims et al. 2008). Here, the change in consecutive depth measurements recorded by shark-attached electronic tags form a time series of move steps suitable for analysis of macroscopic patterns across the long temporal scale (Figure 8.12). A time series of consecutive vertical steps is reminiscent of a Lévy walk rather than a Lévy flight, since vertical move steps are determined across equal time intervals rather than between turns (turning points in a Lévy walk form a Lévy flight; see Shlesinger, Zaslavsky, and Klafter 1993). Using this approach for large, archival datasets (~1.2 million move steps), it was demonstrated that move step frequency distributions of basking shark, bigeye tuna, and Atlantic cod (*Gadus morhua*) tracked on their foraging grounds were consistent with Lévy-like behavior (Sims et al. 2008; Figure 8.13). A modification in the time series was needed for air breathers such as leatherback turtles (*Dermochelys coriacea*) and Magellanic penguins (*Spheniscus magellanicus*), but they too showed move step patterns consistent with Lévy motion. Interestingly, the Lévy exponents for these five species were close to the theoretically optimal $\mu \cong 2$ exponent. Analysis of prey abundance time series and a predator–prey computer simulation was also undertaken and suggested that marine vertebrates in stochastic environments necessitating probabilistic searching may derive benefits from adapting movements described by Lévy processes (Sims et al. 2008).

Further data analysis using maximum likelihood estimation (MLE) to determine Lévy exponents for a further seven species of pelagic sharks (bigeye thresher, *Alopias superciliosus*; blue shark, *Prionace glauca*; shortfin mako, *Isurus oxyrinchus*; porbeagle, *Lamna nasus*; silky, *Carcharhinus falciformis*; oceanic whitetip, *Carcharhinus longimanus*; and whale shark, *Rhincodon typus*) confirm that near-optimal Lévy movement patterns are present in these open ocean foragers (N.E. Humphries and D.W. Sims, unpublished data). However, the Lévy search pattern of all these species was interspersed with other search behavior types, that is, those that have different but characteristic move-step frequency distributions (e.g., an exponential form approximating normal random processes). This observed complexity is predicted by theory, since Lévy searches should be employed where prey is sparse, and random movement should predominate where prey is abundant (Bartumeus et al. 2002). Clearly, these are two prey distribution types sharks are likely to encounter. Although how such patterns arise is not known for certain, it is possible that sharks adjust their search tactics according to the prey field type they encounter (Sims et al. 2008).

These results indicate that archival tag-derived data from sharks may be particularly amenable to movement analysis using the techniques described in various recent studies

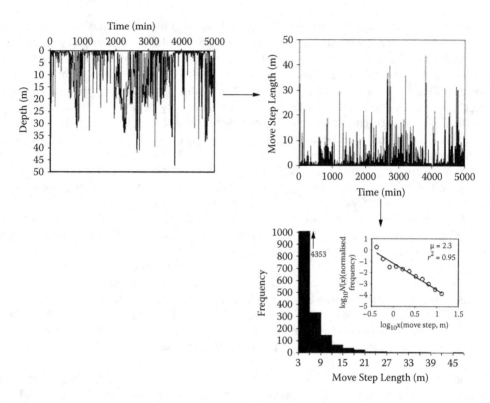

FIGURE 8.12

Analyzing vertical dive displacements for Lévy patterns. Movement time series recorded by electronic tags are analyzed to determine the Lévy exponent to the heavy-tailed distribution. First panel: time series of swimming depths of a 4.5 m long basking shark (*Cetorhinus maximus*). Second panel: The vertical move (dive) steps ($n =$ 5000) for the same shark and time period showing an intermittent structure of longer steps. Third panel: The move step-length frequency distribution for the same data with, inset, the normalized log-log plot of move step frequency versus move step length giving an exponent ($\mu = 2.3$) within ideal Lévy limits ($1 < \mu \le 3$).

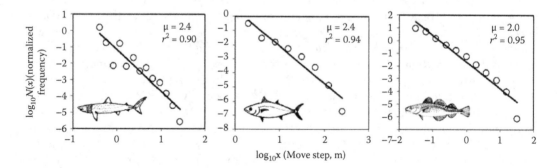

FIGURE 8.13

Lévy-like scaling law among diverse marine vertebrates. Normalized log-log plots of the move-step frequency distributions for (first panel) subadult and adult basking shark ($n =$ 503,447 move steps); (second panel) bigeye tuna (*Thunnus obesus*) ($n =$ 222,282 steps); (third panel) Atlantic cod (*Gadus morhua*) ($n =$ 94,314 steps). Analysis shows Lévy-like movements across species are close to the theoretical optimal for searching ($\mu_{opt} = 2$). (Adapted from Sims DW et al. (2008) Nature 451:1098–1102.)

(Bartumeus et al. 2003; Sims, Righton, and Pitchford 2007; Clauset, Shalizi, and Newman 2007; Sims et al. 2008). Perhaps more importantly though, the recent investigation on marine predators (Sims et al. 2008) showed that in the absence of more direct information (experiments, observations) this approach provides a useful and insightful starting point concerning when and where sharks might be searching, the likely efficiency of such searches, and indeed, why a particular search pattern may be adopted under a particular set of environmental conditions. Determination of the potential movement "rules" sharks (and other animals) have evolved to use their environment will allow a better predictive framework to develop. Hence, Lévy processes may be useful to consider for developing more realistic models of how sharks redistribute themselves in response to environmental changes such as fishing and ocean warming (Sims et al. 2008). Key to this conceptual advance will be the practical one of improvement in accuracy and regularity of locational data during tracking of shark movements so horizontal trajectories are open to rigorous spatial analyses.

8.4 Future Perspectives

The application of cutting-edge tracking and data-logging technology, together with movement analysis and simulation modeling, has recently pointed to potential mechanisms underlying the complex patterns of shark behavior and spatial redistribution patterns. While the discipline known as animal telemetry, or biologging science (to take a more recent name) is reaching some maturity, the field of shark movement ecology is very much in its infancy. The next few years have the potential for a more rapid phase of formal progress. So what lies ahead for us in the study of free-ranging shark behavior? How can new technologies be applied to further enhance our understanding? Of central importance will be the clear identification and testing of behavioral ecological hypotheses combining advanced movement analysis with simulations and modeling to better understand habitat preferences through space and time, and hence to elucidate redistribution patterns. Patterns and dynamics of sexual segregation in sharks will be particularly important to determine, not least because of the potential for biased fishing effects on one sex over another and the implications for deleterious effects on populations (Mucientes et al. 2009).

Linked to this is the need for the next generation of instruments that will provide data of such good quality that much deeper insights into sharks behaving in their natural environments will be possible—we will then be capable of tackling those "why" questions routinely. Some innovative advances will be global positioning system (GPS) tags for sharks for providing more accurate geolocations, the availability of data-logging sensors for signaling prey size and type ingested, and which will be needed for us to begin undertaking natural experiments on the foraging ecology of sharks. Important to this type of analysis is the need for better data on prey abundances and distributions to help interpret movement data through robust testing to quantify habitat selection (e.g., Sims et al. 2006a). The advent of electronic "daily diaries" for sharks that reveal fine-scale data on movement, feeding events, energy expenditure and so on, over long enough time scales to be useful, has potential for making better natural experiments but requires innovative retrieval procedures. In this regard, over the last few years tests have been made of pop-up data-logging tags that instead of transmitting to satellites are capable of downloading data to mobile telephone networks. This has the advantage over satellite transmission, where message lengths are small, in that archival rather than summary data can be transmitted so long as the "call" remains connected.

Movement ecology is also still a young subject (Nathan 2008). However, we can expect rapid advances in this area given the influx of many scientists into the subject from vastly different backgrounds, from plant ecologists and animal behaviorists to mathematical biologists and statistical physicists. This coming together of disciplines to tackle organism movement predicts important advances in theory, methodology, empirical measurements, and modeling. Real insights into behavior, including that of sharks, are likely to occur as a result. The development of ever smaller and more sophisticated tags, together with innovative analysis techniques firmly grounded in behavioral theory will ensure that this knowledge widens yet further to include more species, species interactions, and to identify the links with environmental changes. In so doing I feel this will provide crucial data underpinning better prediction and understanding of shark behavior, that will, ultimately, enhance the possibility for better management and conservation of shark populations.

Acknowledgments

DWS was supported by the UK Natural Environment Research Council (NERC) Oceans 2025 Strategic Research Programme Theme 6 (Science for Sustainable Marine Resources) and a Marine Biological Association Senior Fellowship during the writing of this chapter. V. Wearmouth and N. Humphries are thanked.

References

Arnold GP, Dewar H (2001) Electronic tags in marine research: A 30-year perspective. In Sibert JR, Nielsen JL (ed) Electronic tagging and tracking in marine fisheries, Kluwer, Dordrecht, pp 7–64.

Atkinson RPD, Rhodes CJ, Macdonald DW, Anderson RM (2002) Scale-free dynamics in the movement patterns of jackals. Oikos 98:134–140.

Bartumeus F (2007) Lévy processes in animal movement: An evolutionary hypothesis. Fractals 15:151–162.

Bartumeus F, Catalan J, Fulco UL, Lyra ML, Viswanathan GM (2002) Optimizing the encounter rate in biological interactions: Lévy versus Brownian strategies. Phys Rev Lett 88.

Bartumeus F, Da Luz MGE, Viswanathan GM, Catalan J (2005) Animal search strategies: A quantitative random-walk analysis. Ecology 86:3078–3087.

Bartumeus F, Peters F, Pueyo S, Marrasé C, Catalan J (2003) Helical Lévy walks: Adjusting searching statistics to resource availability in microzooplankton. Proc Natl Acad Sci USA 100:12771–12775.

Baum JK, Myers RA, Kehler DG, Worm B, Harley SJ, Doherty PA (2003) Collapse and conservation of shark populations in the Northwest Atlantic. Science 299:389–392.

Benhamou S (2007) How many animals really do the Lévy walk? Ecology 88:1962–1969.

Berg HC (1993) Random walks in biology. Princeton University Press, Princeton, NJ.

Block BA, Dewar H, Blackwell SB, Williams TD, Prince ED, Farwell CJ, Boustany A, Teo SLH, Seitz A, Walli A, Fudge D (2001) Migratory movements, depth preferences, and thermal biology of Atlantic bluefin tuna. Science 293:1310–1314.

Block BA, Dewar H, Farwell C, Prince ED (1998) A new satellite technology for tracking the movements of Atlantic bluefin tuna. Proc Natl Acad Sci USA 95:9384–9389.

Block BA, Teo SLH, Walli A, Boustany A, Stokesbury MJW, Farwell CJ, Weng KC, Dewar H, Williams TD (2005) Electronic tagging and population structure of Atlantic bluefin tuna. Nature 434:1121–1127.

Bonfil R, Meyer M, Scholl MC, Johnson R, O'Brien S, Oosthuizen H, Swanson S, Kotze D, Paterson M (2005) Transoceanic migration, spatial dynamics, and population linkages of white sharks. Science 310:100–103.

Boustany AM, Davis SF, Pyle P, Anderson SD, Le Boeuf BJ, Block BA (2002) Satellite tagging — Expanded niche for white sharks. Nature 415: 35–36.

Bradshaw CJA, Sims DW, Hays GC (2007) Measurement error causes scale-dependent threshold erosion of biological signals in animal movement data. Ecol Applic 17:628–638.

Brander K (1981) Disappearance of common skate *Raia batis* from Irish Sea. Nature 290:48–49.

Brockmann D, Hufnagel L, Geisel T (2006) The scaling laws of human travel. Nature 439:462–465.

Bruce BD, Stevens JD, Malcolm H (2006) Movements and swimming behaviour of white sharks (*Carcharodon carcharias*) in Australian waters. Mar Biol 150:161–172.

Brunnschweiler JM, Baensch H, Pierce SJ, Sims DW (2009) Deep diving behaviour of a whale shark *Rhincodon typus* during long-distance movement in the western Indian Ocean. J Fish Biol 74:706–714.

Burgess GH, Beerkircher LR, Cailliet GM, Carlson JK, Cortes E, Goldman KJ, Grubbs RD, Musick JA, Musyl MK, Simpfendorfer CA (2005) Is the collapse of shark populations in the Northwest Atlantic Ocean and Gulf of Mexico real? Fisheries 30:19–26.

Burgman MA, Fox JC (2003) Bias in species range estimates from minimum convex polygons: implications for conservation and options for improved planning. Anim Conserv 6:19–28.

Carey FG (1992) Through the thermocline and back again: heat regulation in big fish. Oceanus 35:79–85.

Carey FG, Kanwisher JW, Brazier O, Gabrielson G, Casey JG, Pratt HL (1982) Temperature and activities of a white shark, *Carcharodon carcharias*. Copeia 1982:254–260.

Carey FG, Robison BH (1981) Daily patterns in the activities of swordfish, *Xiphias gladius*, observed by acoustic telemetry. Fish Bull 79:277–292.

Carey FG, Scharold JV (1990) Movements of blue sharks (*Prionace glauca*) in depth and course. Mar Biol 106:329–342.

Casey JM, Myers RA (1998) Near extinction of a large, widely distributed fish. Science 281:690–692.

Chatfield C (1996) The analysis of time series: An introduction, sixth edition. Chapman & Hall, London.

Clarke SC, McAllister MK, Milner-Gulland EJ, Kirkwood GP, Michielsens CGJ, Agnew DJ, Pikitch EK, Nakano H, Shivji MS (2006) Global estimates of shark catches using trade records from commercial markets. Ecol Lett 9:1115–1126.

Clauset A, Shalizi CR, Newman MEJ (2007) Power-law distributions in empirical data. E-Print, arXiv: 0706.1062v1.

Dingle H (1996) Migration: the biology of life on the move. Oxford University Press, Oxford.

Dukas R (ed) (1998) Cognitive ecology: The evolutionary ecology of information processing and decision making. University of Chicago Press, Chicago.

Dulvy NK, Baum JK, Clarke S, Compagno LJV, Cortes E, Domingo A, Fordham S, Fowler S, Francis MP, Gibson C, Martinez J, Musick JA, Soldo A, Stevens JD, Valenti S (2008) You can swim but you can't hide: The global status and conservation of oceanic pelagic sharks and rays. Aquat Conserv 18:459–482.

Eckert SA, Dolar LL, Kooyman GL, Perrin W, Rahman RA (2002) Movements of whale sharks (*Rhincodon typus*) in South-east Asian waters as determined by satellite telemetry. J Zool 257:111–115.

Eckert SA, Stewart BS (2001) Telemetry and satellite tracking of whale sharks, *Rhincodon typus*, in the Sea of Cortez, Mexico, and the north Pacific Ocean. Environ Biol Fish 60:299–308.

Edwards AM, Phillips RA, Watkins NW, Freeman MP, Murphy EJ, Afanasyev V, Buldyrev SV, da Luz MGE, Raposo EP, Stanley HE, Viswanathan GM (2007) Revisiting Lévy flight search patterns of wandering albatrosses, bumblebees and deer. Nature 449:1044–1047.

Fauchald P, Tveraa T (2003) Using first-passage time in the analysis of area-restricted search and habitat selection. Ecology 84:282–288.

Gonzalez MC, Hidalgo CA, Barabasi AL (2008) Understanding individual human mobility patterns. Nature 453:779–782.

Graham RT, Roberts CM, Smart JCR (2006) Diving behaviour of whale sharks in relation to a predictable food pulse. J R Soc Interface 3:109–116.

Guénette S, Pitcher TJ, Walters CJ (2000) The potential of marine reserves for the management of northern cod in Newfoundland. Bull Mar Sci 66:831–852.

Gunn JS, Stevens JD, Davis TLO, Norman BM (1999) Observations on the short-term movements and behaviour of whale sharks (*Rhincodon typus*) at Ningaloo Reef, Western Australia. Mar Biol 135:553–559.

Harden Jones FR (1968) Fish migration. Edward Arnold, London.

Hays GC, Akesson S, Godley BJ, Luschi P, Santidrian P (2001) The implications of location accuracy for the interpretation of satellite-tracking data. Anim Behav 61:1035–1040.

Heithaus MR, Marshall GJ, Buhleier BM, Dill LM (2001) Employing Crittercam to study habitat use and behavior of large sharks. Mar Ecol Prog Ser 209:307–310.

Heithaus MR, Dill LM, Marshall GJ, Buhleier B (2002) Habitat use and foraging behavior of tiger sharks (*Galeocerdo cavier*) in a seagrass ecosystem. Mar Biol 140:237–248.

Heithaus MR, Hamilton IM, Wirsing AJ, Dill LM (2006) Validation of a randomization procedure to assess animal habitat preferences: microhabitat use of tiger sharks in a seagrass ecosystem. Journal of Animal Ecology 75, 666–676.

Heithaus MR, Wirsing AJ, Dill LM, Heithaus LI (2007) Long-term movements of tiger sharks satellite-tagged in Shark Bay, Western Australia. Marine Biology 151: 1455–1461.

Heupel MR, Simpfendorfer CA (2002) Estimation of mortality of juvenile blacktip sharks, *Carcharhinus limbatus*, within a nursery area using telemetry data. Can J Fish Aquat Sci 59:624–632.

Holden C (2006) Inching toward movement ecology. Science 313:779–780.

Holden MJ (1967) Transatlantic movement of a spurdogfish. Nature 214:1140–1141.

Holland KN, Brill RW, Chang RKC, Sibert JR, Fournier DA (1992) Physiological and behavioral thermoregulation in bigeye tuna (*Thunnus obesus*). Nature 358:410–412.

Holland KN, Wetherbee BM, Lowe CG, Meyer CG (1999) Movements of tiger sharks (*Galeocerdo cuvier*) in coastal Hawaiian waters. Mar Biol 134:665–673.

Houston AI, McNamara JM (1999) Models of adaptive behaviour. Cambridge University Press, Cambridge, UK.

Hunter E, Metcalfe JD, Arnold GP, Reynolds JD (2004) Impacts of migratory behaviour on population structure in North Sea plaice. J Anim Ecol 73:377–385.

Jackson JBC, Kirby MX, Berger WH, Bjorndal KA, Botsford LW, Bourque BJ, Bradbury RH, Cooke R, Erlandson J, Estes JA, Hughes TP, Kidwell S, Lange CB, Lenihan HS, Pandolfi JM, Peterson CH, Steneck RS, Tegner MJ, Warner RR (2001) Historical overfishing and the recent collapse of coastal ecosystems. Science 293:629–638.

Johnson AR, Wiens JA, Milne BT, Crist TO (1992) Animal movements and population dynamics in heterogeneous landscapes. Landscape Ecol 7:63–75.

Jouventin P, Weimerskirch H (1990) Satellite tracking of wandering albatrosses. Nature 343:746–748.

Kingman A (1996) Satellite tracking blue sharks. Shark News 7:6.

Klimley AP (1993) Highly directional swimming by scalloped hammerhead sharks, *Sphyrna lewini*, and subsurface irradiance, temperature, bathymetry, and geomagnetic field. Mar Biol 117:1–22.

Klimley AP, Butler SB, Nelson DR, Stull AT (1988) Diel movements of scalloped hammerhead sharks, *Sphyrna lewini* Griffith and Smith, to and from a seamount in the Gulf of California. J Fish Biol 33:751–761.

Klimley AP, Le Boeuf BJ, Cantara KM, Richert JE, Davis SF, Van Sommeran S (2001a) Radio-acoustic positioning as a tool for studying site-specific behavior of the white shark and other large marine species. Mar Biol 138:429–446.

Klimley AP, Le Boeuf BJ, Cantara KM, Richert JE, Davis SF, Van Sommeran S, Kelly JT (2001b) The hunting strategy of white sharks (*Carcharodon carcharias*) near a seal colony. Mar Biol 138:617–636.

Klimley AP, Nelson DR (1984) Diel movement patterns of the scalloped hammerhead shark (*Sphyrna lewini*) in relation to El Bajo Espiritu Santo: a refuging central-position social system. Behav Ecol Sociobiol 15:45–54.

Kohler NE, Turner PA (2001) Shark tagging: A review of conventional methods and studies. Environ Biol Fish 60:191–223.

Kramer DL, Rangeley RW, Chapman LJ (1997) Habitat selection: Patterns of spatial distribution from behavioural decisions. In Godin J-GJ (ed) Behavioural ecology of teleost fishes, Oxford University Press, Oxford, pp 37–80.

Krebs JR, Davies NB (ed) (1997) Behavioural ecology: An evolutionary approach. Blackwell Science, Oxford.

Levin SA (1992) The problem of pattern and scale in ecology. Ecology 73:1943–1967.

Lutcavage ME, Brill RW, Skomal GB, Chase BC, Howey PW (1999) Results of pop-up satellite tagging of spawning size class fish in the Gulf of Maine: do North Atlantic bluefin tuna spawn in the mid-Atlantic? Can J Fish Aquat Sci 56:173–177.

MacArthur RH, Pianka ER (1966) On optimal use of a patchy environment. Amer Nat 100:603–609.

Mackus DL, Boyd CM (1979) Spectral analysis of zooplankton spatial heterogeneity. Science 204:62–64.

Makris NC, Ratilal P, Symonds DT, Jagannathan S, Lee S, Nero RW (2006) Fish population and behavior revealed by instantaneous continental shelf-scale imaging. Science 311:660–663.

Matthews LH (1962) The shark that hibernates. New Scient 280:756–759.

McFarlane GH, Wydoski RS, Prince ED (1990) Historical review of the development of external tags and marks. Am Fish Soc Symp 7:9–29.

Metcalfe JD (2006) Fish population structuring in the North Sea: Understanding processes and mechanisms from studies of the movements of adults. J Fish Biol 69:48–65.

Metcalfe JD, Arnold GP (1997) Tracking fish with electronic tags. Nature 387:665–666.

Metcalfe JD, Fulcher MC, Clarke SR, Challiss MJ, Hetherington S (2009) An archival tag for monitoring key behaviours (feeding and spawning) in fish. In Nielsen JL (ed) Tagging and tracking of marine animals with electronic devices, Reviews: Methods and technologies in fish biology and fisheries 9, Springer, Berlin, pp 1–12.

Morrissey JF, Gruber SH (1993) Home range of juvenile lemon sharks, *Negaprion brevirostris*. Copeia 1993:425–434.

Mucientes GR, Queiroz N, Sousa LL, Tarroso P, Sims DW (2009) Sexual segregation of pelagic sharks and the potential threat from fisheries. Biol Lett 5:156–159.

Myers RA, Baum JK, Shepherd TD, Powers SP, Peterson CH (2007) Cascading effects of the loss of apex predatory sharks from a coastal ocean. Science 315:1846–1850.

Nathan R (2008) An emerging movement ecology paradigm. Proc Natl Acad Sci USA 105:19050–19051.

Nathan R, Getz WM, Revilla E, Holyoak M, Kadmon R, Saltz D, Smouse PE (2008) A movement ecology paradigm for unifying organismal movement research. Proc Natl Acad Sci USA 105:19052–19059.

Neat FC, Wright PJ, Zuur AF, Gibb IM, Gibb FM, Tulett D, Righton DA, Turner RJ (2006) Residency and depth movements of a coastal group of Atlantic cod (*Gadus morhua* L.). Mar Biol 148:643–654.

Nelson DR (1990) Telemetry studies of sharks: A review, with applications in resource management. In Pratt HL, Gruber SH, Taniuchi T (ed) Elasmobranchs as living resources: advances in the biology, ecology, systematics, and status of the fisheries, NOAA Technical Report 90. Washington, DC, NOAA, pp 239–256.

Nielsen A, Bigelow KA, Musyl MK, Sibert JR (2006) Improving light-based geolocation by including sea surface temperature. Fish Oceanogr 15:314–325.

O'Dor RK, Andrade Y, Webber DM, Sauer WHH, Roberts MJ, Smale MJ, Voegeli FM (1998) Applications and performance of radio-acoustic positioning and telemetry (RAPT) systems. Hydrobiologia 372:1–8.

Pade NG, Queiroz N, Humphries NE, Witt MJ, Jones CS, Noble LR, Sims DW (2009) First results from satellite-linked archival tagging of porbeagle shark, *Lamna nasus*: Area fidelity, wider-scale movements and plasticity in diel depth changes. J Exp Mar Biol Ecol 370:64–74.

Pauly D, Christensen V, Guenette S, Pitcher TJ, Sumaila UR, Walters CJ, Watson R, Zeller D (2002) Towards sustainability in world fisheries. Nature 418:689–695.

Pinaud D, Weimerskirch H (2005) Scale-dependent habitat use in a long-ranging central place predator. J Anim Ecol 74:852–863.

Priede IG (1984) A basking shark (*Cetorhinus maximus*) tracked by satellite together with simultaneous remote-sensing. Fish Res 2:201–216.

Priede IG (1992) Wildlife telemetry: An introduction. In Priede IG, Swift SM (ed) Wildlife telemetry: Remote monitoring and tracking of animals, Ellis Horwood, Chichester, UK, pp 3–25.

Priede IG, French J (1991) Tracking of marine animals by satellite. Int J Remote Sensing 12:667–680.

Priede IG, Miller PI (2009) A basking shark (*Cetorhinus maximus*) tracked by satellite together with simultaneous remote sensing. II: New analysis reveals orientation to a thermal front. Fish Res 95:370–372.

Ramos-Fernández G, Mateos JL, Miramontes O, Cocho G, Larralde H, Ayala-Orozco B (2004) Levy walk patterns in the foraging movements of spider monkeys (*Ateles geoffroyi*). Behav Ecol Sociobiol 55:223–230.

Reynolds AM, Smith AD, Reynolds DR, Carreck NL, Osborne JL (2007) Honeybees perform optimal scale-free searching flights when attempting to locate a food source. J Exp Biol 210:3763–3770.

Richardson AJ, Walne AW, John AWG, Jonas TD, Lindley JA, Sims DW, Stevens D, Witt M (2006) Using continuous plankton recorder data. Prog Oceanogr 68:27–74.

Ropert-Coudert Y, Wilson RP (2005) Trends and perspectives in animal-attached remote sensing. Front Ecol Environ 3:437–444.

Royer F, Lutcavage M (2008) Filtering and interpreting location errors in satellite telemetry of marine animals. J Exp Mar Biol Ecol 359:1–10.

Rutz C, Hays GC (2009) New frontiers in biologging science. Biol Lett 5:280–292.

Schaefer KM, Fuller DW (2002) Movements, behavior, and habitat selection of bigeye tuna (*Thunnus obesus*) in the eastern equatorial Pacific, ascertained through archival tags. Fisheries Bulletin 100: 765–788.

Schofield G, Bishop CM, MacLean G, Brown P, Baker M, Katselidis KA, Dimopoulos P, Pantis JD, Hays GC (2007) Novel GPS tracking of sea turtles as a tool for conservation management. J Exp Mar Biol Ecol 347:58–68.

Schuster FL, Levandowsky M (1996) Chemosensory responses of *Acanthamoeba castellanii*: Visual analysis of random movement and responses to chemical signals. J Eukaryotic Microbiol 43:150–158.

Shepard ELC, Ahmed MZ, Southall EJ, Witt MJ, Metcalfe JD, Sims DW (2006) Diel and tidal rhythms in diving behaviour of pelagic sharks identified by signal processing of archival tagging data. Mar Ecol Prog Ser 328:205–213.

Shlesinger MF, Zaslavsky GM, Klafter J (1993) Strange kinetics. Nature 363:31–37.

Sibert J, Hampton J, Kleiber P, Maunder M (2006) Biomass, size, and trophic status of top predators in the Pacific Ocean. Science 314:1773–1776.

Simpfendorfer CA, Heupel MR, Hueter RE (2002) Estimation of short-term centers of activity from an array of omnidirectional hydrophones and its use in studying animal movements. Can J Fish Aquat Sci 59:23–32.

Sims DW (1999) Threshold foraging behaviour of basking sharks on zooplankton: Life on an energetic knife-edge? Proc R Soc B 266:1437–1443.

Sims DW (2003) Tractable models for testing theories about natural strategies: Foraging behaviour and habitat selection of free-ranging sharks. J Fish Biol 63:53–73.

Sims DW (2008) Sieving a living: A review of the biology, ecology and conservation status of the plankton-feeding basking shark *Cetorhinus maximus*. Adv Mar Biol 54:171–220.

Sims DW, Fox AM, Merrett DA (1997) Basking shark occurrence off south-west England in relation to zooplankton abundance. J Fish Biol 51:436–440.

Sims DW, Nash JP, Morritt D (2001) Movements and activity of male and female dogfish in a tidal sea lough: Alternative behavioural strategies and apparent sexual segregation. Mar Biol 139:1165–1175.

Sims DW, Quayle VA (1998) Selective foraging behaviour of basking sharks on zooplankton in a small-scale front. Nature 393:460–464.

Sims DW, Queiroz N, Doyle TK, Houghton JDR, Hays GC (2009a) Satellite tracking of the world's largest bony fish, the ocean sunfish (*Mola mola* L.) in the North East Atlantic. J Exp Mar Biol Ecol 370:127–133.

Sims DW, Queiroz N, Humphries NE, Lima FP, Hays GC (2009b) Long-term GPS tracking of ocean sunfish *Mola mola* offers a new direction in fish monitoring. PLoS ONE 4(10):e7351.

Sims DW, Righton D, Pitchford JW (2007) Minimizing errors in identifying Lévy flight behaviour of organisms. J Anim Ecol 76:222–229.

Sims DW, Southall EJ, Humphries NE, Hays GC, Bradshaw CJA, Pitchford JW, James A, Ahmed MZ, Brierley AS, Hindell MA, Morritt D, Musyl MK, Righton D, Shepard ELC, Wearmouth VJ, Wilson RP, Witt MJ, Metcalfe JD (2008) Scaling laws of marine predator search behaviour. Nature 451:1098–1102.

Sims DW, Southall EJ, Metcalfe JD, Pawson MG (2005a) Basking shark population assessment. Final project report. Global Wildlife Division, Department for Environment, Food and Rural Affairs, London, p 87.

Sims DW, Southall EJ, Richardson AJ, Reid PC, Metcalfe JD (2003) Seasonal movements and behaviour of basking sharks from archival tagging: no evidence of winter hibernation. Mar Ecol Prog Ser 248:187–196.

Sims DW, Southall EJ, Tarling GA, Metcalfe JD (2005b) Habitat-specific normal and reverse diel vertical migration in the plankton-feeding basking shark. J Anim Ecol 74:755–761.

Sims DW, Southward AJ (2006) Dwindling fish numbers already of concern in 1883. Nature 439:660.

Sims DW, Wearmouth VJ, Southall EJ, Hill JM, Moore P, Rawlinson K, Hutchinson N, Budd GC, Righton D, Metcalfe J, Nash JP, Morritt D (2006a) Hunt warm, rest cool: bioenergetic strategy underlying diel vertical migration of a benthic shark. J Anim Ecol 75:176–190.

Sims DW, Witt MJ, Richardson AJ, Southall EJ, Metcalfe JD (2006b) Encounter success of free-ranging marine predator movements across a dynamic prey landscape. Proc R Soc B 273:1195–1201.

Skomal GB, Wood G, Caloyianis N (2004) Archival tagging of a basking shark, *Cetorhinus maximus*, in the western North Atlantic. J Mar Biol Assoc UK 84:795–799.

Smith JNM (1974) The food searching behavior of two European thrushes. II. The adaptiveness of search patterns. Behaviour 49:1–61.

Stephens DW, Krebs JR (1986) Foraging theory. Princeton University Press, Princeton, NJ.

Taillade M (1992) Animal tracking by satellite. In Priede IG, Swift SM (ed) Wildlife telemetry: remote monitoring and tracking of animals, Ellis Horwood, Chichester, UK, pp 149–160.

Tanaka S (1987) Satellite radio tracking of bottlenose dolphins *Tursiops truncatus*. Nippon Suisan Gakkaishi 53:1327–1338.

Teo SLH, Boustany A, Blackwell S, Walli A, Weng KC, Block BA (2004) Validation of geolocation estimates based on light level and sea surface temperature from electronic tags. Mar Ecol Prog Ser 283:81–98.

Turchin P (1998) Quantitative analysis of movement: measuring and modelling population redistribution in animals and plants. Sinauer Associates, Sunderland, MA.

Viswanathan GM, Buldyrev SV, Havlin S, da Luz MGE, Raposo EP, Stanley HE (1999) Optimizing the success of random searches. Nature 401:911–914.

Viswanathan GM, Afanasyev V, Buldyrev SV, Havlin S, da Luz MGE, Raposo EP, Stanley HE (2000) Lévy flights in random searches. Physica A 282:1–12.

Viswanathan GM, Afanasyev V, Buldyrev SV, Murphy EJ, Prince PA, Stanley HE (1996) Lévy flight search patterns of wandering albatrosses. Nature 381:413–415.

Voegeli FA, Smale MJ, Webber DM, Andrade Y, O'Dor RK (2001) Ultrasonic telemetry, tracking and automated monitoring technology for sharks. Environ Biol Fish 60:267–281.

Watkins WA, Daher MA, DiMarzio NA, Samuels A, Wartzok D, Fristrup KM, Gannon DP, Howey PW, Maiefski RR, Spradlin TR (1999) Sperm whale surface activity from tracking by radio and satellite tags. Mar Mamm Sci 15:1158–1180.

Wearmouth VJ, Sims DW (2008) Sexual segregation in marine fish, reptiles, birds and mammals: behaviour patterns, mechanisms and conservation implications. Adv Mar Biol 54:107–170.

Weimerskirch H, Bonadonna F, Bailleul F, Mabille G, Dell'Omo G, Lipp HP (2002) GPS tracking of foraging albatrosses. Science 295:1259–1259.

Welch DW, Rechisky EL, Melnychuk MC, Porter AD, Walters CJ, Clements S, Clemens BJ, McKinley RS, Schreck C (2008) Survival of migrating salmon smolts in large rivers with and without dams. PLoS Biol 6:2101–2108.

Weng KC, Castilho PC, Morrissette JM, Landeira-Fernandez AM, Holts DB, Schallert RJ, Goldman KJ, Block BA (2005) Satellite tagging and cardiac physiology reveal niche expansion in salmon sharks. Science 310:104–106.

West GJ, Stevens JD (2001) Archival tagging of school shark, *Galeorhinus galeus*, in Australia: initial results. Environ Biol Fish 60:283–298.

Wilson RP, Ducamp JJ, Rees WG, Culik BM, Niekamp K (1992) Estimation of location: global coverage using light intensity. In Priede IG, Swift SM (ed) Wildlife telemetry: remote monitoring and tracking of animals, Ellis Horwood, Chichester, UK, pp 131–134.

Wilson RP, Shepard ELC, Liebsch N (2008) Prying into the intimate details of animal lives: use of a daily diary on animals. Endang Species Res 4:123–137.

Wilson RP, Steinfurth A, Ropert-Coudert Y, Kato A, Kurita M (2002) Lip-reading in remote subjects: an attempt to quantify and separate ingestion, breathing and vocalisation in free-living animals using penguins as a model. Mar Biol 140:17–27.

With KA (1994) Using fractal analysis to assess how species perceive landscape structure. Landscape Ecol 9:25–36.

Worm B, Lotze HK, Myers RA (2003) Predator diversity hotspots in the blue ocean. Proc Natl Acad Sci USA 100:9884–9888.

Worton BJ (1987) A review of models of home range for animal movement. Ecol Modelling 38:277–298.

Wray S, Cresswell WJ, Rogers D (1992a) Dirichlet tesselations: a new, non-parametric approach to home range analysis. In Priede IG, Swift SM (ed) Wildlife telemetry: remote monitoring and tracking of animals, Ellis Horwood, Chichester, UK, pp 247–255.

Wray S, Cresswell WJ, White PCL, Harris S (1992b) What, if anything, is a core area? An analysis of the problems of describing internal range configurations. In Priede IG, Swift SM (ed) Wildlife telemetry: remote monitoring and tracking of animals, Ellis Horwood, Chichester, UK, pp 256–271.

9

Sensory Adaptations to the Environment: Electroreceptors as a Case Study

Stephen M. Kajiura, Anthony D. Cornett, and Kara E. Yopak

CONTENTS

9.1 Introduction

Sensory systems are under strong selective pressure to adequately inform organisms about their environment. Since the physical characteristics of the marine environment vary so widely, from the deep sea, to the open ocean, to coastal waters, and reef habitats, it is expected that the sensory systems of organisms in these diverse environments will be differentially optimized to function in these habitats. In this chapter we integrate data on eye size, electroreceptor distribution, and brain morphology to (1) relate differences in sensory structures to the environment and (2) elucidate phylogenetic trends by highlighting patterns between pore distribution and brain morphology within taxa. Because the data are mined from numerous studies across the literature, we are interested only in broad trends rather than a detailed analysis.

The correlation of sensory adaptations to the environment is clearly illustrated by examining the adaptations of the visual system. The visual environment varies dramatically in the ocean, ranging from brightly lit, clear surface waters to the aphotic zone. Deep-sea sharks exhibit a variety of visual system specializations compared to their shallow-water relatives (Hart, Lisney, and Collin 2006). Factors such as eye size, pupil size, lens diameter, visual pigments, and densities of photoreceptor and ganglion cells are modified to accommodate vision in the deep sea (Bozzano 2004; Crescitelli 1991). In addition to these specific adaptations, there are general trends common across taxa. Species that inhabit greater depths tend to be characterized by a greater eye diameter (Figure 9.1). A larger eye will admit a greater number of photons and thus facilitate vision even under low light conditions. These types of ecomorphological adaptations are relatively well documented in visual systems and comparable adaptations should be expected for other senses.

Electroreception is another important, but less well-studied, sensory modality. Electroreception operates in conjunction with vision to inform the shark about the location of prey once the prey item has moved into a blind area close to the head or if vision is occluded, for example, by the nictitating membrane. Thus, electroreceptors serve to position

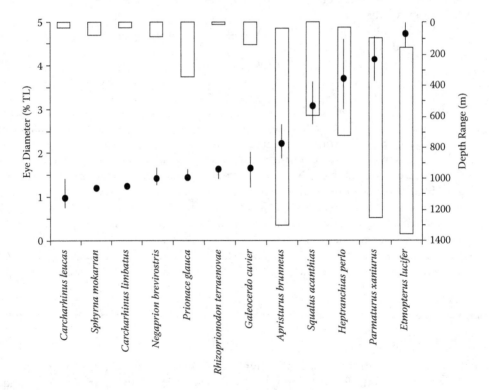

FIGURE 9.1
Eye diameter (filled circles) ± SD and depth range (hollow bars) for selected species. Eye size generally increases with increasing depth.

the shark's mouth over its prey for the final strike. Because the electroreceptors are important for prey detection and localization, they are also subjected to the same selective pressures as vision and likely exhibit similar adaptations to the environment. However, unlike eyes, for which the property of the medium (light) changes dramatically in different marine environments, the electrical properties of seawater are comparatively consistent in a marine environment across a biologically relevant range. Therefore, because electroreceptors continue to function the same in any seawater environment, the electrosensory anatomy would be expected to remain similar. However, the distribution of electroreceptors across the head would likely differ to reflect where the shark primarily searches for prey—ventrally for bottom feeding species and evenly around the head for species that feed in open water.

There is some evidence of morphological differences in electrosensory systems for species that inhabit different environments. In a previous study across 40 skate species, Raschi (1986) determined that the number of electrosensory pores decreased with greater depth. In addition to the decreased numbers of pores at greater depth, the distribution also changed, with greater numbers of pores distributed on the dorsal surface of the disc (Raschi 1986). This provides some evidence of ecomorphological changes in the electrosensory system within the Rajiformes. Like their skate relatives, we would expect shark electroreceptor distributions to reflect adaptations to their environment.

Similarly, there are some detailed data in the literature that have examined brain size and the relative size of the major brain structures across large groups of elasmobranchs from different habitats (Northcutt 1978; Yopak et al. 2007; Yopak and Montgomery 2008; Lisney et al. 2008). In order to compare the morphological data currently available on electroreceptor distributions in sharks to available data on the portions of the brain that receive electroreceptive input and process electrosensory information, brain organization data in a taxonomically and ecologically diverse range of elasmobranchs were compiled from Northcutt (1978), Yopak et al. (2007), and Yopak and Montgomery (2008). These morphometric data were used to ascertain whether adaptations of external morphological characters are similarly reflected in the brains of these species (Figure 9.2).

A confounding factor in the interpretation of all of these trends is phylogenetic relatedness of the species being compared. Do supposed adaptations to the environment merely reflect phylogenetic constraint? For example, if the coastal pelagic species all share a similar electrosensory pore distribution but are also all within the family Carcharhinidae, does the similarity reflect convergence of the electrosensory system to that environment or merely the condition common to members of that family (Figure 9.3)? To tease apart these confounding factors, it is necessary to examine differences in electroreceptors and brain organization among closely related species that inhabit different environments, as well as unrelated taxa that inhabit the same environment. If these characters were merely phylogenetically constrained, we would expect all members of a taxon to share similar pore numbers and distributions regardless of environment (and consequently similar patterns of brain organization). Conversely, if environment does dictate electrosensory adaptations, we would predict similar morphologies among species that inhabit an environment, regardless of their phylogenetic relationship (Figure 9.4).

This chapter is divided into four sections. The first is a general overview of the elasmobranch electrosensory system. In Section 9.2 we examine commonalities in the electrosensory system among species in diverse habitats. This is followed by an examination of the electrosensory system within a few select families to elucidate any phylogenetic patterns. The final section summarizes and discusses the electrosensory adaptations to the environment.

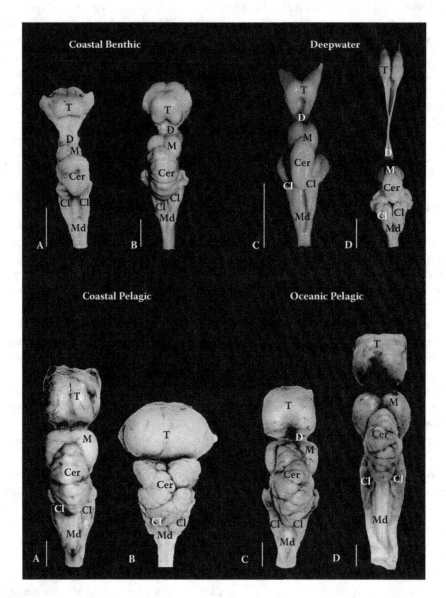

FIGURE 9.2

A representative selection of brains from each of the ecological categories examined in this study in dorsal view, partially adapted from Yopak and Montgomery (2008) and Yopak et al. (2007). (A) *Cephaloscyllium isabellum*, (B) *Mustelus canis*, (C) *Deania calcea*, (D) *Harriotta raleighana*, (E) *Carcharodon carcharias*, (F) *Sphyrna zygaena*, (G) *Alopias vulpinus*, and (H) *Isurus oxyrinchus*. The five main brain structures (T = telencephalon, D = diencephalon, M = mesencephalon, Cer = cerebellum, Cl = cerebellar-like structures, Md = medulla oblongata) are labeled for comparison. Note: In some cases, the diencephalon and/or mesencephalon is obscured in the dorsal view by other brain structures. Scale bar corresponds to 1 cm. (Photos by P. Brown and K. Yopak.)

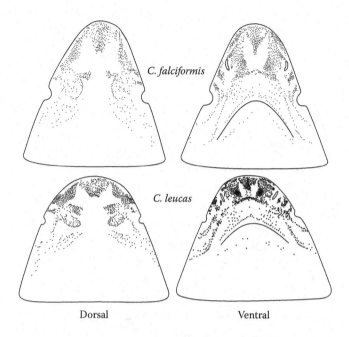

FIGURE 9.3
Pore distribution map for two species in the genus *Carcharhinus* (*C. falciformis*, top, and *C. leucas*, bottom) that share a similar number and distribution of electrosensory pores despite inhabiting dissimilar environments.

9.2 Electrosensory System

9.2.1 Peripheral Anatomy

The elasmobranch electrosensory system is well described and is likely familiar to any student of vertebrate anatomy. An examination of the head of a shark, or the disc of a batoid, reveals well-defined epithelial pores in distinct, bilaterally symmetrical patterns. Gilbert (1967) suggested that the pore distribution patterns were species specific and pore maps have been subsequently generated for several elasmobranch species (Raschi 1984, 1986; Chu and Wen 1979; Kajiura 2001; Tricas 2001; Jordan 2008). The number of pores varies little among individuals within a species and remains constant throughout ontogeny (Kajiura 2001), so quantifying the number of pores for an individual from any developmental stage from juvenile to adult yields a representative count for the species as a whole. Because each pore leads to a single electrosensory ampulla, counting the number of pores provides an exact count of the number of ampullae without necessitating dissection.

Although the pores are the obvious external manifestation of the electrosensory system, the actual electroreceptors are located subdermally. Up to hundreds of individual pores converge in discrete clusters on either side of the head. There are up to five distinct clusters identified in the head of sharks and typically fewer in the batoids (Chu and Wen 1979). Each epithelial pore is continuous with the external environment and leads to a single subdermal canal that can range from several millimeters to many centimeters in length. The cells that comprise the canal walls are connected by tight junctions, which create a high electrical resistance between the lumen of the canal and the internal milieu of the head (New and Tricas 2001). Each canal terminates in a swelling (the ampulla proper) that is comprised of numerous alveoli. The length of the canal and the ampullary lumen contain

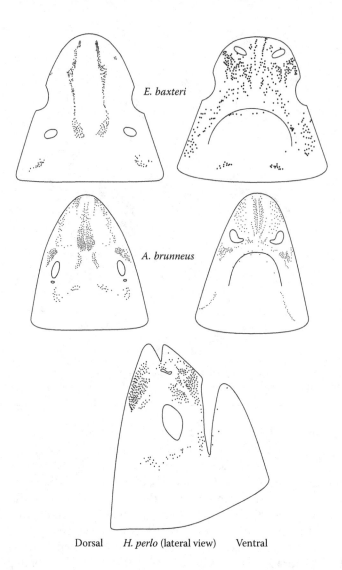

FIGURE 9.4
Pore distribution maps for three deepwater species from different orders that share a similar number and proportion of dorsal:ventral electrosensory pores despite being taxonomically unrelated. Top to bottom: *Etmopterus baxteri* (Squaliformes), *Apristurus brunneus* (Carcharhiniformes), *Heptranchias perlo* (Hexanchiformes).

an ion-rich, glycoprotein gel with an electrical conductivity approximately equal to that of seawater (Waltman 1966). The walls of the alveoli contain the electrosensory epithelium (Waltman 1966). The receptor cells in the sensory epithelium are modified hair cells that are stimulated by a negative charge at the apical membrane (Bullock et al. 1982). Primary afferent neurons at the basal surface of the receptor cells converge and transmit information to the brain via the dorsal branch of the anterior lateral line nerve (ALLN).

9.2.2 Central Anatomy and Pathways

To appreciate the function of the peripheral electrosensory system, it is important to place it in the context of other sensory systems, and the parts of the brain that moderate them.

We offer here a brief overview of basic shark neuroanatomy and sensory systems integration in the elasmobranch brain, followed by a detailed description of the electrosensory brain and its afferents.

9.2.2.1 The Basic Brain

The three main sensory systems of the vertebrate head (olfactory, visual, and octavolateral) are distinctly associated with different brain areas (Gans and Northcutt 1983; Northcutt and Gans 1983; Roberts 1988; Hofmann 1999), and it is the simultaneous integration of these sensory inputs that determines complex behaviors (Bres 1993). In chondrichthyans, the neural structures that receive input from these systems are termed the forebrain (comprised of the telencephalon and diencephalon), the midbrain (also termed the mesencephalon), and hindbrain (which includes the cerebellum and medulla oblongata; Figure 9.2). It is expected that these major brain areas will evolve, to a degree, independently in association with these senses and will be reflected in adapted behaviors or enhanced cognitive capabilities (Barton, Purvis, and Harvey 1995; Barton and Harvey 2000). However, development is also constrained by an animal's evolutionary ancestry (Harvey and Krebs 1990). Recent studies on brain organization and relative development of major neural structures in elasmobranchs have demonstrated evidence that chondrichthyan brains may have evolved as a consequence of behavior, habitat, and lifestyle in addition to phylogeny (Yopak et al. 2007; Yopak and Montgomery 2008; Lisney et al. 2008; Yopak and Frank 2009).

Neural development can often reflect morphological adaptations and sensory specializations (Morita and Finger 1985; Heiligenberg 1987; Kanwal and Caprio 1987; Finger 1988; Kotrschal et al. 1990; Meek and Nieuwenhuys 1997). Previous research on a variety of vertebrate groups, including teleost fishes, birds, and mammals, has shown that similar patterns of brain organization and brain size occur in species that share certain lifestyle characteristics, including diet, habitat complexity, social organization, and/or cognitive capability (e.g., Kotrschal and Palzenberger 1992; Huber et al. 1997; Kotrschal, van Staaden, and Huber 1998; Clark, Mitra, and Wang 2001; de Winter and Oxnard 2001; Iwaniuk and Hurd 2005). Detailed and descriptive illustrations of brain morphology from a number of chondrichthyan species have provided evidence of substantial interspecific variation of component parts (e.g., Garman 1913; Northcutt 1977, 1978; Kruska 1988; Narenda 1991; Smeets 1998), and more recent work has begun to quantify the phylogenetic component of variation in brain size in chondrichthyans (Yopak et al. 2007; Yopak and Montgomery 2008; Lisney et al. 2008). This work shows that brain organization and the relative development of major brain structures (telencephalon, diencephalon, mesencephalon, cerebellum, and medulla oblongata) reflect an animal's ecology, even in phylogenetically unrelated species that share certain lifestyle characteristics (Yopak and Montgomery 2008).

9.2.2.2 The Electrosensory Brain and Its Afferents

The dorsolateral wall of the medulla has distinctive cerebellar-like lobes, which receive electrosensory and lateral line input (Hofmann 1999). The receptor cells of the ampullae of Lorenzini are innervated by five afferent nerve fibers, which project to the dorsal octavolateral nucleus (DON) of the cerebellar-like structures (in addition to other octavolateral afferents) (Montgomery et al. 1995; Bell et al. 1997; Bell 2002) via the dorsal branch of the anterior lateral line nerve (Bell and Maler 2005), similar to the electrosensory lateral line lobe in some teleosts (Bell and Szabo 1986a, 1986b). Neural pathways of the cerebellar-like hindbrain have been well studied and documented (Bodznick and Boord 1986; Schmidt and Bodznick 1987; Bell, Grant, and Serrier 1992; Conley and Bodznick 1994; Bastian 1995; Hjelmstadt, Parks, and Bodznick 1996; Hueter et al. 2004) and will only be reviewed in basic detail here.

The cerebellar-like structures include the DON and the medial octavolateral nucleus (MON) in elasmobranchs, which lies just ventral to the DON on the lateral wall of the medulla. The DON, found in some aquatic amphibians and non-neopterygian fish, including chondrichthyans (Bell and Maler 2005; Bullock et al. 1983; Montgomery et al. 1995), is the site for central termination of primary electrosensory afferents in elasmobranchs. The majority of research on the cerebellar-like structures has been done on the little skate, *Raja erinacea*, where it has been shown that the multipolar principal cells, referred to as ascending efferent neurons (AENs), project to the optic tectum and a portion of the mesencephalic midbrain called the lateral mesencephalic nucleus (LMN) (Bodznick and Boord 1986; Schmidt and Bodznick 1987) in addition to two other midbrain nuclei (Smeets 1982; Bodznick and Boord 1986). These pathways to the mesencephalon are similar to those for other octavolateral afferents (Coombs and Montgomery 2005).

In the elasmobranch brain, electrosensory afferents terminate somatotopically, forming a spatial map of the electroreceptors found subdermally (Bodznick and Schmidt 1984; Bell 2002), likely providing for efficient target location. In batoids, electroreceptors clustered in caudal regions in the skin have been shown to terminate in the DON dorsally, while rostral electroreceptors terminate ventrally (Bodznick and Boord 1986). Similarly in the optic tectum, electrosensory information is aligned spatially with input from the visual field (Bodznick 1991), while electrosensory information outside of the visual field, on the animal's ventral surface, is mapped in the LMN. Interestingly, the lampreys do not show this type of spatial correspondence between electroreceptor positioning on the body and their organization within the brain (Ronan 1986), suggesting that this characteristic has either originated in early cartilaginous fishes or has been lost in specialized lampreys.

Correlations between the medullar development and ecological parameters have been documented in chondrichthyans (Yopak et al. 2007; Yopak and Montgomery 2008; Lisney et al. 2008). Correspondingly, spatial arrangement of electrosensory pores has been linked to ecology and, subsequently, feeding patterns in batoids (Raschi 1986; Raschi and Adams 1988; Raschi, Aadlond, and Keithar 2001; Tricas 2001) and sharks (Kajiura 2001; Tricas 2001), and pore densities have been found to augment ampullae sensitivity (Raschi 1986; Raschi and Adams 1988). This has led to the hypothesis that the spatial array of electroreceptors (and corresponding sensitivity) is likely to also correlate with the development of central anatomy and pathways across this clade, which will be explored in this chapter.

9.3 Adaptations to the Environment

By integrating electrosensory pore distribution and brain morphology data, we attempt to discern general trends across taxa that inhabit ecologically diverse environments. To facilitate this comparison, we categorize each species into one of four habitats. The coastal environment encompasses all waters from the shoreline to 200 m, a depth that generally coincides with the edge of the continental shelf. Coastal benthic species include those sharks that reside on (benthic) or cruise just off (benthopelagic) the bottom. While some of these species do actively swim into the water column, they maintain a strong association with the benthic environment. Deepwater species are those most commonly found at depths greater than 200 m. Coastal pelagic species are those species that actively swim in the water column over continental shelf waters. They may feed on benthic prey, but are not constrained to the seafloor. Oceanic pelagic sharks inhabit the water column over depths greater than 200 m. Species that inhabit pelagic waters may live from the surface to the upper mesopelagic depths (500 to 600 m), but remain in the water column and well off the seafloor.

Previous studies have found strong correlations between brain structure size and ecological factors (e.g., Yopak et al. 2007; Yopak and Montgomery 2008; Lisney et al. 2008; Yopak and Frank 2009). To visually present these patterns, the relative size of each brain area was compared among species using a weighted factor (θ) analysis (Wagner 2001a, 2001b), which has been commonly used to explore variation in brain organization in chondrichthyans (Yopak et al. 2007; Lisney et al. 2008). A brain structure was considered relatively large if $\theta > 0.1$, relatively small if $\theta < -0.1$, and relatively average-sized if $0.1 < \theta < -0.1$. Although the extent to which brain structure size correlates to the specialization of particular sensory modalities is unknown, it is assumed that there are correlations between specific brain areas and functions or behaviors (Kotrschal and Palzenberger 1992).

9.3.1 Coastal Benthic

The coastal benthic environment boasts a large phylogenetic diversity of sharks, with some higher taxonomic groups located exclusively in this environment. This broad category encompasses habitats ranging from turbid estuaries to pristine coral reefs. Organisms in this environment must contend with variation in light levels, turbidity caused by wind and water disturbance, often complex bottom topography, salinity fluctuations from freshwater inputs, and algal blooms. Visually locating prey in this environment is potentially difficult.

The number of electosensory pores varies widely among coastal benthic sharks, with some species possessing three times more pores than others. Despite the wide variation in pore number, the coastal benthic species consistently possess a larger number of pores on the ventral surface. Since most coastal benthic sharks consume benthically associated prey (Compagno 1984a,b; Compagno, Dando, and Fowler 2005), the greater number of pores on the ventral surface reflects an adaptation for prey detection. Although most prey items of coastal benthic sharks are fairly immobile, two coastal benthic species included here consume fishes as their primary prey: the sand tiger shark, *Carcharias taurus* (family Odontaspididae, order Lamniformes) and the lemon shark, *Negaprion brevirostris* (family Carcharhinidae, order Carcharhiniformes). Both species possess similar numbers of electroreceptors but *C. taurus* retains a large ventral pore distribution like other coastal benthic species, whereas *N. brevirostris* possesses an even dorsal:ventral pore distribution, similar to coastal pelagic species. In addition, *N. brevirostris* exhibits the greatest total pore number of any coastal benthic species.

9.3.1.1 Coastal Benthic Nervous System

The relative development of the medulla and telencephalon is the most widespread in the coastal benthic sharks compared to any other ecological group. Interestingly, the coastal benthic sharks with the largest telencephalons, such as *Nebrius ferrugineus* ($\theta = 0.50$) and *Hemiscyllium ocellatum* ($\theta = 0.34$) (Figure 9.5) are species that dwell on or near coral reefs. The relative size of the telencephalon has been previously correlated with an increase in habitat complexity (Yopak et al. 2007), a situation found in other vertebrate groups (Riddell and Corl 1977; Barton, Purvis, and Harvey 1995; Huber et al. 1997; Striedter 2005). Aside

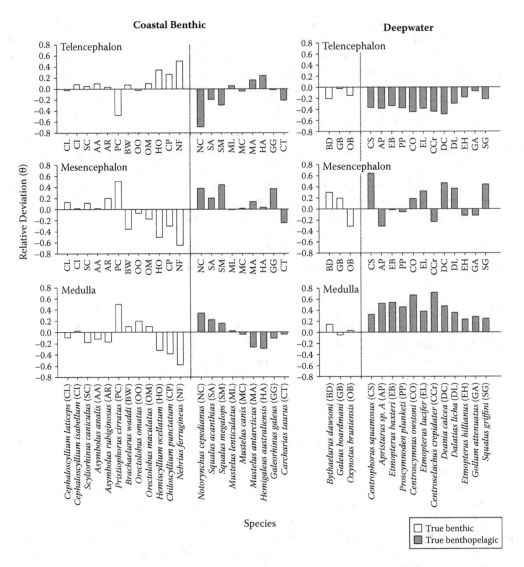

FIGURE 9.5
Weighted factors (θ) for the telencephalon, mesencephalon, and medulla for 21 coastal benthic elasmobranchs and 15 deepwater elasmobranchs, showing the deviation from the average relative volume for each brain structure (Wagner 2001a, 2001b), using data compiled from Northcutt (1978), Yopak et al. (2007), and Yopak and Montgomery (2008). A brain structure was considered relatively large if θ > 0.1, relatively small if θ < –0.1, and relatively average-sized if 0.1 < θ < –0.1. Note: Weighted factors are based on relative structure size data on a total of 61 species (including 7 holocephalans).

from those coastal benthic species that are reef associated, the majority of true benthics in this ecological category do not show high levels of development of the telencephalon (Figure 9.5), or indeed other brain areas (Yopak et al. 2007). Research on cyprinids (Schiemer 1988) has indicated that a brain with no apparent structural enlargement may be correlated with ecological flexibility, lending itself to a more adaptable diet and primary habitat (Brabrand 1985; Lammens, Geursen, and McGillavry 1987). This characteristic has been suggested as a mechanism for true benthic elasmobranchs, where more basic neural development may indicate a more generalized lifestyle (Yopak et al. 2007).

The mesencephalon, which is associated with both visual input and the integrating of electrosensory and mechanosensory information (Bres 1993; Tricas et al. 1997), is average to large in over 70% of the coastal benthic species examined here, ranging from θ = −0.003 in *Mustelus lenticulatus* to θ = 0.51 in *Pristiophorus cirratus* (Figure 9.5). Batoids have been documented to have similar patterns of brain organization, where coastal benthic Torpediniformes and Rajiformes show both enlargement of the mesencephalon and reduction of the telencephalon (Lisney et al. 2008). In contrast, the mesencephalon is noticeably reduced in seven coastal benthic shark species (Figure 9.5; Yopak et al. 2007), all of which are associated with coral reefs (Compagno 1984a, 1984b, 1998, 2001; Compagno and Niem 1998a, 1998b; Musick et al. 2004). The true benthic reef-associated sharks, including *Nebrius*, *Hemiscyllium*, and *Brachaelurus*, and the one true benthopelagic reef-associated shark, *Carcharias taurus*, have reduction of the mesencephalon, in contrast to other coastal benthic species (Figure 9.5). A reduced mesencephalon may reflect eye size differences between non- and reef-associated coastal benthics (Figure 9.6), which is likely associated with visual acuity (Lisney and Collin 2007), though further data are required.

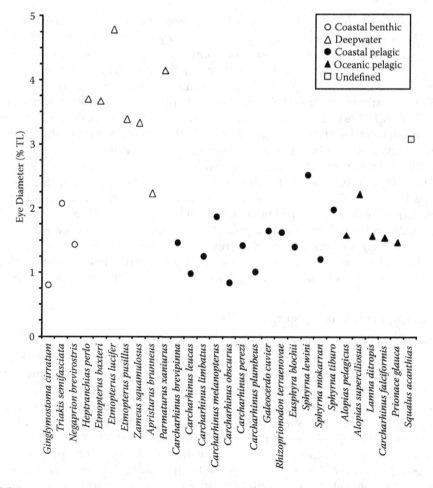

FIGURE 9.6
Eye diameter for each species by environmental classification. The deepwater species generally possess larger eyes than sharks in the other habitats.

The mesencephalon has been associated with behavioral responses to novel visual stimuli (Bodznick 1991), which could be more critical for stationary sharks, and the requirements for responding to rapid changes across their visual field may be influencing the relative hypertrophy of this structure in these species. However, development of the mesencephalon may not wholly reflect visual acuity, as it is unlikely that many of these species depend heavily on vision in their turbid environment. As the mesencephalon is responsible for a variety of functions, enlargement of this structure could result from the need to integrate information from the octavolateralis senses. However, one would expect similarly to see relative enlargement of the medulla and cerebellar-like structures, which receive direct input from the electroreceptors and lateral line organs. Although enlargement of the medulla is seen in some coastal benthic species, such as *Notorynchus cepedianus* ($\theta = 0.34$) and *Orectolobus ornatus* ($\theta = 0.18$) (Figure 9.5), when the cerebellar-like structures are examined, *Pristiophorus cirratus* is the only coastal species to show hypertrophy of this structure ($\theta = 0.50$) (Figure 9.7).

9.3.2 Deep Water

The deep-sea environment is characterized by not only the obvious dramatic reduction in light but also a paucity of potential prey. Although they possess the largest eye size, deepwater sharks have the lowest electrosensory pore number of the four habitats. They also exhibit the widest range in both total pore number (Figure 9.8) and ratio of dorsal:ventral pore distributions (Figure 9.9). Skate species from the same depth range (500 to 1000 m) also exhibit variation in pore distribution, with the proportion of dorsal pores increasing with depth (Raschi 1984). The variation in pore number and distribution may reflect interspecific variation in feeding strategies. For example, some deepwater sharks may hunt off of the bottom, feeding on more active organisms (deepwater teleosts, squid, shrimp), whereas others are benthic invertebrate specialists or ambush predators.

Raschi (1984) hypothesized that if deepwater elasmobranchs swim off the ocean floor, they could maximize their electroreceptive search area. However, this would necessitate the detection of weaker bioelectric fields since field strength decreases dramatically with distance (Kalmijn 1982). Although not quantified, the epithelial pores in deepwater sharks are very large. A large diameter would decrease the electrical impedance along the length of the tubule, which could contribute to increased sensitivity. This adaptation may allow small deepwater species to achieve the electrosensory sensitivity of larger sharks.

9.3.2.1 Deepwater Nervous System

Given the considerable variation in morphological specializations evident in deep-sea chondrichthyans, including their sensory systems, it is not surprising that there are adaptations in neural development as well. Despite their phylogenetic dissimilarity, deepwater species generally have an average to reduced telencephalon ($-0.48 < \theta < -0.2$), an average to large medulla ($-0.04 < \theta < -0.72$) (Yopak and Montgomery 2008), and varied mesencephalic development. Deep-sea sharks show marked hypertrophy of cerebellar-like structures in comparison to other elasmobranchs, with relative deviations as high as $\theta = 0.59$ (*Apristurus* sp. A) and $\theta = 0.55$ (*Centroselachus crepidater*). Previously, the disproportionate size of the dorsal octavolateral nucleus (DON) was noted in the benthic skate, *Raja erinacea* (Bodznick and Schmidt 1984) and the relative size of the cerebellar-like structures was found to be enlarged in deeper dwelling members of Rajidae (Yopak 2007). The deepwater sharks that show reduction of the cerebellar-like structures (Figure 9.7), *Bythaelurus dawsoni* and *Oxynotus bruniensis*,

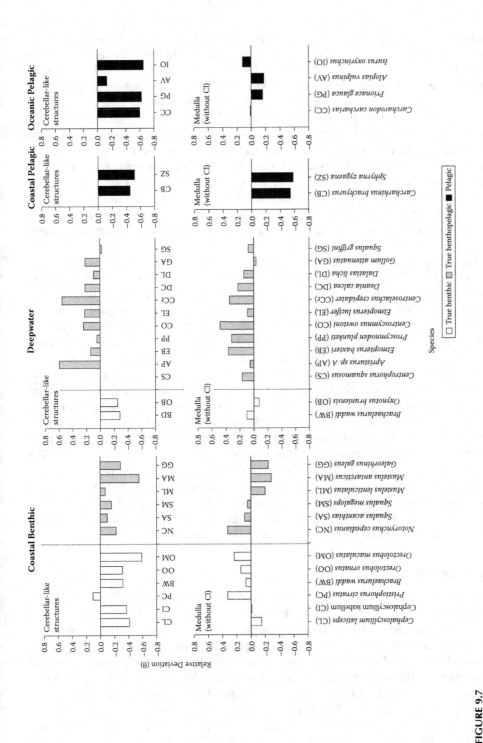

FIGURE 9.7

Weighted factors (θ) for the cerebellar-like structures separate from the rest of the medulla for 12 coastal benthic, 13 deepwater, 2 coastal pelagic, and 4 oceanic pelagic elasmobranchs, showing the average relative volume for each brain structure (Wagner 2001a, 2001b), using data compiled from Northcutt (1978), Yopak et al. (2007), and Yopak and Montgomery (2008). Note: Weighted factors are based on relative structure size data on a total of 37 species, where it was possible to digitally delineate the cerebellar-like tissue from the rest of the medulla.

are the only two true benthics in this ecological category (Yopak and Montgomery 2008), and seldom cruise off the substrate (Compagno 1984b), suggesting that enlargement of this brain area may be related to prey capture for more active deepwater sharks.

Structural enlargement of sensory brain areas has been shown in mesopelagic and abyssal teleosts (Kotrschal, van Staaden, and Huber 1998; Wagner 2001a) as well as blind cavefish (Poulson 1963) and it has been suggested that the brain development of deep-sea fishes reflects the enhancement of nonvisual senses in the deep-sea environment (Ublein et al. 1996; Merrett and Haedrich 1997; Yopak et al. 2007; Yopak and Montgomery 2008). In fact, the greatest development of the brain area associated with the octavolateralis senses is in chondrichthyans that dwell in the mid to lower bathyal zones (500 to 2000 m; Yopak and Montgomery 2008). This type of neural development of the primary acoustico-lateralis areas in bathyal species has been similarly seen in teleost fishes (Fine, Horn, and Cox 1987; Eastman and Lannoo 1995; Kotrschal et al. 1998), as well as the other closely related cartilaginous group, the holocephalans (Yopak et al. 2007; Yopak and Montgomery 2008). This cerebellar-like hypertrophy may be related to the distribution of electrosensory pores extending down the elongated rostrum in some of these animals (Yopak and Montgomery 2008), as this elongated array is then arranged somatotopically in the dorsal nucleus (Bodznick and Boord 1986; Bell 2002).

The relative enlargement of the medulla and, more specifically, the cerebellar-like structures, appears to show an inversely proportional relationship to electrosensory pore number. Deep-sea species possess the lowest number of electrosensory pores (Figure 9.8), and yet the portion of the brain that receives electrosensory input is the most highly developed (Figures 9.5 and 9.7). It is possible that the relative size of the cerebellar-like structures reflects the diameter of the electrosensory pores rather than density, which has resulted in a proposed increased sensitivity. Further, should the foraging strategies of deep-sea sharks, combined with a minimized number of ventrally located pores (Raschi 1984), necessitate the detection of weaker bioelectric fields, the detection of those fields within corrupting, self-generated electroreceptive inputs created through ventilation and body movement (Montgomery 1984; New and Bodznick 1990; Montgomery and Bodznick 1994, 1999; Bodznick, Montgomery, and Carey 1999) becomes more challenging. The cancellation of sensory reafference takes place in the molecular layer of the DON of the cerebellar-like structures (Bodznick, Montgomery, and Carey 1999) and the hypertrophied cerebellar-like structures in the deep-sea sharks may reflect a greater reliance on sensory adaptive filtering in the hindbrain. In the bathyal environment, it becomes more critical for survival for these species to remain sensitive to minute biologically useful signals, particularly at depths where both prey and conspecifics are scarce (Marshall 1979; Herring 2000).

9.3.3 Coastal Pelagic

Sharks in the coastal pelagic environment inhabit the open water near coastlines, typically over the continental shelf at depths <200 m. Because of the strong terrestrial influence, the coastal pelagic environment is characterized by perpetual changes in water conditions. Algal blooms, wind- and wave-driven sediment disturbance, and terrestrial runoff can sometimes result in reduced visibility. This can make prey location challenging, a task compounded by the fast movements of the relatively small fishes that are the dominant prey items of sharks in this environment (Compagno 1984a,b, 2001; Compagno, Dando, and Fowler 2005).

Coastal pelagic species possess the greatest pore number compared to sharks of other environments (Figure 9.8). Their electroreceptive pore numbers are generally consistent within a narrow range of values (Figure 9.8, Table 9.1) and the pores are distributed

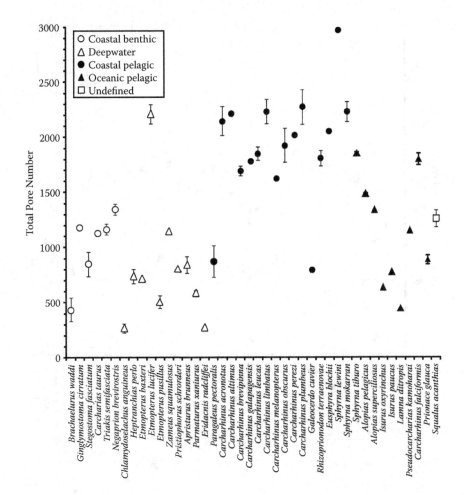

FIGURE 9.8

Total pore number for each species grouped by environmental classification. The coastal pelagic species generally possess more pores than the species in the other habitats.

approximately equally over the dorsal and ventral surfaces of the head (Figure 9.9). Large numbers of evenly distributed pores would enable a shark to accurately resolve the location of its prey, anywhere around its head, even in occasionally turbid coastal waters.

All coastal pelagic species examined are members of the order Carcharhiniformes, with all species (save one) contained in two families (Carcharhinidae and Sphyrnidae) (Table 9.1). The limited phylogenetic scope from one of the four environments complicates direct comparisons and precludes drawing any definitive conclusions. However, carcharhinids and sphyrnids are the dominate sharks in the world's coastal pelagic environments and the species studied accurately represent the taxonomic diversity within this habitat (Compagno, Dando, and Fowler 2005).

9.3.3.1 Coastal Pelagic Nervous System

Coastal pelagic shark species consistently have an enlarged telencephalon and a reduced mesencephalon and medulla (Figure 9.10). Development of the telencephalon in these sharks accounts for over 50% of the brain in these species ($0.29 \leq \theta \leq 0.71$) (Yopak et al. 2007). Species

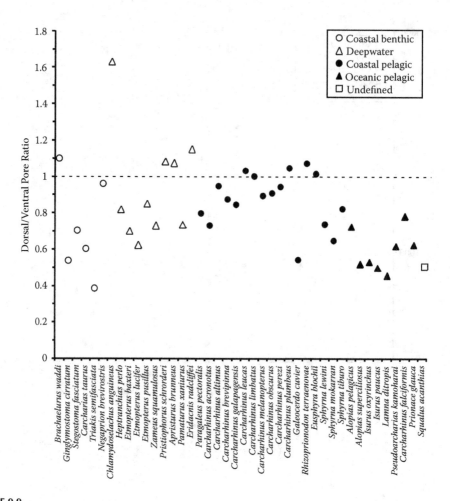

FIGURE 9.9

Dorsal:ventral pore distribution ratio for each species by environmental classification. Most species possess a greater number of pores on the ventral surface of the head (below the dotted line), although some coastal pelagic and deepwater species demonstrate a more even distribution.

in this ecological group similarly have some of the largest brains relative to their body size of all shark species (Yopak et al. 2007; Yopak and Montgomery 2008), a trait that, along with an enlarged telencephalon, has been correlated with "social intelligence" found in species that live in complex habitats (Striedter 2005; Pollen et al. 2007), and is likely linked to the complex intra- and interspecific interactions that occur in species that dwell in these spatially complex environments (Kotrschal, van Staaden, and Huber 1998; Yopak et al. 2007). This pattern has been similarly seen in reef-associated batoids, where the relative size of the telencephalon correlates significantly with lifestyle (Lisney et al. 2008). An increase in brain size and/or telencephalon size has been correlated with spatial complexity and sociality in several other vertebrate groups, including teleosts and chondrichthyans (Bauchot et al. 1977; Northcutt 1977, 1978; Kotrschal, van Staaden, and Huber 1998; Yopak et al. 2007) and birds and mammals (Budeau and Verts 1986; Striedter 2005). Although development of the telencephalon is not as pronounced as in the coastal pelagics, the coastal benthic sharks that show relative enlargement of this structure, such as *Nebrius ferrugineus* and *Chiloscyllium punctatum*, are similarly associated with coral reefs (Yopak et al. 2007; Figure 9.5).

TABLE 9.1

Electrosensory Pore Numbers and Relative Brain Volumes for Various Shark Species

Order[a]	Habitat	D	D%	V	V%	T	T SD	Eye Diameter (% TL)	Common Depth Range (m)	Relative Brain Size					Source
										Tel Vol %	Di Vol %	Mes Vol %	Cer Vol %	Med Vol %	
Hexanchiformes															
Chlamydoselachidae															
Chlamydoselachus anguineus	DW	166	62.3	102	37.7	267	42.4	–	50–1500	–	–	–	–	–	E
Heptranchidae															
Notorhynchus cepedianus	CB	–	–	–	–	–	–	–	1–136	31.31	5.44	17.20	13.78	32.27	C
Heptranchias perlo	DW	333	45.1	406	54.9	739	64.5	3.70	27–720	–	–	–	–	–	E
Squaliformes															
Squalidae															
Squalus acanthias	–	425	33.7	837	66.3	1263	76.4	3.09	0–600	31.26	6.65	15.02	17.66	29.43	C, E
Squalus megalops	CB	–	–	–	–	–	–	–	0–732	27.33	7.17	18.06	19.44	27.99	C
Squalus griffini	DW	–	–	–	–	–	–	–	–	30.53	5.92	18.05	15.36	30.14	C
Centrophoridae															
Centrophorus squamosus	DW	–	–	–	–	–	–	–	230–2400	24.69	6.84	20.53	16.09	31.86	D
Deania calcea	DW	–	–	–	–	–	–	–	400–900	20.06	6.61	18.30	19.67	35.36	D
Dalatiidae															
Dalatias licha	DW	–	–	–	–	–	–	–	200–1800	27.48	8.74	17.14	14.16	32.49	D
Etmopteridae															
Etmopterus baxteri	DW	295	41.2	421	58.8	716	–	3.67	250–1500	25.85	8.19	12.32	16.63	37.00	D, E
Etmopterus hillianus	DW	–	–	–	–	–	–	–	311–695	31.87	9.89	10.99	17.58	29.67	A
Etmopterus lucifer	DW	848	38.4	1362	61.6	2210	87.0	4.78	158–1357	23.92	7.94	16.56	18.45	33.14	D, E
Etmopterus pusillus	DW	232	46.0	273	54.1	504	59.4	3.39	274–1000	–	–	–	–	–	E

Continued

TABLE 9.1 (Continued)

Electrosensory Pore Numbers and Relative Brain Volumes for Various Shark Species

Order[a]	Habitat	D	D%	V	V%	T	T SD	Eye Diameter (% TL)	Common Depth Range (m)	Relative Brain Size					Source
										Tel Vol %	Di Vol %	Mes Vol %	Cer Vol %	Med Vol %	
Oxynotidae															
Oxynotus bruniensis	DW	–	–	–	–	–	–	–	350–650	32.93	9.50	8.58	24.16	24.83	D
Somniosidae															
Centroscymnus owstoni	DW	–	–	–	–	–	–	–	600–1459	21.65	3.32	14.82	19.98	40.22	D
Centroselachus crepidater	DW	–	–	–	–	–	–	–	500–2080	21.81	7.35	9.67	19.85	41.32	D
Proscymnodon plunketi	DW	–	–	–	–	–	–	–	550–732	24.24	5.91	11.85	22.89	35.11	D
Zameus squamulosus	DW	483	42.2	662	57.8	1145	–	3.33	550–1450	–	–	–	–	–	E
Pristiophoriformes															
Pristiophoridae															
Pristiophorus cirratus	CB	–	–	–	–	–	–	–	1–311	20.09	5.25	18.83	19.81	36.01	C
Pristiophorus schroederi	DW	418	52.0	386	48.0	804	–	–	438–952	–	–	–	–	–	E
Orectolobiformes															
Brachaeluridae															
Brachaelurus waddi	CB	229	53.2	208	46.8	437	107.5	–	0–140	41.67	4.39	8.01	19.25	26.67	C, E
Ginglymostomatidae															
Ginglymostoma cirratum	CB	412	34.9	768	65.1	1180	–	0.80	0–12	–	–	–	–	–	E
Nebrius ferrugineus	CB	–	–	–	–	–	–	–	5–30	58.30	5.24	4.39	21.92	10.16	C
Hemiscylliidae															
Chiloscyllium punctatum	CB	–	–	–	–	–	–	–	0–85	49.07	5.56	8.80	21.76	14.81	C
Hemiscyllium ocellatum	CB	–	–	–	–	–	–	–	1–10	52.04	4.59	6.12	20.92	16.33	C
Orectolobidae															
Orectolobus maculatus	CB	–	–	–	–	–	–	–	0–110	42.40	6.15	10.34	14.60	26.50	C
Orectolobus ornatus	CB	–	–	–	–	–	–	–	0–100	37.90	4.86	11.64	16.89	28.82	C

Stegostomatidae															
Stegostoma fasciatum	CB	350	41.1	498	58.9	848	112.4	–	0–62	–	11.42	–	–	–	E
Lamniformes															
Odontaspididae															
Carcharias taurus	CB	425	37.6	705	62.4	1130	–	–	15–25	30.57	11.42	9.47	25.28	23.26	C
Alopiidae															
Alopias pelagicus	OP	623	42.1	860	58.0	1483	11.3	1.59	0–152	–	–	–	–	–	E
Alopias superciliosus	OP	457	34.2	881	65.8	1338	–	2.22	100–500	27.19	3.15	16.19	32.09	21.39	C, E
Alopias vulpinus	OP	–	–	–	–	–	–	–	0–366	26.57	2.75	15.52	30.79	24.38	C
Lamnidae															
Carcharodon carcharias	OP	–	–	–	–	–	–	–	0–1300	38.86	5.57	14.28	17.66	23.63	B, C
Isurus oxyrinchus	OP	220	34.6	415	65.4	635	–	–	0–500	37.70	3.35	18.18	17.03	23.74	C, E
Isurus paucus	OP	258	33.3	516	66.7	774	–	–	–	–	–	–	–	–	E
Lamna ditropis	OP	139	31.3	305	68.7	444	–	1.57	0–225	–	–	–	–	–	E
Pseudocarcharhiidae															
Pseudocarcharias kamoharai	OP	440	38.2	712	61.8	1152	–	–	0–590	33.13	6.88	20.42	16.04	23.54	C, E
Carcharhiniformes															
Syliorhinidae															
Apristurus sp. A	DW	–	–	–	–	–	–	–	940–1290	24.03	6.55	8.59	24.20	36.63	C
Apristurus brunneus	DW	436	51.6	406	48.4	842	75.1	2.23	33–1298	–	–	–	–	–	E
Asymbolus analis	CB	–	–	–	–	–	–	–	40–175	42.55	6.38	12.77	17.02	21.28	C
Asymbolus rubiginosus	CB	–	–	–	–	–	–	–	25–540	40.00	6.25	15.00	18.75	20.00	C
Bythaelurus dawsoni	DW	–	–	–	–	–	–	–	371–420	30.91	4.98	16.21	20.29	27.61	D
Cephaloscyllium isabellum	CB	–	–	–	–	–	–	–	0–400	41.93	5.91	12.72	14.81	24.64	C
Cephaloscyllium laticeps	CB	–	–	–	–	–	–	–	0–60	46.62	5.29	14.12	12.13	21.83	C
Galeus boardmani	DW	–	–	–	–	–	–	–	150–640	37.93	6.90	14.94	17.24	22.99	C
Parmaturus xaniurus	DW	249	42.4	338	57.6	587	20.2	4.14	91–1251	–	–	–	–	–	E
Scyliorhinus retifer	CB	–	–	–	–	–	–	–	73–754	40.70	8.14	13.95	17.44	19.77	A

Continued

TABLE 9.1 (Continued)

Electrosensory Pore Numbers and Relative Brain Volumes for Various Shark Species

Order[a]	Habitat	D	D%	V	V%	T	T SD	Eye Diameter (% TL)	Common Depth Range (m)	Tel Vol %	Di Vol %	Mes Vol %	Cer Vol %	Med Vol %	Source
Proscyllidae															
Eridacnis radcliffei	DW	146	53.5	127	46.5	273	–	–	71-766	–	–	–	–	–	E
Pseudotriakidae															
Gollum attenuatus	DW	–	–	–	–	–	–	–	300-600	36.06	4.79	11.05	17.30	30.80	D
Triakidae															
Galeorhinus galeus	CB	–	–	–	–	–	–	–	2-471	38.19	6.07	17.14	17.19	21.41	C
Mustelus antarcticus	CB	–	–	–	–	–	–	–	0-80	44.99	5.36	14.23	17.77	17.65	C
Mustelus canis	CB	–	–	–	–	–	–	–	0-18	37.21	8.14	12.79	18.60	23.26	A
Mustelus lenticulatus	CB	–	–	–	–	–	–	–	1-220	41.04	4.34	12.47	17.47	24.68	C
Triakis semifasciata	CB	325	27.9	841	72.1	1166	48.1	2.08	0-4	–	–	–	–	–	E
Hemigaleidae															
Hemigaleus australiensis	CB	–	–	–	–	–	–	–	–	48.07	5.71	12.96	16.15	17.12	C
Paragaleus pectoralis	CP	390	44.4	487	55.6	877	142.1	–	0-100	–	–	–	–	–	E
Carcharhinidae															
Carcharhinus acronotus	CP	909	42.4	1238	57.6	2147	131.6	–	18-64	–	–	–	–	–	E
Carcharhinus altimus	CP	1079	48.7	1137	51.3	2216	–	–	90-250	–	–	–	–	–	E
Carcharhinus amblyrhynchos	CP	–	–	–	–	–	–	–	0-140	64.23	4.73	8.30	13.07	9.67	C
Carcharhinus brachyurus	CP	–	–	–	–	–	–	–	0-100	54.90	4.90	12.75	13.11	14.34	C
Carcharhinus brevipinna	CP	791	46.7	903	53.3	1694	44.5	1.46	0-30	–	–	–	–	–	E

Species	Habitat														Source
Carcharhinus falciformis	OP	791	43.9	1010	56.1	1801	53.2	1.54	0–200	63.48	4.99	8.31	12.53	10.68	C,E
Carcharhinus galapagensis	CP	814	45.8	962	54.2	1776	–	–	0–180	–	–	6.07	14.04	13.33	E
Carcharhinus leucas	CP	940	50.8	912	49.2	1852	59.8	0.98	0–30	60.69	5.86	–	–	–	C,E
Carcharhinus limbatus	CP	1117	50.1	1114	50.0	2231	110.5	1.25	0–30	–	–	–	–	–	E
Carcharhinus melanopterus	CP	765	47.2	856	52.8	1621	–	1.87	–	58.34	6.19	7.86	15.84	11.77	C,E
Carcharhinus obscurus	CP	917	47.7	1009	52.3	1926	152.7	0.84	0–400	–	–	–	–	–	E
Carcharhinus perezi	CP	979	48.6	1037	51.4	2016	–	1.42	0–30	–	–	–	–	–	E
Carcharhinus plumbeus	CP	1164	51.2	1112	48.8	2275	157.0	1.01	20–55	54.16	7.23	8.46	14.23	15.92	C,E
Galeocerdo cuvier	CP	280	35.2	518	64.9	798	24.0	1.65	0–140	50.03	7.00	11.79	16.88	14.31	C,E
Negaprion acutidens	CP	–	–	–	–	–	–	–	0–30	50.35	5.56	9.63	13.51	20.95	C,E
Negaprion brevirostris	CB	658	49.0	685	51.0	1343	49.5	1.43	0–92	–	–	–	–	–	E
Prionace glauca	OP	342	38.4	548	61.6	889	40.1	1.47	0–350	49.65	5.96	15.44	10.97	17.98	C,E
Rhizoprionodon terraenovae	CP	937	51.8	873	48.2	1810	70.4	1.63	0–10	–	–	–	–	–	E
Triaenodon obesus	CP	–	–	–	–	–	–	–	0–10	56.84	5.34	8.98	15.89	12.96	C
Sphyrnidae															
Eusphyra blochii	CP	1034	50.4	1018	49.6	2052	–	1.40	–	–	–	–	–	–	E
Sphyrna lewini	CP	1260	42.5	1707	57.5	2967	–	2.52	0–275	53.75	4.44	6.53	23.71	11.57	C,E
Sphyrna mokarran	CP	876	39.3	1352	60.7	2229	90.5	1.21	0–80	66.52	3.39	3.75	18.41	7.93	C,E
Sphyrna tiburo	CP	836	45.1	1016	54.9	1852	14.1	1.98	10–25	–	–	–	–	–	E
Sphyrna zygaena	CP	–	–	–	–	–	–	–	0–20	58.10	3.93	6.08	22.25	9.63	C

Note: Species are organized taxonomically and categorized with their habitat (CB: coastal benthic; DW: deepwater; CP: coastal pelagic; OP: oceanic pelagic) and common depth range (based on Compagno, Dando, and Fowler 2005). The number and percentage of pores on the dorsal and ventral surfaces are indicated along with the total number and standard deviation. The eye diameter expressed as function as total length is also included. The source for each species is indicated in the last column (A: Northcutt 1978; B: Tricas 2001; C: Yopak et al. 2007; D: Yopak and Montgomery 2008; E: Cornett 2006).

a Where both brain data and electroreceptor/eye size data exist, they did not come from the same specimen.

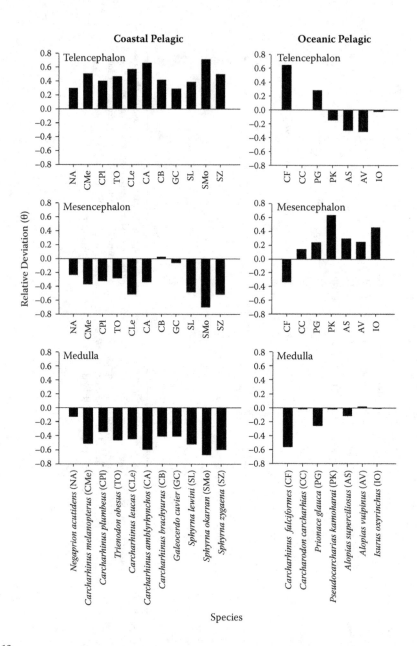

FIGURE 9.10

Weighted factors (θ) for the telencephalon, mesencephalon, and medulla for 11 coastal pelagic elasmobranchs and 7 oceanic pelagic elasmobranchs, showing the deviation from the average relative volume for each brain structure (Wagner 2001a, 2001b), using data compiled from Northcutt (1978), Yopak et al. (2007), and Yopak and Montgomery (2008). A brain structure was considered relatively large if θ > 0.1, relatively small if θ < −0.1, and relatively average-sized if 0.1 < θ < −0.1. Note: Weighted factors are based on relative structure size data on a total of 61 species.

As with the patterns seen above in the reef-associated coastal benthic sharks, there is a general trend for a reduced mesencephalon in species that live on or near coral reefs. In contrast, coastal pelagic species that spend more time in the open water column, such as *Carcharhinus brachyurus*, have an average-sized mesencephalon, which may indicate that this species is a more visual predator (Figure 9.10).

This ecological group contains species that have the most reduced medullas in comparison to other sharks, and although data on the size of the cerebellar-like lobes are only available for two species (Figure 9.7), they similarly demonstrate some of the smallest cerebellar-like structures as well. It is possible, however, that the extreme hypertrophy of the telencephalon in this group overshadowed the development of the acoustico-lateralis areas of the brain, and more data are required to elaborate on these trends. When examining these trends in relation to electrosensory pore numbers, the mean pore number is high in coastal pelagics and electrosensory pores are evenly distributed dorsally and ventrally and do not extend far past the mouth caudally (Figure 9.3). As described above, electroreceptor afferents terminate somatotopically, forming a spatial map within the DON (Bodznick and Boord 1986; Bell 2002), and a more compact pore distribution in these species could correspond to smaller cerebellar-like structures. In contrast, the greatest hypertrophy of both the medulla and the cerebellar-like lobes was observed in the deepwater sharks (Figures 9.5 and 9.7), species that often have an elongated rostrum and, as with *Etmopterus baxteri*, dissimilar pore patterns on the dorsal and ventral surfaces of the head (Figure 9.4), which could necessitate the need for greater cerebellar-like processing in addition to a more complex somatotopic organization within the brain. Further research is required to test whether characteristics, such as more elongated pore patterns, pore width, and varying dorso-ventral distributions, are correlated with cerebellar-like structure development.

9.3.4 Oceanic Pelagic

Like the coastal pelagic environment, prey in the oceanic pelagic environment are free to escape in all dimensions, not constrained to the primarily two-dimensional plane as in the coastal benthic environment. Unlike the coastal benthic and coastal pelagic environments, the oceanic pelagic environment has a paucity of prey items, similar to the deepwater environment. However, unlike the deepwater environment, the oceanic pelagic environment is characterized by well-lit, clear water that provides good to excellent visibility. Oceanic pelagic sharks rely primarily upon vision for prey capture (Compagno 1984a,b; Compagno, Dando, and Fowler 2005), and most likely employ electroreception only in a limited capacity.

Oceanic pelagic species possess the second lowest electrosensory pore number after the deepwater species. All representatives studied have large pore distributions on the ventral surface (Figure 9.9). However, a large range of total pore numbers exists, which reflects the diverse predatory strategies and prey items (Figure 9.8). Pelagic species with the greatest number of pores, such as the thresher sharks (family Alopiidae, order Lamniformes), feed primarily upon smaller schooling fishes and squid and could benefit from finer-scale spatial resolution of electrical signals. In contrast, those taxa with the fewest pores, such as the mackerel sharks (family Lamnidae, order Lamniformes), generally prey upon larger, faster fishes and marine mammals. For adults of these taxa, visual and olfactory cues are important for both prey location and acquisition, with electroreception being relegated to a subordinate role.

9.3.4.1 Oceanic Pelagic Nervous System

Development of the telencephalon varies across the species of this habitat, with some species, such as *Carcharhinus falciformis* (θ = 0.64) and *Prionace glauca* (θ = 0.28), showing enlargement of this structure, while others, such as *Alopias superciliosus* (θ = 0 to 0.30) and *Pseudocarcharias kamoharai* (θ = −0.15), having small forebrains (Figure 9.8). Based on neural development, oceanic pelagic species have been described as relying more on olfaction and vision than other senses (Yopak et al. 2007).

Oceanic pelagic predators living in a three-dimensional environment have been shown to have a more complex cerebellum than species associated with the substrate (Lisney and Collin 2007; Yopak et al. 2007; Figure 9.7). This characteristic has been correlated with agile prey capture and habitat dimensionality in a range of chondrichthyans (Northcutt 1978; Yopak et al. 2007; Yopak and Frank 2007; Yopak and Montgomery 2008; Lisney et al. 2008) and the ability to perform more multifaceted motor tasks (New 2001). Alopiids, in particular, have one of the most highly foliated and large cerebellums of all chondrichthyans (Figure 9.2), a structure comprising almost one-third of the total brain in this group (Yopak et al. 2007). The brain organization of *Alopias* is characterized by an enlarged mesencephalon and a large, highly foliated cerebellum, which has been attributed to the novel method of prey capture documented in these sharks (Yopak et al. 2007; Lisney and Collin 2006). Although there were far fewer numbers of oceanic pelagic species from which to draw comparisons, the overall trend for oceanic pelagic sharks shows greater development of neural structures related to vision, agile prey capture, and movement in three-dimensional space.

9.4 Phylogenetic Considerations

In the preceding sections, we highlighted correlations between pore distributions and brain anatomy for species found in a given environment. In this section, we examine multiple species within a taxon, regardless of which environment they inhabit. Taken together, these two analyses enable us to comment on whether electrosensory adaptations are shared among taxonomically diverse species within an environment, or whether they are specific to a taxon.

Northcutt (1978) proposed a classification scheme for the brains of cartilaginous fishes, based on the phylogenetic relationships of Compagno (1973, 1977). The brains of cartilaginous fishes were grouped into two broad categories: (1) The squalomorph pattern, including the squalomorph and squatinomorph sharks, described small brains comprising a smooth, undifferentiated cerebellar corpus and smaller telencephalon, and (2) the galeomorph pattern, including the advanced galeomorph sharks, which was generalized as a larger brain, with a foliated cerebellum and a hypertrophied telencephalon (Northcutt 1978).

9.4.1 Sharks of the Order Squaliformes

The order Squaliformes (dogfishes) is the second most speciose shark order, with seven families and 130 species (Compagno, Dando, and Fowler 2005). The species in this order reside predominantly in deep water, but can also be coastal benthic. Both the number of electrosensory pores and the dorsal:ventral pore distribution ratios vary widely among the five species examined (Table 9.1). One species, the piked dogfish *Squalus acanthias*, proved to be ecologically unclassifiable. The piked dogfish inhabits a variety of habitats over a substantial depth range (most commonly from the surface to 600 m) and it could be classified as coastal benthic, deepwater, or coastal pelagic. The sensory morphology of the piked dogfish reflects this diversity in its retention of an intermediate to large-sized eye, similar to deepwater species, and a large number of ventral electroreceptors, such as the coastal benthic species. Its broad habitat distribution precluded the piked dogfish from being categorized into any specific environment.

Although we do not distinguish between truly benthic and benthopelagic deepwater predators, the differences in sensory morphology may reflect niche partitioning. For example, the New Zealand lantern shark (*Etmopterus baxteri*) and the smooth lantern shark (*Etmopterus pusillus*) share sensory morphologies similar to other deepwater sharks. Although the stomach contents of *E. baxteri* have not been recorded, analysis of *E. pusillus* has found fish eggs, hake, and lanternfishes, as well as small sharks in the diet (Compagno, Dando, and Fowler 2005). These are predominantly benthic and relatively slow-moving prey items. In contrast to its congeners, *Etmopterus lucifer* has nearly four times the number of electroreceptors and its diet consists of fast-moving prey: squid, shrimp, and small fishes (Compagno, Dando, and Fowler 2005). The large eyes and substantially greater electrosensory pore number may facilitate capture of its fast-moving prey, especially in open water, off the bottom.

9.4.1.1 *Squaliformes Nervous System*

Brain organization of the sharks of the order Squaliformes, including 12 species across six families, is thus far accurately described by Northcutt's original (1978) scheme (Yopak et al. 2007; Yopak and Montgomery 2008). The majority of squaliforms for which brain data are available (including Squalidae, Centrophoridae, Etmopteridae, Somniosidae, Oxynotidae, and Dalatiidae) are deepwater species, with some exceptions (e.g., *S. acanthias*). As mentioned previously, deepwater sharks are characterized by having an enlarged medulla, particularly the cerebellar-like structures, and a reduced telencephalon. Interestingly, the members of the family Squalidae that are not classified as deepwater sharks but rather occupy varying ecological zones, such as *S. acanthias* and *S. megalops*, do not show relative enlargement of the cerebellar-like structures. Sharks of this order also have a smooth, undifferentiated cerebellum (Figure 9.2; Yopak and Montgomery 2008), and it has been suggested that this is linked to lower activity levels and a close association to the substrate (Northcutt 1989; Yopak and Frank 2007; Yopak et al. 2007; Yopak and Montgomery 2008). The squaliforms examined here were spread across four clusters (G, H, I, K; Figure 9.11); 7 of the 12 squaliforms are grouped in cluster K. In general, squaliforms cluster with other deepwater and coastal benthopelagic sharks.

9.4.2 Sharks of the Order Lamniformes

Nearly all lamniforms inhabit the oceanic pelagic environment. Electroreceptor distribution among species in this order is comparable, with all retaining approximately two-thirds of their pores on the ventral surface (Table 9.1). All five species of the family Lamnidae (Figure 9.12) possess relatively few pores and are ecologically remarkably similar as open water predators of large prey items, including large schooling fishes (i.e., salmon, tuna), billfishes, marine mammals, and other sharks (Compagno 2001). For these species, vision

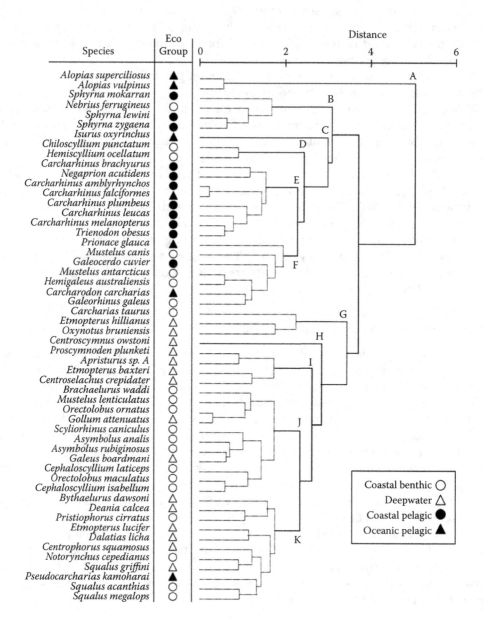

FIGURE 9.11

Multivariate hierarchical cluster analysis dendrogram using Euclidean distances, based on the relative size of each of the five brain areas as a proportion of the total brain in 54 species of elasmobranch, not including holocephalans, from Northcutt (1978), Yopak et al. (2007), and Yopak and Montgomery (2008). Ecological grouping (coastal benthic, deepwater, coastal pelagic, or oceanic pelagic) was coded beside each species. Lighter lines indicate those clusters that are not statistically significant (Primer 6, SIMPROF, $p < 0.05$).

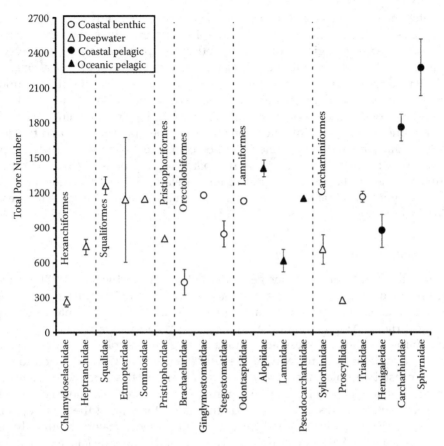

FIGURE 9.12
Mean total pore number (SD) for each family, separated by order. The dominant environment for each family is identified. Sharks in the order Squaliformes show a fairly consistent pore number, whereas sharks in the paraphyletic Carcharhiniformes demonstrate a large range of pore numbers.

and olfaction are the primary sensory modalities used in predation, with electroreception reduced to a tertiary role, or perhaps playing no significant role at all in predation.

An exception to this general rule for large lamniforms is the sand tiger shark, *C. taurus*. Unlike the pelagic lamnids (family Lamnidae), the sand tiger is a coastal benthic species and possesses a greater number of pores than the lamnids while retaining the primarily ventral pore distribution. The small eyes coupled with the greater number of electroreceptors may indicate a greater reliance upon electroreception in this species compared to the lamnids.

Sharks in the family Alopiidae possess more electrosensory pores than any other lamniform species examined, which likely reflects their feeding ecology. Unlike the other lamniform species, thresher sharks prey upon small schooling fishes and squid, using the elongated dorsal lobes of their caudal fin in prey capture (Compagno 2001). While initial prey detection would involve olfaction and vision, the large number of ventral electroreceptors likely provides the spatial resolution to accurately position their relatively small mouth.

The crocodile shark, *Pseudocarcharias kamoharai*, is from the monophyletic family Pseudocarchariidae. This species inhabits the oceanic pelagic environment, commonly at depths of 600 m, making it the single lamniform examined that could be classified as a deepwater shark. Its electrosensory pore patterns reflect a mix of characteristics.

Numerically, the electrosensory pore count of *P. kamoharai* is nearly that of other oceanic pelagic species and the electroreceptors are distributed primarily on the ventral surface as found in both oceanic pelagic sharks and the other lamniforms (Table 9.1). However, the pore openings to both the lateral line and electrosensory systems of *P. kamoharai* are very large, similar to those of the deepwater sharks *Zameus squamulosus* (family Somniosidae, order Squaliformes) and *E. lucifer* (family Etmopteridae, order Squaliformes). These latter two species possess the greatest pore numbers from that environment (Table 9.1). It is possible that these species all inhabit a zone between the upper pelagic and the deepwater environment. This mid-water zone would be completely open water (like the pelagic environment), yet nearly lightless and fairly sparse in prey density (like the deepwater environment). The similar morphologies seen in unrelated lineages (squaloid and galeoid) supports the hypothesis that the environment can exert a selective pressure that results in evolutionary convergence.

9.4.2.1 Lamniformes Nervous System

Members of the Lamniformes order generally fall under Northcutt's (1978) galeomorph pattern for cartilaginous fishes. Brain data are available on lamniforms that occupy the coastal benthic, coastal pelagic, and oceanic pelagic environment (Odontaspidae, Pseudocarchariidae, Alopiidae, and Lamnidae). The majority of lamniforms dwell in the pelagic environment and possess an average to large telencephalon, an enlarged mesencephalon, and an average to reduced medulla (Figure 9.10). One exception to this pattern is the coastal benthic shark, *Carcharias taurus*, which has a reduced telencephalon and mesecephalon. The pattern of brain organization in *C. taurus* is more consistent with other coastal benthic sharks as opposed to other lamniforms, and the small eyes in this species are consistent with the trend for a reduced mesencephalon, commonly seen in reef-associated sharks. Based on cluster analysis, both the alopiids (cluster A) and *Isurus* (cluster C) occupied their own cluster, while *Carcharodon* was grouped with several members of the Carcharhiniformes order, including *Prionace*, another oceanic pelagic species.

The crocodile shark, *Pseudocarcharias kamoharai*, does not group with other lamniforms, nor does it group with other oceanic pelagic species (Figure 9.11). Although several characteristics of the brain organization of this species are similar to other oceanic pelagic sharks, such as a large mesencephalon (Figure 9.10), this species also has little to no foliation of the corpus cerebellum, which is inconsistent with those sharks that dwell in the open water (Yopak et al. 2007) and would likely fall under the squalomorph pattern of brain organization (Northcutt 1978), despite its phylogenetic classification. As the morphological sensory characteristics of *P. kamoharai* are more similar to deepwater sharks, the mesopelagic lifestyle of this species may be influencing its brain development as well, with large eyes (Lisney and Collin 2007), extreme relative enlargement of the mesencephalon, and cerebellar development that is more similar to members of Squalidae (Lisney and Collin 2007; Yopak et al. 2007; Yopak and Montgomery 2008). Large mesencephalons, particularly the optic tectum, have been similarly documented in mesopelagic bony fishes, and vision has been found to be the dominant sense at deeper depths in the open ocean in teleosts, likely linked to the presence of bioluminescent stimuli at depth (Wagner 2001b, 2002). *Pseudocarcharias kamoharai* is grouped into cluster K, and is most closely grouped with the squalids. The morphological similarities of this shark to the squaliform sharks (Last and Stevens 1994), suggest that this shark has very different locomotory and prey capture strategies than other oceanic species (Yopak et al. 2007).

9.4.3 Sharks of the Order Carcharhiniformes

Unlike most orders and families, which have species that reside in primarily one environment, the family Carcharhinidae includes species that can be found in three of the four habitats, a characteristic unique to this family. Families of the order Carcharhiniformes generally possess the greatest number of electroreceptors in any environment in which they reside.

Members of the family Carcharhinidae best demonstrate the effects of phylogeny on electroreceptor pore distribution. All *Carcharhinus* species are morphologically similar, and possess a narrow electrosensory pore range and a relatively even dorsal:ventral pore distribution (Table 9.1). Similarly, all *Carcharhinus* species examined inhabit the coastal pelagic environment, with the exception of the pelagic *C. falciformis*.

Carcharhinus falciformis, the silky shark, is the only oceanic pelagic *Carcharhinus* species. The silky shark has the greatest electroreceptor number of any species from the oceanic pelagic environment, but it has one of the smallest among the *Carcharhinus* species. It also exhibits a ventral pore distribution intermediate between coastal pelagic and oceanic pelagic species (Table 9.1). Its ontogenetic history may account for the intermediate electroreceptor number and distribution. As an adult, *C. falciformis* resides in oceanic pelagic waters and has a body morphology similar to other oceanic pelagic species: long pectoral fins, a sizable but slender body, and a large caudal fin with a long dorsal lobe, which provides for efficient cruising (Compagno 1984a,b; Compagno, Dando, and Fowler 2005). However, juveniles possess shorter pectoral fins, remain within the coastal pelagic environment, and forage closer to shore (Compagno, Dando, and Fowler 2005; Tricas et al. 1997). The intermediate electroreceptor morphology of *C. falciformis* places it between the large number of pores of its congenerics and the small number of pores of exclusively oceanic pelagic species. This suggests that *C. falciformis* may represent a recent immigrant to the oceanic pelagic environment that still carries vestiges of its phylogenetic history. In contrast, *Prionace glauca* (the blue shark) represents the ultimate adaptation to the oceanic pelagic environment. Unlike *C. falciformis*, *P. glauca* is a truly pelagic species at all life stages, with a full suite of morphological adaptations to an open ocean lifestyle, including long pectoral fins, a long, dorsal lobe on its caudal fin for slow cruising, and vastly different electrosensory characteristics. *Prionace glauca* has fewer total electrosensory pores than any other carcharhiniform species in this study (except *G. cuvier*) and maintains a large ventral pore distribution (Figure 9.9), comparable to the ventral distribution for other oceanic pelagic species (Table 9.1).

9.4.3.1 Carcharhiniformes Nervous System

Sharks of the order Carcharhiniformes are the most difficult to describe according to Northcutt's classification schema. This clade contains coastal benthic species (e.g., *Scyliorhinus* and *Cephaloscyllium*), deepwater species (e.g., *Gollum* and *Bythaelurus*), coastal pelagic species (e.g., *Carcharhinus* and *Sphyrna*), as well as oceanic pelagic species (*Prionace* and *C. falciformis*). This ecological breadth allows for the assessment of both phylogenetic and ecological influences on brain organization. This phylogenetic group has the most widespread variation in brain development (Yopak et al. 2007; Yopak and Montgomery 2008), ranging from deep-sea species with the most highly developed cerebellar-like structures to coastal pelagic reef sharks with the largest telencephalons (Figures 9.7 and 9.10).

Within the order Carcharhiniformes are species with some of the largest telencephalons (Carcharhinidae) (Figure 9.10), corresponding with sharks that are primarily associated with reefs. Other species that show relative hypertrophy of the telencephalon are similarly reef-associated, such as the coastal benthic sharks *Nebrius ferrugineus* (Ginglymostomatidae) and *Hemiscyllium ocellatum* (Hemiscylliidae) of the order Orectolobiformes. The majority of other carcharhiniforms, however, that do not live in three-dimensional or complex spatial environments, such as the scyliorhinids and pseudotriakids, have average to small telencephalons (Figure 9.5). The coastal oceanic and oceanic pelagic carcharhiniforms were grouped primarily in clusters B and E, with all members of Carcharhinidae comprising cluster E (Figure 9.11).

Carcharhiniforms also exhibit a range of mesencephalic development. Carcharhinids, sphyrnids, and the pseudotriakid, *Gollum attenuatus*, show reduction of this structure, while the scyliorhinids show average to moderate enlargement. Those with average to large mesencephalons, such as *Cephaloscyllium* and *Scyliorhinus*, are generally grouped into cluster J with other coastal benthic species (Figure 9.11). Both the medulla and the cerebellar-like lobes are reduced across the majority of the Carcharhiniformes, with the exception of the one deepwater member of this order, *Gollum attenuatus*, which is grouped in cluster J, comprising orectolobiforms, lamniforms, and carcharhiniforms. The brain organization of *Gollum* is far more consistent with other deepwater sharks as opposed to members of its order (Figures 9.5 and 9.7). Overall, while brain organization for the members of the same family demonstrates morphological similarity (i.e., Carcharhinidae and Scyliorhinidae), there is extreme variation between orders within the Carcharhiniformes.

9.5 Conclusions

In many cases, it was difficult to classify the habitat of a shark because they are not confined to a single one of the broad ecological categories we assigned each species. For example, the 200 m depth contour generally delineates the margin of the continental shelf and is used to separate coastal from oceanic waters, but many shark species move easily between the two environments. Despite the potential for overlap among the habitats, some broad trends are still apparent. In the oceanic pelagic environment, electroreception usually plays a limited role in predation and oceanic pelagic species have fewer electrosensory pores than coastal pelagic species. Similarly, deepwater species generally possess relatively few pores and the deepwater species with large pore numbers likely forage off the sea floor. Eye diameter generally increases with inhabited depth, and deepwater species possess greater eye diameters than species in other habitats. This trend is consistent across taxa; deepwater representatives from the unrelated Hexanchiformes, Squaliformes, and Carcharhiniformes all demonstrate large eye size. The relative development of these peripheral sensory structures is also reflected in the brain morphology.

Patterns of brain organization are quite consistent across and within the four ecological groups, with similar patterns of development appearing in sharks that live in similar environments. The deepwater species show the greatest hypertrophy of the cerebellar-like structures and a reduced telencephalon. The greatest development of the mesencephalon is found to correspond with species with larger eyes, particularly those in the oceanic pelagic habitats, though this pattern is not consistent through all species, and it is as yet unclear whether size of the mesencephalon and/or optic tectum reflects visual acuity. Although structural enlargement does not necessarily predict ecological patterns (Kotrschal and Palzenberger 1992), relative brain development reflects habitat in addition to phylogeny (Yopak et al. 2007; Yopak and Montgomery 2008). Although there is a phylogenetic component to variation in brain size in sharks (Yopak et al. 2007; Yopak and

Montgomery 2008), holocephalans (Yopak and Montgomery 2008), and batoids (Lisney et al. 2008), much of this variation does not correlate with phylogenetic relationships, but rather reflects ecological correlations, such as the dimensionality of the environment and/ or agile prey capture (Yopak et al. 2007; Yopak and Frank 2007; Yopak and Montgomery 2008; Lisney et al. 2008).

Finally, the correlation of eye size, electroreceptor distribution, and brain morphology with habitat is common across taxa. This indicates that, like the visual system, the electrosensory system also exhibits ecomorphological adaptations to the environment.

Acknowledgments

This work was made possible only with the help of many collaborators. SMK acknowledges the members of the FAU Elasmobranch Research Laboratory. ADC thanks D. Adams, G. Poulakis, B. Flammang, K. Mara, D. Huber, L. Whitenack, B. Casper, and J. Baldwin for specimen collection. J. Finan at the Smithsonian Museum of Natural History and R. Robins and G. Burgess at the Florida Museum of Natural History provided access to rare specimens. KEY would like to thank those who aided in the collection, provision, and/or photography of specimens, past and present, particularly C. Duffy, N. Bagley, P. Brown, J. Montgomery, S. Collin, members of the Leigh Marine Lab (Auckland), members of CSIRO (Hobart), and members of CSCI (San Diego), with particular thanks to T. Lisney, with whom several successful collaborative efforts have been possible. Special thanks to J. Montgomery for his comments and editorial assistance.

References

Barton RA, Harvey PH (2000) Mosaic evolution of brain structure in mammals. Nature 405(6970):1055–1058.

Barton RA, Purvis A, Harvey PH (1995) Evolutionary radiation of visual and olfactory brain systems in primates, bats, and insectivores. Philosophical Translations of the Royal Society, London B 348:381–392.

Bastian J (1995) Pyramidal cell plasticity in weakly electric fish: A mechanism for attenuating responses to reafferent electrosensory inputs. Journal of Comparative Physiology A 176:63–78.

Bauchot R, Bauchot ML, Platel R, Ridet JM (1977) The brains of Hawaiian tropical fishes: Brain size and evolution. Copeia 1(1):42–46.

Bell CC (2002) Evolution of cerebellar-like structures. Brain, Behavior, and Evolution 59(5–6):312–326.

Bell C, Bodznick D, Montgomery J, Bastian J (1997a) The generation and subtraction of sensory expectations within cerebellar-like structures. Brain, Behavior, and Evolution 50(S1):17–31.

Bell CC, Grant K, Serrier J (1992) Sensory processing and corollary discharge effects and sensory processing in the mormyride electrosensory lobe. I. Field potentials and cellular activity in associated structures. Journal of Neurophysiology 68:843–858.

Bell CC, Maler L (2005) Central neuroanatomy of electrosensory systems in fish. In: Bullock TH, Hopkins CD, Popper AN, Fay RR (eds) Electroreception. Springer, New York, pp 68–111.

Bell CC, Szabo T (1986a) Electroreception in mormyrid fish: Central anatomy. In: Bullock TH, Heiliegenberg W (eds) Electroreception. John Wiley & Sons, New York.

Bell CC, Szabo T (1986b) Electroreception: Behavior anatomy and physiology. In: Bullock TH, Heiliegenberg W (eds) Electroreception. John Wiley & Sons, New York, pp 577–612.

Bodznick D (1991) Elasmobranch vision: Multimodal integration in the brain. Journal of Experimental Zoology Supplement 256(S5):108–116.

Bodznick D, Boord RL (1986) Electroreception in Chondrichthyes. In: Bullock TH, Heiligenberg W (eds) Electroreception. John Wiley & Sons, New York, pp 225–256.

Bodznick D, Montgomery JC, Carey M (1999) Adaptive mechanisms in the elasmobranch hindbrain. Journal of Experimental Biology 202:1357–1364.

Bodznick D, Schmidt AW (1984) Somatotopy within the medullary electrosensory nucleus of the little skate, *Raja-Erinacea*. Journal of Comparative Neurology 225:581–590.

Bozzano A (2004) Retinal specializations in the dogfish, *Centroscymus coelolepis*, from the Mediterranean deep-sea. Scientia Marina 68(3):185–195.

Brabrand A (1985) Food of roach (*Rutilus rutilus*) and ide (*Leuciscus idus*): Significance of diet shifts for interspecific competition in omnivorous fishes. Oecologia 66:461–467.

Bres M (1993) The behaviour of sharks. Reviews in Fish Biology and Fisheries 3(2):133–159.

Budeau DA, Verts BJ (1986) Relative brain size and structural complexity of habitats of chipmunks. Journal of Mammalogy 67:579–581.

Bullock TH (1982) Electroreception. Annual Review of Neuroscience. 5:121–170.

Bullock TH, Bodznick DA, Northcutt RG (1983) The phylogenetic distribution of electroreception: Evidence for convergent evolution of a primitive vertebrate sense modality. Brain Research Reviews 6:25–46.

Chu YT, Wen MC (1979) A study of the lateral-line canals system and that of lorenzini ampulae and tubules of elasmobranchiate fishes of China. Science and Technology Press.

Clark DA, Mitra PP, Wang SS-H (2001) Scalable architecture in mammalian brains. Nature 411(6834):189–193.

Cohen JL, Duff TA, Ebbesson SOE (1973) Electrophysiological identification of a visual area in the shark telencephalon. Science 182:492–494.

Compagno LJV (1973) Interrelationships of living elasmobranchs. In: Greenwood PH, Miles RS, Patterson C (eds) Interrelationships of fishes. Academic Press, London, pp 15–61.

Compagno LJV (1977) Phyletic relationships of living sharks and rays. American Zoologist 17:303–322.

Compagno LJV (1984a) FAO Species Catalogue. Sharks of the world: An annotated and illustrated catalogue of shark species known to date. I. Hexanchiformes to Lamniformes. FAO Fisheries Synopsis, Rome.

Compagno LJV (1984b) FAO Species Catalogue. Sharks of the world: An annotated and illustrated catalogue of shark species known to date. II. Carcharhiniformes. FAO Fisheries Synopsis, Rome.

Compagno LJV (1998) Sphyrnidae: Hammerhead and bonnethead sharks. In: Carpenter KE, Niem VH (eds) FAO identification guide for fishery purposes: The living marine resources of the Western Central Pacific. FAO Fisheries Synopsis, Rome, pp 1361–1366.

Compagno LJV (2001) FAO Species Catalogue. Sharks of the world: An annotated and illustrated catalogue of shark species known to date. Bullhead, mackerel, and carpet sharks (Heterodontiformes, Lamniformes and Orectolobiformes). FAO Fisheries Synopsis, Rome.

Compagno LGV, Dando M, Fowler S (2005) Sharks of the world. Princeton University Press, Princeton, NJ.

Compagno LJV, Niem VH (1998a) Carcharhinidae. Requiem sharks. In: Carpenter KE, Niem VH (eds) FAO identification guide for fishery purposes: The living marine resources of the Western Central Pacific. FAO Fisheries Synopsis, Rome, pp 1312–1360.

Compagno LJV, Niem VH (1998b) Squalidae. Dogfish sharks. In: Carpenter KE, Niem VH (eds) FAO identification guide for fishery purposes: The living marine resources of the Western Central Pacific. FAO Fisheries Synopsis, Rome, 1213–1232.

Conley RA, Bodznick D (1994) The cerebellar dorsal grandular ridge in an elasmobranch has proprioceptive and electroreceptive representations and projections homotopically to the medullary electrosensory nucleus. Journal of Comparative Physiology A 174(6):707–721.

Coombs S, Montgomery JC (2005) Comparing octavolateralis systems: What can we learn? In: Bullock TH, Hopkins CD, Popper AN, Fay RR (eds) Electroreception. Springer, New York, pp 318–359.

Cordo PJ, Bell CC, Harnad S (eds) (1997) Motor learning and synaptic plasticity in the cerebellum. Cambridge University Press, Cambridge, UK.

Cornett AD (2006) Ecomorphology of shark electroreceptors. Thesis, Florida Atlantic University.

Crescitelli F (1991) Adaptation of visual pigments to the photic environment of the deep sea. Journal of Experimental Zoology Supplement 5:66–75.

de Winter W, Oxnard CE (2001) Evolutionary radiations and convergences in the structural organization of mammalian brains. Nature 409(6821):710–714.

Eastman JT, Lannoo MJ (1995) Diversification of brain morphology in Antarctic Notothenoid fishes: Basic descriptions and ecological considerations. Journal of Morphology 223:47–83.

Fine ML, Horn MH, Cox B (1987) *Acanthonus armatus*, a deep-sea teleost with a minute brain and large ears. Proceedings of the Royal Society B 230:257–265.

Finger TE (1988) Sensorimotor mapping and oropharyngeal reflexes in goldfish, *Carassius auratus*. Brain, Behavior, and Evolution 31:17–24

Gans C, Northcutt RG (1983) Neural crest and the origin of vertebrates: A new head. Science 220(4594):268.

Garman S (1913) The Plagiostomia (sharks, skates, and rays). Memoirs of the Museum of Comparative Zoology, Volume 36. Harvard University, Cambridge, MA.

Gilbert CR (1967) A revision of the hammerhead shark (family Sphyrnidae). Proceedings of the U.S. Nat. Mus. 119:1–98.

Hart NS, Lisney TJ, Collin SP (2006) Visual communication in elasmobranches. In: Ladich F, Collin SP, Moller P, Kapoor BG (eds) Fish communication. Science Publishers, Enfield, NJ, pp 333–388.

Harvey PJ, Krebs JR (1990) Comparing brains. Science 249(4965):140–146.

Heiligenberg W (1987) Central processing of sensory information in electric fish. Journal of Comparative Physiology A 161:621–631.

Herring PJ (1987) Systematic distribution of bioluminescence in living organisms. Journal of Bioluminescence and Chemoluminescence 1:147–163.

Herring PJ (2000) Species abundance, sexual encounter, and bioluminescent signaling in the deep-sea. Philosophical Transactions of the Royal Society, London B 355:1273–1276.

Highstein SM, Thach WT (2002) The cerebellum: Recent developments in cerebellar research. Annals of the New York Academy of Sciences, New York, Volume 978.

Hjelmstad GO, Parks G, Bodznick D (1996) Motor corollary discharge activity and sensory responses related to ventilation in the skate vestibulolateral cerebellum: Implications for electrosensory processing. Journal of Experimental Biology 199:673–681.

Hofmann MH (1999) Nervous system. In: Hamlet WC (ed) Sharks, skates, and rays: The biology of elasmobranch fishes. Johns Hopkins University Press, Baltimore, pp 273–299.

Huber R, van Staaden MJ, Kaufman LS, Liem KF (1997) Microhabitat use, trophic patterns, and the evolution of brain structure in African cichlids. Brain, Behavior, and Evolution 50:167–182.

Hueter RE, Mann DE, Maruska KP, Sisneros JA, Demski LS (2004) Sensory biology of elasmobranchs. In: Carrier JC, Musick JA, Heithaus MR (eds) Biology of sharks and their relatives. CRC Press, Boca Raton, FL, pp 325–368.

Iwaniuk AN, Hurd PL (2005) The evolution of cerebrotypes in birds. Brain, Behavior, and Evolution 65(4):215–230.

Jordan LK (2008) Comparative morphology of stingray lateral line canal and electrosensory systems. Journal of Morphology 269(11):1325–1339.

Kajiura SM (2001) Head morphology and electrosensory pore distribution of carcharhinid and sphyrnid sharks. Environmental Biology of Fishes 61:125–133.

Kalmijn AJ (1982) Electric and magnetic field detection in elasmobranch fishes. Science 218:916–918.

Kanwal JS, Caprio J (1987) Central projections of the glossopharyngeal and vagal nerves in the channel catfish, *Ictalurus punctatus*: Clues to differential processing of visceral inputs. Journal of Comparative Neurology 264:216–230.

Kotrschal K, Brandstätter R, Gomahr A, Junger H, Palzenberger M, Zaunreiter Z (1990) Brain and sensory systems. In: Winfield J, Nelson JS (eds) The biology of cyprinids. Chapman and Hall, London, pp 284–331.

Kotrschal K, Palzenberger M (1992) Neuroecology of cyprinids: Comparative, quantitative histology reveals diverse brain patterns. Environmental Biology of Fishes 33:135–152.

Kotrschal K, van Staaden MJ, Huber R (1998) Fish brains: Evolution and environmental relationships. Reviews in Fish Biology and Fisheries 8:373–408.

Kruska DCT (1988) The brain of the basking shark (*Cetorhinus maximus*). Brain, Behavior, and Evolution 32:353–363.

Lammens EHRR, Geursen J, McGillavry PJ (1987) Diet shifts, feeding efficiency and coexistence of bream (*Abramis brama*), roach (*Rutilus rutilus*) and white bream (*Blicca bjoercna*) in hypertrophic lakes. In: Kullander S, Fernholm B (eds) Proceedings of the V Congress of European Ichthyologists, Stockholm, pp 153–162.

Last PR, Stevens JD (1994) Sharks and rays of Australia. CSIRO, Melbourne, Australia.

Lisney TJ, Collin SP (2006) Brain morphology in large pelagic fishes: A comparison between sharks and teleosts. Journal of Fish Biology 68:532–554.

Lisney TJ, Collin SP (2007) Relative eye size in elasmobranchs. Brain, Behavior, and Evolution 69(4):266–279.

Lisney TJ, Yopak KE, Montgomery JC, Collin SP (2008) Variation in brain organization and cerebellar foliation in chondrichthyans: Batoids. Brain, Behavior, and Evolution 72:262–282.

Marshall NB (1979) Developments in deep-sea biology. Blandford, Poole, UK.

Meek J, Nieuwenhuys R (1997) Holosteans and teleosts. In: Nieuwenhuys R, Donkelaar HJT, Nicholson C (eds) The central nervous system of vertebrates. Springer, New York, pp 759–937.

Merrett NR, Haedrich RL (1997) Deep-sea demersal fish and fisheries. Chapman and Hall, London.

Montgomery JC (1984) Noise cancellation in the electrosensory system of the thornback ray: Common mode rejection of input produced by the animal's own ventilatory movement. Journal of Comparative Physiology A 155:103–111.

Montgomery JC, Bodznick D (1994) An adaptive filter cancels self-induced noise in the electrosensory and lateral line mechanosensory systems of fish. Neuroscience Letters 174:145–148.

Montgomery JC, Bodznick D (1999) Signals and noise in the elamobranch electrosensory system. Journal of Experimental Biology 202:1349–1355.

Montgomery JC, Carton G, Bodznick D (2002) Error-driven motor learning in fish. Biological Bulletin 203:238–239.

Montgomery JC, Coombs S, Conley RA, Bodznick D (1995) Hindbrain sensory processing in lateral line, electrosensory, and auditory systems: A comparative overview of anatomical and functional similarities. Auditory Neuroscience 1:207–231.

Morita Y, Finger TE (1985) Reflex connections of the facial and vagal gustatory systems in the brainstem of the bullhead catfish, *Ictalurus nebulosus*. Journal of Comparative Neurology 231:547–558.

Musick JA, Harbin MM, Compagno LJV (2004) Historical zoogeography of the Selachii. In: Biology of Sharks and Their Relatives (Carrier JC, Musick JA, Heithaus MR, eds), pp 33–78. London, UK: CRC Press.

Narenda KS (1991) Adaptive control using neural networks. In: Miller WT, Sutton RS, Werbos PJ (eds) Neural networks for control. MIT Press, Cambridge, MA, pp 115–142.

New JG (2001) Comparative neurobiology of the elasmobranch cerebellum: Theme and variations on a sensorimotor interface. Environ Biol Fish 60:93–108.

New JG, Bodznick D (1990) Medullary electrosensory processing in the little skate. II. Suppression of self-generated electrosensory interference during respiration. Journal of Comparative Physiology A 167:295–307.

New JG, Tricas TC (2001) Electroreceptors and magnetoreceptors: Morphology and function. In: Sperlakis N (ed) Cell physiology source book, 3rd ed. Academic Press, San Diego, pp 839–856.

Northcutt RG (1977) Elasmobranch central nervous system organization and its possible evolutionary significance. American Zoologist 17:411–429.

Northcutt RG (1978) Brain organization in the cartilaginous fishes. In: Hodgson ES, Mathewson RF (eds) Sensory biology of sharks, skates, and rays. Office of Naval Research, Arlington, VA, pp 117–194.

Northcutt RG, Gans C (1983) The genesis of neural crest and epidermal placodes: A reinterpretation of vertebrate origins. The Quarterly Review of Biology 58:1–28.

Northcutt RG (1989) Brain variation and phylogenetic trends in elasmobranch fishes. Journal of Experimental Zoology Supplement 1989 252:83–100.

Northcutt RG (2002) Understanding vertebrate brain evolution. Integrative and Comparative Biology 42(4):743–756.

Pollen AA, Dobberfuhl AP, Scace J, Igulu MM, Renn SCP, Shumway CA, Hofmann HA (2007) Environmental complexity and social organization sculpt the brain in Lake Tanganyikan cichlid fish. Brain, Behavior, and Evolution 70(1):21–39.

Poulson TL (1963) Cave adaptations in amblyopsid fishes. American Midland Naturalist 70:257–290.

Raschi WG (1984) Anatomical observations on the ampullae of Lorenzini from selected skates and galeoid sharks of the western north Atlantic. Dissertation, Virginia Institute of Marine Science, The College of William and Mary.

Raschi W (1986) A morphological analysis of the ampullae of Lorenzini in selected skates (Pisces, Rajoidei). Journal of Morphology 189(3):225–247.

Raschi W, Aadlond C, Keithar ED (2001) A morphological and functional analysis of the ampullae of Lorenzini in selected galeoid sharks. In: Kapoor BG, Hara TJ (eds) Sensory biology of jawed fishes: New insights. Science Publishers, Enfield, NJ, pp 297–316.

Raschi W, Adams WH (1988) Depth-related modifications in the electroreceptive system of the eurybathic skate, *Raja radiata* (Chondrichthyes: Rajidae). Copeia 1988(1):116–123.

Riddell WI, Corl KG (1977) Comparative investigation and relationship between cerebral indices and learning abilities. Brain, Behavior, and Evolution 14:305–308.

Roberts BL (1988) The central nervous system. In: Shuttleworth T (ed) Physiology of elasmobranch fishes. Springer-Verlag, New York, pp 49–78.

Ronan MC (1986) Electroreception in cyclostomes. In: Bullock TH, Heiligenberg W (eds) Electroreception. John Wiley & Sons, New York, pp 209–224.

Schiemer F (1988) Gefährdete Cypriniden — Indikatoren für die ökologische Intaktheit von Flußsystemen. Natur und Landschaft 63:370–373.

Schmidt A, Bodznick D (1987) Afferent and efferent connections of the vestibulolateral cerebellum of the little skate, *Raja erinacea*. Brain, Behavior, and Evolution 30:282–302.

Smeets WJAJ (1982) The afferent connections of the tectum mesencephali in two chondrchthyans, the shark, *Scyliorhinus canicula*, and the ray, *Raja clavata*. Journal of Comparative Neurology 205:139–152.

Smeets WJAJ (1998) Cartilaginous fishes. In: Nieuwenhuys R, Roberts BL (eds) The central nervous system of vertebrates. Springer-Verlag, Berlin, pp 551–654.

Striedter GF (2005) Principles of brain evolution. Sinauer Associates, Sunderland.

Tricas TC (2001) The neuroecology of the elasmobranch electrosensory world: Why peripheral morphology shapes behavior. Environmental Biology of Fishes 60:77–92.

Tricas TC, Deacon K, Last P, McCosker JE, Walker TI, Taylor L (1997) Sharks and rays. Time Life Custom Publishing, Sydney.

Uiblein F, Roca JR, Baltanas A, Danielopol DL (1996) Trade-off between foraging and antipredator behaviour in a macrophyte dwelling ostracod. Archiv fur Hydrobiologie 137(1):119–133.

Wagner HJ (2001a) Brain areas in abyssal demersal fishes. Brain, Behavior, and Evolution 57:301–316.

Wagner HJ (2001b) Sensory brain areas in mesopelagic fishes. Brain, Behavior, and Evolution 57(3):117–133.

Wagner HJ (2002) Sensory brain areas in three families of deep-sea fish (slickheads, eels and grenadiers): Comparison of mesopelagic and demersal species. Marine Biology 141(4):807–817.

Waltman B (1966) Electrical properties and fine structure of the ampullary canals of Lorenzini. Acta Physiologica Scandinavica 66(264):3–60.

Yopak KE (2007) Brain organisation in chondrichthyans: Ecological correlations and functional implications. Dissertation, University of Auckland.

Yopak KE, Frank LR (2007) Variation in cerebellar foliation in cartilaginous fishes: Ecological and behavioral considerations. Brain, Behavior, and Evolution 70:210.

Yopak KE, Lisney TJ, Collin SP, Montgomery JC (2007) Variation in brain organization and cerebellar foliation in chondrichthyans: Sharks and holocephalans. Brain, Behavior, and Evolution 69(4):280–300.

Yopak KE, Montgomery JC (2008) Brain organization and specialization in deep-sea chondrichthyans. Brain, Behavior, and Evolution 71:287–304.

Yopak, KE, Frank LR (2009) Brain size and brain organization of the whale shark, Rhincodon typus, using Magnetic Resonance Imaging. Brain, Behavior, and Evolution 74:121–142.

Appendix 9.1

Electrosensory pore distribution maps for several shark species illustrating the number and location of epithelial pores for a representative individual of each species. Pores are distributed on dorsal and ventral surfaces of the head in species-specific patterns.

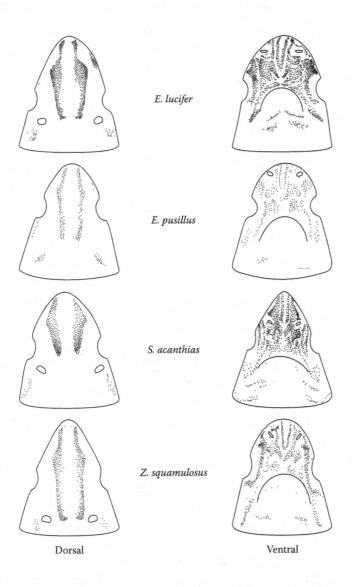

E. lucifer

E. pusillus

S. acanthias

Z. squamulosus

Dorsal   Ventral

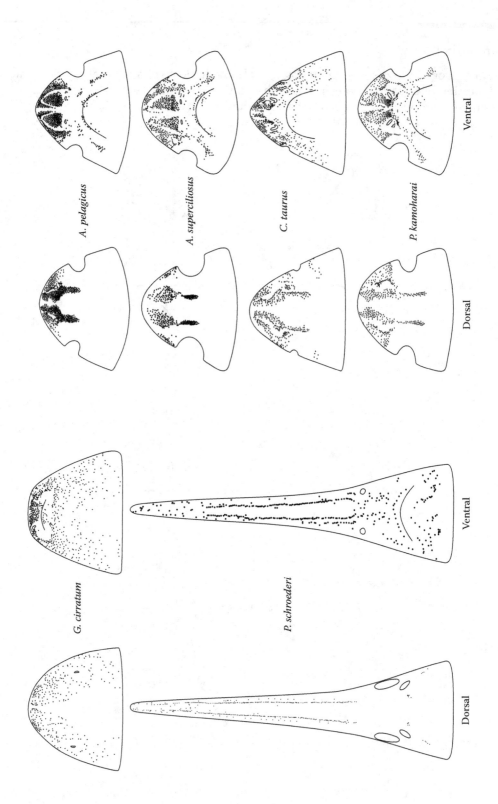

Ventral

A. pelagicus

A. superciliosus

C. taurus

P. kamoharai

Dorsal

G. cirratum

P. schroederi

Ventral

Dorsal

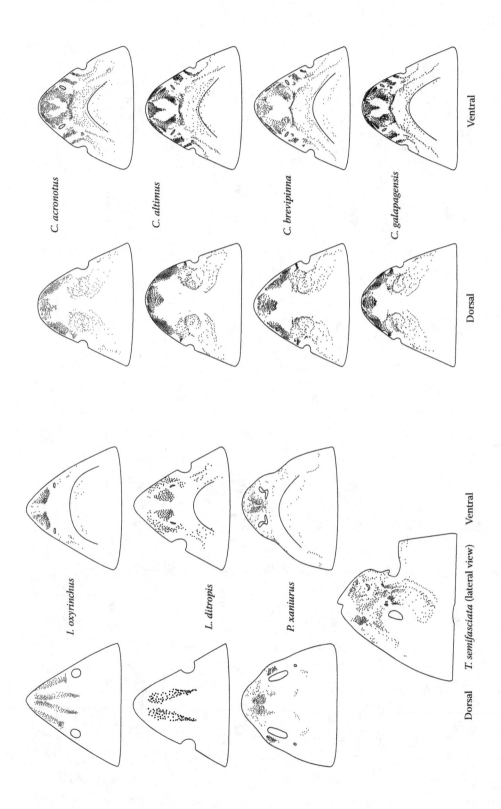

C. acronotus

C. altimus

C. brevipinna

C. galapagensis

Ventral

Dorsal

I. oxyrinchus

L. ditropis

P. xaniurus

T. semifasciata (lateral view)

Ventral

Dorsal

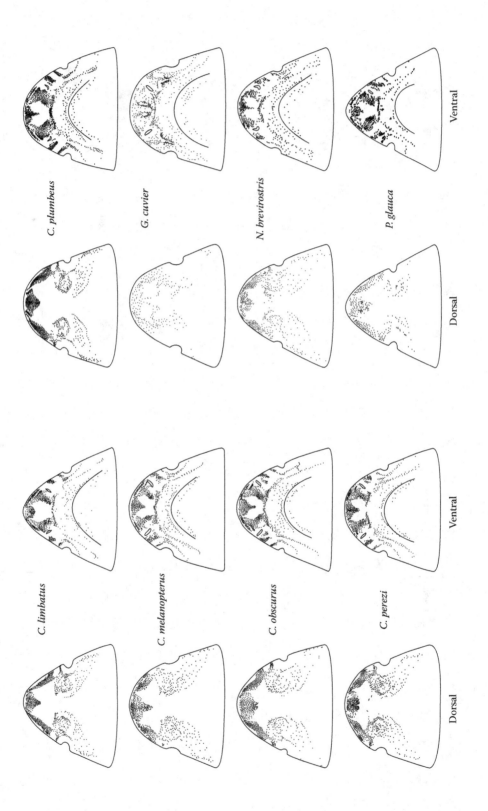

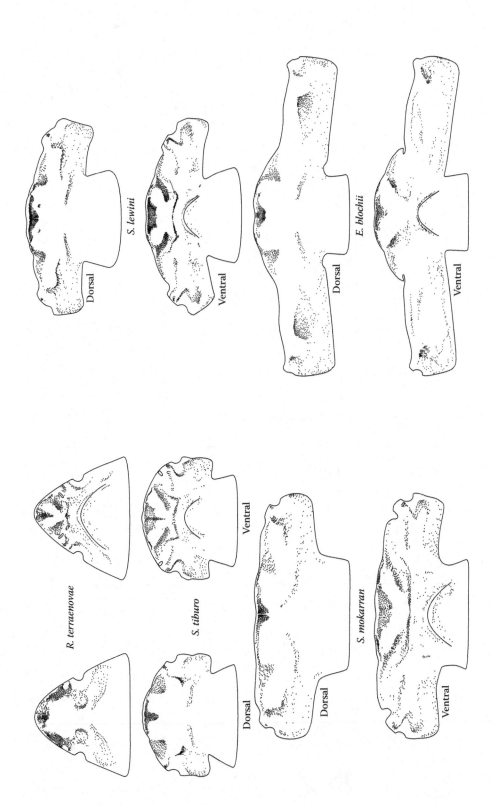

10

Molecular Insights into Elasmobranch Reproductive Behavior for Conservation and Management

David S. Portnoy

CONTENTS

10.1 Introduction

Molecular approaches in elasmobranch research have been used predominantly for phylogenetic inference or to define population structure. However, a variety of high-resolution molecular markers allow for the interpretation of complex patterns of molecular variance as well as the fast and accurate reconstruction of individual molecular profiles. Such techniques can be useful in augmenting current understanding of the reproductive biology of elasmobranchs and, where experimental or observational approaches may be difficult, can provide novel insights. The increased access and ease of utilizing molecular approaches makes the techniques and concepts discussed in this chapter useful for any elasmobranch researcher interested in the study of reproductive biology.

10.2 Molecular Markers

The first step toward utilizing a molecular approach to augment reproductive studies is informed selection of appropriate markers. Many choices exist but the following discussion will be limited to three types: microsatellites, single nuclear polymorphisms (SNPs), and direct DNA sequence data. All three are desirable because they can be highly polymorphic, making them more informative than markers such as restriction fragment length polymorphisms (RFLPs) and allozymes. They also are easier to interpret than amplified fragment length polymorphisms (AFLPs) and more reproducible than randomly amplified polymorphic DNAs (RAPDs). However, there are strengths and weaknesses of each of these marker types and the decision to use one or the other will be based on the type of questions being asked and the resources available to the researcher.

10.2.1 Microsatellites

Microsatellites are short stretches of repetitive DNA found in the nuclear genome, sometimes referred to as short tandem repeats (STRs) or variable number tandem repeats (VNTRs). An individual microsatellite is composed of a motif up to six base pairs that repeats n times, for example CACACACA, which often is expressed as $(CA)_4$, with less repetitive sections of DNA (flanking regions) on either side of the microsatellite repeat. Individual microsatellites and their flanking regions can be amplified using the polymerase chain reaction (PCR) and appropriate primers, which are sequences generally 18 to 24 base pairs long and designed to be complementary to areas in the flanking regions. Individual alleles at a microsatellite are scored as size polymorphisms that generally are differences in the number of individual repeats [e.g., $(CA)_4$ versus $(CA)_6$]. Microsatellite markers have the desirable properties of being highly polymorphic, codominantly inherited, and widely distributed throughout genomes (Weber 1990).

However, use of microsatellite markers relies on the assumption that all size variation is due to changes in the number of repeat units. Changes in repeat number are thought to occur via "slip-strand mispairing" where the repetitive segment aligns improperly during DNA replication, resulting in excision or addition of a repeat unit (Levinson and Gutman 1987; Weber and Wong 1993). This simple model of mutation is known as the stepwise mutational model or SMM (Kimura and Ohta 1978) and is appealing because it makes the relationship between individual alleles easy to understand and model. However, this is generally an oversimplification, as most loci undergo some percentage of mutations that involve more than one repeat unit. The degree to which the SMM is violated depends on the individual microsatellite; consequently, analysis of microsatellite data often requires use of the more complicated two-phase model (TPM) that takes into account both step-wise and larger repeat additions and/or deletions (Di Rienzo et al. 1994; Ellegren 2004). Moreover, many microsatellites feature mutations that do not involve the repetitive unit (Angers and Bernatchez 1997), causing shifts in size that may or may not correspond to the repeat unit. Other mutations may affect the binding affinity of a PCR primer for the target sequence, leading to amplification failure during PCR (Chapuis and Estoup 2007). Loci that are highly polymorphic may feature several of these confounding problems and the risk of homoplasy (where two alleles appear identical but have different genealogical history) may be high (Balloux et al. 2000). Finally, microsatellites frequently exhibit PCR artifacts, such as stutter bands and allele size changes, which further complicate scoring (Figure 10.1). For studies where the goal is individual identification, as in parentage, kinship, or genetic tagging, the nature of mutational events is usually unimportant (as

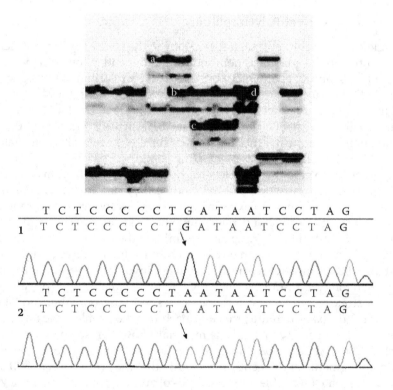

FIGURE 10.1

A color version of this figure follows page 336. A comparison of microsatellite data and SNP data. Top panel is a gel image generated on a LiCor 4200 Global IR² system. Each band is a different allele detected in an array of juvenile *Carcharhinus plumbeus* and each column is one individual. The relationship between alleles marked a, b, and c is assumed to be the result of either the loss or gain of repetitive motif units but is only scored as a size polymorphism so one cannot be sure. Allele d shows a strong stutter band, which is a PCR artifact that may make accurate scoring difficult. Bottom panel is an image of a 20 base pair read that differs by one base pair substitution. In this case the relationship between alleles is clear; 1 has a G at site 10 and 2 has an A at site 10. When scored as an SNP all individuals sampled will show one state or the other for each allele at this locus. This provides less information than a polymorphic microsatellite locus.

mutational rates tend to be low). Genotyping error, however, has the potential to seriously confound results. For other types of studies such as population genetics, where the mutational model may be more important, researchers who are aware of these problems can screen and select loci to avoid them.

Another potential pitfall associated with the use of microsatellites is the expense and time associated with creating enriched libraries and designing species-specific markers. This problem can be circumvented by the use of microsatellites designed for one species on a closely related species. The opportunity to do this is increasing in elasmobranch research, as markers are becoming available for an ever widening range of species across taxa [e.g., *Carcharadon carcharius* (Pardini et al. 2001); *Carcharhinus limbatus* (Keeney and Heist 2003); *Stegostoma fasciatum* (Dudgeon et al. 2006); *Raja clavata* (Chevolot et al. 2007)]. While this type of approach is useful, it is important to note that many microsatellites are most polymorphic in the species for which they have been isolated. For some applications where high levels of polymorphism are important, such as parentage, species-specific markers are likely to be preferable (for a more complete review of microsatellites and their use, see Goldstein and Schlötterer 1999 and Ellegren 2004).

10.2.2 SNPs (Single Nuclear Polymorphisms)

SNPs are single base pair polymorphisms, generally in the nuclear genome. Whereas microsatellite distribution in the genome is nonrandom and dependent on both motif and organism of interest (Chakraborty et al. 1997), SNPs are densely distributed throughout genomes (Vignal et al. 2002). Because SNPs are single base pair substitutions, they have a very simple mutational model and a low probability of homoplasy, due to the low frequency of these substitutions at a given site (Li, Gojobori, and Nei 1981; Martinez-Arias et al. 2001). However, this slow mutational rate means that most SNPs have only two alternate states, making them diallelic. For this reason, datasets must be composed of large numbers of SNPs or panels of linked SNPs to gain the same level of resolution available from a modest number of microsatellites (Glaubitz, Rhodes, and DeWoody 2003; Jones et al. 2009). Finally SNPs are easy to score and interpret, making experimenter error less common (Figure 10.1).

There are many cost-effective and reliable methods for amplifying and scoring SNPs (reviewed in Kwok 2001), but designing and optimizing these markers may require significant sequencing and cloning that can be cost prohibitive (methods of design are reviewed in Vignal et al. 2002). A more cost-effective strategy involves screening large numbers of sequences present on web resources such as GenBank (available at http://www.ncbi.nlm.nih.gov/). This strategy works best for researchers working on either model organisms or intensely studied taxa. Currently, this approach may be problematic for elasmobranch researchers due to the paucity of available materials; however, some exceptions exist (see Nazarian et al. 2007). In addition, much of the available sequence data have originated from functional segments of the genome, meaning that the acquisition of SNPs from neutral parts of the genome (desirable for some types of analysis) may be difficult. While these markers have not yet found wide application in elasmobranch behavioral ecology, as the number of web-available sequences increases, their use may become more common (for a more complete review of SNPs and their use, see Kwok 2001 and Vignal et al. 2002).

10.2.3 DNA Sequence Data

DNA sequence data combines many of the positive aspects of the previously discussed molecular markers and can be nuclear or mitochondrial. Sequences are typically polymorphic, because mutations at multiple sites along the sequenced segment are assayed, and scoring the base changes is easy and accurate. Because the exact mutations that define different haplotypes can be observed directly, modeling relationship(s) between or among sequences is generally straightforward (Goldman 1993). The amount of information available from sequence data, however, is highly dependent on the portion of the genome sequenced. Coding regions (e.g., exons in nuclear genes or protein-coding mitochondrial genes) may be too conserved to be useful in behavioral ecology. Noncoding regions (e.g., introns in nuclear genes or the mitochondrial control region) are generally not constrained by selection and thus are more variable. While these regions may be more useful in behavioral ecology, too much variation may make the relationship(s) between or among sequences problematic for some applications.

There are two major concerns with sequence data. The first involves the design of appropriate primers that amplify the correct target region. This is especially important when amplifying nuclear DNA where there may be multiple, nonorthologous copies of the same gene or pseudogenes (Hillis et al. 1996). In addition, nuclear mitochondrial DNAs (numts; Lopez et al. 1994), mitochondrial DNA that has been transposed into the nuclear genome, can confound analysis of mitochondrial loci. Identification of these artifacts may require significant cloning and sequencing. The second and related concern is price. Although improved technology has made sequencing relatively inexpensive and time efficient, sequencing a large number of individuals is still far more expensive than using either microsatellites or SNPs. In addition, elasmobranchs exhibit an extremely slow rate of mtDNA sequence evolution (Martin, Naylor, and Palumbi 1992). Researchers interested in using mtDNA may therefore need long sequences to capture enough variation, further increasing expense (for a more complete review of sequence data and their use, see Avise 1987 and Ellegren 2004).

For elasmobranch researchers interested in using DNA sequence data, a variety of primers and sequences are available online for both nuclear and mitochondrial DNA. Using these resources to design primers to amplify target sequences in a species of interest may prevent some of the problems (e.g., decreasing the time needed to optimize primers) listed previously. This type of approach has been utilized for multiple phylogeographic studies involving sharks, all of which used the same primer set to amplify the mtDNA control region (Duncan et al. 2006; Keeney and Heist 2006; Schultz et al. 2008; Portnoy 2008).

10.3 Mating System Analysis

10.3.1 Introduction

The use of highly polymorphic molecular markers has revolutionized our understanding of mating systems. Polygynandry (in which both sexes mate with multiple partners) had long been considered the dominant mating system in aggregate spawners with external fertilization and molecular inquiry has confirmed this assumption (DeWoody and Avise 2001; Myers and Zamudio 2004). However, in species with internal fertilization (e.g., mammals and birds), monogamy or polygyny (in which males mate with multiple partners) were considered the dominant mating systems. Studies using high-resolution molecular markers have revealed that polyandry (in which females mate with multiple partners) is common in these species even when observational study suggested female monogamy (Gibbs et al. 1990; Carling, Wiseman, and Byers 2003; Goetz, McFarland, and Rimmer 2003;

Yamaguchi et al. 2004). That females are capable of producing offspring sired by multiple males should not be a surprise, as most animals, even humans, that ovulate multiple eggs have the potential to be inseminated by multiple males (Girella et al. 1997).

There are two major categories of benefits (direct and indirect) that can be used to explain female polyandry. Direct benefits increase the number of offspring a female can produce. These benefits may take the form of nutritive gifts that can be invested in the production of ova, increased sperm volume, or shared parental care (Gray 1997; Avise et al. 2002; Gosselin, Sainte-Marie, and Bernatchez 2005). For species where fecundity is fixed and sperm is not a limiting factor, indirect benefits that increase survivorship and eventual reproductive success of offspring have been invoked to explain polyandrous behavior (Zeh and Zeh 2001). Indirect benefits include increased additive genetic variance in progeny, bet-hedging in unstable environments, precopulatory or postcopulatory trading-up, and postcopulatory defense against genetic incompatibility (Zeh and Zeh 1997; Newcomer, Zeh, and Zeh 1999; Jennions and Petrie 2000; Tregenza and Wedell 2000). While these are appealing explanations for female behavior, they are extremely hard to demonstrate (Byrne and Roberts 2000; Garner and Schmidt 2003). Detecting indirect benefits is even more difficult in long-lived species with late maturity because such benefits can only be measured by reproduction of an individual's offspring. What's more, there is some doubt that indirect benefits can balance costs often associated with mating (Yasui 1998). This has led researchers to suggest that sometimes there may be no benefit to female remating. Instead, females remate to avoid costs associated with resistance, a phenomenon known as convenience polyandry (Alcock et al. 1978).

10.3.2 Methods

Kinship analysis, using highly polymorphic molecular markers, is a straightforward way to assess mating systems (Blouin et al. 1996; Fiumera et al. 2001; Jones and Ardren 2003). The general strategy involves collecting tissues from large numbers of juveniles of known age. Preferably, tissues also will be collected from known parents (for live-bearing animals, this is usually the mother). Arrays of multilocus genotype data can then be collected using microsatellites or SNPs (due to expense, sequencing is likely not practical). When knowledge of parental genotypes is not possible, the basic principle underlying analysis derives from Hamilton's (1964) original formulation of the genetic relatedness between relatives. Put simply, full siblings share one quarter of their mother's genes and one quarter of their father's genes, meaning they have half of their genome in common. Half siblings related maternally still share one quarter of their mother's genes but none of their father's genes. Just like full siblings, offspring and parents share half of their genome. However, offspring directly inherit genes from their parents meaning they will share one allele at every locus examined. Siblings on the other hand may share no alleles at a particular locus. This means that despite the fact both pairings share half of their genome, offspring–parent relationships are always easier to detect than sibling relationships. Individuals who are unrelated also share parts of their genome due to common inheritance in generations past and homplasy. This creates noise that can lead to incorrectly inferred relationships among juveniles. In order to account for this, algorithms must correct for the probability that shared alleles may be identical by state but not identical by descent. Increasing the number of markers and their level of polymorphism further helps increase the accuracy of assignment (see Blouin 2003 for a full review). Individuals can then be grouped by their relatedness or assigned to categories such as full sibling or half-sibling, taking into account allele frequencies, and statistical confidence can be generated for these groupings using Bayesian or

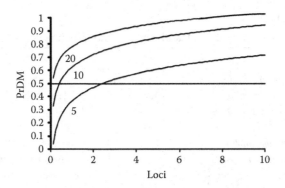

FIGURE 10.2
The probability of detecting multiple paternity (PrDM) increases as the number of loci analyzed increases. The effect is greatest when litter size is small. The three trend lines represent litters of 5, 10, and 20 pups each. PrDM was calculated using PrDM software (Neff and Pitcher 2002). Each locus was assigned five alleles of equal frequency and females were assigned a homozygous condition. If females were heterozygous PrDM would be smaller. A similar affect will occur if the variability of all loci is increased.

maximum likelihood approaches (Goodnight and Queller 1999; Emery et al. 2001). When the maternal genotype is known, reconstructing male genotypes is quite simple. Maternal alleles are identified in progeny and the remaining alleles assigned to males. The algorithms available for this type of analysis either search for the minimum number of sires or the most likely combination of sires based on allele frequencies in the population (Wang 2004; Jones 2005; Kalinowski, Taper, and Marshall 2007). Parental genotypes can also be reconstructed without knowledge of either parent's genotype but these methods rely on the assumption that one sex has greater contribution to the array of progeny than the other (Feldheim, Gruber, and Ashley 2004).

Within a litter, polyandry can be detected by looking for half-siblings that are related by an inferred maternal genotype. This conclusion will be aided if mothers have been genotyped previously. If not, the assumption that maternally related offspring are more common than paternally related offspring must be justified. This approach works well if the number of breeders contributing to offspring in a given location is small and most of the juveniles can be sampled (Feldheim, Gruber, and Ashley 2004; DiBattista et al. 2008b), but will be difficult if the number of breeders is large and only a small fraction of juveniles can be sampled (Portnoy 2008).

Ideally, the maternal multilocus genotypes will be known when litters are analyzed. In such situations, a litter from a monogamous mating will have no more than four alleles present at a given locus. The detection of extra-paternal alleles at one or more loci can be used to infer genetic polyandry. Various algorithms have been constructed that can return the most likely multilocus genotypes of the unsampled parents (Jones 2005; Kalinowski, Taper, and Marshall 2007). When the number of offspring is small, using multiple polymorphic markers (four or more) will increase the probability of detecting extra-pair matings (Figure 10.2). Using multiple markers also ensures that PCR artefacts at a single marker, mistakenly identified as extra paternal alleles, do not lead to a conclusion of polyandry.

10.3.3 Mating Systems in Elasmobranchs

Molecular tools have been used widely in studies of elasmobranch mating systems. So far, all species where multiple litters have been examined have shown genetic polyandry.

TABLE 10.1

Prevalence of Genetic Polyandry in Elasmobranch Species

Species	N	P	L	Study
Ginglymostoma cirratum	2	1.00	28.0	Ohta et al. 2000
Ginglymostoma cirratum	1	1.00	32.0	Saville et al. 2002
Ginglymostoma cirratum	2	1.00	NA	Heist 2004
Triakis scyllium	1	0.00	17.0	Ohta et al. 2000
Carcharhinus plumbeus	20	0.85	9.4	Portnoy et al. 2007
Carcharhinus plumbeus	20	0.40	5.5	Daly-Engel et al. 2007
Carcharhinus galapagensis	1	0.00	7.0	Daly-Engel et al. 2006
Carcharhinus altimus	1	1.00	9.0	Daly-Engel et al. 2006
Sphyrna tiburo	22	0.19	8.5	Chapman et al.2004
Negaprion brevrirosrris	45*	0.86	NA	Feldheim et al. 2004
Negaprion brevrirosrris	46*	0.81	NA	DiBattista et al. 2008
Squalus acanthias	10	0.30	5.0	Lage et al. 2008
Squalus acanthias	29	0.17	5.0	Portnoy unpublished data
Raja clavata	4	1.00	43.0	Chevolot et al. 2007

Note: N is the number of litters examined, P is the percentage of litters that showed polyandry, L is the average litter size calculated from the citation. For studies where polyandry was detected via inferred multilocus parental genotypes (*) no average litter size is presented. Litter sizes were also not reported in Heist (2004) for the two females.

However, the prevalence has differed greatly among species (Table 10.1). The majority of litters examined were genetically monogamous in bonnethead sharks, *Sphyrna tiburo* [19% ($N = 22$), Chapman et al. 2004] and spiny dogfish, *Squalus acanthias* [30% ($N = 10$) and 17% ($N = 29$), Lage at al. 2008; Portnoy, unpublished data]. By contrast, the majority of litters in lemon sharks, *Negaprion brevirostris*, nurse sharks, *Ginglymostoma cirratum*, and sandbar sharks, *Carcharhinus plumbeus*, had multiple sires (Saville et al. 2002; Ohta et al. 2000; Feldheim, Gruber, and Ashley 2004; Portnoy et al. 2007).

These studies were designed to detect genetic polyandry, the presence of multiple sires in a single litter. Females may engage in behavioral polyandry, mating with multiple males, without producing a multiple-sired litter. In addition, in iteroparous species, females may engage in serial monogamy, a type of polyandry in which females mate with one male per reproductive effort but change mates across years (Sugg and Chesser 1994; Karl 2008). The latter is likely a general feature of elasmobranch reproduction even when individual litters are sired by a single male because females almost certainly do not mate with the same male in successive reproductive seasons.

Given the high cost associated with reproduction in elasmobranchs (Pratt and Carrier 2001), the presence of genetic polyandry has led to questions about the benefits of this behavior. Portnoy et al. (2007) found no increase in female fecundity (a direct benefit) for genetically polyandrous female *Carcharhinus plumbeus* in the western North Atlantic. DiBattista et al. (2008a) could not detect indirect benefits, measured as increased survival of pups to age two, in *Negaprion brevirostris*. In both cases, convenience polyandry may be the best explanation for multiple paternity. However, future studies are required as there are multiple types of indirect benefits and they will be difficult to detect in species with long generation times. In *Carcharhinus plumbeus,* the prevalence of genetic polyandry varies between populations with high prevalence in the western North Atlantic (Portnoy et al.

2007) but less than 50% prevalence in Hawaii (Daly-Engel et al. 2007), suggesting that environment or demography may have an effect on female remating rate. In *Negaprion brevirostris*, however, the prevalence of genetic polyandry is consistent among nursery grounds with different environmental characteristics, within the same population (DiBattista et al. 2008a). Our understanding of patterns of genetic polyandry in elasmobranchs is still in its infancy and further study examining possible indirect benefits is necessary. Large comparative studies across multiple populations with different environmental and demographic characteristics as well as long-term studies will be needed to further examine these questions.

Parentage analysis has found an alternate application in examining apparent virgin births in captive sharks. Parthenogenesis had been suggested for both *Sphyrna tiburo* and *Carcharhinus limbatus* when single females of each species gave birth in captivity, apparently without access to males. In both cases molecular analysis failed to turn up paternal alleles, suggesting that these animals produced viable offspring with no paternal contribution (Chapman et al. 2007; Chapman, Firchau, and Shivji 2008). The relevance of these findings in wild populations, however, is unclear.

10.4 Female Philopatry

10.4.1 Introduction

In some animals male and female reproductive strategies may lead to differing dispersal potential. In particular, females, which in live-bearing animals tend to show greater parental investment in offspring than males, may show philopatry to specific areas if it will increase survival of their offspring (Perrin and Mazalov 2000). There are two main reasons to be concerned with this behavior. First, strongly philopatric animals may be at greater risk of localized extinction when exploited (Hueter 1998), making definition of the presence and strictness of philopatry important. Second, adaptive genetic variation is carried in the nuclear genome, so the presence or absence of nuclear gene flow should be used to define populations. When population structure is examined using only mtDNA, misidentification of stocks is likely in highly philopatric animals featuring male-mediated gene flow.

Traditional observational and tagging studies have demonstrated that individual females repeatedly return to the same nursery grounds over multiple years in both *Ginglymostoma cirratum* and *Negaprion brevirostris* (Pratt and Carrier 2001; Feldheim, Gruber, and Ashley 2002). Acoustic monitoring data also suggests that female *Carcharhinus plumbeus* may return to the same nursery ground in multiple years (Portnoy, unpublished data). Tagging studies involving *Carcharhinus limbatus* and *Carcharhinus plumbeus* have shown the tendency for juveniles to return to the same nursery grounds in the summer months (Heupel and Hueter 2001; Grubbs et al. 2007).

Molecular approaches may be important when the ability to tag and resample individuals is compromised by characteristics of the environment or the animal's behavior. In addition, molecular approaches can be used to augment tagging and observational studies, as they allow researchers to examine patterns of philopatry over wider spaces and longer periods of time than tagging or observational studies may allow. The researcher's interests in individual behavior versus patterns of philopatry present over time and space will determine the methodology employed.

10.4.2 Methods

10.4.2.1 Individual Identification

For researchers interested in studies of individual behavior, molecular markers, specifically microsatellites and SNPs, can be used to generate "unique" multilocus genetic identifiers for individuals (genetic fingerprinting). This approach works by genotyping individuals at a modest to large number of markers. As additional markers are assayed, the probability of catching two individuals with identical genotypes rapidly approaches zero. To give a simplified example, consider a study where a researcher employs 10 microsatellites each with 10 equally frequent alleles to genotype individual animals. The probability of sampling an unrelated individual with the same genotype as a previously genotyped individual is $1.0*10^{-17}$. While having a panel of markers that behave this way is highly unlikely, the example demonstrates the power of this methodology. This type of approach can be used to identify and track individuals when traditional tagging is not possible, or if tag shedding is a problem (Palsbøll et al. 1997; Amos, Schlötterer, and Tautz 1993) but requires resampling individuals.

The problem with the above approach is that resampling adult individuals may be inherently difficult, which is why traditional tagging approaches could not be used in the first place. Using the kinship approach, discussed in the previous section, researchers can identify philopatric females by catching and genotyping their progeny. Because this approach can be used to infer maternal genotypes, it negates the need to catch and handle adults. However, the inference of unsampled adult genotypes from juveniles will always require a greater number of markers than the detection of previously sampled adult genotypes from juveniles. In addition, a larger number of SNPs, relative to microsatellites, will be required for this type of analysis because they are diallelic (Glaubitz, Rhodes, and DeWoody 2003).

10.4.2.2 Phylogeographic Approaches

While kinship and genetic fingerprinting are powerful approaches to examining individual philopatry over short periods of time, a phylogeographic approach is needed to examine philopatry over larger geographic areas and across many generations. For this type of application the use of molecular markers with different modes of inheritance is necessary (Karl, Bowen, and Avise 1992; Palumbi and Baker 1994). Specifically, comparisons should be made between patterns of variation across samples with maternally inherited mtDNA versus bi-parentally inherited nuclear genes (microsatellites or SNPs). To accomplish this, the F-statistic or F_{ST}, derived from the inbreeding coefficient (Wright 1965), can be used to detect population structure. This is commonly expressed by examining variance in allele frequency within and between subpopulations as defined by (Weir and Cockerham 1984):

$$\Theta = \frac{s^2}{p(1-p)} \tag{10.1}$$

where p is the average sample frequency of allele A and s^2 is the variance of that allele across populations. The values obtained can be tested against the expectation of the null model (H_0), a population with panmictic mating, using permutation testing (Excoffier, Laval, and Schneider 2005).

Both mtDNA and microsatellite analyses may be enhanced by the use of F_{ST} analogues, Φ_{ST} (Excoffier, Smouse, and Quattro 1992), and R_{ST} (Slatkin 1995), respectively. These measures incorporate assumptions more appropriate for the inheritance patterns and mutational properties of the respective markers when calculating an F_{ST} value. Statistical procedures can examine variance in a pairwise fashion or at different hierarchical levels. When Φ_{ST} estimates, calculated as both the variance within and between groupings, are tested for significance the procedure is analogous to an analysis of variance (ANOVA). For this reason it is referred to as analysis of molecular variance (AMOVA). Φ_{ST} is an unbiased analogue of F_{st} that takes into account both the divergence of alleles as well as their frequencies. It can be defined as (Excoffier, Smouse, and Quattro 1992):

$$\Phi_{st} = \frac{\sigma_a^2 + \sigma_b^2}{\sigma^2} \tag{10.2}$$

where σ_a^2 is the variance component among groups of population, σ_b^2 is the variance component among populations within groups, and σ^2 is the total variance (the sum of the first two elements plus the variance component between individuals within populations). Discordance between results obtained with these markers, in which mtDNA data suggest population structure and nuclear data do not, could be indicative of female philopatric behavior and male-mediated gene flow.

10.4.3 Female Philopatry in Elasmobranchs

The approaches discussed above have been used to explore patterns of female philopatry in elasmobranchs. Genetic tagging, using nine polymorphic microsatellite loci, was used to accompany a traditional tagging study of *Negaprion brevirostris* in Bimini, Bahamas (Feldheim et al. 2002). The results showed that the tagging study failed to detect 12% of the returning animals due to tag shedding. Using the same set of markers, individual adults and juveniles were genotyped across six years of sampling. By assigning juveniles back to sampled females, philopatric behavior was detected in four females (Feldheim, Gruber, and Ashley 2002). These results also were supported by a study in which 32 of 45 females, whose genotypes had been reconstructed from sampled juveniles, gave birth at Bimini in multiple years (Feldheim, Gruber, and Ashley 2004). Similar results were obtained using the same methodology for *Negaprion brevirostris* at Marquesas Key, Florida (DiBattista et al. 2008b). Despite evidence of strong philopatric behavior by individual female *Negaprion brevirostris*, analysis of population structure, using microsatellites, indicated no population structure for this species in the western Atlantic, suggesting male mediated gene-flow (Feldheim, Gruber, and Ashley 2001). For *Carcharhinus plumbeus* pupping in Chesapeake Bay, Delaware Bay, and the Eastern Shore of Virginia, this same approach linked several females of known genotype to their offspring caught at different times (Figure 10.3). However, attempts to infer the genotypes of nonsampled females from juveniles were largely unsuccessful. This may have been caused by a relatively small sample size, about 100 juveniles per annum, compared to a relatively large effective number of breeders, 500 to 1000 per annum (Portnoy et al. 2008a).

Several studies have used a phylogeographic approach to examine sex-biased dispersal. In white sharks, *Carcharodon carcharias*, significant divergence was found in mtDNA sequences between samples from the eastern and western Pacific, while no divergence was detected with microsatellites (Pardini et al. 2001). Likewise in mako sharks, *Isurus oxyrinchus*, population structure across the Atlantic was detected with mtDNA, while nuclear microsatellite allele frequency distributions were homogeneous (Schrey and Heist 2003). Patterns of variation obtained from mtDNA sequence data and microsatellites suggest philopatry in female *Carcharhinus limbatus* to the western North Atlantic, Gulf of Mexico,

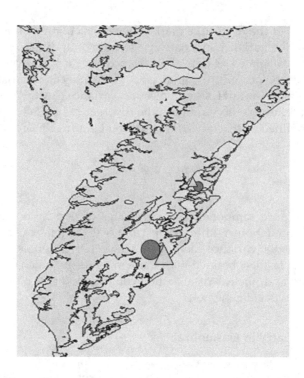

FIGURE 10.3
A color version of this figure follows page 336. The reproductive output of females of known genotype can be detected through the use of multilocus genetic fingerprinting. Here adult and juvenile *Carcharhinus plumbeus* were sampled in the lagoons of the Eastern Shore of Virginia. A mother–offspring relationship was detected using eight microsatellites between a sampled juvenile (small triangle) and a postpartum female (large triangle) sampled two years later. A postpartum female (large circle) and two juveniles (small circles) identified as the female's progeny were caught several months apart in the same year.

and Caribbean Sea, with high levels of male dispersal (Keeney et al. 2005). Long-term female philopatry also has been suggested for *Negaprion brevirostris* in the western Atlantic (Schultz et al. 2008). Similar results with *Carcharhinus plumbeus* were obtained in the Indo-West Pacific, where mtDNA divergence between samples taken from Taiwan and western Australia, and between western Australia and eastern Australia, were significant and more than an order of magnitude greater than the nonsignificant measures of microsatellite divergence (Portnoy et al. 2008a). It is important to remember that because different marker types differ in mode of inheritance, ploidy, and mutation rate, some discordance in results should be expected (see Buonaccorsi, McDowell, and Graves 2001). Sex biased dispersal should only be inferred when the magnitude of this discordance is larger than expected, as in the previous example.

10.5 Effective Population Size

10.5.1 Introduction

Small populations are of concern in conservation and management contexts because they are more susceptible to processes (demographic and genetic) that lead to extinction

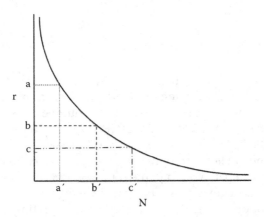

FIGURE 10.4
The size of a population (N) and its relationship with the rate of evolutionary change, in this case drift (r). While the change in N between c′ and b′ equals that between b′ and a′, the resulting change in r between c and b is less than between b and a. As populations shrink, r increases more rapidly, a phenomenon that leads to the fixation of alleles and may rapidly lead to loss of adaptive genetic variation.

(Franklin 1980; Newman and Pilson 1997). In part this is due to the inverse relationship between the number of successful breeders and genetic drift (Figure 10.4). As drift increases so does the probability of the fixation of deleterious (harmful) alleles and loss of additive adaptive variance. In addition, as the gene pool contracts the rate of inbreeding increases. In total these processes may reduce individual fitness across a population that may in turn lead to extirpation. Drift can be simply conceptualized as the sampling variance of gametes during each round of reproduction (Falconer and MacKay 1996):

$$\sigma_p^2 = \frac{pq}{2N_c} \tag{10.3}$$

where p and q are allele frequencies and N_c is the census population size in a Wright–Fisher ideal population (Wright 1931). Since true populations violate most assumptions of an ideal population (equal reproductive success, nonoverlapping generations, equal number of males and females, and random mating; for a more complete review, see Caballero 1994), N_c must be replaced with the measure effective population size (N_e).

It is important to note that N_e and N_c are rarely equal and there is no direct relationship between the two measures. In fact surveys of wild populations have found that the two measures vary greatly across taxa, with N_e/N_c ratios from 10^{-5} in many marine species to nearly 1.0 in some terrestrial vertebrates (Frankham 1995; Hedrick 2005). Therefore, N_e must be estimated from demographic and/or genetic data. Difficulty in obtaining the information required for demographic methods has led to interest in using genetic data to estimate N_e (Caballero 1994; Wang 2005). There are two broad types of N_e that can be estimated from genetic data: historical and contemporary. Understanding the distinction is very important as the meaning and use of the two measures differ greatly.

10.5.1.1 Historical Effective Population Size

Historical or long-term effective size is a backward looking measure that can be defined as:

$$N_e = \frac{\Theta}{4\mu} \quad \text{or} \quad N_e = \frac{\Theta}{2\mu} \tag{10.4}$$

for diploid autosomal loci or haploid mtDNA, respectively (Hartl and Clark 2007). The parameter θ is a function of DNA diversity (sequence data) or multilocus polymorphism (microsatellites, SNPs), which can be calculated in a number of software packages [e.g., MIGRATE (Beerli and Felsenstein 1999); DNASP (Rozas et al. 2003); ARLEQUIN (Excoffier, Laval, and Schneider 2005)]. Put simply, populations that persist over long periods of time at large size will accumulate large amounts of genetic diversity detectable in the present. This, of course, is highly dependent on μ, the mutation rate, because the larger the μ, the more quickly populations of any size accumulate diversity.

There are a number of problems with deriving and interpreting historical N_e estimates. To begin with, using an inappropriate μ will greatly bias the estimate. Because the exact mutation rate at a given locus, let alone across multiple loci, in a species of interest is seldom known, μ must be estimated from the fossil record or from known vicariant events. For many species, this type of information is lacking, so mutation rates estimated for the same locus from closely related species are used. While this does provide a rational "best guess" it does not truly solve the problem. A better way to avert this problem may be to assume that within a species μ is a constant and therefore θ can be left uncorrected. Using this strategy, θ compared across populations within a species will provide information on relative sizes over long periods of time.

A second problem that this strategy cannot overcome is that demographic changes over time, such as fluctuations in population size or past connectivity between or among populations, will lead to differences in historical N_e, making it hard to determine to what time period and population the estimate applies (Crandall, Posada, and Vasco 1999; Schwartz, Tallmon, and Luikart 1999). For example, if one were to estimate historical N_e for a population that is currently isolated but may have had significant immigration from unsampled populations in the past, the N_e estimated might be quite large, as it pertains to all of the populations together instead of the population of interest. Without significant information about connectivity in the past there is no way to tell. Past population bottle necks will have the opposite effect, resulting in historical N_e estimates that are much smaller than contemporary N_e. Finally, from a conservation and management standpoint, historical N_e provides little information about the current number of breeders in a population or that population's evolutionary potential.

10.5.1.2 Contemporary Effective Population Size

Contemporary estimates of N_e apply to generations in the recent past and estimates of the effective number of breeders (N_b) apply directly to the parents of a sampled cohort (Waples 2005). Therefore, the information is more useful for conservation and management of elasmobranchs. There are three major types of contemporary estimates that differ in the type of evolutionary change measured in order to estimate N_e. Variance N_e estimates are based on measures of drift between and within generations and are the most commonly used estimates. Less commonly used are inbreeding N_e estimates, based on the rate of inbreeding within a population, and eigenvalue N_e estimates, based on rates of allele loss within a population. While researchers should be aware of these distinctions (see Crandall, Posada, and Vasco 1999 for review), a more pressing concern is whether one sample (point estimates) or several temporal samples (temporal estimates) are required. Since contemporary estimates are more useful to elasmobranch researchers interested in reproduction and conservation, they will be the focus of the methods section below.

10.5.2 Methods

10.5.2.1 Heterozygote Excess

Heterozygote excess is a point estimate dependent on the idea that when the number of breeders is small, allele frequencies in males and females will tend to be different because of sampling error. This means the number of heterozygote progeny will be greater than expected (Pudovkin, Zaykin, and Hedgecock 1996). N_b can, therefore, be calculated from equations that derive from the simple relationship (Luikart and Cornuet 1999):

$$N_b = \frac{H_{exp}}{2(H_{obs} - H_{exp})} \tag{10.5}$$

where H_{exp} is the expected heterozygosity and H_{obs} is the observed heterozygosity. The method can be used to estimate N_e if adults are sampled. However, heterozygote excess is most accurate and easily differentiated from infinity when the true number of breeders is relatively small, which is unlikely for a whole population (Balloux 2004). Instead, it will be most useful for estimating N_b from juveniles (Zhdanova and Pudovkin 2008). The ability to sample individuals from known cohorts at specific nursery grounds should make this method appealing to elasmobranch researchers interested in the magnitude of breeding effort. Increased sample size and larger numbers of polymorphic loci will increase the accuracy of estimates and shrink confidence intervals.

10.5.2.2 Linkage Disequilibrium

The linkage disequilibrium method is another point estimate based on the premise that when populations are small, alleles at independent loci will appear linked because of drift. It calculates the correlation among alleles at unlinked loci (r), which can be related to N_e by the formula (Hill 1981; Waples 1991):

$$N_e = \frac{1}{3*\left(r^2 - \frac{1}{S}\right)} \tag{10.6}$$

where S is sample size. As with the heterozygote excess method, the linkage disequilibrium method can be used to calculate N_e or N_b but performs best when N_e or N_b is small. Consequently, it also may be useful for estimating N_b from samples of juveniles of known age at specific nursery grounds. In addition, there may be downward bias associated with small sample sizes (England et al. 2006). Though later formulations have accounted for this (Waples 2006), having large sample size ($S > 0.1N_e$) is still advisable.

10.5.2.3 Temporal Estimates

Temporal estimates require multiple samples separated in time and measure changes in gene frequencies caused by genetic drift. The magnitude of these changes will be affected by two parameters: the amount of time between sampling and the effective size of the population. This relationship can be expressed simply as (Krimbas and Tsakas 1971; Nei and Tajima 1981; Pollack 1983):

$$N_e = \frac{T}{2\{F - (1/2S_0 + 1/2S_t)\}} \tag{10.7}$$

where T is the generation time between samples, F is an F_{ST} analogue that employs temporally separated samples instead of spatially separated samples, and S_0 is sample size from the first time period and S_t is the sample from a time period t generations later. In general, this is the most widely used method because it tends to produce both accurate estimates and tight confidence intervals (Waples 1989). Another advantage of this method is that, unlike the previous two methods, which require diploid data (microsatellites, SNPs), one can use mitochondrial sequence data to estimate female effective size (N_{ef}; Laikre, Jorde, and Ryman 1998). This approach requires samples be at least one generation apart to ensure accuracy of the estimate (Waples 1991; Williamson and Slatkin 1999).

This may be problematic for elasmobranch researchers because generation times tend to be long and samples that have been archived in a manner allowing for molecular analysis are rare. Instead, a modified version of the temporal method, which examines shifts in allele frequencies between consecutive cohorts, is preferable (Jorde and Ryman 1995). Drift is then related to N_e by the formula

$$N_e = \frac{C}{2GF'} \tag{10.8}$$

where G is generation time, F' is the F-statistic averaged across cohorts, and C is a parameter used to account for the probability of survival to age (l_i) and reproductive output of each age class (b_i). This methodology requires both samples from juveniles of known age and fairly accurate life history data, but these may be easier to obtain than samples separated by a generation or more. As with other methodologies, accuracy will increase dramatically with increased sample size.

10.5.3 Effective Population Size in Elasmobranchs

Estimates of effective population sizes are generally lacking for elasmobranchs. Long-term N_{ef} (female effective size) was examined in a phylogeographic context in whale sharks, *Rhincodon typus*, scalloped hammerheads, *Sphyrna lewini*, and lemon sharks, *Negaprion brevirostris* and *N. acutidens* (Duncan et al. 2006; Castro et al. 2007; Schultz et al. 2008). In these cases calculations were based on θ_1, which is the long-term effective size after a presumed population expansion. In *Rhincodon typus*, N_{ef} was estimated to be between 119,000 and 238,000 worldwide (Castro et al. 2007) and expanded from an original N_{ef} estimate between 13,000 and 26,000. For populations of *Sphyrna lewini*, N_{ef} estimates varied from 550 up to 31,000,000 females (Duncan et al. 2006). For populations of *N. brevirostris*, N_{ef} estimates varied from 13,000 to 26,000 and from 17,000 to 26,000 for N. *acutidens*. Like other forms of long-term N_e, these estimates require the use of an assumed mutation rate and are greatly affected by demographic processes in the past. In addition, these estimates are highly dependent on the model used (Beaumont 2001), in this case a population expansion model. This means that the actual estimates of N_{ef} must be treated very cautiously.

However, conclusions regarding population increases in *Rhincodon typus* and the relative population sizes in *Sphyrna lewini* and *Negaprion* are robust because they can be based on θ alone and are not affected by uncertainty regarding mutation rates.

Estimates of contemporary N_e and N_b have been made for commercially exploited *Carcharhinus plumbeus*, using linkage disequilibrium and the modified temporal method (Portnoy et al. 2008b). Juvenile sharks of known age were collected from lagoons along the Eastern Shore of Virginia and from the Delaware Bay. N_b was estimated for each cohort at each location and N_e was estimated across cohorts at each location. The results showed fairly consistent estimates of N_b across years within the Delaware Bay and Eastern Shore nursery grounds, with harmonic means of 1059 and 511, respectively. This suggests twice the reproductive effort in the Delaware Bay, likely due to a larger number of females using that nursery ground. In addition, estimate of the census number (N_c) of breeders in the Delaware Bay were compared to N_e and N_b resulting in ratios ($N_{e(b)}/N_c$) near 0.5. This closely approximates the expected relationship between N_e and N_c for species with overlapping generations where reproductive success is even (Nunney 1993). In addition, the estimated ratio is orders of magnitude higher than ratios seen in exploited bony fishes (10^{-3} to 10^{-5}; Hoarau et al. 2005), typical of species that feature large variances in reproductive success (Hedgecock 1994). This suggests that the relationship between N_e and N_c is more affected by life history characteristics than exploitation.

10.6 Summary and Conclusions

The use of molecular techniques to explore the reproductive biology of elasmobranchs has become a more viable option, even for researchers without genetics backgrounds, as the technology and analysis have become easier and more affordable. These investigations require the use of highly polymorphic markers, but each type has pros and cons that must be weighed, not the least of which will be the availability of resources.

With the appropriate markers, multiple aspects of reproductive behavior can be investigated. Kinship and parentage analysis can be used to investigate mating systems or philopatric behavior. In addition, phylogeographic techniques available to investigate philopatry can uncover long-term trends over wide geographic areas. Multiple formulations exist for estimating N_e and N_b. The most useful ones for elasmobranch researchers interested in current reproductive behavior and conservation are contemporary estimates of N_e and N_b and require either one sample or consecutively sampled cohorts.

References

Alcock J, Barrows EM, Gordh G, et al. (1978) The ecology and evolution of male reproductive behavior in the bees and wasps. Zool J Linn Soc 64: 293–326.

Amos B, Schlotterer C, Tautz D (1993) Social structure of pilot whales revealed by analytical DNA profiling. Science 260: 670–672.

Angers B, Bernatchez L (1997) Complex evolution of salmonid microsatellite locus and its consequences in inferring allelic divergence from size information. Mol Biol Evol 14: 230–238.

Avise JC, Arnold J, Ball RM, et al. (1987) Intraspecific phylogeography: the mitochondrial DNA bridge between population genetics and systematics. Annu Rev Ecol Syst 18: 489–522.

Avise JC, Jones AG, Walker D, et al. (2002) Genetic mating systems and reproductive natural histories of fishes: lessons for ecology and evolution. Annu Rev Genet 36: 19–45.

Balloux F (2004) Heterozygote excess in small populations and the heterozygote-excess effective population size. Evolution 58: 1891–1900.

Balloux F, Brünner H, Lugon-Moulin N, et al. (2000) Microsatellites can be misleading: an empirical and simulation study. Evolution 54: 1414–1422.

Beaumont MA (2001) Conservation genetics. In: Balding DJ, Bishop M, Cannings C (eds) Handbook of Statistical Genetics, John Wiley & Sons, New York, pp 779–812.

Beerli P, Felsenstein J (1999) Maximum-likelihood estimation of migration rates and effective population numbers in two populations using a coalescent approach. Genetics 152: 763–773.

Blouin MS (2003) DNA-based methods for pedigree reconstruction and kinship analysis in natural populations. Trends Ecol Evol 18: 503–511.

Blouin MS, Parsons M, Lacaille V, Lotz S (1996) Use of microsatellite loci to classify individuals by relatedness. Mol Ecol 5: 393–401.

Buonacccorsi VP, McDowell JR, Graves JE (2001) Reconciling patterns of inter-ocean molecular variance from four classes of molecular markers in blue marlin (*Makaira nigricans*). Mol Ecol 10: 1179–1196.

Byrne P, Roberts J (2000) Does multiple paternity improve fitness of the frog *Crinia georgiana*. Evolution 54: 968–973.

Caballero A (1994) Developments in the prediction of effective population size. Heredity 73: 657–679.

Carling M, Wiseman PA, Byers JA (2003) Microsatellite analysis reveals multiple paternity in a population of wild pronghorn antelopes. J Mammal 84: 1237–1243.

Castro AL, Stewart BS, Wilson SG, et al. (2007) Population genetic structure of the earth's largest fish, the whale shark (*Rhincodon typus*). Mol Ecol 16: 5183–5192.

Chakraborty R, Kimmel M, Stivers DN, et al. (1997) Relative mutation rates at di-, tri-, and tetranucleotide microsatellite loci. Proc Natl Acad Sci USA 94: 1041–1046.

Chapman DD, Firchau B, Shivji MS (2008) Parthenogenesis in a large-bodied requiem shark, the blacktip *Carcharhinus limbatus*. J Fish Biol 73: 1473–1477.

Chapman DD, Prodohl PA, Gelsleichter J, et al. (2004) Predominance of genetic monogamy by females in a hammerhead shark, *Sphyrna tiburo*: implications for shark conservation. Mol Ecol 13: 1965–1974.

Chapman DD, Shivji MS, Louis E, et al. (2007) Virgin birth in a hammerhead shark. Biol Lett 3: 425–427.

Chapuis MP, Estoup A (2007) Microsatellite null alleles and estimation of population differentiation. Mol Biol Evol 24: 621–631.

Chevolot M, Ellis JR, Rijnsdorp AD, et al. (2007) Multiple paternity analysis in the thornback ray *Raja clavata* L. J Hered 98: 712–715.

Crandall KA, Posada D, Vasco D (1999) Effective population sizes: missing measures and missing concepts. Anim Conserv 2: 317–319.

Daly-Engel TS, Grubbs RD, Bowen BW, Toonen RJ (2007) Frequency of multiple paternity in an unexploited tropical population of sandbar sharks (*Carcharhinus plumbeus*). Can J Fish Aquat Sci 64: 198–204.

Daly-Engel TS, Grubbs RD, Holland K, et al. (2006) Assessment of multiple paternity in single litters from three species of carcharhinid sharks in Hawaii. Environ Biol Fishes 76: 419–424.

DeWoody JA, Avise JC (2001) Genetic perspectives on the natural history of fish mating systems. J Hered 92: 167–172.

DiBattista JD, Feldheim KA, Gruber SH, Hendry AP (2008a) Are indirect genetic benefits associated with polyandry? Testing predictions in a natural population of lemon sharks. Mol Ecol 17: 783–795.

DiBattista J, Feldheim KA, Thibert-Plante X, et al. (2008b) A genetic assessment of polyandry and breeding-site fidelity in lemon sharks. Mol Ecol 17: 3337–3351.

Di Rienzo A, Peterson A, Garza J, et al. (1994) Mutational processes of simple-sequence repeat loci in human populations. Proc Natl Acad Sci USA 91: 3166–3170.

Dudgeon CL, Feldheim KA, Schick M, Ovenden JR (2006) Polymorphic microsatellite loci for the zebra shark *Stegostoma fasciatum*. Mol Ecol Notes 6: 1086–1088.

Duncan KM, Martin AP, Bowen BW, De Couet HG (2006) Global phylogeography of the scalloped hammerhead shark (*Sphyrna lewini*). Mol Ecol 15: 2239–2251.

Ellegren H (2004) Microsatellites: simple sequences with complex evolution. Nature Rev Genet 5: 435–445.

Emery A, Wilson I, Craig S, et al. (2001) Assignment of paternity groups without access to parental genotypes: multiple mating and developmental plasticity in squid. Mol Ecol 10: 1265–1278.

England PR, Cornuet JM, Berthier P, et al. (2006) Estimating effective population size from linkage disequilibrium: severe bias in small samples. Conserv Genet 7: 303–308.

Excoffier L, Laval G, Schneider S (2005) ARLEQUIN (version 3.0): an integrated software package for population genetics data analysis. Evol Bioinform Online 1: 47–50.

Excoffier L, Smouse PE, Quattro JM (1992) Analysis of molecular variance inferred from metric distances among DNA haplotypes: application to human mitochondrial DNA restriction data. Genetics 131: 479–491.

Falconer DS, Mackay TFC (1996) Introduction to quantitative genetics. Pearson Education, Harlow, UK.

Feldheim KA, Gruber SH, Ashley MV (2001) Population genetic structure of the lemon shark (*Negaprion brevirostris*) in the western Atlantic: DNA microsatellite variation. Mol Ecol 10: 295–303.

Feldheim KA, Gruber SH, Ashley MV (2002) The breeding biology of lemon sharks at a tropical nursery lagoon. Proc R Soc London (Biol) 269: 1655–1661.

Feldheim KA, Gruber SH, Ashley MV (2004) Reconstruction of parental microsatellite genotypes reveals female polyandry and philopatry in the lemon shark, *Negaprion brevirostris*. Evolution 58: 2332–2342.

Feldheim KA, Gruber SH, de Marignac JRC, Ashley MV (2002) Genetic tagging to determine passive integrated transponder loss in lemon sharks. J Fish Biol 61: 1309–1313.

Fiumera AC, DeWoody YD, DeWoody JA, et al. (2001) Accuracy and precision of methods to estimate the number of parents contributing to a half-sib progeny array. J Hered 92: 120–126.

Frankham R (1995) Effective population size/adult population size ratios in wildlife: a review. Genet Res 66: 95–106.

Franklin IR (1980) Evolutionary changes in small populations. In: Soule ME, Wilcox BA (eds) Conservation biology: an evolutionary-ecological perspective. Sinauer Associates, Sunderland, MA, pp 135–150.

Garner TWJ, Schmidt BR (2003) Relatedness, body size and paternity in the alpine newt, *Triturus alpestris*. Proc R Soc London (Biol) 270: 619–624.

Gibbs H, Weatherhead P, Boag P, et al. (1990) Realized reproductive success of polygynous red-winged blackbirds revealed by DNA markers. Science 250: 1394–1397.

Girela E, Lorente J, Alvarez J, et al. (1997) Indisputable double paternity in dizygous twins. Fertil Steril 67: 1159–1161.

Glaubitz JC, Rhodes EJ, DeWoody JA (2003) Prospects for inferring pairwise relationships with single nucleotide polymorphisms. Mol Ecol 12: 1039–1047.

Goetz J, McFarland KP, Rimmer CC (2003) Multiple paternity and multiple male feeders in Bicknell's thrush (*Catharus bicknelli*). Auk 120: 1044–1053.

Goldman N (1993) Statistical tests of models of DNA substitution. J Mol Evol 36: 182–198.

Goldstein DB, Schlötterer C (eds) (1999) Microsatellites: evolution and applications. Oxford University Press, New York.

Goodnight KF, Queller DC (1999) Computer software for performing likelihood tests of pedigree relationship using genetic markers. Mol Ecol 8: 1231–1234.

Gosselin T, Sainte-Marie B, Bernatchez L (2005) Geographic variation of multiple paternity in the American Lobster, *Homarus americanus*. Mol Ecol 14: 1517–1525.

Gray EM (1997) Female red-winged blackbirds accrue material benefits from copulating with extra-pair males. Anim Behav 53: 625–639.

Grubbs RD, Musick JA (2007) Spatial delineation of summer nursery areas for juvenile sandbar sharks in Chesapeake Bay, Virginia. Amer Fish Soc Symp 50: 63–85.

Grubbs RD, Musick JA, Conrath C, Romine J (2007) Long-term movements, migration and temporal delineation of a summer nursery for juvenile sandbar sharks in the Chesapeake Bay region. Amer Fish Soc Symp 50: 87–107.

Hamilton WD (1964) The genetical evolution of social behaviour. I. J Theor Biol 7: 1–16.

Hartl DL, Clark AG (2007) Principles of population genetics, 4th ed. Sinauer Associates, Sunderland, MA.

Hedgecock D (1994) Does variance in reproductive success limit effective population sizes of marine organisms? In: Beaumont MA (ed) Genetics and evolution of aquatic organisms. Chapman and Hall, London, pp 122–134.

Hedrick PW (2005) Large variance in reproductive success and the Ne/N ratio. Evolution 59: 1569–1599.

Heist EJ (2004) Genetics of sharks, skates, and rays. In: Carrier JC, Musick JA, Heithaus MR (eds) Biology of sharks and their relatives. CRC Press, Boca Raton, FL, pp 471–486.

Heupel MR, Hueter RE (2001) Use of an automated acoustic telemetry system to passively track juvenile blacktip shark movements. In: Sibert JR, Nielsen JL (eds) Electronic tagging and tracking in marine fisheries. Kluwer Academic Publishers, Boston, pp 7–64.

Hueter RE (1998) Philopatry, natal homing and localized stock depletion in sharks. Shark News 12.

Hill WG (1981) Estimation of effective population size from data on linkage disequilibrium. Genet Res 38: 209–216.

Hillis DM, Mable BK, Larson A, et al. (1996) Nucleic acids, IV: Sequencing and cloning. In: Hillis DM, Moritz C, Mable BK (eds) Molecular systematics. Sinauer Associates, Sunderland, MA, pp 321–381.

Hoarau G, Boon E, Jongma DN, et al. (2005) Low effective population size and evidence for inbreeding in an overexploited flatfish, plaice (*Pleuronectes platessa* L.). Proc R Soc London (Biol) 272: 497–503.

Jennions MD, Petrie M (2000) Why do females mate multiply? A review of the genetic benefits. Biol Rev 75: 21–64.

Jones AG (2005) Gerud 2.0: a computer program for the reconstruction of parental genotypes from half-sib progeny arrays with known or unknown parents. Mol Ecol Notes 5: 708–711.

Jones AG, Ardren WR (2003) Methods of parentage analysis in natural populations. Mol Ecol 12: 2511–2523.

Jones B, Walsh D, Werne L, Fiumera A (2009) Using blocks of linked SNPs as highly polymorphic genetic markers for parentage analysis. Mol Ecol Res 9: 487–497.

Jorde PE, Ryman N (1995) Temporal allele frequency change and estimation of effective size in populations with overlapping generations. Genetics 139: 1077–1090.

Kalinowski S, Taper M, Marshall T (2007) Revising how the computer program CERVUS accommodates genotyping error increases success in paternity assignment. Mol Ecol 16: 1099–1106.

Karl SA (2008) The effect of multiple paternity on the genetically effective size of a population. Mol Ecol 17: 3973–3977.

Karl SA, Bowen BW, Avise JC (1992) Global population genetic structure and male-mediated gene flow in the green turtle (*Chelonia mydas*): RFLP analyses of anonymous nuclear loci. Genetics 131: 163–173.

Keeney DB, Heist EJ (2003) Characterization of microsatellite loci isolated from the blacktip shark and their utility in requiem and hammerhead sharks. Mol Ecol Notes 3: 501–504.

Keeney DB, Heist EJ (2006) Worldwide phylogeography of the blacktip shark (*Carcharhinus limbatus*) inferred from mitochondrial DNA reveals isolation of western Atlantic populations coupled with recent Pacific dispersal. Mol Ecol 15: 3669–3679.

Keeney DB, Heupel M, Hueter RE, Heist EJ (2005) Microsatellite and mitochondrial DNA analyses of the genetic structure of blacktip shark (*Carcharhinus limbatus*) nurseries in the Northwestern Atlantic, Gulf of Mexico, and Caribbean Sea. Mol Ecol 14: 1911–1923.

Kimura M, Ohta T (1978) Stepwise mutation model and distribution of allelic frequencies in a finite population. Proc Natl Acad Sci USA 75: 2868–2872.

Krimbas CB, Tsakas S (1971) The genetics of *Dacas oleae V.* changes of esterase polymorphism in natural population following insecticide control: selection or drift? Evolution 25: 454–460.

Kwok PY (2001) Methods for genotyping single nucleotide polymorphisms. Annu Rev Genomics Hum Genet 2: 235–258.

Lage CR, Petersen CW, Forest D, et al. (2008) Evidence of multiple paternity in spiny dogfish (*Squalus acanthias*) broods based on microsatellite analysis. J Fish Biol 73: 2068–2074.

Laikre L, Jorde PE, Ryman N (1998) Temporal change of mitochondrial DNA haplotype frequencies and female effective size in a brown trout (*Salmo trutta*) population. Evolution 52: 910–915.

Levinson G, Gutman G (1987) Slipped-strand mispairing: a major mechanism for DNA sequence evolution. Mol Biol Evol 4: 203–221.

Li WH, Gojobori T, Nei M (1981) Pseudogenes as a paradigm of neutral evolution. Nature 292: 237–239.

Lopez JV, Yuhki N, Modi W, et al. (1994) Numt, a recent transfer and tandem amplification of mitochondrial DNA in the nuclear genome of the domestic cat. J Mol Evol 39: 171–190.

Luikart G, Cornuet JM (1999) Estimating the effective number of breeders from heterozygote excess in progeny. Genetics 151: 1211–1216.

Martin AP, Naylor GJP, Palumbi SR (1992) Rates of mitochondrial DNA evolution in sharks are slow compared with mammals. Nature 357: 153–155.

Martinez-Arias R, Calafell F, Mateu E, et al. (2001) Sequence variability of human pseudogene. Genome Res 11: 1071–1085.

Myers EM, Zamudio KR (2004) Multiple paternity in an aggregate breeding amphibian: the effect of reproductive skew on estimates of male reproductive success. Mol Ecol 13: 1952–1963.

Nazarian J, Hathout Y, Vertes A, Hoffman EP (2007) The proteome survey of an electricity-generating organ (*Torpedo californica* electric organ). Proteomics 7: 617–627.

Nei M, Tajima F (1981) Genetic drift and estimation of effective population size. Genetics 98: 625–640.

Neff BD, Pitcher TE (2002) Assessing the statistical power of genetic analyses to detect multiple mating in fishes. J Fish Biol 61: 739–750.

Newcomer SD, Zeh JA, Zeh DW (1999) Genetic benefits enhance the reproductive success of polyandrous females. Proc Natl Acad Sci USA 96: 10236–10241.

Newman D, Pilson D (1997) Increased probability of extinction due to decreased genetic effective population size: experimental populations of *Clarkia pulchella*. Evolution 51: 354–362.

Nunney L (1993) The influence of mating system and overlapping generations on effective population size. Evolution 47: 1329–1341.

Ohta Y, Okamura K, Mckinney EC, et al. (2000) Primitive synteny of vertebrate major histocompatibility complex class I and class II genes. Proc Natl Acad Sci USA 97: 4712–4717.

Palsbøll PJ, Allen J, Berube M, et al. (1997) Genetic tagging of humpback whales. Nature 388: 767–769.

Palumbi SR, Baker CS (1994) Contrasting population structure from nuclear intron sequences and mtDNA of humpback whales. Mol Biol Evol 11: 426–435.

Pardini AT, Jones CS, Noble LR, et al. (2001) Sex-biased dispersal of great white sharks. Nature 412: 139–140.

Perrin N, Mazalov V (2000) Local competition, inbreeding, and the evolution of sex-biased dispersal. Amer Nat 155: 116–127.

Pollak, E (1983) A new method for estimating the effective population size from allele frequency changes. Genetics 104: 531–548.

Portnoy DS (2008a) Understanding the reproductive behavior and population condition of the sandbar shark (*Carcharhinus plumbeus*) in the western North Atlantic: a molecular approach to conservation and management. PhD Dissertation, College of William and Mary.

Portnoy DS, McDowell JR, McCandless CT, et al. (2008b) Effective size closely approximates the census size in the heavily exploited western Atlantic population of sandbar sharks, *Carcharhinus plumbeus*. Cons Genet DOI 10.1007/s10592-008-9771-2.

Portnoy DS, Piercy AN, Musick JA, et al. (2007) Genetic polyandry and sexual conflict in the sandbar shark, *Carcharhinus plumbeus*, in the western North Atlantic and Gulf of Mexico. Mol Ecol 16: 187–197.

Pratt HL, Carrier JC (2001) A review of elasmobranch reproductive behavior with a case study on the nurse shark, *Ginglymostoma cirratum*. Environ Biol Fish 60: 157–188.

Pudovkin AI, Zaykin DV, Hedgecock D (1996) On the potential for estimating the effective number of breeders from heterozygote excess in progeny. Genetics 144: 383–387.

Rozas J, Sanchez-DelBarrio JC, Messeguer X, Rozas R (2003) DnaSP, DNA polymorphism analyses by the coalescent and other methods. Bioinformatics 19: 2496–2497.

Saville KJ, Lindley AM, Maries EG, et al. (2002) Multiple paternity in the nurse shark, *Ginglymostoma cirratum*. Environ Biol Fish 63: 347–351.

Schrey AW, Heist EJ (2003) Microsatellite analysis of population structure in the shortfin mako (*Isurus oxyrinchus*). Can J Fish Aquat Sci 60: 670–675.

Schultz JK, Feldheim KA, Gruber SH, et al. (2008) Global phylogeography and seascape genetics of the lemon sharks (genus *Negaprion*). Mol Ecol 17: 5336–5348.

Schwartz MK, Tallmon DA, Luikart G (1999) Using genetics to estimate the size of wild populations: many methods, much potential, uncertain utility. Anim Conserv 2: 321–323.

Slatkin M (1995) A measure of population subdivision based on microsatellite allele frequencies. Genetics 139: 457–462.

Stepien C (1995) Population genetic divergence and geographic patterns from DNA sequences: examples from marine and freshwater fishes. Amer Fish Soc Symp 17: 263–287.

Sugg DW, Chesser RK (1994) Effective population sizes with multiple paternity. Genetics 137: 1147–1155.

Tregenza T, Wedell N (2000) Genetic compatibility, mate choice and patterns of parentage: Invited Review. Mol Ecol 9: 1013–1027.

Vignal A, Milan D, SanCristobal M, Eggen A (2002) A review on SNP and other types of molecular markers and their use in animal genetics. Genet Select Evol 34: 275–305.

Wang J (2004) Sibship reconstruction from genetic data with typing errors. Genetics 166: 1963–1979.

Wang, J (2005) Estimation of effective population sizes from data on genetic markers. Philos Trans R Soc London (Biol) 360: 1395–1409.

Waples RS (1989) A generalized approach for estimating effective population size from temporal changes in allele frequency. Genetics 121: 379–391.

Waples RS (1991) Genetic methods for estimating the effective size of cetacean populations. Rep Int Whal Comm (special issue) 13: 279–300.

Waples RS (2005) Genetic estimates of contemporary effective population size: to what time periods do estimates apply? Mol Ecol 14: 3335–3352.

Waples RS (2006) A bias correction for estimates of effective population size based on linkage disequilibrium at unlinked gene loci. Conserv Genet 7: 167–184.

Weber JL (1990) Informativeness of human (dC-dA)$_n$, (dG-dT)$_n$ polymorphisms. Genomics 7: 524–530.

Weber JL, Wong C (1993) Mutation of human short tandem repeats. Hum Mol Gen 2: 1123–1128.

Weir BS, Cockerham CC (1984) Estimating F-statistics for the analysis of population structure. Evolution 38: 1358–1370.

Williamson EG, Slatkin M (1999) Using maximum likelihood to estimate population size from temporal changes in allele frequencies. Genetics 152: 755–761.

Wright S (1931) Evolution in Mendelian populations. Genetics 16: 97–159.

Wright S (1965) The interpretation of population structure by F-statistics with special regard to systems of mating. Evolution 19: 395–420.

Yamaguchi N, Sarno RJ, Johnson WE, et al. (2004) Multiple paternity and reproductive tactics of free-ranging American minks. J Mammal 85: 432–439.

Yasui Y (1998) The "genetic benefit" of female multiple mating reconsidered. Trends Ecol Evol 13: 246–250.

Zeh JA, Zeh DW (1997) The evolution of polyandry, II: post-copulatory defenses against genetic incompatibility. Proc R Soc Lond (Biol) 264: 69–75.

Zeh JA, Zeh DW (2001) Reproductive mode and the genetic benefits of polyandry. Anim Behav 61: 1051–1063.

Zhdanova OL, Pudovkin AI (2008) Nb_HetEx: a program to estimate the effective number of breeders. J Hered 99: 694–695.

11

Physiological Responses to Stress in Sharks

Gregory Skomal and Diego Bernal

CONTENTS

11.1 Introduction

Despite their long evolutionary history (~400 million years; Grogan and Lund 2004), elasmobranchs are now facing new substantial anthropogenic threats, including habitat degradation and fisheries interactions, which cause acute and chronic stress that may exceed stress levels typically imposed by natural events (e.g., seasonal habitat changes, predator avoidance). Although the physiological stress response of teleosts has been studied for decades (e.g., Adams 1990a), much of this work has centered on salmonids due to the ease of maintaining and manipulating these fishes in captivity as well as the economic interests of the aquaculture industry and the recreational fishing sector. The global expansion of both commercial and recreational fisheries coupled with the mandated release of captured fish due to bag limits, quotas, and minimum sizes, prompted the expansion of stress studies beyond freshwater teleosts to their marine counterparts. However, studies of elasmobranch stress physiology have lagged far behind despite the rapid proliferation of directed fisheries for, and a significant increase in the incidental capture of, these species over the last two decades (NMFS 2008) and the possibility that modifications to coastal habitats, which are important to many elasmobranch species, could increase stress.

In this chapter, we review our understanding of the physiological responses to stress in elasmobranchs and highlight gaps in our current knowledge. Although some elasmobranch stress responses are similar to those exhibited by teleosts, the inherent physiological differences between these two groups create marked differences. Because this chapter centers on the responses of sharks and their relatives to stress (i.e., perturbations to homeostasis), we strongly recommend that the reader be familiar with elasmobranch physiology and homeostasis (e.g., Shuttleworth 1988a; Carlson, Goldman, and Lowe 2004; Evans, Piermarini, and Choe 2004).

11.2 Stress

Numerous definitions of "stress" have emerged in the literature (Adams 1990b), but few have diverged far from the working definition established by Brett (1958): stress is a physiological state produced by an environmental factor that extends the normal adaptive responses of the animal, or disturbs the normal functioning to such an extent that the chances for survival are significantly reduced. In essence, stress is a condition during which the homeostasis of an organism is disrupted by intrinsic or extrinsic stimuli called stressors (Wendelaar-Bonga 1997). The stressor not only threatens or disturbs homeostatic equilibrium, but may also elicit a coordinated response by the animal to compensate for the disruption in an attempt to overcome the threat (Wendelaar-Bonga 1997). The responses may be behavioral (e.g., swimming out of a hypoxic environment), physiological (e.g., a decrease in metabolic rate with declines in ambient oxygen levels), or a combination of both. In general, there are increased metabolic costs associated with the stress response because energy is reallocated from one or several metabolically demanding activities (e.g., growth, reproduction) to others (e.g., respiration, locomotion, tissue repair) in order to restore homeostasis. When compensatory mechanisms (or the energetic reserves necessary to deal with the costs associated with these mechanisms) are not sufficient to correct for large homeostatic perturbations in either the short term (acute) or long term (chronic), the stress response is dysfunctional and reduced fitness and/or mortality is likely (Wendelaar-Bonga 1997).

Although the stress response is inherently organismal, it is often referred to as an integrated response because it encompasses multiple organizational levels ranging from the molecular to those of populations and ecosystems (Adams 1990b; Wendelaar-Bonga 1997). Again, in the case of elasmobranchs, many of these levels have yet to be explored and researchers are just beginning to examine the basic physiological stress response in members of this group. To date, most studies on elasmobranch stress have concentrated on small benthic species that can be readily maintained in captivity [e.g., Port Jackson shark (*Heterodontus portusjacksoni*); Cooper and Morris 2004]. However, work on teleosts has shown that the species-specific characteristics and magnitude of the stress response are tightly linked to overall level of swimming activity (e.g., aerobic and anaerobic scope) and life history (e.g., sluggish benthic vs. active pelagic species; Wood 1991). Given the diversity of elasmobranch activity levels and life history patterns, it is unlikely that there is a "typical" stress response in this group.

A stressor is often classified as acute or chronic, depending on its frequency, duration, and intensity. Acute stress results from the rapid (from minutes to hours) exposure to a stressor or a series of stressors. For example, the capture and handling of a shark may comprise several stressors, including a bout of exhaustive exercise, physical trauma from hooking or netting, and exposure to air. Although acute responses tend to be immediate, the behavioral and physiological consequences may manifest themselves over a period of hours to days (e.g., Skomal 2006). In contrast, chronic stress arises from continuous or periodic exposure to one or more stressors over longer time periods (from days to weeks). For example, sharks held in captivity may exhibit chronic stress from overcrowding, poor water quality, or malnutrition. Chronic stress tends to have long-term sublethal effects on growth, reproduction, and resistance to disease, but acute stress may impact short-term survival (i.e., "lethal stress"; Adams 1990b).

The delineation between acute and chronic stress can be unclear and there are instances when acute stress may cascade into chronic stress. For example, a shark exposed to an acute stressor (e.g., capture) may suffer chronic stress due to the reduced capability to feed or the onset of a secondary infection associated with physical trauma (i.e., wounds). In addition, recent molecular studies on the teleost response to acute stress suggest that cellular damage may result in chronic stress that affects long-term fitness and reduces survivorship (Currie and Tufts 1997; Currie, Tufts, and Moyes 1999; McDonough, Arrell, and Van Eyk 1999; Currie, Moyes, and Tufts 2000; Iwama, Afonso, and Vijayan 2006).

11.2.1 Natural Stressors

Although natural fluctuations in ambient temperature, ionic/osmotic conditions, and dissolved oxygen are not commonly referred to as stressors, homeostasis can be disrupted during encounters with extreme conditions. Such environmental parameters tend to change over seasonal time scales and most fish species, including sharks, are well adapted to moderate environmental fluctuations in their habitat and, therefore, minimize stress through behavioral or physiological compensation. For example, Carlson and Parsons (1999) found that bonnethead sharks (*Sphyrna tiburo*) exhibited seasonal changes in oxygen consumption rates, likely in response to concomitant fluctuations in ambient temperature and, hence, changes in the concentrations of dissolved oxygen available for aerobic metabolic processes. While many shark species exhibit seasonal migratory behavior that is, to varying degrees, driven by moving out of suboptimal environmental conditions, other sharks routinely encounter dramatic and repeated changes in ambient conditions. The blue shark (*Prionace glauca*) is known to make multiple daily dives from the warm, well-mixed surface layers of the ocean to depths in excess of 600 m (well below the thermocline) and may encounter rapid and repeated changes in temperature of up to 19°C (Carey and Scharold 1990), yet the extent to which these sharks are "stressed" is unknown.

Not all natural stressors are chronic. Regional climatic events such as storms, droughts, or volcanic activity can produce significant environmental changes (e.g., hypoxia, increased turbidity, changes in salinity and temperature) that acutely stress aquatic organisms. However, most elasmobranchs are distributed in marine habitats that do not exhibit dramatic and rapid environmental fluctuations in temperature, salinity, pH, and oxygen concentration, and their ability to move may allow them to avoid these adverse stressful conditions. Nonetheless, the physiological effects of acute stress associated with rapid changes in ambient conditions have been investigated in a limited number of smaller elasmobranch species that can be maintained in captivity. Some examples include the effects of: (1) hypoxia on chain dogfish (*Scyliorhinus canicula*; Butler et al. 1978; Butler, Taylor, and Davison 1979) and epaulette sharks (*Hemiscyllium ocellatum*; Routley, Nilsson, and Renshaw 2002; Renshaw, Kerrisk, and Nilsson 2002); (2) lowered salinity on bonnethead sharks (Mandrup-Paulsen 1981), yellow stingrays (*Urolophus jamaicensis*; Sulikowski and Maginniss 2001), and Port Jackson sharks (Cooper and Morris 2004); and (3) hypercapnia (elevated carbon dioxide) in spiny dogfish (*Squalus acanthias*; McKendry, Milsom, and Perry 2001). However, given that most of these species are known to be tolerant of such changes in environmental conditions, it is not likely that their responses to such acute exposures are representative of all elasmobranchs.

11.2.2 Anthropogenic Stressors

Human activities have impacted the marine environment and its inhabitants, including many elasmobranch species and the areas in which they live. These activities include fisheries interactions, habitat destruction and loss, thermal or chemical pollution, and increased runoff and eutrophication associated with coastal development, aquaculture, and agriculture.

11.2.2.1 Ecotoxins

Coastal waters provide essential habitat for many species of elasmobranchs and are subjected to numerous anthropogenic disturbances (McCandless, Kohler, and Pratt 2008). Environmental pollutants such as chemical agents and warm water influx from industrial and agricultural runoff are known to impair overall fish health and cause both acute and chronic stress by depressing immune system response, directly damaging tissues (e.g., gills), and lowering oxygen availability via eutrophication (Adams 1990b; Anderson 1990; Hinton and Lauren 1990). Similar to teleosts, sharks and rays living in essential habitat are exposed to health-threatening levels of aquatic toxins or pollutants and do exhibit signs of pollutant effects; this topic is covered in Chapter 12 of this volume.

11.2.2.2 Capture

Capture and handling cause acute stress responses in elasmobranchs. In the aquarium trade, elasmobranchs also suffer chronic stress associated with captivity. Indeed, much of the early work on stress in this group of fishes arose from an interest in maintaining healthy sharks in captivity (Gruber and Keyes 1981).

Understanding stress responses to capture is important for the conservation of shark populations because many individuals are captured incidentally in fisheries and mortality rates are highly gear- and species-specific (Mandelman et al. 2008). For example, longline-captured blue sharks experience 13% at-vessel mortality (Francis, Griggs, and Baird 2001) while porbeagle shark (*Lamna nasus*) mortalities can be more than three times higher (~40%; G. Skomal, L. Natanson, and D. Bernal, unpublished data). In addition, the overall at-vessel mortality rates in U.S. demersal longline fisheries range widely from 9% for the tiger shark (*Galeocerdo cuvier*) to an alarming 94% for the great hammerhead shark (*Sphyrna mokarran*; Morgan and Burgess 2007). There appears to be no phylogenetic predisposition for at-vessel mortality, as even closely related shark species respond differently to capture. For example, Morgan and Burgess (2007) reported that at-vessel mortality was 36% for the sandbar shark (*Carcharhinus plumbeus*), a value less than half of those for the congeneric blacktip shark (*Carcharhinus limbatus*, 88% mortality) and dusky shark (*Carcharhinus obscurus*, 81% mortality).

Capture causes struggling, muscular fatigue, and time out of water, and is typically accompanied by some level of physical trauma (e.g., tissue damage by hooking). In general, fishes react to the stress of capture, exhaustive exercise, and handling with more exaggerated disruptions to their physiology and biochemistry than in all other higher vertebrates (reviewed by Pickering 1981; Adams 1990b; Wood 1991; Milligan 1996; Wendelaar-Bonga 1997; Keiffer 2000). The myotomal muscle mass (the main muscles used for locomotion) of nearly all species of fishes, including elasmobranchs, is comprised mostly (80% to 95%) of anaerobically powered, fast-twitch, white muscle fibers, which reflect an ability to undergo short-duration, high-intensity, powerful bouts of burst swimming (Driedzic and Hochachka 1978). Moreover, the large mass-specific proportion of white muscle (usually 50% to 60% of body mass) and its intrinsic potential to power short-lived, high energy contractions allows fish to undergo what may be excessive levels of locomotor activity solely powered by anaerobic metabolism. This may ultimately result in acute physiological stress and disruptions to the internal milieu of fish (Wells, Tetens, and Devries 1984).

11.3 Physiological Effects

The integrated stress response in individual fish involves multiple levels of organization, many of which have been traditionally examined by measuring constituents in the blood and muscle tissues. In fishes, physiological disturbances associated with acute stress can be manifested rapidly, with potentially lethal consequences (Black 1958; Strange, Schreck, and Ewing 1978; Wood, Turner, and Graham 1983). Mazeaud, Mazeaud, and Donaldson (1977) characterized the sequence of biochemical and physiological events leading to these disturbances into primary, secondary, and tertiary stress responses. Generally, an acute stressful event elicits an immediate endocrine response (primary), which in turn induces a suite of secondary effects including the rapid mobilization and utilization of energy reserves. Ultimately, the secondary responses may evolve into a more chronic stress event that can have potential sublethal tertiary effects on feeding, growth, reproduction, and the immune system, thereby having consequences at the population level (Cooke et al. 2002).

11.3.1 Primary Stress Response

The primary response to stress in fish is neuroendocrine, with rapid increases in circulating levels of stress hormones (predominantly catecholamines and corticosteroids).

11.3.1.1 Catecholamines

In elasmobranchs, as in most fishes, adrenaline and noradrenaline are the dominant catecholamines released in response to acute stressors, including hypoxia, hypercapnia, and exhaustive exercise during capture and handling (Mazeud, Mazeaud, and Donaldson 1977; Randall and Perry 1992). In response to neural signals, these catecholamines are secreted rapidly (seconds to minutes) from chromaffin tissue, which comprises small clusters of neurosecretory cells on the dorsal surface of the kidneys (Randall and Perry 1992). Although catecholamines are released with the onset of a stressor, they are cleared from the plasma rapidly by the combined effects of tissue accumulation and metabolism and may quickly return to resting levels despite the continuation of the stressor (Randall and Perry 1992). In general, catecholamines act to mobilize and maintain adequate levels of energetic substrates and oxygen supply during periods of stress and metabolic recovery. Some of the numerous physiological effects of catecholamines in fishes include: the activation of hepatic glycogenolysis, which ultimately increases the level of circulating glucose (hyperglycemia); an increase in both gill perfusion and ventilation frequency, which increases branchial blood oxygen uptake; an elevation of cardiac output through changes in stroke volume and beat frequency; and an increase in blood pressure, which promotes blood flow and consequently the delivery of oxygen and energetic substrates to the working tissues (Opdyke, Carroll, and Keller 1982; Butler, Metcalfe, and Ginley 1986; Randall and Perry 1992; Wendelaar-Bonga 1997; Iwama, Afonso, and Vijayan 2006). Therefore, in both teleosts and elasmobranchs, catecholamines impart important changes in cardiorespiratory function. However, unlike teleosts, the heart of elasmobranchs lacks sympathetic innervation of its myocardial cells and, hence, is controlled by the parasympathetic nervous system. Therefore, any catecholamine-mediated change in cardiovascular function will be the result of physiological responses unrelated to the direct control of the heart itself but rather by indirect effectors that may alter its function (e.g., changes in vascular resistance and atrial filling volumes; Taylor and Butler 1982; Farrell 1993; Lai et al. 1997, 2004).

Levels of circulating catecholamines increase significantly in elasmobranchs immediately after the perception of a stressor and reach peak values within several minutes after the cessation of the stress event (Randall and Perry 1992). Hight et al. (2007) found that catecholamine levels increased from 100-fold to as much as up to 1600 times above routine (i.e., baseline, see Section 11.4) swimming levels in blue sharks and shortfin mako sharks (*Isurus oxyrinchus*) after being caught on pelagic longline fishing gear. In such extreme cases, very high catecholamine levels can result in intense vasoconstriction leading to metabolic acidosis or anoxia and potentially to irreversible organ, tissue, and cell damage (Wendelaar-Bonga 1997; Hight et al. 2007). Thus, in highly stressed sharks, the physiological aftermath of an acute but massive release (overload) of catecholamines may have long-term and potentially irreversible consequences on blood and tissue physiology.

11.3.1.2 Corticosteroids

Cortisol is the primary corticosteroid in teleosts and is most often the hormone of choice for quantifying stress in many vertebrates (Romero 2002). However, cortisol is not present in elasmobranchs, but its function is fulfilled by 1-α-hydroxycorticosterone (1α-OHB; Hazon and Balment 1998). This specific corticosterone is produced by and secreted from an adrenocorticoid gland that lies dorsally between two posterior lobes of the kidney, which in contrast to other fishes, is distinct and separate from the catecholamine-releasing chromaffin tissue. The release of 1α-OHB appears to be triggered by the presence of elevated levels of adrenocorticotropic hormone from the pituitary gland as part of the hypothalamic-pituitary-interrenal axis, which is an important feature of the stress response in fish (Hazen and Balment 1998; Gelsleichter 2004). Unfortunately, the hypothalamic-pituitary-interrenal axis and the effects of 1α-OHB secretion are poorly studied in elasmobranchs due largely to the difficulty associated with measuring 1α-OHB (Gelsleichter 2004). Although the detection and quantification of corticosterone has been used as a proxy for 1α-OHB to assess stress in sharks, this hormone plays a key role in the reproductive biology of elasmobranchs. For example, elevated levels of corticosterone in bonnethead sharks and Atlantic stingrays (*Dasyatis sabina*) were associated with reproductive cycles and not to any great degree with either acute or chronic stress (Manire et al. 2007).

In teleosts, cortisol secretion is rapid and plasma concentrations increase dramatically within minutes of the onset of acute stress and remain elevated until the stressor subsides (Skomal 2006). Therefore, in these fishes, plasma cortisol is an excellent indicator of the onset of acute stress and may be used to determine the presence of chronic stress. If elasmobranch 1α-OHB acts similarly to the teleostean cortisol, then it may play a key role during acute stress by mobilizing energy reserves (e.g., glucose) and helping to maintain ionic balance (Hazen and Balment 1998). In elasmobranchs, there is evidence that 1α-OHB plays a role in osmoregulation by maintaining plasma sodium concentrations during osmotic stress (Armour, O'Toole, and Hazon 1993). However, high levels of cortisol in teleosts have been found to inhibit muscle glycogen synthesis (i.e., glyconeogenesis, metabolic recovery) after stress (Pagnotta, Brooks, and Milligan 1994; Milligan 2003) and prolonged periods of elevated cortisol may increase susceptibility to disease, depress growth rates, and interfere with reproduction (Mommsen, Vijayan, and Moon 1999). We know of no comparable work on the extent to which high levels of circulating 1α-OHB may affect elasmobranch recovery from acute stress and, thus, the potential implications of this corticosterone on long-term survivorship warrant further investigation.

11.3.2 Secondary Stress Response

The secondary effects of stress in fish comprise a suite of physiological and biochemical changes that occur as a result of the stressor and are mostly mediated by the primary stress response (i.e., hormonal effects). Secondary effects are realized at all organizational levels, but are typically measured by changes in blood and muscle biochemistry. In general, the secondary effects of stress in fish are tightly linked to the type and duration of the stressor, and commonly include: (1) the rapid increase of circulating levels of glucose; (2) the depletion of intramuscular glycogen, adenosine triphosphate (ATP), and creatine phosphate reserves and the accumulation of lactate (La$^-$) in both white muscle and blood resulting from anaerobic glycolysis; (3) a marked decrease in blood pH (acidemia) resulting from metabolic acidosis [i.e., proton (H$^+$) accumulation] and respiratory acidosis (pCO$_2$ elevation); and (4) the profound disturbance of ionic, osmotic, and fluid volume homeostasis with a change in hemoconcentration and an increase in plasma electrolytes (Black 1958; Driedzic and Hochachka 1978; Beggs, Holeton, and Crossman 1980; Jensen, Nikinmaa, and Weber 1983; Wood, Turner, and Graham 1983; Milligan and Wood 1986; Robertson, Thomas, and Arnold 1988; Thomas and Robertson 1991; Tufts et al. 1991; Wood 1991; Arthur et al.

1992; Ferguson and Tufts 1992; Kieffer et al. 1995). The time for recovery from secondary stress responses is species specific and may take several hours but is highly dependent on the degree of stress (Wood 1991; Arthur et al. 1992; Schulte, Moyes, and Hochachka 1992; Kieffer et al. 1995).

The blood of sharks is mainly composed of red and white blood cells and plasma, with an overall functional role that is similar to that of the blood in most other vertebrates. For example, white blood cells are a critical component of the immune system, whereas red blood cells (RBCs) are involved in the uptake, binding, and transport of oxygen (by the oxygen-binding protein hemoglobin, Hb) from the gills to the rest of the body. The plasma is the main conduit for the extracellular transport of energetic substrates and other meta-bolically important ions and the removal of metabolic waste products (e.g., La^-, CO_2) from active tissues. When a fish undergoes severe stress, the entire blood volume can be sub-jected to immediate and large homeostatic disruptions (e.g., imbalances in ion and acid-base processes, changes in temperature, and fluctuations in the levels of oxygen and CO_2) due to the intimate and continuous contact between the blood, the metabolically active tissues (throughout the body), and the environment (at the gills). The following discussion on specific secondary effects is limited to those that have been most commonly examined in elasmobranchs.

11.3.2.1 Hyperglycemia

Maintaining normal physiological function and recovering from homeostatic disrup-tions are energetically costly. Catecholamines and corticosteroids mobilize hepatic glyco-gen, which is transported by the blood to the metabolically active tissues (McDonald and Milligan 1992). Therefore, hyperglycemia is a common secondary response in all fishes, including elasmobranchs, and blood glucose levels are typically used to characterize and quantify the presence of stress (DeRoos and DeRoos 1978). In sharks, hyperglycemia resulting from capture stress has been reported in dusky sharks (Cliff and Thurman 1984), Atlantic sharpnose sharks (*Rhizoprionodon terraenovae*; Hoffmayer and Parsons 2001), bon-nethead, blacktip, and bull sharks (*Carcharhinus leucus*; Manire et al. 2001), sandbar sharks (Spargo 2001), shortfin mako sharks (Skomal 2006), and spiny dogfish (Mandelman and Farrington 2007). In all these studies, blood glucose increased significantly and remained elevated for hours after the cessation of the stressor.

11.3.2.2 Acid-Base Disturbances

In fishes, disruptions in acid-base homeostasis are typical secondary responses to numer-ous types of stressors. Generally, fish that are exposed to hypoxia (low environmental oxygen) or subjected to bouts of exhaustive exercise show large respiratory perturbations that dramatically alter blood and tissue acid-base chemistry (Wood 1991; Arthur et al. 1992). Even slight modifications in blood and muscle pH can have profound effects on electroneutrality, fluid volume dynamics, and cellular function (Wood 1991; Claiborne, Edwards, and Morrison-Shetlar 2002). A decrease in blood pH by as little as 0.3 represents a doubling of the blood acid load and may lead to an alteration of normal physiological function in most vertebrates (Abelow 1998). Since fish blood and white myotomal muscle comprise from 3% to 6% and from 50% to 60% of body mass, respectively, changes in muscle biochemistry are strongly reflected in the blood (Wells et al. 1986). Therefore, mea-suring changes in blood chemistry provides quantitative information on the magnitude of the stress (Wedemeyer and Yasutake 1977; Wells, Tetens, and Devries 1984). This is particularly true for capture stress.

Acid-base disturbances in fishes have been clinically characterized using methods traditionally applied to higher vertebrate species (Tietz 1987). In short, stress causes a depression in blood pH, called acidemia, which is created by respiratory and/or metabolic mechanisms (reviewed by Pickering 1981; Adams 1990b; Wood 1991; Moyes and West 1995; Milligan 1996; Wendelaar-Bonga 1997; Kieffer 2000; Claiborne, Edwards, and Morrison-Shetlar 2002; Richards, Heigenhauser, and Wood 2003; Skomal 2007). Respiratory failure associated with stress causes blood carbon dioxide (pCO_2) levels to rise significantly resulting in respiratory acidosis. If a stressor alters or deprives the fish from taking up or transporting oxygen in sufficient amounts to maintain normal aerobic metabolism or involves exhaustive exercise (e.g., capture stress) powered by the white myotomal muscle, tissue metabolism switches to anaerobic glycolytic pathways. This quickly depletes large intramuscular glycogen stores and produces large amounts of the metabolic end product lactic acid, which dissociates into the (La^-) anion and hydrogen protons (H^+). In most fishes, the bulk of the La^- is retained in the white muscle for in situ metabolism, but due to very high concentrations of both this metabolic end product and the associated protons, some leaks out of the muscle cells and enters the bloodstream over a protracted time course, resulting in a metabolic acidosis (i.e., lacticacidosis) that usually peaks several hours after the stressor. The rates of La^- and H^+ movement from the muscle into the blood compartment are known to differ among teleost species with some showing very rapid plasma lactate accumulation after an acute bout of exhaustive exercise (i.e., tunas; Arthur et al. 1992), while others take significantly more time (trout; Milligan 1996). Similar information is lacking for sharks and rays, and the overall flux dynamics of lactic acid are poorly known even though the magnitude of the metabolic acidosis can be easily quantified by measuring blood levels of La^- as well as the relative decline in blood bicarbonate (HCO_3^-), which acts to buffer the increased proton load.

To date, there are few studies of acid-base disturbances in shark blood associated with stress, most of which have investigated the effects of exhaustive exercise, capture, and handling. Spotted dogfish (*Scyliorhinus stellaris*) stimulated with electric shocks until fatigued showed a significant drop in blood pH coupled with a rise in blood pCO_2 and La^- (Piiper, Meyer, and Drees 1972; Holeton and Heisler 1978). Work on chain dogfish exposed to long-term, moderate hypoxia showed marked acid-base disturbances and elevated blood La^- levels (Butler, Taylor, and Davison 1979). In general, dusky, Atlantic sharpnose and sandbar sharks subjected to stress induced by rod and reel capture exhibit a decrease in blood pH and HCO_3^- and an increase in blood pCO_2 and La^- (Cliff and Thurman 1984; Hoffmayer and Parsons 2001; Spargo 2001). Richards, Heigenhauser, and Wood (2003) observed similar acid-base disturbances in spiny dogfish exercised to exhaustion. Skomal (2006) found significant acid-base disturbances in rod and reel captured blue, shortfin mako, and spinner (*Carcharhinus brevipinna*) sharks compared to nonstressed individuals. In general, the level of disturbance correlated with stress level (i.e., fight time, Figure 11.1). Manire et al. (2001) quantified serological changes associated with gillnet capture in bonnethead, blacktip, and bull sharks and concluded that differences in mortality were likely associated with species-specific respiratory physiology and the degree of struggling by each species. Recent work on spiny dogfish caught by trawl showed an increase in whole blood pCO_2 and La^- and a decrease in pH (Mandelman and Farrington 2007), a scenario also observed in sandbar, tiger, Atlantic sharpnose, dusky, and blacktip sharks captured by demersal longline gear (Mandelman and Skomal 2009).

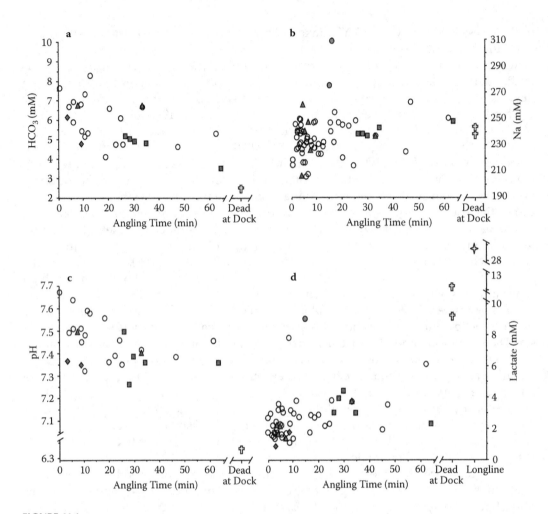

FIGURE 11.1

Changes in the levels of blue shark (*Prionace glauca*) blood. (a) bicarbonate (HCO$_3^-$), (b) sodium (Na), (c) pH, and (d) lactate, relative to angling time. Blue sharks were exposed to rod and reel capture stress, blood sampled, and released (open circles). To assess survivorship, some fish were tracked acoustically for 4.0 to 9.2 hours (filled squares), recaptured after 44 to 215 days at liberty (filled triangles), and satellite tracked (PSAT tags) for 123 days (filled diamonds). Comparative data are presented from Wells et al. (1986) (filled circles), blue sharks sampled dead at the dock (exact angling time unknown) (✝), and blue sharks captured on pelagic longline gear (time on the line 120 to 360 min) (◇). Note the break in scale for pH and lactate. (Modified from Skomal G (2006) Ph.D. dissertation, Boston University; unpublished data from D. Bernal and G. Skomal.)

The extent to which the stress-induced acidemia is of either metabolic or respiratory origin appears to vary among species and, in some cases, by the type of stressor. While high levels of pCO$_2$ and La$^-$ accompanied by low HCO$_3^-$ are indicative of both respiratory and metabolic contributions to acidemia, this is not always the case. The acidemia in spiny dogfish exercised to exhaustion was solely metabolic in origin (Richards, Heigenhauser, and Wood 2003). Similarly, blue sharks exposed to prolonged rod and reel capture exhibited a metabolic acidosis (blood pH decreased by 0.22 units) associated with a slow and significant decline in HCO$_3^-$ and rise in La$^-$ without a concomitant increase in pCO$_2$ (Figure 11.1; Skomal 2006). In addition, Mandelman and Skomal (2009) found that a change in pCO$_2$ levels had the most influence on blood pH in the tiger, sandbar, and blacktip sharks, while the acidemia in dusky and Atlantic sharpnose sharks was found to be exclusively or primarily metabolic in origin. It appears that the overall magnitude of the acidosis measured by Mandelman and Skomal (2009) was tightly linked to the aerobic and anaerobic capacity of each species (see Section 11.5).

11.3.2.3 Electrolyte Levels

Sodium and chloride are the major ions in the blood of all fishes (McDonald and Milligan 1992). In marine teleosts, the levels of sodium and chloride as well as potassium and calcium are influenced by acute stress. Increases in serum ions can result from intramuscular lacticacidosis, which provokes a fluid shift from the extracellular compartment into the tissues resulting in hemoconcentration (i.e., an increase in hematocrit; Turner, Wood, and Clark 1983; McDonald and Milligan 1992). Rises in sodium concentrations can also result from the environmental influx associated with the active branchial exchange of H^+ ions during the recovery from acidemia (Shuttleworth 1988b). In general, electrolyte levels differ between elasmobranchs and teleosts, reflecting differences in osmotic (e.g., high total osmolality through elevated levels of urea and trimethylamines) and ionic regulation (Cliff and Thurman 1984; Wells et al. 1986; Stoskopf 1993; Cooper and Morris 1998; Karnaky 1998; Manire et al. 2001). The effects of stress on changes in electrolyte levels and osmolality have been documented in only a few elasmobranchs, including the blue shark, which exhibited an increase in serum sodium associated with angling stress (Figure 11.1; Skomal 2006).

11.3.2.4 Hematology and Heat Shock Proteins

Unlike mammals, fish RBCs have a nucleus with all the molecular components (e.g., DNA, RNA, ribosomes) needed to complete protein synthesis and are thus capable of expressing a wide array of proteins used in all metabolic processes within the cell. In addition, the fish RBC is capable of expressing proteins that protect the cell's DNA from denaturation during stress (Currie and Tufts 1997; Currie, Moyes, and Tufts 2000). These heat shock proteins (HSPs; also called chaperonins) coat the cell's genetic material and protect it from conformational changes that may irreversibly alter function (reviewed by Iwama, Afonso, and Vijayan 2006). In addition, HSPs protect the critical three-dimensional profiles of many other cellular proteins allowing them to maintain their function under times of stress. Although some HSPs are expressed during normal conditions (constitutive HSPs) other HSPs (induced HSPs) can be rapidly synthesized by the RBCs in response to various types of stress (acute or chronic) and it is believed that they are critical for maintaining normal cellular function. In addition to protecting DNA, it appears that the chaperone-role of HSPs is vital for maintaining the functional structure of Hb (Kihm et al. 2005; Yu et al. 2007). Because any structural alteration to Hb can impact its oxygen binding affinity and the blood's overall oxygen carrying capacity, HSPs may play a critical role in maintaining adequate levels of oxygen transport to the aerobically active tissues during stress. Fishes appear to have an advantage over mammals, as the nucleated RBCs may provide fish with the ability to rapidly (within minutes) synthesize these chaperone molecules in situ in response to a stress event (e.g., exhaustive exercise, changes in ambient temperature) and thus maintain both a high Hb-oxygen affinity and the blood's oxygen carrying capacity. Therefore, stress and the onset of RBC damage in fish may induce or upregulate the synthesis of blood-borne HSPs, which are measurable using either protein or mRNA analyses

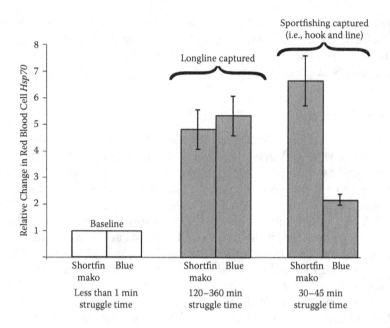

FIGURE 11.2

Change in 70 kDa heat shock protein (HSP70) expression in the red blood cells of shortfin mako (*Isurus oxyrinchus*) and blue (*Prionace glauca*) sharks captured using pelagic longline and conventional recreational fishing techniques. For each capture method, the time on the line is reflective of the level of capture stress. Values shown are mean (+SD) and represent the relative change between the initial, baseline values of unstressed, dipnetted, or quickly (>1 min) landed sharks (relative value = 1) and those after gear-specific landing. Sample size is three for both species. (Unpublished data from D. Bernal and G. Skomal.)

(e.g., Currie, Moyes, and Tufts 2000). This may, relative to mammals, ultimately result in both a quicker and more site-specific stress response.

Preliminary investigations on blue and mako sharks captured by conventional recreational fishing gear (i.e., hook and line) and on commercial pelagic longlines show that, relative to baseline, the level of the 70 kDa HSP (HSP70) expressed in the RBCs appears to increase with exercise-induced capture stress (Figure 11.2). For both species, the molecular response to longline capture (in which the time on the line ranged from 120 to 360 min) was similar, with values of HSP70 between five and six times higher than those of baseline (i.e., both species were highly stressed). However, relative to blue sharks, makos showed a much larger (threefold) molecular stress response when captured with conventional sportfishing gear, during which the time on the line ranged from 30 to 45 min. This difference may be attributed to the higher intensity struggling by makos when hooked (often fighting to complete exhaustion). Although limited to a single family of chaperone proteins in just two species, these results suggest the important role that shark RBCs may play during a stress response. However, further studies are needed on the effects of stress on HSP expression and the potential presence of other indices of RBC stress (e.g., antioxidant levels, the degree of oxidatively damaged proteins) to better understand the molecular mechanisms initiated by a stress event and their role during recovery.

11.3.3 Tertiary Stress Response

Although poorly studied in fishes, stress can negatively impact individual fitness and possibly have impacts on populations (Cooke et al. 2002). Both acute and chronic stress can reduce growth rates, reproductive output or investment, and resistance to disease (Adams 1990b) because of the energetic costs of restoring homeostasis (Iwama, Afonso, and Vijayan 2006). The tertiary effects of stress have yet to be explored in elasmobranchs.

11.4 Baseline

To understand the physiological effects of stress, it is essential to have baseline or "stress-free" estimates of the parameters being measured. This is problematic for elasmobranchs because many species are difficult, if not impossible, to maintain in captivity and/or are too large to sample without causing stress during capture and handling. Several techniques, however, have been used to approximate baseline (i.e., unstressed) conditions. In vitro manipulation of blood or muscle allows controlled experiments to assess the physiological response from steady-state conditions (e.g., shortfin mako shark; Wells and Davie 1985). Working with captive animals provides the optimal setting for experimentation and many in vivo studies on elasmobranchs have been conducted on small individuals or species that can be maintained in captivity (e.g., sandbar shark; Spargo 2001; Brill et al. 2008).

Secondary stress responses have generally been quantified by comparing the relative change in an indicator of interest (e.g., pCO_2, La^-, HCO_3^-, pH) between a stressed state and an estimated value from a "minimally stressed" individual. To approximate baseline conditions, researchers have applied the "grab and stab" caudal venipuncture blood sampling technique in captive elasmobranchs (e.g., spiny dogfish; Mandelman and Skomal 2009), free-swimming sharks (e.g., blue and shortfin mako sharks; Skomal 2006; Figures 11.1 and 11.2), and sharks taken with reduced capture and handling times (e.g., grey reef sharks, *Carcharhinus amblyrhynchos*; Skomal, Lobel, and Marshall 2008). Presumably, the blood biochemistry of minimally stressed animals (i.e., those exposed to brief handling times) approximates baseline conditions because many of the parameters have not yet been significantly disrupted during the short handling period. While this may be the case for many secondary effects, including blood pH, the primary effects (stress hormones) are initiated within seconds of the onset of stress and, thus, current measurements may not be representative of baseline data.

Using an alternative method, Skomal (2006) derived baseline estimates in several large pelagic teleosts and elasmobranchs using a mathematical approach. Baseline blood biochemistry was calculated from the statistically significant relationship between the duration of the capture event and blood parameters. For example, blood gas data were collected from blue sharks captured on rod and reel over a broad range of angling times, yielding a significant negative relationship between blood pH and angling time (Figure 11.1C). This mathematical relationship may then be used to extrapolate the blood pH to the intercept (i.e., time = 0), a theoretical baseline. This approach may be useful for other types of stressors, but only if a relationship between the stressor and the physiological effect can be quantified (e.g., handling time). As an alternative, Manire et al. (2001) used a subjective behavioral scale to categorize the level of stress in bull, blacktip, and bonnethead sharks taken in gillnets and found significant differences in blood biochemistry between those qualified as minimally stressed relative to those qualified as heavily stressed.

11.5 Intra- and Interspecific Variation in Stress Responses

In teleosts, there are inter- and intraspecific differences in physiological responses to stress that appear to be linked to variation in life history, activity level, individual health, size, thermal physiology, and environmental conditions. Because of the paucity of studies on the physiological response of elasmobranchs to stress and the relatively narrow phylogenetic and ecological range of species studied to date, factors influencing variation in stress responses cannot be adequately resolved. Also, the diverse methodologies (and stressors) applied to date makes interspecific comparisons difficult. Nonetheless, the magnitude and nature of the stress response in sharks is likely linked to both their metabolic capacity and thermal biology.

11.5.1 Metabolic Scope

The majority of studies on shark metabolism have investigated the standard metabolic rate (SMR, the metabolic rate at rest; Brill 1987), usually of small species easily held in captivity [e.g., spiny dogfish (Brett and Blackburn 1967), bamboo shark (*Chiloscyllium plagiosum*; Tullis and Baillie 2005), and sandbar shark (Dowd et al. 2006)]. Little work has focused on fragile open-ocean, continuously swimming species (Graham et al. 1990; reviewed by Carlson, Goldman, and Lowe 2004; Sepulveda, Graham, and Bernal 2007). Recent work suggests that SMR is species specific and highest in actively swimming species with the cardiovascular and respiratory specializations that allow for high rates of energy turnover (e.g., large gill surface areas, high mitochondrial densities in the locomotor muscles; Bernal et al. 2003a; Sepulveda, Graham, and Bernal 2007). More biologically relevant than the absolute value of SMR is the difference between the maximum metabolic rate and SMR, the "aerobic scope" (Priede 1985), because it represents the potential capacity to handle multiple simultaneous aerobic demands (e.g., continuous swimming, gonadal and somatic growth, and digestion), including recovery from acute stress (Priede 1985; Brill and Bushnell 1991a; Brill 1996; Korsmeyer et al. 1996). Aerobic scope in sharks appears to correlate with the level of swimming activity (e.g., benthic, sluggish species have a lower aerobic scope when compared to fast moving, pelagic species; Graham et al. 1990; reviewed by Carlson, Goldman, and Lowe 2004; Sepulveda, Graham, and Bernal 2007), suggesting that the intensity and magnitude of both primary and secondary stress responses in sharks could be directly linked to their aerobic scope. Thus, actively swimming pelagic sharks must be able to recover from a stress event in the absence of refuge and without compromising any of the aerobically demanding metabolic processes critical to their routine activity patterns (e.g., swimming).

Comparative studies on the metabolic biochemical capacities of locomotor muscles in fishes have shown that sharks have aerobic and anaerobic capacities comparable to ecologically similar teleosts species (Dickson et al. 1993; Dickson 1996; Bernal et al. 2001, 2003a, 2003b). Since muscle biochemistry is strongly reflected in the blood (Wells et al. 1986), higher muscle biochemical metabolic capacities will also be manifested in the secondary effects of stress present in the blood. The magnitude and intensity of the stress response in each species is likely linked to: (1) its metabolic scope as it pertains to the capacity for cruise and burst swimming; (2) the ability to physiologically respond to stress (e.g., enhanced oxygen delivery); and (3) the capacity to recover from physiological perturbation (e.g., acid-base disturbances). Mandelman and Skomal (2009) found significant interspecific differences between five closely related carcharhinids, including three species of

Carcharhinus, exposed to capture stress (Figure 11.3). Relative to the other species, the tiger shark and, to a lesser extent, the sandbar shark were less impacted by longline capture. Conversely, the blood acid-base chemistry of the dusky, Atlantic sharpnose, and blacktip sharks was highly disturbed, displaying the most depressed pH and HCO_3^- values, the most elevated blood La$^-$ levels, and high blood CO_2 tensions. In addition, plasma La$^-$ values for shortfin mako and porbeagle sharks, two highly active species, are more than twice as elevated relative to those of the most affected carcharhinids (Figure 11.3). This suggests that the degree of homeostatic disruption may be inherently linked to interspecific variation in metabolic scope.

11.5.2 Endothermy

In contrast to most fishes, the lateral circulatory system of lamnid sharks and the common thresher shark (*Alopias vulpinus*) gives rise to a network of red muscle-associated vessels that act as countercurrent heat exchangers, which effectively conserve heat produced metabolically by this tissue during sustained swimming. Consequently, these species maintain the temperature of their aerobic, red myotomal muscle significantly above that of the ambient water (i.e., regional endothermy; Carey 1969, 1973; Carey and Teal 1969; Carey et al. 1985; Bernal et al. 2001) and warms, to some degree, via conductive heat transfer the white muscle in close proximity to the red muscle. The resultant more thermally stable operating environment of the locomotor muscles has been linked to higher in vivo anaerobic and aerobic metabolic biochemical capacities in the white and red muscle of regionally endothermic fishes (Dickson et al. 1993; Dickson 1996; Bernal et al. 2001, 2003a, 2003b). The maintenance of elevated muscle temperatures may result in a higher metabolic scope and, therefore, may allow these sharks to swim faster and to recover more rapidly from anaerobic activity (reviewed by Bernal et al. 2001). Taken together, their capacity to undergo intense and repeated bouts of anaerobically powered burst swimming and to recover more rapidly (see Section 11.7) from these events suggests that both the nature and magnitude of the stress response in these fishes is likely to be greater than in their ectothermic counterparts. For example, Skomal (2006) found that blood acid-base disturbances associated with capture stress were lower in the ectothermic blue shark than in the sympatric endothermic shortfin mako.

Some species of elasmobranchs experience rapid changes in water temperature that can induce stress responses affecting a suite of physiological processes. In most vertebrates, the affinity between hemoglobin (Hb) and oxygen is strongly temperature dependent. For example, warming of the blood generally lowers Hb-O_2 affinity (i.e., higher pO_2 is required to saturate the Hb molecule) and cooling of the blood generally increases Hb-O_2 affinity (Powers 1983, 1985; Nikinmaa and Salama 1998). In teleosts, this change in blood temperature modifies the quaternary structure of Hb (Rossi-Fanelli and Antonini 1960) and affects CO_2 transport and acid-base regulation (Nikinmaa 2006). What does this mean for sharks? In the case of warm-water species (e.g., sandbar sharks) that inhabit a very stable thermal environment and rarely encounter a large decrease in blood temperature, their Hb has potentially evolved to have high oxygen affinity even at warm temperatures. A similar scenario (little or no change in blood temperature) can be predicted for a cold-water species (e.g., spiny dogfish) as they remain at either high latitudes or at depth throughout the year and their Hb-O_2 affinity will be high under cold conditions. However, there is evidence that if the blood of fishes is warmed, it will potentially decrease Hb-O_2 affinity and deleteriously affect CO_2 transport and acid-base regulation (Cech, Laurs, and Graham 1984; Perry et al. 1985; Brill and Bushnell 1991b; Brill et al. 1992; Lowe, Brill, and Cousins

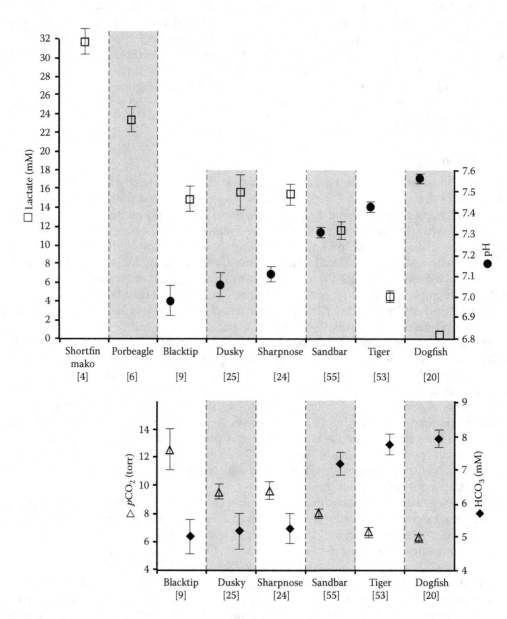

FIGURE 11.3
Interspecific differences in blood biochemistry of sharks exposed to longline-capture stress. Shortfin mako (*Isurus oxyrinchus*) and porbeagle (*Lamna nasus*) sharks were caught on pelagic longline gear (time of capture stress 120 to 360 min). All carcharhinid sharks [blacktip (*Carcharhinus limbatus*), dusky (*C. obscurus*), Atlantic sharpnose (*Rhizoprionodon terraenovae*), sandbar (*C. plumbeus*), and tiger shark (*Galeocerdo cuvier*)] were exposed to standardized demersal longline capture (time of capture stress ≤330 min). Dogfish (*Squalus acanthias* and *Mustelus canis*) are control, nonstressed specimens held in captivity. Values shown are mean (+SD). Sample size shown in brackets. (Modified from Mandelman and Skomal 2009; unpublished data from D. Bernal.)

2000). For sharks that are capable of undergoing extensive and rapid vertical movements between the warm-surface layers of the ocean and the cold water below the thermocline (e.g., mako and blue sharks), their blood will experience rapid cooling and rapid warming, which may affect CO_2 transport and acid-base regulation. A shark that rapidly ascends from cooler depths will have regions of the body remote from the gills that are colder than ambient temperature. The inverse is also observed in sharks that rapidly descend from the warm surface layers to depths below the thermocline, as their body usually remains temporally warmer than ambient due to thermal inertia. In either case, this thermal gradient in the systemic circulation can be expected to alter Hb kinetics and, thus, the solubilities and partial pressures of O_2 and CO_2. For example, the warming of arterial (i.e., oxygenated) blood while en route to an aerobically active tissue (e.g., cold arterial blood from the gills is warmed before it reaches the locomotor muscles) will result in a lower Hb-O_2 affinity and drive bound O_2 off Hb and into the plasma and cause an increase in arterial blood pO_2 (Lowe, Brill, and Cousins 2000; Nikinmaa 2006). Saturation changes of this nature, termed closed-system changes because they occur in the absence of either a source or sink for O_2 and CO_2 equilibration or proton equivalents (i.e., the total content of these gases in the blood and plasma remains unchanged), affect both blood chemistry and respiratory gas diffusion (Brill and Bushnell 2006).

For regionally endothermic sharks, the more thermally stable operating environment of the locomotor muscles may, to some degree, allow muscle performance to remain high even when these sharks penetrate cold waters and, thus, may provide these species with the ability to undergo rapid and repeated vertical movements across thermal fronts (reviewed by Bernal et al. 2009). This raises the question of to what extent can diving sharks protect their blood-O_2 equilibria and, hence, defend their overall oxygen transport capacity from the consequences of these changes in ambient temperature?

Since shark RBCs have a nucleus capable of protein synthesis, during times of thermal stress they should be capable of inducing or upregulating metabolic pathways or synthesizing HSPs that allow for cell function to be maintained until complete recovery to pre-stress conditions. The finding that HSPs play an important role in maintaining the tertiary structure of Hb (Kihm et al. 2005; Yu et al. 2007) suggests that HSPs may participate in preserving adequate levels of oxygen transport to the aerobically active tissues under times of stress. Moreover, because shark erythrocytes should be capable of rapidly (within minutes) synthesizing these chaperone molecules in response to a stress event (e.g., exhaustive exercise, changes in ambient temperature, metabolic acidosis), sharks should be able to maintain both a high Hb-O_2 affinity and high blood oxygen carrying capacity. Unfortunately, to date there is no information on the capacity of the blood of sharks to constitutively express or induce HSPs in response to thermal stress and whether this response is species specific and reflected by their thermal biology (i.e., the role of HSPs in the maintenance of proper Hb function during times of large and rapid changes in temperature).

Studies on the Hb-O_2 binding affinity of regionally endothermic fishes have shown a correlation between a decreased Hb thermal sensitivity and ambient temperature changes (Cech, Laurs, and Graham 1984; Lowe, Brill, and Cousins 2000; Brill and Bushnell 1991b, 2006). This decreased thermal sensitivity may thus be a mechanism that allows for adequate O_2 delivery across thermal gradients within the body resulting from metabolic heat conservation. In lamnids, Andersen et al. (1973) found that the 10°C warming of mako and porbeagle Hb solutions only minimally affected O_2 affinity, while Larsen, Malte, and Weber (2004) reported that, in the presence of the Hb-affinity modulator ATP, a 16°C warming of purified porbeagle Hb actually increased O_2 affinity (a reverse temperature effect). However, little is known on how untreated blood (i.e., fresh, unfrozen, whole blood preparations) of sharks with red muscle endothermy is able to mitigate the effects of a continuous thermal stress that can expose it to changes of up to 20°C between the gills and the red muscle. Recent work on blue and mako shark blood showed that a rapid closed-system temperature increase under arterial conditions (low pCO_2) markedly decreased the Hb-O_2 affinity in the blue shark (a predicted result) but had little or no effect (i.e., decreased thermal sensitivity) in makos (Figure 11.4). This suggests that regionally endothermic sharks may have evolved the capacity to defend their Hb-O_2 affinity when faced with the closed-

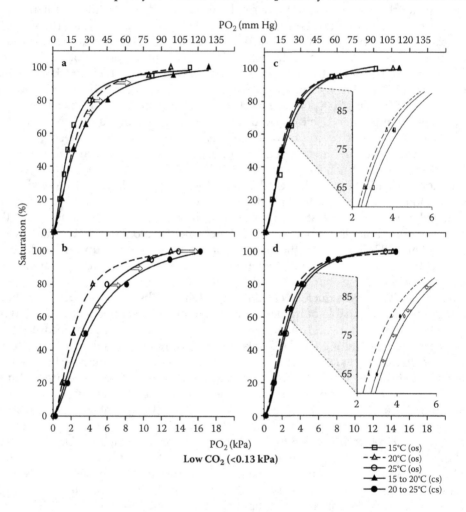

FIGURE 11.4

Oxygen equilibrium curves for blue (*Prionace glauca*) and shortfin mako (*Isurus oxyrinchus*) shark blood under arterial conditions (i.e., low CO_2: < 0.13 kPa) undergoing either open-system (*os*) or closed-system (*cs*) temperature changes (see below for definitions). (a) Blue shark, *os*: 15°C and 20°C, *cs*: from 15°C to 20°C. (b) Blue shark, *os*: 20°C and 25°C, *cs*: from 20°C to 25°C. (c) Mako shark, *os*: 15°C and 20°C, *cs*: from 15°C to 20°C. (d) Mako shark, *os*: 20°C and 25°C, *cs*: from 20°C to 25°C. Arrows illustrate the direction of change in curve positions between *os* at 20°C and *cs* warming from 15°C to 20°C or *os* at 25°C and *cs* warming from 20°C to 25°C. Note the presence of a thermal effect (right shift) in blue shark curves (a, b) and the lack of such an effect in the mako sharks (c, d). Inserts in c and d are enlargements of saturation values between 60% and 80%. *os* conditions: O_2 and CO_2 partial pressure, variable O_2 and CO_2 content (e.g., blood at the gills); *cs* conditions: instant warming, without sinks or reservoirs for gases and proton equivalents, constant O_2 and CO_2 content, variable O_2 and CO_2 partial pressure (e.g., blood in vessels remote from the gills). (Unpublished data from D. Bernal, J. Cech, and J. Graham.)

system warming of their blood, a scenario that in the blue sharks would significantly impact their capacity to deliver O_2 to their working tissues. It is possible that blue sharks (and other ectothermic sharks) may not require this Hb thermal adaptation because, when compared to makos, they have a lower metabolic scope (Bernal et al. 2003a, 2003b), which provides them with the ability to tolerate a temporary decrease in O_2 delivery to their working tissues during times of thermal stress. Clearly, additional work is needed over a broader phylogenetic range to elucidate closed system thermal effects under conditions that reflect a stressed state (low pH, high pCO_2, high La^-) in sharks and related species.

11.6 Compensatory Mechanisms

When faced with stress, fishes, like all vertebrates, employ a variety of physiological mechanisms to compensate for the homeostatic disruptions. Compensatory mechanisms include, for example, the responses by the cardio-respiratory and renal systems to restore acid-base chemistry. In general, compensation is tightly linked to restoring acid-base balance, but this should not be confused with the ability to buffer such changes, which include bicarbonate and nonbicarbonate compounds that mitigate for acidemia by binding metabolically produced protons.

Teleosts are able to make cardio-respiratory adjustments to compensate for metabolic and respiratory acidoses associated with stress (see Farrell 1993, 1996; Brill and Bushnell 2001). These mechanisms include increases in hematocrit due to splenic contraction to increase blood O_2 carrying capacity (McDonald and Milligan 1992) and catecholamine activation of RBC Na^+-H^+ exchangers, which restore intracellular pH and promote RBC swelling (Nikinmaa 1992). In concert, these processes result in the rapid return of Hb-O_2 affinity despite significant acidemia.

It has been thought that the cardio-respiratory system of elasmobranchs generally lacks these compensatory mechanisms. Bushnell et al. (1982) found a Bohr shift (decrease in O_2 affinity with drop in pH) in the blood of the lemon shark (*Negaprion brevirostris*), which exhibited only a nominal increase in hematocrit during exercise. When exposed to exhaustive exercise, the shovelnosed ray (*Glaucostegus typus*) did not make adjustments to O_2 carrying capacity and, therefore, required more time to recover (Lowe, Wells, and Baldwin 1995). However, this may not be the case for all elasmobranchs. A study by Wells and Davie (1985) on mako shark blood collected from capture-stressed specimens showed a high hematocrit (mean = 32.4) and no significant Bohr effect despite high concentrations of blood lactate. [Generally, an increase in blood-associated lactate and protons results in a decrease of the Hb-O_2 binding affinity that is observed as a right shift in the O_2-equilibrium curve (Nikinmaa and Salama 1998).] However, a recent study on mako sharks using fresh, whole blood samples showed that the Bohr effect was significantly greater than that in blue sharks, indicating a higher capacity to offload oxygen at the working tissues and suggesting a higher aerobic demand in makos (D. Bernal, J. Cech, and J. Graham, unpublished data). Additionally, unlike the high levels of lactate in the Wells and Davie (1985) study, lactate levels remained low during all experimental treatments, and most of the principal changes in O_2-binding affinity were attributed to an increase in CO_2. Thus, unlike the previous report of an absence of a Bohr effect, indicating the capacity of mako Hb to maintain function in the face of acidosis, the more recent findings suggest that both pH and CO_2 play a significant role in the blood-O_2 binding properties under normal acid-base conditions. Work on sandbar sharks by Brill et al. (2008) found that capture stress significantly increased RBC volume, hematocrit, and Hb concentration. These mechanisms, coupled with the alkalization of intracellular pH, resulted in no loss of blood oxygen affinity despite the 0.2 to 0.3 drop in blood pH. Therefore, the cardio-respiratory system of some elasmobranchs is better adapted for responding to stress, but additional studies are needed given the diversity of this group.

11.7 Recovery and Survivorship

Recovery from stress typically requires not only the continued delivery of oxygen to sustain normal metabolic functions, but the repayment of the oxygen debt associated with the removal of lactate and the return of acid-base and osmotic balance. This is particularly important in elasmobranchs that are obligate ram ventilators, which must continue to swim to facilitate respiration. For example, thresher sharks captured using recreational fishing gear are commonly hooked in the tail, which prevents adequate gill ventilation during the capture event. Under these conditions, excessive time on the line (e.g., more than 60 min) usually results in the shark reaching complete exhaustion that leads to death.

Although the immediate and delayed mortality associated with stress is poorly known for most shark species, in many fishes, including a limited number of elasmobranchs, the acute stress associated with capture elicits a suite of potentially lethal physiological changes (Wood, Turner, and Graham 1983; Cliff and Thurman 1984; Hoffmayer and Parsons 2001; Manire et al. 2001; Skomal 2007). Capture mortality, both immediate and delayed (weeks to months later), appears to be tightly linked to the nature, severity, and duration of the stress imposed, as well as the metabolic capacity (aerobic and anaerobic) of the species and its ability to recover from homeostatic disruption (Wood, Turner, and Graham 1983; Davis 2002; Skomal 2007). These latter attributes vary widely among elasmobranch species and even among closely related sharks (Mandelman and Skomal 2009). Thus, the assessment of species-specific physiological responses to capture stress is vital to fisheries conservation.

Unfortunately, the physiological consequences of stress are poorly understood in elasmobranchs and, given the logistical problems associated with measuring the suite of primary, secondary, and tertiary effects of stress in these fishes, restoration of acid-base balance is the most commonly used measure of recovery. However, when assessing recovery from a particular stressor, it is important to consider all potential sources of stress. For example, chronic stress caused by captivity may confound the effects of acute stress imposed experimentally by a researcher. Moreover, in addition to causing physiological stress, capture and handling in the laboratory or in the field cause physical trauma to tissue and organs that may exacerbate the physiological stress response and/or impede recovery and survivorship (Skomal 2007). For example, hook damage from capture events alters post-release behavior in the grey reef shark (Skomal, Lobel, and Marshall 2008).

11.7.1 Laboratory Studies

Although many species of elasmobranchs are difficult to maintain in captivity, several smaller species have proven resilient and adapt well to life within an enclosed and controlled environment. Captive studies of elasmobranchs allows: (1) the controlled application of a single or multiple stressors; (2) the measurement of the effects of the stressor(s); and (3) the characterization of recovery from the stress event. To date, in vivo and in vitro studies in captive settings have shown that the time needed to recover from stress induced by exhaustive exercise is species specific. For example, the time required to recover physiologically from exhaustive exercise may be prolonged in some sharks, ranging from 12 hours in the spiny dogfish (Richards, Heigenhauser, and Wood 2003) to up to 24 hours in the spotted dogfish (Piiper, Meyer, and Drees 1972; Holeton and Heisler 1978), smooth dogfish (*Mustelus canis*; Barham and Schwartz 1992), and dusky shark (Cliff and Thurman 1984). By contrast, other species appear to recover more rapidly. For example, acid-base blood chemistry in the sandbar shark returned to prestress levels in less than six hours (Spargo 2001; Skomal, unpublished data; Figure 11.5). These species-specific differences suggest that sharks with increased cardio-respiratory performance may be able to preserve oxygen affinity during stress (Brill et al. 2008) and, hence, remove or reconvert metabolic end products and mobilize energetic substrates to the working tissues, resulting in a decrease in the time needed to return to baseline, pre-stress values.

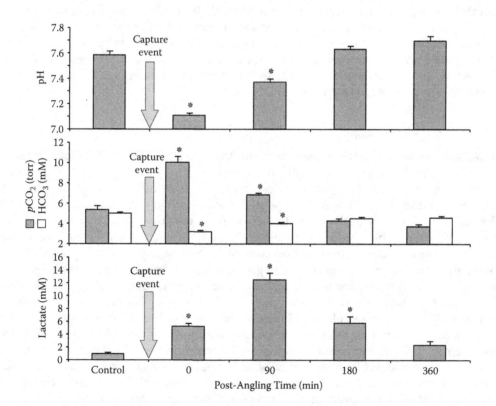

FIGURE 11.5

Postexercise recovery of blood biochemical parameters in juvenile sandbar sharks (*Carcharhinus plumbeus*) exposed to ten minutes of angling stress (i.e., capture event) and held in captivity. Blood acid-base disturbances associated with the capture event are both respiratory (elevated pCO_2) and metabolic (elevated La^- and depressed HCO_3^-) in origin. Although blood lactate remains elevated, acid-base status is restored to prestress (i.e., control) levels 3 to 6 hours after the stress event. Values shown are mean (+SD) and $n = 7$ for each group. Asterisks show a significant (a = 0.05) difference between baseline, nonstressed (i.e., control) values and each post-angling time (i.e., recovery). (Unpublished data from G. Skomal and N. Kohler.)

11.7.2 Field Studies

For those species that cannot be studied in captivity, work onboard a research vessel allows for simple, short-duration experiments (Graham et al. 1990; Lai et al. 1997). Alternatively, sharks can be outfitted with devices that record or transmit movement, environmental, and/or physiological data. Of the limited field studies conducted to date on sharks, most have centered on temporal and spatial movement patterns, with few focusing on the physiological consequences of capture stress on post-release survivorship. Conventional tags provide low resolution data on survival (i.e., two data points: capture and recapture) over broad temporal and spatial scales. However, this methodology is fisheries-dependent and recapture rates do not necessarily reflect survival rates. More detailed information on shark survivorship comes from tracking studies, which offer high-resolution data on fine-scale movements for hours to days post-release. Unfortunately, this method is not able to address the question of long-term survival. Recent technological advancements in satellite-based archival tagging provide the tools to collect high-resolution behavioral, and hence survivorship, data over broad spatial and temporal scales. Even though this new technology has allowed researchers to examine habitat use, ecology, and movements of several shark species, it has seldom been used to link the level of physiological stress at the time of release to the post-release behavior and survivorship. For the most part, snapshot blood samples at the time of capture can be used to quantify the level of stress and these, in turn, can be linked to the nature and duration of the capture event (Skomal 2007). Data collected by the tag can then be used to examine the consequences of the capture event given a known level of stress (Moyes et al. 2006).

11.7.2.1 Conventional Tagging

While tag recapture rates, averaging 5.1% in sharks (Kohler and Turner 2001), provide some indication of mortality associated with stress, other factors, including tag shedding, emigration, population size, tag reporting rates, and fishing/natural mortality, may affect estimates of recapture rates. However, there are indications that conventional tag-recapture rates reflect the capacity of the shark to recover from the stress associated with capture. Francis (1989) found that the recapture rate of the gummy shark (*Mustelus lenticulatus*) was lower in sharks captured in trawls when compared to those captured using set nets, suggesting that trawl-caught fish experienced greater capture stress. Skomal (2006) correlated blood biochemistry to angling time in blue sharks in order to quantify physiological stress and used tag-recaptures to assess survivorship (Figure 11.6). Recaptures of blue sharks with severely compromised blood pH may be indicative of the capacity of this species to recover from the physiological stress associated with angling times in excess of 30 minutes. Hight et al. (2007) compared tag-recapture information with blood catecholamine levels in live and moribund longline-caught shortfin mako sharks to derive a catecholamine threshold for post-release survivorship. Mandelman and Skomal (2009) found a correlation between species-specific levels of physiological stress, tag-recapture rates, and at-vessel mortality rates in tiger, sandbar, dusky, Atlantic sharpnose, and blacktip sharks exposed to capture stress.

11.7.2.2 Acoustic Tracking

Survivorship and post-release recovery of sharks have been directly observed with acoustic telemetry (reviewed in Sundstrom et al. 2001). These studies indicate that the stress

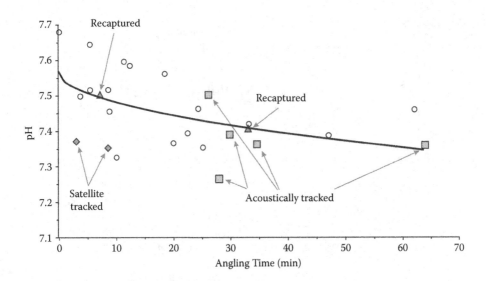

FIGURE 11.6
Changes in blood pH relative to angling time in 27 blue sharks (*Prionace glauca*) captured using conventional recreational fishing gear. Blue sharks were exposed to rod and reel capture stress, blood sampled, and released (open circles). To assess survivorship, some fish were acoustically tracked for 4.0 to 9.2 hours (filled squares), recaptured after 44 to 215 days at liberty (filled triangles), and satellite tracked (PSAT tags) for 123 days (filled diamonds). Line shows the best fit relationship between angling time and blood pH (i.e., pH = 7.57 to $0.0278*T^{-0.5}$; where T is time in minutes, $r = -0.52$, $p < 0.01$). (Modified from Skomal G (2006) Ph.D. dissertation, Boston University.)

associated with capture and handling may not compromise survivorship. For example, most sharks released from fishing gear and subsequently tracked using ultrasonic telemetry showed what appeared to be normal vertical and horizontal movements over the duration of the track (Sundstrom et al. 2001). However, most of the tracking studies to date have not had the objective to assess the effects of stress on post-release survivorship. Rather, most have taken steps to minimize capture-related stress so as to increase the probability that post-release movement data reflect more natural swimming behavior. For this reason, specimens that are in particularly poor health (show obvious signs of complete exhaustion) are not usually considered for further study, which reduces the utility of these studies for assessing post-release mortality. Furthermore, even if fish survive a few days after being released, in some cases, their immediate swimming behavior differs from the expected in displaying a decreased selectivity for specific water temperatures or depths, and appearing to ignore topographic features (e.g., Holts and Bedford 1993; Carey and Sharold 1990; Figure 11.7). For example, captured and released salmonids, which normally show a stereotyped migration pattern, sometimes move in directions other than expected (Walker et al. 2000).

Acoustic tracking studies coupled with blood sampling can provide insights into physiological recovery from capture stress. For example, acoustic tracking of blue sharks combined with blood chemistry showed that physiological recovery required <3.5 hours based on vertical behavior, which was limited during this period (Skomal 2006; Figures 11.6 and 11.7), a scenario typically seen in other shark tracking studies (Holts and Bedford 1993; Carey and Scharold 1990; Sepulveda, Graham, and Bernal 2007). Further technological advances in animal-borne instruments should greatly enhance stress-related research (e.g., Skomal, Lobel, and Marshall 2008). Passive acoustic tracking systems significantly

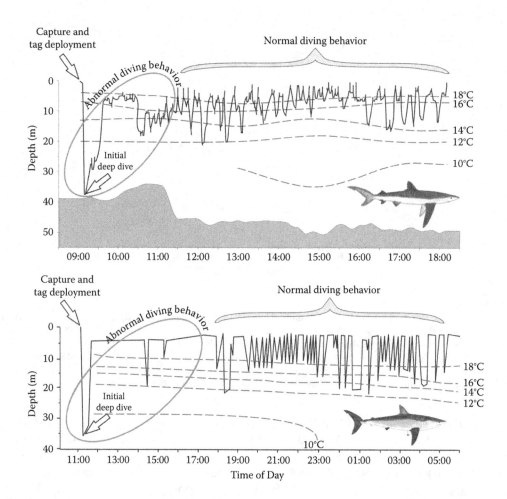

FIGURE 11.7

Representative vertical movement patterns (solid lines) as determined by acoustic telemetry of blue (*Prionace glauca*, top) and shortfin mako (*Isurus oxyrinchus*, bottom) sharks after exposure to capture stress on conventional recreational fishing gear. Note the deep initial dives of both sharks after being released and the subsequent abnormal diving behavior during the following 2 to 4 hours. Recovery is implied from the return to normal swimming behavior after this initial period. Dashed lines represent isotherms. (Modified from Skomal G (2006) Ph.D. dissertation, Boston University; Holts DB, Bedford DW (1993) Aust J Mar Freshw Res 44:45–60.)

extend tracking periods to several months and have been used to examine natural mortality in sharks (e.g., Heupel and Simpfendorfer 2002). For elasmobranch species that remain in discrete areas, this technology has tremendous potential for investigating the long-term effects of stress on behavior and recovery.

11.7.2.3 Archival Tagging

The advent of new tagging technologies has allowed for the assessment of short- and long-term survivorship after exposure to stress. Archival tags and pop-up satellite archival transmitting (PSAT) tags are ideal for investigating long-term (up to one year) post-release behavior, recovery, and survivorship. Despite the growing use of PSAT tags on large sharks (e.g., white sharks, *Carcharodon carcharias*; Bonfil et al. 2005, and basking sharks,

Cetorhinus maximus; Skomal, Wood, and Caloyianis 2004), we know of only two studies, both on blue sharks, that have attempted to correlate at-capture blood biochemistry data with PSAT tagging results. In addition to conventional and acoustic tagging, Skomal (2006) deployed PSAT tags on blood-sampled blue sharks to show that angling stress did not compromise post-release survivorship (Figure 11.6). Similarly, Moyes et al. (2006) coupled blood biochemistry data with PSAT tagging results to model post-release survivorship of blue sharks caught on pelagic longline gear and concluded that blood lactate and Mg^{2+} levels influenced post-release survival.

The use of archival tags to assess the effects of stress on survivorship has shortcomings. Although implanted archival tags, which collect depth, temperature, and light level data, are less expensive than PSAT tags, more fish must be tagged because fish must be recaptured and the tag returned. Moreover, implantation surgery creates additional stress that may exacerbate the effects of capture stress. For these reasons, satellite-based archival (PSAT) tagging is the better tool to assess post-release survivorship in highly migratory pelagic fish species (e.g., Graves, Luckhurst, and Prince 2002; Moyes et al. 2006). However, given the high cost associated with the purchase and deployment of these tags, their broadscale use with the specific objective of assessing the effects of capture on physiological stress is likely to be limited.

11.8 Conclusions

There is a paucity of information on the physiological effects of stress in elasmobranchs, but anthropogenic activities, including habitat degradation and fisheries exploitation, are impacting sharks worldwide and elevating the importance of stress-related research. Of the limited number of studies conducted to date, most have centered on sharks; batoids are poorly represented. Like teleosts, sharks exhibit primary and secondary responses to stress that are manifested in their blood biochemistry. The former is characterized by immediate and profound increases in circulating catecholamines and corticosteroids, which are thought to mobilize energy reserves and maintain oxygen supply and osmotic balance. Mediated by these primary responses, the secondary effects of stress in elasmobranchs include hyperglycemia, acidemia resulting from metabolic (lacticacidosis) and respiratory (pCO_2 elevation) acidoses, and the profound disturbance of ionic, osmotic, and fluid volume homeostasis. The nature and magnitude of these secondary effects are species specific and may be tightly linked to metabolic scope and thermal physiology as well as the type and duration of the stressor.

The extent to which elasmobranchs can recover from stress is poorly understood and has only been characterized in a few species. Laboratory and field-based tagging (conventional, acoustic, and PSAT) studies have shown that some species require longer periods to recover from stress (12 to 24 hours), whereas others have cardiorespiratory adaptations that allow them to compensate for the secondary effects of stress and recover within a few hours. Clearly, given the diversity of sharks and their relatives, additional studies that characterize the nature, magnitude, and consequences of physiological stress over a broad spectrum of stressors are essential for the development of conservation measures. Such studies should include the coupling of stress indicators with quantifiable aspects of the stressor, which will allow researchers to test hypotheses on survivorship and, ultimately, derive models that effectively link physiology to mortality. Given the unrelenting increase of anthropogenic stressors, studies of this nature are essential for decision making that will result in the effective management and conservation of sharks and their relatives.

Acknowledgments

The authors acknowledge support through MA Marine Fisheries and Federal Aid in Sportfish Restoration, the Large Pelagics Research Center (sub-award 06-125), and NSF IOS-0617403. The views expressed herein are those of the authors and do not necessarily reflect the views of their agencies. We would like to thank B. Chase and J. Chisholm for technical assistance and W. Skomal, J. Valdez, S. Bernal, S. Adams, P. Bernal, and T. Tazo for logistical support. This is Massachusetts Division of Marine Fisheries Contribution no. 25.

References

Abelow B (1998) Understanding acid–base. Williams and Wilkins, Baltimore.

Adams SM (ed) (1990a) Biological indicators of stress in fish. Am Fish Soc Symp 8.

Adams SM (1990b) Status and use of biological indicators for evaluating the effects of stress on fish. Am Fish Soc Symp 8:1–8.

Anderson DP (1990) Immunological indicators: effects of environmental stress on immune protection and disease outbreaks. Am Fish Soc Symp 8:38–50.

Andersen ME, Olson JS, Gibson QH, Carey FG (1973) Studies on ligand binding to hemoglobins from teleosts and elasmobranchs. J Biol Chem 248:331–341.

Armour KJ, O'Toole LB, Hazon N (1993) The effect of dietary protein restriction on the secretory dynamics of 1-hydroxycorticosterone and urea in the dogfish, *Scyliorhinus canicula*: a possible role for 1-hydroxycorticosterone in sodium retention. J Endocrinol 1993:275–282.

Arthur PG, West TG, Brill RW, Schulte PM, Hochachka PW (1992) Recovery metabolism of skipjack tuna (*Katsuwonus pelamis*) white muscle: rapid and parallel changes in lactate and phosphocreatine after exercise. Can J Zool 70:1230–1239.

Barham WT, Schwartz FJ (1992) Physiological responses of newborn smooth dogfish, *Mustelus canis*, during and following temperature and exercise stress. J Elisha Mitchell Sci Soc 108(2):64–69.

Beggs GL, Holeton GF, Crossman EJ (1980). Some physiological consequences of angling stress in muskellunge, *Esox masquinongy* Mitchell. J Fish Biol 17:649–659.

Bernal D, Dickson KA, Shadwick RE, Graham JB (2001) Review: analysis of the evolutionary convergence for high performance swimming in lamnid sharks and tunas. Comp Biochem Physiol 129A:2–3.

Bernal D, Sepulveda C, Mathieu-Costello O, Graham JB (2003a) Comparative studies of high performance swimming in sharks. I. Red muscle morphometrics, vascularization, and ultrastructure. J Exp Biol 206:2831–2843.

Bernal D, Sepulveda C, Musyl M, Brill R (2009) The eco-physiology of swimming and movement patterns of tunas, billfishes, and large pelagic sharks. In Domenici P, Kapoor D (eds) Fish locomotion: an etho-ecological perspective. Science Publishers, Enfield, NH, pp 436–483.

Bernal D, Smith D, Lopez G, Weitz D, Grimminger T, Dickson K, Graham JB (2003b) Comparative studies of high performance swimming in sharks. II. Metabolic biochemistry of locomotor and myocardial muscle in endothermic and ectothermic sharks. J Exp Biol 206:2845–2857.

Black EC (1958) Hyperactivity as a lethal factor in fish. J Fish Res Bd Can 15(4):573–586.

Bonfil R, Meÿer M, Scholl MC, Johnson R, O'Brien S, Oosthuizen H, Swanson S, Kotze D, Paterson M (2005) Transoceanic migration, spatial dynamics, and population linkages of white sharks. *Science* 310(5745):100–103.

Brett JR (1958) Implications and assessments of environmental stress. In Larkin PA (ed) Investigations of fish-power problems. H.R. MacMillan Lectures in Fisheries, University of British Columbia, Vancouver, pp 69–83.

Brett JR, Blackburn JM (1978) Metabolic rate and energy expenditure of the spiny dogfish *Squalus acanthias*. J Fish Res Bd Can 35:816–821.

Brill RW (1987) On the standard metabolic rates of tropical tunas including the effect of body size and acute temperature change. Fish Bull 85:25–36.

Brill RW (1996). Selective advantages conferred by the high performance physiology of tunas, bill-fishes, and dolphin fish. Comp Biochem Physiol A 113:3–15.

Brill RW, Bushnell PG (1991a) Metabolic and cardiac scope of high energy demand teleosts: the tunas. Can J Zool 69:2002–2009.

Brill RW, Bushnell PG (1991b) Effects of open and closed-system temperature changes on blood oxygen dissociation curves of skipjack tuna (*Katsuwonus pelamis*) and yellowfin tuna (*Thunnus albacares*). Can J Zool 69:1814–1821.

Brill RW, Bushnell PG (2001) The cardiovascular system of tunas. In Block BA, Stevens ED (eds) Fish physiology, vol. 19, tuna: physiology, ecology and evolution, Academic Press, San Diego, CA, pp 79–120.

Brill RW, Bushnell PG (2006) Effects of open- and closed-system temperature changes on blood O_2-binding characteristics of Atlantic bluefin tuna (*Thunnus thynnus*). Fish Physiol Biochem 32:283–294.

Brill RW, Bushnell PG, Jones DR, Shimizu M (1992) Effects of acute temperature change *in-vivo* and *in-vitro* on the acid-base status of blood from yellowfin tuna (*Thunnus albacares*). Can J Zool 70:654–662.

Brill R, Bushnell P, Schroff S, Seifert R, Galvin M (2008) Effects of anaerobic exercise accompanying catch-and-release fishing on blood-oxygen affinity of the sandbar shark (*Carcharhinus plumbeus*, Nardo). J Exp Mar Biol Ecol 354(1):132–143.

Bushnell PG, Brill RW (1992) Oxygen transport and cardiovascular responses in skipjack tuna (*Katsumonus pelamis*) and yellowfin tuna (*Thunnus albacares*) exposed to acute hypoxia. J Comp Physiol 162:131–143.

Bushnell PG, Lutz PL, Steffensen JF, Oikari A, Gruber SH (1982) Increases in arterial blood oxygen during exercise in the lemon shark (*Negaprion brevirostris*). J Comp Physiol 147:41–47.

Butler PJ, Taylor EW, Davison W (1979) The effect of long term, moderate hypoxia on acid-base balance, plasma catecholamines and possible anaerobic end products in the unrestrained dogfish *Scyliorhinus canicula*. J Comp Physiol 132:297–303.

Butler PJ, Metcalfe JD, Ginley SA (1986) Plasma catecholamines in the lesser spotted dogfish and rainbow trout at rest and during different levels of exercise. J Exp Biol 123:409–421.

Butler PJ, Taylor EW, Copra MF, Davison W (1978) The effect of hypoxia on the levels of circulating catecholamines in the dogfish *Scyliorhinus canicula*. J Comp Physiol 127:325–330.

Carey FG (1969) Warm fish. Oceanus 15.

Carey FG (1973) Fishes with warm bodies. Sci Am 228:36–44.

Carey FG, Casey JG, Pratt HL Jr, Urquhart D, McCosker JE (1985) Temperature, heat production and heat exchange in lamnid sharks. Mem S Cal Acad Sci 9:92–108.

Carey FG, Scharold JV (1990) Movements of blue sharks, *Prionace glauca*, in depth and course. Mar Biol 106:329–342.

Carey FG, Teal JM (1969) Mako and porbeagle are warm-bodied sharks. Comp Biochem Physiol 28:199–204.

Carlson JK, Goldman KJ, Lowe CG (2004) Metabolism, energy demand, and endothermy. In Carrier JC, Musick JA, Heithaus MR (eds) Biology of sharks and their relatives, CRC Press, Boca Raton, FL, pp 203–224.

Carlson JK, Parsons GR (1999) Seasonal differences in routine oxygen consumption rates of the bonnethead shark. J Fish Biol 55:876–879.

Cech JJJ, Laurs RM, Graham JB (1984) Temperature induced changes in blood gas equilibria in the albacore (*Thunnus alalunga*) a warm bodied tuna. J Exp Biol 109:21–34.

Claiborne JB, Edwards SL, Morrison-Shetlar AI (2002) Acid-base regulation in fishes: cellular and molecular mechanisms. J Exp Zool 293:302–319.

Cliff G, Thurman GD (1984) Pathological effects of stress during capture and transport in the juvenile dusky shark, *Carcharhinus obscurus*. Comp Biochem Physiol 78A:167–173.

Cooke SJ, Schreer JF, Dunmall KM, Philipp DP (2002) Strategies for quantifying sublethal effects of marine catch-and-release angling: insights from novel freshwater applications. In Lucy JA, Studholme AL (eds) Catch and release in marine recreational fisheries, Am Fish Soc Symp 30, pp 121–134.

Cooper AR, Morris S (1998) The blood respiratory, haematological acid–base and ionic status of the port jackson shark, *Heterodontus portusjacksoni*, during recovery from anaesthesia and surgery: a comparison with sampling by direct caudal puncture. Comp Biochem Physiol 119A:895–903.

Currie S, Moyes CD, Tufts BL (2000) The effects of heat shock and acclimation temperature on hsp70 and hsp30 mRNA expression in rainbow trout: *in vivo* and *in vitro* comparisons. J Fish Biol 56:398–408.

Currie S, Tufts BL (1997) Synthesis of stress protein 70 (Hsp70) in rainbow trout (*Oncorhynchus mykiss*) red blood cells. J Exp Biol 200:607–614.

Currie S, Tufts BL, Moyes CD (1999) Influence of bioenergetic stress on heat shock protein gene expression in nucleated red blood cells of fish. Am J Physiol 276:R990–R996.

Davis MW (2002) Key principles for understanding fish bycatch discard mortality. Can J Fish Aquat Sci 59:834–1843.

DeRoos R, DeRoos CC (1978) Elevation of plasma glucose levels by catecholamines in elasmobranch fish. Gen Comp Endocrin 34:447–452.

Dickson KA (1996) Locomotor muscle of high performance fishes: what do comparisons of tunas with ectothermic sister taxa reveal? Comp Biochem Physiol 113A:39–49.

Dickson KA, Gregorio MO, Gruber SJ, Loefler KL, Tran M, Terrell C (1993) Biochemical indices of aerobic and anaerobic capacity in muscle tissues of California elasmobranch fishes differing in typical activity level. Mar Biol 117:185–193.

Dowd WW, Brill RW, Bushnell PG, Musick JA (2006) Standard and routine metabolic rates of juvenile sandbar sharks (*Carcharhinus plumbeus*), including the effects of body mass and acute temperature change. Fish Bull 104:323–331.

Driedzic WR, Hochachka PW (1978) Metabolism in fish during exercise. In Hoar WS, Randall DJ (eds) Fish physiology, vol. VII, locomotion, Academic Press, New York, pp 503–543.

Evans DH, Piermarini PM, Choe KP (2004) Homeostasis: osmoregulation, pH regulation, and nitrogen metabolism. In: Carrier JC, Musick JA, Heithaus MR (eds) Biology of sharks and their relatives, CRC Press, Boca Raton, FL, pp 247–268.

Farrell AP (1993) Cardiovascular system. In Evans DH (ed) The physiology of fishes, CRC Press, Boca Raton, FL, pp 219–250.

Farrell AP (1996) Features heightening cardiovascular performance in fishes, with special reference to tunas. Comp Biochem Physiol A 113:61–67.

Ferguson RA, Tufts BL (1992) Physiological effects of brief air exposure in exhaustively exercised rainbow trout (Oncorhynchus mykiss): implications for "catch and release" fisheries. Can J Fish Aquat Sci 49:1157–1162.

Francis MP (1989) Exploitation rates of rig (*Mustelus lenticulatus*) around the South Island of New Zealand. NZ J Mar Freshw Res 23:239–245.

Francis MP, Griggs LH, Baird SJ (2001) Pelagic shark bycatch in the New Zealand tuna longline fishery. Mar Freshw Res 52:165–178.

Gelsleichter J (2004) Hormonal regulation of elasmobranch physiology. In Carrier JC, Musick JA, Heithaus MR (eds) Biology of sharks and their relatives, CRC Press, Boca Raton, FL, pp 287–323.

Graham JB, Dewar H, Lai NC, Lowell WR, Arce SM (1990) Aspects of shark swimming performance determined using a large water tunnel. J Exp Biol 151:175–192.

Graves JE, Luckhurst BE, Prince ED (2002) An evaluation of pop-up satellite tags for estimating postrelease survival of blue marlin (*Makaira nigricans*) from a recreational fishery. Fish Bull 100:134–142.

Grogan ED, Lund R (2004) The origin and relationships of early Chondrichthyes. In Carrier JC, Musick JA, Heithaus MR (eds) Biology of sharks and their relatives, CRC Press, Boca Raton, FL, pp 3–31.

Gruber SH, Keyes RS (1981) Keeping sharks for research. In Hawkins AD (ed) Aquarium systems, Academic Press, New York, pp 373–402.

Hazen N, Balment RJ (1998) Endocrinology. In Evans DE (ed) The physiology of fishes, second edition, CRC Press, Boca Raton, FL, pp 441–464.

Heupel MR, Simpfendorfer CA (2002) Estimation of mortality of juvenile blacktip sharks, *Carcharhinus limbatus*, within a nursery area using telemetry data. Can J Fish Aquat Sci 59:624–632.

Hight BV, Holts D, Graham JB, Kennedy BP, Taylor V, Sepulveda CA, Bernal D, Ramon D, Rasmussen R, Lai NC (2007) Plasma catecholamine levels as indicators of the post-release survivorship of juvenile pelagic sharks caught on experimental drift longlines in the Southern California Bight. Mar Freshw Res 58:145–151.

Hinton DE, Lauren DJ (1990) Integrative histopathological approaches to detecting effects of environmental stressors in fishes. Am Fish Soc Symp 8:51–66.

Hochachka PW, Hulbert WC, Guppy M (1978) The tuna power plant and furnace. In Sharp GD, Dizon AE (eds) The physiological ecology of tunas, Academic Press, New York, pp 153–174.

Hoffmayer ER, Parsons GR (2001) The physiological response to capture and handling stress in the Atlantic sharpnose shark, *Rhizoprionodon terraenovae*. Fish Physiol Biochem 25:277–285.

Holeton G, Heisler N (1978) Acid-base regulation by bicarbonate exchange in the gills after exhausting activity in the larger spotted dogfish *Scyliorhinus stellaris*. Physiol 21:56.

Holts DB, Bedford DW (1993) Horizontal and vertical movements of the shortfin mako, *Isurus oxyrinchus*, in the southern California Bight. Aust J Mar Freshw Res 44:45–60.

Iwama GK, Afonso LOB, Vijayan MM (2006) Stress in fishes. In Evans DE, Claiborne JB (eds) The physiology of fishes, third edition, CRC Press, Boca Raton, FL, pp 319–342.

Jensen FB, Nikinmaa M, Weber RE (1983) Effects of exercise stress on acid-base balance and respiratory function in blood of the teleost *Tinca tinca*. Resp Physiol 51:291–301.

Karnaky KL (1998) Osmotic and ionic regulation. In Evans DE (ed) The physiology of fishes, second edition, CRC Press, Boca Raton, FL, pp 157–176.

Kieffer JD (2000) Limits to exhaustive exercise in fish. Comp Biochem Physiol 126A:161–179.

Kieffer JD, Kubacki MR, Phelan FJS, Philipp DP, Tufts BL (1995) Effects of catch and release angling on nesting male smallmouth bass. Trans Am Fish Soc 124:70–76.

Kihm AJ, Kong Y, Hong W, Russell JE, Rouda S, Adachi K, Simon MC, Blobel GA, Weiss MJ (2005) An abundant erythroid protein that stabilizes free-haemoglobin. Ann NY Acad Sci 1054:103–117.

Kohler NE, Turner PA (2001) Shark tagging: a review of conventional methods and studies. Env Biol Fishes 60:191–223.

Korsmeyer KE, Dewar H, Lai NC, Graham JB (1996) The aerobic capacity of tunas: adaptation for multiple metabolic demands. Comp Biochem Physiol A 113:17–24.

Lai NC, Dalton N, Lai YY, Kwong C, Rasmussen R, Holts D, Graham JB (2004) A comparative echocardiographic assessment of ventricular function in five species of sharks. Comp Biochem Physiol A 137:505–521.

Lai NC, Korsmeyer KE, Katz S, Holts DB, Laughlin LM, Graham JB (1997) Hemodynamics and blood properties of the shortfin mako shark (*Isurus oxyrinchus*). Copeia 2:424–428.

Larsen C, Malte H, Weber RE (2004) ATP-induced reverse temperature effect in isohemoglobins from the endothermic porbeagle shark, *Lamna nasus*. J Biol Chem 278:30741–30747.

Lowe TE, Brill RW, Cousins KL (2000) Blood oxygen-binding characteristics of bigeye tuna (*Thunnus obesus*), a high-energy-demand teleost that is tolerant of low ambient oxygen. Mar Biol 136:1087–1098.

Lowe TE, Wells RMG, Baldwin J (1995) Absence of regulated blood-oxygen transport in response to strenuous exercise by the shovelnosed ray, *Rhinobatos typus*. Mar Freshw Res 46:441–446.

Mandelman JW, Cooper PW, Werner TB, Lagueux KM (2008) Shark bycatch and depredation in the U.S. Atlantic pelagic longline fishery. Rev Fish Biol Fisheries18:427–442.

Mandelman JM, Farrington MA (2007) The physiological status and mortality associated with otter trawl capture, transport, and captivity of an exploited elasmobranch *Squalus acanthias*. ICES J Mar Sci 64:122–130.

Mandelman and Skomal (2009) Differential sensitivity to capture stress assessed by blood acid–base status in five carcharhinid sharks. J Comp Physiol B 79(3):267–277.

Mandrup-Paulsen J (1981) Changes in selected blood serum constituents, as a function of salinity variations, in the marine elasmobranch, *Sphyrna tiburo*. Comp Biochem Physiol 70A:127–131.

Manire C, Heuter R, Hull E, Spieler R (2001) Serological changes associated with gillnet capture and restraint in three species of sharks. Trans Am Fish Soc 130:1038–1048.

Manire CA, Rasmussen LEL, Maruska KP, Tricas TC (2007) Sex, seasonal, and stress-related variations in elasmobranch corticosterone concentrations. Comp Biochem Physiol 148A:926–935.

Mazeaud MM, Mazeaud F, Donaldson EM (1977) Primary and secondary effects of stress in fish: some new data with a general review. Trans Am Fish Soc 106(3):201–212.

McCandless CT, Kohler NE, Pratt HL Jr (eds) (2008) Shark nursery grounds of the Gulf of Mexico and the east coast waters of the United States. Am Fish Soc Symp 50, Bethesda. 390 pp.

McDonald DG, Milligan CL (1992) Chemical properties of the blood. In Hoar WS, Randall DJ, Farrell AP (eds) Fish physiology, vol 12, part B: the cardiovascular system, Academic Press, San Diego, CA, pp 56–133.

McDonough JL, Arrell DK, Van Eyk JE (1999) Troponin I degradation and covalent complex formation accompanies myocardial ischemia/reperfusion injury. Circ Res 84:9–20.

McKendry JE, Milsom WK, Perry SF (2001) Branchial CO_2 receptors and cardiorespiratory adjustments during hypercarbia in Pacific spiny dogfish (*Squalus acanthias*). J Exp Biol 204:1519–1527.

Milligan CL (1996) Metabolic recovery from exhaustive exercise in rainbow trout. Comp Biochem Physiol 113A:51–60.

Milligan CL (2003) A regulatory role for cortisol in muscle glycogen metabolism in rainbow trout *Oncorhynchus mykiss* Walbaum. J Exp Biol 206:3167–3173.

Milligan CL, Wood CM (1986) Intracellular and extracellular acid-base status and H^+ exchange with the environment after exhaustive exercise in the rainbow trout. J Exp Biol 123:93–121.

Mommsen TP, Vijayan MM, Moon TW (1999) Cortisol in teleosts: dynamics, mechanisms of action, and metabolic regulation. Rev Fish Biol. Fish 9:211–268.

Morgan A, Burgess GH (2007) At-vessel fishing mortality for six species of sharks caught in the Northwest Atlantic and Gulf of Mexico. Gulf Carib Res 19:123–129.

Moyes CD, Fragoso N, Musyl MK, Brill RW (2006) Predicting post-release survival in large pelagic fish. Trans Am Fish Soc 135(5):1389–1397.

Moyes CD, West TG (1995) Exercise metabolism in fish. In Hochachka PW, Mommsen TP (eds) Metabolic biochemistry, Elsevier, New York, pp 368–392.

Nikinmaa M (1992) Membrane transport and the control of haemoglobin oxygen-affinity in nucleated erythrocytes. Physiol Rev 72:301–321.

Nikinmaa M (2006) Gas transport. In Evans DE, Claiborne JB (eds) The physiology of fishes, third edition, CRC Press, Boca Raton, FL, pp 153–174.

Nikinmaa M, Salama A (1998) Oxygen transport in fish. In: Perry SF, Tufts B (eds) Fish physiology, vol. 17, Academic Press, San Diego, CA, pp 141–184.

NMFS (National Marine Fisheries Service) (2008) Final amendment 2 to the consolidated Atlantic highly migratory species fishery management plan. U.S. Department of Commerce, Washington, DC.

Opdyke DF, Carroll RG, Keller NE (1982) Catecholamine release and blood pressure changes induced by exercise in dogfish. Am J Physiol 242:R306–R310.

Pagnotta A, Brooks L, Milligan L (1994) The potential regulatory roles of cortisol in recovery from exhaustive exercise in rainbow trout. Can J Zool 72:2136–2146.

Perry SF, Daxboeck C, Emmett B, Hochachka PW, Brill RW (1985) Effects of temperature change on acid-base regulation in skipjack tuna (*Katsuwonus pelamis*) blood. Comp Biochem Physiol A 81:49–54.

Pickering AD (ed) (1981) Stress and fish. Academic Press, New York.

Piiper J, Meyer M, Drees F (1972) Hydrogen ion balance in the elasmobranch *Scyliorhinus stellaris* after exhausting exercise. Resp Physiol 16:290–303.

Powers DA (1983) Adaptation of erythrocyte function during changes in environmental oxygen and temperature. In Cossins AR, Sheterline P (eds) Cellular acclimatization to environmental change. Society for experimental biology seminar series, vol 17, Cambridge University Press, New York, pp 227–244.

Powers DA (1985) Molecular and cellular adaptations of fish hemoglobin-oxygen affinity to environmental changes. In Lamy J, Truchot JP, Gilles R (eds) Respiratory pigments in animals: relation structure-function, Springer-Verlag, New York, pp 97–124.

Priede IG (1985) Metabolic scope in fishes. In Tytler P, Calow TP (eds) Fish energetics: new perspectives, Johns Hopkins University Press, Baltimore, MD, pp 28–46.

Randall DJ, Perry SF (1992) Catecholamines. In Hoar WS, Randall DJ (eds) Fish physiology, vol.12B, Academic Press, New York, pp 255–300.

Renshaw GMC, Kerrisk CB, Nilsson GE (2002) The role of adenosine in the anoxic survival of the epaulette shark, *Hemiscyllium ocellatum*. Comp Biochem Physiol B 131:133–141.

Richards JG, Heigenhauser GJF, Wood CM (2003) Exercise and recovery metabolism in the pacific spiny dogfish (*Squalus acanthias*). J Comp Physiol 173B: 463–474.

Robertson L, Thomas P, Arnold CR (1988) Plasma cortisol and secondary stress responses of cultured red drum (*Sciaenops ocellatus*) to several transportation procedures. Aquaculture 68:115–130.

Romero LM (2002) Seasonal changes in plasma glucocorticoid concentrations in free-living vertebrates. Gen Comp Endocrin 128:1–24.

Rossi-Fanelli A, Antonini E (1960) Oxygen equilibrium of hemoglobin from *Thunnus thynnus*. Nature 186:895–896.

Routley MH, Nilsson GE, Renshaw GMC (2002) Exposure to hypoxia primes the respiratory and metabolic responses of the epaulette shark to progressive hypoxia. Comp Biochem Physiol A 131:313–321.

Schulte PM, Moyes CD, Hochachka PW (1992) Integrating metabolic pathways in post-exercise recovery of white muscle. J Exp Biol 166:181–195.

Sepulveda CA, Graham JB, Bernal D (2007) Aerobic metabolic rates of swimming juvenile mako sharks, *Isurus oxyrinchus*. Mar Biol 152:1087–1094.

Sepulveda CA, Kohin S, Chan C, Vetter R, Graham JB (2004) Movement patterns, depth preferences, and stomach temperatures of free-swimming juvenile mako sharks, *Isurus oxyrinchus*, in the Southern California Bight. Mar Biol 145:191–199.

Shuttleworth TJ (1988a) Physiology of elasmobranch fishes. Springer-Verlag, Berlin.

Shuttleworth TJ (1988b) Salt and water balance: extrarenal mechanisms. In Shuttleworth TJ (ed) Physiology of elasmobranch fishes, Springer-Verlag, Berlin, pp 171–199.

Skomal G (2006) The physiological effects of capture stress on post-release survivorship of sharks, tunas, and marlin. Ph.D. dissertation, Boston University.

Skomal G (2007) Evaluating the physiological and physical consequences of capture on post-release survivorship in large pelagic fishes. Fish Manage Ecol 14:81–89.

Skomal G, Lobel PS, Marshall G (2008) The use of animal-borne imaging to assess post-release behavior as it relates to capture stress in grey reef sharks, *Carcharhinus amblyrhynchos*. J Mar Tech Soc 41(4):44–48.

Skomal G, Wood G, Caloyianis N (2004) Archival tagging of a basking shark, *Cetorhinus maximus*, in the western North Atlantic. J Mar Biol Assoc UK 84:795–799.

Spargo A (2001) The physiological effects of catch and release angling on the post-release survivorship of juvenile sandbar sharks (*Carcharhinus plumbeus*). MS Thesis, University of Rhode Island.

Stoskopf MK (1993) Fish medicine. WB Saunders, Philadelphia.

Strange RJ, Schreck CB, Ewing RD (1978) Cortisol concentrations in confined juvenile chinook salmon (*Oncorhynchus tshawytscha*). Trans Am Fish Soc 107:812–819.

Strange RJ, Schreck OB, Golden JT (1977) Corticoid stress responses to handling and temperature in salmonids. Trans Am Fish Soc 106(3):213–218.

Sulikowski JA, Maginniss LA (2001) Effects of environmental dilution on body fluid regulation in the yellow stingray, *Urolophus jamaicensis*. Comp Biochem Physiol A 128:223–232.

Sundstrom LF, Gruber SH, Clermont SM, Correia JPS, de Marignac JRC, Morrissey JF, Lowrance CR, Thomassen L, Oliveira MT (2001) Review of elasmobranch behavioral studies using ultrasonic telemetry with special reference to the lemon shark, *Negaprion brevirostris*, around Bimini Island, Bahamas. Env Biol Fishes 60:225–250.

Taylor EW, Butler PJ (1982) Nervous control of heart rate: activity in the cardiac vagus of the dogfish. J Appl Physiol 53:1330–1335.

Thomas P, Robertson L (1991) Plasma cortisol and glucose stress responses of red drum (*Sciaenops ocellatus*) to handling and shallow water stressors and anaesthesia with MS-222, quinaldene sulfate, and metomidate. Aquaculture 96:69–86.

Tietz NW (1987) Fundamentals of clinical chemistry, WB Saunders, Philadelphia.

Tufts BL, Tang Y, Tufts K, Boutilier RG (1991) Exhaustive exercise in "wild" Atlantic salmon (*Salmo salar*): acid-base regulation and blood gas transport. Can J Fish Aquat Sci 48:868–874.

Turner JD, Wood CM, Clark D (1983) Lactate and proton dynamics in the rainbow trout (*Salmo gairdneri*). J Exp Biol 104:247–268.

Tullis A, Baillie M (2005) The metabolic and biochemical responses of tropical whitespotted bamboo shark *Chiloscyllium plagiosum* to alterations in environmental temperature. J Fish Biol 67:950–968.

Walker RV, Myers KW, Davis N, Aydin KY, Friedland FD, Carlson HR, Boehlert GW, Urawa S, Ueno Y, Anma G. (2000) Diurnal variation in thermal environment experienced by salmonids in the North Pacific as indicated by data storage tags. Fish Ocean 9:171–186.

Wang Y, Heigenhauser GJF, Wood CM (1994) Integrated responses to exhaustive exercise and recovery in rainbow trout white muscle: acid-base, phosphogen, carbohydrate, lipid, ammonia, fluid volume and electrolyte metabolism. J Exp Biol 195:227–258.

Wedemeyer GA, Yasutake WT (1977) Clinical methods for the assessment of the effects of environmental stress on fish health. Tech Paper US Fish Wildl Serv 89:1–18.

Wendelaar-Bonga SE (1997) The stress response in fish. Physiol Rev 77:591–625.

Wells RMG, Davie PS (1985) Oxygen binding by the blood and hematological effects of capture stress in two big gamefish: mako shark and striped marlin. Comp Biochem Physiol 81A:643–646.

Wells RMG, McIntyre RH, Morgan AK, Davie PS (1986) Physiological stress responses in big gamefish after capture: observations on plasma chemistry and blood factors. Comp Biochem Physiol 84A:565–571.

Wells RMG, Tetens V, Devries AL (1984) Recovery from stress following capture and anaesthesia of antarctic fish: hematology and blood chemistry. J Fish Biol 25:567–576.

Wood CM (1991) Acid-base and ion balance, metabolism, and their interactions after exhaustive exercise in fish. J Exp Biol 160:285–308.

Wood CM, Turner JD, Graham MS (1983) Why do fish die after severe exercise? J Fish Biol 22(2):189–201.

Yu X, Kong Y, Dore LC, Abdulmalik O, Katein AM, Zhou S, Choi JK, Gell D, Mackay JP, Gow AJ, Weiss MJ (2007) An erythroid chaperone that facilitates folding of α-globin subunits for hemoglobin synthesis. J Clin Invest 117:1856–1865.

12

Pollutant Exposure and Effects in Sharks and Their Relatives

James Gelsleichter and Christina J. Walker

CONTENTS

12.1 Introduction

Due to their relative longevity, moderate to large size, slow metabolism, lipid-rich livers, and high trophic position, many cartilaginous fish are capable of accumulating significant concentrations of environmental pollutants. This is due to the chemical behavior of these compounds, many of which are lipid soluble and have the tendency to *bioconcentrate* in aquatic organisms, or accumulate at concentrations greater than ambient levels; *bioaccumulate* over time, or increase in concentrations as animals grow in size; and *biomagnify* in aquatic food webs, or increase in concentrations with rising trophic position. The accumulation of environmental pollutants in sharks and their relatives may also be associated with the high use of frequently polluted near-shore areas as nursery and/or breeding grounds in some coastal species, as well as the large-scale migratory patterns of some pelagic species, which may expose them to compounds banned in most developed countries but still in use in developing nations. Because of this, environmental pollutants may pose risks to the health and sustainability of some elasmobranch populations, particularly those that have already been threatened by other, typically more immediate impacts, including commercial and recreational exploitation and habitat loss. The accumulation of environmental pollutants in shark and ray tissues also poses health risks to humans who consume or use elasmobranch-based products, especially since many of these pollutants can be easily absorbed by the human gastrointestinal system and even traverse the placental barrier in pregnant females and impact the developing fetus.

This chapter provides a comprehensive overview of what is known regarding the accumulation and potential health risks of major classes of environmental pollutants in sharks and their relatives. Some attention is also given to the limited body of data concerning the biotransformation of environmental pollutants in cartilaginous fish, especially since many of these physiological mechanisms can be used to identify biological indicators, or biomarkers, of pollutant effects in these animals. Last, the potential risks that pollutant levels in elasmobranch tissues pose to human health are discussed via consideration of regulatory guidelines concerning safe limits of these compounds in seafood intended for human consumption.

12.2 Organic Compounds

The major organic contaminants examined in cartilaginous fish to date include pesticides, polychlorinated dibenzo-*p*-dioxins, polychlorinated dibenzofurans, polychlorinated biphenyls, polycyclic aromatic hydrocarbons, and some personal care products. Although organometals are by definition organic compounds, they are discussed along with metals and metalloids from which they are derived.

12.2.1 Pesticides

The large and diverse group of chemicals classified as pesticides include naturally produced and synthetic compounds that are used to control the propagation of insects (insecticides), unwanted plant species (herbicides), fungi (fungicides), and a variety of other target taxa. These chemicals are released into the environment through both residential and agricultural use, the latter of which accounts for the large majority (i.e., >70%) of total pesticide use in most developed countries (Kiely, Donaldson, and Grube 2004). Lower, nonagricultural use of pesticides is still substantial, and is likely underestimated due to difficulties in accurately measuring residential application rates (Kiely, Donaldson, and Grube 2004). Following their release into the environment, pesticides from both agricultural and residential sources can enter aquatic habitats through a variety of point and nonpoint sources, including industrial or municipal wastewater discharge, stormwater runoff, atmospheric transport and deposition, and groundwater intrusion (Costa 2008). Historically, pesticide use increased dramatically between the 1950s and 1980s due largely to the spread of modern agricultural practices that occurred during the Green Revolution (Pimental 1996), as well as efforts to reduce the transmission of vector-borne diseases (Rogan and Chen 2005). Despite public misconceptions, pesticide use has largely stabilized since the 1980s because of the development of more effective compounds and a movement toward the use of integrated pest management approaches and organic farming (Costa 2008). Nonetheless, the total amount of pesticides used on a worldwide basis remains high and represents a significant health risk to aquatic wildlife, including sharks and their relatives.

To date, the only published studies that have examined pesticide exposure and uptake in elasmobranchs have focused on organochlorine pesticides (OCPs), a large group of synthetic compounds that includes the well-known insecticide DDT (dichlorodiphenyltrichloroethane) and its derivatives; the cyclodiene pesticides, such as chlordane and dieldrin; the hexachlorocyclohexanes, such as lindane; and caged structures, such as mirex (Costa 2008). These compounds were used extensively in agriculture and vector control from the 1940s to the 1970s and 1980s, but were largely banned for use in most developed countries because of their slow breakdown in the environment, tendency to accumulate in wildlife, and potential for causing chronic effects on reproduction. For example, the high use of DDT in the United States from the 1950s to the 1970s has been linked with major declines in bird populations, apparently resulting from reductions in eggshell thickness and egg viability caused by exposure to the DDT metabolite dichlorodiphenyldichloroethylene (DDE; Lundholm 1997). Despite bans on many of these compounds and their replacement by other insecticides, OCPs continue to be common pollutants in aquatic habitats due to their high persistence, as well as continued or reintroduced use of some OCPs as vector control agents in some less developed countries.

Concentrations of DDT and its primary metabolites DDE and dichlorodiphenyldichloroethane (DDD) have been measured in at least 20 species of cartilaginous fish, representing several locations throughout the world and a variety of life history traits likely to influence the uptake and accumulation of organic contaminants (Table 12.1). In addition, a small number of studies have also examined concentrations of other widely used OCPs, particularly in coastal species (Table 12.1). Although no studies have reported muscle OCP concentrations in sharks that exceed estimated safe limits for human consumption (e.g., 5 μg/g wet weight for total DDTs in the United States; ATSDR 2002), the results of these studies demonstrate that OCPs can often accumulate at elevated and potentially hazardous levels in some chondrichthyan species. This has been best illustrated in the white shark *Carcharodon carcharias* (Schlenk, Sapozhnikova, and Cliff 2005) and several deepwater shark species, such as the Greenland shark *Sominosus microcephalus* (Fisk et al. 2002), gulper shark *Centrophorus granulosus*, and longnose spurdog *Squalus blainvillei* (Storelli and Marcotrigiano 2001), which have been shown to accumulate total DDT concentrations in liver comparable with if not greater than those observed in other top marine predators (e.g., toothed whales, polar bears). In contrast, most coastal species, such as the bonnethead *Sphyrna tiburo* (Gelsleichter et al. 2005, 2008), leopard shark *Triakis semifasciata* (Davis et al. 2002), black dogfish *Centroscyllium fabricii* (Berg et al. 1997), and the bamboo shark *Chiloscyllium plagiosum* (Cornish et al. 2007) exhibited much lower concentrations of OCPs in comparison with pelagic and deepwater sharks. This is likely because they occupy lower trophic positions, are smaller, and have shorter lifespans and/or perhaps lower lipid content. However, moderately elevated OCP concentrations have been reported in some small and young sharks, such as <3-month-old neonate blacktip sharks *Carcharhinus limbatus* from the northwest Atlantic Ocean (Gelsleichter, Szabo, and Morris 2007). Since OCP concentrations in these animals were far greater than those later observed in larger and older *S. tiburo* from the same region (Gelsleichter et al. 2008), they may reflect maternal transfer of OCPs to offspring via yolk reserves or perhaps the placenta. This is supported by observations of DDT concentrations higher than those found in some maternal tissues (i.e., muscle) in yolk of other shark species, such as *C. granulosus* and *S. blainvillei* (Storelli and Marcotrigiano 2001).

TABLE 12.1

Concentrations of Organochlorine Pesticides Observed in Organs or Tissues of Cartilaginous Fish*

Species	Area/Tissue	Total DDTs	Total Cyclodienes or Chlordanes	Reference
Alopias vulpinus	Italian coast (adipose fat)	16–300	—	Corsolini et al. 1995
Carcharhinus limbatus	NW Atlantic	451 ± 83	146 ± 43	Gelsleichter, Szabo, and Morris 2007
	E. Gulf of Mexico	57 ± 10–95 ± 32	65 ± 7–73 ± 22	Gelsleichter, Szabo, and Morris 2007
Carcharhinus plumbeus	NW Atlantic	413 ± 80–563 ± 167	135 ± 29–162 ± 42	Gelsleichter, Szabo, and Morris 2007
Carcharodon carcharias	South Africa (muscle)	600–2626 7.6–20.1	—	Schlenk, Sapozhnikova, and Cliff 2005
Centrophorus granulosus	Mediterranean Sea (muscle)	4481 ± 961	—	Storelli and Marcotrigiano 2001
	(eggs)	49 ± 13	—	
		1018 ± 186	—	
Centroscyllium fabricii	Davis Strait	111 ± 81	—	Berg et al. 1997[a]
Chiloscyllium plagiosum	Hong Kong (muscle)	0.6–24	—	Cornish et al. 2007
Dalatias licha	Mediterranean Sea	4554 ± 2046	—	Storelli, Storelli, and Marcotrigiano 2005
Dasyatis sabina	E. Gulf of Mexico	15 ± 8	49 ± 4	Gelsleichter et al. 2006
	Central Florida lakes	20 ±4–80 ± 10	31 ±2–88 ± 3	Gelsleichter et al. 2006
Isurus oxyrincus	Brazilian coast (muscle)	2.1	—	Azevedo-Silva et al. 2009

Species	Location / tissue			Reference
Mustelus henlei	U.S. Pacific coast (muscle)	—	5–16	Fairey et al. 1997
Prionace glauca	Mediterranean Sea	—	2392 ± 1439	Storelli, Storelli, and Marcotrigiano 2005
	Brazilian coast (muscle)	—	0.4–2.1	Azevedo-Silva et al. 2007
	Italian coast (adipose fat)	—	14–35	Corsolini et al. 1995
Scoliodon sorrakowah	Arabian Sea (muscle)	—	1.39	Shailaja and Nair 1997
Scyliorhinus canicula	Mediterranean Sea	—	1170 ± 471	Storelli et al. 2006[a]
Scyliorhinus torazame	Japan	—	77–1,600	Horie, Tanaka, and Tanaka 2004[a]
Somniosus microcephalus	Davis Strait	1,815 ± 273	7159 ± 1271	Fisk et al. 2002[a]
Sphyrna tiburo	NW Atlantic	11 ± 2–24 ± 7	25 ±3–66 ± 12	Gelsleichter et al. 2008
	E. Gulf of Mexico	23 ± 4–104 ± 9	19 ±7–158 ± 49	Gelsleichter et al. 2005, 2008
Sphyrna zygaena	Brazilian coast (muscle)	—	2.13–2.96	Azevedo-Silva et al. 2009
Squalus blainvillei	Mediterranean Sea	—	1625 ± 439	Storelli and Marcotrigiano 2001
	muscle	—	17 ± 9	
	eggs	—	238 ± 53	
Triakis semifasciata	U.S. Pacific coast (muscle)	0.2–2.5 ± 0.9	4.5 ± 0.7–8 ± 2	Davis et al. 2002

* Values presented are for single measurements, means ± SE, range of single measurements, or range of means ± SE for total DDTs and total cyclodienes or chlordanes in liver specifically unless otherwise indicated. Concentrations are presented in ng g[-1] wet weight unless otherwise indicated.

[a] Values presented in ng/g lipid weight.

Given their low diversity, it is not surprising that few studies have examined OCP concentrations in freshwater elasmobranchs. However, as demonstrated in a recent study on pollutant exposure in freshwater and estuarine populations of the Atlantic stingray *Dasyatis sabina* (Gelsleichter et al. 2006), it is possible that freshwater sharks and rays may face greater risks of exposure to these and other organic contaminants than their marine and/or euryhaline counterparts. For example, Gelsleichter et al. (2006) reported that hepatic total DDT and chlordane concentrations in *D. sabina* from an agriculture-impacted freshwater lake in Florida (Lake Jesup) were two to five times higher than those in rays from a Florida estuary generally viewed as moderately to highly contaminated by OCPs (Tampa Bay). Although the actual OCP concentrations observed in Lake Jesup rays were not particularly high (e.g., they were comparable to the levels observed in small coastal sharks, such as *S. tiburo*; Gelsleichter et al. 2005, 2008), this may be a function of the low trophic position of *D. sabina*. Therefore, larger, predatory elasmobranchs that frequent similarly polluted freshwater sites (e.g., the bull shark *Carcharhinus leucas*) could accumulate far greater and potentially hazardous concentrations of OCPs. Given this, future studies on pollutant burden in freshwater sharks and rays are warranted.

Despite the number of studies on OCP concentrations in cartilaginous fish, there is little information available on the possible effects of these compounds in sharks and their relatives. Whereas liver concentrations of total DDTs in some elasmobranchs exceed threshold values for biological effects in juvenile and adult fish (estimated at 0.6 µg/g wet weight; Beckvar, Dillon, and Read 2005), these values are believed to be provisional at best due to limited data on the sublethal effects of these compounds. A more realistic screening value for adverse effects of total DDTs may be ~3 µg/g wet weight, which has been shown to be associated with reproductive effects (e.g., increased fry mortality, eggshell thinning) in species most sensitive to these compounds, such as the brown pelican *Pelecanus occidentalis* and the lake trout *Salvelinus namaycush* (Blus 2003). Based on this, the only chondricthyan species studied to date for which DDT-related effects may be expected are the Greenland shark *Sominosus microcephalus* (Fisk et al. 2002) and gulper shark *Centrophorus granulosus* (Storelli and Marcotrigiano 2001). DDT-related effects may also occur in adult white sharks *Carcharodon carcharias*, given that juveniles have been shown to accumulate close to 3 µg total DDTs/g wet weight liver at only a third of the total maximum length for this species (Schlenk, Sapozhnikova, and Cliff 2005).

Although some OCPs continue to be used even in developed countries (e.g., endosulfan), they were largely replaced for agricultural and household applications in the 1980s by less persistent but more acutely toxic pesticides, such as the organophophorus insecticides (OPs) and carbamates. More recently, due to the high toxicity of these compounds, OPs and carbamates have been mostly replaced for residential use by the less toxic and minimally persistent pyrethroid insecticides. To the best of the authors' knowledge, no published studies have examined exposure levels and uptake of these "current use" insecticides in elasmobranchs. However, given recent evidence for their accumulation and potential health effects in fish despite their expectedly low environmental persistence (Dutta and Meijer 2003; Soumis et al. 2003; Henry and Kishimba 2006; Srivastava, Yadav, and Trivedi 2008), future studies on this topic are needed. Future research should also consider the potential uptake of herbicides in sharks and their relatives, because these compounds generally account for a far greater proportion of the total pesticides used worldwide than insecticides and because they are the pesticides most commonly found in aquatic habitats. Whereas most herbicides function by altering physiological processes specific to plants, some of the most commonly used compounds (e.g., 2,4-D, atrazine) are capable of causing health effects in at least some fishes (Ateeq, Abul Farah, and Ahmad 2005; Spano et al.

2004). Although some studies have described the biotransformation and excretion of certain herbicides in sharks (Guarino, James, and Bend 1977; Koschier and Pritchard 1980), the accumulation of these compounds in elasmobranchs has been largely unexplored.

12.2.2 Polychlorinated Dibenzo-*p*-Dioxins and Polychlorinated Dibenzofurans

Polychlorinated dibenzo-*p*-dioxins (PCDDs or dioxins) and polychlorinated dibenzofurans (PCDFs) are two groups of structurally similar compounds that are produced naturally by the incomplete combustion of organic material (e.g., wood, coal), and unintentionally formed as by-products in a number of industrial processes, including the manufacture of chlorinated phenols (e.g., chlorophenoxy herbicides) and paper bleaching. Dioxins comprise a group of 75 alternate forms or congeners including the compound 2,3,7,8-tetra-chlorodibenzo-*p*-dioxin (TCDD), which is well known as the major toxicant found in the defoliant Agent Orange (a combination of two chlorophenoxy herbicides, 2,4-D and 2,4-T), used extensively in the Vietnam War. The lesser known PCDFs include 135 congeners that differ little from PCDDs and are formed in many of the same processes. Dioxins and PCDFs enter the environment through a number of sources, including naturally occurring fires, waste incineration, pesticide application, and air and sewage emissions from various industries, such as paper mills and pesticide manufacturing plants. Like OCPs, PCDDs/Fs are highly persistent in aquatic ecosystems and tend to bioaccumulate and biomagnify in top predators. A total of 17 of these compounds (7 PCDDs and 10 PCDFs) are known to be toxic and capable of causing a variety of health effects in vertebrates, including immuno-suppression, fetal or embryonic abnormalities, reproductive dysfunction, carcinogenesis, and a severe loss of body mass known as "wasting syndrome" (Schecter et al. 2006). In fact, TCDD in particular is commonly viewed as one of the most harmful substances known because of its high acute toxicity in animal models and purported effects in exposed human populations (Grassman et al. 1998; Schecter et al. 2006). Because of this, the toxicity of PCDDs/Fs, and similarly acting compounds (i.e., dioxin-like polychlorinated biphenyls, as discussed later in this chapter) are typically expressed as a fraction of TCDD's potency using a measure known as the toxic equivalency factor (TEF). Using this approach, the concentrations of these compounds in environmental matrices or animal tissues can be multiplied by their respective TEFs and the resulting values summed to obtain a single value known as the toxic equivalency quotient (TEQ), which can be used to express the overall toxicity of a complex chemical mixture.

To date, only one published study has used analytical procedures to measure concentrations of PCDDs/Fs in an elasmobranch. In this study, Strid et al. (2007) examined muscle and liver concentrations of six PCDDs and nine PCDFs in Greenland sharks *Somniosus microcephalus* from northeast Atlantic waters near Iceland. Liver from *S. microcephalus* was found to contain sizeable quantities of total PCDDs/Fs that, on a lipid-weight (mean = 530 pg/g) and TEQ (mean = 100 pg TEQ/g lipid weight) basis, were higher than those observed in some other large marine predators from Arctic and northeast Atlantic waters (e.g., polar bears; Kumar et al. 2002). While liver concentrations of PCDDs/Fs alone do not exceed threshold levels for biological effects in aquatic vertebrates (i.e., 160 to 1400 pg TEQ/g lipid weight; Kannan et al. 2000), combined levels of these compounds and the similarly acting dioxin-like polychlorinated biphenyls did surpass these limits, suggesting the potential for health effects in Greenland shark populations. Similarly, while muscle concentrations of PCDDs/Fs alone in *S. microcephalus* (mean = 13 pg/g and 3.1 pg TEQ/g lipid weight) did not exceed levels recommended for fish consumption (e.g., 8 pg TEQ/g wet weight in

Europe, 50 pg TCDD/g wet weight in the United States), combined levels of PCDDs/Fs and dioxin-like PCBs did in most cases, suggesting the potential for human health risks.

Using a different approach known as the Chemical-Activated Luciferase Gene Expression (CALUX) bioassay, which measures the presence of dioxin-like compounds in a sample extract on the basis of its ability to induce "dioxin receptor" (i.e., aryl hydrocarbon receptor or AhR) mediated responses in cells transfected with the firefly luciferase gene, Baeyens et al. (2007) also measured dioxin-like activity in muscle of unidentified shark and ray samples from a Belgium fish market. While dioxin-like activity was observed in these samples (mean = 0.71 and 0.53 pg CALUX-TEQ/g lipid weight for ray and shark, respectively), it was low in comparison to most other fish species examined and could have been a function of exposure to other compounds, such as dioxin-like polychlorinated biphenyls.

12.2.3 Polychlorinated Biphenyls

Polychlorinated biphenyls (PCBs) are a large group of synthetic organic compounds with 1 to 10 chlorine atoms attached to a biphenyl core structure. Since there are 10 possible positions for chlorine atoms to bind to the core molecule, there are 209 theoretically possible PCB congeners. Only about 130 of these compounds have been produced for commercial use, always in the form of PCB mixtures labeled with the trade name Aroclor in the United States and numbered on the basis of the type of molecule (i.e., "12" for PCBs) and the percentage of chlorine by mass in the mixture (e.g., Aroclor 1248, which is a PCB mixture containing 48% chlorine by mass). Because of their unique physiochemical properties, such as high stability, low flammability, and low electrical conductivity, PCB mixtures were used extensively from the 1930s to the 1970s as insulating agents in electrical capacitators and transformers and for a diverse variety of other industrial purposes (e.g., as lubricating oils, hydraulic oils, fire retardants, and as additives in plastics, paints, varnishes, and pesticide formulations). However, the manufacture and use of these compounds were banned in most industrialized countries in the late 1970s due to concerns about their persistence in the environment and potential toxicity to human and wildlife populations. These concerns were largely motivated by studies that first demonstrated the growing presence of PCBs in animal and human tissues worldwide (see review in Erickson 2001), as well as a mass human poisoning by PCB-contaminated rice-bran oil, which occurred in the western part of Japan in 1968 (Yoshimura 2003). This event resulted in the death of approximately 400,000 birds that ingested the contaminated oil because of its use as a supplement in animal feed and adverse health effects known as "Yusho disease" in approximately 14,000 Japanese citizens. A similar incident occurred in Taiwan in 1979, resulting in a condition known as "Yu-cheng disease" in approximately 2000 Taiwanese individuals. Both of these events were caused by leakage of PCBs from heating coils used to deodorize oil during its processing for human consumption. The health effects observed in Yusho and Yu-cheng disease patients as a result of acute PCB exposure included abnormal pigmentation, swelling and increased discharge from the upper eyelids, an acne-like eruption of skin pustules commonly known as chloracne, various subjective symptoms (e.g., nausea, numbness of extremities, loss of appetite) and in some cases, death (Yoshimura 2003; Guo et al. 2004). Transgenerational effects have also been observed in offspring born to women with Yusho and Yu-cheng disease, such as hyperpigmentation, eyelid swelling, chloracne, poor tooth development, reduced growth, and alterations in neurological and sexual development.

During the heyday of their use, large amounts of PCBs were released to aquatic habitats via industrial discharge, particularly from facilities involved in the manufacture or disposal of electrical equipment. Because of the tendency of PCBs to associate with organic material, this resulted in extensive contamination of aquatic sediments in a number of industrialized areas that persists to this day; a testament to the longevity and slow environmental breakdown of these compounds. In addition to these "legacy" reservoirs of PCB contamination, there is also new input of these compounds into aquatic habitats as a result of environmental redistribution of PCBs from terrestrial stores. Therefore, despite restrictions on their production and use, PCBs continue to pose significant pollution risks to aquatic ecosystems. Like OCPs and PCDDs/Fs, PCBs also biomagnify in aquatic food webs and tend to accumulate in top predators. This is well illustrated by the high number of fish consumption advisories in the United States due to PCB contamination, which is second only to that for mercury.

Concentrations of PCBs have been examined in at least 36 species of cartilaginous fish (Table 12.2), making these compounds one of the most extensively studied groups of pollutants in the field of chondrichthyan ecotoxicology. It is likely that even more data concerning PCB levels in sharks and their relatives are available, but are either unpublished or included as unidentified components of broad, multispecies studies of PCB contamination in regional seafood resources. Although several aspects of published studies such as methodology, detection limits, sample type (e.g., liver or muscle), and reported units (i.e., wet weight or lipid weight) vary, the results of these surveys generally suggest that only certain, high trophic level sharks are likely to accumulate PCBs at potentially hazardous concentrations. In fact, of the 33 species for which published data are available, only two have been shown to exhibit PCB concentrations above the threshold for organism-level effects in fish and aquatic mammals (e.g., growth and reproduction, which are impaired at PCB concentrations >50 µg/g; Kannan et al. 2000; Giesey et al. 2006): the Greenland shark *Somniosus microcephalus* and bull shark *Carcharhinus leucas*. In some cases, the PCB concentrations measured in individuals of these species even exceeded levels used to classify materials as "hazardous waste" in some countries (e.g., 50 µg/g in Canada). These findings are not particularly surprising because both *S. microcephalus* and *C. leucas* are large, long-lived, and terminal consumers in their respective habitats. While lower, total PCB concentrations reported in three other predatory shark species, the Atlantic sharpnose shark *Rhizoprionodon terranovae*, tiger shark *Galeocerdo cuvier*, and smooth hammerhead *Sphyrna zygaena*, are also of toxicological relevance because they are within the range of threshold levels for some cell- and molecular-level effects in aquatic vertebrates, such as immunosuppression and hormonal alterations (~10 to 30 ppm; Kannan et al. 2000; Giesey et al. 2006). Although some biochemical or cell-level effects may be experienced by species in which low parts-per-million-range PCB concentrations (i.e., 1 to 4 µg/g) have been observed, the majority of other chondrichthyan species are likely to be relatively unimpacted by PCB exposure. However, these predications are based solely on studies on other taxa and the threshold PCB concentrations at which detrimental effects may occur in cartilaginous fish are virtually unknown.

Due to their ability to adopt a planar configuration and induce cellular responses similar to those elicited by toxic PCDDs/Fs, a group known as the coplanar or dioxin-like PCBs (DL-PCBs) are believed to be largely responsible for the toxicity of technical PCB mixtures. This group is composed of two subgroups, both of which contain four or more chlorine atoms at both *para* and *meta* positions in the core molecule, but differ based on the presence of a chlorine atom in *ortho* positions: the highly toxic non-*ortho* chlorine substituted DL-PCBs, such as PCB-77, PCB-126, and PCB-169; and the moderately toxic mono-*ortho* DL-PCBs, such as PCB-105, PCB-118, and PCB-156. Given the differences in the toxicity of DL-PCBs and other PCB congeners, congener-specific analysis of PCB residues in aquatic wildlife are

TABLE 12.2

Concentrations and Toxic Equivalence Quotients (TEQ) of Total Polychlorinated Biphenyls (PCBs) Observed in Organs or Tissues of Cartilaginous Fish*

Species	Area/Tissue	Total PCBs	TEQ	Reference
Alopias superciliosus	Brazilian coast (muscle)	8.29	—	Azevedo-Silva et al. 2009
Alopias vulpinus	Italian coast	160–4400	—	Corsolini et al. 1995
Carcharhinus albimarginatus	Japan (immature)	120–520	—	Haraguchi et al. 2009
Carcharhinus leucas	Florida coast (1993–94)	3930–8310	—	Johnson-Restrepo et al. 2005[a]
	Florida coast (2002–04)	2930–327,000	—	Johnson-Restrepo et al. 2005[a]
	Japan	35,000	—	Haraguchi et al. 2009
Carcharhinus limbatus	NW Atlantic (neonate)	2926 ± 577	—	Gelsleichter, Szabo, and Morris 2007
	E. Gulf of Mexico (neonate)	185 ± 33–417 ± 87	—	Gelsleichter, Szabo, and Morris 2007
Carcharhinus plumbeus	NW Atlantic (immature)	1311 ± 275–2054 ± 465	—	Gelsleichter, Szabo, and Morris 2007
	Japan (immature)	280	—	Haraguchi et al. 2009
Centrophorus granulosus	Mediterranean Sea	1741 ± 531	197	Storelli and Marcotrigiano 2001
	muscle	28 ± 11	—	
	eggs	416 ± 85	—	
Centrophorus squamosus	NE Atlantic	2097–3570	17–287	Serrano et al. 1997, 2000
Centroscyllium fabricii	Davis Strait	545 ± 258	—	Berg et al. 1997[a]
Centroscymnus coelolepis	NE Atlantic	102 ± 64–4723 ± 3670	1.2–56	Serrano et al. 1997, 2000
Centroscymnus cryptacanthus	NE Atlantic	920–1426	68–88	Serrano et al. 1997, 2000
Chiloscyllium plagiosum	Hong Kong (muscle)	1.05–4.77	—	Cornish et al. 2007
Chimaera monstrosa	Mediterranean Sea	280 ± 78	0.33	Storelli et al. 2004
Dalatias licha	NE Atlantic	2304	13.5	Serrano et al. 2000
	Mediterranean Sea	1827 ± 349	1.46	Storelli, Storelli, and Marcotrigiano 2005[a]
Dasyatis sabina	E. Gulf of Mexico	97 ± 32	—	Gelsleichter et al. 2006
	Florida coast	60–3160	—	Johnson-Restrepo et al. 2005[a]
	Central Florida lakes	52 ± 5–121 ± 51	—	Gelsleichter et al. 2006
Deania histricosa	NE Atlantic	39–1532	0.15–0.42	Serrano et al. 1997, 2000

Deania produndorum	NE Atlantic	392–433	69–118	Serrano et al. 1997, 2000
Etmopterus princeps	NE Atlantic	6.5–2290	2.4–38	Serrano et al. 1997, 2000
Galeocerdo cuvier	Japan (juvenile)	72–2800	—	Haraguchi et al. 2009
	Japan (adult)	3600–11,000	—	Haraguchi et al. 2009
Galeus melastomus	Mediterranean Sea	853 ± 471–1072 ± 387	0.55–0.83	Storelli et al. 2003
Hydrolagus affinis	NE Atlantic	956	4.2	Serrano et al. 2000
Isurus oxyrincus	Brazilian coast (muscle)	9.32	—	Azevedo-Silva et al. 2009
Mustelus henlei	U.S. Pacific coast (muscle)	16.5–148	—	Fairey et al. 1997
Prionace glauca	Mediterranean Sea	2482 ± 1020	2.51	Storelli, Storelli, and Marcotrigiano 2005[a]
	Brazilian coast (muscle)	1.53–6.88	—	Azevedo-Silva et al. 2007
	Italian coast	70–110	—	Corsolini et al. 1995
Raja spp. (3 spp.)	Mediterranean Sea	314 ± 200	0.48	Storelli et al. 2004
Rhizoprionodon terranovae	Florida coast	1770–13,400	—	Johnson-Restrepo et al. 2005[a]
Scyliorhinus canicula	Mediterranean Sea	1292 ± 577	0.40	Storelli et al. 2006[a]
Scyliorhinus torazame	Japan	400–4400	—	Horie, Tanaka, and Tanaka 2004[a]
Somniosus microcephalus	Davis Strait	3442 ± 650	—	Fisk et al. 2002[a]
	NE Atlantic	990–10,000	39–820	Strid et al. 2007[a]
Sphyrna tiburo	NW Atlantic	113 ± 73–520 ± 64	—	Gelsleichter et al. 2008
	E. Gulf of Mexico	29 ± 7–367 ± 67	—	Gelsleichter et al. 2005, 2008
Sphyrna zygaena	Mediterranean Sea	0–17,390	4843	Storelli, Storelli, and Marcotrigiano 2003
	Brazilian coast (muscle)	4.5–18.6	—	Azevedo-Silva et al. 2009
Squalus acanthias	Florida coast	605–982	—	Johnson-Restrepo et al. 2005[a]
Squalus blainvillei	Mediterranean Sea	958 ± 658	166	Storelli and Marcotrigiano 2001
	muscle	11 ± 7	—	
	eggs	214 ± 41	—	
Triakis semifasciata	U.S. Pacific coast (muscle)	0–27 ± 10	—	Davis et al. 2002

* Values presented are for single measurements, means ± SE, range of single measurements, or range of means ± SE for total DDTs and total cyclodienes or chlordanes in liver specifically unless otherwise indicated. Concentrations are presented in ng g^{-1} wet weight unless otherwise indicated. TEQ are presented in pg TEQ g^{-1} unless otherwise indicated.

[a] Values presented in ng g^{-1} lipid weight.

typically conducted to gain a better understanding of the toxic potential of these compounds. When this is performed, concentrations of individual DL-PCBs can be adjusted by
their respective TEFs and resulting values summed to obtain a TEQ specifically for these
compounds, as described in the section above on PCDDs/Fs. This has been conducted for a
number of cartilaginous fish (Table 12.3) and, like measurements of total PCB residues, they
show high potential for PCB effects in some high trophic level species. This is true for both
S. microcephalus and *S. zygaena*, two species for which TEQ levels greatly exceed toxicity
reference values for organism-level effects in some aquatic vertebrates. For example, maximum TEQ levels found in both of these species were several-fold greater than threshold
TEQs for reproductive effects in aquatic mammals, which range from 160 to 1400 pg TEQ/g
lipid weight (Kannan et al. 2000). Use of this approach also suggests some potential for PCB
effects in both gulper shark *Centrophorus granulosus* and longnose spurdog *Squalus blanvillei*
from the Mediterranean Sea, a conclusion also supported to some extent by data on total
PCB residues in these species (Storelli and Marcotrigiano 2001). In contrast, TEQ values
observed for all other species for which this approach has been employed were generally
low, suggesting limited potential for PCB effects in these species.

12.2.4 Polycyclic Aromatic Hydrocarbons

Polycyclic aromatic hydrocarbons (PAHs) are a group of over 100 compounds with two or
more fused aromatic rings, which are present in natural fossil fuel deposits and enter the
environment by the incomplete combustion of these materials (i.e., pyrolytic sources) and/
or their physical release and breakdown (i.e., petrogenic sources). PAH contamination in
terrestrial and aquatic ecosystems can result from naturally occurring processes, such as
volcanic activity or geological seepage of buried fuel deposits, as well as from a number of
human-related activities including the combustion of coal, oil, and fuel for both industrial
and residential needs (e.g., power generation, heating, automobile usage, coke and coal
tar production, waste incineration) and the accidental or purposeful discharge of petroleum (e.g., oil spills) or other PAH-containing substances (e.g., the wood preservative creo-
sote). Like most of the other organic contaminants described in this chapter, PAHs tend
to associate with organic material and as a result are persistent contaminants in aquatic
sediments. However, since many organisms (most vertebrates and some invertebrates) are
capable of rapidly transforming most PAHs into more excretable metabolites, the parent
compounds generally do not biomagnify in aquatic food webs and usually only accumulate in species that lack the ability to biotransform them (e.g., mollusks). As demonstrated
in studies on fish populations inhabiting creosote-contaminated sites, PAH exposure can
have significant effects on fish, such as the formation of toxicopathic lesions, including
neoplasms (Vogelbein et al. 1990; Myers et al. 2008). Such effects likely result from the
tendency of some higher molecular weight PAHs, such as benzo(*a*)pyrene, to be biotransformed to metabolites capable of forming adducts with DNA and altering the expression
of tumor suppression and/or tumor promoting genes (Varanasi et al. 1987; Baumann 1998;
Reichert et al. 1998).

 Due to their rapid metabolism in many organisms, PAHs are typically present at only
trace levels in most marine vertebrates and are rarely measured as reliable indicators of
exposure to these compounds. Nonetheless, several studies have examined PAH accumulation in marine wildlife populations, and levels ranging from 1 to 4 µg/g wet weight have
been observed in lipid-rich tissues of species residing in moderately to highly contaminated habitats (Albers and Loughlin 2003; Kannan and Perrotta 2008). Even higher PAH
concentrations have been reported in some top marine predators, such as Mediterranean

TABLE 12.3

Concentrations of Three Groups of Brominated Fire Retardants in Cartilaginous Fish: Polybrominated Diphenyl Ethers (PBDEs), Hexabromocyclododecanes (HBCDs), and Tetrabromobisphenol A (TBBPA)*

Species	Area	Total PBDEs	Total HBCD	TBBPA	Reference
Carcharhinus albimarginatus	Japan (immature)	8–30	—	—	Haraguchi et al. 2009
Carcharhinus leucas	Florida coast (1993–94)	16–287	9.2–413	4.2–8.1	Johnson-Restrepo et al. 2005; Johnson-Restrepo, Adams, and Kannan 2008
	Florida coast (2002–04)	12–4190	16.6–310	0.035–35.6	Johnson-Restrepo et al. 2005; Johnson-Restrepo, Adams, and Kannan 2008
	Japan	850	—	—	Haraguchi et al. 2009
Carcharhinus plumbeus	Japan (immature)	17	—	—	Haraguchi et al. 2009
Dasyatis sabina	Florida coast	2.3–83	—	—	Johnson-Restrepo et al. 2005
Galeocerdo cuvier	Japan (juvenile)	4–180	—	—	Haraguchi et al. 2009
	Japan (adult)	110–650	—	—	Haraguchi et al. 2009
Rhizoprionodon terranovae	Florida coast	45–1930	1.8–156	0.5–1.4	Johnson-Restrepo et al. 2005; Johnson-Restrepo, Adams, and Kannan 2008
Squalus acanthias	Florida coast	16–60	—	—	Johnson-Restrepo et al. 2005
Triakis semifasciata	California coast	1.4[a]	—	—	Holden et al. 2003

* Values presented are ranges in ng g^{-1} lipid. Tissues examined were liver (Haraguchi et al. 2009) and muscle (Johnson-Restrepo et al. 2005; Johnson-Restrepo, Adams, and Kannan 2008).

a Defined as "edible parts" and presented in ng g^{-1} wet weight.

Sea striped dolphins, in which total blubber PAH concentrations of up to 198 µg/g wet weight have been observed (Marsili et al. 2001). To the best of the authors' knowledge, only one published study has used analytical procedures to measure PAH residues in elasmobranchs. In this study, Al-Hassan et al. (2000) reported total liver PAH concentrations ranging from 0.13 to 33 µg/g wet weight in a multispecies sample of sharks from the Arabian Gulf. These levels were comparable to those observed in liver of teleosts from the same area, which ranged from 1.69 to 10.79 µg/g wet weight (Al-Hassan et al. 2003). These results are consistent with a high degree of oil pollution in this area, particularly due to events such as the Kuwait oil fires, in which 6 to 10 million barrels of crude oil were intentionally discharged into the local marine environment (Al-Hassan 1992). However, whether exposure to such high levels of PAHs impacts local shark populations remains to be studied.

A more reliable and commonly used approach for assessing exposure of marine fish to PAHs is to examine the presence of PAH metabolites in bile, where they are concentrated and stored prior to excretion (Vuorinen et al. 2006). This is commonly assessed by measuring the concentration of fluorescent aromatic compounds (FACs) in bile, which is indicative of PAH content, but less complicated and expensive to perform than more traditional analytical procedures (e.g., high-performance liquid chromatography or HPLC). Fuentes-Rios et al. (2005) examined biliary fluorescence in redspotted catsharks *Schroederichthys chilensis* obtained from pollutant-impacted and reference sites on the south Pacific coast of Chile. The results of this study suggest that FAC analysis may also be a useful measure of PAH exposure in elasmobranchs. Sharks sampled from more contaminated sites exhibited a threefold higher concentration of bile FACs than those from the reference location.

12.2.5 Brominated Flame Retardants

The group of chemicals known collectively as brominated flame retardants (BFRs) includes a variety of bromine-containing synthetic compounds that are added to various combustible products, such as electronic equipment and household consumer goods, to reduce their flammability and meet fire safety standards (Janssen 2005). Although diverse in structure, all BFRs function in a similar fashion; that is, by releasing bromine and forming hydrogen bromide gas, which acts as a free radical scavenger and interferes with the chemical reactions that create and spread flames (Rahman et al. 2001). "First-generation" BFRs known as polybrominated biphenyls (PBBs) were the predominant flame retardants used in commercial goods in the early 1970s, but have been banned for manufacture and use in most developed countries because of concerns about their persistence and toxicity. In particular, restrictions on PBB production and use in the United States were prompted by a highly publicized event in which animal feed for dairy cattle, livestock, and poultry in the state of Michigan was accidentally contaminated by these chemicals, resulting in exposure of over nine million individuals to PBBs. Current-use BFRs include three classes of chemicals: the polybrominated diphenyl ethers (PBDEs), tetrabromobisphenol A (TBBPA), and hexabromocyclodecanes (HBCDs). Due to their greater efficiency, wider applications, and lower cost, these compounds are more commonly used for commercial applications than alternative flame retardants. All BFRs are released into the environment from a number of point sources including BFR production sites, manufacturing facilities that incorporate BFRs into commercial products, and recycling plants. In addition, BFRs that are not covalently bound into synthetic polymers (i.e., PBDEs, HBCDs) can diffuse out of the polymer matrix over time and enter the environment via atmospheric transport.

PBDEs are a large group of compounds that differ from one another based on the degree of bromination of a diphenyl ether core structure (Siddiqi, Laessig, and Reed 2003; Jansen 2005). Since there are 10 possible positions for bromine atoms on the core molecule, there are 209 theoretically possible PBDE congeners, which are divided into 10 groups from mono- to deca-BDE. Commercial formulations of PBDEs do not consist solely of single congeners, but are rather combinations of multiple compounds named for the most abundant group of congeners in the mixture. There are three commercially available PBDE mixtures used in industry: penta-bromodiphenyl ether (penta-BDE), octa-bromodiphenyl ether (octa-BDE), and deca-bromodiphenyl ether (deca-BDE). Currently, only deca-BDE remains in high use in most developed countries, particularly in the United States, where it is the most commonly used flame retardant in commercial products. Both penta-BDE and octa-BDE have been banned for use in Europe and certain regions in North America in recent years, and the major U.S. manufacturer of these mixtures voluntarily agreed to end production of them in 2005. These events were largely motivated by concerns about rising levels of PBDEs in the environment, as well as in wildlife and human populations, in addition to their ability to cause a variety of toxic effects (e.g., alterations in thyroid function, neurotoxicity, and liver tumors) in exposed organisms. Since deca-BDE is believed to be less persistent, bioavailable, and toxic than other formulations, it has generally been viewed as safer for commercial use. However, since recent studies have demonstrated that deca-BDE is capable of bioaccumulating in some aquatic organisms and may be degraded to more toxic and persistent lower brominated congeners in both the environment and exposed animals, its safety has been understandably questioned.

A recent study examined muscle concentrations of PBDEs in four elasmobranch species from coastal waters off Florida (the Atlantic stingray *Dasyatis sabina*, Atlantic sharpnose shark *Rhizoprionodon terranovae*, spiny dogfish *Squalus acanthais*, and bull shark *Carcharhinus leucas*), and compared these measurements to those observed in bony fish and cetaceans sampled from the same regions (Johnson-Restrepo et al. 2005). The results of this study illustrate the reasons for recent concerns about these compounds that, because of their growing presence in the environment, have been referred to as "the PCBs of the future" (Siddiqi, Laessig, and Reed 2003). First, PBDE concentrations in Florida fish and cetaceans were strongly associated with trophic level (Table 12.3), demonstrating the tendency of these compounds to biomagnify in aquatic ecosystems and bioaccumulate in top predators (Yogui and Sericano 2009). PBDE concentrations in elasmobranchs in particular followed this same general trend, with low average concentrations detected in *D. sabina* and *S. acanthias*, 10-fold higher concentrations observed in *R. terranovae*, and 30-fold higher levels detected in *C. leucas*. In fact, lipid-normalized PBDE concentrations detected in some individual bull sharks were among the highest levels ever measured in fish species and rivaled those observed in Florida cetaceans (Holden et al. 2003; Yogui and Sericano 2009). Second, a comparison of PBDE concentrations in Florida *C. leucas* sampled from 1993 to 1994 with those observed in sharks obtained in 2002 to 2004 demonstrate the exponential rise in environmental PBDE concentrations that has occurred in the past two decades, and has been previously observed in wildlife and human populations alike (Janssen 2005; Yogui and Sericano 2009). Another interesting point about this study is that the deca-PBDE congener BDE-209 was the most abundant PBDE observed in all shark species. This finding supports the premise that deca-PBDE is available for uptake into at least some aquatic organisms; however, its abundance in sharks also suggests that it may not be metabolized to more toxic congeners in these fish.

A more recent study reported total PBDE concentrations in liver of tiger shark *Galeocerdo cuvier*, bull shark *Carcharhinus leucas*, silvertip shark *Carcharhinus albimarginatus*, and sandbar shark *Carcharhinus plumbeus* from southern Japan waters (Haraguchi et al. 2009) (Table 12.3). Last, a single measurement of total PBDEs in edible parts from leopard shark *Triakis semifasciata* was performed in a multispecies study on contamination of fishery resources in San Francisco Bay, California (Holden et al. 2003). The concentrations of PBDEs in these species were comparable with those reported in sharks with similar life history traits by Johnson-Restrepo et al. (2005).

Using the same samples examined for PBDEs in Johnson-Restrepo et al. (2005), the same authors recently measured muscle concentrations of the two other major classes of BFRs, TBBPA and HBCDs, in Florida *R. terranovae* and *C. leucas* (Johnson-Restrepo, Adams, and Kannan 2008; Table 12.3). Although both of these classes are used less extensively than deca-BDE in the United States, they still represent a sizeable amount of BFRs manufactured in this country and are in fact the more dominant flame retardants used in Europe and Asia (Janssen 2005). Interestingly, lipid-normalized concentrations of both TBBPA and HBCDs in shark muscle were in most cases far greater than those observed in cetaceans from the same area, suggesting species-specific differences in the metabolism of these compounds. In the case of TBBPA, this is not reason for concern because the overall concentrations of this compound in aquatic wildlife are generally low. This likely reflects lower release of TBBPA into the environment despite greater use of this compound than HBCDs, which in itself is probably because TBBPA is generally covalently bound to synthetic polymers and less likely to diffuse or leach from them over time. While the higher levels of HBCDs in shark muscle raise some questions about their potential toxicity in elasmobranchs, they are still far lower than the maximum HBCD concentrations observed in aquatic wildlife (i.e., 21 µg/g lipid weight in harbor porpoises; Law et al. 2006) and much remains to be learned about the toxicological effects of these chemicals (Darnerud 2003).

12.2.6 Pharmaceuticals and Personal Care Products

There is growing concern about the presence of human pharmaceuticals and personal care products (PPCPs) in the aquatic environment (Daughton and Ternes 1999). These chemicals include the active compounds and metabolites of prescription and nonprescription drugs and cosmetic goods, such as sun-screen products, fragrances, soaps, makeup, skin care products, and similar items. PPCPs are continually introduced into coastal eco-systems, primarily via wastewater discharge, but also through direct release to coastal waters as a result of human recreation. Although research on the uptake and effects of PPCPs in wildlife populations is still in its infancy, a number of studies have demonstrated that aquatic organisms can accumulate some of these compounds, such as antidepressant drugs, human contraceptives, synthetic musk fragrances, ultraviolet light-filtering agents, and the bactericide commonly used in antibacterial soaps, triclosan (Balmer et al. 2004; Brooks et al. 2005; Gibson et al. 2005; Mottaleb et al. 2009; Ramirez et al. 2009). Since virtu-ally all PPCPs that have been tested to date have been shown to be capable of altering ani-mal physiology in some manner (Smital et al. 2004; Dussault et al. 2008; Harada et al. 2008), they pose emerging risks to the well-being of aquatic communities. For example, recent field-based studies using a Canadian Experimental Lakes system have demonstrated that exposure to environmentally relevant levels of the estrogen-mimicking contraceptive agent 17α-ethynylestradiol can result in a variety of reproductive effects (e.g., alterations in gonadal development) that can eventually lead to reductions in wild fish populations (Kidd et al. 2007).

To the best of the authors' knowledge, only two published studies have examined PPCP concentrations in cartilaginous fish. Both of these studies focused on synthetic musks, compounds commonly used to add a pleasing odor to a wide variety of personal care products including perfumery, cosmetics, soaps, and laundry detergents. These compounds include three majors groups: aromatic nitro musks, polycyclic musks, and macrocyclic musks, the first two of which are of special toxicological relevance because they accumulate in biological systems and are capable of disrupting endocrine function and inhibiting cellular defenses against other toxic chemicals (Schmeiser, Gminski, and Mersch-Sundermann 2001; van der Burg et al. 2008). Both Kallenborn et al. (2001) and, later, Nakata (2005) detected the presence of two polycyclic musks, galaxolide (1,3,4,6,7,8,8-hexamethylcyclopental[g]-2-benzopyran or HHCB) and tonalide (7-acetyl-1,1,3,4,4,6-hexamethyltetrahydronaphthalene or AHTN), and two nitro musks, musk ketone and musk xylene, in liver of the thornback ray *Raja clavata* and the scalloped hammerhead *Sphyrna lewini*. Liver concentrations of these compounds were generally lower in elasmobranchs in comparison with those observed in other aquatic vertebrates examined in previous studies (Nakata 2005). However, while these data may suggest lower accumulation of synthetic musks in cartilaginous fish, they may also reflect lower use of some of these compounds in the study areas.

12.3 Metals, Metalloids, and Organometals

Metals and metalloids are naturally occurring elements that are ubiquitous in the environment and redistribute between different compartments of the biosphere by geological and biological processes, such as volcanic activity, the weathering of rocks, transport in the atmosphere and ocean, and uptake and incorporation into biological systems. Because of their useful properties, including high electrical and thermal conductivity, malleability, and strength, metals and metalloids are used extensively for human applications and become environmental pollutants due to actions that shorten their residence time in ore deposits and increase their concentration in the atmosphere, soil, and/or aquatic habitats. These activities include mining, smelting, combustion of fossil fuels, waste incineration, and processing of metals for use in industrial and household products (e.g., fertilizers, batteries, alloys, pigments, cement, and medical products). Although most metals are released from anthropogenic sources in their elemental or ionic form, they can be transformed into organometals, that is, metal atoms covalently attached to organic molecules, by a variety of natural processes (e.g., microbial biomethylation). When this occurs, the addition of organic groups to a metal may drastically increase its uptake and persistence in aquatic or terrestrial organisms, increasing the potential for bioaccumulation and toxic effects. Organometals can also be intentionally synthesized for use in commercial products, such as paints, pesticides, polymer stabilizers, wood preservatives, and catalysts. If used extensively, such products can lead to extensive contamination of organometals in the environment (e.g., the use of the antiknock agent tetraalkyllead in gasoline resulted in significant contamination of lead in the atmosphere until restrictions were placed on leaded gasoline in the 1970s).

Metals and metalloids can be subdivided into those that are essential for life due to their necessity in various biological processes, and those that have no known biological role in at least most species and are viewed primarily as nonessential, toxic compounds. Despite these designations, essential metals can also be toxic to organisms at concentrations above physiologically optimal levels, whereas nonessential, toxic metals are generally tolerable at only very low concentrations. The critical need for essential metals can influence the uptake of nonessential metals in biological systems because toxic metals can make use of physiological processes intended to promote the absorption of essential metals via mimicry. Since many of these uptake mechanisms are increased by essential metal deficiency, the nutritional status of an organism can have a significant influence on its sensitivity to toxic metals. However, physiological mechanisms that have evolved for transporting and storing essential metals in a nonreactive form (i.e., by binding to extracellular and intracellular proteins) may also provide organisms with some resistance to the effects of low concentrations of toxic metals because these same processes generally sequester nonessential metals in similarly inactive forms.

A large number of studies have examined the presence of nonessential and essential metals in cartilaginous fish. The following summary focuses primarily on three naturally occurring toxic nonessential metals/metalloids that have been well studied in elasmobranchs, mercury, cadmium, and arsenic (Table 12.4), and one synthetic organometal, tributyltin (TBT). However, measurements of two other naturally occurring nonessential toxic metals, lead and silver, have been made in cartilaginous fish in several studies and readers are recommended to consult these references for observed values (Stevens and Brown 1974; Vas 1991; Vas and Gordon 1993; Gibbs and Miskiewicz 1995; Turoczy et al. 2000; Powell and Powell 2001; Storelli et al. 2003; Cornish et al. 2007; Malek et al. 2007; McMeans et al. 2007). Concentrations of seven essential metals/metalloids (selenium, iron, copper, chromium, zinc, manganese, and nickel) in elasmobranch tissues have also been provided along with references for these studies for readers interested in the potential toxicity of these compounds (Table 12.5). Readers are also directed to recent articles on the uptake of radioisotopic metals in oviparous elasmobranchs, which have demonstrated that cartilaginous fish may be more susceptible to toxic metal accumulation than bony fish because of higher rates of metal uptake (Jeffree et al. 2006b). These same studies have also shown that the egg case of oviparous sharks may act as a reservoir for radioisotopic metals, and therefore a potential source of radiation exposure to developing embryos (Jeffree et al. 2006a, 2006b; Jeffree, Oberhansli, and Teyssie 2008). Last, the authors also recommend a recent article on high toxic metal concentrations in cestode parasites from the whitecheek shark *Carcharhinus dussumieri* that suggests that these organisms may benefit the health of hosts by serving as heavy metal filters or sinks (Malek et al. 2007).

12.3.1 Mercury

The nonessential toxic metal mercury (Hg) becomes concentrated in the environment largely due to anthropogenic activities, particularly the combustion of Hg-rich coal (Wiener et al. 2003). Through this and other processes, Hg is released into the atmosphere primarily as elemental mercury (Hg^0), which can be distributed by atmospheric transport and eventually oxidized to its water-soluble inorganic form (Hg^{2+}) and returned to soil or surface waters in rainwater. Although it can be returned to the atmosphere through bacterial reduction back to Hg^0, Hg^{2+} is also capable of being transformed via bacterial methylation to the organometal monomethylmercury (MeHg) in aquatic ecosystems. This process occurs largely due to the actions of anaerobic sulfate-reducing bacteria and therefore occurs primarily, although not exclusively, under the hypoxic and anoxic conditions present in organic-rich aquatic sediments. Sediment bacteria are also capable of demethylating MeHg back to Hg^{2+} and eventually Hg^0, particularly under more oxygenated conditions, potentially resulting in rapid turnover of MeHg in aquatic habitats.

TABLE 12.4

Concentrations of the Nonessential Toxic Metals Mercury (Hg), Cadmium (Cd), and Arsenic (As) Observed in Muscle (Top Row) or Liver (Bottom Row in Parentheses, if Available) of Cartilaginous Fish*

Species	Area	MeHg Mean ± SD	MeHg Range	Hg Mean ± SD	Hg Range	Cd Mean ± SD	Cd Range	As Mean ± SD	As Range	References
Carcharhinus acronotus	Florida	0.53 ± 0.13 (NA)	—	—	—	—	—	—	—	Hueter et al. 1995
Carcharhinus albimarginatus	W. Pacific	—	—	1.80 ± 0.45 (0.70 ± 0.42)	1.19–2.35 (0.26–1.41)	NA (0.26 ± 0.46)	NA (0.11–1.40)	—	—	Endo et al. 2008
Carcharhinus amblyrhynchoides	Australia	—	—	1.90 (NA)	0.55–3.30 (NA)	—	—	—	—	Lyle 1986
Carcharhinus amboinensis	Australia	—	—	1.51 (NA)	0.55–3.30 (NA)	—	—	—	—	Lyle 1986
Carcharhinus brachyurus	Australia	—	—	0.6 (NA)	0.23–1.20 (NA)	—	—	—	—	Walker 1988
Carcharhinus brevipinna	Australia	—	—	0.14 (NA)	0.06–0.23 (NA)	—	—	—	—	Lyle 1986
	Australia	—	—	1.48 (NA)	0.61–2.10 (NA)	—	—	—	—	Walker 1988
	Florida	0.59 ± 0.53	0.09–0.75 (NA)	—	—	—	—	—	—	Hueter et al. 1995[3]
Carcharhinus cautus	Australia	—	—	1.14 (NA)	0.34–2.30 (NA)	—	—	—	—	Lyle 1986
Carcharhinus dussumieri	Australia	—	—	0.35 (NA)	0.15–0.51 (NA)	—	—	—	—	Lyle 1986
	Persian Gulf	—	—	—	—	5.0E-5 ± 3.0E-5 (2.0E-4 ± 2.0E-4)	NA	—	—	Malek et al. 2007
	Persian Gulf	—	—	—	—	5E-5 ± 3E-5 (6E-5 ± 4E-5)	—	—	—	Malek et al. 2007
Carcharhinus falciformis	Florida	0.96 ± 0.53 (NA)	0.90–0.97 (NA)	—	—	—	—	—	—	Hueter et al. 1995[3]
Carcharhinus fitzroyensis	Australia	—	—	0.90 (NA)	0.15–1.60 (NA)	—	—	—	—	Lyle 1986

Continued

TABLE 12.4 (Continued)

Concentrations of the Nonessential Toxic Metals Mercury (Hg), Cadmium (Cd), and Arsenic (As) Observed in Muscle (Top Row) or Liver (Bottom Row in Parentheses, if Available) of Cartilaginous Fish*

Species	Area	MeHg Mean ± SD	MeHg Range	Hg Mean ± SD	Hg Range	Cd Mean ± SD	Cd Range	As Mean ± SD	As Range	References
Carcharhinus leucas	Florida	—	—	0.77 ± 0.32 (NA)	0.24–1.70 (NA)	—	—	—	—	Adams and McMichael 1999
	Florida	1.03 ± 0.42 (NA)	0.45–1.27 (NA)	—	—	—	—	—	—	Hueter et al. 1995[3]
	W. Pacific	—	—	3.65 (28.1)	—	NA (2.97)	—	—	—	Endo et al. 2008
Carcharhinus limbatus	Australia	—	—	1.5 (NA)	0.48–2.50 (NA)	—	—	—	—	Lyle 1986
	Florida	—	—	0.77 ± 0.71 (NA)	0.16–2.3 (NA)	—	—	—	—	Adams and McMichael 1999
	Florida	1.30 ± 0.83 (NA)	0.34–2.14 (NA)	—	—	—	—	—	—	Hueter et al. 1995[3]
	Papua New Guinea	—	—	—	—	0.01 (NA)	—	—	—	Powell and Powell 2001
Embryo	Florida	—	—	0.69 ± 0.07 (NA)	0.63–0.78 (NA)	—	—	—	—	Adams and McMichael 1999
Carcharhinus macloti	Australia	—	—	0.25 (NA)	0.09–0.50 (NA)	—	—	—	—	Lyle 1986
Carcharhinus melanopterus	Australia	—	—	1.59 (NA)	0.36–3.10 (NA)	—	—	—	—	Lyle 1986
Carcharhinus obscurus	Florida	1.47 ± 0.13 (NA)	—	—	—	—	—	—	—	Hueter et al. 1995
Carcharhinus perezi	Florida	2.25 ± 0.15 (NA)	—	—	—	—	—	—	—	Hueter et al. 1995
Carcharhinus plumbeus	Florida	0.77 ± 0.40 (NA)	0.63–0.86 (NA)	—	—	—	—	—	—	Hueter et al. 1995[3]
	W. Pacific	—	—	1.66 (3.62)	—	NA (0.73)	—	—	—	Endo et al. 2008
Carcharhinus signatus	Atl. Equator	—	—	1.74 (NA)	0.33–3.48 (NA)	—	—	—	—	Ferreira et al. 2004
	Brazil	—	—	1.77 (NA)	1.09–2.57 (NA)	—	—	—	—	de Pinho 1998

Species	Location									Reference
Carcharhinus sorrah	Brazil	—	—	1.77 ± 0.56 (NA)	—	—	—	—	—	de Pinho et al. 2002
	Australia	—	—	0.33 (NA)	0.06–0.68 (NA)	—	—	—	—	Lyle 1986
Carcharodon carcharias	Indian Ocean	—	—	—	—	—	—	—	5.20 (NA)	Hanaoka et al. 1986
Centrophorus granulosus	Albania	—	—	9.09 ± 0.83 (NA)	7.90–10.0 (NA)	—	—	—	—	Storelli et al. 2002
Centroscymnus crepidator	Mediterranean	—	—	—	—	0.06 (1.76)	—	—	—	Hornung et al. 1993
	Tasmania	—	—	—	—	0.003 (NA)	—	—	—	Vas and Gordon 1993
Centroscymnus oustonii	Tasmania	—	—	—	—	0.013 (NA)	—	—	—	Turoczy et al. 2000
Chimaera monstrosa	Mediterranean	2.67 ± 1.25 (NA)	1.14–4.56 (NA)	3.14 ± 1.39 (NA)	1.30–5.16 (NA)	—	—	—	—	Storelli et al. 2002
Chiloscyllium plagiosum	W. Pacific	—	—	—	—	0.01 (0.24)	<0.001–0.17 (0.04–1.09)	—	—	Cornish et al. 2007
Dalatias licha	Ionian Sea	—	—	3.81 ± 0.69 (NA)	3.24–5.0 (NA)	—	—	—	—	Storelli et al. 2002
Dania calcea	Tasmania	—	—	—	—	0.01 (NA)	—	—	—	Vas and Gordon 1993
Etmopterus spinax	Ionian Sea	—	—	0.58 ± 0.26 (NA)	0.17–0.97 (NA)	—	—	—	—	Storelli et al. 2002
	Mediterranean	—	—	—	—	0.08 (0.79)	—	—	—	Hornung et al. 1993
	NE Atlantic	—	—	—	—	0.25 (1.76)	—	—	—	Vas 1991
Galeocerdo cuveri	Australia	—	—	0.77 (NA)	0.39–1.10 (NA)	—	—	—	—	Lyle 1986
	Florida	0.24 ± 0.14 (NA)	0.20–0.26 (NA)	—	—	—	—	—	—	Hueter et al. 1995[3]
	W. Pacific	—	—	0.78 ± 0.29 (1.17 ± 3.14)	0.38–1.34 (0.11–20.1)	NA (0.15 ± 0.24)	NA (0.02–1.45)	—	—	Endo et al. 2008
Galeorhinus australis	Australia	—	—	1.175 (NA)	0.60–2.20 (NA)	—	—	13.58 (NA)	5.00–23.00 (NA)	Glover 1979
	Tasmania	—	—	—	—	<0.05 (NA)	—	—	—	Estace 1974
Galeorhinus galeus	Australia	—	—	0.64 (NA)	0.01–4.9 (NA)	—	—	—	—	Walker 1988

Continued

TABLE 12.4 (Continued)

Concentrations of the Nonessential Toxic Metals Mercury (Hg), Cadmium (Cd), and Arsenic (As) Observed in Muscle (Top Row) or Liver (Bottom Row in Parentheses, if Available) of Cartilaginous Fish*

Species	Area	MeHg		Hg		Cd		As		References
		Mean ± SD	Range	Mean ± SD	Range	Mean ± SD	Range	Mean ± SD	Range	
	Celtic Sea	—	—	1.1 ± 0.6 (NA)	0.5–1.9 (NA)	0.4 ± 0.2 (NA)	0.2–0.8 (NA)	—	—	Domi, Bouquegneau, and Das 2005?
	NE Atlantic	—	—	—	—	<0.02 (<0.02)	—	—	—	Vas 1991
Galeorhinus viaminicus	Argentina	—	—	0.34 ± 0.17 (NA)	—	—	—	—	—	Scapini, Andrade, and Marcovecchio 1993
Galeus melastomus	Aegean Sea	0.40 ± 0.61 (NA)	0.07–1.89 (NA)	1.02 ± 1.22 (NA)	0.17–4.09 (NA)	—	—	—	—	Storelli and Marcotrigiano 2002
	Aegean Sea	—	—	1.55 ± 1.23 (NA)	0.58–4.32 (NA)	—	—	—	—	Storelli et al. 2002
	Albania	0.15 ± 0.12 (NA)	0.06–0.42 (NA)	0.32 ± 0.22 (NA)	0.12–1.00 (NA)	—	—	—	—	Storelli and Marcotrigiano 2002
	Albania	—	—	—	0.20–1.21 (NA)	—	—	—	—	Storelli et al. 1998
	Albania	—	—	1.01 ± 0.58 (NA)	0.23–1.99 (NA)	—	—	—	—	Storelli et al. 2002
	Celtic Sea	—	—	2.1 ± 1.4 (NA)	1.0–4.4 (NA)	0.7 ± 0.6 (NA)	<0.18–1.5 (NA)	—	—	Domi, Bouquegneau, and Das 2005?
	Ionian Sea	0.04 ± 0.02 (NA)	0.02–0.11 (NA)	0.13 ± 0.08 (NA)	0.04–0.26 (NA)	—	—	—	—	Storelli and Marcotrigiano 2002
	Ionian Sea	—	—	0.74 ± 0.52 (NA)	0.25–2.20 (NA)	—	—	—	—	Storelli et al. 2002
	Italy	0.56 ± 0.47 (NA)	0.13–1.71 (NA)	1.08 ± 1.14 (NA)	0.16–3.57 (NA)	—	—	—	—	Storelli and Marcotrigiano 2002

Species	Location								Reference
	Italy	—	—	0.14–3.39 (NA)	—	—	—	—	Storelli et al. 1998
	Italy	—	2.11 ± 0.96 (NA)	0.47–3.70 (NA)	—	—	—	—	Storelli et al. 2002
Halaelurus bivius	Mediterranean	—	—	—	0.07 (1.57)	—	—	—	Hornung et al. 1993
	NE Atlantic	—	—	—	0.08 (0.07)	—	—	—	Vas 1991
	Argentina	—	2.51 ± 0.30 (NA)	—	—	—	—	—	Marcovecchio, Moreno, and Perez1991
Heptranchias perlo	Ionian Sea	—	1.20 ± 0.17 (NA)	1.00–1.41 (NA)	—	—	—	—	Storelli et al. 2002
Heterodontus portojacksoni	Australia	—	—	—	0.005 (NA)	—	—	—	Gibbs and Miskiewicz 1995
Hexanchus griseus	Mediterranean	—	—	—	0.04 (1.06)	—	—	—	Hornung et al. 1993
Isurus oxyrhinchus	Brazil	—	0.384 ± 0.246 (NA)	0.12–0.69 (NA)	—	—	—	—	Marsico et al. 2007
	South Africa	—	2.12 (NA)	0.59–5.58 (NA)	—	—	—	—	Watling, McClurg, and Stanton1981
Lamna nasus	NE Atlantic	—	—	—	NA (<0.05)	—	—	—	Stevens and Brown 1974
	NE Atlantic	—	—	—	0.79 (NA)	—	—	—	Vas 1991
	New Zealand	—	—	—	—	—	—	—	Vlieg, Murray, and Body 1993
Mustelus antarticus	Australia	—	0.73 (NA)	0.30–1.40 (NA)	0.026 (NA)	<0.01–0.08 (NA)	17.08 (NA)	7.00–30.00 (NA)	Glover 1979
	Tasmania	—	—	—	<0.05 (NA)	—	—	—	Estace 1974
Mustelus asterias	Celtic Sea	—	2.2 ± 0.5 (NA)	1.7–3.1 (NA)	<0.17 (NA)	—	—	—	Domi, Bouquegneau, and Das 2005[2]
Mustelus canis	Brazil	0.20 ± 0.04 (NA)	0.33 ± 0.05 (NA)	—	—	—	—	—	de Pinho et al. 2002
Mustelus higmani	Brazil	—	0.055 (NA)	0.013–0.163 (NA)	—	—	—	—	Lacerda et al. 2000
Mustelus manazo	NA	—	—	—	—	—	17.3 (NA)	—	Hanaoka et al. 1987[4]

Continued

TABLE 12.4 (Continued)

Concentrations of the Nonessential Toxic Metals Mercury (Hg), Cadmium (Cd), and Arsenic (As) Observed in Muscle (Top Row) or Liver (Bottom Row in Parentheses, if Available) of Cartilaginous Fish*

Species	Area	MeHg Mean ± SD	Range	Hg Mean ± SD	Range	Cd Mean ± SD	Range	As Mean ± SD	Range	References
Mustelus mustelus	Italy	—	—	0.23 ± 0.05 (NA)	0.18–0.28 (NA)	—	—	—	—	Storelli et al. 2002
Mustelus norrisi	Brazil	—	—	0.36 ± 0.28 (NA)	—	—	—	—	—	de Pinho et al. 2002
Mustelus schmitti	Argentina	—	—	0.89 ± 0.29 (NA)	—	0.14 (NA)	—	—	—	Marcovecchio, Moreno, and Perez 1991
	Argentina	—	—	0.45 ± 0.30 (NA)	—	—	—	—	—	Scapini, Andrade, and Marcovecchio 1993
Myliobatis aquila	Mediterranean	0.63 ± 0.22 (NA)	0.40–0.84 (NA)	0.83 ± 0.17 (NA)	0.67–1.01 (NA)	—	—	—	—	Storelli et al. 2002
Myliobatis goodei	Argentina	—	—	0.43 ± 0.14 (NA)	—	—	—	—	—	Marcovecchio, Moreno, and Perez 1988a
Negaprion acutidens	Australia	—	—	0.50 (NA)	0.46–0.55 (NA)	—	—	—	—	Lyle 1986
Notorhynchus sp.	Argentina	—	—	2.99 ± 0.18 (NA)	—	—	—	—	—	Marcovecchio, Moreno, and Perez 1991
Prionace glauca	Adriatic Sea	—	—	0.38 (NA)	—	—	—	—	—	Storelli et al. 2001
	Atl. Equator	—	0.65–1.95 (0.077–0.83)	—	0.68–2.5 (0.15–2.2)	—	—	—	—	Branco et al. 2007
	Azores	—	—	—	0.16–1.20 (NA)	—	—	—	—	Branco et al. 2004
	Azores	—	0.18–1.2 (0.01–0.80)	—	0.22–1.30 (0.032–0.96)	—	—	—	—	Branco et al. 2007
	Brazil	—	—	0.398 ± 0.29 (NA)	0.012–1.15 (NA)	—	—	—	—	Marsico et al. 2007
	Canary Is.	—	—	—	0.16–1.84 (NA)	—	—	—	—	Branco et al. 2004

Species	Location									Reference
	Italy	—	—	—	—	—	—	—	—	Kannan et al. 1996
	NE Atlantic	—	—	—	—	—	<0.05 (<0.05–8.4)	—	—	Stevens and Brown 1974
	NE Atlantic	—	—	—	—	0.45 (0.25)	—	—	—	Vas 1991
	Tasmania	—	—	—	0.27–1.20 (NA)	—	—	—	—	Davenport 1995
Rhizoprionodon acutus	Australia	—	—	1.01 (NA)	0.25–1.50 (NA)	—	—	—	—	Lyle 1986
	Papua New Guinea	—	—	—	—	0.01 (NA)	—	—	—	Powell and Powell 2001
Rhizoprionodon lalandei	Brazil	—	—	0.0746 (NA)	0.0215–0.28 (NA)	—	—	—	—	Lacerda et al. 2000
Rhizoprionodon porosus	Brazil	—	—	0.0422 (NA)	0.008–0.091 (NA)	—	—	—	—	Lacerda et al. 2000
Rhizoprionodon taylori	Australia	—	—	0.56 (NA)	0.07–1.10 (NA)	—	—	—	—	Lyle 1986
Rhizoprionodon terraenovae	Florida	—	—	1.06 ± 0.63 (NA)	0.11–2.3 (NA)	—	—	—	—	Adams and McMichael 1999
Embryo	Florida	—	—	0.22 ± 0.02 (NA)	0.17–0.29 (NA)	—	—	—	—	Adams and McMichael 1999
Scyliorhinus canicula	Celtic Sea	—	—	1.9 ± 0.8 (NA)	0.8–3.1 (NA)	1.0 ± 0.4 (NA)	0.4–1.7 (NA)	—	—	Domi, Bouquegneau, and Das 2005[2]
	Italy	—	—	1.23 ± 0.49 (NA)	0.68–2.00 (NA)	—	—	—	—	Storelli et al. 2002
	Mediterranean	1.01 ± 0.58 (NA)	0.23–1.99 (NA)	1.10 ± 0.62 (NA)	0.26–2.06 (NA)	—	—	8.62 ± 2.84 (NA)	5.75–14.49 (NA)	Storelli, Busco, and Marcotrigiano 2005
Scyliorhinus stellaris	NE Atlantic	—	—	—	—	0.78 (<0.02)	—	—	—	Vas 1991
	NE Atlantic	—	—	—	—	<0.02 (<0.02)	—	—	—	Vas 1991
Scymnorhinus licha	NE Atlantic	—	—	—	—	<0.02 (<0.02)	—	—	—	Vas 1991
Somniosus microcephalus	Cumberland Sound	—	—	NA (0.49 ± 0.06)	—	NA (3.91 ± 0.44)	—	NA (9.82 ± 0.70)	—	McMeans et al. 2007[1]
Somniosus pacificus	Prince William Sound	—	—	NA (0.12 ± 0.01)	—	NA (2.63 ± 0.35)	—	NA (5.36 ± 0.15)	—	McMeans et al. 2007[1]

Continued

TABLE 12.4 (Continued)

Concentrations of the Nonessential Toxic Metals Mercury (Hg), Cadmium (Cd), and Arsenic (As) Observed in Muscle (Top Row) or Liver (Bottom Row in Parentheses, if Available) of Cartilaginous Fish*

Species	Area	MeHg		Hg		Cd		As		References
		Mean ± SD	Range	Mean ± SD	Range	Mean ± SD	Range	Mean ± SD	Range	
Somniosus rostratus	Mediterranean	—	—	—	—	0.07 (3.97)	—	—	—	Hornung et al. 1993
Sphyrna blochii	Australia	—	—	0.58 (NA)	0.14–1.70 (NA)	—	—	—	—	Lyle 1986
Sphyrna lewini	Australia	—	—	1.21 (NA)	0.25–2.80 (NA)	—	—	—	—	Lyle 1986
	Papua New Guinea	—	—	—	—	0.02 (NA)	—	—	—	Powell and Powell 2001
Sphyrna mokarran	Australia	—	—	1.52 (NA)	0.19–3.70 (NA)	—	—	—	—	Lyle 1986
Sphyrna tiburo	Florida	—	—	0.50 ± 0.36 (NA)	0.13–1.50 (NA)	—	—	—	—	Adams and McMichael 1999
Embryo	Florida	—	—	0.16 ± 0.07 (NA)	0.08–0.35 (NA)	—	—	—	—	Adams and McMichael 1999
Sphyrna zygaena	Brazil	—	—	0.443 ± 0.299 (NA)	0.015–0.704 (NA)	—	—	—	—	Marsico et al. 2007
	Ionian Sea	—	—	16.06 ± 0.04 (NA)	—	—	—	—	—	Storelli et al. 2002
	Ionian Sea	14 ± 4.40 (23.77 ± 3.20)	7.45–19.57 (19.10–26.30)	12.15 ± 4.60 (35.89 ± 3.58)	8.55–21.07 (32.31–39.48)	0.03 ± 0.005 (19.77 ± 1.29)	0.02–0.03 (18.43–21.00)	18.00 ± 8.57 (44.22 ± 2.22)	15.65–20.21 (42.0–46.43)	Storelli et al. 2003
Squalina argentina	Argentina	—	—	0.48 ± 0.23 (NA)	—	—	—	—	—	Marcovecchio, Moreno, and Perez 1988a

Species	Location						Reference
	Argentina	0.41 ± 0.26 (NA)	—	—	—	—	Marcovecchio, Moreno, and Perez 1988b
Squalus acanthias	Celtic Sea	0.2 ± 0.06 (NA)	0.2–0.3 (NA)	<0.16 (NA)	—	—	Domi, Bouquegneau, and Das 2005[2]
	Canada	—	17.70 (NA)	—	—	—	Goessler et al. 1998[2,4]
Squalus blainvillei	Albania	4.05 ± 1.29 (NA)	3.22–7.24 (NA)	—	—	—	Storelli et al. 2002
Squalus megalops	Brazil	1.90 ± 0.58 (NA)	—	—	—	—	de Pinho et al. 2002
Squalus mitsukurii	Brazil	1.75 ± 0.19 (NA)	—	—	—	1.71 ± 0.97 (NA)	de Pinho et al. 2002
	Japan	—	—	0.007 (NA)	—	—	Taguchi et al. 1979
Sympterygia bonapartei	Argentina	0.18 ± 0.06 (NA)	—	—	—	—	Marcovecchio, Moreno, and Perez 1988a
	Argentina	0.10 ± 0.05 (NA)	—	—	—	—	Marcovecchio, Moreno, and Perez 1988b
Torpedo nobiliana	Mediterranean	2.42 ± 0.86 (NA)	1.65–3.59 (NA)	—	—	1.90 ± 0.57 (NA) / 1.15–2.76 (NA)	Storelli et al. 2002

* Metal concentrations are presented in µg g^{-1} wet weight unless otherwise indicated. NA = not available. When possible, concentrations of the organometal methylmercury (MeHg) is provided in place of or in addition to total Hg measurements.

[1] mean ± standard error; [2]dry weight; [3]range of means; [4]ppm.

TABLE 12.5

Concentrations of the Essential Toxic Metals Selenium (Se), Iron (Fe), Copper (Cu), Zinc (Zn), Chromium (Cr), Manganese (Mn), and Nickel (Ni) Observed in Muscle (Top Row) or Liver (Bottom Row in Parentheses, if Available) of Cartilaginous Fish*

Species	Area	Se	Fe	Cu	Zn	Cr	Mn	Ni	Reference
Apristurus spp.	NE Atlantic	—	—	0.41 (0.38)	—	—	0.04 (<0.02)	0.36 (0.08)	Vas and Gordon 1993
Carcharhinus albimarginatus	W. Pacific	—	3.26 ± 1.93 (35.1 ± 17.3)	NA (1.24 ± 0.43)	3.40 ± 0.81 (4.28 ± 1.07)	—	—	—	Endo et al. 2008
Carcharhinus amblyrhynchoides	Australia	0.84 (NA)	—	—	—	—	—	—	Lyle 1986
Carcharhinus amboinensis	Australia	0.66 (NA)	—	—	—	—	—	—	Lyle 1986
Carcharhinus brevipinna	Australia	0.61 (NA)	—	—	—	—	—	—	Lyle 1986
Carcharhinus cautus	Australia	1.12 (NA)	—	—	—	—	—	—	Lyle 1986
Carcharhinus dussumieri	Australia	1.86 (NA)	—	—	—	—	—	—	Lyle 1986
Carcharhinus fitzroyensis	Australia	0.60 (NA)	—	—	—	—	—	—	Lyle 1986
Carcharhinus leucas	W. Pacific	—	3.51 (69.7)	NA (2.38)	4.36 (5.02)	—	—	—	Endo et al. 2008
Carcharhinus limbatus	Australia	0.65 (NA)	—	—	—	—	—	—	Lyle 1986
	Papua New Guinea	—	—	0.31 (NA)	3.11 (NA)	—	—	—	Powell and Powell 2001
Carcharhinus macloti	Australia	0.69 (NA)	—	—	—	—	—	—	Lyle 1986
Carcharhinus melanopterus	Australia	0.70 (NA)	—	—	—	—	—	—	Lyle 1986
Carcharhinus plumbeus	W. Pacific	—	1.99 (102)	NA (4.07)	3.35 (6.16)	—	—	—	Endo et al. 2008
Carcharhinus sorrah	Australia	0.61 (NA)	—	—	—	—	—	—	Lyle 1986
Centrophorus granulosus	Mediterranean	—	—	0.36 (1.35)	3.13 (8.72)	—	0.12 (0.55)	—	Hornung et al. 1993
Centroscyllium fabricii	NE Atlantic	—	—	0.07 (<0.02)	—	—	0.24 (1.46)	0.43 (0.32)	Vas and Gordon 1993
Centroscymnus coelepis	NE Atlantic	—	—	0.19 (0.26)	—	—	0.03 (0.07)	0.98 (0.16)	Vas and Gordon 1993
Centroscymnus crepidator	NE Atlantic	—	—	1.91 (2.72)	—	—	0.08 (0.06)	<0.02 (<0.02)	Vas and Gordon 1993
	Tasmania	—	—	0.065 (NA)	2.2 (NA)	—	0.25 (NA)	<0.01 (NA)	Turoczy et al. 2000
Centroscymnus owstonii	Tasmania	—	—	0.007 (NA)	2.4 (NA)	—	0.19 (NA)	<0.01 (NA)	Turoczy et al. 2000
Chiloscyllium plagiosum	W. Pacific	—	—	0.15 (1.07)	7.42 (8.72)	0.21 (0.12)	0.09 (0.21)	<0.001 (<0.001)	Cornish et al. 2007

Species	Location								Reference
Dalatias licha	Ionian Sea	—	—	—	—	—	—	—	Storelli et al. 2002
	NE Atlantic	—	—	3.8 (5.42)	—	—	0.13 (<0.02)	<0.02 (<0.02)	Vas and Gordon 1993
Deania calcea	NE Atlantic	—	—	<0.02 (<0.02)	—	—	<0.02 (<0.02)	2.77 (0.20)	Vas and Gordon 1993
Etmopterus princeps	Tasmania	—	—	0.058 (NA)	2.0 (NA)	—	0.12 (NA)	<0.1 (NA)	Turoczy et al. 2000
Etmopterus spinax	NE Atlantic	—	—	0.13 (0.51)	—	—	0.1 (0.04)	0.37 (0.26)	Vas and Gordon 1993
	Mediterranean	—	—	0.58 (2.58)	4.93 (10.3)	—	0.18 (0.75)	NA	Hornung et al. 1993
	NE Atlantic	—	—	5.31 (3.24)	—	—	0.44 (0.19)	1.54 (<0.02)	Vas and Gordon 1993
	NE Atlantic	—	—	—	<0.05 (4.6)	—	0.86 (0.26)	—	Vas 1991
Galeocerdo cuvieri	Australia	0.48 (NA)	—	—	—	—	—	—	Lyle 1986
	W. Pacific	—	3.10 ± 1.47 (20.2 ± 6.5)	NA (1.67 ± 0.73)	4.72 ± 3.28 (2.88 ± 0.90)	—	—	—	Endo et al. 2008
Galeorhinus australis	Australia	0.0275 (NA)	—	—	—	—	0.41 (NA)	—	Glover 1979
	Tasmania	—	—	—	5.0 (NA)	0.5 (NA)	<0.5 (NA)	—	Estace 1974
Galeorhinus galeus	Celtic Sea	2.2±0.5 (NA)	14.5 ± 3.4 (NA)	1.1 ± 0.3 (NA)	16.5 ± 2.4 (NA)	—	—	—	Domi, Bouquegneau, and Das 2005[2]
	NE Atlantic	—	—	0.44 (0.45)	2.12 (1.44)	—	0.03 (<0.02)	1.79 (0.78)	Vas 1991
Galeus melastomus	Aegean Sea	0.77±0.16 (NA)	—	—	—	—	—	—	Storelli and Marcotrigiano 2002
	Albania	1.04±0.25 (NA)	—	—	—	—	—	—	Storelli and Marcotrigiano 2002
	Celtic Sea	1.1±0.2 (NA)	15.6 ± 2.3 (NA)	1.2 ± 0.3 (NA)	14.5 ± 1.6 (NA)	—	—	—	Domi, Bouquegneau, and Das 2005[2]
	Ionian Sea	0.80±0.30 (NA)	—	—	—	—	—	—	Storelli and Marcotrigiano 2002
	Italy	1.21±0.51 (NA)	—	—	—	—	—	—	Storelli and Marcotrigiano 2002
Galeus murinus	Mediterranean	—	—	0.45 (2.55)	3.63 (14.0)	—	0.2 (NA)	NA (1.03)	Hornung et al. 1993
	NE Atlantic	—	—	0.44 (0.25)	—	—	<0.02 (0.04)	1.41 (1.88)	Vas 1991
	NE Atlantic	—	—	<0.02 (0.02)	—	—	<0.02 (<0.02)	<0.02 (<0.02)	Vas and Gordon 1993
Heterodontus portojacksoni	Australia	—	—	0.22 (NA)	5.52 (NA)	0.14 (NA)	—	—	Gibbs and Miskiewicz 1995

Continued

TABLE 12.5 (*Continued*)

Concentrations of the Essential Toxic Metals Selenium (Se), Iron (Fe), Copper (Cu), Zinc (Zn), Chromium (Cr), Manganese (Mn), and Nickel (Ni) Observed in Muscle (Top Row) or Liver (Bottom Row in Parentheses, if Available) of Cartilaginous Fish*

Species	Area	Se	Fe	Cu	Zn	Cr	Mn	Ni	Reference
Hexanchus griseus	Mediterranean	—	—	0.31 (4.38)	4.28 (15.0)	—	0.18 (1.21)	—	Hornung et al. 1993
Isurus oxyrinchus	New Zealand	—	—	0.35 (NA)	4.0 (NA)	—	0.05 (NA)	—	Vlieg, Murray, and Body 1993
Lamna nasus	NE Atlantic	—	—	NA (18.1)	NA (41)	—	—	—	Stevens and Brown 1974
	NE Atlantic	—	—	—	7.2 (NA)	—	0.93 (NA)	—	Vas 1991
	New Zealand	—	—	0.4 (NA)	4.0 (NA)	—	0.06 (NA)	—	Vlieg, Murray, and Body 1993
Mustelus antarticus	Australia	0.30 (NA)	—	0.28 (NA)	3.92 (NA)	—	0.31 (NA)	—	Glover 1979
	Tasmania	—	—	0.25 (NA)	4.6 (NA)	—	<0.5 (NA)	—	Estace 1974
Mustelus asterias	Celtic Sea	3.0±1.0 (NA)	12.0 ± 4.1 (NA)	0.4 ± 0.4 (NA)	13.9 ± 1.8 (NA)	—	—	—	Domi, Bouquegneau, and Das 2005[2]
Mustelus schmitti	Argentina	—	—	—	16.9 (NA)	—	—	—	Marcovecchio, Moreno, and Perez 1991
Negaprion acutidens	Australia	0.37 (NA)	—	—	—	—	—	—	Lyle 1986
Prionace glauca	Atl. Equator	0.23–0.46 (0.82–3.0)	—	—	—	—	—	—	Branco et al. 2007
	Azores	0.084–0.30 (0.47–3.0)	—	—	—	—	—	—	Branco et al. 2007
	NE Atlantic	—	—	4.4 (5.7)	35 (39)	—	—	—	Stevens and Brown 1974
	NE Atlantic	—	—	0.24 (0.65)	—	—	1.55 (0.37)	2.58 (3.23)	Vas 1991
Rhizoprionodon acutus	Australia	0.72 (NA)	—	—	—	—	—	—	Lyle 1986
	Papua New Guinea	—	—	0.29 (NA)	3.13 (NA)	—	—	—	Powell and Powell 2001
Rhizoprionodon taylori	Australia	0.46 (NA)	—	—	—	—	—	—	Lyle 1986

Scyliorhinus canicula	Celtic Sea	1.7±0.8 (NA)	20.1 ± 6.5 (NA)	3.3 ± 5.2 (NA)	49.0 ± 26.0 (NA)	—	—	—	—	Domi, Bouquegneau, and Das 2005[2]
	NE Atlantic	—	—	0.39 (2.22)	—	—	—	2.07 (0.70)	1.70 (0.67)	Vas 1991
Scyliorhinus stellaris	NE Atlantic	—	—	0.56 (7.85)	—	—	—	<0.02 (2.06)	<0.02 (NA)	Vas 1991
Scymnorhinus licha	NE Atlantic	—	—	—	<0.05 (<0.05)	—	—	0.24 (<0.02)	—	Vas 1991
Somniosus microcephalus	Cumberland Sound	NA (0.52±0.03)	—	NA (1.75 ± 0.20)	NA (6.89 ± 0.34)	—	—	NA (0.18±0.01)	—	McMeans et al. 2007[1]
Somniosus pacificus	Prince William Sound	NA (0.54±0.03)	—	NA (0.83 ± 0.07)	NA (10.39 ± 0.71)	—	—	NA (0.43 ± 0.02)	—	McMeans et al. 2007[1]
Somniosus rostratus	Mediterranean	—	—	0.64 (2.23)	4.53 (15.90)	—	—	—	—	Hornung et al. 1993
Sphyrna blochii	Australia	1.13 (NA)	—	—	—	—	—	—	—	Lyle 1986
Sphyrna lewini	Australia	0.81 (NA)	—	—	—	—	—	—	—	Lyle 1986
	Papua New Guinea	—	—	0.37 (NA)	3.8 (NA)	—	—	—	—	Powell and Powell 2001
Sphyrna mokarran	Australia	0.79 (NA)	—	—	—	—	—	—	—	Lyle 1986
Sphyrna zygaena	Ionian Sea	3.24±0.36 (8.09±1.46)	—	1.45 ± 0.41 (6.06 ± 1.20)	6.97 ± 0.17 (26.66 ± 1.85)	0.18±0.04 (0.53±0.03)	—	—	—	Storelli et al. 2003
Squalus acanthias	Celtic Sea	1.1±0.2 (NA)	8.8 ± 1.4 (NA)	<0.3 (NA)	10.1 ± 1.2 (NA)	—	—	—	—	Domi, Bouquegneau, and Das 2005[2]
Squalus mitsukurii	Japan	—	—	0.3 (NA)	2.5 (NA)	—	—	—	—	Taguchi et al. 1979

* Metal concentrations are presented in mean ± SD or range when mean was not available (μg g^{-1} wet weight) unless otherwise indicated. NA = not available.

[1] mean ± standard error; [2] dry weight

Although Hg^{2+} makes up the majority of total Hg (THg) found in the environment and can be rapidly absorbed by fish via the gill, MeHg generally makes up a much larger proportion (>90%) of THg in these animals and as a result, it is considered to be the most toxicologically relevant form of this metal. Whereas it also capable of being absorbed by gill, most studies have demonstrated that MeHg is primarily absorbed by the gastrointestinal system in fish, where its uptake is approximately five times more efficient than that of Hg^{2+}. Following its uptake, MeHg is rapidly distributed and redistributed in fish tissues because of its lipid-soluble nature, which makes it easy for MeHg to cross biological membranes in virtually all tissues and organs. Eventually, MeHg accumulates in muscle, in which it associates with thiol-containing amino acids because of its high affinity for sulfhydryl groups. As a result, MeHg concentrations in fish muscle can range from a million to 10 million times greater than that present in the aquatic medium in which they reside.

Mercury concentrations have been measured in muscle and/or liver of at least 75 species of cartilaginous fish, making it the most extensively studied environmental toxicant in these organisms (Table 12.4). This is primarily due to concerns about human consumption of Hg-contaminated shark meat, which are well founded based on the results of these surveys. Of the large number of investigations on Hg concentrations in sharks from various habitats, approximately 70% have observed muscle Hg concentrations that fell within or exceeded the range of maximum allowable limits for human consumption in most developed countries (e.g., 0.5 and 1.0 µg/g wet weight for Canada and the United States, respectively). This is largely due to extensive bioaccumulation of Hg in these animals, which is well illustrated by the strong associations between body length and muscle Hg concentrations that have been observed for several shark species. In addition to increased uptake of Hg as a result of ontogenetic shifts in dietary habits and biomagnification, high Hg concentrations also may be associated with the slow clearance of MeHg in elasmobranchs. Experimental studies on biliary elimination of ^{203}Hg-labeled MeHg in little skate *Leucoraja erinacea* and spiny dogfish *Squalus acanthias* have demonstrated slow excretion of this compound in both species, presumably due to a slow rate of bile formation (Ballatori and Boyer 1986). Hg concentrations as high as 60% of those found in maternal tissue have been observed in some embryonic sharks, suggesting that transgenerational exposure of Hg via yolk, uterine secretions, and/or placental transfer may also contribute to Hg accumulation in viviparous elasmobranchs (Adams and McMichael 1999). Recent studies have also demonstrated that Hg^{2+} present in the water column can be absorbed by developing embryos of oviparous elasmobranchs, both by transfer across the egg case as well as by accumulation in the egg case itself followed by its release and uptake in embryonic tissues. This has been demonstrated in embryos of the oviparous spotted dogfish *Scyliorhinus canicula*, in which ^{203}Hg was shown to be capable of traversing the egg case prior to the opening of seawater apertures and accumulating in developing offspring (Jeffree, Oberhansli, and Teyssie 2008). This same study also determined that substantial amounts of ^{203}Hg that accumulate in the egg case during this period can undergo depuration following exposure, making it bioavailable for uptake in the embryo.

Although most studies on Hg accumulation in elasmobranchs have focused on the possible human health risks of consuming Hg-contaminated shark meat, there is also some potential for Hg-related health effects in a small number of shark populations that have been surveyed to date. For example, muscle Hg concentrations exceeding thresholds for sublethal effects in adult fish (e.g., 6 to 20 µg/g wet weight in axial muscle, Wiener et al. 2003) have been observed in at least three shark species, the gulper shark *Centrophorus granulosus*, longnose spurdog *Squalus blainvillei*, and smooth hammerhead *Sphyrna zygaena*. While no studies have reported health abnormalities in these animals, it is possible that they may experience similar effects that have been observed in bony fish exposed to comparable Hg levels, such as behavioral alterations, emaciation, cerebral lesions, and impaired gonadal development (Wiener et al. 2003). Some evidence for reproductive effects of Hg exposure in sharks has been observed in studies using the spiny dogfish *Squalus acanthais* in vitro testis model, in which mercuric chloride has been shown to be capable of dramatically inhibiting DNA synthesis, a measure of mitotic activity (Redding 1992). These effects are similar to those observed in most other vertebrate groups (e.g., teleosts; Crump and Trudeau 2009) and, as shown in these taxa, could potentially lead to significant declines in male shark fertility. Toxic effects of Hg also have been observed in the spiny dogfish rectal gland, in which mercuric chlorine can inhibit chloride secretion (Silva, Epstein, and Solomon 1992; Kinne-Saffron and Kinne 2001; Ratner et al. 2006). However, notwithstanding these possible effects, some researchers have hypothesized that slow accumulation of Hg may allow some fish to tolerate high tissue concentrations of this metal without experiencing negative effects (Wiener et al. 2003). This is presumably due to binding of Hg to proteins in muscle, which reduces Hg exposure in other, more sensitive organs, including the testis and brain.

12.3.2 Cadmium

Although it is released to the environment in large quantities due to a number of natural processes (e.g., volcanic activity, the weathering and erosion of sedimentary rocks), the nonessential toxic metal cadmium (Cd) becomes concentrated in the biosphere primarily due to anthropogenic activities, particularly its extensive use in commercial products (e.g., batteries and pigments), the mining and smelting of nonferrous metal ores, and the disposal of Cd-rich gypsum produced during the manufacture of phosphate fertilizers. Following its release from these and other sources, Cd enters aquatic ecosystems via atmospheric and waterborne transport and may persist in either its hydrated ionic form (Cd^{2+}) or more commonly as waterborne or sediment-bound inorganic or organic complexes. Cd can be absorbed by fish in both the gill and gastrointestinal system, and the relative contribution that these organ systems make to total Cd uptake depends greatly on the relative Cd concentrations of the water and food to which fish are exposed. For example, although Cd^{2+} can be rapidly and efficiently absorbed by fish via the gill, it represents only a small proportion (1% to 3%) of the total Cd present in most aquatic ecosystems (Neff 2002). As a result, gastrointestinal uptake of Cd via diet is likely to be more important than branchial uptake in fish exposed to environmentally realistic levels of Cd. Following its absorption, Cd tends to accumulate in visceral organs, such as liver, kidney, and the gut, to a far greater extent than in muscle. Compared to Hg, Cd does not consistently biomagnify in aquatic food webs or bioaccumulate in marine vertebrates. However, because of its high bioavailability and efficient uptake, bioconcentration can occasionally lead to elevated levels of Cd in some vertebrate organs that probably pose greater risks to the organism in question than to human seafood consumers.

Cd concentrations have been measured in liver and/or muscle of at least 33 species of cartilaginous fish (Table 12.4). In general, Cd concentrations in shark tissues and organs are comparable with those observed in other aquatic vertebrates (e.g., fish to marine mammals), which range from geometric means of 0.10 to 0.78 and 2.6 to 3.26 µg/g for muscle and liver, respectively (Neff 2002). The similarities between Cd concentrations in sharks and those in a diverse array of other marine taxa are consistent with the observation that Cd does not biomagnify in aquatic food webs in a consistent and predictable manner. However, this hypothesis has been challenged by recent studies (Croteau, Luoma, and Stewart 2005). No correlations between body length and Cd concentrations have been observed in sharks (McMeans et al. 2007; Storelli et al. 2003), also supporting the premise that Cd may not bioaccumulate in at least some marine vertebrates. This is likely to be in part due to Cd metabolism; for example, Jeffree et al. (2006a) found that bioconcentration of the radioisotope [109]Cd was reduced as encased spotted dogfish *Scyliorhinus canicula* embryos grew in size, suggesting increased clearance of Cd in larger, more metabolically active individuals. As demonstrated in other vertebrates, Cd appears to accumulate in visceral organs more so than in muscle in cartilaginous fish, a point also supported by the greater distribution of [109]Cd in kidney and liver of *S. canicula* in comparison to that in muscle following waterborne exposure to this radioisotope (Jeffree et al. 2006b). None of the cartilaginous fish examined to date have been shown to possess muscle Cd concentrations above estimated safe limits for human consumption (2.7 µg/g wet weight; Neff 2002).

Although Cd accumulation in sharks appears to pose little risk to human health, at least three shark species, the smooth hammerhead *Sphyrna zygaena*, Greenland shark *Somniosus microcephalus*, and little sleeper shark *Somniosus rostratus*, can possess liver Cd concentrations within the range of tissue residue levels associated with toxic effects in marine vertebrates (3.5 to 33,077 µg/g wet weight; Neff 2002). This suggests that some sharks may be susceptible to sublethal effects of Cd exposure, many of which impact vertebrate reproduction, such as reductions in the maturation and viability of gametes and overall declines in animal fertility (Thompson and Bannigan 2008). Depending on Cd exposure levels, comparable effects are likely to occur in elasmobranchs based on experimental studies. For example, Redding (1992) observed dose-dependent reductions in DNA synthesis reflecting impaired mitotic activity in spiny dogfish *Squalus acanthias* spermatocysts cultured in the presence of $CdCl_2$. More recently, McCluskey (2006) reported that in vivo treatment with $CdCl_2$ can induce apoptosis, or programmed cell death, specifically in premeiotic (PrM) spermatogonia in male *S. acanthias* as well as disrupt the function of the blood-testis barrier. McCluskey (2008) later confirmed these observations as well as those by Redding (1992) regarding Cd-related disruption of mitosis in PrM-stage tissue using the *Squalus* in vitro testis model. Interestingly, the stage-specific effects of Cd on PrM spermatogonia in the *Squalus* testis appear to be a function of stage-dependent accumulation of Cd in these and germinal zone cells, which has been shown to occur following both in vitro and in vivo exposure to [109]Cd (Betka and Callard 1999; McCluskey 2008).

12.3.3 Arsenic

The toxic nonessential metalloid arsenic (As) becomes concentrated in the environment due to both natural and human activities, such as volcanic activity, smelting, the combustion of As-rich coal, and its past and present use in pesticides, wood preservatives, and the electronics industry (Neff 2002). Following its release from these sources, As enters aquatic habitats primarily via atmospheric deposition and may exist in a variety of inorganic or organic forms. Inorganic As is generally only bioavailable to marine bacteria and plants, which are capable of transforming it to organoarsenic compounds that can be absorbed by fish and other marine vertebrates via diet. Although accumulation of As in marine fish is often high, a large proportion of the total As in these animals is typically in the form of the virtually nontoxic organoarsenic compound arsenobetaine. Therefore, even elevated concentrations of total As in marine fish generally pose little risk to their health or that of human seafood consumers.

As concentrations have been measured in muscle and/or liver of at least nine species of cartilaginous fish (Table 12.4). Muscle As concentrations in sharks and their relatives have been shown to regularly exceed estimated safe limits for human consumption of As-contaminated fishery products, which are 0.04 and 0.81 µg/g wet weight for carcinogenic and noncarcinogenic effects, respectively. However, since arsenobetaine accounts for virtually all of the soluble As in sharks (e.g., 94% in white shark *Carcharodon carcharias*; Hanaoka et al. 1986), As present in elasmobranch tissues is unlikely to pose a threat to human health. There is also limited potential for toxic effects in sharks themselves based on the low toxicity of this organoarsenic compound.

12.3.4 Tributyltin

Tributyltin (TBT) compounds are a group of synthetic organic derivatives of tin commonly used as antifouling agents for ocean-going vessels. TBT compounds enter aquatic ecosystems by leaching from marine paints and can bind strongly to organic material, persisting in sediments for extended periods of time (Antizar-Ladislao 2008). These compounds have been shown to be highly toxic to most marine organisms, particularly mollusks, in which TBT exposure can result in masculinized features in genetic females (i.e., the condition known as "imposex") as a result of chemically induced increases in testosterone levels. Because of concerns that these effects were linked with marked declines in coastal mollusk populations throughout the world, the use of TBT-based antifouling paints on small vessels has been banned in most developed countries, while its use on larger vessels is regulated on the basis of leaching rate (Champ 2000). Whereas such restrictions have reduced the overall input of TBTs into coastal habitats, their persistence in aquatic sediments and eventual desorption from these reservoirs may still threaten wildlife populations (Ruiz et al. 2008). Since TBTs accumulate in marine organisms and have been shown to be toxic to human immune cells in vitro, they also pose potential health risks to human seafood consumers.

In the only study to examine TBT accumulation in cartilaginous fish, Kannan et al. (1996) reported subcutaneous fat, liver, and kidney TBT concentrations in blue shark *Prionace glauca* ranging from 1 to 9, 20 to 36, and 75 to 220 ng/g wet weight, respectively. These values were one to two times less than those observed in bottlenose dolphin *Tursiops truncatus* and bluefin tuna *Thunnus thynnus thynnus* examined in the same study, suggesting greater potential for health risks in these species. TBT concentrations observed in *P. glauca* also fell well below maximum levels observed in other studies, which have been reported to be as high as >2000 ng/g wet weight in some marine mammals (Antizar-Ladislao 2008). Although limited, these data may quell concerns about TBT effects in elasmobranchs or human consumers of shark meat. This is fortunate as experimental exposure to acute waterborne concentrations of tributyltin oxide (50 to 4000 ng/L) has been shown to be highly toxic to the yellow stingray *Urolophus jamaicensis*, causing lipid peroxidation in the gill epithelium and resulting in membrane degradation, cell loss, and tissue exfoliation (Dwivedi and Trombetta 2006).

12.4 Biotransformation, Pollutant Tolerance, and Biomarkers of Pollutant Exposure and Effects

Although few studies have investigated the ability of cartilaginous fish to biotransform organic pollutants, physiological mechanisms comparable to those employed by other vertebrates appear to be present in this group. This generally involves a two-step process in which reactive groups such as carboxyl, hydroxyl, amino, or sulfhydryl groups are first added to lipid-soluble compounds to slightly increase their water solubility (phase I reactions), typically followed by conjugation of phase I metabolites with endogenous groups (e.g., the triapeptide glutathione) to significantly increase their hydrophilicity and promote their elimination (phase II reactions).

One of the most common and toxicologically important phase I reactions involves the addition of oxygen to organic contaminants, a process that is catalyzed by a superfamily of enzymes known as the cytochrome P450 (CYP)-dependent monooxygenases. Members of the cytochrome P450 1A (CYP1A) subfamily, which are highly involved in the biotransformation of PCDDs/Fs, DL-PCBs, and PAHs, have been shown to be present in the little skate *Leucoraja erinacea* and smooth dogfish *Mustelus canis* and presumably are involved in the metabolism of these compounds (Hahn et al. 1998). As in other vertebrates, CYP1A activity in sharks and their relatives appears to be induced by exposure to PCDDs/Fs, DL-PCBs, PAHs, and related compounds by their noncovalent interactions with cytosolic aryl hydrocarbon receptors (AHR), also known as dioxin receptors (Hahn et al. 1997, 1998). After binding to these ligands, AHR combines with AHR nuclear translocator (ARNT) to produce an active transcription factor that regulates the production of CYP1A enzymes (particularly CYP1A1). Although this generally results in detoxification and clearance of many CYP1A1-inducing pollutants, the actions of CYP1A enzymes can also result in production of pollutant metabolites more toxic than the parent compound; the process known as toxication. For example, oxidation of the PAH benzo(*a*)pyrene (BaP) by CYP1A1 produces the highly reactive metabolite benzo(*a*)pyrene diol epoxide, which is capable of binding covalently to nitrogenous bases in nucleic acids and causing tumorigenic misreplication of DNA to occur.

A second group of phase I enzymes known as the flavin-containing monooxygenases (FMOs) have also been shown to be present in some cartilaginous fish (Schlenk and Li-Schlenk 1994). Although less research has been performed on these enzymes than on CYP1A1, they may play an important role in the oxidation of certain organometals, such as organoselenides.

Very few studies have been performed on the occurrence and actions of phase II enzymes in sharks and their relatives. However, these studies suggest that several major phase II enzymes are present in these animals and likely perform roles comparable to those reported in other vertebrates. This includes the enzymes glutathione-*S*-transferase (GST), sulfotransferase (SULT), and uridinediphospho glucuronosyltransferase (UDP-GT), which catalyze the attachment of glutathione (GSH), sulfate, and glucuronic acid to parent toxicants or their phase I metabolites, respectively (Chowdhury, Chowdhury, and Arias 1982; Foureman et al. 1987; Macrides et al. 1994). More research must be conducted to clarify the role that these compounds play in the biotransformation of endogenous and exogenous substrates in sharks and their relatives.

Although organometals can be enzymatically biotransformed to their inorganic form in vertebrate tissues (e.g., MeHg can be demethylated to Hg^{2+}), the primary manner by which metal ions can be detoxified is by their binding to extracellular or intracellular proteins, which generally sequesters them in a nonreactive state. This can involve the use of plasma-associated proteins, such as albumin, as well as a class of small intracellular sulfhydryl-rich proteins known as metallothioneins (MTs), which are individually capable of binding up to six to seven metal atoms and play an important role in regulating essential metal homeostasis. The production of MTs can be induced by exposure to toxic as well as nontoxic metals, providing a form of cellular resistance against their harmful effects at least until a point at which metal concentrations exceed the production rate and/or binding-capacity of MTs. Although albumin is either absent or in low concentrations in elasmobranch plasma and unlikely to function in sequestering harmful metal ions, the presence of MTs has been reported in several elasmobranch species (i.e., spiny dogfish *Squalus acanthias*, thornback ray *Raja clavata*, cloudy catshark *Scyliorhinus torazame*) and these proteins appear to contribute to toxic metal resistance in these fish (Hildalgo, Tort, and Flos 1985; Hildago and Flos 1986a, 1986b; Hylland, Haux, and Hogstrand 1995; Cho et al. 2005). As shown in other vertebrates, production of MTs or MT-like proteins can be induced in elasmobranchs by exposure to a number of nonessential and essential metals, including Cd, copper, and zinc (Planas et al. 1991; Betka and Callard 1999; Cho et al. 2005).

In addition to mechanisms of pollutant biotransformation or sequesterization, several other physiological processes shown to be important in reducing the toxic effects of environmental pollutants in many vertebrates also appear to be present in cartilaginous fish. For example, the presence of various cell membrane transporters capable of promoting uptake of organic pollutants into hepatocytes for their metabolism (i.e., organic-anion transporting peptides or OATPs), as well as others involved in transporting intracellular pollutants and their metabolites into bile for their clearance (i.e., P-glycoprotein or multi-drug resistant protein 1) have been identified in the liver of the little skate *Leucoraja erinacea* (Jacquemin et al. 1995; Cai et al. 2002, 2003; Meier-Abt et al. 2007). As in other vertebrates, these transporters likely play an important role in the biliary excretion of both endogenous and exogenous substrates in cartilaginous fish. The presence of certain antioxidants (e.g., catalase) capable of reacting with and/or reducing the activity of harmful oxyradicals, common by-products of pollutant biotransformation, has also been observed in some elasmobranchs (Gorbi et al. 2004). Last, cellular mechanisms that can repair molecular damage of environmental toxicants are also likely to contribute to pollutant tolerance in sharks and their relatives. For example, Dwivedi and Trombetta (2006) observed induction of heat shock proteins (HSPs), chaperones involved in the disaggregation and refolding of denatured proteins following chemically induced damage, in yellow stingray *Urolophus jamaicensis* exposed to acute levels of TBT.

The induction of various biotransformation enzymes, MT, antioxidants, and repair proteins following exposure to certain environmental toxicants provides researchers with valuable diagnostic biomarkers for detecting pollutant exposure and effects in wild fish populations. Although these tools have been used extensively in pollution studies on bony fish, they have rarely been employed in research on elasmobranch ecotoxicology (Gorbi et al. 2004; Fuentes-Rios et al. 2005). However, since these studies have demonstrated some evidence for relationships between pollutant exposure levels and biomarker responses in field-sampled sharks, the continued development and application of these tools is recommended. For example, Fuentes-Rios et al. (2005) observed associations between CYP1A1 activity [as measured using the 7-ethoxyresorufin-*O*-deethylase dealkylation (EROD) assay] and PAH exposure (as measured via FAC analysis) in redspotted catsharks *Schroederichthys chilensis*, suggesting that this biomarker would be useful in detecting exposure and biological effects of these and other CYP1A1-inducing pollutants in sharks and their relatives.

12.5 Conclusions and Future Directions

As predicted based on their life history traits and illustrated by a number of studies reviewed in this chapter, many cartilaginous fish are capable of accumulating environmental pollutants that pose potential risks to their health as well as that of human seafood consumers. However, with the exception of studies on MeHg, most reports of pollutant concentrations in elasmobranch tissues that exceed safe limits for animal health and/or human consumption are restricted to a small number of large upper trophic level sharks. Given this, future work should examine potential health consequences associated with pollutant exposure in these and similar species, as well as the implications of MeHg accumulation in all elasmobranchs. These studies should make special use of pollutant biomarkers as diagnostic tools for linking chemical exposure with biological effects, when possible. Last, while self-enforced restrictions on the consumption of shark meat due primarily to MeHg accumulation are well founded (especially by pregnant women), it is important to remember that modest consumption of shark meat and other seafood products provides a variety of health benefits (e.g., reductions in coronary heart disease) that have been shown to outweigh the risks of occasional, low MeHg exposure (Mozaffarian and Rimm 2006). Nonetheless, continued monitoring of pollutant concentrations in shark meat as well as other shark-related products (e.g., shark liver oil supplements, which were recently shown to possess only low concentrations of PBDEs and PCBs; Akutsu, Tanaka, and Hayakawa 2006) is recommended, especially since some of these items may originate from species known to show high levels of pollutant accumulation.

Acknowledgments

The authors thank the editors for their patience and efforts in including this chapter in the volume, and the University of North Florida for providing the time and resources needed to prepare this chapter.

References

Adams DH, McMichael RH Jr (1999) Mercury levels in four species of sharks from the Atlantic coast of Florida. Fish Bull 97:372–379.

Agency for Toxic Substances and Disease Registry (ATSDR) (2002) Toxicological profile for DDT, DDE, DDD. Atlanta, GA: U.S. Department of Health and Human Services, Public Health Service.

Akutsu K, Tanaka Y, Hayakawa K (2006) Occurrence of polybrominated diphenyl ethers and polychlorinated biphenyls in shark liver oil supplements. Food Addit Contam 23:1323–1329.

Albers PH, Loughlin TR (2003) Effects of PAHs on marine birds, mammals and reptiles. In: Douben PET (ed) PAHs: an ecotoxicological perspective. John Wiley & Sons, West Sussex, England, pp 243–261.

Al-Hassan JM (1992) Iraqi invasion of Kuwait: an environmental catastrophe. Fahad Al-Marzouk Publishers, Kuwait.

Al-Hassan JM, Afzal M, Rao CV, Fayad S (2000) Petroleum hydrocarbon pollution in sharks in the Arabian Gulf. Bull Environ Contam Toxicol 65:391–398.

Al-Hassan JM, Afzal M, Rao CV, Fayad S (2003) Polycyclic aromatic hydrocarbons (PAHs) and aliphatic hydrocarbons (AHs) in edible fish from the Arabian Gulf. Bull Environ Contam Toxicol 70:205–212.

Alvarado NE, Quesada I, Hylland K, Marigomez I, Soto M (2006) Quantitative changes in metallo-thionein expression in target cell-types in the gills of turbot (*Scophthalmus maximus*) exposed to Cd, Cu, Zn and after a depuration treatment. Aquat Toxicol 77:62–77.

Antizar-Ladislao B (2008) Environmental levels, toxicity and human exposure to tributyltin (TBT)-contaminated marine environment: a review. Environ Int 34:292–308.

Ateeq B, Abul Farah M, Ahmad W (2005) Detection of DNA damage by alkaline single cell gel elec-trophoresis in 2,4-dichlorophenoxyacetic-acid- and butachlor-exposed erythrocytes of *Clarias batrachus*. Ecotoxicol Environ Saf 62:348–354.

Azevedo-Silva CE, Azeredo A, Cássia Lima Dias A, Costa P, Lailson-Brito J, Malm O, Guimarães JRD, Torres JPM (2009) Organochlorine compounds in sharks from the Brazilian coast. Mar Pollut Bull 58:294–298.

Azevedo-Silva CE, Azeredo A, Lailson-Brito J, Torres JPM, Malm O (2007) Polychlorinated biphenyls and DDT in swordfish (*Xiphias gladius*) and blue shark (*Prionace glauca*) from Brazilian coast. Chemosphere 67:S48–S53.

Baeyens W, Leermakers M, Elskens M, Van Larebeke N, De Bont R, Vanderperren H, Fontaine A, Degroodt JM, Goeyens L, Hanot V, Windal I (2007) PCBs and PCDD/FS in fish and fish prod-ucts and their impact on the human body burden in Belgium. Arch Environ Contam Toxicol 52:563–571.

Ballatori N, Boyer JL (1986) Slow biliary elimination of methyl mercury in the marine elasmobranchs, *Raja erinacea* and *Squalus acanthias*. Toxicol Appl Pharmacol 85:407–415.

Balmer ME, Poiger T, Droz C, Romanin K, Bergqvist KP, Muller MD, Buser HR (2004) Occurrence of methyl triclosan, a transformation product of the bactericide triclosan, in fish from various lakes in Switzerland. Environ Sci Technol 38:390–395.

Baumann PC (1998) Epizootics of cancer in fish associated with genotoxins in sediment and water. Mutat Res 411:227–233.

Beckvar N, Dillon TM, Read LB (2005) Approaches for linking whole-body fish tissue residues of mercury or DDT to biological effects thresholds. Environ Toxicol Chem 24:2094–2105.

Berg V, Ugland KI, Hareide NR, Aspholm PE, Polder A, Skaare JU (1997) Organochlorine contamina-tion in deep-sea fish from the Davis Straight. Mar Environ Res 44:135–148.

Berntssen M, Aspholm O, Hylland K, Bonga S, Lundebye A (2001) Tissue metallothionein, apoptosis and cell proliferation responses in Atlantic salmon (*Salmo salar* L.) parr fed elevated dietary cadmium. Comp Biochem Physiol 128C:299–310.

Betka M, Callard GV (1999) Stage-dependent accumulation of cadmium and induction of metallothion-ein-like binding activity in the testis of the dogfish shark, *Squalus acanthias*. Biol Reprod 60:14–22.

Blus RJ (2003) Organochlorine pesticides. In: Hoffman DJ, Rattner BA, Burton HA Jr, Cairns J Jr (eds) Handbook of toxicology, 2nd edition. CRC Press, Boca Raton, FL, pp 313–340.

Branco V, Canário J, Vale C, Raimundo J, Reis C (2004) Total and organic mercury concentrations in muscle tissue of the blue shark (*Prionace glauca* L. 1758) from the Northeast Atlantic. Mar Pollut Bull 49:854–874.

Branco V, Vale C, Canário J, dos Santos MN (2007) Mercury and selenium in blue shark (*Prionace glauca*, L. 1758) from two areas of the Atlantic Ocean. Environ Pollut 150:373–380.

Brooks BW, Chambliss CK, Stanley JK, Ramirez A, Banks KE, Johnson RD, Lewis RJ (2005) Determination of select antidepressants in fish from an effluent-dominated stream. Environ Toxicol Chem 24:464–469.

Bustamante P, Caurant F, Fowler SW, Miramand P (1998) Cephalopods as a vector for the trans-fer of cadmium to top marine predators in the north-east Atlantic Ocean. Sci Total Environ 220:71–80.

Cai SY, Soroka CJ, Ballatori N, Boyer JL (2003) Molecular characterization of a multidrug resis-tance-associated protein, Mrp2, from the little skate. Am J Physiol Regul Integr Comp Physiol 284:R125–R130.

Cai SY, Wang W, Soroka CJ, Ballatori N, Boyer JL (2002) An evolutionarily ancient Oatp: insights into conserved functional domains of these proteins. Am J Physiol Gastrointest Liver Physiol 282:G702–G710.

Champ MA (2000) A review of organotin regulatory strategies, pending actions, related costs and benefits. Sci Total Environ 258:21–71.

Cho YS, Choi BN, Ha EM, Kim KH, Kim SK, Kim DS, Nam YK. 2005. Shark (*Scyliorhinus torazame*) metallothionein: cDNA cloning, genomic sequence, and expression analysis. Mar Biotechnol 7:350–362.

Chowdhury NR, Chowdhury JR, Arias IM (1982) Bile pigment composition and hepatic UDP-glucuronyl transferase activity in the fetal and adult dogfish shark, *Squalus acanthias*. Comp Biochem Physiol 73B:651–653.

Cornish AS, Ng WC, Ho VC, Wong HL, Lam JC, Lam PK, Leung KM (2007) Trace metals and organo-chlorines in the bamboo shark *Chiloscyllium plagiosum* from the southern waters of Hong Kong, China. Sci Total Environ 376:335–345.

Corsolini S, Focardi S, Kannan K, Tanabe S, Borrell A, Tatsukawa R (1995) Congener profile and toxicity assessment of polychlorinated biphenyls in dolphins, sharks and tuna fish from Italian waters. Mar Environ Res 40:33–53.

Costa DL (2008) Toxic effects of pesticides. In Klaassen CD (ed) Casarett & Doull's toxicology: the basic science of poisons, 7th edition. McGraw-Hill, New York, pp 883–930.

Croteau M-N, Luoma SN, Stewart AR (2005) Trophic transfer of metals along freshwater food webs: evidence of cadmium biomagnification in nature. Limnol Oceanogr 50:1511–1519.

Crump KL, Trudeau VL (2009) Mercury-induced reproductive impairment in fish. Environ Toxicol Chem 28:895–907.

Darnerud PO (2003) Toxic effects of brominated flame retardants in man and in wildlife. Environ Int 29:841–853.

Daughton CG, Ternes TA (1999) Pharmaceuticals and personal care products in the environment: agents of subtle change? Environ Health Perspect 107 (Suppl 6):907–938.

Davenport S (1995) Mercury in blue sharks and deepwater dogfish from around Tasmania. Austr Fish 54(3):20–22.

Davis JA, May MD, Greenfield BK, Fairey R, Roberts C, Ichikawa G, Stoelting MS, Becker JS, Tjeerdema RS (2002) Contaminant concentrations in sport fish from San Francisco Bay, 1997. Mar Pollut Bull 44:1117–1129.

De Pinho AP (1998) Mercurio total em elasmobranquios e teleosteos da costa leste do Brasil. Dissertation, Universidade Federal do Rio de Janeiro.

De Pinho AP, Guimarães JRD, Martins AD, Costa PAS, Olavo G, Valentin J (2002) Total mercury in muscle tissue of five shark species from Brazilian offshore waters: effects of feeding habit, sex, and length. Environ Res Sect A 89:250–258.

Denton GRW, Breck WG (1981) Mercury in tropical marine organisms from North Queensland. Mar Pollut Bull 12:116–121.

Domi N, Bouquegneau JM, Das K (2005) Feeding ecology of five commercial shark species of the Celtic sea through stable isotope and trace metal analysis. Mar Environ Res 60:551–569.

Dussault E, Balakrishnan V, Sverko E, Solomon K, Sibley P (2008) Toxicity of human pharmaceutical and personal care products to benthic invertebrates. Environ Toxicol Chem 27:425–432.

Dutta HM, Meijer HJ (2003) Sublethal effects of diazinon on the structure of the testis of bluegill, *Lepomis macrochirus*: a microscopic analysis. Environ Pollut 125:355–360.

Dwivedi J, Trombetta LD (2006) Acute toxicity and bioaccumulation of tributyltin in tissues of *Urolophus jamaicensis* (yellow stingray). J Toxicol Environ Health A 69:1311–1323.

Endo T, Hisamichi Y, Haraguchi K, Kato Y, Ohta C, Koga N (2008) Hg, Zn and Cu levels in the muscle and liver of tiger sharks (*Galeocerdo cuvier*) from the coast of Ishigaki Island, Japan: relationship between metal concentrations and body length. Mar Pollut Bull 56:1774–1780.

Erickson MD (2001) Introduction: PCB properties, uses, occurrence, and regulatory history. In Robertson LW, Hansen LG (eds) PCBs: recent advances in environmental toxicology and health effects. University Press of Kentucky, Lexington, pp xi–xxx.

Estace IJ (1974) Zinc, cadmium, copper and manganese in species of finfish and shellfish caught in Derwent estuary, Tasmania. Aust J Mar Freshw Res 25:209–220.

Fairey R, Taberski K, Lamerdin S, Johnson E, Clark RP, Downing JW, Newman J, Petreas M (1997) Organochlorines and other environmental contaminants in muscle tissues of sportfish collected from San Francisco Bay. Mar Pollut Bull 34:1058–1071.

Ferreira AG, Faria VV, de Carvalho CEV, Lessa RPT, da Silva FMS (2004) Total mercury in the night shark, *Carcharhinus signatus* in the western equatorial Atlantic Ocean. Braz. Arch Biol Tech 47:629–634.

Figdor E (2004) Reel danger: power plant mercury pollution and the fish we eat. FLPE Fund. August, i–iii:1–55.

Fisk AT, Tittlemier SA, Pranschke JL, Norstrom JR (2002) Using anthropogenic contaminants and stable isotopes to assess the feeding ecology of Greenland sharks. Ecology 83:2162–2172.

Foureman GL, Hernandez O, Bhatia A, Bend JR (1987) The stereoselectivity of four hepatic glutathione S-transferases purified from a marine elasmobranch (*Raja erinacea*) with several K-region polycyclic arene oxide substrates. Biochem Biophys Acta 914:127–135.

Fuentes-Rios D, Orrego R, Rudolph A, Mendoza G, Gavilan JF, Barra R (2005) EROD activity and biliary fluorescence in *Schroederichthys chilensis* (Guichenot 1848): biomarkers of PAH exposure in coastal environments of the South Pacific Ocean. Chemosphere 61:192–199.

Gelsleichter J, Manire CA, Szabo NJ, Cortes E, Carlson J, Lombardi-Carlson L (2005) Organochlorine concentrations in bonnethead sharks (*Sphyrna tiburo*) from four Florida estuaries. Arch Environ Contam Toxicol 48:474–483.

Gelsleichter J, Szabo NJ, Belcher CN, Ulrich GF (2008) Organochlorine contaminants in bonnethead sharks (*Sphyrna tiburo*) from Atlantic and Gulf estuaries on the U.S. east coast. Mar Pollut Bull 56:359–363.

Gelsleichter J, Szabo NJ, Morris JJ (2007) Organochlorine contaminants in juvenile sandbar (*Carcharhinus plumbeus*) and blacktip (*Carcharhinus limbatus*) sharks from major nursery areas on the east coast of the United States. In: McCandless C, Kohler N, Pratt HL Jr (eds) Shark nursery grounds of the Gulf of Mexico and east coast waters of the United States. Am Fish Soc Symp 50:153–164.

Gelsleichter J, Walsh CJ, Szabo NJ, Rasmussen LE (2006) Organochlorine concentrations, reproductive physiology, and immune function in unique populations of freshwater Atlantic stingrays (*Dasyatis sabina*) from Florida's St. Johns River. Chemosphere 63:1506–1522.

Gibbs PJ, Miskiewicz AG (1995) Heavy metals in fish near a major primary treatment sewage plant outfall. Mar Pollut Bull 30:667–674.

Gibson R, Smith MD, Spary CJ, Tyler CR, Hill EM (2005) Mixtures of estrogenic contaminants in bile of fish exposed to wastewater treatment works effluents. Environ Sci Technol 39:2461–2471.

Giesy JP, Kannan K, Blankenship AL, Jones PD, Newsted JL (2006) Toxicology of PCBs and related compounds. In: Norris DP, Carr JA (eds) Endocrine disruption: biological bases for health effects in wildlife and humans. Oxford University Press, New York, pp 245–331.

Glover JW (1979) Concentrations of arsenic, selenium and ten heavy metals in school shark, *Galeorhinus australis* (Macleay), and gummy shark, *Mustelus antarcticus* Gunther, from southeastern Australian waters. Austr J Mar Freshw Res 30:505–510.

Goessler W, Kuehnelt D, Schlagenhaufen C, Slejkovec Z, Irgolic KJ (1998) Arsenobetaine and other arsenic compounds in the National Research Council of Canada Certified Reference Materials DORM 1 and DORM 2. J Anal At Spectrom 13:183–187.

Gorbi S, Pellegrini D, Tedesco S, Regoli F (2004) Antioxidant efficiency and detoxification enzymes in spotted dogfish *Scyliorhinus canicula*. Mar Environ Res 58:293–297.

Grassman JA, Masten SA, Walker NJ, Lucier GW (1998) Animal models of human response to dioxins. Environ Health Perspect 106 (Suppl 2):761–775.

Guarino AM, James MO, Bend JR (1977) Fate and distribution of the herbicides 2,4-dichlor-phenoxyacetic acid (2,4-D) and 2,4,5-trichlorophenoxyacetic acid (2,4,5-T) in the dogfish shark. Xenobiotica 7:623–631.

Guo YL, Lambert GH, Hsu CC, Hsu MM (2004) Yucheng: health effects of prenatal exposure to polychlorinated biphenyls and dibenzofurans. Int Arch Occup Environ Health 77:153–158.

Hahn ME, Karchner SI, Shapiro MA, Perera SA (1997) Molecular evolution of two vertebrate aryl hydrocarbon (dioxin) receptors (AHR1 and AHR2) and the PAS family. Proc Natl Acad Sci USA 94:13743–13748.

Hahn ME, Woodin BR, Stegeman JJ, Tillitt DE (1998) Aryl hydrocarbon receptor function in early vertebrates: inducibility of cytochrome P450 1A in agnathan and elasmobranch fish. Comp Biochem Physiol C Pharmacol Toxicol Endocrinol 120:67–75.

Hanaoka K, Matsuda H, Kaise T, Tagawa S (1986) Identification of arsenobetaine as a major arsenic compound in the muscle of a pelagic shark, *Carcharodon carcharias*. J Shimon Univ Fish 35(1):37–40.

Hanaoka K, Fujita T, Matsuura M, Tagawa S, Kaise T (1987) Identification of arsenobetaine as a major arsenic compound in muscle of two demersal sharks, shortnose dogfish *Squalus brevirostris* and starspotted shark *Mustelus manazo*. Comp Biochem Physiol B 86: 681–682.

Harada A, Komori K, Nakada N, Kitamura K, Suzuki Y (2008) Biological effects of PPCPs on aquatic lives and evaluation of river waters affected by different wastewater treatment levels. Water Sci Technol 58:1541–1546.

Haraguchi K, Hisamichi Y, Kotaki Y, Kato Y, Endo T (2009) Halogenated bipyrroles and methoxylated tetrabromodiphenyl ethers in tiger shark (*Galeocerdo cuvier*) from the southern coast of Japan. Environ Sci Technol 43:2288–2294.

Henry L, Kishimba MA (2006) Pesticide residues in Nile tilapia (*Oreochromis niloticus*) and Nile perch (*Lates niloticus*) from Southern Lake Victoria, Tanzania. Environ Pollut 140:348–354.

Hidalgo J, Flos R (1986a) Dogfish metallothionein. I. Purification and characterization and comparison with rat metallothionein. Comp Biochem Physiol C 83:99–103.

Hidalgo J, Flos R (1986b) Dogfish metallothionein. II. Electrophoretic studies and comparison with rat metallothionein. Comp Biochem Physiol C 83:105–109.

Hidalgo J, Tort L, Flos R (1985) Cd-, Zn-, Cu-binding protein in the elasmobranch *Scyliorhinus canicula*. Comp Biochem Physiol C 81:159–165.

Holden A, She J, Tanner M, Lunder S, Sharp R, Hooper K (2003) PBDEs in San Francisco Bay Area: measurements in fish. Organohalogen Compd 61:255–258.

Hornung H, Krom MD, Cohen Y, Bernhard M (1993) Trace metal content in deep-water sharks from the eastern Mediterranean sea. Mar Biol 115:331–338.

Horie T, Tanaka H, Tanaka S (2004) Bioaccumulation of PCBs and DDE in cloudy catshark, *Scyliorhinus torazame*, caught in four locations around Japan. J Sch Mar Sci Technol 2:33–43.

Hueter RE, Fong WG, Henderson G, French MF, Manire CA (1995) Methylmercury concentrations in shark muscle by species, size and distribution of sharks in Florida coastal waters. Water Air Soil Pollut 80:893–899.

Hylland H, Haux C, Hogstrand C (1995) Immunological characterization of metallothionein in marine and freshwater fish. Mar Environ Res 39:111–115.

Jacquemin E, Hagenbuch B, Wolkoff AW, Meier PJ, Boyer JL (1995) Expression of sodium-independent organic anion uptake systems of skate liver in *Xenopus laevis* oocytes. Am J Physiol 268:G18–G23.

Janssen S (2005) Brominated flame retardants: rising levels of concern. For Healthcare without Harm. Available via http://www.noharm.org/details.cfm?typedocument&ID=1095. Accessed December, 2006.

Jones-Otazo HA, Clarke JP, Diamond ML, Archbold JA, Ferguson G, Harner T, et al. (2005) Is house dust the missing exposure pathway for PBDEs? An analysis of the urban fate and human exposure to PBDEs. Environ Sci Technol 39:5121–5130.

Jeffree RA, Oberhansli F, Teyssie JL (2007) Accumulation and transport behaviour of [241]americium, [60]cobalt and [134]cesium by eggs of the spotted dogfish *Scyliorhinus canicula*. Mar Pollut Bull 54:912–920.

Jeffree RA, Oberhansli F, Teyssie JL (2008) The accumulation of lead and mercury from seawater and their depuration by eggs of the spotted dogfish *Scyliorhinus canicula* (Chondrichthys). Arch Environ Contam Toxicol 55:451–461.

Jeffree RA, Warnau M, Oberhansli F, Teyssie J-L (2006a) Bioaccumulation of heavy metals and radio-nuclides from seawater by encased embryos of the spotted dogfish *Scyliorhinus canicula*. Mar Pollut Bull 52:1278–1286.

Jeffree RA, Warnau M, Teyssie J-L, Markich SJ (2006b) Comparison of the bioaccumulation from seawater and depuration of heavy metals and radionuclides in the spotted dogfish *Scyliorhinus canicula* (Chondrichthys) and the turbot *Psetta maxima* (Actinopterygii: Teleostei). Sci Total Environ 368:839–852.

Johnson-Restrepo B, Adams DH, Kannan K (2008) Tetrabromobisphenol A (TBBPA) and hexabromo-cyclododecanes (HBCDs) in tissues of humans, dolphins, and sharks from the United States. Chemosphere 70:1935–1944.

Johnson-Restrepo B, Kannan K, Addink R, Adams DH (2005) Polybrominated diphenyl ethers and polychlorinated biphenyls in a marine foodweb of coastal Florida. Environ Sci Technol 39:8243–8250.

Kallenborn R, Gatermann R, Nygard T, Knutzen J, Schlabach M (2001) Synthetic musks in Norwegian marine fish samples collected in the vicinity of densely populated areas. Fresenius Environmental Bulletin 10: 832–842.

Kannan K, Blankenship AL, Jones PD, Giesy JP (2000) Toxicity reference values for the toxic effects of polychlorinated biphenyls to aquatic mammals. Human Ecol Risk Assess 6:181–201.

Kannan K, Corsolini S, Focardi S, Tanabe S, Tatsukawa R (1996) Accumulation pattern of butyltin compounds in dolphin, tuna, and shark collected from Italian coastal waters. Arch Environ Contam Toxicol 31:19–23.

Kannan K, Perrotta E (2008) Polycyclic aromatic hydrocarbons (PAHs) in livers of California sea otters. Chemosphere 71:649–655.

Karouna-Renier NK, Rao KR, Lanza JJ, Rivers SD, Wilson PA, Hodges DK, Levine KE, Ross GT (2008) Mercury levels and fish consumption practices in women of child-bearing age in the Florida panhandle. Environ Res 108:320–326.

Kidd KA, Blanchfield PJ, Mills KH, Palace VP, Evans RE, Lazorchak JM, Flick RW (2007) Collapse of a fish population after exposure to a synthetic estrogen. Proc Natl Acad Sci USA 104:8897–8901.

Kiely T, Donaldson D, Grube A (2004) Pesticide industry sales and usage. 2000 and 2001 market estimates. U.S. Environmental Protection Agency. Office of Prevention, Pesticides, and Toxic Substances, Office of Pesticide Programs, Biological and Economic Analysis Division. Available via http://www.epa.gov/oppbead1/pestsales/01pestsales/market_estimates2001.pdf.

Kinne-Saffran E, Kinne RK (2001) Inhibition by mercuric chloride of Na-K-2Cl cotransport activity in rectal gland plasma membrane vesicles isolated from *Squalus acanthias*. Biochem Biophys Acta 1510:442–451.

Koschier FJ, Pritchard JB (1980) Renal handling of 2,4-D by the dogfish shark (*Squalus acanthias*). Xenobiotica 10:1–6.

Kumar KS, Kannan K, Corsolini S, Evans T, Giesy JP, Nakanishi J, Masunaga S (2002) Polychlorinated dibenzo-p-dioxins, dibenzofurans and polychlorinated biphenyls in polar bear, penguin and south polar skua. Environ Pollut 119:151–161.

Lacerda LD, Paraquetti HHM, Marins RV, Rezende CE, Zalmon IR, Gomes MP, Farias V (2000) Mercury content in shark species from the south-eastern Brazilian coast. Rev Bras Biol 60:571–576.

Law RJ, Bersuder P, Allchin CR, Barry J (2006) Levels of the flame retardants hexabromocyclodo-decane and tetrabromobisphenol A in the blubber of harbor porpoises (*Phocoena phocoena*) stranded or bycaught in the U.K., with evidence for an increase in HBCD concentrations in recent years. Environ Sci Technol 40:2177–2183.

Lundholm CD (1997) DDE-induced eggshell thinning in birds: effects of p,p'-DDE on the calcium and prostaglandin metabolism of the eggshell gland. Comp Biochem Physiol C Pharmacol Toxicol Endocrinol 118:113–128.

Lyle JM (1986) Mercury and selenium concentrations in sharks from Northern Australian waters. Aust J Mar Freshw Res 37:309–321.

Macrides TA, Faktor DA, Kalafatis N, Amiet RG (1994) Enzymic sulfation of bile salts. Partial purification and characterization of an enzyme from the liver of the shark *Heterodontus portusjacksoni* that catalyses the sulfation of the shark bile steroid 5 beta-scymnol. Comp Biochem Physiol Biochem Mol Biol 107:461–469.

Malek M, Haseli M, Mobedi I, Ganjali MR, Mackenzie K (2007) Parasites as heavy metal bioindicators in the shark *Carcharhinus dussumieri* from the Persian Gulf. Parasitology 134:1053–1056.

Marcovecchio JE, Moreno VJ, Perez A (1988a) Determination of heavy metal concentrations in biota of Bahia Blanca, Argentina. Sci Tot Environ 75:181–190.

Marcovecchio JE, Moreno MJ, Perez A (1988b) Total mercury levels in marine organisms of the Bahia Blanca estuary food web, Argentina. In: Seeliger U, Lacerda LD, Patchineelam SR (eds) Metals in coastal environments of Latin America. Springer Verlag, Berlin, pp 122–129.

Marcovecchio JE, Moreno VJ, Perez A (1991) Metal accumulation in tissues of sharks from the Bahia Blanca Estuary, Argentina. Mar Environ Res 31:263–274.

Mársico ET, Machado MES, Knoff M, São Clemente SC (2007) Total mercury in sharks along the southern Brazilian coast. Arq Bras Med Vet Zootec 59:656–662.

Marsili L, Caruso A, Fossi MC, Zanardelli M, Politi E, Focardi S (2001) Polycyclic aromatic hydrocarbons (PaHs) in subcutaneous biopsies of Mediterranean cetaceans. Chemosphere 44:147–154.

McClusky LM (2006) Stage-dependency of apoptosis and the blood-testis barrier in the dogfish shark (*Squalus acanthias*): cadmium-induced changes as assessed by vital fluorescence techniques. Cell Tissue Res 325:541–553.

McClusky LM (2008) Cadmium accumulation and binding characteristics in intact Sertoli/germ cell units, and associated effects on stage-specific functions in vitro: insights from a shark testis model. J Appl Toxicol 28:112–121.

McMeans BC, Borga K, Bechtol WR, Higginbotham D, Fisk AT (2007) Essential and non-essential element concentrations in two sleeper shark species collected in arctic waters. Environ Pollut 148:281–290.

Meier-Abt F, Hammann-Hanni A, Stieger B, Ballatori N, Boyer JL (2007) The organic anion transport polypeptide 1d1 (Oatp1d1) mediates hepatocellular uptake of phalloidin and microcystin into skate liver. Toxicol Appl Pharmacol 218:274–279.

Mottaleb MA, Usenko S, O'Donnell JG, Ramirez AJ, Brooks BW, Chambliss CK (2009) Gas chromatography-mass spectrometry screening methods for select UV filters, synthetic musks, alkylphenols, an antimicrobial agent, and an insect repellent in fish. J Chromatogr A 1216:815–823.

Mozaffarian D, Rimm EB (2006) Fish intake, contaminants, and human health: evaluating the risks and the benefits. JAMA 296:1885–1899.

Myers MS, Anulacion BF, French BL, Reichert WL, Laetz CA, Buzitis J, Olson OP, Sol S, Collier TK (2008) Improved flatfish health following remediation of a PAH-contaminated site in Eagle Harbor, Washington. Aquat Toxicol 88:277–288.

Nakata H (2005) Occurrence of synthetic musk fragrances in marine mammals and sharks from Japanese coastal waters. Environ Sci Technol 39:3430–3434.

Neff JM (2002) Bioaccumulation in marine organisms: effects of contaminants from oil well produced water. Elsevier, Oxford.

Pimental D (1996) Green revolution agriculture and chemical hazards. Sci Total Environ 188 (Suppl 1):S86–S98.

Piscator M (1985) Dietary exposure to cadmium and health effects: impact of environmental changes. Environ Health Persp 63:127–132.

Planas J, Tort L, Torres P, Flos R (1991) Cadmium induction of metallothioneins in several dogfish organs. Rev Esp Fisiol 47:75–80.

Poulin J, Gibb H, Pruss-Ustan A (2008) Mercury: assessing the environmental burden of disease at national and local levels. WHO Environ Burden Disease 16, i–vii:1–60.

Powell JH, Powell RE (2001) Trace elements in fish overlying subaqueous tailings in the tropical west pacific. Water Air Soil Pollut 125:81–104.

Rahman F, Langford KH, Scrimshaw MD, Lester JN (2001) Polybrominated diphenyl ether (PBDE) flame retardants. Sci Tot Environ 275:1–17.

Ramirez AJ, Brain RA, Usenko A, Mottaleb MA, O'Donnell JG, Stahl LL, Wathen JB, Snyder BD, Pitt JL, Perez-Hurtado P, Dobbins LL, Brooks BW, Chambliss CK (2009) Occurrence of pharmaceuticals and personal care products (PPCPs) in fish: results of a national pilot study in the U.S. Environ Toxicol Chem 25:1 [Epub ahead of print.]

Ratner MA, Decker SE, Aller SG, Weber G, Forrest JN Jr (2006) Mercury toxicity in the shark (*Squalus acanthias*) rectal gland: apical CFTR chloride channels are inhibited by mercuric chloride. J Exp Zool A Comp Exp Biol 305:259–267.

Redding JM (1992) Effects of heavy metals on DNA synthesis in the testis of dogfish (*Squalus acanthias*). Bull Mt Desert Is Biol Lab 31:42–43.

Reichert WL, Myers MS, Peck-Miller K, French B, Anulacion BF, Collier TK, Stein JE, Varanasi U (1998) Molecular epizootiology of genotoxic events in marine fish: linking contaminant exposure, DNA damage, and tissue-level alterations. Mutat Res 411:215–225.

Rogan WJ, Chen A (2005) Health risks and benefits of bis(4-chlorophenyl)-1,1,1-trichloroethane (DDT). Lancet 366:763–773.

Ruiz JM, Barreiro R, Couceiro L, Quintela M (2008) Decreased TBT pollution and changing bioaccumulation pattern in gastropods imply butyltin desorption from sediments. Chemosphere 73:1253–1257.

Scapini EM, Andrade S, Marcovecchio JE (1993) Total mercury distribution in two shark species from Buenos Aires province coastal waters, in Argentina. Proc Intern Conf Heavy Metals in the Environ, Toronto 1:82–85.

Schecter A, Birnbaum L, Ryan JJ, Constable JD (2006) Dioxins: an overview. Environ Res 101:419–428.

Schlenk D, Li-Schlenk R (1994) Characterization of liver flavin-containing monooxygenase of the dogfish shark (*Squalus acanthias*) and partial purification of liver flavin-containing monooxygenase of the silky shark (*Carcharhinus falciformis*). Comp Biochem Physiol B Biochem Mol Biol 109:655–664.

Schlenk D, Sapozhnikova Y, Cliff G (2005) Incidence of organochlorine pesticides in muscle and liver tissues of South African great white sharks *Carcharodon carcharias*. Mar Pollut Bull 50:208–211.

Schmeiser HH, Gminski R, Mersch-Sundermann V (2001) Evaluation of health risks caused by musk ketone. Int J Hyg Environ Health 203:293–299.

Serrano R, Fernandez MA, Hernandez LM, Hernandez M, Pascual P, Rabanal RM, Gonzalez MJ (1997) Coplanar polychlorinated biphenyl congeners in shark livers from the north-western African Atlantic Ocean. Bull Environ Contam Toxicol 58:150–157.

Serrano R, Fernandez M, Rabanal R, Hernandez M, Gonzalez MJ (2000) Congener-specific determination of polychlorinated biphenyls in shark and grouper livers from the northwest African Atlantic Ocean. Arch Environ Contam Toxicol 38:217–224.

Shailaja MS, Nair M (1997) Seasonal differences in organochlorine pesticide concentrations of zooplankton and fish in the Arabian Sea. Mar Environ Res 44:263–274.

Siddiqi MA, Laessig RH, Reed KD (2003) Polybrominated diphenyl ethers (PBDEs): new pollutants–old diseases. Clin Med Res 1:281–290.

Silva P, Epstein FH, Solomon RJ (1992) The effect of mercury on chloride secretion in the shark (*Squalus acanthias*) rectal gland. Comp Biochem Physiol C 103:569–575.

Smital T, Luckenbach T, Sauerborn R, Hamdoun AM, Vega RL, Epel D (2004) Emerging contaminants: pesticides, PPCPs, microbial degradation products and natural substances as inhibitors of multixenobiotic defense in aquatic organisms. Mutat Res 552:101–117.

Soumis N, Lucotte M, Sampaio D, Almeida DC, Giroux D, Morais S, Pichet P (2003) Presence of organophosphate insecticides in fish of the Amazon river, Acta Amazon 3:325–338.

Spano L, Tyler CR, van Aerle R, Devos P, Mandiki SN, Silvestre F, Thome JP, Kestemont P (2004) Effects of atrazine on sex steroid dynamics, plasma vitellogenin concentration and gonad development in adult goldfish (*Carassius auratus*). Aquat Toxicol 66:369–379.

Srivastava RK, Yadav KK, Trivedi SP (2008) Devicyprin induced gonadal impairment in a freshwater food fish, *Channa punctatus* (Bloch). J Environ Biol 29:187–191.

Stevens JD, Brown BE (1974) Occurrence of heavy metals in the blue shark *Prionace glauca* and selected pelagic in the NE Atlantic Ocean. Mar Biol 26:287–293.

Storelli MM, Barone G, Santamaria N, Marcotrigiano GO (2006) Residue levels of DDTs and toxic evaluation of polychlorinated biphenyls (PCBs) in S*cyliorhinus canicula* liver from the Mediterranean Sea (Italy). Mar Pollut Bull 52:696–700.

Storelli MM, Busco VP, Marcotrigiano GO (2005) Mercury and arsenic speciation in the muscle tissue of *Scyliorhinus canicula* from the Mediterranean Sea. Bull Environ Contam Toxicol 75:81–88.

Storelli MM, Ceci E, Storelli A, Marcotrigiano GO (2003) Polychlorinated biphenyl, heavy metal and methylmercury residues in hammerhead sharks: contaminant status and assessment. Mar Pollut Bull 46:1035–1039.

Storelli MM, Giacominelli-Stuffler, Marcotrigiano GO (1998) Total mercury in muscle of benthic and pelagic fish from the South Adriatic Sea (Italy). Food Addit Contam 15: 876–883.

Storelli MM, Giacominelli-Stuffer R, Marcotrigiano GO (2001) Total mercury and methylmercury in tuna fish and sharks from the South Adriatic Sea. Italian J Food Sci 13(1):101–106.

Storelli MM, Giacominelli-Stuffler R, Marcotrigiano G (2002) Mercury accumulation and speciation in muscle tissue of different species of sharks from Mediterranean Sea, Italy. Bull Environ Contam Toxicol 68:201–210.

Storelli MM, Marcotrigiano GO (2001) Persistent organochlorine residues and toxic evaluation of polychlorinated biphenyls in sharks from the Mediterranean Sea (Italy). Mar Pollut Bull 42:1323–1329.

Storelli MM, Marcotrigiano GO (2002) Mercury speciation and relationship between mercury and selenium in liver of *Galeus melastomus* from the Mediterranean sea. Bull Environ Contam Toxicol 69:516–522.

Storelli MM, Storelli A, D'Addabbo R, Barone G, Marcotrigiano GO (2004) Polychlorinated biphenyl residues in deep-sea fish from Mediterranean Sea. Environ Int 30:343–349.

Storelli MM, Storelli A, Marcotrigiano GO (2003) Coplanar polychlorinated biphenyl congeners in the liver of *Galeus melastomus* from different areas of the Mediterranean Sea. Bull Environ Contam Toxicol 71:276–282.

Storelli MM, Storelli A, Marcotrigiano GO (2005) Concentrations and hazard assessment of polychlorinated biphenyls and organochlorine pesticides in shark liver from the Mediterranean Sea. Mar Pollut Bull 50:850–855.

Strid A, Jorundsdottir H, Papke O, Svavarsson J, Bergman A (2007) Dioxins and PCBs in Greenland shark (*Somniosus microcephalus*) from the north-east Atlantic. Mar Pollut Bull 54:1514–1522.

Taguchi M, Yasuda K, Toda S, Shimizu M (1979) Study of metal contents of elasmobranch fishes. I. Metal concentration in the muscle tissues of a dogfish, *Squalus mitsukurii*. Mar Environ Res 2:239–249.

Thompson J, Bannigan J (2008) Cadmium: toxic effects on the reproductive system and the embryo. Reprod Toxicol 25:304–315.

Turoczy NJ, Laurenson LJ, Allinson G, Nishikawa M, Lambert OF, Smith C, Cottier JP, Irvine SB, Stagnitti F (2000) Observations on metal concentrations in three species of shark (*Deania calcea*, *Centroscymnus crepidater*, and *Centroscymnus owstoni*) from southeastern Australian waters. J Agric Food Chem 48:4357–4364.

van der Burg B, Schreurs R, van der Linden S, Seinen W, Brouwer A, Sonneveld E (2008) Endocrine effects of polycyclic musks: do we smell a rat? Int J Androl 31:188–193.

Varanasi U, Stein JE, Nishimoto M, Reichert WL, Collier TK (1987) Chemical carcinogenesis in feral fish: uptake, activation, and detoxication of organic xenobiotics. Environ Health Perspect 71:155–170.

Vas P (1991) A field guide to the sharks of British coastal waters. FSC Publications, Shrewsbury.

Vas P, Gordon JDM (1993) Trace metals in deep-sea sharks from the Rockall Trough. Mar Pollut Bull 26:400–402.

Vlieg P, Murray T, Body DR (1993) Nutritional data on six oceanic pelagic fish species from New Zealand waters. J Food Compos Anal 6:45–54.

Vogelbein WK, Fournie JW, Van Veld PA, Huggett RJ (1990) Hepatic neoplasms in the mummichog *Fundulus heteroclitus* from a creosote-contaminated site. Cancer Res 50:5978–5986.

Vuorinen PJ, Keinanen M, Vuontisjarvi H, Barsiene J, Broeg K, Forlin L, Gercken J, Kopecka J, Kohler A, Parkkonen J, Pempkowiak J, Schiedek D (2006) Use of biliary PAH metabolites as a biomarker of pollution in fish from the Baltic Sea. Mar Pollut Bull 53:479–487.

Walker TI (1976) Effects of species, sex, length and locality on the mercury content of school shark *Mustelus antarticus* (Guenther) from south-eastern Australian waters. Aust J Mar Freshw Res 27:603–616.

Walker TI (1988) Mercury concentration in edible tissues of elasmobranchs, teleosts, crustaceans and mollusks from south-eastern Australian waters. Austr J Mar Freshw Res 39:39–49.

Watling RJ, McClurg TP, Stanton RC (1981) Relation between mercury concentration and size in the mako shark. Bull Environ Contam Toxicol 26:352–358.

WHO (2003) UN committee to recommend new dietary intake limits for mercury. World Health Organization: Notes for the Press 20.

Wiener JG, Krabbenhoft DP, Heinz GH, Scheuhammer AM (2003) Ecotoxicology of mercury. In: Hoffman DJ, Rattner BA, Burton GA Jr, Cairns J Jr (eds) Handbook of toxicology, second edition. CRC Press, Boca Raton, FL, pp 409–463.

Yogui GT, Sericano JL (2009) Polybrominated diphenyl ether flame retardants in the U.S. marine environment: a review. Environ Int 35:655–666.

Yoshimura T (2003) Yusho in Japan. Ind Health 41:139–148.

Section III

Conservation

13

Factors Contributing to Shark Attacks on Humans: A Volusia County, Florida, Case Study

George H. Burgess, Robert H. Buch, Felipe Carvalho,
Brittany A. Garner, and Christina J. Walker

CONTENTS

13.1 Introduction

Why sharks attack humans is not fully understood. However, it is believed that most unprovoked attacks are cases of misidentification wherein human-produced sensory stimuli are interpreted by sharks as being from natural prey items, triggering predatory attack (Burgess 1990; Myrberg and Nelson 1991). International Shark Attack File (ISAF) investigations of such interactions document an average of 64 unprovoked shark attacks,

TABLE 13.1

Regions of the World Having the Highest Recorded
Number of Shark Attacks in the Last Decade, 1999–2008

Geographic Region	No. of Shark Attacks	% of World Total
World	639	100
United States	428	66.9
Florida	275	43
Volusia County	135	21.1
Hawaii	40	6.3
South Carolina	32	5
California	26	4.1
North Carolina	22	3.4
Texas	14	2.2
Australia	79	12.4
New South Wales	33	5.2
Western Australia	16	2.5
Queensland	14	2.2
South Australia	11	1.7
South Africa	35	5.5
Eastern Cape	18	2.8
Western Cape	15	2.3
Brazil	23	3.6
Pernambuco	14	2.2
Bahamas	11	1.7

Source: Data from ISAF.

resulting in an average of 4.6 fatalities, per year worldwide in the most recent 10-year period (1999 to 2008). Most incidents are recorded in waters of the United States, with Florida and Volusia County annually having the highest number of incidents (Table 13.1). Although shark attack as a phenomenon is statistically uncommon given the billions of hours spent by humans in the sea and adjacent littoral and estuarine regions annually, and fatalities yet rarer, even a single attack may profoundly influence the public's view of the risks involved with aquatic recreation. Shark attacks deleteriously impact tourism (Gaiser 1976), resulting in severe economic losses in affected communities, a situation recorded from such geographically and culturally diverse areas as Australia (Dudley 1997), Brazil (Hazin, Burgess, and Carvalho 2008), South Africa (Davies 1995), and the United States (Bullion 1976).

Located on the north-central east coast of Florida in the southeastern United States, Volusia County has 47 miles of north–south-oriented Atlantic Ocean coastline and highly utilized beaches that have helped make tourism a cornerstone of this region's economy. Volusia County has a resident population of more than 500,000 (http://www.volusia.org/about/intro.htm) and about 10 million tourists visit the county's beaches each year (http://www.volusia.org/beach/shark_facts.htm). Average summer air temperature is 81°F and winter air temperatures average 61.5°F (http://volusia.org/weather.htm), fostering year-round bathing, surfing, and other aquatic recreational activities.

Intense recreational use of Volusia County beaches has occurred since the mid-1950s. The ISAF has records of 231 unprovoked attacks (all nonfatal) from 1956 to 2008, a total that represents 10.2% of worldwide incidents during that period. More recently, the number of unprovoked attacks in Volusia County has represented an even larger portion of the world's attack number, averaging 21.1% during the past decade (1999 to 2008) (Table 13.1). Because this is a high tourism area and shark attack is a headline-maker of great journalistic interest, shark attacks historically have been heavily covered by local, national, and international media, who have dubbed the area the "Shark Attack Capital of the World." In this study, we examine environmental and anthropological factors associated with shark attacks in Volusia County, document historical attack trends, and relate the Volusia County experience to other regions.

13.2 Methodological Approaches

Studies using ISAF shark attack statistics cannot fully follow standards of experimental design and scientific protocol because attack data gathering is, by definition, an opportunistic process that relies on collaboration and cooperation of multiple human parties, and often "luck." As such, ISAF investigations are in essence "accident reports" that are uncovered largely by word of mouth, pen, or keyboard. Therefore data is gathered in a nonrandomized, nonstratified manner. While the ISAF has a long-established (since 1958) reputation as the definitive scientific repository of shark attack data, which opens many investigative doors, and ISAF staff and their many worldwide collaborators work diligently to uncover and investigate all shark attacks on humans, it is clear that these investigations are in essence targets of opportunity that generate data that does not lend itself to nor is intended to be part of a carefully developed experimental design. Therefore, while some data categories can be subjected to statistical analyses, other data cannot and will never be able to be subjected to statistical review because of the lack of control data. Historically (Baldridge 1974; Burgess and Callahan 1996; Hazin, Burgess, and Carvalho 2008) such information as victim demographics (sex, height, skin color/race), coloration of victim clothing/gear, victim activity, depth of water at point of attack, and a large suite of other potentially contributing factors are presented for informational purposes and accepted for what they are—observations that may point to key causes but are simply statistically untestable (see Burgess and Callahan 1996 for further discussion). Owing to the close proximity of the region to the home base of the ISAF, the "high profile" nature of Daytona Beach as an international tourist attraction, and the historically well-developed local press, Volusia County probably has the most complete ISAF data set for any given geographic region in the world, yet even here we must acknowledge that the pre-1990 data is surely incomplete. Thus this study is fundamentally based on *statistics*, not an elegant *statistical treatment* of experimentally derived data. The inability to provide controls for virtually all data categories is a stumbling block and of particular concern is the inability to quantify baseline historic human use and activity patterns, apparel type and coloration, etc.—the "denominators" needed to relate the phenomenon to the norms of the day.

13.2.1 Study Area

Volusia County's Atlantic Ocean coastline is bisected by the Ponce de Leon Inlet (commonly referred to as Ponce Inlet), bounded north and south by manmade rock jetties (Figure 13.1). Located north of the inlet is Daytona Beach, and New Smyrna Beach lies to the south. Dredging associated with inlet channel maintenance and jetty placement affect

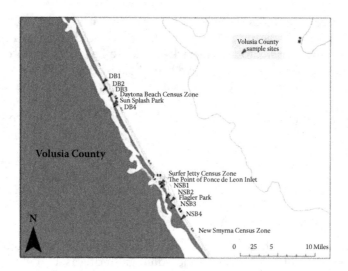

FIGURE 13.1
A color version of this figure follows page 336. Volusia County, Florida, beaches and aquatic user visual census stations used in study.

the natural north–south pattern of movement, producing sandbar development south of the inlet in New Smyrna Beach that produces surf conditions highly favored by surfers.

In order to understand how humans use the beaches in Volusia County, visual censuses were performed at five locations north of Ponce Inlet and five stations located south of the inlet (Figure 13.1). An additional 11th site (The Point at Ponce Inlet Jetty) was set aside for surfer utilization analysis. From north to south, the monitored stations were:

DB1 (29°15′32.81″N, 81°01′21.70″W): A stretch of beach that is car accessible but easily closed off due to high tides. Relatively low human water use, with the majority of people walking, biking, or wading; northernmost waypoint north of the jetty.

DB2 (29°14′39.72″N, 81°00′55.34″W): Located behind a few hotels and time-shares, this waypoint is relatively secluded and has no easy car access.

DB3 (29°13′59.10″N, 81°00′35.80″W): Located in the heart of tourist attractions in Daytona Beach, this point is not easily car accessible but is largely visited by guests of large resorts; includes rental facilities along the beach for umbrellas, food, surf-boards, etc.

Sunsplash Park (29°13′06.59″N, 81°00′09.74″W): Heavily used waypoint north of the jetty, this point is a city park and has a free parking lot, bathrooms, showers, and picnic tables for beach users.

DB4 (29°12′48.79″N, 81°00′01.82″W): Southernmost waypoint north of the jetty, this point has easy human and car access; located behind a few smaller hotels and time-shares.

The Point at Ponce Inlet Jetty (29°04′10.49″N, 80°54′34.47″W): A span of beach beginning at NSB1 and going north until the land hits the jetty, this area is known for its high swells and excellent surfing conditions. The majority of Volusia County shark attacks on record in the ISAF have occurred at this almost exclusively surfer-utilized location.

NSB1 (29°03′46.80″N, 80°54′26.23″W): Located adjacent to the Point, this spot is located behind a vast high-security condominium complex and is accessible by car or by guests of the condominium. This is the northern-most swimmer-utilized section of beach south of Ponce Inlet.

NSB2 (29°02′36.21″N, 80°53′52.71″W): Found behind time-share condominiums near the northern part of New Smyrna Beach, this waypoint has a relatively small beach space that hosts both swimmers and surfers regularly.

Flagler Park (29°02′15.21″N, 80°53′40.47″W): A very large span of heavily utilized beach, this is a city park that has a free parking lot, bathrooms, showers, picnic tables, and a nearby lifeguard facility.

NSB3 (29°01′23.03″N, 80°53′11.35″W): A secluded stretch of beach located behind smaller homes of beach residents, accessible by a wooden walkway.

NSB4 (29°00′13.59″N, 80°52′27.72″W): The southernmost waypoint south of the jetty, this beach has no easy car access and is visited mainly by people walking to and from their beach condominiums or guest residences.

13.2.2 Attack Data Collection

Most of the Volusia County shark attack data in the ISAF originates from the voluntary submissions of ISAF questionnaires and interviews conducted by ISAF staff of cooperating victims, witnesses, physicians, and beach safety personnel. The ISAF staff works closely with Volusia County Beach Patrol personnel in order to obtain accurate data for each Volusia County incident. Local physicians and hospital emergency room workers also are cooperative in connecting victims with the ISAF. In addition, mutual interest in the subject matter has fostered the development of a close relationship between the regional working press and the ISAF, thereby ensuring that news of virtually all Volusia County shark-related incidents are brought to the attention of the ISAF. Since beach patrol and medical personnel often are in direct contact with victims immediately after a shark-bite incident, they frequently promote completion of an ISAF attack questionnaire (available online at http://www.flmnh.ufl.edu/fish/sharks/isaf/questionintro.htm), and, if approved by the victim, provide contact information to the ISAF so an investigation can be launched. With the development of the Internet and e-mail, it is not unusual to receive an unsolicited submission from a victim who used an Internet search engine and "found" the ISAF on the web (http://www.flmnh.ufl.edu/fish/sharks/ISAF/ISAF.htm) after a low-grade interaction, typically requiring no immediate care or professional medical treatment, went undiscovered by the media.

Older case data are derived from available documentation sources, most often newspaper, magazine, and/or book accounts. Whenever possible such accounts are verified by interviews with surviving victims or witnesses to an attack. However information on attacks many decades old often cannot be augmented by additional investigation. Data submitted to the ISAF is screened, coded, and computerized in a SQL database. Hard copy documentation, including original notes, press clippings, photographs, audio/video tapes, digital archival information, and medical/autopsy reports, is permanently archived.

In this study we analyze cases of confirmed unprovoked shark attacks, ignoring provoked attacks, attacks on boats, and "scavenge" cases in which the victim was dead prior to suffering shark-inflicted damage. "Unprovoked attacks" are defined as incidents where an attack on a live human by a shark occurs in its natural habitat without human provocation of the shark. See Burgess and Callahan (1996) and Hazin, Burgess, and Carvalho (2008) for additional details of methods.

As in earlier analyses of shark attack (Schultz 1963; Baldridge 1974; Burgess and Callahan 1996; McCosker and Lea 2006; Hazin, Burgess, and Carvalho 2008; Trape 2008), the present study relates attacks to many biotic and abiotic factors, such as time of year, attack location, time of the day, victim activity, injury location, tidal phase, moon phase, and type of injury. Activities of victims were divided into three categories: (1) board riders (surfing, body boarding, partial-emersion kayaking, and rafting), (2) swimmers (swimming, bodysurfing, floating, and wading), and (3) divers (free, hard hat, hookah, pearl, scuba, and snorkeling).

13.2.3 Beach Use Data

Visual censuses were performed between May 25 and July 28, 2008. Data collection days were drawn randomly until each day of the week had been selected twice. This was designed to ideally collect data from 14 days from each month, although there was slight deviation because of random circumstances and logistics that occasionally prevented scheduled field work. Using GPS equipment, all beaches were split evenly into four quadrants according to city limits, and one waypoint for data collection was plotted evenly within each quadrant. The areas were split geographically without prior knowledge or experience of the location, thus acting as a random way of selection for waypoints that still spanned the geographic area in question.

Using markers on the beach, each waypoint was sectioned to approximately 100 meters. Upon data collection at a waypoint, a digital camera was used to take four to six pictures of the waypoint beginning at the leftmost start of the 100 meter section and ending at the right of it. The photographer stood near the center of the waypoint facing east toward the Atlantic Ocean. After collection, pictures were analyzed using J Image computer software with the cell counter application (Sheffield 2007). Scanning within the marked 100 meters, counts were made of all people on the beach and in the water, the latter differentiating surfers and swimmers/waders.

Monitoring of people on the beach and in the water occurred during daylight hours from 0800 to 1800. Each waypoint was visited approximately six times daily (between 0600 and 0800; 0800 and 1000; 1000 and 1200; 1200 and 1400; 1400 and 1600; 1600 and 1800), resulting in a total of approximately 1000 data points for summer human use of the beaches. Data collection days were split between the beaches, with equal amounts of time spent at each location, morning and afternoon location drawn randomly for each collection day. This method was chosen over a full day/one location method in order to allow a more random and variable collection of data accounting for changes in weather, temperature, traffic, and other factors.

13.2.4 Statistical Analysis

A priori review of historic attack location frequency data and observations of differing patterns of human use of waters north and south of the jetty indicated that Daytona Beach and New Smyrna Beach, the two locations most often hosting such interactions, were ideal study sites for comparison in this study.

The number of shark attacks per unit of effort (APUE) is used in considering factors possibly contributing to shark–human interactions in Volusia County. This relative abundance index is expressed in terms of the number of shark attacks per 100,000 inhabitants of Volusia County. County population data is derived from the latest U.S. census. The formula used to calculate the APUE is:

$$APUE = \Sigma \text{ attacks} * 10{,}000/100{,}000 \text{ habitants}$$

To verify significant differences in the location of the injury on the body, time of the attack, and activity of the victim, a Chi-square (χ^2) test was used. The mean values of the APUE were calculated as related to attack variables including decade, month, lunar phase, and tidal phase, and for the two designated geographical regions north and south of Ponce de Leon Inlet. The variables were checked for normality and homoscedasticity. In the case that the ANOVA normality and homoscedasticity assumptions did not apply, the Kruskal–Wallis test was used to test for the equality of population medians among groups.

13.3 Results

13.3.1 Attack Location

The first unprovoked shark attack in Volusia County was recorded in 1956. During the 1956 to 2008 period, 231 unprovoked shark attacks occurred in the waters of Volusia County. Of the 220 that have detailed locality information, 79.1% took place in New Smyrna Beach and 15.5% in Daytona Beach (Table 13.2).

13.3.2 Attack Timing: Month of the Year

The number of attacks north of Ponce Inlet was particularly high for swimmers in the month of July (Kruskal–Wallis, $F = 4.45$; $P = 0.01$). Attacks on surfers north of the inlet were found to be significantly higher during September (Kruskal–Wallis, $F = 1.32$; $P = 0.04$) (Figure 13.2). The number of attacks south of the inlet were high during the months of September for swimmers (Kruskal–Wallis, $F = 4.97$; $P = 0.03$), and in August for surfers in that region (Kruskal–Wallis, $F = 5.12$; $P = 0.02$) (Figure 13.3).

13.3.3 Attack Timing: Day of the Week

Review of the daily frequency of attacks during the study period revealed a pattern of increasing number of attacks starting Thursday and peaking on Sunday, followed by declines through Wednesday, when the lowest number was recorded. The number of attacks on Sunday was significantly different from the number of attacks that occurred on other days of week for both surfers (Kruskal–Wallis, $F = 3.08$; $P = 0.02$) and swimmers (Kruskal–Wallis, $F = 1.57$; $P = 0.03$) in attacks occurring north of Ponce Inlet (Figure 13.4). In incidents south of the inlet, day of the week had a significant influence on attacks on swimmers (Kruskal–Wallis, $F = 2.77$; $P = 0.01$); however, it did not demonstrate significant influence on surfer attacks (Kruskal–Wallis, $F = 3.58$; $P = 0.09$) (Figure 13.5).

13.3.4 Attack Timing: Time of Day

Although we found no statistical differences between time of attacks north of Ponce Inlet, surfers suffered more bites early and late in the day (0800 to 1000 and 1600 to 1800 hours;

TABLE 13.2

Location of 221 Shark Attacks off Volusia County, Florida, 1956–2008

| Year | South of Ponce de Leon Inlet | | North of Ponce de Leon Inlet | | Total |
	New Smyrna Beach	Cape Canaveral	Ormond	Daytona	Volusia County
1956	—	—	—	—	1
1957	—	—	1	—	1
1962	1	—	—	—	1
1963	1	—	—	—	1
1969	1	—	—	—	1
1971	—	—	—	1	1
1974	1	—	—	3	4
1975	2	—	—	1	3
1977	2	—	—	1	3
1981	—	—	—	2	2
1982	3	—	—	—	3
1984	—	—	—	1	1
1985	—	—	—	1	1
1986	1	—	—	1	2
1987	1	—	—	—	1
1988	1	—	—	—	1
1991	2	—	—	—	2
1992	1	—	—	—	1
1993	1	—	—	1	2
1994	9	1	—	2	12
1995	9	—	1	2	12
1996	7	—	—	—	7
1997	7	—	1	2	10
1998	6	—	1	1	8
1999	7	—	—	1	8
2000	8	1	2	3	14
2001	22	—	—	—	22
2002	17	—	1	1	19
2003	16	—	—	4	20
2004	—	1	1	1	3
2005	8	—	1	1	10
2006	7	—	—	2	9
2007	13	—	—	2	15
2008	20	—	—	—	20
Total	174	3	9	34	221

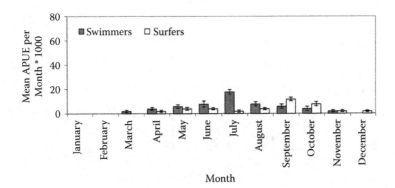

FIGURE 13.2
Mean number (±SE) of shark attacks per month in Volusia County, Florida, north of Ponce de Leon Inlet between 1957 and 2008.

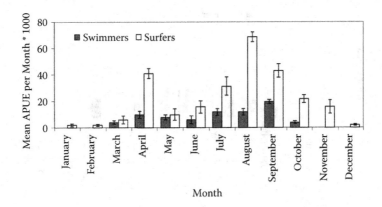

FIGURE 13.3
Mean number (±SE) of shark attacks per month in Volusia County, Florida, south of Ponce de Leon Inlet between 1957 and 2008.

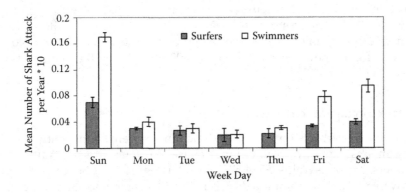

FIGURE 13.4
Mean number (±SE) of attacks per day of week, Volusia County, Florida, north of Ponce de Leon Inlet between 1957 and 2008.

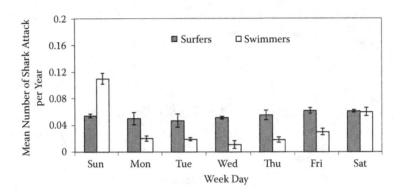

FIGURE 13.5
Mean number (±SE) of attacks per day of week, Volusia County, Florida, south of Ponce de Leon Inlet between 1957 and 2008.

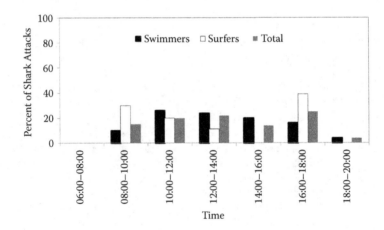

FIGURE 13.6
Percentage of shark attacks on swimmers and surfers in Volusia County, Florida, north of Ponce de Leon Inlet between 1957 and 2008 as a function of time of the day.

$\chi2 = 1.01$; df = 2; $P = 0.07$) and swimmers were most frequently attacked during the late morning between 1000 and 1200 hours ($\chi2 = 3.45$; df = 2; $P = 0.06$) (Figure 13.6). South of the inlet, surfers were also bitten more often early and late in the day, while swimmers were attacked more often during the mid-afternoon hours (1400 to 1600 hours). No significant difference was found, however (surfers: $\chi2 = 4.00$; df = 2; $P = 0.1$; swimmers: $\chi2 = 5.23$; df = 2; $P = 0.07$) (Figure 13.7).

13.3.5 Attack Timing: Moon Phase

North of Ponce Inlet attacks on surfers and swimmers were significantly more common during new moons than during other lunar phases (surfers: Kruskal–Wallis, $F = 3.71$; $P = 0.01$; swimmers: Kruskal–Wallis, $F = 4.57$; $P = 0.007$) (Figure 13.8). South of the inlet attacks on surfers were much more common during new and full moons than during first and last quarters (Kruskal–Wallis, $F = 4.61$; $P = 0.008$). Attacks were significantly higher on swimmers south of Ponce Inlet during the new moon phase (Kruskal–Wallis, $F = 5.11$; $P = 0.03$) (Figure 13.9).

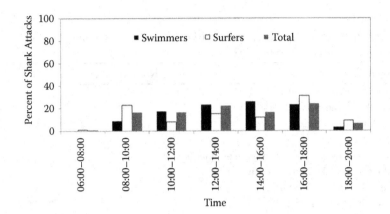

FIGURE 13.7
Percentage of shark attacks on swimmers and surfers in Volusia County, Florida, south of Ponce de Leon Inlet between 1957 and 2008 as a function of time of the day.

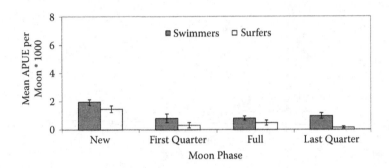

FIGURE 13.8
Mean number (±SE) of shark attacks per lunar phase in Volusia County, Florida, north of Ponce de Leon Inlet between 1957 and 2008.

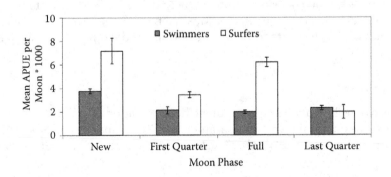

FIGURE 13.9
Mean number (±SE) of shark attacks per lunar phase in Volusia County, Florida, south of Ponce de Leon Inlet between 1957 and 2008.

TABLE 13.3

Habitats in which Sharks Attacked Humans in Volusia County, Florida, between 1957 and 2008 ($n = 182$)

Habitat	No. of Attacks	% of Attacks
Just beyond breaker, surf line	1	0.5
Mouth of river, creek	1	0.5
Offshore reef, bar, bank	2	1.1
Open seas	2	1.1
Waters between sandbar and shore or other bar	8	4.4
Alongside breaker, jetty, dock, wharf	9	4.9
Surf zone (no specifics)	27	14.8
Inside breaker, surf line	54	29.7
Near-shore waters, beaches (no specifics)	78	42.9

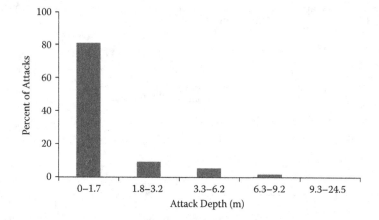

FIGURE 13.10
Depth at which sharks attacked humans in Volusia County, Florida, between 1957 and 2008 ($n = 52$).

13.3.6 Attack Depth and Total Water Depth at Attack Site

Volusia County shark attacks were overwhelmingly near-shore and near-surface occurrences (Table 13.3, Figure 13.10). More than 80% of incidents occurred in water depths of less than 1.8 meters. Waters close to shore, namely the surf zone associated with beaches, hosted virtually all (96.7%) of the attacks. Most victims (79%) were at the water surface at the time of attack, 19% were standing on the bottom, and 2% of victims were engaged in underwater activity.

13.3.7 Water Clarity

In the absence of technical testing equipment, determination of water clarity is subjective. Despite difficulties in qualifying this parameter, 74.7% of shark attacks were reported to occur in "turbid," "murky," or "muddy" waters and 25.3% in "clear" waters.

13.3.8 Victim Demographics and Apparel

Most (53.1%) shark attack victims were under 21 years of age (Table 13.4) and, overwhelmingly, the race and sex of the victims were white (97.7%, Table 13.5) and male (89.5%,

TABLE 13.4

Ages of 196 Victims Involved in Shark
Attacks in Volusia County, Florida,
between 1958 and 2008

Age	No. of Attacks	% of Attacks
<11	3	1.5
11–15	46	23.5
16–20	55	28.1
21–25	27	13.8
26–30	22	11.2
31–35	9	4.6
36–40	8	4.1
41–45	6	3.1
46–50	11	5.6
>50	9	4.6

TABLE 13.5

Race of Victims Involved in Shark
Attacks in Volusia County, Florida,
between 1958 and 2008 ($n = 171$)

Race	No. of Attacks	% of Attacks
White	167	97.7
Black	4	2.3

TABLE 13.6

Sex of Victims Involved in Shark Attacks
in Volusia County, Florida, between 1958
and 2008 ($n = 220$)

Sex	No. of Attacks	% of Attacks
Male	197	89.5
Female	23	10.5

Table 13.6). The primary, secondary, and tertiary colors of victims' gear and clothes were assigned to one of nine major color groups. When examined in combination, the primary and secondary color grouping most often seen included the color black (78%) (Table 13.7).

13.3.9 Victim Activity

The primary (66.7%) activity of the victims south of Ponce Inlet was surfing ($\chi2 = 5.21$; df = 2; $P < 0.001$), while in attacks north of the inlet, swimmers represented the majority (62.8%) of the victims ($\chi2 = 3.11$; df = 2; $P < 0.001$) (Table 13.8). Since only one attack on a diver was recorded in Volusia County waters (in 1999), we have excluded this user group from the further analyses. When we examine data for the entire county, combining both north and south of the inlet, surfers (60.9%) were the most frequent victims ($\chi2 = 4.03$; df = 2; $P < 0.001$) (Table 13.8). The effect of fishing at the attack site was investigated, but

TABLE 13.7

Color Combinations (Primary and Secondary) of the Gear Worn by Victims Involved in Shark Attacks in Volusia County, Florida, between 1958 and 2008 ($n = 95$)

Color	Attacks (%)
Black and white	22.1
Black and yellow	18.9
Black and red	11.6
Black and orange	9.5
Black and green	8.4
Black and blue	7.4
Blue and yellow	5.3
Blue and black	5.3
Blue and orange	5.3
Yellow and red	3.2
Yellow and green	2.1
Brown and green	1.1

TABLE 13.8

Activity of Victims Involved in Shark Attacks in Volusia County, Florida, between 1958 and 2008 ($n = 220$)

Activity	South of Ponce de Leon Inlet		North of Ponce de Leon Inlet		Volusia County
	New Smyrna Beach	Cape Canaveral	Ormond	Daytona	Total (%)
Divers	1	0	0	0	1 (0.5)
Swimmers	56	2	4	23	85 (38.7)
Surfers	117	1	5	11	134 (60.9)
Total	174	3	9	34	220

fishing by others in the area was reported in only 2% of the attacks. See Section 13.3.14 Beach Use for complementary data that dovetails with these results.

13.3.10 Victim Injury

There have been no fatal unprovoked shark attacks in Volusia County. Surfers and swimmers were mainly bitten on the legs both north and south of Ponce Inlet (surfers north of the inlet: $\chi^2 = 2.67$; df = 2; $P < 0.001$; surfers south of the inlet: $\chi^2 = 1.41$; df = 2; $P = 0.03$; swimmers north of the inlet: $\chi^2 = 3.44$; df = 2: $P < 0.01$; swimmers south of the inlet: $\chi^2 = 2.10$; df = 2; $P = 0.02$). Volusia County attack victims were bitten on the legs significantly more often (71.8%) when compared with the other anatomical injury locations ($\chi^2 = 4.03$; df = 2; $P < 0.001$) (Table 13.9).

13.3.11 Attacking Shark Species

Identification of attacking shark species is a difficult task, even to those trained in the task, so it is not surprising that there were only 16 cases in which the shark was identified.

TABLE 13.9

Bite Location on Body of Victims Subjected to Traumatic Injury from Shark Attack in Volusia County, Florida, between 1958 and 2008 (*n* = 220)

	South of Ponce de Leon Inlet			North of Ponce de Leon Inlet			Total No. of Victims
Injury Location	Swimmers	Surfers	Divers	Swimmers	Surfers	Divers	
Arms	7	16	0	5	6	0	34
Legs	40	90	1	16	11	0	158
Torso	5	5	0	5	0	0	15
Multiple	6	7	0	0	0	0	13
Total	58	118	1	26	17	0	220

FIGURE 13.11
A color version of this figure follows page 336. Blacktip shark (*Carcharhinus limbatus*) jumping out of the water in a spinning motion near surfer in New Smyrna Beach, Volusia County, Florida, 2008. (Photo © KemMcNair. com. All rights reserved.)

Spinner sharks, *Carcharhinus brevipinna* (14), and bull sharks, *C. leucas* (2), were the identified species. However, since the spinner and blacktip (*C. limbatus*) sharks are frequently confused by fishers (and some scientists) and the blacktip is regionally abundant in the surf zone environment (Dodrill 1977) and often jumps out of the water in a spinning motion (Figure 13.11) equal to its aptly named congener, the spinner identifications likely include both species. Circumstantial evidence from the east coast of Florida (ISAF unpublished data) suggests that the blacktip and spinner sharks are the major culprits in this region, with bull sharks less often involved.

13.3.12 Attacking Shark Behavior

Shark behavior was recorded before, during, and after a strike on 89 occasions. Single bites were the norm; repetitive bites occurred in only 18% of attacks (Table 13.10). In 83% of recorded incidents, it was presumed that the attacking shark left the immediate area following the initial bite; in 17% of the cases, the shark remained nearby (Table 13.11).

TABLE 13.10

Number of Individual Bites made by the Attacking
Shark on Its Victim and/or Associated Gear in Volusia
County, Florida, between 1958 and 2008 (n = 89)

Number of Bites	Attacks (%)
None	10.1
One	71.9
More than one	17.9
"Two"	6.7
"Three"	5.6
"Several"	3.4
"Many"	1.1
"Too many to specify"	1.1

TABLE 13.11

Behavior of the Attacking Shark after Final Strike on Its Victim
in Volusia County, Florida, between 1958 and 2008 ($n = 82$)

Shark Behavior	Attacks (%)
Presumably left area	82.9
Not seen after final strike	67.1
Seen leaving area	15.9
Remained in area	17.1
Remained in immediate area	9.8
Followed victim/rescuers to shore	6.1
Remained attached to the victim, had to be forcibly removed	1.2

13.3.13 Long-Term Attack Trend

The shark attack rate on swimmers has been increasing significantly both north and south
of Ponce Inlet in recent decades (Kruskal–Wallis: north of the inlet, $F = 3.54$; $P = 0.04$; south
of the inlet, $F = 2.61$; $P = 0.002$). There has been a 16-fold increase between the mean APUE
per year south of the inlet between the decades of the 1980s and 1990s (Figure 13.12) and,
although at time of writing there is still one year left in the first decade of the twenty-first

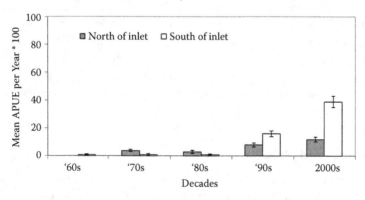

FIGURE 13.12
Mean APUE (±SE) of shark attack per decade for swimmers north and south of Ponce de Leon Inlet.

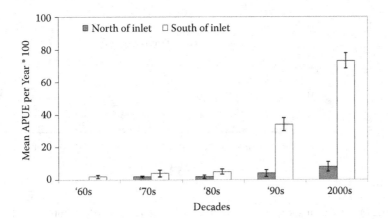

FIGURE 13.13
Mean APUE (±SE) per year per decade for surfers north and south of Ponce Inlet, Florida.

century, it is apparent that this trend is continuing. This pattern is similarly marked in the attack rates on surfers in both areas (Kruskal–Wallis: north of the inlet, $F = 4.88$; $P < 0.001$; south of the inlet, $F = 5.21$; $P = 0.001$) (Figure 13.13).

13.3.14 Beach Use

Results from the visual censuses showed that the mean number of total aquatic users (= surfers plus swimmers) in Daytona Beach waters is significantly higher on Sundays (Kruskal–Wallis, $F = 4.11$; $P = 0.01$), surfers (Kruskal–Wallis, $F = 2.87$; $P = 0.02$), and swimmers (Kruskal–Wallis, $F = 1.48$; $P = 0.04$) (Figure 13.14). The Point at Ponce Inlet Jetty did not show any statistically significant differences for mean number of aquatic users (Kruskal–Wallis, $F = 5.37$; $P = 0.09$), surfers (Kruskal–Wallis, $F = 4.03$; $P = 0.1$), and swimmers (Kruskal–Wallis, $F = 2.08$; $P = 0.06$) (Figure 13.15), in the water. However, unlike the beaches located north of the inlet, the mean number of surfers at the Point was higher than mean number of swimmers. Pooled New Smyrna Beach station utilization data showed the mean number of aquatic users (Kruskal–Wallis, $F = 4.11$; $P = 0.01$), surfers (Kruskal–Wallis, $F = 2.87$; $P = 0.02$), and swimmers (Kruskal-Wallis, $F = 1.48$; $P = 0.04$) in the water is significantly higher on Sundays (Figure 13.16). For the overall study period Daytona Beach

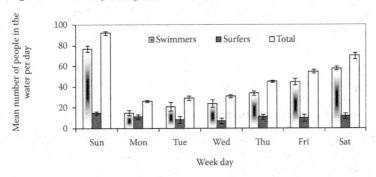

FIGURE 13.14
Mean number (±SE) of people in the water per day at Daytona Beach, Florida, obtained through visual censuses between May 25 and July 28, 2008.

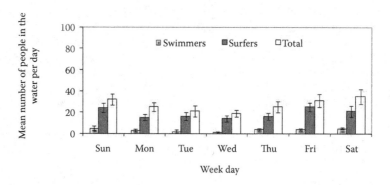

FIGURE 13.15
Mean number (±SE) of people in the water per day at Ponce de Leon Inlet, Florida, south jetty, obtained through visual censuses between May 25 and July 28, 2008.

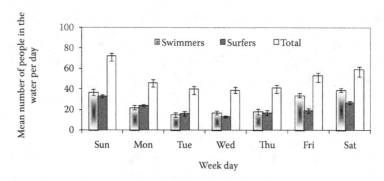

FIGURE 13.16
Mean number (±SE) of people in the water per day at New Smyrna Beach, Florida, obtained through visual censuses between May 25 and July 28, 2008.

showed a significantly higher mean number of swimmers in the water (Kruskal–Wallis, $F = 3.46$; $P = 0.01$) and New Smyrna Beach had a significantly higher mean for surfers (Kruskal–Wallis, $F = 3.61$; $P < 0.001$) (Figure 13.17). Although there were no statistical differences found between the percent of aquatic users north of Ponce Inlet, both surfers and swimmers were more numerous between 1000 and 1400 hours ($\chi^2 = 3.21$; df = 2; $P = 0.09$) in the counting census (Figure 13.18). South of the inlet, surfers were more often present early in the day and late afternoon; however, no significant difference was found ($\chi^2 = 5.10$; df = 2; $P = 0.07$) (Figure 13.19).

13.4 Analysis and Interpretations

This is the third in a series of studies in which shark attacks are analyzed by location and/or type of shark involved. Worldwide patterns of white shark (*Carcharodon carcharias*) attacks were reviewed by Burgess and Callahan (1996), characterizing the type of attack that is most common in cool-water regions of the world, most notably the southern coasts of Australia and Africa and Pacific coast of the United States. Tiger (*Galeocerdo cuvier*) and bull sharks, the other two-thirds of the "big three" of most dangerous sharks, were

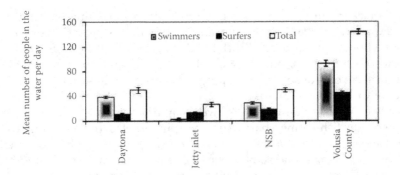

FIGURE 13.17
Mean number (±SE) of people in the water per day in Volusia County, Florida, obtained through visual censuses between May 25 and July 28, 2008.

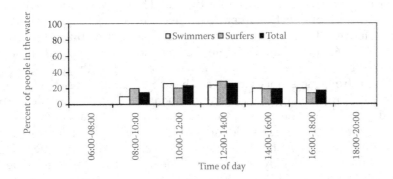

FIGURE 13.18
Recreational user group utilization of Volusia County, Florida, waters north of Ponce de Leon Inlet as function of time of the day, obtained through visual censuses between May 25 and July 28, 2008.

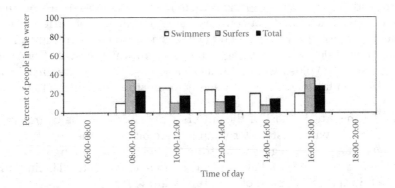

FIGURE 13.19
Recreational user group utilization of Volusia County, Florida, waters south of Ponce de Leon Inlet as function of time of the day, obtained through visual censuses between May 25 and July 28, 2008.

the primary attackers implicated in Recife, Brazil, attacks examined at length by Hazin, Burgess, and Carvalho (2008). White, tiger, and bull sharks are large (3 m or longer) species that seek large prey items as adults (Cliff, Dudley, and Davis 1989; Cliff and Dudley 1991; Simpfendorfer 1992; Burgess, unpublished observations) and possess serrated dentition effective in shearing off portions of their prey, often resulting in serious traumatic injuries, and occasionally deaths, in humans.

In this study of Volusia County attacks we see a different style of attack primarily perpetrated by smaller fish-eating sharks such as spinner, blacktip, and perhaps blacknose (*C. acronotus*) and sharpnose (*Rhizoprionodon terraenovae*) sharks, with most attackers being two meters or less in length and some as small as a meter. Because of their size, these species normally seek smaller prey items and additionally have more erect teeth effective at grabbing and holding. The types of injuries inflicted on victims are correspondingly less severe, usually ranging from lacerations to tooth impression wounds. There seldom is tissue loss and reduction of function is largely confined to cases where tendons of the foot or hand have been damaged. All attacks have been nonfatal.

Burgess (1990) placed shark attacks in three broad categories, "sneak," "bump and bite," and "hit and run" attacks. Attacks by white, tiger, and bull sharks tend to fall into the two former categories, both of which frequently involve repeat attacks or continued threat by the shark and often result in high severity of trauma to the victim. By contrast, most attacks occurring in Volusia County waters fall into the "hit and run" class. "Hit and run" attacks are characterized by an initial bite or raking slash followed by the apparent prompt exit of the perpetrator from the immediate area. The combination of relatively small sharks, reduced water visibility in a high-energy environment (breaking surf, undercurrents, tidal currents), and the lack of repeat attacks or observations of the attacking shark remaining in the area suggest that these incidents are primarily cases of misidentification. As most attacks occur at or near the water surface, it seems likely that the splashing and movements of humans, especially the kicking of feet and hand/arm movements associated with swimming or maintaining stability, are attractive to sharks seeking normal prey items such as schooling mullet (*Mugil* spp.) or herrings (Clupeidae) that occupy the same habitat. This area is a nursery for blacktip and spinner sharks (Aubrey and Snelson 2007), and it appears that young (approximately 1 m total length), inexperienced, individuals are responsible for some attacks.

The in-water behavior of surfers—including spending greater time in the water and venturing farther from shore (Cliff 1991), therefore becoming isolated, and the provocative nature of arm and leg movements—render them more vulnerable to an encounter with a shark than swimmers. It should not come as a surprise, then, that surfers were the most affected Volusia County user group and that attacks occurred in significantly higher numbers south of Ponce Inlet, where surfing is the main in-water activity. The most heavily used surfing location in Volusia County is near the jetty south of Ponce Inlet. In addition to favorable environmental conditions, another factor in attracting sharks to this area could be the presence of recreational fishers along the jetty.

At the time of the attack, most of the surfers were sitting on the board waiting for a wave, which explains why most surfer victims were bitten on the legs. It is also reasonable to assume that a surfer's leg hanging from the board and moving in deeper water is more attractive to a shark than a bather standing in shallow water (Hazin, Burgess, and Carvalho 2008). Surfers spend most of their time waiting for waves sitting on their boards with their legs and feet dangling in turbulent, murky water and, often, in the midst of bait fishes. It also is no coincidence that the majority of surfers receive injuries to their legs and feet during times of reported increased baitfish activity.

Swimmers and bathers, by contrast, were the dominant recreational group using the waters north of the Ponce Inlet and were less likely to experience a shark attack. They are also more likely to heed the warnings from lifeguards about increased shark bite risk than surfers (personal observation). This surely helps in reducing the risk to swimmers. Like surfers, swimmers were also most often bitten on the legs. This likely is because most unprovoked shark attacks on swimmers occur in shallow water (less than 2 m) where the legs are most often submerged while walking or maintaining balance in the surf zone.

It is unremarkable that the dominant color of clothing/gear worn by attack victims was black (Burgess and Callahan 1996). Although there is no data on color choice in aquatic recreational apparel, either historic or current, black is the dominant color of choice in wet suits (worn by surfers) and frequently is featured as a primary, secondary, or tertiary color in swimsuits. Lacking solid utilization data reflective of changing fashion trends over the years, it is unlikely we will ever be able to determine if sharks were selectively choosing one color over another.

Temporal attack descriptor categories—time of day, day of the week, and month of the year—all appear to influence the probability that an attack will occur, in many cases because of their influences on the number of people in the water rather than influences on shark behavior. Indeed, the number of attacks is directly related to the times of day when the most people are in the water. Surfers tend to surf and be bitten more in the early morning (prior to 0800–1000) and late afternoon (after 1600–1800) hours when the waves are highest. The early morning hours, before the wind picks up, as well as the late afternoon, after the winds drop, usually produce the best surfing conditions, more surfers, and more shark–surfer interactions. Times of greatest swimmer use also corresponded to the times with the greatest number of attacks. Beach use and the number of shark attacks peak between 1100 and 1900. The greatest numbers of attacks and swimmers occurred prior to and after the "lunch break" with a dip between 1200 and 1400.

There was also a relationship between attack numbers and beach use with day of the week. The highest number of people and attacks occurred over the weekend and the lowest on Tuesday and Wednesday. As expected based on the number of people using beaches, the number of attacks on Sunday was significantly higher than the number of attacks that occurred on other days of week.

Similarly, the same relationship would be expected to exist between month of the year, the number of people engaged in aquatic recreation, and number of shark attacks. Because only a portion of the visual beach use census was completed prior to preparation of this chapter (to be addressed in a subsequent study), we are unable to compare numbers of people on the beach by month. However, in northern Florida generally water temperatures begin to warm in late March–April and remain warm enough for mainstream recreational activity until October–November. "Spring Break" for various U.S. and Canadian universities occurs in March and April, bringing increased numbers of student tourists from northern latitudes to Daytona Beach and other seaside tourism communities during a time period when most Florida residents are reluctant to enter the still (by their standards) cool waters. Daytona Beach is one of the most highly selected Spring Break locations in North America. Secondary schools and many universities end their academic years in late May–June and start the new academic year in late August–September. During the intervening period, Volusia County hosts family vacations and increased numbers of teenagers and young adults, many engaged in the popular pastime of surfing. Historically the attack numbers mirror use patterns, with the highest number of attacks recorded in August and September when the air and water temperatures are, on average, highest and aquatic recreation at its highest levels. During the early summer the waves are not particularly good for surfing in Volusia County but build in size in late August and September, leading to better surfing conditions. This attraction could explain why areas south of the inlet had their highest number of attacks in these months. An April peak in attacks occurring south of Ponce Inlet, chiefly on surfers, corresponds to a period of northward shark movements apparently triggered by warming water temperatures that also is marked by a south-to-north wave of shark–human interactions (ISAF, unpublished data).

Temporal factors affect swimmers more than surfers, who tend to frequent the water when waves are good and ignore the water when waves are poor, regardless of time of day or day of week. Because the predominant surfer demographic is young, there is reduced surfer participation during late fall and winter months when school is in session. However, because early morning and late afternoon hours produce desirable surfing conditions, surfing is a year round activity for aficionados, regardless of academic or employment status.

The significantly higher frequency of attacks on surfers south of the inlet and on bathers north of the inlet (the predominant user group in that region) during a new moon may be related to the higher tides occurring during this moon phase, which may facilitate the immigration of sharks into normally shallow near-shore areas. Past studies on lunar phase and marine life have shown that lunar phase can significantly affect the behavior of many marine organisms, including fish (Naylor 1999), invertebrates (Gliwicz 1986), and mammals (Wright 2005). The results of this study, which demonstrates that shark bites occur more frequently around a new moon than they do around a full moon, supports this idea.

Although no definitive reason has been pinpointed for the increase in shark attacks during new moons, looking at the behavior of other pelagic shark species as well as the behavior of their prey may provide an answer. School sharks (*Galeorhinus galeus*) have been reported to inhabit significantly deeper water during full moons than they do during any other moon phase (West and Stevens 2001). Similarly, Lowry, Williams, and Metti (2007) found that catch rates of blue sharks (*Prionace glauca*) and shortfin mako sharks (*Isurus oxyrinchus*) increased during new moons, suggesting that they were inhabiting surface waters during that period. Both studies concluded that the vertical distribution of sharks reflected the prey distribution as related to the increase in lunar illumination and the deep scattering layer (DSL). As light increases and penetrates deeper into the water column, light-sensitive prey will swim deeper and, in many cases, be followed by predators (Lowry, Williams, and Metti 2007). Although this phenomenon has not been observed in all species of sharks, the sharks inhabiting the waters off the coast of Volusia County may be following similar cues, coming into shallow, near-shore waters to feed when light is low and moving seaward to deeper water when light is high.

Lunar phase may also affect the migratory patterns and spawning patterns of baitfish off Volusia County. "Mullet runs" occur frequently up and down the coast of New Smyrna Beach and lifeguards take extra precaution in warning bathers and surfers about entering the water during the "runs." Katselis et al. (2007) found that seaward migration of numerous fish, including striped mullet (*Mugil cephalus*), increased during full moons. There often is an upturn in shark–human interactions in October–November "mullet runs" as humans, primarily surfers, mullets, and sharks intermingle in the surf zone.

13.5 Conclusions and Prospects

Shark populations along the U.S. East Coast have been in decline for at least the past two decades, primarily as a result of overfishing (National Marine Fisheries Service 2006), yet there has been a steady increase in the number of shark attacks since the 1980s in Volusia County (and in many areas of the world; ISAF, unpublished data). While the population decline of some coastal shark species, most notably the blacktip, apparently has been stabilized (National Marine Fisheries Service 2006), the continuous rise in number of attacks throughout this time period suggests that human use patterns likely played a more important role in dictating the absolute number of interactions. Indeed, Volusia County's resident human population and visitor numbers increased markedly during this time period. According to the U.S. Bureau of Census (http://quickfacts.census.gov/qfd/states/12/12127. html), the population in Volusia County increased from 74,229 people in 1950 to 496,575 people in 2006. Volusia County is an extremely popular tourist location, averaging about eight million visitors a year over the last decade (Daytona Beach Area Convention and Visitors Bureau, personal communication). However, although the population grew 3.7% from 1990 to 2000, likely with a proportional increase in the number of people in Volusia County waters, this rise is not sufficient to explain the sharp increase in shark attacks in the last decade.

A key factor influencing the increased number of recorded Volusia County attacks has been the increased level of effort expended by ISAF staff and its resultant success in documenting these incidents. The U.S. Navy's initial support of the ISAF was terminated in 1968, and efforts to secure continued support from traditional funding sources were unsuccessful (Burgess 1990). The ISAF was transferred from Mote Marine Laboratory, where it had been sent for David Baldridge's (1974) seminal analysis, to the University of Rhode Island, where it resided at the National Undersea Safety Program until it was ultimately moved to the Florida Museum of Natural History, University of Florida, in 1988. As a result, there have been unequal amounts of effort dedicated to discovering and investigating shark attacks, resulting in incomplete coverage during the fiscally unsupported 1968 to 1988 period. Since transfer to the Florida Museum of Natural History, ISAF staffing has increased and important relationships developed with both beach safety and medical personnel in Volusia County, thereby ensuring increased efficiency in documenting incidents from the region.

As Volusia County's resident and tourist populations have grown, so has the presence of the media. The combination of increased human presence at the beach and a highly developed press has ensured that nearly all shark incidents in Volusia County are witnessed and recorded by ISAF. Consequently, the increase in number of recorded attacks in the past 20 years is the result of increased human use, an increased and more efficient effort by the current ISAF staff, and increased public and media attention to the phenomenon in the region.

The elevated number of shark attacks in Volusia County has become a *cause célèbre* that transcends local public safety concerns. Widespread international media coverage, ranging from newspaper stories to television documentaries, not only affects the regional tourism-based economy but also influences the global view of sharks, their threat to humans, and indirectly fishery management and conservation of sharks on both local and international scales.

Despite the relatively minor nature of injuries occurring in Volusia County attacks—many of which are no more severe than a dog bite—there is a lingering perception fueled by widespread and sometimes inflammatory media attention that equates these incidents with those involving larger and more dangerous species such as white, tiger, and bull sharks. Region- and user group-specific beach safety measures derived from scientific analyses should be implemented to decrease the risk to humans. Such measures ultimately serve a larger purpose by minimizing negative publicity, which influences the health of the local economy and the public's view of sharks. Further, outreach effort must continuously be maintained to educate the media and their target audiences regarding the roles human demographics and behavior play in influencing attack numbers, thereby keeping the phenomenon in perspective.

References

Arnold, EL, Thompson JR (1958) Offshore spawning of the striped mullet (*Mugil cephalus*) in the Gulf of Mexico. Copeia 1958:130–132.

Aubrey C, Snelson Jr F (2007) Life history of spinner shark in Florida waters. In McCandless CT, Kohler NE, Pratt HL Jr (eds) Shark nursery grounds of the Gulf of Mexico and the East Coast waters of the United States, American Fisheries Society, Bethesda, MD, pp 175–189.

Baldridge HD (1974) Shark attack: a program of data reduction and analysis. Contributions from the Mote Marine Laboratory 1:98.

Bullion J (1976) How tourist center reacts to shark attack publicity. In Seaman W (ed) Sharks and man: a perspective. Florida Sea Grant Program Publication. 10:9.

Burgess GH (1990) Shark attack and the International Shark Attack File. In Gruber SH (ed) Discovering sharks, American Littoral Society, Sandy Hook, pp 101–105.

Burgess GH, Callahan M (1996) Worldwide patterns of white shark attacks on humans. In Klimley P, Ainley D (eds) Great white sharks: the biology of *Carcharodon carcharias*. Academic Press, San Diego, CA, pp 457–470.

Cliff G (1991) Shark attacks on the South African coast between 1960 and 1990. S Afr J Sci 87:513–518.

Cliff G, Dudley SFJ (1991) Sharks caught in the protective gill nets off Natal, South Africa. IV. The bull shark. *Carcharhinus leucas* (Valenciennes). S Afr J Sci 10:253–270.

Cliff G, Dudley SFJ, Davis B (1989) Sharks caught in the protective gill nets off Natal, South Africa. II. *Carcharodon carcharias* (Linnaeus). S Afr J Sci 8:131–144.

Collins MR, Stender BW (1989) Larval striped mullet (*Mugil cephalus*) and white mullet (*Mugil curema*) off the southeastern United States. Bull Mar Sci 45(3):580–589.

Davis C (1995, July 1) Shark attacks diver. The Monterey County Herald, pp 1A, 4A.

Dodrill J (1977) Hook and line survey of the shark found within five hundred meters off shore along Melbourne Beach, Brevard County, Florida. Master's thesis, Florida Institute of Technology.

Dudley, S. F. J. (1997). A comparison of shark control programs of NSW and Queensland (Australia) and KwaZulu-Natal (South Africa). Ocean and Coastal Management 34(1):1–27.

Gaiser D (1976) Impact of sharks on tourism. In: Seaman W (ed) Sharks and man: a perspective, Florida Sea Grant Program Publication 10:8.

Gliwicz ZM (1986) A lunar cycle in zooplankton. Ecology 67:883–897.

Hazin F, Burgess GH, Carvalho F (2008) Shark attack outbreak of Recife, Pernambuco, Brazil: 1992–2006. Bull Mar Sci 89(2):199–212.

Katselis G, Koukou K, Dimitriou E, Koutsikopoulos C (2007) Short-term seaward fish migration in the Messolonghi-Etoliko Lagoons (western Greek coast) in relation to climatic variables and the lunar cycle. Estuar Coast Shelf Sci 73:571–582.

Lowry M, Williams D, Metti Y (2007) Lunar landings: relationship between lunar phase and catch rates for an Australian gamefish-tournament fishery. Fish Res 88:15–23.

McCosker JE, Lea RN (2006) White shark attacks upon humans in California and Oregon, 1993–2003. Proc Calif Acad Sci 57:479–501.

Myrberg A, Nelson DF (1991) The behavior of sharks: what have we learned? Underw Nat 20 (1991):92–100.

National Marine Fisheries Service (2006) Final consolidated Atlantic highly migratory species Fishery Management Plan. U.S. Department of Commerce, NOAA, NMFS, Silver Spring, MD, p 118.

Naylor E (1999) Marine animal behavior in relation to lunar phase. Earth Moon Planets 85–86:291–302.

Schultz L (1963) Attacks by sharks as related to the activities of man. In: Gilbert PW (ed) Sharks and survival, D.C. Heath and Company, Boston, pp 425–452.

Sheffield JB (2007) Image J, a useful tool for biological image processing and analysis. Microsc Microanal 13:200–201.

Simpfendorfer C (1992) Biology of tiger sharks (*Galeocerdo cuvier*) caught by the Queensland shark meshing program off Townville, Australia. Aust J Mar Freshw Res 43:33–43.

Trape S (2008) Shark attacks in Dakar and the Cap Vert Peninsula, Senegal: low incidence despite high occurrence of potentially dangerous species. PLoS ONE 3(1):e1495. doi:10.1371/journal.pone.0001495.

West G, Stevens J (2001) Archival tagging of school shark, *Galeorhinus galeus*, in Australia: initial results. Environ Biol Fish 60:283–298.

Wright AJ (2005) Lunar cycles and sperm whales (*Physeter macrocephalus*) strandings on the North Atlantic coastlines of the British Isles and eastern Canada. Mar Mamm Sci 2:145–149.

14

Shark Control: Methods, Efficacy, and Ecological Impact

Sheldon F.J. Dudley and Geremy Cliff

CONTENTS

14.1 Introduction

Shark attacks attract considerable publicity and have affected a number of regions around the world. In only a small number of locations, however, has public reaction led to socio-economic pressures on the local tourism and hospitality industry that resulted in physical intervention to reduce the risk of shark attack. One of the earliest examples was in the early 1930s off the beaches of Sydney, Australia, where shark attacks had led to a lack of confidence in bathing, and in 1934 the New South Wales Government set up a Shark Menace Advisory Committee to investigate the best methods of protecting bathers (Coppleson and Goadby 1988). It has long been acknowledged that the psychological impact of shark attack outweighs that of other forms of accident, such as those involving motor vehicles, despite the fact that the latter greatly outnumber the former (Davies 1961). In South Africa's infamous "Black December" of 1957 and during the subsequent Easter holidays, shark attacks had a substantial effect on the south coast of KwaZulu-Natal (KZN), a 160 km stretch of coastline between Durban and Port Edward, where the local economy is almost entirely dependent on tourism. Three attacks on bathers in waist-deep water occurred in 11 days spanning Christmas 1957 at the adjacent holiday resorts of Uvongo and Margate (Davies 1964; Wallett 1983). The first two incidents were fatal, while in the third a teenage girl lost her arm at the shoulder. The following April a young widow was fatally bitten at Uvongo. Wallett (1983) describes the consequences:

> Midday news broadcasts announced details of the incident to a nation grown sensitive to shark attack. During the twenty-four hour period following these broadcasts, the mass hysteria which mounted and spread through the resorts has not been equalled since. Thousands of cars streamed away from the resorts, blocking roads for kilometres at the single-carriage bridges characterizing the South Coast at the time. Hoteliers could not believe what was happening as the little resorts became virtual ghost towns within hours. South Coast residents were directly affected and during the ensuing months many hotels went bankrupt.

This is one of the strongest indications of how a series of shark attacks can affect a tourism industry. While shark attack led to the introduction of protective measures off the swimming beaches of places such as Sydney and KZN, however, this has not occurred in certain other parts of the world where beach tourism is important. In the United States "a single attack is front page news across the nation" (Randall 1963) but, although Florida has the highest number of shark attacks of any region in the world (22 per annum; 1990 to 2007; International Shark Attack File: http://www.flmnh.ufl.edu/fish/sharks/ISAF/ISAF.htm), no protective measures exist there or anywhere else on the mainland United States.

14.2 Methods for Reducing Risk of Shark Attack

The focus of this section is on methods of reducing risk of shark attack at bathing beaches, as distinct from protection of the individual. Reviews of devices developed for use by individuals are provided by Nelson (1983) and Johnson (1999). Intervention in response to shark attack has varied in its nature, often determined by the prevailing sea conditions. Types of response have included the use of shark exclusion methods (Section 14.2.1), shark detection methods (Section 14.2.2), and shark fishing devices (Section 14.2.3), as well as programs to educate the public about behavioral measures that would reduce the risk of attack.

14.2.1 Shark Exclusion Devices

Shark exclusion methods (barriers) provide complete protection but are suited to sheltered conditions where there is little or no surf. They do not involve the capture of sharks. A barrier was built off a Durban surf beach in 1907. The semicircular enclosure, 600 ft in diameter, was constructed of steel piles with vertical steel grids and designed to provide safe bathing as well as protection from shark attack. This sturdy structure was demolished in 1928 due to the damage it had suffered through wave action and corrosion and the high cost of maintenance (Davies 1964; Figure 14.1). In 1929 an enclosure was constructed at a surf beach at Coogee, New South Wales, and it was maintained until at least 1935 (Shark Menace Advisory Committee 1935). It was subjected to continual damage by heavy seas and was eventually dismantled (Coppleson and Goadby 1988). In the late 1950s a variety of physical barriers were erected at popular beaches on the KZN south coast but proved impractical to maintain in heavy surf and were short lived (Coppleson and Goadby 1988;

FIGURE 14.1
Shark-proof bathing enclosure on Durban beachfront, early 20th century. (Photo courtesy of Local History Museums Collection, Durban, South Africa.)

Davies 1964). Barrier nets are in still in use in some sheltered environments such as at Sydney Harbor (Figure 14.2), Queensland's Gold Coast marinas, and Hong Kong (Dudley and Gribble 1999).

Attempts to develop electrical repellents to protect bathing beaches (e.g., Smith 1974) have not yet resulted in a practical solution (Cliff and Dudley 1992b). In South Africa, the National Physical Research Laboratory of the Council for Scientific and Industrial Research began investigating electrical repellents in 1958 (Davies 1964) and continued, despite several hiatuses, until the late 1980s when attempts to evaluate the efficacy of a

FIGURE 14.2
A color version of this figure follows page 336. Shark-proof bathing enclosure at Manly, Sydney Harbor, 2008.

FIGURE 14.3
A color version of this figure follows page 336. Shark spotter overlooking Fish Hoek Beach, Cape Town, 2009.
(Photo by Darryl Colenbrander.)

third electrical cable that had been installed at Margate were abandoned. Although electrical repellents designed for individual use reduce substantially the probability of a white shark *Carcharodon carcharias* being able to access a bait (Smit and Peddemors 2003), the challenge remains to surround a bathing area, particularly at a surf beach, with a suitable electrical field. Given that species such as the bull shark *Carcharhinus leucas* are able to swim in very shallow water, the field would need to traverse the surf zone.

14.2.2 Surveillance Schemes

Lookout towers were used on certain NSW surf beaches, with the lookout sounding a bell or siren to warn bathers to leave the water, but their success was dependent on water that was both clear and sufficiently calm (Whitley 1963). These beaches are now protected by shark fishing devices. In 2004, subsequent to a number of shark attacks (Cliff 2006), the use of "spotters" as a form of shark detection was formally introduced at some of the beaches of Cape Town, South Africa (Oelofse and Kamp 2006). Most of these are beaches that have a mountainside close to the water's edge, thereby providing an elevated vantage point to the spotters, who alert bathers when a shark is sighted (Figure 14.3). No incidents have taken place at beaches where spotters have been operational. Nevertheless, there were concurrent changes in the patterns of recreational beach use, with fewer surfski paddlers and junior lifeguards (G. Oelofse, City of Cape Town, personal communication).

14.2.3 Shark Control (Fishing) Devices

Shark fishing using either anchored, large-mesh gillnets (Figure 14.4) or baited drumlines (Figure 14.5) has been effective in reducing the risk of shark attack at surf beaches but carries an environmental cost in terms of catches both of sharks and other animals (Dudley 2006). These shark control "fisheries" are either fully or heavily subsidized, with the justification being enhancement of coastal tourism (Dudley and Gribble 1999). "As early as 1929

FIGURE 14.4
A color version of this figure follows page 336. Shark net off Durban.

FIGURE 14.5
A color version of this figure follows page 336. Drumline with a shark net in the background off Cairns, Queensland.

it was suggested that 'regular and systematic netting affords a cheap and effective way of greatly minimising the shark peril' (New South Wales Shark Menace Committee 1929, 1). Later, it was suggested that the NSW public would be reassured by a demonstration of declining shark numbers (Shark Menace Advisory Committee 1935). These shark control measures achieve their function by reducing shark numbers locally, and thereby the probability of an encounter between a shark and a bather (Cliff and Dudley 1992b; Davies 1961, 1964; Paterson 1979; Last and Stevens 1994; Springer and Gilbert 1963). Despite education programs, however, some members of the beach-going public retain the misperception that shark nets function as a physical barrier.

The three major active shark control (or "meshing") programs are in New South Wales (NSW) and Queensland, Australia, and KZN, South Africa (Figure 14.6). In addition, large-mesh shark nets provide bather protection off the beaches of Dunedin, New Zealand, during the summer (Francis 1998). Shark nets of an unknown type were in use at the main bathing beach in Qingdao (Shantung Province, China) in 1982 (R.B. Clark, University of Newcastle, personal communication). Whether such nets still exist there or elsewhere in China is unknown (Dudley and Gribble 1999).

Six shark control programs with varying fishing intensity were conducted between 1959 and 1976 by the state government of Hawaii (Wetherbee, Lowe, and Crow 1994). Over $300,000 was spent in deploying near-shore longlines, which caught 4668 sharks. The reasons for the termination of the program are not clear, but Wetherbee, Lowe, and Crow (1994, 109) concluded that "the shark control programs and the associated reduction in coastal shark populations do not appear to have had a dramatic effect on the rate of shark attacks in Hawaii." In 2004 at Recife, Brazil, after shark attacks over the previous decade had had a substantial effect on tourism revenue, 20 drumlines and two 100-hook longlines were introduced to fish for potentially dangerous sharks (Dudley 2006).

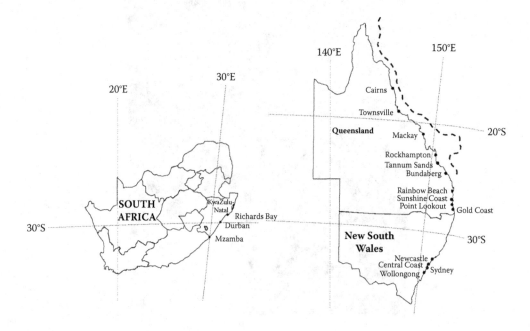

FIGURE 14.6
The netted regions of KwaZulu-Natal, Queensland, and New South Wales. (From Dudley SFJ (1997) Ocean Coast Manag 34:1–27. Reprinted with permission.)

Given that shark fishing systems achieve their effectiveness through reducing the probability of an encounter between a shark and a bather, rather than through the physical exclusion of sharks from bathing areas, they would not be expected to provide absolute protection. The three major programs have, however, reduced the rate of shark attack at protected beaches (number of attacks per year) by between 88% and 91% (Dudley 1997). Sceptics in KZN have suggested that the cessation of whaling in KZN in 1975 may have had more to do with the reduction in shark attacks than the introduction of shark nets. Large sharks were known to follow whale carcasses to a whaling station at Durban (Davies 1964). Shark attack at Durban's beaches ended with the introduction of nets in 1952, while whaling continued until 1975. Also, Davies (1964) pointed out that most shark attacks occurred in the summer months, whereas whaling took place primarily in winter.

The bull shark, the white shark, and the tiger shark *Galeocerdo cuvier* occur in New South Wales, Queensland, and KZN and have probably been responsible for most of the incidents in all three regions (Dudley 1997).

14.3 Population Effects of Shark Control Mechanisms

14.3.1 Sharks

The effect of shark safety programs on shark populations has been investigated in NSW (Reid and Krogh 1992), Queensland (Paterson 1990; Simpfendorfer 1993), and KZN (Dudley and Cliff 1993a; Dudley and Simpfendorfer 2006).

Dudley and Simpfendorfer (2006) assessed the population status of the 14 species that are caught most commonly in the protective shark nets off KZN, based on an analysis of catch rate and shark length data collected over the 26-year period 1978 to 2003. They also used a demographic modeling approach, together with the catch information, to assess the potential effect of the nets on the populations (Table 14.1). Catch rates (numbers of sharks caught per kilometer of net per year) of four species declined, nine species showed no trend, and one species, the tiger shark, showed an increase (Figure 14.7). Catch rate is assumed to be proportional to total population size off the KZN coast but, given the probable existence of localized stock depletion (see Section 14.3.1.1), this assumption may not apply to all species. The four species showing a decline were the bull shark, blacktip shark *C. limbatus*, scalloped hammerhead *Sphyrna lewini*, and great hammerhead *S. mokarran*.

The bull (Zambezi) shark is the species that was probably responsible for the majority of surf-zone attacks on bathers in KZN before shark nets were introduced (Davies 1964) and it is believed to be primarily through the capture of locally resident bull sharks that the nets have brought about a substantial reduction in risk of attack (Cliff and Dudley 1991a; Wallett 1983). Some uncertainty exists about the identification of species caught in the early years of shark netting but catch rates of bull sharks appear to have declined rapidly with the installation of nets at Durban in 1952 (Davies 1963) and at other beaches in the 1960s (Cliff and Dudley 1991a). Catches of bull sharks over the period 1978 to 2003 (45 year⁻¹) were not particularly high but were evidently sufficient to maintain pressure on the population, although Dudley and Simpfendorfer (2006) noted that additional factors may have played a role, such as an extended period of closure of Lake St. Lucia, the major nursery for the species in the region. Also, increased riverine sediment loads resulting from catchment mismanagement have been largely accountable for loss of feeding grounds

TABLE 14.1

Summary of Status of 14 Shark Species Commonly Caught by the KZN Beach Protection Program and Potential Effect of This Program on These Populations

Species	CPUE	Size	R	Potential Effect of Program	Reasoning for Program Effect
Carcharhinus amboinensis	Stable	**Declining**	—	Low	Very small catch
Carcharhinus brachyurus	Stable	Stable	—	Low	Most catch avoided by removing nets for sardine run
Carcharhinus brevipinna	Stable	Stable/**increasing**	Low	Low	Very high $F_{r=0}$
Carcharhinus leucas	**Declining**	Stable	Low	Medium	Low r, moderate catch, low $F_{r=0}$
Carcharhinus limbatus	**Declining**	**Declining**	Low	Low	Very high $F_{r=0}$
Carcharhinus obscurus	Stable	Stable	Very low	High	Very low r, large catch, low $F_{r=0}$
Carcharhinus plumbeus	Stable	Stable	Low	Low	Small catch
Carcharias taurus	Stable	Stable	Very low	High	Very low r, small population, large catch
Carcharodon carcharias	Stable	Stable/**declining**	Low	Medium	Low r, small population, moderate catch
Galeocerdo cuvier	**Increasing**	Stable	Medium	Low	Medium r, increasing CPUE
Isurus oxyrinchus	Stable	Stable	—	Low	Very small catch
Sphyrna lewini	**Declining**	Stable/**increasing**	Low	Medium	Low r, large catch, moderate $F_{r=0}$
Sphyrna mokarran	**Declining**	Stable	—	Low	Very small catch
Sphyrna zygaena	Stable	Stable	—	Low	Only take juveniles

Source: After Dudley SFJ, Simpfendorfer CA (2006) Mar Freshw Res 57:225–240.
Note: r, intrinsic rate of population increase (from Musick et al. 2000); high >0.5, medium 0.5–0.16, low 0.15–0.05, very low <0.05; $F_{r=0}$, fishing mortality rate at which population growth is zero.

available to fish species associated with southern African estuaries (Whitfield 1998), and this loss of habitat may have affected bull sharks either directly or indirectly. In terms of possible localized depletion, Cliff and Dudley (1991a) described an occasion where a new net installation, at Mbango on the KZN south coast, was established in February 1991. The installation was situated near the mouth of the large Umzimkulu River, about 5 km south of an existing installation at Umtentweni and 10 km north of another installation, at St. Michael's-on-Sea. In the first six weeks, 11 bull sharks, 10 of which were larger than 2 m PCL (precaudal length), were caught at Mbango. The typical annual catch of large bull sharks at each of Umtentweni and St. Michael's was less than one for the period 1978 to 1990. They concluded that an "isolated" group of large bull sharks had survived at Mbango despite more than two decades of netting to both north and south. Similarly, shark net installations are situated about 400 m from shore and their effect on bull sharks occupying reefs farther offshore appears to have been limited. St. Michael's and adjacent netted beaches are situated inshore of one such reef system, Protea Banks, and today bull

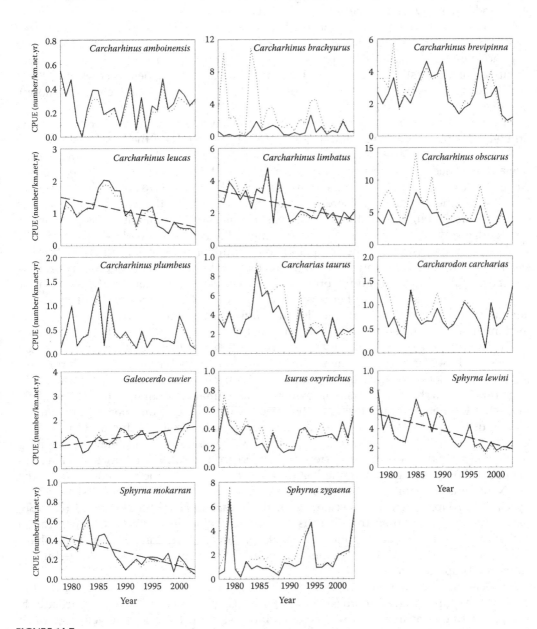

FIGURE 14.7
Annual catch per unit effort (CPUE) of 14 species of sharks caught in the KwaZulu-Natal beach protection program nets between 1978 and 2003 for all installations excluding Richards Bay, Mzamba, and the months of June and July (solid line). The effects of the "sardine run" in winter months are illustrated by the CPUE of all installations (except for Richards Bay) in all months (dotted lines). Where the slope of a linear regression fitted to the CPUE data was significant, a dashed line is shown. (From Dudley SFJ, Simpfendorfer CA (2006) Mar Freshw Res 57:225–240. Reprinted with permission.)

sharks are seldom caught in those nets, yet bull sharks are one of the species that, in the 1990s and 2000s, divers have visited Protea to see.

Although the shark nets have probably had a substantial effect on bull shark numbers in the vicinity of the netted beaches, there are no nets on the KZN coast north of Richards Bay and there is negligible fishing pressure on bull sharks in that region, which constitutes about 42% of the coastline (Figure 14.6). Bull sharks are using estuaries to the south of KZN as nursery grounds (S. Wintner, KwaZulu-Natal Sharks Board, personal communication); there are no nets in that region.

With regard to the other three species that showed a decline in catch rate, it seems unlikely that the nets have been a major contributory factor and they may instead have acted as a fisheries independent monitoring device. The blacktip shark is relatively productive and is fished commercially elsewhere in the world, including off the U.S. east coast. An annual catch of 104 sharks in the shark nets over the period 1978 to 2003 is not large in terms of commercial fishing. There is some concern, however, that the nursery grounds of this species, thought to be the shallow coastal waters of Mozambique (Bass, D'Aubrey, and Kistnasamy 1973), may be subject to artisanal gillnetting (Sousa, Marshall, and Smale 1997). Also, notwithstanding the nonquantitative nature of anecdotal information, this is a species encountered frequently by both the recreational shark diving community and the boat-based linefishing community off the KZN coast, suggesting that blacktip sharks may be more abundant slightly farther offshore than is indicated by the shark net data. The annual catch of largely immature scalloped hammerheads is only about 140 individuals, in comparison to the estimated 3000 neonates of this species taken annually as bycatch in the demersal prawn trawl fishery on the Tugela Banks, which lie off the netted region (Fennessey 1994). Finally, catch rates of the great hammerhead have declined alarmingly. KZN is at the southern limit of the range of this species and it has not been caught in large numbers in the shark nets (approximately 10 animals per annum); hence it is likely that what the nets are monitoring is the effect of IUU (illegal, unreported, and unregulated) fishing in the Mozambique Channel. A report to the IOTC (Indian Ocean Tuna Commission) in 2005 indicated that approximately 120 longline vessels may be operating illegally in coastal waters of the Western Indian Ocean. The KZN time series of catch rates of both scalloped and great hammerhead sharks contributed to the recent revision of the IUCN Red List categories of these species by the Species Survival Commission's Shark Specialist Group, to which the scientists of the KwaZulu-Natal Sharks Board (KZNSB) belong. It could be argued that there is an irony in belonging both to an organization that is responsible for catching sharks as well as to an organization responsible for shark conservation, but the KZNSB attempts to maximize the value of the data collected in the course of its operations.

A species upon which one might expect the KZN shark nets to have had a substantial effect is the spotted ragged-tooth (sand tiger) shark *Carcharias taurus*, given its particularly low productivity (intrinsic rate of population increase, $r = 0.043$) and relatively high catch in the nets (194 sharks per annum, 1978 to 2003), although 30% to 40% of those caught survive capture and are released. Dudley and Simpfendorfer (2006) found no evidence of a population decline, however, suggesting that the catch is not high relative to the total population size. Dicken, Booth, and Smale (2008) found that the population size for both juvenile and adult sharks over the previous decade appears to have remained constant. Reported sightings have declined at the Aliwal Shoal, however, a popular local dive site, and a Ph.D. student has investigated whether this may be due to diver disturbance (M. van Tienhoven, personal communication). Encouragingly, photographs taken on the Shoal in July 2008 showed large numbers of ragged-tooth sharks, with one group estimated at almost 100. In 2007 the KZNSB was invited to participate in a collaborative project with colleagues from the New South Wales Department of Primary Industries to investigate methods of enhancing recruitment in NSW waters, where the conservation status of the species is critical. Dicken, Booth, and Smale (2008), using mark-recapture techniques, estimated the adult ragged-tooth population on South Africa's east coast to be 16,700 animals, whereas the population for NSW and southern Queensland was estimated to be about 1000 animals only (Otway, Bradshaw, and Harcourt 2004).

Another species for which the potential effect of the nets might be expected to be high, given the relatively high catch (234 sharks per annum, 1978 to 2003), comprising immature and mature animals, and the very low intrinsic rate of population increase ($r = 0.041$), is the dusky shark *Carcharhinus obscurus*, yet Dudley and Simpfendorfer (2006) found no decline in catch rate or length. Again this may indicate that catches are not high relative to population size. There is a commercial and recreational fishery for young dusky sharks on the KZN coast, with catches in the commercial fishery estimated to be between 5000 and 6000 sharks in 2004 (Dudley, unpublished data) and the competition shore angling fishery catching approximately 2300 sharks per annum in 1977 to 2001, albeit that, since about 1997, much of this catch has been released (Pradervand, Mann, and Bellis 2007).

A species that has attained somewhat iconic status in shark conservation circles is the white shark. The white shark received legal protection in South Africa in 1991 but, unusually, this was a preemptive measure that was not based on evidence of declining stocks (Compagno 1991). An assessment of South African abundance indices (FAO 2004) concluded that catches in recent decades appear to be sustainable. There was no trend in catch rate in the KZN shark nets over the period 1978 to 2003 or in the mean or median size of males or the median size of females (Dudley and Simpfendorfer 2006). Mean size of females did show a decline, however, suggesting some change in this component of the population. Genetic research has indicated that females are more philopatric than males (Pardini et al. 2001) and hence may be more susceptible to fishing pressure in South African waters (Dudley and Simpfendorfer 2006), although more recent research using satellite tracking technology has demonstrated transoceanic movement of a philopatric female (Bonfil et al. 2005).

Catch rates of the tiger shark exhibited an increase over the period 1978 to 2003, suggesting a trend of increasing local abundance (Dudley and Simpfendorfer 2006). Similar increases have been reported in the Australian programs (Dudley 1997; Reid and Krogh 1992; Paterson 1990). Of the 14 species assessed by Dudley and Simpfendorfer (2006), *G. cuvier* was the only one with a moderate level of r (intrinsic rate of population increase), all the others being low or very low, and it may have a competitive advantage in a multi-species shark "fishery" (Wintner and Dudley 2000; see Section 14.4). This may be enhanced by the fact that tiger sharks have a high survival rate in the nets and approximately 42% of the catch is released (Dudley and Simpfendorfer 2006). Given that the tiger shark has been implicated in a number of attacks on humans (West 1993; Wetherbee et al. 1994) this apparent increase in abundance is of concern in terms of bather safety.

14.3.1.1 Localized Stock Depletion

Holden (1977) described initial trends in catch rates at Durban, where nets were first installed in 1952, and at Brighton Beach, where nets were first installed in 1961. These installations are only about 10 km apart, but trends in catch rates of various species appear to be independent. Although Holden (1977) speculated that the pattern might be due to territoriality, it more likely is due to individuals showing a high degree of site fidelity or philopatry, which is common even in wide-ranging species (see Hueter et al. 2005 for a review). Dudley and Simpfendorfer (2006) noted that such localized depletion would not only increase the effectiveness of beach protection programs but also the likelihood that observed declines in CPUE represent local changes rather than population-level changes in abundance (e.g., bull sharks, see Section 14.3.1). In addition, anecdotal support for the possibility that distribution of a number of species is heterogeneous is provided by reports from the commercial linefishing community in KZN. On numerous occasions in the past decade, line fishers have voiced their frustration at the loss of hooked teleosts to sharks, implying a high level of shark activity on fishing grounds on the continental shelf. Activity appears to be greatest in water of 21 to 30 m, to some 70 m deep, whereas the nets are set inshore of that, in water 12 to 14 m deep. In recent years, commercial linefishers targeting geelbek *Atractoscion aequidens* have reported losing over 200 hooks per night to sharks in season (B. Mann, personal communication), yet catches in the shark nets closest to the fishing grounds have shown no evidence of elevated shark activity.

14.3.2 Bycatch

All three shark control programs generate bycatch (Table 14.2; Dudley and Cliff 1993a; Krogh 1996; Gribble, McPherson, and Lane 1998a; Patterson 1979, 1990), principally dolphins, sea turtles, batoids, teleosts, and nonthreatening sharks. All live animals are released in KZN, but survival rates range widely from <5% in dolphins and teleosts to 34% in sea turtles and about 70% in batoids (Dudley and Cliff 1993a). An analysis of catch rates of each of the eight most commonly caught ray species in the KZN program, African angelshark *Squatina africana*, five species of turtle and bottlenose dolphin *Tursiops aduncus*, common dolphin *Delphinus capensis*, and humpback dolphin *Sousa plumbea* over the period 1981 to 2006 revealed a decline in only one species, the giant guitarfish *Rhynchobatus djiddensis* (KZNSB, unpublished data). Nesting populations of both turtle species within a monitored zone on the KZN coast are either increasing (the loggerhead *Caretta caretta*) or stable (the leatherback *Dermochelys coriacea*) (C. Mulqueeny, Ezemvelo KwaZulu-Natal Wildlife, personal communication). Common dolphin catches are small relative to the total stock size (Cockcroft and Peddemors 1990). Despite concern that mortalities of the bottlenose dolphin and the humpback dolphin may be unsustainable (Cockcroft 1990; Ross et al. 1989), catch rates of both species have shown no decline between 1981 and 2006 (KZNSB, unpublished data). Commercially manufactured 10 kHz "pingers" (sound-emitting devices) as well as reflectors of dolphin sonar, in the form of small, hollow floats, have been deployed in selected net installations in KZN as a means of alerting dolphins to the presence of the nets, with the objective of reducing catches of these animals. Evaluation of the effectiveness of these devices is ongoing. In NSW two such pingers are deployed on every net (V. M. Peddemors, NSW Department of Primary Industries, personal communication) and three in Queensland (T. Ham, Queensland Department of Primary Industries, personal communication).

In all three shark control programs entanglement of baleen whales has proved an increasing problem over the last decade, the most common species being the humpback whale *Megaptera novaeanglidae*, followed by the southern right whale *Eubalaena australis*. Whale populations in the southern hemisphere had been severely depleted by decades of commercial whaling but the total ban on hunting humpback whales implemented in 1963 and all other fin whales in 1976 (Best and Ross 1989) has been followed by recovery. In the case of humpback whales off South Africa's east coast, for example, population

TABLE 14.2

Average Annual Catches in the KZN Shark Nets, 2000–2007, Inclusive of Animals Other Than Large Predatory Sharks

Species	Common Name	No. Caught	Released (%)	Species	Common Name	No. Caught	Released (%)
Birds				*Gymnura natalensis*	Backwater butterflyray	22.8	63
Sula capensis	Cape gannet	4.6	0	*Himantura gerrardi*	Sharpnose stingray	0.5	25
Turtles				*Himantura uarnak*	Honeycomb stingray	0.8	67
Eretmochelys imbricata	Hawksbill	1.6	23	Myliobatoidei	Unidentified ray	2.5	35
Lepidochelys olivacea	Olive ridley	0.3	50	*Torpedo sinuspersici*	Marbled electric ray	0.3	50
Caretta caretta	Loggerhead	37.6	61	Torpediniformes	Electric ray	0.3	100
Chelonia mydas	Green	8.6	28	*Rhina ancylostoma*	Bowmouth guitarfish	0.9	29
Cheloniidae	Unidentified turtle	0.4	100	*Rhynchobatus djiddensis*	Giant guitarfish	43.5	68
Dermochelys coriacea	Leatherback	3.8	37				
				Teleosts			
Cetaceans				*Lichia amia*	Garrick	6.9	0
Sousa plumbea	Indo-Pacific humpbacked dolphin	6.6	2	*Scomberoides* spp.	Queenfish	0.9	0
Delphinus delphis	Common dolphin	24.4	1	*Caranx ignobilis*	Giant kingfish	0.1	0
Tursiops truncatus	Bottlenose dolphin	28.3	3	*Caranx heberi* (formerly *C. sem*)	Blacktip (yellowtail) kingfish	0.3	0
Stenella coeruloealba	Striped dolphin	0.1	0	*Seriola lalandi*	Giant yellowtail	0.3	0
Delphinidae	Unidentified dolphin	0.8	0	*Alectis indicus*	Indian mirrorfish	0.3	0
Balaenoptera acutorostrata	Minke whale	0.4	67	*Parastromateus niger*	Black pomfret	0.1	0
Megaptera novaeangliae	Humpback whale	3.3	77	*Thunnus albacares*	Yellowfin tuna	3.5	0
Eubalaena australis	Southern right whale	1.3	90	*Euthynnus affinis*	Eastern little tuna	1.1	0
Sharks				*Sarda orientalis*	Striped bonito	2.5	0
Rhizoprionodon acutus	Milk	6.5	6	Scombridae	Unidentified tuna, bonito	6.3	0
Mustelus mustelus	Smooth-hound	1.0	25	*Rachycentron canadum*	Prodigal son	0.4	0
Mustelus palumbes	Whitespotted smooth-hound	0.1	100	*Argyrosomus japonicus*	Kob	1.3	0

Continued

TABLE 14.2 (*Continued*)

Average Annual Catches in the KZN Shark Nets, 2000–2007, Inclusive of Animals Other Than Large Predatory Sharks

Species	Common Name	No. Caught	Released (%)	Species	Common Name	No. Caught	Released (%)
Mustelus sp.	Unidentified houndshark	0.9	86	*Atractoscion aequidens*	Geelbek	33.8	1
Rhincodon typus	Whale shark	1.1	56	*Makaira indica*	Black marlin	0.1	0
Squatina africana	African angel	15.5	36	*Tetrapturus audax*	Striped marlin	0.1	0
Carcharhinus sealei	Blackspot shark	1.1	22	*Istiophorus platypterus*	Sailfish	0.3	0
Rays				*Elops machnata*	Ladyfish (springer)	1.1	0
Aetobatus narinari	Spotted eagleray	20.9	75	*Epinephelus lanceolatus*	Brindlebass	0.5	0
Myliobatis aquila	Eagleray	2.4	53	*Sparodon durbanensis*	Brusher (white musselcracker)	0.8	0
Pteromylaeus bovinus	Bullray	40.1	40	*Tripterodon orbis*	Spadefish	0.1	0
Rhinoptera javanica	Flapnose ray	13.9	45	*Pomadasys kaakan*	Javelin grunter	0.1	0
Manta birostris	Manta	49.9	59	*Pomatomus saltatrix*	Shad (elf)	0.5	0
Mobula eregoodootenkee	Longhorned devilray	0.5	0	*Galeichthys feliceps*	White sea-catfish	0.1	0
Mobula kuhlii	Shorthorned devilray	18.1	8		Unidentified fish	0.8	0
Mobula thurstoni	Smoothtail (or bentfin) devilray	0.1	0				
Mobula spp.	Devilray	21.9	50				

growth rates of 7% to 10% per annum have been estimated (Best 2007 and references therein), and they are commonly seen during the winter months. The larger individuals invariably break their way out of a shark net but younger individuals, particularly new-born calves, lack the necessary strength and, without the timely intervention of trained release teams, soon die. The mortalities of these animals are low, numbering just over one per annum in KZN from 2003 to 2007 (KZNSB, unpublished data), and as such will have minimal impact on whale population recovery. In New South Wales the shark nets are out of the water during the winter months (May to August) when whale abundance is greatest. Since 1994, there have been no recorded southern right whale entanglements and one humpback whale mortality, where the animal may already have been dead before washing into the nets (V. M. Peddemors, NSW Department of Primary Industries, per-sonal communication). In Queensland there have been four humpback whale mortalities between 1992 and 2008 (T. Ham, Queensland Department of Primary Industries, personal communication). Such captures attract adverse publicity and are used by lobby groups to campaign for the closure of shark control programs, but are unlikely to have population-level impacts.

14.4 Ecosystem Effects

A number of studies have addressed the ecosystem effects of fishing for sharks and rays (e.g., Stevens et al. 2000; Bascompte, Melián, and Sala 2005; Myers et al. 2007; see Chapter 17 of this volume). In the context of the KZN shark control programs, the apparent stability of most shark populations (Dudley and Simpfendorfer 2006) may reduce the potential for such effects. Alternatively, the possibility that depletion of stocks of certain species, such as the bull shark, may be localized (see sections 14.3.1 and 14.3.1.1) may suggest that investigation of ecosystem effects should focus on the immediate vicinity of protected beaches. Identification of any such effects may be difficult to distinguish from the effects of other anthropogenic factors such as fishing. It is possible, however, that the increased catch rate of tiger sharks reported in the shark control programs of KZN, Queensland, and NSW (see Section 14.3.1) may represent species replacement, a phenomenon discussed by Stevens et al. (2000). Of all the species commonly caught in the KZNSB shark control program, the tiger shark has the highest intrinsic rate of population increase (Dudley and Simpfendorfer 2006). This relatively high productivity and a cosmopolitan diet potentially enable it to capitalize on any reduction in competition such as increased prey availability.

Van der Elst (1979) showed that from 1968 until the mid-1970s there was a sharp increase in the CPUE of small sharks, especially the juvenile dusky shark and the milk shark, *Rhizoprionodon acutus*, in KZN's shore-based sport fishery and he concluded that "reduction in numbers of the large inshore shark species along the Natal coast, as achieved by intensive gillnetting, has resulted in reduced predation on small sharks and is the cause of their proliferation" (p. 358). This conclusion has been given prominence by various authors (Castro 1987; Compagno 1987; Compagno, Ebert, and Smale 1989; Gruber and Manire 1989; Paterson 1990; Stevens et al. 2000). Van der Elst based his conclusion partly on qualitative feeding data for various shark species (Bass, D'Aubrey, and Kistnasamy 1973) and on the observation that captive bull and spotted ragged-tooth sharks fed preferentially on small sharks in comparison with teleosts. He predicted a 40-fold increase in the number of adult dusky sharks by the early 1980s. Dudley and Cliff (1993a), while agreeing with van der Elst that a reduction in predation had occurred, published a partial rebuttal of his work, making the points (1) that bull and ragged-tooth sharks formed only 21% of the annual shark catch of 14 species, each species having its own dietary characteristics, (2) that small sharks are less important in the diet of wild caught bull and ragged-tooth sharks than they appeared to be in captivity, and (3) that there is no evidence from catch rates in the shark nets that the 40-fold increase in adult dusky shark numbers has taken place. More recent research by Dudley and Simpfendorfer (2006) showed that there was no trend in the catch rate of dusky sharks in the shark nets between 1978 and 2003.

14.5 Catch Reduction Initiatives

Total effort in the KZN program, expressed as kilometers of netting, increased from 1.6 km (in a single installation) in 1952 to a peak of just over 44 km (distributed across 43 to 44 installations) in the late 1980s and early 1990s. A comparison of the programs of KZN, New South Wales, and Queensland concluded that fishing effort in KZN could be reduced

(Dudley 1997) to reduce captures while maintaining adequate bather safety. Between 1999 and 2004 the total length of netting was reduced by about one-third. In addition, prior to and during this same period, several installations, including that with the highest annual shark catch, were removed completely for socioeconomic or political reasons.

Since the late 1990s the KZNSB has, through timely removal of the shark nets ahead of approaching sardine shoals each winter, managed to bring about a major reduction in the catch of those sharks that accompany the run (Dudley and Cliff, in press). These measures collectively have reduced by about 50% the annual shark catch, from approximately 1200 in the 1990s to approximately 600 sharks in the subsequent decade. As an additional measure to reduce mortalities, since the late 1980s all sharks found alive have been released, with most being tagged first.

A drumline is a more selective shark fishing device than a shark net (Dudley et al. 1998; Gribble, McPherson, and Lane 1998b), and consists of a single hook suspended from a large, anchored float. Following an extended period of experimentation on the KZN coast, and with support of a 45-year dataset from off the beaches of Queensland, Australia, KZN introduced drumlines as a partial replacement for nets. The objective was to reduce bycatch while continuing to catch potentially dangerous sharks. In 2005 some netting at Richards Bay was replaced with three drumlines as a measure to reduce bycatch of humpback dolphins. In early 2007 approximately half of the nets in the 18 southernmost net installations were replaced with drumlines. After this process was complete the total quantity of gear on the KZN coast was 23 km of netting and 79 drumlines.

In the first year of fishing the total shark catch taken by the combination of nets and drumlines on the Hibiscus Coast (236 sharks, 16% released) was very similar to the mean annual catch taken in nets alone over the previous five years (227 sharks, SE 24.7, 20% released, Table 14.3). The species composition was different, however, with the mixed-gear installations catching more dusky sharks *Carcharhinus obscurus* and fewer blacktip, spinner *C. brevipinna*, and spotted ragged-tooth sharks. Mixed-gear installations had similar catch rates to net-only installations for dangerous species, including tiger, great white, and bull sharks. Mixed-gear installations caught only 17 sharks of small species, including 13 houndsharks *Mustelus* spp. that were all caught on drumlines. Finally, the mixed-gear installations caught 47 nonshark animals, 66% of which were released, compared with a mean annual catch of 106.0 (SE 11.7, 42% released) in nets alone in the preceding five years. Of the 47 animals, 43 were caught in nets; the four drumline catches were two turtles, both of which were released alive, and two sea catfishes of the family Arridae.

14.6 Shark Control Mechanisms as a Research Platform

In the mid-1960s, the first director of the KZNSB, Beulah Davis, began documenting the catches of animals in the shark nets on the KZN coast, although species identifications by contractors responsible for servicing the nets were not completely reliable. In addition, Davis recognized the potential value for research purposes of animals captured in the shark nets and the KZNSB appointed its first research officer in January 1972. In the late 1980s the number of staff scientists increased in order to produce scientific publications with a focus on the life history and catch trends of sharks captured in the nets (Allen and Cliff 2000; Cliff 1995; Cliff and Dudley 1991a, 1991b, 1992a; Cliff, Dudley, and Davis 1988, 1989, 1990; Cliff, Dudley, and Jury 1996; Cliff et al. 1996; de Bruyn et al. 2005; Dudley

TABLE 14.3

Effect of Drumline Introduction on Catch Composition, South Coast, KwaZulu-Natal*

Species	Catch in Nets Only[a]	Catch in Mixed Gear[b]
Carcharhinus limbatus	**24** (median PCL 163 cm)	**10** (median PCL 150 cm)
C. obscurus	**39** (170 cm)	**116** (130 cm)
C. plumbeus	4 (129 cm)	1 (143 cm)
Sphyrna lewini	25 (119 cm)	25 (102 cm)
S. zygaena	**33** (105 cm)	**9** (114 cm)
Carcharhinus brevipinna	**29** (165 cm)	**16** (176.5 cm)
Galeocerdo cuvier	24 (190 cm)	26 (184 cm)
Carcharhinus leucas	4 (170 cm)	3 (170 cm)
C. amboinensis	1 (132 cm)	0 (-)
Carcharias taurus	**21** (187 cm)	**8** (210 cm)
Carcharodon carcharias	9 (200 cm)	8 (215 cm)
Carcharhinus brachyurus	10 (178 cm)	3 (192 cm)
Isurus oxyrinchus	2 (208 cm)	**5** (210 cm)
Subtotal (large shark species)	**227** ± 24.7SE (20% released)	**236** (16% released)
Small shark species	**12** (27% released)	**17** (53 % released)
Nonshark bycatch	**108** ± 10.5SE (42% released)	**47** (66% released)

* Figures in bold type indicate a significant difference between the catch taken in nets only and that taken in the mixed gear (1-sample *t*-test, $p < 0.05$). PCL, precaudal length.

[a] 44 nets, annual mean number of animals February 20, 2002–February 19, 2007.

[b] 21 nets and 76 drumlines, number of animals February 20, 2007–February 19, 2008.

and Cliff 1993b; Dudley et al. 2005; Shelmerdine and Cliff 2006). Other foci have included shark age and growth (Allen and Wintner 2002; Wintner 2000; Wintner and Cliff 1996, 1999; Wintner and Dudley 2000; Wintner et al. 2002), matters pertaining directly to shark attack and the reduction in risk thereof (Cliff 1991; Cliff and Dudley 1992b; Davis, Cliff, and Dudley 1989; Dudley 1997; Dudley and Gribble 1999; Dudley et al. 1998; Smit and Peddemors 2003), population effects of the shark control program (Dudley and Cliff 1993a; Dudley and Simpfendorfer 2006), the annual influx of sardines *Sardinops sagax* into KZN waters and the relationship with shark distribution (Armstrong et al. 1991; Dudley and Cliff, in press), and the local distribution and abundance of whale sharks (Beckley et al. 1997; Cliff et al. 2007). Opportunistic research has included documenting the occurrence of plastic debris in shark stomachs (Cliff et al. 2002) and scavenging by white and tiger sharks on a whale carcass (Dudley et al. 2000).

The program in KZN differs from those in NSW and Queensland in two major aspects (Dudley and Gribble 1999). In KZN all live sharks, including potentially dangerous species, are released and tagged with external dart tags. Some of these sharks are also injected with oxytetracycline for aging studies. Recaptures of these tagged sharks have provided insight into movement patterns (e.g., Dudley et al. 2005), and recaptures of animals that have been injected with oxytetracycline have been used to validate growth estimates (Wintner and Cliff 1999; Wintner and Dudley 2000) and contributed to population estimates (Cliff et al. 1996; Dicken, Booth, and Smale 2008). In the Australian programs only those sharks considered harmless or endangered are released (Dudley and Gribble 1999). Second, sharks caught in the Australian programs are examined on board the vessel used to service the equipment, with the result that limited data are recorded from each capture. In NSW this situation is changing and the contractor who is responsible for maintaining the shark control gear is encouraged to return catches to fisheries officials (V.M Peddemors, NSW Department of Primary Industries, personal communication). In KZN all catches, apart from live animals and those in an advanced state of decomposition, are returned to shore for examination by the research department of the KZNSB. This has not only enabled the KZNSB to collect and publish data on the life histories of the common species but also to respond to a growing number of requests from colleagues in various parts of the world for access to biological material from sharks and other animals caught in the nets. Every effort is made to accede to these requests in order to maximize the research potential of this source of material (Table 14.4).

14.7 The Future of Shark Control Mechanisms

All three shark control programs are under pressure to reduce mortalities, particularly with regard to nontarget species (e.g., Gribble, McPherson, and Lane 1998a; Paxton 2003). Both Australian programs have been subjected to periodic governmental review. Queensland Ministerial Enquiries were conducted in 1992 and 1996, resulting in several initiatives to reduce bycatch. Both enquiries established, however, that any modifications must be made with due regard to the primary responsibility of protecting human life (Gribble, McPherson, and Lane 1998a). More recently, the Queensland government reaffirmed its commitment "to minimise the risk of shark attack on specific beaches while reducing inadvertent impacts on non target species" (B.H. Lane, Queensland Department of Primary Industries, personal communication in Dudley 2006, 104).

The New South Wales shark meshing program has been declared a "key threatening process" because of the capture of nontargeted species such as the grey nurse shark and threatened marine mammals (Anon 2005). The view of the NSW state government, however, is that public safety takes precedence over shark conservation (D.D. Reid, NSW Department of Primary Industries, personal communication in Dudley 2006). In the last decade the number of protected beaches in both Queensland and NSW has remained constant.

In KZN a commitment to reducing environmental impact is now listed in its defining legislation (KwaZulu-Natal Sharks Board Act 2008). Historically there was no such stipulation (Anti-Shark Measures Control Ordinance, Natal Ordinance No. 10, 1964). Further replacement of nets with drumlines is planned, provided sufficient suitable bait can be sourced. No new installation has been established since 1996. Any future requests for new protected beaches will be considered by a committee comprising representatives from sectors such as tourism, economic development, and environmental affairs, rather than by the KZNSB alone, as was the case in the past.

Although the risk of shark attack is extremely low, a questionnaire survey of beachgoers in Queensland, Australia, revealed that beach safety is a critical element of a beach experience and that beach safety is associated with protection from shark attack (Richards 1997). Similar findings for KZN have been obtained from focus groups (J. Seymour, Tourism KwaZulu-Natal, personal communication). During the public comment period for the new KZNSB Bill, very few comments questioned the need for bather protection but there was a call for an ongoing commitment to keeping environmental impact to a minimum. Environmental pressures aside, the future of all these programs is dependent on government willingness to fund them.

TABLE 14.4

Collaborative Research on Sharks Undertaken Using Material or Information Obtained from the KZN Shark Control Program

Subject	Species	Reference
Population size and status	*Carcharodon carcharias*	Cliff et al. (1996)
	Various species	Dudley and Simpfendorfer (2006)
Genetics and population structure	*Carcharias taurus*	Feldheim et al. (2007)
	Carcharodon carcharias	Pardini et al. (2001)
Occurrence, movement, and distribution	*Carcharias taurus*	Dicken et al. (2007)
	Carcharodon carcharias	Cliff et al. (2000)
	Rhincodon typus	Beckley et al. (1997)
	Carcharhinus obscurus	Hussey et al. (2009a)
Age and growth	*Carcharhinus leucas*	Wintner et al. (2002)
	Isurus oxyrinchus	Natanson et al. (2006)
	Cetorhinus maximus	Natanson et al. (2008)
Reproduction and shark condition	*Isurus oxyrinchus*	Mollet et al. (2000)
	Carcharhinus obscurus	Hussey et al. (2009b)
Shark predation and diet	Various species	Cockcroft et al. (1989)
	Sphyrna spp; *Galeocerdo cuvier*	Smale and Cliff (1998)
Liver lipids	Various species	Davidson and Cliff (2002)
	Various species	Davidson et al. (2007)
Ectoparasites	*Carcharodon carcharias*	Dippenaar et al. (2008)
	Galeocerdo cuvier	Dippenaar et al. (2009)
Sharks and environmental conditions	*Carcharodon carcharias*	Cliff et al. (1996)
	Various species	Vennemann et al. (2001)
Shark attack	Various species	Woolgar et al. (2001)
	Various species	Mokoena and Cliff (1997)
Shark control	Various species	Dudley and Gribble (1999)
	Various species	Dudley et al. (1998)

Following a series of shark attacks in Cape Town in the late 1990s and early 2000s, a workshop was held to discuss the problem and possible solutions (Nel and Peschak 2006). The use of shark control measures was rejected, with one of the reasons being the potential for high catches of white sharks, a fully protected species. Hence it was decided to expand the spotter program, which has proven successful (see Section 14.2.2; Oelofse and Kamp 2006).

Exclusion nets continue to be deployed at 32 Hong Kong beaches (G. Ng, Hong Kong Government 1823 Citizen's Easy Link, personal communication, in Dudley 2006) but they cannot withstand anything other than low surf action. Like shark control measures, these devices require regular maintenance, but they come at very little environmental cost.

The potential usefulness of electrical shark barriers remains uncertain, especially at surf beaches. The high costs of research and development have restricted attempts by the KZNSB to pursue this form of protection against shark attack.

References

Allen BR, Cliff G (2000) Sharks caught in the protective gill nets off KwaZulu-Natal, South Africa. IX. The spinner shark *Carcharhinus brevipinna* (Müller and Henle). S Afr J Mar Sci 22:199–215.

Allen BR, Wintner SP (2002) Age and growth of the spinner shark *Carcharhinus brevipinna* (Müller and Henle, 1839) off the KwaZulu-Natal coast, South Africa. S Afr J Mar Sci 24:1–8.

Anon (2005) Shark meshing. http://www.dpi.nsw.gov.au/fisheries/species-protection/species-conservation/what-current/key/shark-meshing.

Armstrong MJ, Chapman P, Dudley SFJ, Hampton I, Malan PE (1991) Occurrence and population structure of pilchard *Sardinops ocellatus*, round herring *Etrumeus whiteheadi* and anchovy *Engraulis capensis* off the east coast of South Africa. S Afr J Mar Sci 11:227–249.

Bascompte J, Melián CJ, Sala E (2005) Interaction strength combinations and the overfishing of a marine food web. Proc Natl Acad Sci USA 102:5443–5447.

Bass AJ, D'Aubrey JD, Kistnasamy N (1973) Sharks of the east coast of southern Africa. l. The genus *Carcharhinus* (Carcharhinidae). Investig Rep Oceanogr Res Inst, Durban 33:1–168.

Beckley LE, Cliff G, Smale MJ, Compagno LJV (1997) Recent strandings and sightings of whale sharks in South Africa. Environ Biol Fishes 50:343–348.

Best P (2007) Whales and dolphins of the Southern Africa Subregion. Cambridge University Press, New York.

Best PB, Ross GJB (1989) Whales and whaling. In Payne AIL, Crawford RJM (eds) Oceans of life off Southern Africa, Vlaeberg Publishers, Cape Town, pp 315–338.

Bonfil R, Meÿer M, Scholl MC, Johnson R, O'Brien S, Oosthuizen H, Swanson S, Kotze D, Paterson M (2005) Transoceanic migration, spatial dynamics, and population linkages of white sharks. Science 310:100–103.

Castro JI (1987) The position of sharks in marine biological communities: an overview. In Cook S (ed.) Sharks: an inquiry into biology, behavior, fisheries, and use, Oregon State University Extension Service, pp 11–17.

Cliff G (1991) Shark attacks on the South African coast between 1960 and 1990. S Afr J Sci 87:513–518.

Cliff G (1995) Sharks caught in the protective nets off Natal, South Africa. VIII. The great hammerhead shark *Sphyrna mokarran* (Rüppell). S Afr J Mar Sci 15:105–114.

Cliff G (2006) A review of shark attacks in False Bay and the Cape Peninsula between 1960 and 2005. In Nel DC, Peschak TP (eds) Finding a balance: white shark conservation and recreational safety in the inshore waters of Cape Town, South Africa. Proceedings of a specialist workshop. WWF S Afr Rep Ser 2006/Marine/001: 20–31.

Cliff G, Anderson-Reade MD, Aitken AP, Charter GE, Peddemors VM (2007) Aerial census of whale sharks (*Rhincodon typus*) on the northern KwaZulu-Natal coast, South Africa. Fish Res 84:41–46.

Cliff G, Compagno LJV, Smale MJ, van der Elst RP, Wintner SP (2000) First records of white sharks, *Carcharodon carcharias*, from Mauritius, Zanzibar, Madagascar and Kenya. S Afr J Sci 96:365–367.

Cliff G, Dudley SFJ (1991a) Sharks caught in the protective nets off Natal, South Africa. IV. The bull shark *Carcharhinus leucas* Valenciennes. S Afr J Mar Sci 10:253–270.

Cliff G, Dudley SFJ (1991b) Sharks caught in the protective nets off Natal, South Africa. V. Java shark *Carcharhinus amboinensis* (Müller and Henle). S Afr J Mar Sci 11:443–453.

Cliff G, Dudley SFJ (1992a) Sharks caught in the protective gill nets off Natal, South Africa. VI. The copper shark *Carcharhinus brachyurus* (Günther). In: Payne AIL, Brink KH, Mann KH, Hilborn R (eds) Benguela trophic functioning. S Afr J Mar Sci 12:663–674.

Cliff G, Dudley SFJ (1992b) Protection against shark attack in South Africa, 1952 to 1990. In: Pepperell JG (ed) Sharks: biology and fisheries. Aust J Mar Freshw Res 43:263–272.

Cliff G, Dudley SFJ, Davis B (1988) Sharks caught in the protective gill nets off Natal, South Africa. I. The sandbar shark *Carcharhinus plumbeus* (Nardo). S Afr J Mar Sci 7:255–265.

Cliff G, Dudley SFJ, Davis B (1989) Sharks caught in the protective gill nets off Natal, South Africa. II. The great white shark *Carcharodon carcharias* (Linnaeus). S Afr J Mar Sci 8:131–144.

Cliff G, Dudley SFJ, Davis B (1990) Sharks caught in the protective nets off Natal, South Africa. III. The shortfin mako shark *Isurus oxyrinchus* Rafinesque. S Afr J Mar Sci 9:119–126.

Cliff G, Dudley SFJ, Jury MR (1996) Catches of white sharks in KwaZulu-Natal, South Africa and environmental influences. In Klimley AP, Ainley DG (eds) Great white sharks: the biology of *Carcharodon carcharias*, Academic Press, San Diego, CA, pp 351–362.

Cliff G, Dudley SFJ, Ryan PG, Singleton N (2002) Large sharks and plastic debris in KwaZulu-Natal, South Africa. Mar Freshw Res 53:575–581.

Cliff G, Van der Elst RP, Govender A, Witthuhn TK, Bullen EM (1996) First estimates of mortality and population size of white sharks on the South African coast. In: Klimley AP, Ainley DG (eds) Great white sharks: the biology of *Carcharodon carcharias*, Academic Press, San Diego, CA, pp 393–400.

Cockcroft VG (1990) Dolphin catches in the Natal shark nets, 1980 to 1988. S Afr J Wildl Res 20:44–51.

Cockcroft VG, Cliff G, Ross GJB (1989) Shark predation on Indian Ocean bottlenose dolphins *Tursiops truncatus* off Natal, South Africa. S Afr J Zool 24:305–310.

Cockcroft VG, Peddemors VM (1990) Seasonal distribution and density of common dolphins *Delphinus delphis* off the south-east coast of southern Africa. S Afr J Mar Sci 9:371–377.

Compagno LJV (1987) Shark attack in South Africa. In: Stevens JD (ed) Sharks: an illustrated encyclopedic survey by international experts, Intercontinental Publishing Corporation, Hong Kong, pp 134–147.

Compagno LJV (1991) Government protection for the great white shark (*Carcharodon carcharias*) in South Africa. S Afr J Sci 87:284–285.

Compagno LJV, Ebert DA, Smale MJ (1989) Guide to the sharks and rays of Southern Africa. Struik, Cape Town.

Coppleson V(M), Goadby P (1988) Shark attack: how, why, when and where sharks attack humans. Angus and Robertson, Sydney.

Davidson B, Cliff G (2002) The liver lipid fatty acid profiles of seven Indian Ocean shark species. Fish Physiol Biochem 26:171–175.

Davidson BC, Rottanburg D, Prinz W, Cliff G (2007) The influence of shark liver oils on normal and transformed mammalian cells in culture. In Vivo 21:333–337.

Davies DH (1961) The shark problem. Bull S Afr Assoc Mar Biol Res 2:23–27.

Davies DH (1963) Shark attack and its relationship to temperature, beach patronage and the seasonal abundance of dangerous sharks. Investig Rep Oceanogr Res Inst, Durban 6:1–43.

Davies DH (1964) About sharks and shark attack. Shuter and Shooter, Pietermaritzburg.

Davis B, Cliff G, Dudley SFJ (1989) Sharks and the Natal Sharks Board. Part 2: The Natal Sharks Board. In: Payne AIL, Crawford RJM (eds) Oceans of life off Southern Africa, Vlaeberg Publishers, Cape Town, pp 209–213.

De Bruyn P, Dudley SFJ, Cliff G, Smale MJ (2005) Sharks caught in the protective gill nets off KwaZulu-Natal, South Africa. XI. The scalloped hammerhead shark *Sphyrna lewini* (Griffith and Smith). Afr J Mar Sci 27:517–528.

Dicken ML, Booth AJ, Smale MJ (2008) Estimates of juvenile and adult raggedtooth shark (*Carcharias taurus*) abundance along the east coast of South Africa. Can J Fish Aquat Sci 65:621–632.

Dicken ML, Booth AJ, Smale MJ, Cliff G (2007) Spatial and seasonal distribution patterns of juvenile and adult raggedtooth sharks (*Carcharias taurus*) tagged off the east coast of South Africa. Mar Freshw Res 58:127–134.

Dippenaar SM, van Tonder RC, Wintner SP (2009) Is there evidence of niche restriction in the spatial distribution of *Kroyeria dispar* Wilson, 1935, *K. papillipes* Wilson, 1932 and *Eudactylina pusilla* Cressey, 1967 (Copepoda: Siphonostomatoida) on the gill filaments of tiger sharks *Galeocerdo cuvier* off KwaZulu-Natal, South Africa? Hydrobiologia 619:89–101.

Dippenaar S, van Tonder R, Wintner S, Zungu P (2008) Spatial distribution of *Nemesis lamna* Risso 1826 (Copepoda: Siphonostomatoida: Eudactylinidae) on the gills of white sharks *Carcharodon carcharias* off KwaZulu-Natal, South Africa. Afr J Mar Sci 30:143–148.

Dudley SFJ (1997) A comparison of the shark control programs of New South Wales and Queensland (Australia) and KwaZulu-Natal (South Africa). Ocean Coast Manag 34:1–27.

Dudley SFJ (2006) International review of responses to shark attack. In: Nel DC, Peschak TP (eds) Finding a balance: white shark conservation and recreational safety in the inshore waters of Cape Town, South Africa. Proceedings of a specialist workshop. WWF S Afr Rep Ser 2006/ Marine/001: 95–108.

Dudley SFJ, Anderson-Reade MD, Thompson GS, McMullen PB (2000) Concurrent scavenging off a whale carcass by great white sharks, *Carcharodon carcharias*, and tiger sharks, *Galeocerdo cuvier*. Fish Bull 98:646–649.

Dudley SFJ, Cliff G (1993a) Some effects of shark nets in the Natal nearshore environment. Env Biol Fishes 36:243–255.

Dudley SFJ, Cliff G (1993b) Sharks caught in the protective gill nets off Natal, South Africa. VII. The blacktip shark *Carcharhinus limbatus* (Valenciennes). S Afr J Mar Sci 13:237–254.

Dudley SFJ, Cliff G (in press). The influence of the annual sardine run on catches of large sharks in the protective gill nets off KwaZulu-Natal, South Africa, and the occurrence of *Sardinops sagax* in shark diet. Afr J Mar Sci.

Dudley SFJ, Cliff G, Zungu MP, Smale MJ (2005) Sharks caught in the protective gill nets off KwaZulu-Natal, South Africa. X. The dusky shark *Carcharhinus obscurus* (Lesueur 1818). Afr J Mar Sci 27:107–127.

Dudley SFJ, Gribble NA (1999) Management of shark control programmes. In: Shotton R (ed) Case studies of the management of elasmobranch fishes, FAO Fish Tech Pap, Food and Agriculture Organization, Rome, pp 819–859.

Dudley SFJ, Haestier RC, Cox KR, Murray M (1998) Shark control: experimental fishing with baited drumlines. Mar Freshw Res 49:653–661.

Dudley SFJ, Simpfendorfer CA (2006) Population status of 14 shark species caught in the protective gillnets off KwaZulu-Natal beaches, South Africa, 1978–2003. Mar Freshw Res 57:225–240.

FAO (Food and Agriculture Organization of the United Nations) (2004) Report of the FAO Ad Hoc Expert Advisory Panel for the Assessment of Proposals to Amend Appendices I and II of CITES Concerning Commercially-Exploited Aquatic Species. FAO Fisheries Report 748. FAO, Rome.

Feldheim KA, Stow AJ, Ahonen H, Chapman DD, Shiviji M, Peddemors V, Wintner S (2007) Polymorphic microsatellite markers for studies of the conservation and reproductive genetics of imperilled sand tiger sharks (*Carcharias taurus*). Mol. Ecol Notes 7:1366–1368.

Fennessy ST (1994) Incidental capture of elasmobranchs by commercial prawn trawlers on the Tugela Bank, Natal, South Africa. S Afr J Mar Sci 14:287–296.

Francis MP (1998) New Zealand shark fisheries: development, size and management. Mar Freshw Res 49: 579–592.

Gribble NA, McPherson G, Lane B (1998a) Effect of the Queensland Shark Control Program on non-target species: whale, dugong, turtle and dolphin—a review. Mar Freshw Res 49:645–651.

Gribble NA, McPherson G, Lane B (1998b) Shark control: a comparison of meshing with set drumlines. In: Gribble NA, McPherson G, Lane B (eds) Shark and management conservation. Proceedings of the Sharks and Man Workshop of the Second World Fisheries Congress, Department of Primary Industries, Queensland, Brisbane, Australia, pp 98–124.

Gruber SH, Manire CA (1989) Challenge of the chondrichthyans. Chondros 1(1):1–3.

Holden MJ (1977) Elasmobranchs. In: Gulland JA (ed) Fish population dynamics, John Wiley & Sons, New York, pp 187–215.

Hueter RE, Heupel MR, Heist EJ, Keeney DB (2005) Evidence of philopatry in sharks and implications for the management of shark fisheries. J Northwest Atl Fish Sci 35:239–247.

Hussey NE, McCarthy ID, Dudley SFJ, Mann BQ (2009a). Movement patterns and growth rates of dusky sharks, *Carcharhinus obscurus* (Lesueur, 1818); a long-term tag and release study in South African waters. Mar Freshw Res 60:571–583.

Hussey NE, Cocks DT, Dudley SFJ, Wintner SP, McCarthy ID (2009b). The condition conundrum: application of multiple condition indices to the dusky shark, *Carcharhinus obscurus* (Lesueur, 1818). Mar Ecol Prog Ser 380:199–212.

Johnson CS (1999) Repelling sharks. In: Stevens JD (ed.) Sharks, Second Edition, Checkmark Books, New York, pp 196–205.

Krogh M (1996) Bycatch in the protective shark meshing programme off south-eastern New South Wales, Australia. Biol Conserv 77:219–226.

Last PR, Stevens JD (1994) Sharks and rays of Australia. CSIRO Division of Fisheries, Australia.

Mokoena TR, Cliff G (1997) Injury from bites. In: Cooper GJ, Dudley HAF, Gann DS, Little RA, Maynard RL (eds) Scientific Foundations of Trauma. Butterworth Heinemann, Oxford, pp. 347–364.

Mollet HF, Cliff G, Pratt HL Jr, Stevens JD (2000) Reproductive biology of the female shortfin mako shark *Isurus oxyrinchus* Rafinesque, 1810. Fish Bull 98:299–318.

Musick JA, Harbin MM, Berkeley SA, Burgess GH, Eklund AM, Findley L, Gilmore RG, Golden JT, Ha DS, Huntsman GR, McGovern JC, Parker SJ, Poss SG, Sala E, Schmidt TW, Sedbery GR, Weeks H, Wright SG et al. (2000) Marine, estuarine, and diadromous fish stocks at the risk of extinction in North America (exclusive of Pacific salmonids). Fisheries 25:6–30.

Myers RA, Baum JK, Shepherd TD, Powers SP, Peterson CH (2007) Cascading effects of the loss of apex predatory sharks from a coastal ocean. Science 315:1846–1850.

Natanson LJ, Kohler NE, Ardizzone D, Cailliet GM, Wintner SP, Mollet HF (2006) Validated age and growth estimates for the shortfin mako, *Isurus oxyrinchus*, in the North Atlantic Ocean. Env Biol Fishes 77:367–383.

Natanson LJ, Wintner SP, Johansson F, Piercy A, Campbell P, De Maddalena A, Gulak SJB, Human B, Fulgosi FC, Ebert DA, Hemida F, Mollen FH, Vanni S, Burgess GH, Compagno LJV, Wedderburn-Maxwell A (2008) Ontogenetic vertebral growth patterns in the basking shark *Cetorhinus maximus*. Mar Ecol Prog Ser 361:267–278.

Nel DC, Peschak TP (eds) Finding a balance: white shark conservation and recreational safety in the inshore waters of Cape Town, South Africa. Proceedings of a specialist workshop. WWF S Afr Rep Ser 2006/Marine/001.

Nelson DR (1983) Shark attack and repellency research: an overview. In: Zahuranec BJ (ed) Shark repellents from the sea: new perspectives, Westview Press, Boulder, CO, pp. 1–74.

New South Wales Shark Menace Committee (1929) Summary of New South Wales Menace Committee's Report. Government Printer, Sydney.

Oelofse G, Kamp Y (2006) Shark spotting as a water safety programme in Cape Town. In Nel DC, Peschak TP (eds) Finding a balance: white shark conservation and recreational safety in the inshore waters of Cape Town, South Africa. Proceedings of a specialist workshop. WWF S Afr Rep Ser 2006/Marine/001: 121–129.

Otway NM, Bradshaw CJA, Harcourt RG (2004) Estimating the rate of quasi-extinction of the Australian grey nurse shark (*Carcharias taurus*) population using deterministic age- and stage-classified models. Biol Cons 119:341–350.

Pardini AT, Jones CS, Noble LR, Kreiser B, Malcolm H, Bruce BD, Stevens JD, Cliff G, Scholl MC, Francis M, Duffy CAJ, Martin AP (2001) Sex-biased dispersal of great white sharks. Nature 412:139–140.

Paterson R (1979) Shark meshing takes a heavy toll of harmless marine animals. Aust Fish 38:17–23.

Paterson RA (1990) Effects of long-term anti-shark measures on target and non-target species in Queensland, Australia. Biol Conserv 52(2):147–159.

Paxton J (2003) Shark nets in the spotlight. Nat Aust Spring, 84.

Pradervand P, Mann BQ, Bellis, MF (2007) Long-term trends in the competitive shore fishery along the KwaZulu-Natal coast, South Africa. Afr Zool 42:216–236.

Randall JE (1963) Dangerous sharks of the Western Atlantic. In Gilbert PW (ed) Sharks and survival, D.C. Heath and Company, Boston, pp 339–361.

Reid DD, Krogh M (1992) Assessment of catches from protective shark meshing off New South Wales beaches between 1950 and 1990. In: J.G. Pepperell (ed) Sharks: biology and fisheries. Aust J Mar Freshw Res 43:283–296.

Richards F (1997) Beach visitation and health and safety awareness report. Department of Tourism, James Cook University, Townsville.

Ross GJB, Cockcroft VG, Melton DA, Butterworth DS (1989) Population estimates for bottlenose dolphins *Tursiops truncatus* in Natal and Transkei waters. S Afr J Mar Sci 8:119–129.

Shark Menace Advisory Committee (1935) Report on suggested methods of protecting bathers from shark attack. Legislative Assembly, New South Wales. Government Printer, Sydney.

Shelmerdine RL, Cliff G (2006) Sharks caught in the protective gill nets off KwaZulu-Natal, South Africa. XII. The African angel shark *Squatina africana* (Regan). Afr J Mar Sci 28:581–588.

Simpfendorfer C (1993) The Queensland Shark Meshing Program: analysis of the results from Townsville, north Queensland. In: Pepperell JG, West J, Woon P (eds) Shark conservation. Proceedings of an international workshop on the conservation of elasmobranchs, Sydney, Australia, February 24, 1991. Zoological Parks Board of NSW, Mosman, pp 71–85.

Smale MJ, Cliff G (1998) Cephalopods in the diets of four shark species (*Galeocerdo cuvier*, *Sphyrna lewini*, *S. zygaena* and *S. mokarran*) from KwaZulu-Natal, South Africa. In Payne AIL, Lipinski MR, Clarke MR, Roeleveld MAC (eds) Cephalopod biodiversity, ecology and evolution. S Afr J Mar Sci 20:241–253.

Smit CF, Peddemors V (2003) Estimating the probability of a shark attack when using an electric repellent. S Afr Stat J 37:59–78.

Smith ED (1974) Electro-physiology of the electrical shark-repellent. Trans S Afr Inst Electr Eng 68:166–180.

Sousa MI, Marshall NT, Smale MJ (1997) The shark trade in Mozambique. In: Marshall NT, Barnett R (eds) The trade in sharks and shark products in the western Indian and southeast Atlantic Oceans, TRAFFIC East/Southern Africa, Nairobi, pp 67–79.

Springer S, Gilbert PW (1963) Anti-shark measures. In: Gilbert PW (ed) Sharks and survival, D.C. Heath and Company, Boston, pp 465–476.

Stevens JD, Bonfil R, Dulvy NK, Walker PA (2000) The effects of fishing on sharks, rays, and chimaeras (chondrichthyans), and the implications for marine ecosystems. ICES J Mar Sci 57:476–494.

Van der Elst RP (1979) A proliferation of small sharks in the shore-based Natal sport fishery. Env Biol Fishes 4:349–362.

Vennemann TW, Hegner E, Cliff G, Benz GW (2001) Isotopic composition of recent shark teeth as a proxy for environmental conditions. Geochim Cosmochim Acta 65:1583–1599.

Wallett TS (1973) Analysis of shark meshing returns off the Natal coast. M Sc thesis, University of Natal, Durban.

Wallett T (1983) Shark attack in southern African waters and treatment of victims, C. Struik Publishers, Cape Town.

West JG (1993). The Australian Shark Attack File with notes on preliminary analysis of data from Australian waters. In: Pepperell JG, West J, Woon P (eds) Shark conservation. Proceedings of an international workshop on the conservation of elasmobranchs, Sydney, Australia, February 24, 1991. Zoological Parks Board of NSW, Mosman, pp 93–101.

Wetherbee BM, Lowe CG, Crow GL (1994) A review of shark control in Hawaii with recommendations for future research. Pac Sci 48:95–115.

Whitfield AK (1998) Biology and ecology of fishes in southern African estuaries, J.L.B. Smith Institute of Ichthyology, Grahamstown.

Whitley GP (1963) Shark attacks in Australia. In: Gilbert PW (ed) Sharks and survival, D.C. Heath and Company, Boston, pp 329–338.

Wintner SP (2000) Preliminary study of vertebral growth rings in the whale shark, *Rhincodon typus*, from the east coast of South Africa. Env Biol Fishes 59:441–451.

Wintner SP, Cliff G (1996) Age and growth determination of the blacktip shark, *Carcharhinus limbatus*, from the east coast of South Africa. Fish Bull 94:135–144.

Wintner SP, Cliff G (1999) Age and growth determination of the white shark, *Carcharodon carcharias*, from the east coast of South Africa. Fish Bull 97:153–169.

Wintner SP, Dudley SFJ (2000) Age and growth estimates for the tiger shark, *Galeocerdo cuvier*, from the east coast of South Africa. Mar Freshw Res 51:43–53.

Wintner SP, Dudley SFJ, Kistnasamy N, Everett B (2002) Age and growth estimates for the Zambezi shark, *Carcharhinus leucas*, from the east coast of South Africa. Mar Freshw Res 53:557–566

Woolgar JD, Cliff G, Nair R, Hafez H, Robbs JV (2001) Shark attack: review of 86 consecutive cases. J Trauma Inj Infect Crit Care 50:887–891.

15

DNA Forensic Applications in Shark Management and Conservation

Mahmood S. Shivji

CONTENTS

15.1 Introduction

Declining shark populations coupled with extensive, global-scale industrial fishing to satisfy the market demand for shark fins have resulted in widespread recognition that sustainable fishing for these species is unlikely under current exploitation levels (FAO 2000; Musick et al. 2000; Baum et al. 2003; P. Ward and Myers 2005; Dulvy et al. 2008). Furthermore, there are concerns about the impacts of such large-scale apex predator removal on marine ecosystem function and stability (Stevens et al. 2000; Bascompte, Melián, and Sala 2005; P. Ward and Myers 2005; Ferretti et al. 2008; Heithaus et al. 2008). These issues in the face of increasing shark fin trade (growing by more than 5% per year; Clarke 2004) have resulted in a consensus that prompt implementation of substantially improved conservation and management measures is essential if shark populations are to recover and be fished sustainably (FAO 1998, 2000; Musick et al. 2000; NMFS 1999, 2001; Baum et al. 2003; Dulvy et al. 2008).

In light of these concerns, the U.N. Food and Agriculture Organization (FAO) and at least seven nations have developed International and National, respectively, Plans of Action for the Conservation and Management of Sharks (FAO 1998, 2000; NMFS 2001). A major, recurring recommendation in these plans is the collection of data on shark landings and trade on a species and also, whenever possible, on a geographic population- or stock-specific basis. This recommendation has its roots in (1) the historical absence of reliable data on shark catch and trade on a species- and population-specific basis, making robust stock status assessments nearly impossible in most cases, and (2) the realization that individual shark species can differ substantially in their life history characteristics and therefore their susceptibility to exploitation (Smith, Au, and Show 1998; Cortés 2002). There is also the recognition that recording landings of shark catch for expediency in grossly imprecise categories (e.g., "sharks," or "small vs. large" sharks), or even in the slightly more precise categories (i.e., "large coastal" vs. "small coastal" vs. "pelagic sharks"; NMFS 1999), prevents the early recognition of fundamental changes in community structure and dangerous reductions in threatened populations and individual species, and precludes effective management intervention.

Implementing the sensible and overarching recommendation that shark catches be identified to the species level and their geographic origin recorded, however, is fraught with difficulties. Many fished shark species have widespread (often circumglobal) distributions, are highly migratory, often crossing national management boundaries, and are exploited by multinational fisheries. To be effective, the management and conservation of such sharks needs to be considered in an internationally cooperative context (FAO 2000), a socio-politically complicated undertaking. Compounding the problems of international shark management are more basic difficulties in identifying many exploited shark species, either as whole animals or as dismembered body parts (e.g., "logs," carcasses, detached fins, meat), that are typically landed or found in international trade (NMFS 1999; FAO 2000; Shivji et al. 2002). Even though several shark species are protected from landings and trade by various well-meaning national and international regulations, enforcing these regulations has often been impractical due to the difficulties with species identification. Furthermore, although progress is being made (see Section 15.4) there is currently insufficient detailed information on population genetic structure for most shark species to allow reliable tracking of the population origin of landed and traded shark body parts. Given the poor status of many shark populations and the international management scope of the problem, there is need for widely applicable, efficient, and accurate forensic methods to track the species and population origin of landed and traded shark body parts for comprehensive management and conservation efforts.

15.2 DNA Forensic Approaches

Genetic methods have proven powerful and accurate for species-level identification of organism body parts and even processed animal products in diverse taxa. The predominant DNA-based approaches to marine vertebrate species identification for ecological, conservation, management, and law enforcement purposes have been either polymerase chain reaction (PCR) amplification of a specific mitochondrial or nuclear genome region followed by restriction enzyme analysis to generate a form of DNA fingerprint (i.e., the PCR-RFLP approach; e.g., Innes, Grewe, and Ward 1998; Heist and Gold 1999; Asensio et al. 2000; Gharrett, Gray, and Heifetz 2001; Vandersea et al. 2008), or a combination of direct sequence comparison and clustering approaches to determine similarities and group affiliations between DNA sequences from samples of unknown identity and reference species (i.e., the phylogenetic approach; e.g., Baker and Palumbi 1994; Malik et al. 1997; Palumbi and Cipriano 1998; Dizon et al. 2000; Roman and Bowen 2000, R.D. Ward et al. 2008; Wong and Hanner 2008). The above approaches are generally effective for species identification, except in a minority of cases (depending on the species group) where a DNA fingerprint or sequence derived from the sample is unavailable in the reference species database it is being compared to.

Despite their overall utility for species diagnosis, the widespread adoption of these DNA forensic approaches for routine marine fish conservation and management practice, especially in developing countries, remains in its infancy. Even in developed countries, use of these DNA methods is still largely limited to specialized situations usually related to law enforcement cases and ecological and taxonomic research. One important reason for the lack of routine implementation in practical fisheries management contexts is that these methods are still relatively time consuming and expensive. Both DNA approaches described above require further analysis of the amplified products after the PCR step by either restriction endonuclease analysis or DNA sequencing, thereby involving more reagent and personnel time costs. Although it has been legitimately argued that the cost of DNA sequencing continues to decrease drastically, the infrastructure costs of the technology (automated sequencers and maintenance) and personnel training required are still high enough to be a deterrent for routine use in monitoring the species identity of landed and traded fishery products.

The above caveats notwithstanding, DNA technology has enormous promise for aiding the management and conservation of the world's fisheries, including sharks. Here, I review the DNA-based forensic tools currently being used to determine the species identity of shark body parts, and their application in shark fisheries trade characterization and law enforcement. I also provide an overview of analytical approaches that show great promise for improving shark management by allowing tracking of the population origin of shark products. I conclude by sharing some perspectives on the beneficial impact that routine implementation of these DNA tools can have on helping the recovery of shark species and populations.

15.2.1 Multiplex PCR Approach to Shark Species Identification

The PCR has become the basis for rapidly detecting nucleic acids of specific organisms, ranging from viral and microbial pathogens to animal body parts. The streamlined and comparatively inexpensive PCR-based approach developed for shark species identification relies on the experimentally controllable, exquisite specificity of synthetic primer annealing to unique shark DNA sequences in a PCR reaction. The basic approach as applied to sharks utilizes species-specific primers, and is detailed in several publications (Shivji et al. 2002; Chapman et al. 2003; Abercrombie, Clarke, and Shivji 2005; Magnussen et al. 2007) and overviewed here.

DNA identification methods for most marine vertebrates, including sharks rely on mitochondrial genome sequence differences as the target for species differentiation (Heist and Gold 1999; Greig et al. 2005; R.D. Ward et al. 2008; and see the next section). In contrast, the multiplex PCR (i.e., a PCR reaction involving more than two primers simultaneously) forensic approach developed for sharks utilizes a noncoding region of the nuclear genome, the ribosomal internal transcribed spacer 2 (ITS2). This locus has proven exceptionally useful for developing species-specific PCR primers to distinguish shark species for several reasons:

1. The DNA sequence of the ITS2 is highly conserved within shark species, showing very little population differentiation even within species that have global distributions.

2. The ITS2 sequence is sufficiently variable even between closely related species (e.g., members of the speciose and highly exploited genus *Carcharhinus*) to provide species-diagnostic nucleotide substitutions.

3. This locus occurs in high copy number in vertebrate genomes, providing an abundant target for amplification even from very limited amounts of shark tissue.

4. The ITS2 is bi-parentally inherited, circumventing potential identification errors that could result from maternally inherited mitochondrial markers in the event of hybridization between shark species.

The multiplex PCR approach for sharks utilizes species-specific PCR primers designed to anneal to regions of fixed nucleotide differences (one or more bases), that occur among shark species, in the ITS2 (Shivji et al. 2002). The species-specific primer(s) are used in combination with two shark "universal" primers in a single tube PCR reaction. The universal primers are designed to amplify the entire ITS2 region from all sharks, and are both included in the multiplex PCR reaction to provide greater accuracy to the technique by preventing false negative results, that is, the absence of any amplification, which can occur for several technical reasons, from being incorrectly interpreted as the absence of the target species (Shivji et al. 2002). Further streamlining of the multiplex PCR diagnostic process is achieved by combining several species-specific primers together, allowing efficient discrimination among several shark species simultaneously with just a single-tube reaction (Figure 15.1).

Advantages of this forensic approach are (1) it requires just three steps: DNA extraction from the unknown sample, the multiplex PCR reaction, and running an agarose gel of the reaction products to derive the identity of the sample; (2) it is easy to perform and the results simple to interpret; and (3) the main equipment necessary is limited to a thermal cycler and gel electrophoresis and visualization apparatus. These are relatively low-cost and very low-maintenance pieces of equipment that are standard in just about every genetics laboratory worldwide. Furthermore, this equipment is also compact and sturdy enough to be easily portable—that is, deployable in a mobile van-lab or on board a fisheries management or law-enforcement vessel. Limitations of this approach are mainly that the suite of species-specific primers currently available is limited to about 28 species,

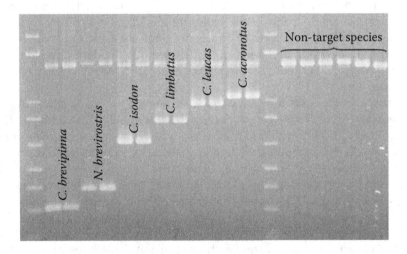

FIGURE 15.1

Agarose gel showing the results of an eight primer multiplex PCR reaction. Diagnostic sized amplicons for each species (two individuals each) are indicated. Species-specific primers for the six species shown plus two universal primers were multiplexed in a single PCR reaction. (Adapted from Nielsen J (2004) Master's thesis, Nova Southeastern University, Florida.)

although more primers are in development, and that unlike DNA barcode sequencing (see below) no species information is available from the multiplex primer assay when the target species is absent.

15.2.2 DNA Barcoding for Shark Species Identification

DNA barcoding is a concept that relies on using the nucleotide sequence of a short segment of a mitochondrial gene to distinguish animal species (Hebert et al. 2003; Stoeckle and Hebert 2008). The underlying premise is that the sequence of an ~650 base pair portion of the mitochondrial cytochrome oxidase I gene (COI) is much more conserved within than between species, thus allowing species to be distinguished based on fixed nucleotide differences that occur among them. DNA barcoding is showing considerable promise in being able to distinguish a diversity of animal species, and this efficacy has led to the establishment of a large, international effort to obtain COI reference barcodes for all animals. The barcoding movement is growing rapidly under the oversight of the Consortium for the Barcode of Life (CBOL; www.barcoding.si.edu/), and includes 150 institutions from 45 countries (Stoeckle and Hebert 2008). A major part of the Barcode of Life initiative is also making the reference barcodes for each species and associated species identification tools available via a public database with an easy-to-use web interface (see Barcode of Life Data Systems [BOLD]; www.barcodinglife.org). The BOLD database is a searchable library of COI sequences from reference species that can be queried with a COI sequence from a sample of unknown species origin. Species identification may be achieved by comparing the barcode sequence from the unknown specimen to the library of reference barcode sequences in the database (Ratnasingham and Hebert 2007).

The most extensive investigations thus far of the efficacy of COI barcoding for identification of sharks have been conducted by R.D. Ward et al. (2005; 2007; 2008) and Wong, Shivji, and Hanner (2009). The findings suggest that COI barcodes may be very promising in distinguishing most shark species and chondricthyians generally. Ward et al. (2008) compared barcodes for 210 chondricthyan species from 36 families (including 123 shark species), and found that 99% of these species could be discriminated based on their COI sequences. Clear separation of only very few shark species that likely diverged recently (e.g., a few species in the genus *Carcharhinus*) or may be potentially cryptic species (e.g., in the genus *Mustelus* and *Orectolobus*) proved uncertain based on their COI barcode (Ward et al. 2008; Holmes, Steinke, and Ward 2009).

Despite its promise for shark species discrimination, the robustness of COI barcodes for universal application awaits further confirmation. The investigations thus far have examined shark species from restricted parts of their distribution (mainly Australasia) and relatively small numbers of individuals per species. This limited survey is likely to have produced an underestimate of intraspecific variation (Dasmahapatra and Mallet 2006). With many of the commercially exploited shark species possessing extensive and often global distributions, the extent of geographic intraspecific sequence variation in the COI barcode using at least 10 individuals per species (Stoeckle and Hebert 2008) and preferably more for globally distributed species, needs to be examined to ensure that such variation does not exceed the interspecific variation that accurate species discrimination hinges on (Moritz and Cicero 2004).

Reliance on an exclusively mitochondrial genome barcode for species discrimination is not without controversy (Moritz and Cicero 2004; Rubinoff 2006; Rubinoff, Cameron, and Will 2006), some of which may be applicable to sharks: The mitochondrial genome is largely maternally inherited in vertebrates, sections of mitochondrial DNA have been found to reside as psuedogenes with slightly different sequences in the nuclear genome of many taxa (Bensasson et al. 2001; Buhay 2009), and mitochondria not uncommonly show heteroplasmy (i.e., multiple copies with slightly different sequence variants; Kmiec, Woloszynska, and Janska 2006). These factors may contribute to some as yet unknown level of identification errors. For example, hybrids will be identified as the maternal species contributor of the mitochondrial genome. There is some evidence for hybridization between shark species (laboratory of M. Shivji, unpublished data), although this may not be a frequent occurrence. The occurrence of mitochondrial COI-derived psuedogenes in shark genomes has not been investigated to date, so its potential impact on barcode accuracy is unknown. Heteroplasmy has been documented in the white shark, *Carcharodon carcharias*, mitochondrial genome control region locus (Pardini et al. 2001) but not specifically investigated for COI.

The above issues notwithstanding, the DNA barcoding approach shows considerable promise for discriminating shark species based on the pilot studies conducted thus far. The number of shark species with DNA barcodes is growing rapidly and can be tracked on the FISH Barcode of Life database (FISHBOL; http://www.fishbol.org/progress.php). The major current caveat with its application as a tool for routine shark fisheries management and conservation worldwide, however, is that barcoding is still dependent on DNA sequencing, a not trivial endeavor in terms of infrastructure costs.

15.3 Applying DNA Forensics to Study the Shark Fin Trade and Its Impacts

The demand for shark fins in the global market is arguably the major driver of shark exploitation worldwide, with an estimated minimum market value of US$400 to $550 million per year (Clarke et al. 2006a; Clarke, Milner-Gulland, and Bjørndal 2007). This lucrative trade has contributed to fins being sourced from worldwide origins to the main trading centers in Asia (Clarke 2004). Despite this demand and the aforementioned concerns that many shark populations appear to have declined or are declining at rates incongruent with sustainable fisheries, most shark fisheries still remain unregulated. Even for those countries practicing some regulation, the incidence of illegal and mis- or unreported shark catch is believed substantial (Clarke et al. 2006b; Dulvy et al. 2008). These factors have introduced tremendous uncertainty in assessments of the true level of shark mortality. Furthermore, the near absence of shark landings data by species has made stock assessments and dependent management efforts imprecise (ICCAT 2005; Clarke, Milner-Gulland, and Bjørndal 2007).

With the absence of reliable landings data for most shark fisheries globally, surveying the main shark fin trade centers is an alternative approach to derive estimates of the number of sharks, by species, killed worldwide to supply the fin trade, and understand fin market dynamics generally to assess its impacts on shark populations (Clarke et al. 2006a, 2006b). Shelley Clarke has performed a series of pioneering studies in this regard, with a focus on the Hong Kong market, the world's largest fin trading center (Clarke 2004). Earlier surveys of the Hong Kong fin market (e.g., Parry-Jones 1996; Vannuccini 1999; Fong and Anderson 2000) provided a very general characterization of the trade. However, the utility of information from these earlier surveys for assessing the impacts of the fin trade on shark species has been limited by insufficient knowledge about the precise relationship between the many trader Chinese market names for the fins and actual shark taxa, and the lack of statistically robust market survey designs to avoid incorrect inferences.

Trader Chinese market names (e.g., *ya jian*) for the fins appear to be based primarily on the quality of fin needles and other distinguishing features (e.g., fin size and position)—factors that control the price of the dried fins (Clarke, McAllister, and Michielsens 2004). Randomized survey sampling of the market is made difficult by the sensitive and covert nature of the trade. To determine which shark species contributed the highest weight to the Hong Kong market, Clarke et al. (2006a) designed a survey strategy that would obtain statistically sufficient number of fins sampled across traders, trade names (categories), and geographic source regions, under the conditions of constrained market access. They focused their survey on 11 trade categories that, based on preceding trader interviews, contained the most common fin types. DNA forensic testing using the multiplex PCR method outlined above provided the level of concordance between these market categories and scientific taxa. This concordance information was then incorporated into a stochastic model used to estimate fin weights in each category based on trader auction records (Clarke, McAllister, and Michielsens 2004) to derive biomass estimates of the market fins by species.

By taking a multidisciplinary approach combining DNA forensics with statistically based sampling design and stochastic modeling of trader auction records, Clarke et al. (2006a) provided the first quantitative view of the composition and weight of dried fins by shark species in the world's largest fin-trading center. The 11 Chinese trade categories containing the most common fin types in the market comprised approximately 40% of the auctioned fins and were surprisingly made up of only 14 shark species (Table 15.1). Fins from blue sharks (*Prionace glauca*) made up by far the largest proportion of the fins (~17%), with other species/species groups comprising much lower proportions (~0.1% to 4.4%) of the trade. It is not clear if the species composition and proportions simply reflect global fisheries catch abundance of particular species or a market preference for those types of fins. The fact that the highest proportions by single species belong to three

TABLE 15.1

Means and 95% Probability Intervals for Estimated Proportion (Percent) of the Hong Kong Shark Fin Trade Composed of Each Species/Species Group

Market Category	Species	Mean[a]	Lower 95% P.I.	Upper 95% P.I.
Ya jian	*P. glauca*	17.3	15.5	19.1
Qing lian	*I. oxyrinchus*	2.7	2.3	3.1
Wu yang	*C. falciformis*	3.5	3.1	4.0
Hai hu	*C. obscurus*	1.4	1.2	1.7
Bai qing	*C. plumbeus*	2.4	1.9	2.8
Ruan sha	*G. cuvier*	0.13	0.08	0.19
Chun chi	*S. zygaena* or *S. lewini*	4.4	3.9	4.9
Gu pian	*S. mokarran*	1.5	1.2	1.8
Wu gu	*Alopias* spp.	2.3	2.0	2.7
Sha qing	*C. leucas*	2.2	1.8	2.6
Liu qiu	*C. longimanus*	1.8	1.6	2.1

Source: Adapted from Clarke SC, Magnussen JE, Abercrombie DL, McAllister M, and Shivji MS (2006a) Cons Biol 20:201–211.

[a] The possible presence of additional fins of these taxa within the portion of the trade that was not characterized implies that these estimates are minimums.

pelagic species (*P. glauca*, *Carcharhinus falciformis*, and *Isurus oxyrhinchus*; Table 15.1) often caught incidentally in tuna and swordfish longline fisheries is suggestive, at least partly, of the former.

Management and conservation decisions for individual fishery species are much better informed if robust population (stock) assessments exist. The robustness of these assessments is in turn dependent on accurate information on the number or biomass of fish individuals harvested over time. To obtain this information for sharks by species, Clarke and colleagues built on their studies of the Hong Kong market by combining trade data with DNA forensic information for the 11 trade categories in a Bayesian statistical framework to derive minimum estimates of the number and biomass of sharks killed worldwide to supply the major global fin markets (Clarke et al. 2006b). By converting species-specific fin weights to whole sharks, Clarke and colleagues also provided the first quantitative, fishery-independent assessment of the scale of shark catches worldwide by species. Assuming the fin trade is the major driver of shark mortality, these estimates provide an approximation of the bulk of exploitation pressure on individual species. Additionally, comparison of these trade-derived, species-specific catch estimates to stock assessment sustainability reference points for individual species (where they exist) can serve to guide the prioritization of the limited management and conservation resources typically available.

Clarke et al.'s (2006b) overall findings raise concerns about the level of shark exploitation even further. Their analyses showed that shark biomass just in the global fin trade was three to four times higher than the total shark landings reported by fishing nations to the United Nations FAO database in 2000. This large difference confirms that most shark catches worldwide fall into the illegal, unreported, or unregulated (IUU) category. Further fueling concerns about the future of shark resources is the strong likelihood that the fin trade-derived biomass calculations are underestimates of the true scale of the global catches supplying the fin trade. Underestimates are likely because domestic consumption of fins by fishing nations was not accounted for in the methodology used by Clarke and colleagues.

To demonstrate how the trade-based estimates of shark removals might be helpful for assessing impacts of the fin trade on individual species, Clarke et al. (2006b) compared their estimates of the *number* of sharks killed per year for the most common species in the trade (blue sharks) to the global maximum sustainable yield (MSY) for this species extrapolated from a stock assessment available for blue sharks in the North Pacific. Similarly, they compared their trade based estimates of blue shark *biomass* removed per year to the global MSY biomass extrapolated from a stock assessment available for the North and South Atlantic basins. The trade-based median number of blue sharks killed (10,740,000 year^{-1}; 95% PI = 4,640,000–15,760,000 year^{-1}) was very similar to the global MSY (7.26–12.60 m year^{-1}) extrapolated from the North Pacific. Their trade-based estimate of blue shark median biomass removed (0.36 m tonnes year^{-1}; 95% PI = 0.20–0.62 m tonnes year^{-1}) compared to the Atlantic-based, global MSY biomass reference point (0.73–1.09 m tonnes year^{-1}) was lower. However, taking into consideration margins of error and that shark exploitation is likely underestimated by the trade-based analyses, it is likely that catches exceeded the MSY.

Given the global nature of Clarke et al.'s (2006b) assessment of the fin trade, the application of their approach to estimating sustainable or unsustainable use of species other than blue sharks will require more detailed stock assessment reference points on a broad geographic scale. In the absence of more direct measures of shark catch by species, trade-based analyses in combination with DNA forensics may provide the best opportunity for assessing the level of exploitation and sustainability of individual shark species.

15.4 Tracking the Population Origin of Sharks in International Fisheries and Trade

A central goal of most fisheries and wildlife management and conservation is to manage the resources not only by species but also by their individual populations/stocks (hereafter populations) (Dizon et al. 1992). The biological rationale underlying this goal is that the reproductive and often geographic isolation that occurs among components of a widespread species can result in adaption to local environments due to selection. Maintaining the health of a species as a whole therefore requires safeguarding the diversity and evolutionary potential of its component populations. Although the same rationale applies to sharks, their management on a population basis has been rare for two reasons. First, shark fisheries management was historically largely ignored, being considered a low priority compared to management of teleost fisheries. Second, although it is now widely acknowledged that sharks urgently need strong management focus, population-level management is dependent on first identifying populations, information that has been sparse for most shark species on ocean basin and worldwide scales. Fortunately, with increased conservation awareness and concomitant research attention, this situation is quickly improving. The pioneering DNA-based studies of shark population structure by Heist, Graves, and Musick (1995) and Heist, Musick, and Graves (1996a, 1996b) have been followed up by several recent studies using higher resolution mitochondrial sequence and nuclear microsatellite markers to investigate the population structure of some protected and common fishery species from regional to worldwide scales (e.g., Feldheim, Gruber, and Ashley 2001; Pardini et al. 2001; Schrey and Heist 2003; Keeney et al. 2003, 2005; Duncan et al. 2006; Hoelzel et al. 2006; Keeney and Heist 2006; Stow et al. 2006; Castro et al. 2007; Schultz et al. 2008; Ahonen, Harcourt, and Stow 2009; Chabot and Allen 2009; Schmidt et al. 2009; Chapman, Pinhal, and Shivji 2009).

Despite the large size and long-distance movement potential of many sharks, the published (above references) and ongoing studies (Bernard et al. 2008; Horn et al. 2008; Testerman et al. 2007; Testerman, Prodohl, and Shivji 2008; laboratory of M. Shivji, unpublished data) on shark populations are providing evidence of surprising levels of population genetic differentiation. At the minimum, interocean basin or hemispheric population differentiation based on mitochondrial or nuclear markers, or both, has been found in nearly all species that have global distributions (e.g., *Carcharhinus falciformis, C. obscurus, C. altimus, C. albimarginatus, Carcharias taurus, Carcharodon carcharias, Galeocerdo cuvier, Galeorhinus galeus, Isurus oxyrhinchus, Lamna nasus, Negaprion brevirostris, Rhincodon typus, Sphyrna mokarran, S. lewini, S. zygaena*). For some species, statistically significant genetic population structure has even been detected within ocean basins [*S. lewini* (Duncan et al. 2006; Chapman, Pinhal, and Shivji 2009); *G. cuvier* (Bernard et al. 2008); *C. amblyrhynchos* (Horn et al. 2008); *N. brevirostris* (Schultz et al. 2008); *S. zygaena* (Testerman, Prodohl, and Shivji 2008); *G. galeus* (Chabot and Allen 2009)]. In some cases this structure is present over relatively small geographic scales (1000 to 2500 km) and even along continuous coastlines [*Carcharhinus limbatus* (Keeny et al. 2003); *S. lewini* (Chapman, Pinhal, and Shivji 2009); *G. cuvier* (Bernard et al. 2008); *C. amblyrhynchos* (Horn et al. 2008)].

Although more detailed information on shark population structure is still necessary (i.e., studies with many more sampling locations for each species and more statistical power obtained with more markers and/or larger sample sizes), the growing evidence that shark populations can be genetically strongly differentiated bodes well for being able to track the population origin of shark products in fisheries and trade. This ability will be particularly beneficial to international shark management and conservation efforts given that (1) sharks commonly cross national boundaries and are thus subject to multinational fisheries, and (2) shark management and regulatory efforts are regionally "patchy," a result of some nations practicing active shark management while others practice little to no management despite indulging in extensive fisheries. Furthermore, some nations prohibit domestic landings of specific species, but allow importation of shark products from other countries that have the same species in their fisheries. This patchwork of national prohibitions is likely to result in illegal domestic landings being passed off as legal imports (see Section 15.5). Globally patchy management efforts are also unlikely to detect overexploitation of individual shark populations.

If population genetic differentiation exists, several approaches are available to determine the population origin of individuals. The power and resolution for determining the population origin of samples is greatest if source populations are unambiguously distinguished by fixed, diagnostic genetic characters or are phylogenetically distinct. This circumstance, although not common at the population level, does exist in some shark species if their populations are considered on a broad geographic scale. For example, *Carcharhinus limbatus*, considered on a global scale, falls into two monophyletic mitochondrial DNA lineages composed of northwestern Atlantic vs. eastern Atlantic/Indo-Pacific haplotypes (Keeney and Heist 2006). If this clear lineage separation is confirmed with larger sample sizes from the eastern Atlantic, it will be straightforward to assign unknown origin *C. limbatus* individuals (taxonomic issues notwithstanding) to one of these broad geographic regions by simple phylogenetic analysis tools.

Similarly, with the caveat of possible reference errors due to sample size limitations, *Carcharodon carcharias* and *Carcharias taurus* from South Africa and Australia each also separate almost completely into monophyletic mitochondrial DNA lineages (Pardini et al. 2001; Stowe et al. 2006), suggesting that samples of these species obtained from Indo-Pacific fisheries and originating from one of these regions may be directly genetically traceable back to their region of origin. Just-published studies also suggest that further global phylogeographic divisions are present in these species (Ahonen et al. 2009; Jorgensen et al. 2009). *Lamna nasus* assessed on a global scale based on mitochondrial DNA control region sequences also form two monophyletic lineages corresponding to the northern and southern hemispheres (Testerman et al. 2007). The northern hemisphere *L. nasus* are severely overfished (IUCN Endangered listing in the western Atlantic and Critically Endangered listing in the eastern Atlantic; IUCN 2008 Red List of Threatened Species. www.iucnredlist.org; downloaded on March 1, 2009). The clear genetic separation of northern and southern hemisphere populations therefore provides a robust phylogenetic basis for tracking the hemispheric origins of unknown *L. nasus* samples obtained in international trade. Furthermore, northern and southern hemisphere *L. nasus* populations possess fixed nucleotide differences in the control region locus, thereby allowing the development of population-specific PCR primers that can distinguish animals from the two hemispheres with a single-step PCR assay (Testerman and Shivji, unpublished data).

In contrast to the above situation, genetic differentiation especially over smaller geographic scales is mostly manifested as nuclear allele or mitochondrial haplotype frequency differences among populations. In these cases, several genetic information-based, probabilistic approaches broadly known as assignment methods are available to estimate population membership of individuals or groups of individuals (reviewed in Manel, Gaggiotti, and Waples 2005). One type of assignment method, known more specifically as an assignment test, requires multilocus genetic data (e.g., microsatellite genotypes) from both the potential source populations and the sample whose population of origin is in question. The sample of unknown origin is then "assigned" to one of the potential source populations based on its multilocus genotype and the expected probability of that genotype occurring in each potential source population. These traditional assignment tests, however, usually perform best when all potential source populations are predefined, a condition that is not always met. Newly developed statistical frameworks that require some but not all potential sources to be predefined (e.g., the "smoothed continuous assignment technique"; Wasser et al. 2004, 2007) may be applicable over relatively small areas (e.g., the north or south Atlantic). In some situations, particularly when sampling of potential source populations is limited for practical reasons, a related assignment test can also be used to *exclude* a particular population as the origin of the sample (Cornuet et al. 1999; Primmer, Koskinen, and Piironen 2000; Manel, Berthier, and Luikart 2002).

In cases where information on potential source populations is lacking or incomplete, alternative assignment methods based on individual clustering are also available to estimate the origin of specific samples. These clustering methods decompose undefined mixtures of individuals based on their multilocus genotypes into groups within which linkage disequilibrium is minimized. The resulting groups are considered populations, with individuals in the starting mixture becoming assigned to the specific groups. Thus, if a sample whose origin is being questioned is included in the starting mixture, it will also be assigned to its population of origin, assuming its source population is present within the mixture (Manel, Gaggiotti, and Waples 2005).

Sometimes in fisheries management the issue being investigated is not the population origin of an individual, but rather the population composition of a mixed fishery to ensure that a particular population is not being disproportionately harvested. Genetic analysis can be especially useful for such assessments, assuming genetic population differentiation exists (Shaklee and Currens 2003). This approach, commonly known as mixed population/stock analysis, is commonly used for teleost fishery management. An extension of this approach was used by Baker et al. (2002), who, in an elegant and conservation relevant forensic application, applied it to track the population origin of North Pacific minke whale products in Japanese markets. Using mixed stock analysis based on mitochondrial DNA haplotype frequency differences between protected and legal to harvest populations, Baker and colleagues showed that ~31% of the minke whale market products were derived from the protected population. Furthermore, this market information integrated into a population dynamics model indicated that continued exploitation of the protected population at levels consistent with the market survey would result in a decline of the protected stock toward extinction in just a few decades (Baker et al. 2002).

There are numerous examples of various genetic assignment methods productively employed for tracking the population origin of wildlife and fish individuals (e.g., Primmer, Koskinen, and Piironen 2000; Manel, Gaggiotti, and Waples 2005; Frantz et al. 2006; Wasser et al. 2007, 2008; Baker 2008). These methods have yet to be applied to address similar questions for sharks. However, the rapidly accumulating genetic population structure information becoming available for sharks with evidence for strong population differentiation in many cases indicates great promise for these methods to play an important role in shark conservation and management.

To utilize the full promise of these analytical tools that continue to increase in power and speed as statistical frameworks and computing capabilities improve (Manel et al. 2007; Wasser et al. 2004, 2007), researchers must obtain increased sample sizes across a broader spatial distribution when assessing the population structure of shark species. An illustration of the importance of improved sampling is provided by the following case: although

Duncan et al. (2006) did not detect evidence for population structure in *S. lewini* along the continuous coastline of North and South America, at least three clear populations were found over even shorter distances by Chapman, Pinhal, and Shivji (2009), who used the same DNA marker but larger sample sizes. In fact, the haplotype frequency differences among the three populations were large enough to theoretically permit the use of mixed stock analysis to determine the contribution of these individual populations to *S. lewini* fin mixtures obtained from trade. In cases where only small sample sizes are available for some locations and population structure not detected, it is advisable to incorporate power analyses of the marker and sample sets used in the analyses.

15.5 DNA Applications in Shark Fisheries Law Enforcement

With increasing facilitation of global commerce, illegal international trade in wildlife products is burgeoning and estimated to be worth over US$20 billion a year worldwide (see Wasser et al. 2008). The global volume of IUU trade in fisheries products is extremely difficult to estimate because of the enormous spatial scope of exploitation efforts, which range from coastal to high seas fisheries. Surveying the major market end points of the global supply chain may thus be the best option for determining the true volume of the overall and IUU trade (Clarke 2004; Baker 2008). By using this approach and comparing their trade-based estimates to shark landings reported by fishing nations to the UN FAO, Clarke et al. (2006b) estimated that ~65% to 75% of shark landings globally fell into the IUU category. This proportion is expected to differ substantially at the level of individual countries, depending on the degree to which they regulate their shark fisheries and invest in regulation enforcement. The United States likely has one of the most highly regulated shark fisheries in the world, with 19 shark species from U.S. Atlantic EEZ waters prohibited from landings (NMFS 1999). We have employed the DNA multiplex PCR and barcoding forensic methods outlined above in approximately 21 cases since 2003 to assist the NOAA's Office for Law Enforcement (NOAA OLE) in determining the species composition of confiscated, suspect shark fins. Most of these cases have demonstrated the presence of at least some fins from prohibited species (most frequently *Carcharhinus obscurus*, *C. signatus*, and *C. altimus*, among others), and even occasionally CITES listed species (Shivji et al. 2005; Magnussen et al. 2007). Several of these DNA methodology-assisted cases have resulted in successful prosecutions and collection of fines from the responsible parties (P. Raymond, Assistant Special Agent in Charge, NOAA OLE, personal communication).

The fact that landings of some widespread shark species are prohibited by some nations but not others can present challenges for law enforcement. For example, landings of the dusky shark, *Carcharhinus obscurus*, from the U.S. Atlantic EEZ are prohibited by U.S. law. However, in some cases (depending on U.S. state law), it may not be an offense for a U.S. trader to import and market *C. obscurus* fins from a country where their landing is legal (P. Raymond, Assistant Special Agent in Charge, NOAA OLE, personal communication). These differences in national regulations provide the opportunity for poached fins to be passed off as legally obtained fins of the same species. In such cases, the relevant law enforcement issue is centered on whether the fins (or some proportion) originate from the protected population or not. With sufficient knowledge of the genetic makeup of most (but not necessarily all; Wasser et al. 2004, 2007) potential source populations, assignment tests could be used to determine the geographic source of the fins.

Alternatively, if this information was not available for most potential source populations but was available for the protected population of interest, law enforcement efforts could still be productively informed if the protected population was statistically excluded (by exclusion assignment tests; Cornuet et al. 1999; Primmer, Koskinen, and Piironen 2000) as the source of the fins.

15.6 Concluding Perspectives

The power of DNA tools to aid in broad forensic applications ranging from human crime cases to identification of pathogens in medical cases is well established. These tools are also finding increasing application for solving wildlife crimes and informing wildlife management and conservation efforts. These tools can be of equal value in aiding management and conservation efforts for sharks. To date, however, the power and utility of these tools outside the research arena has not been widely recognized by managers of shark fisheries. This likely results from the general lack of cross-disciplinary communication between geneticists and fishery managers and policy makers. It is telling that until recently, conference presentations reporting genetic data and fisheries data have typically been scheduled in different topical sessions, even though many genetic studies use management and conservation goals as their applied justification, and many fisheries studies are limited by the lack of population structure information. In this review I have attempted to illustrate how DNA analysis tools can be highly informative if adopted in practical shark management efforts. For example, DNA tools for shark species identification are sufficiently well developed to be of immediate use for monitoring shark landings by species, especially in developed nations. The ability to monitor shark landings by population origin is foreseeable in the near future, awaiting only more refined delineation of shark populations, which is occurring at a rapid pace. Adoption of these powerful tools to aid traditional management efforts is even more pressing in light of numerous indications that sharks are experiencing unsustainable levels of exploitation pressure.

Acknowledgments

Many thanks are due to my former and current students (D. Abercrombie, A. Bernard, D. Chapman, M. Debiasse, M. Henning, J. Hester, R. Horn, J. Nielsen, J. Magnussen, L. Murphy, M. Pank, V. Richards, and C. Testerman) who have been instrumental in developing and assisting in various aspects of the shark forensic research described here. I am indebted to numerous colleagues and collaborators who have kindly provided tissue samples over the years for my lab's research. Funding for the research described herein has been provided by grants from the Save Our Seas Foundation, the Institute for Ocean Conservation Science (founded as the Pew Institute for Ocean Science), the Hai Stiftung/ Shark Foundation, the Florida Sea Grant Program, the National Science Foundation, the National Marine Fisheries Service, the Wildlife Conservation Society, the Munson Foundation, and operational funds for the Guy Harvey Research Institute from Guy Harvey Incorporated, the Guy Harvey Ocean Foundation, and Nova Southeastern University.

References

Abercrombie DL, Clarke SC, Shivji MS (2005) Global-scale genetic identification of hammerhead sharks: application to assessment of the international fin trade and law enforcement. Cons Gen 6:775–788.

Ahonen H, Harcourt RG, Stow AJ (2009) Nuclear and mitochondrial DNA reveals isolation of imperilled grey nurse shark populations (*Carcharias taurus*). Mol Ecol 18:4409–4421.

Asensio L, Gonzalez I, Fernandez A, Cespedes A, Hernandez P, Garcia T, Martin R (2000) Identification of Nile perch (*Lates noliticus*), grouper (*Epinephelus guaza*), and wreck fish (*Polyprion americanus*) by polymerase chain reaction-restriction fragment length polymorphism of a 12S rRNA gene fragment. J Food Protec 63(9):1248–1252.

Baker CS (2008) A truer measure of the market: the molecular ecology of fisheries and wildlife trade. Mol Ecol 17:3985–3998.

Baker CS, Lento GM, Cipriano F, Palumbi SR (2002) Predicted decline of protected whales based on molecular genetic monitoring of Japanese and Korean markets. Proc Roy Soc London B Biol Sci 267:1191–1199.

Baker CS, Palumbi SR (1994) Which whales are hunted? A molecular genetic approach to monitoring whaling. Science 265:1538–1539.

Bascompte J, Melián CJ, Sala E (2005) Interaction strength combinations and the overfishing of a marine food web. Proc Natl Acad Sci USA 102:5443–5447.

Baum JK, Myers RA, Kehler DG, Worm B, Harley SJ, Doherty PA (2003) Collapse and conservation of shark populations in the Northwest Atlantic. Science 299:389–392.

Bensasson D, Zhang D-X, Hartl D, Hewitt GM (2001) Mitochondrial pseudogenes: evolution's misplaced witnesses. Trends Ecol Evol 16:314–321.

Bernard A, Feldheim K, Howey L, Wetherbee B, Heithaus M, Shivji M (2008) Defining management units of a migratory species: the global genetic population structure of the tiger shark (*Galeocerdo cuvier*). Presented at the American Elasmobranch Society 24th Annual Meeting, July 23–28, Montreal.

Blanco M, Pérez-Martín RI, Sotelo CG (2008) Identification of shark species in seafood products by forensically informative nuceotide sequencing (FINS). J Agric Food Chem 56:9868–9874.

Buhay JE (2009) "COI-like" sequences are becoming problematic in molecular systematic and DNA barcoding studies. J Crust Biol 29:96–110.

Castro ALF, Stewart BS, Wilson SG, Hueter RE, Meekan MG, Motta PJ, Bowen BW, Karl SA (2007) Population genetic structure of earth's largest fish, the whale shark (*Rhincodon typus*). Mol Ecol 16:5183–5192.

Chabot CL, Allen LG (2009) Global population structure of the tope (*Galeorhinus galeus*) inferred by mitochondrial control region sequence data. Mol Ecol 18:545–552.

Chapman D, Abercrombie D, Douady C, Pikitch E, Stanhope M, Shivji M (2003) A streamlined, bi-organelle, multiplex PCR approach to species identification: application to global conservation and trade monitoring of the great white shark, *Carcharodon carcharias*. Cons Gen 4:415–425.

Chapman DD, Pinhal D, Shivji MS (2009) Tracking the fin trade: genetic stock identification in western Atlantic scalloped hammerhead sharks, *Sphyrna lewini*. Endng Spec Res. doi:10.3354/esr00241.

Clarke S (2004) Understanding pressures on fishery resources through trade statistics: a pilot study of four products in the Chinese dried seafood market. Fish and Fisheries 5:53–74.

Clarke SC, Magnussen JE, Abercrombie DL, McAllister M, Shivji MS (2006a) Identification of shark species composition and proportion in the Hong Kong shark fin market using molecular genetics and trade records. Cons Biol 20:201–211.

Clarke S, McAllister M, Michielsens C (2004) Estimates of shark species composition and numbers associated with the shark fin trade based on Hong Kong auction data. J Northwest Atl Fish Sci 35:453–465.

Clarke SC, McAllister MK, Milner-Gulland EJ, Kirkwood GP, Michielsens CGJ, Agnew DJ, Pikitch EK, Nakano H, Shivji MS (2006b) Global estimates of shark catches using trade records from commercial markets. Ecol Lett 9:1115–1126.

Clarke S, Milner-Gulland EJ, Bjørndal T (2007) Social, economic and regulatory drivers of the shark fin trade. Mar Res Econ 22:305–327.

Cornuet J-M, Piry S, Luikart G, Estoup A, Solignac M (1999) New methods employing multilocus genotypes to select or exclude populations as origins of individuals. Genetics 153:1989–2000.

Cortés E (2002) Incorporating uncertainty into demographic modeling: application to shark populations and their conservation. Cons Biol 16:1048–1062.

Dasmahapatra KK, Mallet J (2006) DNA barcodes: recent successes and future prospects. Heredity 97:254–255.

Dizon A, Baker CS, Cipriano F, Lento G, Palsboll P, Reeves R (eds) (2000). Molecular genetic identification of whales, dolphins, and porpoises. Proceedings of a workshop on the forensic use of molecular techniques to identify wildlife products in the marketplace. U.S. Department of Commerce, NOAA Tech. Memorandum, NOAA-TM-NMFS-SWFSC-286. U.S. Department of Commerce, Washington, DC.

Dizon AE, Lockyer C, Perrin WF, DeMaster DP, Sisson J (1992) Rethinking the stock concept: a phylogeographic approach. Cons Biol 6:24–36.

Dulvy NK, Baum J, Clarke SC, et al. (2008) You can swim but you can't hide: the global status and conservation of oceanic pelagic sharks and rays. Aquatic Conserv Mar Freshw Ecosyst 18:459–482.

Duncan KM, Martin AP, Bowen BW, De Couet HG (2006) Global phylogeography of the scalloped hammerhead shark (*Sphyrna lewini*). Mol Ecol 15:2239–2251.

FAO (1998) International Plan of Action for the conservation and management of sharks. Document FI:CSS/98/3, Oct. 1998. Food and Agriculture Organization, Rome.

FAO (2000) Fisheries Management. 1. Conservation and Management of Sharks. FAO Technical Guidelines for Responsible Fisheries. No. 4, Suppl. 1. Food and Agriculture Organization, Rome.

Feldheim KA, Gruber SH, Ashley MV (2001) Population genetic structure of the lemon shark (*Negaprion brevirostris*) in the western Atlantic: DNA microsatellite variation. Mol Ecol 10:295–303.

Ferretti F, Myers RA, Serena F, Lotze HK (2008) Loss of large predatory sharks from the Mediterranean Sea. Cons Biol 22:952–964.

Fong QSW, Anderson JL (2002) International shark fin markets and shark management: an integrated market preference-cohort analysis of the blacktip shark (*Carcharhinus limbatus*). Ecol Econ 40:117–130.

Frantz AC, Pourtois JT, Heuertz M, Schley L, Flamand MC, Krier A, Bertouille S, Chaumont F, Burke T (2006) Genetic structure and assignment tests demonstrate illegal translocation of red deer (*Cervus elaphus*) into a continuous population. Mol Ecol 15:3191–3203.

Gharrett AJ, Gray AK, Heifetz J (2001) Identification of rockfish (*Sebastes* spp.) by restriction site analysis of the mitochondrial ND-3/ND-4 and 12S/16S rRNA gene regions. Fish Bull 99:49–62.

Greig TW, Moore KM, Woodley CM, Quattro JM (2005) Mitochondrial gene sequences useful for species identification of western North Atlantic Ocean sharks. Fish Bull 103:516–523.

Hebert PDN, Cywinska A, Ball SL, Dewaard JR (2003) Biological identifications through DNA barcodes. Proc Roy Soc London B Biol Sci 27:313–321.

Heist EJ, Gold JR (1999) Genetic identification of sharks in the U.S. Atlantic shark longline fishery. Fish Bull 97:53–61.

Heist EJ, Graves JE, Musick JA (1995) Population genetics of the sandbar shark (*Carcharhinus plumbeus*) in U.S. coastal waters. Copeia 1995:555–562.

Heist EJ, Musick JA, Graves JE (1996a) Genetic population structure of the shortfin mako (*Isurus oxyrinchus*) inferred from restriction fragment length polymorphism analysis of mitochondrial DNA. Can J Fish Aquat Sci 53:583–588.

Heist EJ, Musick JA, Graves JE (1996b) Mitochondrial DNA diversity and divergence among sharpnose sharks, *Rhizoprionodon terraenovae*, from the Gulf of Mexico and Mid-Atlantic Bight. Fish Bull 94:664–668.

Heithaus MR, Frid A, Wirsing AJ, Worm B (2008) Predicting the consequences of declines in marine top predators. Trend Ecol Evol 23:202–210.

Hoelzel AR, Shivji MS, Magnussen J, Francis MP (2006) Low worldwide genetic diversity in the basking shark (*Cetorhinus maximus*). Biol Lett 2:639–642.

Holmes BH, Steinke D, Ward RD (2009) Identification of shark and ray fins using DNA barcoding. Fish Res 95:280–288.

Horn R, Robbins W, McCauley D, Shivji M (2008) Genetic structure of the gray reef shark (*Carcharhinus amblyrhynchos*), based on microsatellite and mitochondrial DNA analyses with implications for management. Presented at the American Elasmobranch Society 24th Annual Meeting.

ICCAT (2005) Report of the 2004 Inter-sessional meeting of the ICCAT Subcommittee on By-catches: Shark stock assessment. Col Vol Sci Pap ICCAT 58:799–890.

Innes BH, Grewe PM, Ward RD (1998) PCR-based genetic identification of marlin and other billfish. Mar Freshwat Res 49(5):383–388.

Jorgensen SJ, Reeb CA, Chapple TK, Anderson S, Perle C, Van Sommeran S, Fritz-Cope C, Brown AC, Klimley AP, Block BA (2009) Philopatry and migration of Pacific white sharks. Proc Roy Soc B doi:10.1098/rspb.2009.1155.

Keeney DB, Heist EJ (2006) Worldwide phylogeography of the blacktip shark (*Carcharhinus limbatus*) inferred from control region sequences. Mol Ecol 15:3669–3679.

Keeney DB, Heupel M, Hueter RE, Heist EJ (2003) Genetic heterogeneity among blacktip shark, *Carcharhinus limbatus*, continental nurseries along the U.S. Atlantic and Gulf of Mexico. Mar Biol 143:1039–1046.

Keeney DB, Heupel MR, Hueter RE, Heist EJ (2005) Microsatellite and mitochondrial DNA analyses of the genetic structure of blacktip shark (*Carcharhinus limbatus*) nurseries in the northwestern Atlantic, Gulf of Mexico, and Caribbean Sea. Mol Ecol 14:1911–1923.

Kmiec B, Woloszynska M, Janska H (2006) Heteroplasmy as a common state of mitochondrial genetic information in plants and animals. Curr Gen 50:149–159.

Magnussen JE, Pikitch EK, Clarke SC, Nicholson C, Hoelzel AR, and Shivji MS (2007) Genetic tracking of basking shark (*Cetorhinus maximus*) products in international trade. Anim Cons 10:199–207.

Malik S, Wilson PJ, Smith RJ, Lavigne DM, White BN (1997) Pinneped penises in trade: a molecular-genetic investigation. Cons Biol 11(6):1365–1374.

Manel S, Berthier P, Luikart G (2002) Detecting wildlife poaching: identifying the origin of individuals with Bayesian assignment tests and multilocus genotypes. Cons Biol 16:650–659.

Manel S, Berthoud F, Bellemain E, Gaudeul M, Luikart G, Swenson JE, Waits LP, Taberlet P, and Intrabiodiversity Consortium (2007) A new individual-based spatial approach for identifying genetic discontinuities in natural populations. Mol Ecol 16:2031–2043.

Manel S, Gaggiotti OE, Waples RS (2005) Assignment methods: matching biological questions with appropriate techniques. Trends Ecol Evol 20:136–142.

Moritz C, Cicero C (2004) DNA barcoding: promise and pitfalls. PLoS Biol 2(10):1529–1530.

Musick JA, Burgess G, Caillet G, Camhi M, Fordham S (2000) Management of sharks and their relatives (Elasmobranchii). Fisheries 25(3):9–13.

Nielsen J (2004) Molecular genetic approaches to species identification and delineation in elasmobranchs. Masters thesis, Nova Southeastern University, Florida.

NMFS (1999) Final Fishery Management Plan for Atlantic Tuna, Swordfish, and Sharks. NMFS Report. U.S. Department of Commerce, National Oceanic and Atmospheric Administration, Washington, DC.

NMFS (2001) Final United States National Plan of Action for the Conservation and Management of Sharks. NMFS Report. U.S. Department of Commerce, NOAA, Washington, DC.

Palumbi SR, Cipriano F (1998) Species identification using genetic tools: the value of nuclear and mitochondrial gene sequences in whale conservation. J Hered 89:459–464.

Pardini AT, Jones CS, Noble LR, Kreiser B, Malcolm H, Bruce BD, Stevens JD, Cliff G, Scholl MC, Francis M, Duffy CAJ, Martin AP (2001) Sex-biased dispersal of great white sharks. Nature 412:139–140.

Parry-Jones R (1996) TRAFFIC report on shark fisheries and trade in Hong Kong. In The world trade in sharks: a compendium of TRAFFIC's regional studies. TRAFFIC International, Cambridge, UK, pp 87–143.

Primmer CR, Koskinen MT, Piironen J (2000) The one that did not get away: individual assignment using microsatellite data detects a case of fishing competition fraud. Proc Roy Soc London B 267:1699–1704.

Ratnasingham S, Hebert PDN (2007) BOLD: The Barcode of Life Datasystem. Mol Ecol Notes 7:355–364.

Roman J, Bowen BW (2000) The mock turtle syndrome: genetic identification of turtle meat purchased in the south-eastern United States of America. Anim Cons 3:61–65.

Rubinoff D (2006) Utility of mitochondrial DNA barcodes in species conservation. Cons Biol 20:1026–1033.

Rubinoff D, Cameron S, Will K (2006) A genomic perspective on the shortcomings of mitochondrial DNA for "barcoding" identification. J Hered 97:581–594.

Schmidt JV, Schmidt CL, Ozer F, Ernst RE, Feldheim KA, Ashley MV, Levine M (2009) Low genetic differentiation across three major ocean populations of the whale shark, *Rhincodon typus*. PLoS ONE 4(4): e4988. doi:10.1371/journal.pone.0004988.

Schrey A, Heist EJ (2003) Microsatellite analysis of population structure in the shortfin mako (*Isurus oxyrinchus*). Can J Fish Aquat Sci 60:670–675.

Schultz JK, Feldheim KA, Gruber SH, Ashley MV, Mcgovern TM, Bowen BW (2008) Global phylogeography and seascape genetics of the lemon sharks (genus *Negaprion*). Mol Ecol 17:5336–5348.

Shaklee JB, Currens KP (2003) Genetic stock identification and risk assessment. In Hallerman EM (ed) Population genetics: principles and applications for fisheries scientists. American Fisheries Society, Bethesda, MD, pp 291–328.

Shivji MS, Chapman DD, Pikitch EK, Raymond PW (2005) Genetic profiling reveals illegal international trade in fins of the great white shark, *Carcharodon carcharias*. Cons Gen 6:1035–1039.

Shivji MS, Clarke SC, Pank M, Natanson L, Kohler N, Stanhope M (2002) Genetic identification of pelagic shark body parts for conservation and trade monitoring. Cons Biol 16:1036–1047.

Smith SE, Au DW, Show C (1998) Intrinsic rebound potentials of 26 species of Pacific sharks. Mar Freshw Res 49:663–678.

Stevens JD, Bonfil R, Dulvy NK, Walker PA (2000) The effects of fishing on sharks, rays, and chimaeras (chondrichthyans), and the implications for marine ecosystems. ICES J Mar Sci 57:476–494.

Stoeckle MY, Hebert PDN (2008) Barcode of Life. Scient Amer October, 82–88.

Stow A, Zenger K, Briscoe D, Gillings M, Peddemors V, Otway N, Harcourt R (2006) Isolation and genetic diversity of endangered grey nurse shark (*Carcharias taurus*) populations. Biol Lett 2:308–311.

Testerman C, Prodohl P, Shivji M (2008) Global phylogeography of the great (*Sphyrna mokarran*) and smooth (*Sphyrna zygaena*) hammerhead sharks. Presented at the American Elasmobranch Society 24th Annual Meeting, July 23–28, Montreal.

Testerman C, Richards V, Francis M, Pade N, Jones C, Noble L, Shivji M (2007) Global phylogeography of the porbeagle shark (*Lamna nasus*) reveals strong genetic separation of northern and southern hemisphere populations. Presented at the American Elasmobranch Society 23rd Annual Meeting, July 12–16, St. Louis, MO, USA.

Vandersea MW, Litaker RW, Marancik KE, et al. (2008) Identification of larval sea basses (*Centropristis* spp.) using ribosomal DNA-specific molecular assays. Fish Bull 106:183–193.

Vannuccini S (1999) Shark utilization, marketing and trade. FisheriesTechnical Paper 389. Food and Agriculture Organization, Rome.

Ward P, Myers RA (2005) Shifts in open-ocean fish communities coinciding with the commencement of commercial fishing. Ecology 86:835–847.

Ward RD, Holmes BH, White WT, Last PR (2008) DNA barcoding Australasian chondrichthyans: results and potential uses in conservation. Mar Freshw Res 59:57–71.

Ward RD, Holmes BH, Zemlak TS, Smith PJ (2007) DNA barcoding discriminates spurdogs of the genus *Squalus*. In: Last PR, White WT, Pogonoski JJ (eds) Descriptions of new dogfishes of the genus *Squalus* (Squaloidea: Squalidae). CSIRO Marine and Atmospheric Research Paper 014. CSIRO, Hobart, Australia, pp 117–130.

Ward RD, Zemlak TS, Innes BH, Last PR, Hebert PDN (2005) DNA barcoding Australia's fish species. Philos Trans R Soc London B Biol Sci 360:1847–1857.

Wasser SK, Clark WJ, Drori O, Kisamo ES, Mailand C, Mutayoba B, Stephens M (2008) Combating the illegal trade in African elephant ivory with DNA forensics. Cons Biol 22:1065–1071.

Wasser SK, Mailand C, Booth R, Mutayoba B, Kisamo E, Clark B, Stephens M (2007) Using DNA to track the origin of the largest ivory seizure since the 1989 trade ban. Proc Natl Acad Sci USA 104:4228–4233.

Wasser SK, Shedlock AM, Comstock K, Ostrander EA, Mutayoba B, Stephens M (2004) Assigning elephant DNA to geographic region of origin: applications to the ivory trade. Proc Natl Acad Sci USA 101:14847–14852.

Wong EH-K, Hanner RH (2008) DNA barcoding detects market substitution in North American seafood. Food Res Int 41:828–837.

Wong EH-K, Shivji MS, Hanner RH (2009) Identifying sharks with DNA barcodes: assessing the utility of a nucleotide diagnostic approach. Mol Ecol Res 9:243–256.

16

Unraveling the Ecological Importance of Elasmobranchs

Michael R. Heithaus, Alejandro Frid, Jeremy J. Vaudo,
Boris Worm, and Aaron J. Wirsing

CONTENTS

16.1 Introduction

Ongoing and rapid declines in populations of many large-bodied sharks throughout the world are now widely recognized as a critical conservation challenge (e.g., Musick, Branstetter, and Colvocoresses 1993; Baum et al. 2003; Baum and Myers 2004; Clarke et al. 2006; Robbins et al. 2006; Myers et al. 2007; Heithaus et al. 2007a; Ferretti et al. 2008). Despite conservation efforts, shark bycatch and a growing demand for shark fins and meat have increased exploitation rates in recent years and many large sharks are listed as Endangered, Vulnerable, or Threatened on the IUCN Red List (but see Okey, Wright, and Brubaker 2007 for a possible exception). Populations of many other elasmobranchs, including rays, skates, and smaller-bodied sharks are responding directly to fishing pressure and indirectly to the removal of large sharks and other marine top predators. Many species that are captured in targeted fisheries or as bycatch are declining, while those exempt from significant fishing mortality may be increasing in locations where their predators or competitors have declined (Shephard and Myers 2005; Okey, Wright, and Brubaker 2007). Further, ongoing climate change is predicted to modify distributions of elasmobranchs and other apex predators (Cairns, Gaston, and Huettmann 2008). Understanding and predicting the broader ecological consequences of these changes in elasmobranch populations is important for marine conservation. To date, however, few studies have provided detailed insights into the ecological role of these fishes.

It is likely that, as upper trophic level predators, elasmobranchs have influenced the structure of marine communities over geological time. Indeed, the Chondrichthyes evolved as large, mobile predators around 350 million years ago, and species that resembled modern sharks survived all major mass extinctions. Extant large sharks, at the level of order and genus, have existed much longer than other major marine predators (Sepkowski 2002). Even in the Mesozoic, some sharks occupied apex trophic positions and preyed on other large-bodied predators (e.g., mosasaurs, Rothschild, Martin, and Schulp 2005). Further suggesting the importance of elasmobranchs as a structuring force over geological time, the unique invertebrate fauna of present day Antarctica is thought to exist because of the dramatic reduction in predation that resulted from the loss of skates and benthic-foraging sharks that occurred during the Eocene (Aronson and Blake 2001).

Given their long history in the oceans, elasmobranchs likely have influenced the ecology and evolution of marine life for millions of years. But do sharks and their relatives still play this role today? What are the ecological consequences of large changes in their abundance? Elasmobranchs can be difficult to study in the wild, and we have only begun to answer these questions using a variety of techniques (see the Appendix). Here, we summarize our current understanding of the ecological importance of elasmobranchs and suggest directions for future research. In particular we strive to stimulate studies of the ecological contexts in which elasmobranchs can influence community dynamics.

16.2 The Complete Role of Predators

At the most basic level, the ecological role of a species is defined by its position in the food web, that is, by what it eats and what eats it. A species' ecological influence, however, extends beyond trophic position and encompasses a broader range of interspecific interactions and indirect effects on the wider community. Ultimately, we are concerned with a species' ecological importance, and the consequences of a substantial change in its abundance. Therefore, it is important to consider the mechanisms through which consumers might influence populations of their prey and the community at large. These effects may manifest through predation rates inflicted on prey, behavioral changes induced in prey, or mechanisms outside predator–prey interactions. Before discussing what we know about the ecological importance of elasmobranchs, we briefly review general mechanisms through which predators may influence their prey and communities.

16.2.1 Direct Predation Effects

The most studied mechanism through which predators affect their communities is the consumption of prey ("direct predation"), which may depress populations of prey and competitors. The influence of direct predation on prey populations can be estimated from the diet composition, metabolic rate, and population size of the predator combined with demographic analyses of its prey (Williams et al. 2004). The first three parameters determine the rate at which prey are removed from the population while the fourth is necessary for estimating the impact of these removals on the prey population. Thus, estimating the ecological role of direct predation by elasmobranchs requires studies on the population biology of their prey, especially because predators may consume many individuals with little or no effect on equilibrium population sizes of prey. For example, predation on adults may have little long-term effect on the population sizes of species whose abundance is determined by larval or juvenile mortality (e.g., Piraino, Fanelli, and Boero 2002). Also, predation may result in increased prey population densities if it alters patterns of population fluctuations, relaxes competition, thereby enhancing productivity among survivors, or if there is temporal separation between mortality and density dependence ("hydra effect"; Abrams and Quince 2005; Abrams and Matsuda 2005; Abrams 2009).

16.2.2 Risk Effects

Predators may influence prey through mechanisms other than direct predation. One of the most important of these mechanisms is antipredator behavior by prey—hiding, habitat shifts, vigilance, and others—that lowers the risk of predation but at the cost of reduced rates of acquiring food or other resources (see Lima and Dill 1990 for a general review; see Heithaus 2004 for elasmobranch examples). Thus, predators can functionally depress carrying capacity because prey that invest in antipredator behavior cannot make full use of available resources (Creel and Christianson 2008). Estimating the importance of these "risk effects" (also called nonconsumptive effects or nonlethal effects) is difficult in large-scale marine settings. Theory (e.g., Abrams 1995; Frid et al. 2007), small-scale experiments in diverse settings (e.g., Preisser, Bolnick, and Benard 2005), and empirical studies in large-scale terrestrial systems suggest that risk effects can be substantial. For example, empirical data suggest that in Yellowstone National Park, 67% to 74% of the predator-induced decline in elk (*Cervus elaphus*) calf recruitment following the 1995 reintroduction of wolves (*Canis lupis*) was due to antipredator responses by elk that reduced rates of energy intake (Creel, Christianson, and Winnie 2007; Creel and Christianson 2008). Importantly, antipredator behavior can be so effective that prey organisms may experience substantial risk effects, even when predator-inflicted mortality is rare (Lima and Dill 1990; Creel and Christianson 2008; Heithaus et al. 2008a, 2008b). Further, a growing body of evidence suggests that risk effects can be equally or more important to community dynamics than direct predation (e.g., Werner and Peacor 2003; Preisser, Bolnick, and Benard 2005).

16.2.3 Cascading Effects of Predator–Prey Interactions

Both direct predation and risk effects can cause trophic cascades in marine ecosystems (see Pace et al. 1999; Heithaus et al. 2008a). In the case of direct predation, a predator reduces the population size or density of its prey (a "mesoconsumer"), which leads to increases in the "resource species" eaten by the mesoconsumer. Conversely, where predators decline, mesoconsumer populations can expand and reduce populations of resource species. Risk-induced cascades occur when mesoconsumers alter their behavior (e.g., foraging rates on resource species and changes in spatial and temporal patterns of resource exploitation) in response to predation risk and this changes the abundance or dynamics of resource species populations (Abrams 1995; Werner and Peacor 2003; Dill, Heithaus, and Walters 2003; Heithaus et al. 2008a). Consequently, predators may indirectly increase the mesoconsumer's exploitation rate of resources in safer habitats or during safer times while decreasing their exploitation rates in more dangerous contexts (Heithaus et al. 2008a). Through these behavioral mechanisms, which might involve diet or habitat switching by the mesoconsumer, upper level predators may indirectly influence mesoconsumer impacts on resource species without necessarily having population-level effects on mesoconsumers ("leapfrog effect"; Frid, Baker, and Dill 2008; Heithaus et al. 2008a). Importantly, observed changes in mesoconsumer and resources species populations in response to changes in predator populations are the result of both risk and direct-predation effects, and these effects may act synergistically because decreased body condition as a result of antipredator behavior can lead to increased risk taking and higher rates of predation (see Anholt and Werner 1995; Sinclair and Arcese 1995; Abrams 2008; Heithaus et al. 2008a).

Although the strength and frequency of trophic cascades in marine systems has been controversial (e.g., Kitchell et al. 2002; Bascompte, Melián, and Sala 2005), several well-documented cases do exist, several of them involving elasmobranchs (Estes et al. 1998; Pace et al. 1999; Heithaus et al. 2008a; Baum and Worm 2009; see Section 16.3.3). Most of these interactions became apparent when the removal of a top predator was followed by changes to community structure. While these cascades often are linked implicitly to direct predation (Baum and Worm 2009) they are the cumulative effect of direct predation and risk effects. Indeed, risk effects may drive, or contribute to, many classic examples of predator–prey interactions traditionally thought to involve direct predation exclusively (Peckarsky et al. 2008). What remains unclear for many observed cascades is the relative contribution of each mechanism. Unraveling these mechanisms is an important task because the magnitude and timing of ecological effects may vary depending on which pathway predominates. For example, because predators inflict risk effects on many more individual prey than they actually kill, a higher density of predators may be necessary to maintain effects resulting from direct predation as opposed to those involving risk effects. Further, direct predation involves energy flow to the elasmobranch population while risk effects do not. Thus, when risk effects predominate over direct predation, aware prey become more difficult to catch and predators may experience a functional reduction of carrying capacity (see Brown, Laundré, and Gurung 1999). In general, direct predation is expected to predominate over risk effects when mesoconsumers are energetically stressed and have less scope for antipredator behavior, while risk effects should predominate when mesoconsumers have abundant resources and are not energetically stressed (Heithaus et al. 2008a).

Community-level effects of predators need not propagate vertically through communities. Predators may enhance community diversity by reducing interspecific competition at lower trophic levels through species-specific changes in prey behavior or by regulating densities of competitive dominants. Predators may also mediate indirect interactions in communities. For example, apparent competition is a negative relationship between the abundances or distributions of two or more prey species that is driven indirectly by a shared predator (Holt 1977). In a direct mortality context, high abundance of a preferred prey type may keep the predator population at higher density, thereby indirectly increasing predator-inflicted mortality on secondary prey. In a risk-effect context, prey may segregate spatially because the predator preferentially targets either one prey type or, if given a chance, the aggregation of several prey types (e.g., Heithaus and Dill 2002; Dill, Heithaus, and Walters 2003; James et al. 2004).

16.3 The Ecological Importance of Elasmobranchs: Knowns, Unknowns, and Best Estimates

16.3.1 Elasmobranchs as Predators

Elasmobranchs generally occupy high trophic levels. Based on standardized estimates of trophic level (see Cortés 1999; Ebert and Bizzaro 2007 for methods), sharks are typically third-order consumers (above trophic level 4), which is similar to marine mammals and significantly higher than the trophic level of seabirds (Cortés 1999). There are, however, notable exceptions, including the orectolobiforms (e.g., nurse, carpet, and zebra sharks), which are secondary consumers (trophic level <4), and zooplanktivorous species (e.g., whale shark, basking shark) that occupy lower trophic levels, more similar to baleen whales. In general, the trophic level of sharks increases with greater body size, and the relationship strengthens if zooplanktivorous species are analyzed as a separate functional group (Cortés 1999). Skates occupy slightly lower trophic levels than sharks—often between trophic level 3.5 and 4.2—but their trophic level also tends to increase with body size (Ebert and Bizzaro 2007). The standardized trophic levels of other batoids have yet to be studied, but available data suggests that their diets are similar to those of skates. Higher proportions of mollusks (primarily bivalves) in the diets of myliobatids and rhinopterids (Gray, Mulligan, and Hannah 1997; Yamaguchi, Kawahara, and Ito 2005; Collins et al. 2007) and filter feeding habits of mobulids (Wetherbee and Cortés 2004), however, suggest lower trophic levels for these groups. Despite their generally high trophic levels, most elasmobranchs, including sharks, also are prey during at least some portion of their lives (see Section 16.3.3). During these vulnerable stages, therefore, they are likely to mediate indirect effects of their predators (see Heithaus 2004) and the spatial and temporal pattern of their own ecosystem effects will be influenced by their need to manage their risk of predation (Heithaus et al. 2008a).

Many elasmobranch studies have analyzed stomach contents (see Cortés 1999; Wetherbee and Cortés 2004; Ebert and Bizzaro 2007) or trophic position based on stable isotopes and other techniques (e.g., Domi, Bouquegneau, and Das 2005; Fisk et al. 2002; see Appendix). Nonetheless, our understanding of how elasmobranch predation impacts prey populations is limited by the scarcity of studies on prey species and on how diets of many species of elasmobranchs might vary in space and time. The importance of a particular prey item in a predator's diet, however, does not necessarily reflect the importance of direct predation or risk effects generated by an elasmobranch predator for that prey species (e.g., Piraino, Fanelli, and Boero 2002; Creel and Christianson 2008).

16.3.1.1 Effects on Prey

It is plausible that large elasmobranchs at high trophic levels provide top-down control of mesopredators. For example, analyses by Myers et al. (2007) linked declines in populations of large-bodied sharks to changes in mesopredator populations. Time-series data from 17 independent research surveys conducted between 1970 and 2005 on the U.S. eastern seaboard showed increasing catch rates for 12 of 14 small elasmobranch mesopredators including skates, rays, and small-bodied sharks. These increases coincided with declining catches of large sharks at higher trophic levels, leading Myers et al. (2007) to suggest predatory release as a mechanism. Other analyses are consistent with this hypothesis. In the tropical Pacific, increases in catches of pelagic stingrays (*Pteroplatytrygon violacea*) and other small mesoconsumers coincided with declining catch rates of large sharks and other large predators (Ward and Myers 2005). Similarly, in the Gulf of Mexico, there were declines in catches of large coastal sharks and increases in catches of two deep-water, but not shallow-water, elasmobranchs (Shepherd and Myers 2005). Finally, off the coast of South Africa, declines in the abundance of large sharks may have resulted in increases in smaller sharks (van der Elst 1979).

The predatory release hypothesis, however, is controversial. In some cases, existing diet data do not support trophic links between declining sharks and increasing mesopredators and in other cases distributions of declining predators do not overlap with increasing mesopredators. Also, the speed of some mesopredator population increases does not appear to be consistent with their slow life history parameters (e.g., cownose rays, *Rhinoptera bonasus*; age of maturity 8 to 10 years, one pup produced annually; Smith and Merriner 1986, 1987). Finally, there are other possible mechanisms that might explain increasing catches of mesopredators that are not mutually exclusive with predatory release, including population redistribution similar to that by skates in the northwest Atlantic Ocean (Frisk et al. 2008), increases in population densities at higher latitudes of their range in response to climatic variation as seen for reef fishes off the coast of North Carolina (Parker and Dixon 1998), competitive release in response to overfishing of previously dominant mesopredators (Link 2007), and changes in the timing of migration relative to sampling periods. Further studies of large shark diets and mesopredator populations should help to elucidate the relative importance of predatory and competitive release and other mechanisms in regulating populations of elasmobranch, and other, mesopredators.

In some situations, sharks can exert top-down control of marine mammal populations. For example, off Sable Island, in the northwest Atlantic Ocean, direct predation by sharks was estimated to contribute to at least 50% of the decline in harbor seal (*Phoca vitulina*) pup production from 1995 to 1997 (Lucas and Stobo 2000; Bowen et al. 2003). For Prince William Sound, Alaska, a combination of theoretical models and empirical data suggest that harbor seals and stellar sea lions (*Eumetopias jubatus*) underutilize deep water resources (i.e., use less than expected based on their potential profitability) to reduce the risk of predation from Pacific sleeper sharks (*Somniosus pacificus*; Frid et al. 2007, 2009; Frid, Baker, and Dill 2008). Theoretical models predict that declines of near-surface resources, such as Pacific herring, could force seals and sea lions to increase deep foraging, thereby indirectly raising rates of shark predation on these pinnipeds (Frid, Baker, and Dill 2006; Frid et al. 2009).

Studies in Shark Bay, Western Australia, have revealed strong risk effects of tiger sharks (*Galeocerdo cuvier*) on a variety of mesoconsumers (see Heithaus et al. 2008a, 2008b; Wirsing et al. 2008 for summaries). Briefly, almost all potential prey of tiger sharks that have been investigated—piscivorous dolphins (*Tursiops aduncus*), cormorants (*Phalacrocorax varius*) and seasnakes (*Hydrophis elegans*), and herbivorous green turtles (*Chelonia mydas*) and dugongs (*Dugong dugon*)—shift habitats in response to changes in tiger shark predation risk (Heithaus and Dill 2002, 2006; Heithaus 2005; Heithaus et al. 2007b, 2009; Wirsing, Heithaus, and Dill 2007a, 2007b; Wirsing et al. 2008; Kerford et al. 2008). Prey species respond to increased risk, which is highest in more productive shallow habitats, by shifting into safer but less productive deep habitats. In addition, dugongs change their foraging behavior from excavating more nutritious seagrass rhizomes during periods of low shark abundance to cropping less nutritious seagrass leaves when shark abundance is high. This is thought to occur in order to avoid creating sediment plumes during excavation that would interfere with dugongs' ability to detect sharks (Wirsing, Heithaus, and Dill 2007c). By reducing rates of resource acquisition, tiger sharks likely reduce prey populations below the levels that might be reached in the absence of sharks, even if direct predation is not substantial (Creel, Christianson, and Winnie 2007; Creel and Christianson 2008; Heithaus et al. 2008b).

One important result from the Shark Bay studies is that risk effects of tiger sharks may vary with individual condition. Green turtles in good condition respond strongly to the risk of tiger shark predation presumably because they can afford to give up more foraging opportunities to be safe. In contrast, green turtles in poor condition take greater risks because the foraging costs of predator avoidance may be less affordable to them (Heithaus et al. 2007b). State-dependent behaviors of this sort provide the mechanisms via which resource availability and predation can synergistically regulate prey populations (e.g., Anholt and Werner 1995; Sinclair and Arcese 1995; Heithaus et al. 2008a; Frid et al. 2009).

Another interesting finding is that not all prey species respond to the presence of tiger sharks in the same way (Heithaus et al. 2009). Despite higher shark encounter rates, species with maneuverability superior to that of tiger sharks (e.g., dolphins, dugongs) shift into edge microhabitats, which appear to facilitate enhanced escape probabilities, as tiger shark abundance increases. In contrast, species that are not able to modify their escape probability by shifting habitats (e.g., cormorants) move into interior habitats where shark encounter rates are lower. These behavioral contrasts between prey types suggest that predator abundance alone does not necessarily provide a good measure of overall predation risk, which is determined by the combined probabilities of encountering predators, being attacked given an encounter, and escaping to safety given an attack (see also Lima and Dill 1990; Hebbelwhite, Merrill, and McDonald 2005) and indirect effects of predator presence may be the reverse of that expected based on predator distributions (Heithaus et al. 2009).

Risk effects imposed by large sharks may play an important role in regulating population sizes of other shark species, including their own. Young of many species of sharks are found in restricted nursery areas to reduce the risk of predation and cannibalism (see Heithaus 2007; Heupel, Carlson, and Simpfendorfer 2007 for reviews). By staying within these nurseries, the resources available to young sharks are constrained relative to their overall availability in the environment (see Heithaus 2004). Thus, although populations may seem to be regulated within nurseries by competition, this is a direct result of high risk of predation outside nurseries ("foraging arena hypothesis"; see Walters and Juanes 1993). Direct predation by large sharks on larger juveniles outside of nurseries, however, might also be important in regulating population sizes of some shark species. Elasticity/sensitivity analyses of demographic models for a variety of shark species suggest that intrinsic rates of increase are most heavily influenced by mortality rates of older juveniles that reside outside of nurseries (see Kinney and Simpfendorfer 2009 for a review) and are prey for some large sharks.

Most studies of the ecological role of batoids have focused on their overall effects on invertebrate community structure rather than on specific prey taxa (VanBlaricom 1982; Thrush et al. 1994). Thrush et al. (1994) found that excluding predators (sea birds and eagle rays together) allows increases in population sizes of some potential prey, including clams and sea cucumbers. Exclosures that only excluded birds (there were no exclosures that only excluded rays) were not significantly different from reference plots, suggesting that ray predation can lead to decreased population sizes of benthic invertebrates. Indeed, Hines et al. (1997) estimated that in this system eagle rays (*Myliobatis tenuicaudatus*) consumed 1.6% of the *Macomona* clam population during the 31-day study, with foraging focused on areas of high clam density where rays inflicted 4% mortality. Such selective foraging in areas of high prey densities may lead to rays stabilizing prey populations and reducing spatial heterogeneity in prey distribution (Hines et al. 1997). In some locations, batoid predation can be stronger. Cownose rays completely removed bay scallops (*Argopecten irradians*) from productive habitats, turning these habitats into population sinks (Peterson et al. 2001). These population sinks can have dramatic effects on scallop population dynamics because cownose predation and the resulting scallop population crash precede the annual spawning of the scallops. Although relatively few adults are required to create substantial larval supply, enhanced cownose ray predation may decrease recruitment and population sizes (Myers et al. 2007).

While the above studies suggest strong effects of elasmobranchs on their prey in some situations, in others elasmobranch predation may be much weaker. For example, mako sharks (*Isurus oxyrinchus*) consume 4% to 14% of bluefish populations along a stretch of the U.S. east coast, but appear not to influence equilibrium population sizes substantially (Stillwell and Kohler 1982). Similarly, Ellis and Musick (2007) concluded that the dietary flexibility of juvenile sandbar sharks (*Carcharhinus plumbeus*) limits their potential impacts on prey populations through direct predation, and an ecosystem model predicted little influence of direct predation by small coastal sharks on populations of their prey (Carlson 2007).

16.3.1.2 Cascading Effects

Recent research has provided insights into how predation by elasmobranchs might influence community dynamics. For example, the Myers et al. (2007) study discussed in Section 16.3.1.1 proposed that declines in populations of large sharks led to increases in catches of cownose rays and that this larger ray population caused the collapse of the North Carolina bay scallop population. There is little argument that large shark populations have declined or that cownose rays can depress scallop populations. Indeed, cownose rays clearly have been causing major disturbances to benthic habitats in Chesapeake Bay for years (Orth 1975; see this section below) and exclosure studies show that cownose rays can collapse scallop populations (Peterson et al. 2001; Myers et al. 2007). However, the relative importance of predatory release in driving increases in ray catches remains unclear (see Section 16.3.1.1) and critics have argued that the timing of increases in ray catches (1990s) is not consistent with the timing of collapse in bay scallop fisheries (1980s). Because of the magnitude of changes in the abundance of large shark populations in the northwest Atlantic, and around the world, elucidating the importance of shark predation in inducing trophic cascades is a high research priority.

Studies in Shark Bay suggest that tiger sharks likely play important roles as transmitters of indirect effects among their prey species (e.g., mediate apparent competition) and as initiators of indirect effects within the community. First, seasonal changes in dugong density appear to help initiate fluctuations in the abundance of large (>3.5 m total length) sharks; Wirsing, Heithaus, and Dill 2007d), which prefer shallow habitats where the abundance of prey, including dugongs, is highest (Heithaus et al. 2002). Shallow habitats also have the highest abundance of prey for dolphins, but high shark densities induce most dolphins to abandon these habitats to reduce their risk of predation (Heithaus and Dill 2002). Thus, tiger sharks mediate an indirect effect between herbivorous dugongs and piscivorous dolphins in a behavioral analog of apparent competition (Heithaus and Dill 2002; Dill, Heithaus, and Walters 2003). More importantly, antipredator responses by prey mediate indirect effects of tiger sharks on seagrass communities; the spatial pattern of nutrient composition of seagrass leaves is consistent with spatial shifts in herbivory observed in response to tiger shark predation risk (Heithaus et al. 2007b). Based on experimental studies elsewhere that show strong top-down effects of turtle and dugong grazing on seagrass communities (e.g., Preen 1995; Masini, Anderson, and McComb 2001), it is likely that shark-induced spatial shifts in herbivory and changes in dugong foraging behavior in Shark Bay have cascading effects on seagrass community composition and the structure of faunal communities associated with them (Heithaus et al. 2008a, 2008b). Data from Bermuda support the possibility that tiger shark removals would have significant consequences for seagrass communities (Heithaus et al. 2008b). Along with steep declines in shark abundance (Baum et al. 2003) there have been increases in herbivorous green turtle populations and losses of seagrass (Murdoch et al. 2007). Ongoing exclosure experiments will help to test whether seagrass declines in Bermuda and spatial patterns of nutrient composition in Shark Bay are driven by cascading top-down effects of tiger sharks.

The relative strength of top-down control and the presence of trophic cascades and keystone species vary considerably among ecosystems (Power et al. 1996; Piraino, Fanelli, and Boero 2002). Unsurprisingly, ecosystem models suggest an important role for elasmobranchs for some but not all ecosystems (Libralato, Christensen, and Pauly 2006). In general, these models suggest that sharks play key roles in determining ecosystem dynamics when (1) large-bodied sharks are major predators on other large-bodied species, or (2) where large teleost predators don't overlap with sharks in their habitat use. For example, ecosystem models suggest that removal of salmon sharks (*Lamna ditropis*) in the Alaska Gyre (north Pacific), tiger sharks in French Frigate Shoals (central Pacific), Galapagos (*Carcharhinus galapagensisi*) and white-tipped reefs sharks (*Triaenodon obesus*) in the Galapagos Islands, and of several species of small-bodied sharks in the Northeast Venezuela Shelf could disrupt trophic cascades (Stevens et al. 2000; Okey et al. 2004).

Similar to the above ecosystem models, the sleeper shark–harbor seal behavioral optimization model of Frid, Baker, and Dill (2008) suggests that the removal of sharks would have cascading effects on the ecosystem. Seal avoidance of sharks is predicted to reduce seal predation rates on profitable deep-strata fish while increasing seal predation on shallow-strata fish. In response to simulated removals of sharks, seals reversed their foraging preferences, leading to increased seal predation on deep-strata fish and relaxed predation of shallow-water fishes. Because harbor seals are generally abundant in the northeast Pacific (though declining in some areas; Small, Pendleton, and Pitcher 2003), the release of shark predation risk for seals could potentially cause considerable changes to the teleost community.

In contrast to the above examples, ecosystem models do not predict a keystone role for sharks in tropical pelagic systems where large teleosts have similar trophic levels. Models based on the current trophic structure of the central North Pacific ecosystem and simulated trophic dynamics with greater shark predation (e.g., higher shark trophic levels, higher shark abundance) that might have occurred before fishing reduced their popula-tions (Kitchell et al. 2002) suggest that the community would not change dramatically due to shark declines (despite potentially large changes within the shark components of the food web) due to compensatory responses of billfishes and tunas. Kitchell et al. (2002) con-cluded that it is unlikely that sharks would play a keystone role in tropical and subtropical pelagic ecosystems where scombrid and xiphoid fishes occur because the food habitats and life histories of these teleosts make them capable of rapid compensatory responses following shark population declines.

Ecosystem models of coral reef habitats differ in their predictions. In the French Frigate Shoals model, simulated removal of reef sharks has little community-level effect, because reef fish appear not to be regulated by predation (Stevens et al. 2000). An ecosystem model of the Caribbean, however, suggests that although strong interactions and the possibil-ity of cascades was generally low, sharks were involved in 48% of the strong tri-trophic interactions, potentially influencing community stability (Bascompte, Melián, and Sala 2005). Intraguild predation (IGP), an interaction in which predator and prey compete for basal resources, can buffer trophic cascades. Sharks (tiger sharks, Caribbean reef sharks *Carcharhinus perezi*, and silky sharks *Carcharhinus falciformis*) were major IGP players in the Caribbean model (Kondoh 2008), but also were involved in strong tri-trophic interac-tions not buffered by IGP (Bascompte, Melián, and Sala 2005). The role of sharks in IGP interactions warrants further investigation since it is predicted to have major implications for the strength of trophic cascades and ecosystem stability and IGP appears to be a com-mon aspect of trophic interactions involving large sharks (Kitchell et al. 2002; Bascompte, Melián, and Sala 2005; Kondoh 2008).

Batoid foraging can influence the structure of benthic communities. Meiofaunal species disrupted by batoid excavations are found at lower densities in foraging pits and heavy ray foraging can reduce densities over larger areas (Cross and Curran 2000, 2004). Disturbance, coupled with higher rates of organic matter accumulation can allow for the persistence of early colonist species such as small amphipods that can exploit the organic matter found in ray pits. These early colonists are not found in structurally similar systems that lack small-scale disturbance, such as ray feeding events (VanBlaricom 1982). As a result, disturbance by rays coupled with rapid recolonization may be important in maintaining dominance patterns in soft bottom communities (Thrush et al. 1991).

Batoid foraging also may structurally change an ecosystem. For example, schools of cownose rays have fragmented large seagrass beds in Chesapeake Bay (Hovel and Lipcius 2001) and, in an extreme case, destroyed approximately 90 hectares of seagrass beds and substantially eroded sediments (Orth 1975). By altering these habitats, the rays almost certainly modify the distribution and abundance of seagrass-associated species. While myliobatids and rhinopterid rays likely disrupt seagrass beds heavily, dasyatid rays prob-ably have lesser effects on habitat structure. In enclosure experiments, Valentine et al. (1994) found no significant destruction of seagrasses even when foraging densities of large (>90 cm disc width) dasyatid rays were elevated experimentally.

Collectively, the above studies suggest that elasmobranchs can play important roles in structuring communities. These effects could be particularly pronounced for invertebrate communities impacted by batoid foraging and for large-bodied sharks in coastal areas (away from scombrid and xiphoid fishes). Effects may be more pronounced for species that consume or threaten prey with slow life histories that are capable of considerable ecosystem effects themselves (e.g., marine mammals, other elasmobranchs). However, we are only beginning to comprehend the complex ways that elasmobranchs might influence their communities and ecosystems.

16.3.2 Elasmobranch Roles in Nutrient Dynamics

Consumers can influence nutrient dynamics within ecosystems and the transfer of nutrients among ecosystems. In oligotrophic ecosystems, spatial and temporal patterns of primary productivity can be controlled by inputs of nutrients from outside the ecosystem ("spatial subsidies" or "allochthonous inputs"; Polis and Strong 1996; Persson et al. 1996; Heck et al. 2008) that often are driven by consumer movements across ecosystem boundaries. For example, Pacific salmon migrating from the ocean to spawn in rivers and streams add oceanic nitrogen to freshwater and terrestrial habitats (Helfield and Naiman 2001). Also, consumers associated with coral reefs and mangroves import productivity from adjacent seagrass habitats where they forage (Valentine and Heck 2005; Heck et al. 2008; Valentine et al. 2008). Within ecosystems, consumers can create hotspots of nutrient recycling through excretion in habitats where they are most abundant (McIntyre et al. 2008; Herbert and Fourqurean 2008) or through bioturbation and the resuspension of sediments. In fact, bioturbation is a key driver of the high biological activity found in marine sediment of coastal zones (see Kogure and Wada 2005 for a review); even small-scale resuspension events caused by benthic fish (Yahel, Yahel, and Genin 2002; Yahel et al. 2008) can release buried nutrients into the water column, where they become available to microorganisms and primary producers. The ensuing increase in microbial activity can significantly reduce the amount of organic carbon sequestered in sediments (Yahel et al. 2008).

Bioturbation by batoids provides a clear example of elasmobranch roles in nutrient dynamics. Many batoids feed on infaunal species by excavating prey with water jets from the mouth or gills (Gregory et al. 1979; Sasko et al. 2006), possibly aided by flapping or undulation of the pectoral fins (Howard, Mayou, and Heard 1977). Using these methods, batoids can create feeding pits up to 40 cm deep (Smith and Merriner 1985) and can cause a great deal of sediment turnover. For example, on a New Zealand intertidal sandflat, Thrush et al. (1991) estimated that foraging by eagle rays could completely turn over a 700 to 800 m² area in 70 days. Similarly high rates of bioturbation were found in the sandflats of Bahia La Choya, Mexico, where batoids were estimated to rework the sandflats area to a depth of 20 cm in 72 days (Myrick and Flessa 1996). Several studies in known batoid foraging grounds report that, on average, approximately 1% (and up to about 5%) of a given study site is covered by new foraging pits per day (Reidenauer and Thistle 1981; VanBlaricom 1982; Sherman et al. 1983) and up to 30% of some areas were covered with foraging pits of various ages (Grant 1983). In addition to foraging pits, batoids also turn over sediment during their burial and emergence from the substrate. Work with flatfish indicates that these behaviors can be a significant source of sediment resuspension (Yahel et al. 2008). Unfortunately, the direct effects of batoid bioturbation on nutrient fluxes have not been examined. In areas of high tidal reworking the volume of sediment reworked by rays may be <1% of that of ripple migration (Grant 1983). However, areas where foraging pits remain for extended periods tend to accumulate organic matter faster than adjacent areas (VanBlaricom 1982). High levels of organic matter settling in ray pits can cause rapid remineralization of the organic matter in the sediments below the pits and can decrease the impact of tidally driven advective porewater flux, the primary agent for enhancing remineralization rates and carbon cycling in sands (D'Andrea, Aller, and Lopez 2002).

Have changes in shark biomass altered energy and nutrient transfer rates in some ecosystems? Empirical data directly addressing this question are lacking, but consider the following. The top predator biomass in relatively undisturbed Pacific reefs far exceeds that of nearby fished reefs, suggesting that top predators, including sharks, were historically much more abundant. For example, in the northern Line Islands, at the unfished reefs of Kingman and Palmyra atolls, top predators made up 85% and 44% of total fish bio-mass, respectively, compared to 19% at nearby heavily fished Kiritamati (Sandin et al. 2008; Stevenson et al. 2007). Reef sharks accounted for 74% of top predator biomass at Kingman and 57% at Palmyra, but virtually absent from Kiritimati (Sandin et al. 2008) where the top predator guild consisted of smaller groupers (Sandin et al. 2008). Similarly, large predator (sharks, jacks, and large snapper) biomass was 54% of fish biomass in the remote Northwest Hawaiian Islands (Friedlander and DeMartini 2002) compared to only 3% to 7% off the more heavily fished Main Hawaiian Islands and Line Island Atolls (Christmas and Fanning Island). Replacing such a large biomass of long-lived and possibly wide-ranging predators with smaller short-lived species (e.g., small teleosts and invertebrates) suggests not only strong implications for community structure, but also for nutrient cycling and energy flow (e.g., Gascuel et al. 2008).

Elasmobranchs may transport nutrients across ecosystem boundaries or modify pat-terns of consumer-mediated nutrient transport by their prey. For example, many species of teleosts take shelter from predators, including elasmobranchs, among mangrove prop roots or in reefs. Yet, many of these species forage in adjacent seagrass habitats and then transport seagrass-derived nutrients and energy into reef and mangrove habitats where they are safer (e.g., Valentine and Heck 2005; Valentine et al. 2008). Declines in elasmo-branch predators, therefore, might reduce cross-ecosystem transfers of nutrients in coastal systems because of a relaxation in predation risk to mesoconsumers in seagrass habitats, which might then reduce their use of refuge habitats like reefs and mangroves. Whether elasmobranch declines might enhance or reduce nutrient flow is difficult to predict, how-ever, and is likely to depend on the trophic position of elasmobranchs relative to species important to cross-boundary transfers (Estes et al. 1998). Furthermore, such effects are unlikely to manifest unless the entire guild of large predators declines. Indeed, despite reductions in many (but not all) upper trophic level predators in the Florida Keys, con-sumer-mediated linkages between coral reef and seagrass habitats have been maintained (Valentine et al. 2008).

Finally, coastal sharks may import nutrients into near-shore nursery areas. Juvenile mortality may approach 90% in the first several weeks (Heupel and Simpfendorfer 2002), adding nutrients in a process analogous, though of lower magnitude, to that of salmon carcasses in Alaskan streams. Surviving young may also play a role in nutrient dynamics if they forage in areas away from where they spend the majority of their time and deposit nutrients through excretion. Movements of blacktip sharks support this possibility; young sharks primarily select habitats within their coastal nurseries to avoid predators (larger sharks) and make relatively short excursions from core areas (Heupel and Hueter 2002) possibly to forage.

16.3.3 Elasmobranchs as Prey

Because of their large body size, most elasmobranchs generally are preyed upon only by larger elasmobranchs and, in some areas, by killer whales (*Orcinus orca*) (Heithaus 2004). Killer whales off New Zealand's North Island, for instance, may specialize in eating sharks and batoids; whales made successful attacks on elasmobranchs during 80% of researcher encounters with whales (Visser 1999, 2005). Several species of large elasmobranchs prey largely on smaller sharks and batoids, but the extent to which these prey items dominate their diet is variable (Cortés 1999; Wetherbee and Cortés 2004). Whether population sizes of large elasmobranchs depend on the availability of smaller elasmobrach prey items is unknown.

16.3.4 Elasmobranchs as Facilitators

Few studies have focused on the role of elasmobranchs as facilitators. However, sharks and rays have been documented at cleaning stations (Sazima and Moura 2000) and several species of teleosts that maintain close proximity with batoids and sharks (e.g., cobia, *Rachycentron canadum*, remora, "pilot fishes") might benefit from reduced risk of predation or access to food (Smith and Merriner 1982). Also, benthic foraging by batoids and some sharks might facilitate other predators that can take advantage of injured or disturbed prey. For example, bat rays (*Myliobatus californica*) and round stingrays (*Urobatis halleri*) facilitate foraging by sand dabs (*Citharichthys stigmaeus*) by ejecting infaunal prey that are otherwise unavailable to sand dabs (VanBlaricom 1982). Such facilitation of other predators by ray foraging is likely to occur in diverse systems. Elasmobranch modification of benthic habitats also may play a facilitating role in communities. For example, cownose rays appear to facilitate juvenile crab survival (Hovel and Lipcius 2001). In mid- to late summer, cownose rays forage heavily in seagrass habitats fragmenting otherwise continuous beds. This process leads to enhanced survival of juvenile crabs because the density of their chief predator—adult blue crabs—is lower in the fragmented habitat. Also, the removal of species (including prey and nonprey) from foraging areas makes habitats available for early colonizing invertebrate species (VanBlaricom 1982).

16.4 Toward a Deeper Understanding of the Ecological Importance of Elasmobranchs

Elasmobranchs can play crucial roles in marine ecosystems through diverse mechanisms, yet their relative ecological importance varies among species, ecosystems, and contexts (Table 16.1). Like other marine predators, elasmobranchs likely will play a greater role in marine communities when

1. They prey upon, or threaten, longer-lived species that are more likely to invest in antipredator behavior or are more susceptible to predation.
2. They are the primary predator on a limited number of prey (more specialized diets) and, therefore, could have larger impacts on equilibrium population sizes of these species through direct predation.
3. They prey upon or threaten species that are likely to play an important ecological role themselves due to trophic position (e.g., herbivores) or have the potential to influence populations of their own prey (e.g., high metabolic rates of marine mammals).
4. They are capable of inflicting major structural changes to environments.
5. Their predation is focused on life history stages of prey where density-dependent selection occurs (but see section 16.2.3 for how predators can induce trophic cascades in the absence of effects on mesoconsumer population sizes).
6. There are no trophically similar predators capable of exerting similar direct predation and effects on shared prey (see Piraino, Fanelli, and Boero 2002; Heithaus et al. 2008a for reviews of these concepts).

TABLE 16.1

Studies Addressing the Ecological Importance of Elasmobranchs

Species	Location and Habitat	Large Role?	Proposed Mechanism	Methods	References
Tiger shark	Shark Bay, Australia; seagrass ecosystem	Yes	Risk effects on key mesoconsumers induces changes in seagrass nutrient dynamics, possibly community composition; potential risk-induced reductions in population sizes of mesoconsumers	Field observations under temporal variation in shark abundance; seagrass nutrient analysis	Reviews in Heithaus et al. 2007c, 2008a, 2008b; Wirsing et al. 2008
Tiger shark	Bermuda, seagrass ecosystem	Possible	Possible mesoconsumer release of green turtles with shark decline leading to overgrazing	Time-series analysis (sharks); seagrass surveys	Murdoch et al. 2007; Heithaus et al. 2008b
Large coastal sharks	Northwest Atlantic Ocean	Possible	Declines in large sharks may cause release of ray mesoconsumers and declines in shellfish	Exclosure studies; time-series analyses	Myers et al. 2007
Pelagic sharks	Pacific Ocean	Possible	Declines in sharks accompanied with mesoconsumer release; other large predators declined concurrently; relative effect of shark loss unclear	Time-series analysis	Ward and Myers 2005
Pelagic sharks	Central Pacific	No	Direct predation; predicted compensatory response in large teleosts mitigates effects of shark declines	Ecosystem model	Kitchell et al. 2002
Reef sharks	Pacific Islands	No	Direct predation but reef fish not regulated by shark predation	Ecosystem model	Stevens et al. 2000
Reef sharks	Caribbean	Yes	Strong predation on key species	Ecosystem model	Bascompte et al. 2005
Galapagos and white-tipped reef sharks	Galapagos Islands	Yes	Direct predation	Ecosystem model	Okey et al. 2004
Small coastal sharks	Northern Gulf of Mexico	No	Direct predation	Ecosystem model	Carlson 2007
Mako shark	NW Atlantic	No	4%–14% of bluefish consumed; little effect on bluefish population predicted	Bioenergetic model; prey population dynamics estimates	Stillwell and Kohler 1982
Greenland shark and white sharks	NW Atlantic Ocean	Yes	Direct predation led to declines in harbor seals	Observed shark kills, seal population surveys	Lucas and Stobo 2000; Bowen et al. 2003

Group	Location	Effect	Description	Method	Reference
Sleeper sharks	NE Pacific Ocean	Yes	In computer experiments, risk of sleeper shark predation induces pinnipeds to reduce foraging on profitable prey at depth, enhancing pinniped predation on shallow-water fishes; could influence population sizes of pinnipeds through direct predation and risk effects	Behavioral optimization model; empirical data on pinniped diving	Frid, Baker, and Dill 2006, 2008; Frid et al. 2007, 2009
Large coastal sharks	South Africa	Possible	Small sharks proliferated with decline of large sharks; effect may not be as extreme as initially thought	Time-series data	Van der Elst 1979
Bat rays, round stingrays	California; subtidal sand community	Yes	Ray foraging changed benthic invertebrate community allowing persistence of species that are disturbance-adapted	Exclosures	VanBlaricom 1982
Eagle rays	New Zealand intertidal sandflats	Yes	Create more homogeneous densities of bivalves across sand flats by foraging selectively in high bivalve densities; depress populations of some invertebrate prey	Exclosures; field observations	Thrush et al. 1991, 1994; Hines et al. 1997
Cownose rays (normal population sizes and foraging patterns)	North Carolina; shallow coastal bank	Moderate	Can extirpate scallops from high-quality habitat but scallops from low-quality habitat can compensate reproductively	Exclosures	Peterson et al. 2001
Cownose rays (increased populations)	North Carolina; shallow coastal banks	Yes	Induce declines in scallop populations	Exclosure studies; timeseries analyses	Myers et al. 2007
Cownose rays	NW Atlantic; seagrass ecosystem	Yes	Rhizomes broken, large areas of seagrass habitat fragmented or destroyed	Field observation	Orth 1975; Hovel and Lipcius 2001
Southern stingray	Gulf of Mexico; seagrass ecosystem	No	No major effects on seagrass cover	Enclosure study	Valentine et al. 1994
Batoids	NW Atlantic; high tidal reworking of sediment	No	Enhanced nutrient cycling through bioturbation minimal due to tidal effects	Field study	Grant 1983
Batoids	NW Atlantic; low tidal reworking of sediment	Yes	Enhanced nutrient cycling through bioturbation	Field study	Grant 1983

From these general concepts it should be possible to predict when and where elasmobranchs influence community and ecosystem processes. Further work, however, clearly is required to fill gaps in our understanding. First, adequate data on trophic interactions are available for only few species, and the extent of intraspecific variation in diet preferences (individual specialization) is generally unknown. Studies of individual specialization are important to our understanding of ecological and evolutionary dynamics and may also contribute to conservation strategies (e.g., Bolnick et al. 2003, 2007). Second, studies of elasmobranchs tend to focus on particular life history stages, but an integrated understanding of how elasmobranch populations are regulated and how their ecological role changes with body size is important. For example, if large individuals prey on unique prey items not consumed by smaller individuals, as is the case for copper sharks (*Carcharhinus brachyurus*), removal of the largest individuals may lead to the loss of unique ecological functions, even if the overall population persists (Lucifora et al. 2009). Third, we need studies addressing the extent to which trophic redundancy buffers trophic cascades. Although in some contexts predators may fill similar roles (e.g., tuna, billfishes, and sharks; Kitchell et al. 2002), top predators with similar diets are not necessarily redundant and the presence of multiple predators can still induce complex direct effects on shared prey and strong cascading effects (Siddon and Witman 2004; Bruno and Cardinale 2008). Fourth, further investigations are needed on the importance of intraguild predation in buffering communities from trophic cascades and the possibility that the loss of strongly interacting intraguild predators, like sharks in the Caribbean, could leave ecosystems more susceptible to further perturbations. Finally, and perhaps most critically, we need studies of species that elasmobranchs prey on and of the communities in which these prey species are embedded. In particular, the effect of direct predation and of risk effects imposed by elasmobranchs on their prey must be better understood in order to understand how elasmobranchs indirectly influence lower trophic levels.

Recent advances in methods for studying food webs, general ecological theory, comparative analyses of the wider literature, and experimental techniques can potentially advance research on the ecological role of sharks and their relatives. Potentially fruitful avenues include exclosure experiments and comparisons of communities across temporal or spatial gradients of shark densities. In particular, studies in relatively undisturbed ecosystems are needed as baselines against which more exploited ecosystems are compared. Studies in reef and open ocean habitats, where the effects of shark removals are particularly unclear, should be a priority.

Hopefully, the upcoming years will see a proliferation of studies on the ecological roles of elasmobranchs. We particularly hope to see the formulation of rigorous predictions on how communities might change under various scenarios of resource exploitation and climate change, and of empirical research programs addressing these predictions. The ensuing body of work could guide priorities for conservation and management. We can only hope that these priorities will eventually move from halting population declines to restoring populations to the ecologically meaningful densities that maintain the trophic interactions inherent to community diversity and ecosystem resilience (Soule et al. 2005; Chesson and Kuang 2008).

Acknowledgments

During the preparation of this chapter, AF was supported by an NSERC-Canada IRFD fellowship. Work by MH was supported by NSF grants OC0745606 and DBI0620409. BW acknowledges funding by the Lenfest Ocean Program and NSERC. Thanks to Jack Musick and Dean Grubbs for helpful comments on this chapter.

References

Abrams PA (1995) Implications of dynamically variable traits for identifying, classifying, and measuring direct and indirect effects in ecological communities. Am Nat 146: 112–134.

Abrams PA (2008) Measuring the impact of dynamic antipredator traits on predator-prey-resource interactions. Ecology 89: 1640–1649.

Abrams PA (2009) When does greater mortality increase population size? The long history and diverse mechanisms underlying the hydra effect. Ecol Lett 12: 462–474.

Abrams PA, Matsuda H (2005) The effect of adaptive change in the prey on the dynamics of an exploited predator population. Can J Fish Aquat Sci 62: 758–766.

Abrams PA, Quince C (2005) The impact of mortality on predator population size and stability in systems with stage-structured prey. Theor Popul Biol 68: 253–266.

Anholt BR, Werner EE (1995) Interaction between food availability and predation mortality mediated by adaptive behavior. Ecology 76: 2230–2234.

Aronson RB, Blake DB (2001) Global climate change and the origin of modern benthic communities in Antarctica. Am Zool 41: 27–39.

Bascompte JC, Melián CJ, Sala E (2005) Interaction strength combinations and the overfishing of a marine food web. Proc Natl Acad Sci USA 102: 5443–5447.

Baum JK, Myers RA (2004) Shifting baselines and the decline of pelagic sharks in the Gulf of Mexico. Ecol Lett 7: 135–145.

Baum JK, Myers RA, Kehler DG, Worm B, Harley SJ, Doherty PA (2003) Collapse and conservation of shark populations in the Northwest Atlantic. Science 299: 389–392.

Baum JK, Worm B (2009) Cascading top-down effects of oceanic predators. J Anim Ecol 78: 699–714.

Beck CA, Iverson SJ, Bowen WD (2005) Blubber fatty acids of gray seals reveal sex differences in the diet of a size-dimorphic marine carnivore. Can J Zool 83: 377–388.

Bergstrom U, Englund G, Leonardsson K (2006) Plugging space into predator-prey models: an empirical approach. Am Nat 167: 246–259.

Bolnick DI, Scanbäck R, Araújo MS, Persson L (2007) Comparative support for the niche variation hypothesis that more generalized populations also are more heterogeneous. Proc Natl Acad Sci USA 104: 10075–10079.

Bolnick DI, Scanbäck R, Fordyce JA, Yang LH, Davis JM, Hulsey CD, Forister ML (2003) The ecology of individuals: incidence and implications of individual specialization. Am Nat 161: 1–28.

Bowen WD, Ellis SL, Iverson SJ, Boness DJ (2003) Maternal and newborn life-history traits during periods of contrasting population trends: implications for explaining the decline of harbour seals, *Phoca vitulina*, on Sable Island. J Zool (London) 261: 155–163.

Brown JS, Kotler B (2004). Hazardous duty pay and the foraging cost of predation. Ecol Lett 10: 999–1014.

Brown JS, Laundré JW, Gurung M (1999) The ecology of fear: optimal foraging, game theory, and trophic interactions. J Mammal 80: 385–399.

Bruno JF, Cardinale BJ (2008) Cascading effects of predator richness. Front Ecol Environ 6: 539–546.

Cairns DK, Gaston AJ, Huettmann F (2008) Endothermy, ectothermy and the global structure of marine vertebrate communities. Mar Ecol Prog Ser 356: 239–250.

Carlson JK (2007) Modeling the role of sharks in the trophic dynamics of Apalachicola Bay, Florida. Am Fish Soc Symp 50: 281–300.

Chasar LC, Chanton JP, Koenig CC, Coleman FC (2005) Evaluating the effect of environmental disturbance on the trophic structure of Florida Bay, U.S.A.: multiple stable isotope analyses of contemporary and historical specimens. Limnol Oceanogr 50: 1059–1072.

Chesson P, Kuang JJ (2008). The interaction between predation and competition. Nature 456: 235–238.

Christensen V, Walters CJ (2004) ECOPATH with ECOSIM: methods, capabilities, and limitations. Ecol Model 172: 109–139.

Church MR, Ebersole JL, Rensmeyer KM, Couture RB, Barrows FT, Noakes DLG (2009) Mucus: a new tissue fraction for rapid determination of fish diet switching using stable isotope analysis. Can J Fish Aquat Sci 66: 1–5.

Clarke SC, McAllister MK, Milner-Gulland EJ, Kirkwood GP, Michielsens CGJ, Agnew DJ, Piktich EK, Nakano H, Shivji MS (2006) Global estimates of shark catches using trade records from commercial markets. Ecol Lett 9: 1115–1126.

Collins AB, Heupel MR, Hueter RE, Motta PJ (2007) Hard prey specialists or opportunistic generalists? An examination of the diet of the cownose ray, *Rhinoptera bonasus*. Mar Freshw Res 58: 135–144.

Cortés E (1997) A critical review of methods of studying fish feeding based on analysis of stomach contents: application to elasmobranch fishes. Can J Fish Aquat Sci 54: 726–738.

Cortés E (1999). Standardized diet compositions and trophic levels of sharks. ICES J Mar Sci 56: 707–717.

Creel S, Christianson D (2008) Relationships between direct predation and risk effects. Trend Ecol Evol 23: 194–201.

Creel S, Christianson D, Winnie JA Jr (2007) Predation risk affects reproductive physiology and demography of elk. Science 315: 960.

Cross RE, Curran MC (2000) Effects of feeding pit formation by rays on an intertidal meiobenthic community. Estuar Coast Shelf Sci 51: 293–298.

Cross RE, Curran MC (2004) Recovery of meiofauna in intertidal feeding pits created by rays. Southeast Nat 3: 219–230.

D'Andrea AF, Aller RC, Lopez GR (2002) Organic matter flux and reactivity on a South Carolina sandflat: the impacts of porewater advection and macrobiological structures. Limnol Oceanogr 47: 1056–1070.

Dill LM, Heithaus MR, Walters CJ (2003) Behaviorally mediated indirect interactions in marine communities and their conservation implications. Ecology 84: 1151–1157.

Domi N, Bouquegneau M, Das K (2005) Feeding ecology of five commercial shark species of the Celtic Sea through stable isotope and trace metal analysis. Mar Environ Res 60: 551–569.

Ebert DA, Bizzaro JJ (2007) Standardized diet compositions and trophic levels of skates (Chondrichthyes: Rajiformes: Rajoidei). Environ Biol Fish 80: 221–237.

Ellis JK, Musick JA (2007) Ontogenetic changes in the diet of the sandbar shark, *Carcharhinus plumbeus*, in lower Chesapeake Bay and Virginia (USA) coastal waters. Environ Biol Fish 80: 51–60.

Englund G (1997) Importance of spatial scale and prey movements in predator caging experiments. Ecology 78: 2316–2325.

Englund G (2005) Scale dependent effects of predatory fish on stream benthos. Oikos 111: 19–30.

Englund G, Cooper SD, Sarnelle O (2001) Application of a model of scale dependence to quantify scale domains in open predation experiments. Oikos 92: 501–514.

Estes JA, Tinker MT, Williams TM, Doak DF (1998) Killer whale predation on sea otters linking oceanic and nearshore ecosystems. Science 282: 473–476.

Ferretti F, Myers RA, Serena F, Lotze HK (2008) Loss of large predatory sharks from the Mediterranean Sea. Conserv Biol 22: 952–964.

Ferry LA, Cailliet GM (1996) Sample size and data: are we characterizing and comparing diet properly? In: Mackinlay D, Shearer K (eds) Feeding ecology and nutrition in fish. Proceedings of the Symposium on the Feeding Ecology and Nutrition in Fish, International Congress on the Biology of Fishes, San Francisco, California. American Fisheries Society, San Francisco, CA, pp 70–81.

Fisk AT, Tittlemier SA, Pranschke JL, Norstrom RJ (2002) Using anthropogenic contaminants and stable isotopes to assess the feeding ecology of Greenland sharks. Ecology 83: 2162–2172.

Frid A, Baker GG, Dill LM (2006) Do resource declines increase predation rates on North Pacific harbor seals? A behavior-based plausibility model. Mar Ecol Prog Ser 312: 265–275.

Frid A, Baker GG, Dill LM (2008). Do shark declines create fear-released systems? Oikos 117: 191–201.

Frid A, Burns J, Baker GG, Thorne RE (2009). Predicting synergistic effects of resources and predators on foraging decisions by juvenile Steller sea lions. Oecologia 158: 775–776.

Frid A, Dill LM, Thorne RE, Blundell GM (2007) Inferring prey perception of relative danger in large-scale marine systems. Evol Ecol Res 9: 635–649.

Friedlander AM, DeMartini EE (2002) Contrasts in density, size, and biomass of reef fishes between the northwestern and the main Hawaiian islands: the effects of fishing down apex predators. Mar Ecol Prog Ser 230: 253–264.

Frisk MG, Miller TJ, Martell SJD, Sosebee K (2008) New hypothesis helps explain elasmobranch "OutBurst" on Georges bank in the 1980s. Ecol Appl 18: 234–245.

Gascuel D, Morissette L, Palomares MLD, Christensen V (2008) Trophic flow kinetics in marine ecosystems: toward a theoretical approach to ecosystem functioning. Ecol Model 217: 33–47.

Grant J (1983) The relative magnitude of biological and physical sediment reworking in an intertidal community. J Mar Res 41: 673–689.

Gray AE, Mulligan TJ, Hannah RW (1997) Food habits, occurrence, and population structure of the bat ray, *Myliobatis californica*, in Humboldt Bay, California. Environ Biol Fish 49: 227–238.

Gregory MR, Ballance PF, Gibson GW, Ayling AM (1979) On how some rays (Elasmobranchia) excavate feeding depressions by jetting water. J Sedimen Petrol 49: 1125–1130.

Hebblewhite M, Merrill E, McDonald T (2005) Spatial decomposition of predation risk using resource selection functions: an example in a wolf-elk predator-prey system. Oikos 111: 101–111.

Heck KL Jr, Carruthers TJB, Duarte CM, Hughes AR, Kendrick G, Orth RJ, Williams SW (2008) Trophic transfers from seagrass meadows subsidize diverse marine and terrestrial consumers. Ecosystems 11: 1198–1210.

Heithaus MR (2004) Predator-prey interactions. In: Carrier JC, Musick JA, Heithaus MR (eds) The biology of sharks and their relatives. CRC Press, Boca Raton, FL, pp 487–521.

Heithaus MR (2005) Habitat use and group size of pied cormorants (*Phalacrocorax varius*) in a seagrass ecosystem: possible effects of food abundance and predation risk. Mar Biol 147: 27–35.

Heithaus MR (2007) Nursery areas as essential shark habitats: a theoretical perspective. Am Fish Soc Symp 50: 3–13.

Heithaus MR, Burkholder D, Hueter RE, Heithaus LI, Pratt HW Jr, Carrier JC (2007a) Spatial and temporal variation in shark communities of the lower Florida Keys and evidence for historical population declines. Can J Fish Aquat Sci 64: 1302–1313.

Heithaus MR, Dill LM (2002) Food availability and tiger shark predation risk influence bottlenose dolphin habitat use. Ecology 83: 480–491.

Heithaus MR, Dill LM (2006) Does tiger shark predation risk influence foraging habitat use by bottlenose dolphins at multiple spatial scales? Oikos 114: 257–264.

Heithaus MR, Dill LM, Marshall GJ, Buhleier B (2002) Habitat use and foraging behavior of tiger sharks (*Galeocerdo cuvier*) in a seagrass ecosystem. Mar Biol 140: 237–248.

Heithaus MR, Frid A, Wirsing AJ, Dill LM, Fourqurean J, Burkholder D, Thomson J, Bejder L (2007b) State-dependent risk-taking by green sea turtles mediates top-down effects of tiger shark intimidation in a marine ecosystem. J Anim Ecol 76: 837–844.

Heithaus MR, Frid A, Wirsing AJ, Worm B (2008a) Predicting ecological consequences of marine top predator declines. Trend Ecol Evol 23: 202–210.

Heithaus MR, Wirsing AJ, Burkholder D, Thomson J, Dill LM (2009). Interaction of landscape features and antipredator behavior can amplify or reverse indirect effects of predation. J Anim Ecol 78: 552–562.

Heithaus MR, Wirsing AJ, Frid A, Dill LM (2007c) Species interactions and marine conservation: lessons from an undisturbed ecosystem. Isr J Ecol Evol 53: 355–370.

Heithaus MR, Wirsing AJ, Thomson JA, Burkholder DA (2008b) A review of lethal and non-lethal effects of predators on adult marine turtles. J Exp Mar Biol Ecol 356: 43–51.

Helfield JM, Naiman RJ (2001) Effects of salmon-derived nitrogen on riparian forest growth and implications for stream productivity. Ecology 82: 2403–2409.

Herbert DA, Fourqurean JW (2008) Ecosystem structure and function still altered two decades after short-term fertilization of a seagrass meadow. Ecosystems 11: 688–700.

Heupel MR, Carlson JK, Simpfendorfer CA (2007) Shark nursery areas: concepts, definition, characterization and assumptions. Mar Ecol Prog Ser 337: 287–297.

Heupel MR, Hueter RE (2002). The importance of prey density in relation to the movement patterns of juvenile sharks within a coastal nursery area. Mar Freshw Res 53: 543–550.

Heupel MR, Simpfendorfer CA (2002) Estimation of mortality of juvenile blacktip sharks, *Carcharhinus limbatus*, within a nursery area using telemetry data. Can J Fish Aquat Sci 59: 624–632.

Hines AH, Whitlatch RB, Thrush SF, Hewitt JE, Cummings VJ, Dayton PK, Legendre P (1997). Nonlinear foraging response of a large marine predator to benthic prey: eagle ray pits and bivalves in a New Zealand sandflat. J Exp Mar Biol Ecol 216: 191–210.

Holt R (1977) Predation, apparent competition, and the structure of prey communities. Theor Popul Biol 12: 197–229.

Hovel KA, Lipcius RN (2001) Habitat fragmentation in a seagrass landscape: patch size and complexity control blue crab survival. Ecology 82: 1814–1829.

Howard JD, Mayou TV, Heard RW (1977) Biogenic sedimentary structures formed by rays. J Sedimen Petrol 47: 339–346.

Iverson SJ, Field C, Bowen WD, Blanchard W (2004) Quantitative fatty acid signature analysis: a new method of estimating predator diets. Ecol Monogr 74: 211–235.

James AR, Boutin S, Herbert DM, Rippin AB (2004) Spatial separation of caribou from moose and its relation to predation by wolves. J Wildl Manage 68: 799–809.

Kerford M, Wirsing AJ, Heithaus MR, Dill LM (2008) Danger on the rise: habitat use by bar-bellied sea snakes in Shark Bay, Western Australia. Mar Ecol Prog Ser 358: 289–294.

Kinney MJ, Simpfendorfer CA (2009) Reassessing the value of nursery areas to shark conservation and management. Conserv Lett 2: 53–60.

Kitchell JF, Essington TE, Boggs CH, Schindler DE, Walters CJ (2002) The role of sharks and longline fisheries in a pelagic ecosystem of the central Pacific. Ecosystems 5: 2002–2016.

Kogure K, Wada M (2005) Impacts of macrobenthic bioturbation in marine sediment on bacterial metabolic activity. Microb Environ 20: 191–199.

Kondoh M (2008) Building trophic modules into a persistent food web. Proc Natl Acad Sci USA 105: 16631–16635.

Libralato S, Christensen V, Pauly D (2006) A method for identifying keystone species in food web models. Ecol Model 195: 153–171.

Lima SL, Dill LM (1990) Behavioral decisions made under the risk of predation: a review and prospectus. Can J Zool 68: 619–640.

Link LS (2007) Underappreciated species in ecology: "ugly fish" in the northwest Atlantic Ocean. Ecol Appl 17: 2037–2060.

Lucas Z, Stobo WT (2000) Shark-inflicted mortality on a population of harbor seals (*Phoca vitulina*) at Sable Island, Nova Scotia. J Zool (London) 252: 405–414.

Lucifora LO, Garcia VB, Menni RC, Escalante AG, Hozbor NM (2009) Effects of body size, age and maturity stage on diet in a large shark: ecological and applied implications. Ecol Res 24: 109–118.

MacNeil MA, Skomal GB, Fisk AT (2005) Stable isotopes from multiple tissues reveal diet switching in sharks. Mar Ecol Prog Ser 302: 199–206.

Masini RJ, Anderson PK, McComb AJ (2001) A *Halodule*-dominated community in a subtropical embayment: physical environment, productivity, biomass, and impact of dugong grazing. Aquat Bot 71: 179–197.

McIntyre PB, Flecker AS, Vanni MJ, Hood JM, Taylor BW, Thomas SA (2008) Fish distributions and nutrient cycling in streams: can fish create biogeochemical hotspots? Ecology 89: 2335–2346.

Murdoch TJT, Glasspool AF, Outerbridge M, Ward J, Manuel S, Gray J, Nash A, Coates KA, Pitt J, Fourqurean JW, Barnes PA, Vierros M, Holzer K, Smith SR (2007) Large-scale decline of offshore seagrass meadows in Bermuda. Mar Ecol Prog Ser 339: 123–130.

Musick JA, Branstetter S, Colvocoresses JA (1993) Trends in shark abundance 1974–1991 for the Chesapeake Bight of the U.S. Mid-Atlantic coast. NOAA Tech Rep NMFS 115: 1–18.

Myers RA, Baum JK, Shepherd TD, Powers SP, Peterson CH (2007) Cascading effects of the loss of apex predatory sharks from a coastal ocean. Science 315: 1846–1850.

Myers RA, Worm B (2003). Rapid worldwide depletion of predatory fish communities. Nature 423: 280–283.

Myrick JL, Flessa KW (1996) Bioturbation rates in Bahia La Choya, Sonora, Mexico. Ciencias Marinas 22: 23–46.

Okey TA, Banks S, Born AF, Bustamante RH, Calvopiña M, Edgar GJ, Espinoza E, Fariña JM, Garske LE, Recke GK, Salazar S, Shepherd S, Toral-Granda V, Wallem P (2004) A trophic model of a Galápagos subtidal rocky reef for evaluating fisheries and conservation strategies. Ecol Model 173: 383–401.

Okey TA, Wright BA, Brubaker MY (2007) Salmon shark connections: North Pacific climate change, indirect fisheries effects, or just variability? Fish and Fisheries 8: 359–366.

Orth RJ (1975) Destruction of eelgrass, *Zostera marina*, by the cownose ray, *Rhinoptera bonasus*, in the Chesapeake Bay. Chesapeake Sci 16: 205–208.

Pace ML, Cole JJ, Carpenter SR, Kitchell JF (1999) Trophic cascades revealed in diverse ecosystems. Trends Ecol Evol 14: 483–488.

Paraino S, Fanelli G, Boero F (2002) Variability of species' roles in marine communities: change of paradigms for conservation priorities. Mar Biol 140: 1067–1074.

Parker RO Jr, Dixon RL (1998) Changes in a North Carolina reef fish community after 15 years of intense fishing: global warming implications. Trans Am Fish Soc 127: 908–920.

Peckarsky BL, Abrams PA, Bolnick DI, Dill LM, Grabowski JH, Luttbeg B, Orrock JL, Peacor SD, Preisser EL, Schmitz OJ, Trussell GC (2008). Revisiting the classics: considering nonconsumptive effects in textbook examples of predator-prey interactions. Ecology 89: 2416–2425.

Persson L, Bengtsson J, Menge BA, Power ME (1996) Productivity and consumer regulation: concepts, patterns, and mechanisms. In: Polis G, Winemiller KO (eds) Food webs. Chapman & Hall, New York, pp 396–434.

Peterson CH, Fodrie FJ, Summerson HC, Powers SP (2001) Site-specific and density-dependent extinction of prey by schooling rays: generation of a population sink in top-quality habitat for bay scallops. Oecologia 129: 349–356.

Piraino S, Fanelli G, Boero F (2002) Variability of species' roles in marine communities: change of paradigms for conservation priorities. Mar Biol 140: 1067–1074.

Polis GA, Strong DR (1996) Food web complexity and community dynamics. Am Nat 147: 813–846.

Post DM (2002) Using stable isotopes to estimate trophic position: models, methods, and assumptions. Ecology 83: 703–718.

Power ME, Tilman D, Estes JA, Menge BA, Bond WJ, Mills LS, Daily G, Castilla JC, Lubchenco J, Paine RT (1996) Challenges in the quest for keystones. BioScience 46: 609–620.

Preen A (1995) Impacts of dugong foraging on seagrass habitats: observational and experimental evidence for cultivation grazing. Mar Ecol Prog Ser 124: 201–213.

Preisser EL, Bolnick DI, Benard MF (2005) Scared to death? The effects of intimidation and consumption in predator-prey interactions. Ecology 86: 501–509.

Reidenauer JA, Thistle D (1981) Response of a soft-bottom harpacticoid community to stingray (*Dasyatis sabina*) disturbance. Mar Biol 65: 261–267.

Ripple WJ, Beschta RL (2004) Wolves and the ecology of fear: can predation risk structure ecosystems? Bioscience 54: 755–766.

Robbins WD, Hisano M, Connolly SR, Choat JH (2006) Ongoing collapse of coral-reef shark populations. Curr Biol 16: 2314–2319.

Rothschild BM, Martin LD, Schulp AS (2005) Sharks eating mosasaurs, dead or alive? Neth J Geosci 84: 335–340.

Rubenstein DR, Hobson KA (2004) From birds to butterflies: animal movement patterns and stable isotopes. Trends Ecol Evol 19: 256–263.

Sandin SA, Smith JE, DeMArtini EE, Dinsdale EA, Donner SD, Friedlander AM, Knonotchick T, Malay M, Maragos JE, Obura D, Pantos O, Paulay G, Richie M, Rohwer F, Schroeder RE, Walsh S, Jackson JBC, Knowlton N, Sala E (2008) Baselines and degradation of coral reefs in the northern Line Islands. PLoS ONE 3: e1548. doi: 10.1371/journal.pone.0001548.

Sasko DE, Dean MN, Motta PJ, Hueter RE (2006) Prey capture behavior and kinematics of the Atlantic cownose ray, *Rhinoptera bonasus*. Zoology 109: 171–181.

Sazima I, Moura RL (2000) Shark (*Carcharhinus perezi*), cleaned by the goby (*Elacatinus randalli*), at Fernando de Noronha Archipelago, western south Atlantic. Copeia 2000: 297–299.

Sepkowski JJ (2002) A compendium of fossil marine animal genera. Bull Am Paleontol 363: 1–560.

Shepherd TD, Myers RA (2005) Direct and indirect fishery effects on small coastal elasmobranchs in the northern Gulf of Mexico. Ecol Lett 8: 1095–1104.

Sherman KM, Reidenauer JA, Thistle D, Meeter D (1983) Role of a natural disturbance in an assemblage of marine free-living nematodes. Mar Ecol Prog Ser 11: 23–30.

Sibert J, Hampton J, Kleiber P, Maunder M (2006) Biomass, size, and trophic status of top predators in the Pacific Ocean. Science 314: 1773–1776.

Siddon CE, Witman JD (2004) Behavioral indirect interactions: multiple predator effects and prey switching in the rocky subtidal. Ecology 85: 2938–2945.

Sinclair ARE, Arcese P (1995) Population consequences of predation-sensitive foraging: the Serengeti wildebeest. Ecology 76: 882–891.

Small RJ, Pendleton GW, Pitcher KW (2003) Trends in abundance of Alaska harbor seals, 1983–2001. Mar Mamm Sci 19: 344–362.

Smith JW, Merriner JV (1982) Association of cobia, *Rachycentron canadum*, with cownose ray, *Rhinoptera bonasus*. Estuaries 5: 240–242.

Smith JW, Merriner JV (1985) Food habits and feeding behavior of the cownose ray, *Rhinoptera bonasus*, in lower Chesapeake Bay. Estuaries 8: 305–310.

Smith JW, Merriner JV (1986) Observations on the reproductive biology of the cownose ray, *Rhinoptera bonasus*, in Chesapeake Bay. Fish Bull US 84: 871–877.

Smith JW, Merriner JV (1987) Age and growth, movements and distribution of the cownose ray, *Rhinoptera bonasus*, in Chesapeake Bay. Estuaries 10: 153–164.

Soulé ME, Estes JA, Miller B, Honnold DL (2005) Strongly interacting species: conservation policy, management, and ethics. BioScience 55: 168–176.

Stevens JD, Bonfil R, Dulvy NK, Walker PA (2000) The effects of fishing on sharks, rays, and chimeras (chondrichthyans), and the implications for marine ecosystems. ICES J Mar Sci 57: 476–494.

Stevenson C, Katz LS, Micheli F, Block B, Heiman KW, Perle C, Weng K, Dunbar R, Witting J (2007) High apex predator biomass on remote Pacific Islands. Coral Reefs 26: 47–51.

Stillwell CE, Kohler NE (1982) Food, feeding habits, and estimates of daily ration in the shortfin mako (*Isurus oxyrinchus*) in the Northwest Atlantic. Can J Fish Aquat Sci 39: 407–414.

Theimann GW, Iverson SJ, Stirling I (2008) Polar bear diets and arctic marine food webs: insights from fatty acid analysis. Ecol Monogr 78: 591–613.

Thrush SF, Pridmore RD, Hewitt JE, Cummings VJ (1991) Impact of ray feeding disturbances on sandflat macrobenthos: do communities dominated by polychaetes or shellfish respond differently? Mar Ecol Prog Ser 69: 245–252.

Thrush SF, Pridmore RD, Hewitt JE, Cummings VJ (1994) The importance of predators on a sandflat: interplay between seasonal changes in prey densities and predator effects. Mar Ecol Prog Ser 107: 211–222.

Valentine JF, Heck KL Jr (2005) Interaction strength at the coral reef-seagrass interface: has overfishing diminished the importance of seagrass habitat production for coral reef food webs? Coral Reefs 24: 209–213.

Valentine JF, Heck KL Jr, Blackmon D, Goecker ME, Christian J, Kroutil RM, Peterson BJ, Vanderklift MA, Kirsch KD, Beck M (2008) Exploited species impacts on trophic linkages along reef-seagrass interfaces in the Florida Keys. Ecol Appl 18: 1501–1515.

Valentine JF, Heck KL, Harper P, Beck M (1994) Effects of bioturbation in controlling turtlegrass (*Thalassia testudinum* Banks ex Konig) abundance: evidence from field enclosures and observations in the northern Gulf of Mexico. J Exp Mar Biol Ecol 178: 181–192.

Van der Elst (1979) A proliferation of small sharks in the shore-based Natal sports fishery. Environ Biol Fish 4: 349–362.

van der Stap I, Vos M, Tollrian R, Mooij W (2008) Inducible defenses, competition and shared predation in planktonic food chains. Oecologia 157: 697–705.

VanBaricom GR (1982) Experimental analyses of structural regulation in a marine sand community exposed to oceanic swell. Ecol Mongr 52: 283–305.

Visser I (1999) Benthic foraging on stingrays by killer whales (*Orcinus orca*) in New Zealand waters. Mar Mamm Sci 15: 220–227.

Visser I (2005) First observations of feeding on thresher (*Alopias vulpinus*) and hammerhead (*Sphyrna zygaena*) sharks by killer whales (*Orcinus orca*) specialising on elasmobranch prey. Aquat Mamm 31: 83–88.

Walters C, Christensen V, Pauly D (1997) Structuring dynamic models of exploited ecosystems from trophic mass-balance assessments. Rev Fish Biol Fish 7: 139–172.

Walters C, Juanes F (1992) Recruitment limitation as a consequence of natural selection for use of restricted feeding habitats and predation risk taking by juvenile fishes. Can J Fish Aquat Sci 50: 2058–2070.

Ward P, Myers RA (2005) Shifts in open-ocean fish communities coinciding with the commencement of commercial fishing. Ecology 86: 835–847.

Werner E, Peacor S (2003). A review of trait-mediated indirect interactions in ecological communities. Ecology 84: 1083–1100.

Wetherbee BM, Cortés E (2004) Food consumption and feeding habits. In: Carrier JC, Musick JA, Heithaus MR (eds) The biology of sharks and their relatives. CRC Press, Boca Raton, FL, pp 224–246.

Williams TM, Estes JA, Doak DF, Springer AM (2004) Killer appetites: assessing the role of predators in ecological communities. Ecology 85: 3373–3384.

Wirsing AJ, Heithaus MR, Dill LM (2007a) Living on the edge: dugongs prefer to forage in microhabitats that allow escape from rather than avoidance of predators. Anim Behav 74: 93–101.

Wirsing AJ, Heithaus MR, Dill LM (2007b) Fear factor: do dugongs (*Dugong dugon*) trade food for safety from tiger sharks (*Galeocerdo cuvier*)? Oecologia 153: 1031–1040.

Wirsing AJ, Heithaus MR, Dill LM (2007c) Can you dig it? Use of excavation, a risky foraging tactic, by dugongs is sensitive to predation danger. Anim Behav 74: 1085–1091.

Wirsing AJ, Heithaus MR, Dill LM (2007d) Can measures of prey availability improve our ability to predict the abundance of large marine predators? Oecologia 153: 563–568.

Wirsing AJ, Heithaus MR, Frid A, Dill LM (2008) Seascapes of fear: methods for evaluating sublethal predator effects experienced and generated by marine mammals. Mar Mamm Sci 24: 1–15.

Worm B, Myers RA (2003) Meta-analysis of cod-shrimp interactions reveals top-down control in oceanic food webs. Ecology 84: 162–173.

Yahel R, Yahel G, Genin A (2002) Daily cycles of suspended sand at coral reefs: a biological control. Limnol Oceanogr 47: 1071–1083.

Yahel G, Yahel R, Katz T, Lazar B, Herut B, Tunnicliffe V (2008) Fish activity: a major mechanism for sediment resuspension and organic matter remineralization in coastal marine sediments. Mar Ecol Prog Ser 372: 195–209.

Yamaguchi A, Kawahara I, Ito S (2005) Occurrence, growth and food of longheaded eagle ray, *Aetobatus flagellum*, in Ariake Sound, Kyushu, Japan. Environ Biol Fish 74: 229–238.

Appendix 16.1

Approaches to Understanding the Role of Elasmobranchs: Promise and Pitfalls

For many elasmobranchs we lack even basic information on their biology, let alone their ecological importance. Indeed, because large-scale changes in elasmobranch populations occurred prior to the initiation of most research programs (e.g., Baum and Myers 2004; Heithaus et al. 2007a) elucidating the role of elasmobranchs in natural communities is difficult. However, analyses of time-series data from multiple sources, empirical investigations employing new tools, and mathematical modeling efforts can shed further light on this question. Here we briefly outline how these different approaches can be used.

A.1 Time-Series Analysis

Time-series data from dedicated research surveys and fisheries catches are important for estimating the sizes of elasmobranch populations and structure of their communities. Careful comparisons through time in multiple surveys can also help to reveal the effects of changes in population sizes of key species. There are several long-term scientific surveys of shark populations in particular areas, such as the Virginia Institute of Marine Sciences longline survey that has measured catch rates of elasmobranch populations of Chesapeake Bay and Virginia coastal waters every year since 1973. Such long-term records for elasmobranchs are few, but their importance for determining rates of change in shark and other elasmobranch populations can hardly be overestimated. However, high sampling variability and biases in any particular sampling gear often make the interpretation of single time series challenging. This is why some researchers have favored a meta-analytic approach whereby multiple datasets are statistically combined, and population parameters are estimated across all observations. Typically, a mixed model approach is favored, where it is assumed that there is an overall rate of change (fixed effect) that is modified at each locale by a random variable, which pools site-specific environmental and sampling variation. For example, Myers and Worm (2003) used this approach to estimate the magnitude of decline in large predatory fishes worldwide. Their analysis reported both individual estimates for the residual proportion of biomass (ranging from 5.3% to 21.5%), as well as the mixed model mean (10.3%). This means that there is substantial variation in the estimated residual biomass, but that on average 90% of biomass has been lost. While this method is widely accepted, these particular conclusions have been disputed, based largely on uncertainties in the Japanese longline data for tuna and billfish (see for example Sibert et al. 2006).

Similar meta-analytic times-series analyses have been employed to estimate the strength of predator–prey interactions in oceanic food webs (Worm and Myers 2003). This method relies on correlating predator and prey biomass across multiple systems and testing for consistent outcomes. Correlation, of course, cannot prove causation and therefore needs to be backed up with a clear understanding of mechanism; for example, through diet studies or experimental manipulations (see below). A meta-analytic approach can make inferences more robust, since different regions can serve as replicate "plots" across which these interactions are examined. For detailed discussion on the meta-analysis of species interactions refer to Worm and Myers (2003).

A.2 Ecological Modeling

Mathematical models have become an important tool for investigating marine community dynamics. The most widely used ecosystem models, ECOSIM with ECOPATH and ECOSPACE, use existing data on species diets, abundances, energetics, and demographics to model the dynamics of communities. Using simulations it is then possible to make predictions, based on explicit assumptions, about how a community might change with the removal of a particular species or guild (e.g., Kitchell et al. 2002). These models are useful because of their ability to account for the complexity of interspecific interactions in food webs and their ability to be optimized using our understanding of current ecosystems. They often require strong assumptions, for example, about the strength of top-down control and the proportion of prey populations vulnerable to predation. Testing of these assumptions can generate important new research questions. See Christensen and Walters (2004) for a discussion of ECOPATH with ECOSIM and ECOSPACE.

Another approach that has recently been used to investigate the possible ecological importance of elasmobranchs is behavioral optimization theory. This framework is based on the understanding that prey generally give up foraging opportunities to reduce the risk of predation (e.g., Lima and Dill 1990), but individuals that are closer to starvation will take greater risks to acquire food. Thus, behavioral optimization models can highlight behavior-mediated indirect effects of predators. They also can partition the relative contribution of direct predation effects and risk effects on mesoconsumers and to trophic cascades (see Heithaus et al. 2008a).

A.3 Estimating Trophic Position

Stomach contents analysis is the primary method for assessing a species' diet and trophic position because it is relatively inexpensive and provides a record of what an individual has consumed over a relatively short time period. When samples from a large number of individuals are available across a variety of locations, seasons, age, and sex classes, then stomach contents analysis can provide highly detailed information on the trophic level as well as geographic and ontogenetic variation in diets. A review of stomach contents analysis and particular methodological considerations is beyond our scope (see Cortés 1997), but several considerations warrant mention here. First, the power of stomach contents analysis lies in its ability to provide fine resolution of diets (i.e., specific prey species and size classes) and it is possible to use rarefaction curves to determine if the diet for a particular population (or subset of the population) has been fully captured in the analyses (see Ferry and Calliet 1996; Lucifora et al. 2009). However, attaining large sample sizes is unlikely for many wide-ranging elasmobranch species that often show marked ontogenetic and geographic variation in diets unless they are commonly captured in fisheries. Furthermore, unless individuals can be repeatedly sampled, which is unlikely because elasmobranchs generally are sampled destructively or are unlikely to be recaptured, stomach contents analysis provides only a snapshot of an individual's diet. Thus, it is more difficult to understand how variable diets might be both within and among individuals.

To overcome some of the drawbacks of stomach contents analysis, researchers have increasingly turned to stable isotopic studies (e.g., Post 2002). Stable isotopic studies take advantage of natural variation in the ratio at which heavy and light isotopes of specific elements are incorporated into the tissues of primary producers and consumers to gain insights into the trophic level at which an individual is foraging (using the ratio of $^{15}N:^{14}N$ or $\delta^{15}N$) and the sources of primary productivity in an individual's diet and foraging location (using the ratio of $^{12}C:^{13}C$ or, less often, $^{33}S:^{34}S$; Post 2002; Rubenstein and Hobson 2004; Chasar et al. 2005). As with stomach contents analysis, a review of stable isotopic techniques is beyond our scope, but we highlight several considerations for studies of elasmobranchs. Despite the lower resolution of data from stable isotopic studies relative to stomach contents analysis, there are several benefits. First, stable isotopes are incorporated into the tissues of consumers at a relatively slow rate and thus provide a longer-term picture of foraging by an individual. The rate at which isotopic signatures are incorporated into various body tissues varies, however, allowing for insights into how an individual's diet varies over a range of time scales from days to months or years if multiple tissues (e.g., mucus, muscle, blood plasma, and whole blood) are analyzed (Rubenstein and Hobson 2004; MacNeil, Skomal, and Fisk 2005; Church et al. 2009). Our understanding of tissue turnover rates in elasmobranchs, however, is limited to one study of freshwater rays (MacNeil, Skomal, and Fisk 2005). Finally, this method can be applied nondestructively, which is important for the study of declining and endangered populations.

There are several important issues that must be considered when applying stable isotopic methods. First, it is critical that stable isotopic studies achieve adequate sample sizes. Especially in species that may exhibit a broad feeding niche, small sample sizes could lead to inaccurate conclusions about the foraging ecology of a population. The use of rarefaction curves or other statistical methods to estimate adequate sample sizes should be employed. Second, stable isotopes cannot usually resolve specific prey taxa of a predator because potential prey species may have identical isotopic signatures or different diet combinations can lead to the same isotopic signature in a consumer. Third, variation in isotopic signatures in a consumer population (e.g., across seasons or space) could be the result of shifts in the signatures of their prey, necessitating isotopic studies of at least community modules to interpret signatures of a consumer of interest (see Post 2002). Finally, although $\delta^{15}N$ is an indicator of trophic position, its utility in determining absolute trophic level is questionable because fractionation rates can vary among taxa of a community, across a landscape, or with variation in environmental conditions (e.g., salinity; Post 2002). Furthermore, whether urea retention in elasmobranchs may lower $\delta^{15}N$ and result in underestimates of trophic level (Fisk et al. 2002) remains to be examined.

Other emerging techniques, like fatty-acid analysis (Iverson et al. 2004) may ultimately provide more detailed information on the foraging ecology of elasmobranches. They have been applied to studies of other marine predators (e.g., gray seals, Beck, Iverson, and Bowen 2005; polar bears, Thiemann, Iverson, and Stirling 2008) but require extensive libraries of the signatures of potential prey and are subject to some of the same caveats applied to stable isotopic analysis.

A.4 Empirical Estimates of Elasmobranch Impacts on Their Prey and Communities

Determining the ecological importance of elasmobranchs—or how communities might respond to changes in elasmobranch population sizes—is particularly difficult. Both observational and experimental methods may be useful. Comparative studies of prey populations or communities in areas with relatively intact and heavily impacted predator populations can provide information on the importance and role of large predators While it is preferable to make comparisons among similar ecosystems or areas that are geographically similar, studies of elasmobranchs and their communities in any relatively pristine location are of great value for elucidating the dynamics of communities with intact elasmobranch populations and setting benchmarks for conservation and restoration. Another observational technique that has been used effectively in terrestrial systems is to document community changes that occur with the decline or reintroduction of predators. For example, the ecological importance of wolves has been determined based on changes in prey populations and wider communities following their extirpation and subsequent reintroduction (Ripple and Beschta 2004). Finally, assessing changes in prey behavior (e.g., habitat use and foraging behavior) and community dynamics in locations with temporal variation in predator abundance serves as a "natural experiment" and can be useful for determining how prey populations and communities might respond to long-term changes in population sizes of predators through either climate change or overfishing (Heithaus et al. 2008a). Risk effects of predators can be quantified as the amount of resources (e.g., food) mesoconsumers are willing to give up in order to stay safe (e.g., giving-up densities; Brown and Kotler 2004), which in marine habitats, can be estimated by measuring differences in mesoconsumer distributions relative to food resources during periods of predator abundance and scarcity (Heithaus et al. 2007c; Wirsing et al. 2008).

Experimental studies (exclosures and enclosures) have been used in studies of batoid impacts on benthic communities, but such studies have not yet been widely used to elucidate the ecological importance of sharks. Although exclosure studies are likely to be of great value, extrapolating results for mobile prey species to scales beyond that of the experimental plot must be done with care (e.g., Englund 1997, 2005; Englund, Cooper, and Sarnelle 2001). Responses of sessile consumers and primary producers are likely to scale-up because changes in their abundance and biomass within an exclosure are due to changes in mortality rates, growth rates, and settlement/reproductive rates with the absence of excluded consumers. In contrast, mobile species may accumulate in exclosures due to movement to a refuge. These species may exert strong top-down effects on their own resources within an exclosure and accumulate in densities considerably higher than would be expected for a habitat at large if predators were removed from the system. Therefore, it is important to consider the size of exclosures relative to movement scale of prey of interest (Englund 1997, 2005; Englund, Cooper, and Sarnelle 2001) as well as the effects of spatial heterogeneity (Bergstrom, Englund, and Leonardsson 2006) when designing experimental studies.

17

Life Histories, Population Dynamics, and Extinction Risks in Chondrichthyans

Nicholas K. Dulvy and Robyn E. Forrest

CONTENTS

17.1 Introduction

The greatest threat to chondrichthyan (sharks, rays, and chimaeras) populations and species is fishing mortality. This can come from directed fisheries targeting sharks (Bonfil 1994; T.I. Walker 1998; Punt et al. 2005), mortality imposed as bycatch in more valuable crustacean fisheries (Stobutzki, Miller, and Brewer 2001), demersal fish trawling (Graham, Andrew, and Hodgson 2001; Ellis et al. 2005), pelagic trawling (Zeeberg, Corten, and de Graaf 2006), pelagic line fishing (Gilman et al. 2008), recreational fisheries (Anderson 2002) or through finning of sharks captured, mainly as bycatch, in pelagic fisheries (Clarke et al. 2006).

The response of a chondrichthyan population or species to elevated mortality, and its risk of achieving threatened status or a raised risk of extinction, depends largely on the intrinsic life history of the population. Life histories of chondrichthyans vary widely, particularly their reproductive traits—indeed they could arguably be among the most diverse of all vertebrates. Chondrichthyans exhibit considerable interspecific life history variation: gestation period (2 to 42 months), egg hatching period (1 to 27 months), ovum diameter (0.5 to 600 mm), reproductive mode (egg-laying, live-bearing), maternal investment (yolk-only versus uterine milk, oophagosity, uterine cannibalism, placentation), fecundity (1 to 400 offspring), offspring size (20 to 1800 cm long), age at maturity (1.5 to 30+ years), and longevity (5 to 50+ years) (Compagno 1990; Dulvy and Reynolds 1997; Cortés 2000; Goodwin, Dulvy, and Reynolds 2002, 2005).

Such life history information can provide considerable insight into the response of shark populations to exploitation. For example, different life history strategies give rise to very different responses to fishing in two similarly sized sharks in the family Triakidae (*Mustelus antarcticus* and *Galeorhinus galeus*), which are targeted in the same Australian fishery. One species (*M. antarcticus*) matures relatively early, living to about 16 years, and consequentially has a fairly high rate of population growth. The fishery for this species has been assessed as sustainable. In contrast, *G. galeus* matures later, grows slowly, lives for around 60 years, and has a much lower rate of population growth. This species had been consistently overexploited, despite being subject to similar fishing pressure (Stevens 1999).

There have been a large number of studies in recent years, linking life history to risk of overexploitation and extinction in chondrichthyans (Hoenig and Gruber 1990; Kirkwood, Beddington, and Rossouw 1994; Cortés 1998, 2002; S.E. Smith, Au, and Show 1998; Heppell, Crowder, and Menzel 1999; Musick 1999; Gedamke et al. 2007; Au, Smith, and Show 2008; Forrest and Walters, in press). One of the main reasons for the strong interest in using life history approaches to inform management of chondrichthyans is the extreme lack of data worldwide for conventional stock and risk assessments (Bonfil 1994; T.I. Walker 1998; FAO 2000; Stevens et al. 2000). Reliable time series of catch, catch per unit effort, or other indices of abundance are usually unavailable because sharks are caught as bycatch or are otherwise of low management priority (Bonfil 1994). Life history data describing growth, fecundity, age at maturity, and maximum age are, however, routinely collected in many parts of the world. Here, we review advances in understanding links between these life history data and chondrichthyan population dynamics and discuss implications for management. We then summarize evidence for extirpation, local and regional extinction, and the likelihood of impending global extinction of chondrichthyan populations, based on the IUCN Red List assessments. Finally, we consider the relative vulnerability of chondrichthyans to climate change.

17.2 Life Histories and Population Dynamics

Predictive population models can be used to help gain a more formal understanding of how life history characteristics contribute to risks of overfishing and extinction. One of the simplest models linking life histories to population dynamics is the logistic model of population growth. In this model, the change in population size (dN) over a period of time (dt) is modeled as a function of the intrinsic rate of population increase (r) and the carrying capacity of the population (K), determined by size, productivity, or quality of ecological habitat (Jennings et al. 2008). The change in numbers of a population over time (t) is described by

$$dN_t/dt = rN_t\,(1 - N_t/K)$$

Under the assumptions of this model, species or populations with higher values of r recover more rapidly from small population sizes and reach carrying capacity more quickly than those with lower rates of population increase. There are two parts to the equation that capture the two key determinants of a species' productivity and resilience to fishing: the intrinsic rate of population increase r and the strength of density dependence in population growth rate, represented here by $1 - N_t/K$.

For fished populations, the model is modified by subtraction of annual yield or catch C_t, (Schaefer 1954).

$$dN_t/dt = rN_t\,(1 - N_t/K) - C_t$$

For any exploited species, there is a theoretical constant long-term harvest rate (U) that would achieve long-term maximum sustainable yield (MSY). Here the long-term sustainable catch rate U_{MSY} is expressed as the proportion of the population killed each year. Under the assumptions of this model, U_{MSY} is equal to half the intrinsic population growth rate ($r/2$). Annually, killing a proportion of the population greater than the intrinsic rate of population increase r would lead to eventual extinction of the population. The parameter r is therefore representative of the intrinsic productivity of a population and also a direct determinant of its resilience to fishing. While the logistic model has now largely been replaced by the use of fully age-structured models that account for age schedules of survival and maturity, it provides a valid and useful means of illustrating linkages between life history, productivity, and impacts of fishing on different types of species (Figure 17.1).

In the following sections, we first review the links between life histories and r. Second we explore the role of density dependence in population dynamics in relation to chondrichthyans. Third we explore life history strategies, and finally the comparative demography of chondrichthyans and recent modeling approaches for determining linkages between sustainable exploitation rates and life history.

17.3 Life History Strategies

The logistic growth equation underlies the concept of an r-K life history continuum (MacArthur and Wilson 1967; Pianka 1970). Under this scheme, species living in highly

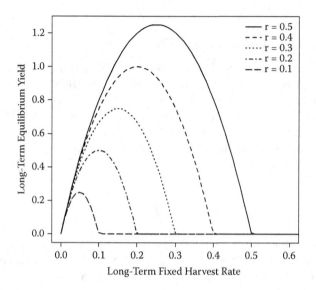

FIGURE 17.1

Equilibrium yield curves predicted from the logistic surplus production model (Schaefer 1954) for five hypothetical species with different intrinsic rates of growth *r*. Note that maximum sustainable yield (MSY) occurs at the peak of each curve at a value of *r*/2. In reality, these curves tend to be asymmetric with peak yield occurring to the left of center at less than *r*/2 (Fowler 1981; Sutherland and Gill 2001).

variable, unpredictable environments and suffering repeated catastrophic mortality events (hence unlikely to reach carrying capacity) were termed *r*-selected species. These *r*-selected species tended to be characterized by frequent colonization and recolonization, with broad niches, small body size, early reproduction, high fecundity and short lifespans, high production to biomass (P/B) ratios, and a high degree of density-independent mortality. At the other end of continuum *K*-selected species were more typically found in more stable, predictable habitats and exhibited narrower niches, larger body sizes, low fecundity, long life spans, and a predominance of intrinsic density-dependent mortality. While this conceptual framework energized ecology in the 1960s and 1970s, it is now viewed as incomplete, particularly since it overlooks high fecundity bet-hedging strategies exhibited by many broadcast spawning fishes and plants (Stearns 1977; Reznick, Bryant, and Bashey 2002). In recent years, the idea has been extended by various authors, in recognition of the limitations of the one-dimensional *r-K* continuum. Following Grime's (1974) classification of plant life histories, a triangular life history continuum consisting of three strategies (opportunistic; periodic, and equilibrium) has been described based largely on teleost fishes (Winemiller and Rose 1992; Winemiller 2005). In this realm, opportunistic strategists are categorized by short generation time, small body-size, and high lifetime reproductive output, with low batch fecundity and low parental investment per offspring. Periodic strategists are characterized by long generation time, large body size, and moderate reproductive output, with large batch size and low investment per offspring. Finally, equilibrium strategists are characterized by long generation time and low reproductive output, with low batch fecundity and high parental investment per offspring. Equilibrium strategists conform most closely to the idea of *K*-selected species and would typically include most chondrichthyans. They are expected to exhibit relatively low interannual variability in recruitment and, rather, to respond in consistent density-dependent manners to changes in habitat quality or resource availability (Winemiller 2005; Goodwin et al. 2006). In general, longer-lived species tend to have evolved mechanisms, such as large body size and fast growth through smaller size classes, to reduce adult mortality and have age at maturity and reproductive rates that reflect life history strategies either dependent on strong iteroparity (repeated breeding events), as in many teleosts (Roff 1984; Heppell, Crowder, and Menzel 1999) or high juvenile survival rates, as in many chondrichthyans (Branstetter 1990; Hoenig and Gruber 1990; Gruber, de Marignac, and Hoenig 2001).

17.4 Life Histories and the Intrinsic Rate of Population Increase *r*

Life history traits are static measures of fishes' life history that can provide considerable insight into the response of populations and species to exploitation. The key life history traits include the von Bertalanffy growth completion rate (K), age at maturity (a_α), lifespan (a_{max}), and natural mortality rate (M) or survival rate (e^{-M}). These traits form the backbone of the demography and population dynamics, and also contribute to risk of population decline or eventual extinction. It turns out that surprisingly few metrics describe and limit the range of possible population dynamics for chondrichthyans. This can be understood in terms of the trade-offs among life history traits. It is commonly said that there is no "free lunch," that is, there is no such thing as a fast-growing, highly fecund animal that matures late and lives for a long time (Law 1979). Such a "Darwinian Demon" cannot exist anywhere in our universe because the laws of thermodynamics constrain metabolic processes. Organisms survive and reproduce by acquiring energy through foraging and feeding and transforming it by somatic (body) growth, metabolism, excretion, and reproduction. Because energy cannot be created or destroyed, the transformation of energy imposes fundamental constraints or trade-offs on the possible combination of life histories. These could almost be considered as the "rules of life":

The faster you grow, the quicker you die: $M = K \cdot 1.65$ to 2

The faster you grow, the smaller your maximum size: $L_\infty = K^{-0.33}$

The quicker you die, the shorter your lifespan: $M \approx 1/a_{max}$ or a_{max}^{-1}

The shorter your lifespan, the earlier you must breed: $L_\alpha = L_\infty \cdot 0.66$ to 0.73 and $M = 1.65/a_\alpha$

where L_∞ is the asymptotic maximum length and L_α is length at maturity. It follows that the shorter your active reproductive life, the more offspring you must produce each year, and vice versa. It also follows that greater reproductive investment this year may limit future investments (Beverton and Holt 1957; Charnov 1993; Jensen 1996).

The rules of life were originally discovered by Ray Beverton and Sidney Holt (1957, 1959), who noticed that ratios of these life history traits greatly simplified the mathematics of fisheries catch models (Jennings and Dulvy 2008). These ratios appear to be robust across a wide range of taxa and are now known as dimensionless ratios or life history invariants, which form the foundations of life history theory (Beverton 1987, 1992; Charnov 1993). Despite their widespread acceptance and use, there have been relatively few estimates of life history invariant ratios for elasmobranchs (see Chapter 6), although this gap has recently been addressed by Frisk, Miller, and Fogarty (2001). Their results suggest that invariant ratios for elasmobranchs (especially large pelagic species) may differ considerably from teleosts and reptiles. The rules of chondrichthyan life so far are:

The length at maturity is typically 70% of maximum size: $L_\alpha \sim 0.70 \cdot L_{max}$

Age at maturity occurs at around 38% of maximum age: $a_\alpha \sim 0.38 \cdot a_{max}$

Natural mortality rate is 42% of the growth rate: $M \sim K \cdot 0.42$

Ultimately trade-offs among life history traits tend to result in large-bodied species having lower intrinsic rates of population increase. The intrinsic rate of population increase is difficult to measure. However, a simple approximation can be used to show the relationship between rate of population increase and body size. In this approach, potential population increase r' is calculated as $r' = \ln(\text{fecundity})/t_{mat}$ (Jennings, Reynolds, and Mills 1998; Frisk, Miller, and Fogarty 2001). The potential rate of population increase of 18 shark and skate species has been shown to be negatively related to maximum size (L_{max}) with slope -0.53 ± 0.13 SE (see Figure 17.2 and Frisk, Miller, and Fogarty 2001). This slope was slightly steeper than expected from metabolic theory of ecology that predicts the intrinsic rate of population increase should scale with body mass B as $r \sim B^{-1/4}$ (Savage et al. 2004). This discrepancy was probably because of the use of an indirect measure of r and because temperature was not controlled for (Brown et al. 2004). An analysis of the intrinsic rate of population increase derived from 63 European marine teleost stock-recruit relationships showed that $r \sim B^{-0.308}$. This was not significantly different from the -0.25 scaling predicted from metabolic theory (Denney, Jennings, and Reynolds 2002; Maxwell and Jennings 2005). Hoenig and Gruber (1990) present results from several empirical studies showing a strong negative relationship between r and body size and between r and generation time in chondrichthyans. Studies such as these illustrate the role of constraints and trade-offs in producing predictable negative relationships between

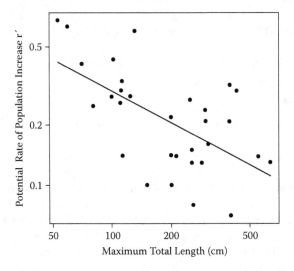

FIGURE 17.2
The potential rate of population increase is negatively related to maximum body size in elasmobranchs. Body size is measured as total length in centimeters. The tiger shark *Trianodon obesus* outlier was removed; however, this does not affect the overall result. The significance of the fit is improved but the estimated parameters change little. The line is a robust regression model, $r^2 = 0.34$, $F_{1,29} = 16.1$, $P < 0.001$, $\ln(r') = 0.54 + 0.533*\ln(\text{maximum length})$.

intrinsic rate of population increase and body size (see also Jensen 1996). Further analytical and empirical evidence suggests that smaller elasmobranch species may have greater resilience to fishing and/or rebound faster from depleted states than larger species (van der Elst 1979; Dulvy et al. 2000; Myers et al. 2007; Au, Smith, and Show 2008). There are, of course, exceptions to this general rule, notably Australian school shark (*G. galeus*) and several species of dogshark in the order Squaliformes (S.E. Smith, Au, and Show 1998; Cortés 2002; Braccini, Gillanders, and Walker 2006b; Forrest and Walters, in press). Other factors contributing to overexploitation and extinction risk in these species include spatial effects resulting from species distribution and vulnerability to fishing gear (Stevens 1999) and extremely low fecundity in some smaller species such as dogsharks (Daley, Stevens, and Graham 2002; Forrest and Walters, in press).

17.5 Density-Dependent Mortality and Productivity of Shark Populations

Most animal populations exhibit some form of compensatory density-dependent population regulation resulting from improvement in rates of growth, fecundity, or survival of young as the population size is reduced (Myers 2001; Rose et al. 2001; Brook and Bradshaw 2006; Goodwin et al. 2006).* Compensatory density-dependent population regulation forms the ecological basis for sustainable fishing. Without such negative feedback control, any fishing regime removing a constant proportion of the population would eventually lead to extinction of the population (Hilborn and Walters 1992). In low fecundity elasmobranchs, density-dependent increases in fecundity might seem to be the most important mechanism conferring resilience to increased fishing mortality (Holden 1973, 1977). However, simulation approaches have shown that density-dependent improvement in fecundity is unlikely to be sufficient to offset increased mortality due to fishing in many shark populations (Wood, Ketchen, and Beamish 1979; Brander 1981; Bonfil 1996). In sharks, as in most exploited fish populations, measurable compensatory effects are most likely to be realized as improvement in the survival rate of juveniles at lower densities (Wood, Ketchen, and Beamish 1979; Brander 1981; Hoenig and Gruber 1990; Gruber, de Marignac, and Hoenig 2001; Gedamke et al. 2007).

Mechanisms for improved juvenile survival at lower population sizes include decreased territorial behavior, reduced competition for resources, and decreased vulnerability to predation or cannibalism at lower densities (Branstetter 1990; Walters and Korman 1999; Gruber, de Marignac, and Hoenig 2001; Rose et al. 2001; Heupel and Simpfendorfer 2002). Predation (by other sharks) is therefore likely to be the most important source of mortality in young sharks, although such effects may be reduced in species that employ nursery grounds for their young (Gruber, de Marignac, and Hoenig 2001; Heupel and Simpfendorfer 2002). At least one study has shown that juvenile sharks in nursery areas

---

* Here, we will only discuss compensatory processes, where population growth rates decrease with increasing population size. Depensatory processes, where population growth rates decrease with decreasing population size appear to be less common in nature (but see Liermann and Hilborn 1997) or only occur at extremely low population sizes. Mechanisms for depensation include the "Allee effect," where low density of adults results in inability to find mates; and predatory effects, where predation rates increase as juvenile numbers decrease. This last effect is exacerbated if predators have benefited from a reduction in the number of their own predators due to fishing (Rudstam et al. 1994; Walters and Kitchell 2001). Depensatory effects such as these can lead to population biomass becoming trapped at low levels and, in the worst cases, lead to local extinction.

may have difficulty capturing enough food to satisfy metabolic requirements, suggesting that food limitation may also be a source of juvenile mortality affected by density in some populations (Bush and Holland 2002).

The magnitude of compensatory improvement in juvenile survival is variable among populations and, because it is one of the main determinants of the resilience of fish populations to fishing, is of principal concern in management of fisheries. Density dependence in juvenile survival is usually represented using standard stock-recruit functions that plot the average number of surviving recruits against average spawning stock biomass or eggs produced (Ricker 1954; Beverton and Holt 1957). These average relationships are typically asymptotic, representing the near-ubiquitous observation in exploited fish populations that average number of surviving recruits is stable over a wide range of population sizes (Myers 2002). Asymptotic stock-recruit relationships arise directly from the assumption of linear increase in natural mortality with population density (Beverton and Holt 1957). A fundamental assumption of stock-recruit relationships in assessment models is that density-dependent effects occur before individuals are first vulnerable to fisheries, although this assumption may not always be true (Heupel and Simpfendorfer 2002; Gedamke et al. 2007; Gazey et al. 2008).

The slope of a straight line fitted through a stock-recruit relationship at any given stock size represents the average rate of juvenile survival at that stock size. It follows, therefore, that if the number of recruits is stable over a wide range of stock sizes, the rate of juvenile survival (surviving recruits per egg) must increase as stock size is reduced (see Figure 17.3). The maximum rate of juvenile survival therefore occurs at very low stock sizes where density dependence is minimal. The slope of the stock-recruit function near the origin (i.e., maximum juvenile survival rate, α), is proportional to the maximum intrinsic rate of population increase (Myers, Mertz, and Fowlow 1997; Myers, Brown, and Barrowman 1999). The strength of density dependence can be measured as the compensation ratio (CR), defined as the ratio of α to the unfished juvenile survival rate (Goodyear 1993). This unitless ratio represents the maximum possible improvement in juvenile survival as population size is reduced (Myers, Brown, and Barrowman 1999). The equilibrium unfished juvenile survival rate (shown as line (ii) in Figure 17.3) can be calculated from life history data alone, as it occurs in the absence of fishing. It is given by the inverse of the equilibrium eggs per recruit summation across ages, i.e.,

$$\left(\sum_{a=1}^{amax} e^{-M(a-1)} f_a \right)^{-1}$$

where f_a is fecundity at age, the term $e^{-M(a-1)}$ represents survivorship at age, and all density-dependent processes are assumed to occur during the first year of life. All other parameters being equal, the unfished juvenile survival rate is inversely proportional to fecundity—the biological interpretation being that, for an unfished population to maintain itself at equilibrium, production of fewer eggs must be accompanied by greater survival rates of those eggs. Accurate estimation of α and CR requires long time series of catch and abundance data that reflect the rate of change of population growth over a wide range of population densities (Hilborn and Walters 1992). The key challenge is that there are few such datasets available for chondrichthyans. We are aware of only five published stock-recruit relationships—all populations of spurdog or piked dogfish (*Squalus acanthias*) and the barndoor skate (da Silva 1993; Myers, Bridson, and Barrowman 1995; Gedamke et al. 2009). However,

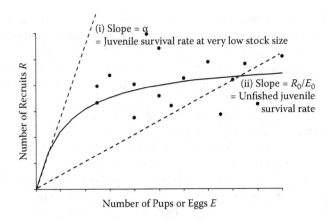

FIGURE 17.3

Stock recruitment relationship for a hypothetical fished population. Points represent observed number of recruits R plotted against number of pups/eggs E. The solid line shows a fitted Beverton–Holt stock recruitment curve. Dashed lines represent juvenile survival rate: (i) close to the origin; and (ii) at unfished (maximum) production of eggs (i.e., E_0 where the 0 subscript indicates fishing mortality $F = 0$). The maximum juvenile survival rate, that is, slope of dashed line (i) is called α and occurs at the fishing mortality rate F_{ext}, which, if applied consistently, would cause extinction of the population. The ratio of slopes (i) and (ii) is called the recruitment compensation ratio, CR (Goodyear 1993; Myers RA, Brown KG, Barrowman NJ (1999) Can J Fish Aquat Sci 56:2404–2419) and represents the maximum possible improvement in juvenile survival as stock size is reduced. Note that R_0/E_0 is the inverse of unfished eggs per recruit, and, therefore, CR = α (E_0/R_0) or αSPR$_0$, where SPR$_0$ is unfished spawners per recruit.

life history strategies exhibited by many chondrichthyan populations may provide constraints to the magnitude of the compensatory response that make the unavailability of time series data less of an issue than for many teleost populations.

As far as we are aware, there have been only three studies that have attempted to directly measure the survival rate of juvenile elasmobranchs (Manire and Gruber 1990; Gruber, de Marignac, and Hoenig 2001; Heupel and Simpfendorfer 2002). Gruber, de Marignac, and Hoenig (2001) estimated the survival rate of a population of age 0 lemon sharks (*Negaprion brevirostris*) in a lagoonal nursery area to be between 38% and 65% over a four-year study. A fifth year of data, consistent with the original observations, has since been added (Gedamke et al. 2007). Preliminary results suggested that the survival rate of age 0 sharks was almost linearly related to the density of juveniles, with the highest survival rate (65%) occurring at the lowest density, although only five years of data were available. Although densities in the lagoon were unlikely to have resulted from changes in adult population size over such a short period of time, results were nonetheless consistent with the assumption of a linear relationship between mortality and density that underpins conventional stock-recruitment theory (Beverton and Holt 1957; Walters and Korman 1999; Walters and Martell 2004). The equilibrium unfished juvenile survival rate for lemon sharks has been estimated from a demographic model to be 39% (Hoenig and Gruber 1990). In this example, it is easy to see that there is only limited room for improvement in the juvenile survival rate as density of juveniles is reduced from its maximum (only a 2.5-fold improvement on a 40% unfished survival rate would result in a maximum of 100% survival).

In a recent simulation study, equilibrium unfished juvenile survival rates were calculated, accounting for uncertainty in life history parameters, for 12 species of dogshark caught in trawl fisheries on the continental slope of southeastern Australia (Forrest and Walters, in press). Many of these species have been depleted by fishing (Graham, Andrew, and Hodgson 2001), and are believed to have very low fecundity and late maturity (Daley, Stevens, and Graham 2002), resulting in mean estimates of unfished juvenile survival rates ranging from between around 0.05 to 0.2 for the different species (Forrest and Walters, in press). The maximum possible compensation ratio for these species (i.e., that which results in all individuals surviving at low population densities) would therefore range between around 20 and 5.

It is easy to see that a life history strategy dependent on high rates of survival of few large young places a fundamental constraint on the possible magnitude of the compensatory response, thereby reducing the amount of uncertainty in stock assessment that is due to uncertainty in recruitment. This is in contrast with teleosts, where the magnitude of compensatory response may be very large in some populations; although increases greater than 100-fold, compared to the unfished state, are rare even in teleosts (Myers, Brown, and Barrowman 1999; Goodwin et al. 2006). The low fecundities and high juvenile survival rates exhibited by many shark species have led a number of authors to suggest that density dependence in recruitment can be ignored in sharks, especially with regard to giving management advice such as sustainable harvest rates (Branstetter 1990). However, while the magnitude of any compensatory response to change in population size is undoubtedly extremely low for many chondrichthyan species, especially those that produce few live young, it may still play a key role in determining the response of a population to fishing and its rate of recovery from depletion, even if the magnitude is low (Hoenig and Gruber 1990; Cortés 2007; Gedamke et al. 2007). Also many smaller or more fecund shark species may have unfished juvenile survival rates more comparable with large teleosts (Au, Smith, and Show 2008) and, therefore, greater potential scope for compensatory effects that it would be unwise to ignore.

17.6 Comparative Demographic Studies of Chondrichthyan Populations

Comparative demographic approaches aim to indicate relative responses of populations to perturbations, such as fishing, by providing methods to estimate r (Simpfendorfer 2005a). These types of studies have proved particularly important for gauging the impacts of fishing and climate change on data-limited chondrichthyan populations and have now been applied in a large number of studies (Hoenig and Gruber 1990; Au and Smith 1997; Cortés 1998, 2002, 2008; S.E. Smith, Au, and Show 1998; Heppell, Crowder, and Menzel 1999; McAllister, Pikitch, and Babcock 2001; Mollet and Cailliet 2002; Frisk, Miller, and Dulvy 2005; Gedamke et al. 2007; Au, Smith, and Show 2008). Essentially, demographic models are age-structured models that enable estimation of the rate of population increase r under a fixed set of parameters, assuming no density dependence (but see Au and Smith 1997; Smith, Au, and Show 1998; Gedamke et al. 2007; Au, Smith, and Show 2008). Demographic models that have been applied to chondrichthyans have been reviewed by Simpfendorfer (2005a) and Gedamke et al. (2007) and readers are referred to these papers and the references above for a full description of the approach. Briefly, there are two general approaches for estimating r from demographic models: (1) life tables and (2) matrix models. Both approaches provide similar results if used in comparable ways and, therefore, the choice of which method to use is a matter of preference, although matrix models are more common in the literature (Simpfendorfer 2005a). One advantage of matrix models is they allow calculation of the elasticity (i.e., proportional sensitivity) of estimates r

to changes in individual parameters (Heppell, Crowder, and Menzel 1999; Simpfendorfer 2005a). They can therefore be used to identify which part of the life cycle has the greatest contribution to *r* and, therefore, where best to direct data-collection and management efforts (Heppell, Crowder, and Menzel 1999; Cortés 2002; Frisk, Miller, and Dulvy 2005; Braccini, Gillanders, and Walker 2006b). From these studies, the population growth rate appears to be relatively insensitive to fecundity (in agreement with Wood, Ketchen, and Beamish 1979; Brander 1981; Bonfil 1996). Instead, the most sensitive part of the life history tends to be the survival of juveniles to maturity rather than the survival of neonates (age 0 to 1), particularly for longer-lived sharks (Cortés 2002; Frisk, Miller, and Dulvy 2005; Kinney and Simpfendorfer 2009).

There are five advantages of demographic approaches: (1) they incorporate the best biological information available; (2) they can be used to develop biological characteristics compared to those obtained from alternative stock assessment approaches (e.g., aggregated surplus production models); (3) they allow examination of constraints imposed by life history traits; (4) they can be used to evaluate the effects of harvesting; and (5) they allow for species-specific assessment and management (Cortés 1998). Life table approaches, particularly those that incorporate life history information, tend to produce more conservative and realistic estimates of *r* than aggregated surplus production models (Cortés 1998). However, the two approaches can be combined in a Bayesian framework, where a surplus production estimate is improved by incorporating prior probabilities of *r* derived from a demographic model (McAllister, Pikitch, and Babcock 2001). For example, McAllister, Pikitch, and Babcock (2001) found that estimates of *r* for the sandbar shark (*Carcharhinus plumbeus*) were an order of magnitude lower than those obtained without demographic information using this approach.

Despite the advantages of demographic approaches, a shortcoming of most demographic models is they do not account for density dependence in juvenile survival (Heppell, Crowder, and Menzel 1999; Gedamke et al. 2007). This is especially a problem with the many demographic models that do not include a fishing component to the mortality because the resulting estimate of *r* represents the unfished population growth rate and fails to account for the likelihood of increased population growth rates under increased mortality rates associated with fishing (Cortés 1998; Gedamke et al. 2007). These approaches are therefore unable to identify sustainable fishing mortality rates, which are necessary for successful management of sharks in targeted fisheries or in multispecies fisheries where they are an unavoidable bycatch. This assessment problem may be worse for smaller, more fecund species than for low-fecundity, live bearing species, where the compensation ratio is highly constrained and there is, therefore, less uncertainty in the magnitude of density-dependent effects on population growth rates (see above).

One approach to address density dependence in juvenile survival in demographic models is to include an estimate of fishing mortality at maximum sustainable yield MSY (assuming $F = M$) and incorporate it into the model (Au and Smith 1997; Smith, Au, and Show 1998; Au, Smith, and Show 2008). The intrinsic "rebound potential" r_{2M} is then estimated as the rate at which the population rebounds from the MSY state after the fishing mortality is removed. The approach is based on the assumptions of the surplus production model, where $F = M$, although in recent updates this has been revised to a level considered more appropriate for sharks, $F = 0.5M$ (Au, Smith, and Show 2008; Cortés 2008; Smith, Au, and Show 2008). While the intrinsic rebound method is unable to produce an estimate of the maximum rate of intrinsic increase (and therefore the maximum sustainable fishing mortality that the population can withstand), the method still provides a logical framework for directly comparing relative productivities of different populations with different life histories, accounting for density dependence (Gedamke et al. 2007). In one of the first major applications of the approach to chondrichthyans, Smith, Au, and Show (1998) calculated r_{2M} for 26 species of shark. The study suggested that the most important parameter determining "rebound" potential for sharks is age at maturity; that is, those with the lowest expected resilience to fishing were those that matured late.

17.7 Age-Structured Models Incorporating Density Dependence in Juvenile Survival

A recently developed approach has avoided the assumption of density independence in age-0 juvenile survival by using fully age-structured models that incorporate a Beverton–Holt (1957) stock-recruit function, therefore assuming that all density-dependent mortality occurs in the first year of life (Forrest and Walters, in press). The approach was based on work by Forrest et al. (2008), who presented an analytical relationship between maximum juvenile survival rate α and U_{MSY} and showed that the relationship between the compensation ratio and U_{MSY} is strongly influenced by life history (notably natural mortality, growth rate, and maximum age) and selectivity parameters (age at first capture). Therefore, the degree to which density dependence determines sustainable harvest rate is unique to an individual population under a given selectivity regime. Under some parameter combinations, and assuming Beverton–Holt recruitment, Forrest and Walters (in press) showed that the range of plausible hypotheses for U_{MSY} approaches an asymptotic maximum value as the compensation ratio (CR) increases, with the maximum possible value of U_{MSY} constrained by the particular combination of life history (and selectivity) parameters of the population. For species with very slow life histories and low fecundity, the upper limit to U_{MSY} could be shown to be very small indeed. This is illustrated in Figure 17.4, which shows the relationship between the CR and U_{MSY} for Harrisson's dogshark (*Centrophorus harrissoni*) under three different ages at 50% first capture. Figure 17.4 illustrates that the maximum possible U_{MSY} value of occurs at around 0.04 when age at 50% first capture is 1 and increases to 0.09 if first capture is delayed until sharks are 15 years old (note that this incorporates an assumption, based on known length at maturity, that these sharks mature at around 18 years old). Application of this model to 12 species of Australian dogshark suggested that the maximum possible hypothesis for U_{MSY} for deepwater dogfishes is very low (5% to 10%), especially when individuals are caught at very young ages (Forrest and Walters, in press). These authors were also able to systematically show that later-maturing, slower-growing less-fecund species have a smaller range of possible values of U_{MSY} than shorter-lived, faster-growing species (Figure 17.5). The main advantage of the approach is that it explicitly accounts for the degree to which density dependence in juvenile survival determines sustainable harvest rates and shows that there are cases (e.g., in slow-growing, live bearing species) where U_{MSY} is so highly constrained by factors such as low fecundity and slow growth, even under the highest possible recruitment compensation (100% survival of juveniles at low population density), that knowledge of density-dependent effects would have a relatively minor effect on management decisions. Since the upper limit to U_{MSY} can be estimated using life history

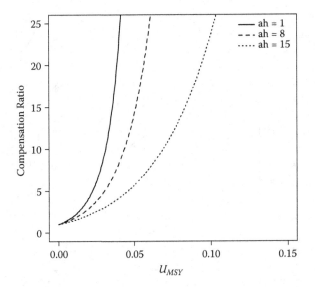

FIGURE 17.4
Curves showing relationship between the long-term sustainable catch rate U_{MSY} and the compensation ratio CR for Harrisson's dogshark (*C. harrissoni*), under three hypothesized values of *ah* (age in years at 50% first harvest). Values were calculated using an analytical relationship between U_{MSY} and CR (Forrest et al. 2008) assuming a Beverton–Holt stock-recruit relationship. Parameter values for this species can be found in Forrest and Walters (in press). The parameters representing growth rate, age at maturity, maximum age, litter size, and the ratio of the growth rate to natural mortality were treated as uncertain and drawn randomly from distributions given in Forrest and Walters (in press). The curves therefore represent the mean U_{MSY}-CR relationships from 100 Monte Carlo simulations. Curves are truncated at the average maximum possible compensation ratio for this species under this set of parameters (i.e., 100% juvenile survival at very low stock size, $\alpha = 1$; see text). Note that a recruit is here defined as an age 1 individual, regardless of the age at entry to the fishery. All density-dependent mortality is therefore assumed to occur at age 0.

and selectivity data alone, this approach is appropriate for data-limited species. However, the method estimates the upper limit of U_{MSY}, not U_{MSY} itself, because the true magnitude of compensation remains unknown. For more fecund, faster-growing species, the upper limit may be quite high, and uncertainty in the compensation ratio will become a more important concern, as it is in most teleost assessments. A key advantage of this method compared to demographic approaches is that it allows for explicit consideration of vulnerability to fishing gear and can therefore be used to search for selectivity schedules that would allow enough individuals to reproduce for fishing mortality to become insignificant (Myers and Mertz 1998). Another advantage is that it is flexible to a wide variety of assumptions about the adult mortality schedule—another limitation of demographic approaches (T.I. Walker 1998; Cortés 2007).

It is worth noting here that U_{MSY} is regaining popularity as a limit reference point for use in fisheries management, that is, as a threshold to fishing mortality that should not be exceeded (Mace 2001; Punt and Smith 2001). It represents a biologically valid threshold to exploitation that will prevent both growth and recruitment overfishing if successfully implemented (Sissenwine and Shepherd 1987; Mace 1994; R.M. Cook, Sinclair, and Stefansson 1997; Punt 2000). Therefore, while achieving MSY is rarely a goal in management of shark populations, knowledge of U_{MSY} is still important for sustainable management, especially when capture of sharks is unavoidable in multispecies fisheries.

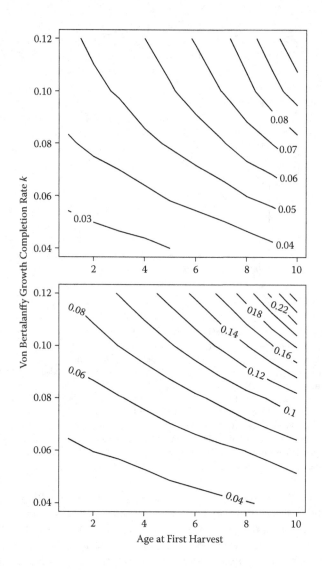

FIGURE 17.5

Life history, selectivity, and maximum possible harvest rate U_{MSY}. Contour plots showing maximum possible U_{MSY} over a range of tested values of age at 50% first harvest and von Bertalanffy growth rate k, for two fecundity scenarios—litter size of one (upper panel) and ten (lower panel) and holding all other parameter values constant. The two plots show the effect of increasing litter size on the maximum possible value of U_{MSY}, which occurs at the maximum possible hypothesis for α (i.e., 100% juvenile survival at very low stock size, $\alpha = 1$; see text and Figure 17.3). Here we assume a maximum age of 30 years and an age at maturity of 10. (Adapted from Forrest RE, Walters CJ (in press) Can J Fish Aquat Sci. See their paper for detailed methods.)

17.8 Management Implications of Life Histories and Demography

The above discussions have hopefully shown that one of the most important determinants of population regulation in chondrichthyans, and therefore of risk to overexploitation and extinction, is density dependence in survival of very young individuals. The above studies discussed mostly live bearing species, although similar arguments could apply to egg-laying species (e.g., Frisk, Miller, and Dulvy 2005; Gedamke et al. 2007). Over the last decade, a number of innovative modeling approaches have greatly improved understanding of the impacts of fishing on chondrichthyan populations. In particular, approaches based on readily obtainable life history information help to overcome some of the problems of extreme data limitation in most of the world's fished chondrichthyan populations. A key recommendation from these approaches is that management should focus on maintaining reserves of reproducing adults and protection of relatively abundant juveniles and young reproductive adults that have survived the first year of high mortality (Au, Smith, and Show 2008). In coastal species, this may be achievable by creation of reserves where juveniles are known to occur, although a recent review has shown that protection of juvenile nursery areas alone will likely be insufficient and management plans must also include older age classes (Kinney and Simpfendorfer 2009). In many cases the greatest gains may be achieved by modifying fishing gear or fishing practices so that these portions of the population are not vulnerable to capture [i.e., gauntlet fisheries (Prince 2005; Kinney and Simpfendorfer 2009)]. T.I. Walker (1998) discussed effects of size selectivity in gillnets for sharks but noted that there have been few selectivity studies of sharks in trawl nets. Bycatch reduction devices (BRDs), such as escape panels and grids, may be effective at reducing catches of sharks (Brewer et al. 1998). In a global study of pelagic longline fisheries, Gilman et al. (2008) found that longline fishers employed a range of methods to decrease shark catches, although these tended only to be employed when there were legislative disincentives to catch sharks. Shark-repellent technologies, involving magnets or chemicals, may also be effective in the future (Gilman et al. 2008; Kaimmer and Stoner 2008).

A key lesson from recent modeling is that while high adult and juvenile survival rates may suggest large reservoirs of biomass and, therefore, high potential returns for harvesting, the slow growth rates and long generation times exhibited by many shark species imply that even very strong compensatory responses in recruitment would not be enough to offset high harvest rates (Heppell, Crowder, and Menzel 1999; Forrest and Walters, in press). Therefore high harvests of low-productivity species achieved in the initial years of a fishery are analogous to the mining of a nonrenewable resource; that is, large biomass reserves are fished down but are not replaced at fast enough rates for the fishery to remain sustainable, resulting in a "boom and bust" fishery. Such fisheries are exhibited by many sharks, such as the Californian soupfin shark and Norwegian spurdog fisheries (Ripley 1946; Holden 1979; Koslow and Tuck 2001). In fisheries where low-productivity species are bycatch (many of which catch chondrichthyans), it is an inevitability that these species simply cease to form a significant part of the catch or become extremely rare (Brander 1981; Dulvy et al. 2000; Graham, Andrew, and Hodgson 2001). In sum, the emerging life history and demographic theory is rapidly catching up with the increasing weight of empirical evidence to suggest that many chondrichthyan populations and species are declining and are threatened due to fisheries. Next we highlight some case studies of decline, extirpation, local and regional extinction, and regional rates of threat in chondrichthyans.

17.9 Decline, Extirpation, and Extinction of Sharks and Rays

17.9.1 Documented Population Declines of Sharks and Rays in the Mediterranean Sea

Over the past decade, there has been an increasing number of studies documenting declines of coastal and oceanic sharks and rays. Increasing scientific awareness of elasmobranch vulnerability and high rates of fishing mortality are leading scientists to develop innovative methods to infer population trends. For example, a combination of sightings records, and commercial and recreational catch data was used to reconstruct nine time series of abundance indices for parts of the Mediterranean Sea (Ferretti et al. 2008). Large pelagic sharks with adequate data, including hammerheads (*Sphyrna* spp.), blue shark (*Prionace glauca*), porbeagle shark (*Lamna nasus*), shortfin mako (*Isurus oxyrhinchus*), and thresher shark (*Alopias vulpinus*) had declined between 96% and 99.99% relative to former abundance (Ferretti et al. 2008). These rates of decline would be consistent with an IUCN Red Listing of Critically Endangered (Ferretti et al. 2008). However, at the time of the last IUCN Red Listing exercise (see section below on the IUCN Red List assessment process), these data were not available and it was only then defensible to assign some of these species (smooth hammerhead *S. zygaena*, blue and thresher sharks) with a lesser threat status of Vulnerable. The porbeagle and shortfin mako were assigned Critically Endangered status listings, which were subsequently confirmed by these new trends in abundance indices (Cavanagh and Gibson 2007). The subsequent more detailed analyses confirm that the IUCN Red List categorization process is, if anything, conservative in the sense that commercially exploited species are usually assigned a lower threat status than can be defended with more detailed retrospective analyses. IUCN Red List assessments do not raise false alarms—a comparison of IUCN threat status and fisheries management status (inside or outside safe biological limits) demonstrates that exploited northeast Atlantic teleost fishes have always been designated as overexploited (outside safe biological limits) before a threatened status criterion is triggered (Dulvy et al. 2005).

17.9.2 Steep Declines of Australian Deepwater Sharks

Around 12 species of dogshark (Order Squaliformes) are caught on the southeastern Australian continental shelf and slope. One species (*Centrophorus harrissoni*) has been listed as Critically Endangered on the IUCN Red List of Threatened Species (Pogonoski and Pollard 2003). Its congenerics, *C. zeehaani** and *C. moluccensis,* are listed as Data Deficient. All three species have recently been added to the Australian federal government's Priority Assessment List, which could see them listed as threatened species in Australia (DEWHA 2008). Should they be listed, the government will be required to develop a comprehensive management plan to reduce further risks.

These sharks can be considered particularly prone to risk of overfishing and extinction because of life history strategies that place them at the lower end of the shark productivity spectrum (Daley, Stevens, and Graham 2002; Forrest and Walters, in press). For example, *C. harrissoni* is thought to live for more than 40 years, does not reach maturity until close to its maximum length, and has only one or two pups every two years (Daley, Stevens, and Graham 2002). Growth parameters are not available, but other dogsharks have been reported to grow very slowly (C.D. Wilson and Seki 1994; Braccini, Gillanders, and Walker 2006a; Irvine, Stevens, and Laurenson 2006). Like other dogsharks, *C. harrissoni* is live bearing with yolk-only provisioning (ovoviviparous), giving birth to large (~40 cm) pups that are potentially immediately vulnerable to trawl nets or longline hooks (Daley, Stevens, and Graham 2002).

During the 1970s, Australia's fisheries were considered underexploited and, with the impending 1979 declaration of the 200 nautical mile Australian Fishing Zone (Rothwell

---

* This species was formerly thought to be the more widely distributed *C. uyato* but has now been reclassified as a separate species endemic to Australia (White, Ebert, and Compagno 2008).

and Haward 1996) the Australian government provided considerable funding for exploratory surveys of the waters of the southeast Australian slope to assess potential commercial opportunities (Tilzey and Rowling 2001). This led to a set of surveys by the Fisheries Research Vessel *Kapala*. The initial, exploratory upper-slope surveys were done in 1976 to 1977 and were not fully replicated until 20 years later in 1996 to 1997. This allowed for some striking comparisons of the abundance of many species (Andrew et al. 1997; Graham, Wood, and Andrew 1997; Graham, Andrew, and Hodgson 2001). In the 20 years between surveys, there had been significant declines in the abundance of many demersal sharks, skates, and several teleost species. Notable declines were reported for deepwater dogsharks (*Centrophorus* spp., *Squalus* spp., and *Deania* spp.), as well as sawsharks (Pristiophoridae), angel sharks (Squatinidae), school sharks (*Galeorhinus galeus*), and skates (Rajidae). Mean catch rates of *Centrophorus* spp. had declined by more than 99% in the period between the two surveys. The surveys were partially replicated in 1979, indicating that large declines in populations of some species may have occurred in the early years of the fishery, almost undoubtedly due to fishing (Andrew et al. 1997; Graham, Andrew, and Hodgson 2001).

Commercial trawling on the slope began with two vessels in 1968, followed by rapid expansion of the fishery during the 1980s (Graham, Andrew, and Hodgson 2001). Most vessels fishing in dogshark habitat (300 to 650 m) target valuable teleosts such as blue grenadier (*Macruronus novaezelandiae*), blue-eye (*Hyperoglyphe antarctica*), and pink ling (*Genypterus blacodes*). For these operators, dogsharks are bycatch, although they have some commercial value. Dogshark flesh is sold as "flake," a generic term for shark fillets in Australia, popular in fish and chips because they are boneless. Also, livers of *Centrophorus* spp. (and, to a lesser extent, *Deania* and *Centroscymnus* spp.) have a high content of squalene, an oil that is extracted, refined, and exported for use in cosmetics, sometimes fetching very high prices per kilogram (Summers 1987; Deprez, Volkman, and Davenport 1990; Daley, Stevens, and Graham 2002). At its peak, the price obtained for *Centrophorus* livers was around $7 per kilogram, although the price has fallen in recent years due to declining catches, improvements in the profitability of synthetic squalene, and other economic factors affecting the cosmetics industry (Daley, Stevens, and Graham 2002). Despite dogsharks being caught and marketed in southeastern Australia for more than three decades, large gaps exist in the catch and effort data that limit their usefulness for assessment purposes (Daley, Stevens, and Graham 2002; T.I. Walker and Gason 2007). These types of problems are common in fisheries around the world, where there is a general lack of reliable data and low priority given to sharks (Bonfil 1994; FAO 2000). The lack of reliable data for dogsharks in Australia compromises the ability to perform risk assessments to determine threatened species status for three species of *Centrophorus* under consideration (DEWHA 2008). Recent work has placed credible limits on productivity parameters for these species, showing that extremely low fecundity, slow growth, and late maturity imply very low sustainable harvest rates (Forrest and Walters, in press). Demographic analyses of *Squalus* species have come to similar conclusions (Cortés 2002; Braccini, Gillanders, and Walker 2006a). In southeastern Australia, three spatial closures have been announced off the states of New South Wales, South Australia, and Tasmania, aimed at protecting populations of *C. moluccenis*, *C. zeehaani*, and *C. harrissoni*, respectively. The success of spatial refugia as a harvest control measure depends upon spatial distribution and movement of the population, that is, how much of the population is protected from fishing and how far outside the refuge do individuals move on foraging or mating excursions (Gerber and Heppell 2004; Gerber et al. 2005). Very little is known about Australian deepwater dogsharks in these respects, although a recently launched tagging program, in collaboration with the fishing industry, may provide some answers (R. Daley, CSIRO, personal communication).

17.9.3 Extirpation of the British Columbia Basking Shark

Extirpation is a term usually reserved to describe extinction from part of a species' former range or to convey some degree of uncertainty of the disappearance of the species. This usage, while widespread, is incorrect. Instead the use of the word extinction along with a sense of the spatial scale of the extinction, such as local extinction or regional extinction, might be preferred (Dulvy, Sadovy, and Reynolds 2003). Strictly speaking extirpation is defined as the *intentional* eradication of a species. This usage is pejorative and directly implies the conscious proactive intention to eliminate a population or a species from part of its geographic range. Numerous populations of sharks and rays have become locally extinct and have not recovered or returned to the area even after several decades (Dulvy, Pinnegar, and Reynolds 2009). These populations and species have disappeared; however, fishers and fisheries managers would claim it unfair to blame them by describing local and regional extinctions as extirpations, as it would imply that fishers and or fisheries management agencies have actively chosen to eradicate populations or species. Fishermen more often than not are motivated by the need for financial security and have a high regard for the biodiversity and ecosystems that underpin their livelihoods. It seems unjust to suggest resource users willingly choose to extirpate populations and species, unless there is evidence of the intention of resource users to eradicate species. To illustrate our point we summarize the extirpation of the basking shark off the coast of British Columbia, Canada, which to our knowledge is the first extirpation, in the true sense of the word, of a marine species.

Sea monsters have been reported from the coast of British Columbia, Canada, and the west coast of Vancouver Island for over a century. These sea monsters were probably basking sharks (*Cetorhinus maximus*). They were frequently entangled in set nets targeting the vast runs of Pacific salmon (*Onchorhynchus* spp.) as they returned to their spawning grounds in coastal rivers and lakes. The basking sharks were attracted, not by the Pacific salmon, but by the rich and locally abundant phytoplankton blooms in coastal bays and estuaries of salmon spawning streams. The entanglement of basking sharks resulted in damaged gear and lost fishing time. In 1949, basking sharks—like black bears, wolves, seals, sea lions, merganser ducks, and kingfishers—were officially classified as "destructive pests" by the federal Department of Fisheries and Oceans (DFO). This list reflected the perceived need for the control and eradication of this species. The local branch of DFO hunted and killed basking sharks by ramming them using a specially modified patrol vessel called the *Comox Post*. The prow of this patrol boat was fitted with a forward-pointing, large-curved blade and the intention was to ram and kill the shark. On April 24, 1956, the newly modified vessel put to sea, whereupon it rammed and killed 34 basking sharks in Pachena Bay, Vancouver Island (Wallace and Gisborne 2006). According to DFO annual reports, the *Comox Post* killed 413 basking sharks in 14 years in the central west coast of Vancouver Island. Three other DFO vessels rammed any basking sharks encountered on their patrols. One vessel, *Laurier*, was estimated to have killed 200 to 300 individuals. DFO reports and newspapers covered only a small fraction of basking shark kills; entanglements with fishing gear are thought to have killed the greatest number of basking sharks (Wallace and Gisborne 2006). A single gill-netter caught seven basking sharks in the 1952 season alone. The true number killed by entanglement in fishing gears is unknown, but based on the extent of documented basking shark–gillnet interactions it has been conservatively estimated that several hundred sharks were killed this way (Darling and Keogh 1994; Wallace and Gisborne 2006). Spearfishing for sport and harassment may have been responsible for the death of several hundred more individuals. Sharks were harassed by motor boaters who would use the basking sharks as "ski jumps," whereas others were killed with harpoons, and by shooting and ramming.

"For many coastal residents, harassing basking sharks was simply a way of life in the 1950s and 1960s" (Wallace and Gisborne 2006). Based on newspaper reports, anecdotes, and the reports of DFO, "it is likely that several thousand sharks may have been killed in British Columbia between 1920–1970" (Wallace and Gisborne 2006). During the last decade, only a handful of basking sharks have been sighted or caught. The British Columbia trawl fleet has had comprehensive observer coverage since 1996 and only four basking sharks have been captured, three off the Queen Charlotte Islands and one in Rennel Sound, in Northern British Columbia. There have been no recent reports of capture in salmon gillnets, though fishing effort has decreased markedly in recent years. Now the likelihood of spotting a basking shark in the eastern Pacific Ocean and the Californian and British Columbian coastlines is vanishingly small. This extirpated population is currently being considered for legal protection under the Canadian Species at Risk Act.

17.9.4 Local Extinctions of North Atlantic Skates

The rapid decline in fisheries landings of the common skate (*Dipturus batis*) from the Irish Sea warned fisheries scientists to check the status of the majority of fished species that were not typically subject to stock assessments or scientific scrutiny (Brander 1981). The common skate and other large species of skate were found to have disappeared and declined from both the Irish and North Seas. At least two of the largest species in the Irish Sea, the common skate, white skate (*D. alba*), and possibly the long-nosed skate (*D. oxyrhinchus*) have disappeared virtually unnoticed (Brander 1981; Dulvy et al. 2000). Uncertainty remains over the long-nose skate, as it is unclear whether this species previously existed in the Irish Sea, though it is documented in the older taxonomies and species lists (Dulvy et al. 2000). An analysis of annual research survey data revealed that of the remaining five species, the two largest had declined and the two smallest had increased, with the intermediate-sized species remaining moderately stable over time (Dulvy et al. 2000). Fishermen tend to target larger individuals and species and this pattern remained even when the rate of fishing mortality was controlled for. A detailed demographic analysis of the North Sea skates (Rajidae) demonstrated that demersal fishing mortality, typically of otter and beam trawlers, was 10% to 20% greater than the rate of replacement of the four largest species (P.A. Walker and Heessen 1996). The replacement rate of the starry ray (*Amblyraja radiata*), the smallest skate in the North Sea, was greater than the (high) rate of fishing mortality and this species is now one of the most abundant large-bodied fishes in the North Sea (P.A. Walker and Heessen 1996; Ellis et al. 2005). This study provided a more detailed mechanistic link between the rates of fishing and the demographic capacity of each species to replace numbers killed by fishing. This study also showed how the replacement rates of these skates were sufficient to explain the current distribution and abundance of the skates remaining in the North Sea. The common skate is now only rarely caught on the northern fringes of the North Sea, the geographic distribution of the largest remaining skate (thornback ray) is now largely restricted to the Thames Estuary in the southwest North Sea (Rogers and Ellis 2000; Ellis et al. 2005; Hunter et al. 2006). It is increasingly recognized that population trajectories, threat status, and extinction risk result from the interaction of the intrinsic vulnerability of species and extrinsic fishing mortality. While an increasing number of studies have explored intrinsic vulnerability we are aware of only two that have explicitly considered both (P.A. Walker and Heessen 1996; Dulvy et al. 2000).

The disappearance of the largest skates and increases in the abundance and distribution of smaller skates has been repeated elsewhere—in the northwest Atlantic shelf seas a large skate had also disappeared (Casey and Myers 1998). The barndoor skate (*Dipturus laevis*), the second largest skate species after the common skate, was found to have been fished out across the shelf seas. This species remains on deep slope waters >450 meters deep, and appears to be recovering in the southern part of its range particularly in and around closed no-take areas on the Georges Bank and the Southern New England Shelf (Kulka 1999; Frisk, Miller, and Fogarty 2002; Kulka, Frank, and Simon 2002; Simon, Frank, and Kulka 2002; Gedamke et al. 2008). More generally one wonders whether declines of large skates and increases in the smaller species are occurring in other temperate shelf seas fisheries.

17.9.5 Regional Extinction of the Angel Shark

The angel shark (*Squatina squatina*) is a large benthic sit-and-wait predator, and in the northeast Atlantic shelf it was originally caught as bycatch in demersal trawl fisheries. This species was originally marketed and sold as "monkfish"—so-called because the head of the angel shark resembled the cowl worn by monks. The decline and disappearance of this species throughout its range went undetected because as angel shark catches declined they were supplanted by catches of anglerfish (*Lophius piscatorius* and *L. budegassa*) that were marketed under the same "monkfish" brand. While previously the subject of large fisheries, by the 1980s they were virtually absent in the Irish Sea—they were sufficiently rare and unusual that specimens were more often brought to public aquaria for display rather than sold on the market. One of us (NKD) saw a single captive specimen in an aquarium in St. David's, SW Wales, in the mid-1990s. Aside from these anecdotal reports of previous abundance followed by modern rarity, until recently there was little scientific evidence of the status of this species. A recent analysis of more than 29,000 research trawl surveys over the past three decades across most of the northeast Atlantic range of this species (except the Mediterranean Sea) failed to uncover a single individual. This compilation spanned from the Bay of Biscay in the south to the Barents Sea in the North and from around 1980 to 2005 (ICES WGFE 2006). A voluntary tagging program of *Squatina squatina* captured by recreational anglers was carried out in Tralee and Clew Bays on the Atlantic coast of western Ireland (Fitzmaurice and Green 2009). A total of 1107 individuals were marked between 1970 and 2001, with most captured in Tralee Bay (939). To date 187 individuals (18.3%) have been recaptured, with most (179) recaptured around western Ireland, and five recaptured in French waters, two captured in the Western English Channel, and one captured off the North Coast of Spain. Almost half were recaptured by angling (47.6%), while 19.3% were caught by trawling, 21% by tangle and gillnets, and five tags were washed ashore. There has been a "dramatic fall-off" in the numbers caught from 1977 onward: "in the five year period 1987–1991, 320 angel sharks were tagged whereas in the period 1997–2001 only 16 individuals have been tagged despite the angling effort being relatively constant" (Fitzmaurice and Green 2009). In 2006, this species was taken off the official listings of the Irish Specimen Fish Committee as a precautionary measure, in recognition that they "are under serious threat due to commercial fishing pressure" (Irish Specimen Fish Committee 2009).

Further details of the decline come from a retrospective comparison of historic and recent trawl surveys, standardized by swept area, in two locations around the British Isles (west central Irish Sea and Start Bay in the western English Channel; Rogers and Ellis 2000). Historically, moderate catch rates of between 2 (Irish Sea) and 19 (English Channel) individuals were captured per 24 hours of trawl survey between 1901 and 1907. More recently, none were caught in comparable modern surveys from 1989 to 1997, although the modern survey must undoubtedly have had higher fishing power (Rogers and Ellis 2000). Angel shark comprised 2% of the catch in Start Bay, English Channel, prior to the First World

War and angel shark was as abundant, at least in Start Bay, as adult North Sea cod (*Gadus morhua*) are presently!

It is possible that some angel sharks might remain in the Mediterranean Sea; however, this is looking increasingly unlikely. The MEDITS trawl survey, which consists of around 1000 hauls each year in depths ranging from 10 to 800 m in the West, North, and Eastern Mediterranean captured angel sharks only in two out of a total of 9095 hauls carried out between 1994 and 1999 (Baino et al. 2001). These angel sharks were caught around the Balearic Islands in the Western Mediterranean in depths between 50 and 100 meters. However, a more recent and comprehensive trawl survey of the Balearic Islands, consisting of 143 hauls from 46 to 1713 meters from 1996 to 2001, failed to capture a single angel shark. Consequently, this species has been listed as Critically Endangered globally by the IUCN Red List in 2006 (Cavanagh and Gibson 2007). Remaining hope for the continued existence of this species lies with unsurveyed habitats in the southern North African coast of the Mediterranean and possibly in the Canary Islands were there are reports that they have been observed by individuals while SCUBA diving (S. Fowler, personal communication). Without urgent action to uncover and protect any remaining viable populations of this species, we are concerned that the angel shark could become one of the first species of fish to be driven to global extinction (Cavanagh and Gibson 2007).

17.9.6 Regional Extinctions of Guitarfishes and Sawfishes

In addition to angel shark and skates, other coastal shark species have declined or disappeared from large parts of their former geographic range. Guitarfishes (Rhinobatidae) and sawfishes (Pristidae) are highly sensitive to fishing pressure as they are large bodied, and presumably have a low intrinsic rate of population increase. They are also highly exposed to fishing mortality and have relatively high catchability. Sawfishes are easily entangled in nets. They tend to be restricted to shallow depths and consequently most of their depth range lies within reach of inshore and coastal fisheries. The Brazilian guitarfish (*Rhinobatos horkelli*) is endemic to the southwest Atlantic and has undergone severe declines >80% since 1986 following intensive exploitation by fisheries and is consequently listed as Critically Endangered (Lessa and Vooren 2007).

Similarly, sawfishes appear to be in trouble worldwide—all are listed as Critically Endangered on the IUCN Red List. Sawfishes were once common in the Mediterranean Sea but are now absent. None had been captured within the living memory of the Mediterranean scientists present at the IUCN Red List Mediterranean Sea workshop in San Marino in September 2003 (Cavanagh and Gibson 2007) and none have been caught in the Mediterranean-wide MEDITS annual trawl survey. It seems highly likely that two species, common sawfish (*Pristis pristis*) and smalltooth sawfish (*P. pectinata*), are regionally extinct from the Mediterranean Sea and northeast Atlantic (S.F. Cook and Compagno 2000; Cavanagh and Gibson 2007). These sawfishes may also be close to global extinction. They were formerly found along the West African coast. Large specimens were regularly captured by Russian trawl surveys in the 1950 to 1960s, but none were observed in more recent surveys in the 1970s and 1980s (F. Litvanov, personal communication). This anecdotal evidence is corroborated by Norwegian surveys conducted by RV *Fritjov Nansen*; over the last decade these surveys have failed to capture a single individual sawfish. The most recent catches of sawfishes occurred in Guinea-Bissau and Sierra Leone according to questionnaire surveys undertaken at fisheries landing sites (Robillard and Seret 2006). Possibly the last remaining population of common and smalltooth sawfishes in the eastern Atlantic is found around the Bijagos Islands, Guinea-Bissau (Mika Diop, CSRP, SICAP

AMITIE 3, VILLA 4430, BP 25485, Dakar Sénégal; personal communication). Here the saw-fish is revered as totem of the indigenous people and recent landings surveys and questionnaire surveys hint that sawfishes are still present and occasionally captured (Robillard and Seret 2006). However, only there have been only three catches of individuals of either common or smalltooth sawfishes there since early 2008. While the Bijagos Islands are a UNEP Biosphere reserve, Guinea-Bissau is the fifth poorest country in the world and is politically highly unstable—the president and head of the army were assassinated while this chapter was being written—making the conservation of the last populations of large sawfishes in the eastern Atlantic a major challenge.

The largetooth sawfish (*Pristis perotteti*) was the subject of pioneering biological studies by Thomas Thorson in the 1960s and 1970s (Thorson 1982). It was distributed in the western Atlantic Ocean and previously found in large numbers in Lake Nicaragua. This migratory lake-dwelling population is now close to extinction as are any adjacent Caribbean and Meso-American populations due to capture, probably as bycatch, in commercial and artisanal fisheries. The most likely location for the remnant populations may be in the northern coastal region of South America (Charvet-Almeida et al. 2007).

The smalltooth sawfish (*Pristis pectinata*) was similarly formerly widely distributed in the western central Atlantic. Large catches of large individuals were historically taken by U.S. recreational fishers in the 1930s to 1950s. Their distribution is currently over a small fraction (<5%) of their former range (National Marine Fisheries Service 2000). Large numbers were known from the Gulf of Mexico, but this species is locally extinct along the eastern U.S. coast, mainly due to incidental capture in commercial fisheries and recreational fisheries (National Marine Fisheries Service 2000; Simpfendorfer 2005a). Habitat loss may have contributed to the decline and may hamper recovery efforts as mangroves and other shallow coastal habitats are used as a juvenile nursery habitat (Simpfendorfer 2007). A small population of smalltooth sawfishes remains in coastal Florida, which is currently monitored and protected by the U.S. Endangered Species Act (Simpfendorfer 2005b; Carlson, Osborne, and Schmidt 2007). Anglers in this region now return sawfishes alive (National Marine Fisheries Service 2000; Simpfendorfer 2005b).

17.10 Global Threat Status of Chondrichthyans

17.10.1 A Brief Summary of the IUCN Red List Process

The Shark Specialist Group, under the auspices of the World Conservation Union for Nature and Natural Resources (www.iucnredlist.org), has undertaken a comprehensive evaluation of the threat status of all chondrichthyans since 1991. Global threat evaluations will have been completed for all species by the end of 2009. This collaborative effort has drawn upon the vast expertise of elasmobranch researchers, fisheries scientists, and the staff of nongovernmental organizations worldwide—and includes many of the writers and readers of this book.

17.10.2 The Global Status of Chondrichthyans

By the end of 2007, almost half (591) of all chondrichthyans had been evaluated at a global scale and 126 species or 21.3% of the known chondrichthyans were threatened. A small proportion has been assigned the highest threat status (Critically Endangered). Two species (3.7%) were Critically Endangered, 29 (4.9%) Endangered, and 75 (12.7%) Vulnerable (Dulvy et al. 2008a). A further 117 species (18%) were listed as Near Threatened, largely on the basis of the ongoing or increasing degree of potential threat faced by these species. It may or may not be a surprise that there are a large number of species for which little is known—201 species (34%) were listed as Data Deficient.

17.10.3 Regional Variation in Chondrichthyan Threat Status

This global picture does not capture considerable regional variation in the degree of threat faced by chondrichthyans. To date, regional Red List assessments have been published for three regions, the Northeast Atlantic Ocean, the Mediterranean Sea, and Australia and Oceania (Cavanagh et al. 2003; Cavanagh and Gibson 2007; Gibson et al. 2008). The greatest proportion of threatened species is found in the Mediterranean Sea, followed by the Northeast Atlantic then Australia and Oceania. In the Mediterranean there are 80 known chondrichthyan species, 30 out of the 71 assessed species (42%) are Threatened (Critically Endangered, Endangered, Vulnerable) and over half of all species 42 (60%) are Threatened or Near Threatened. In the Northeast Atlantic all 116 chondrichthyan species were assessed and a similar number of species are threatened in the Northeast Atlantic (30 species or 25%), and 53 species (45%) are Near Threatened.

A similar number of species (34) are threatened in Australia and Oceania as in the other two regions; however, the higher regional diversity brings down the percentages. The Australia Oceania region has around a third of the world's chondrichthyan diversity with an estimated 350 species and a large number of endemic species—118, comprising 94 endemic sharks and 14 endemic batoids (Last and Stevens 1994; Cavanagh et al. 2003). So far 175 have been assessed in this region, and 34 species (16%) are Threatened, with a total of 86 species (40%) classed as Threatened or Near Threatened.

Scientific knowledge of a large proportion of the chondrichthyan faunas of the three regions remains poor; around a quarter of the chondrichthyans from all three regions were listed as Data Deficient (Cavanagh et al. 2003; Cavanagh and Gibson 2007; Gibson et al. 2008). The true lack of knowledge may be underestimated because the majority of Australian and Oceania species have yet to be evaluated at the regional scale. A large number of species have recently been described, but many remain to be named (Last, White, and Pogonoski 2008a, 2008b; Last et al. 2008). Many of these Data Deficient species may be threatened at smaller spatial scales. For example, the manta ray (*Manta birostris*) is Vulnerable in the South China Sea and Sulu Seas but Data Deficient regionally. Conversely there are regionally Data Deficient species that are locally Least Concern: this pertains particularly to Australia, where there is considerably higher scientific capacity for monitoring and management than in the rest of the region. This includes some large carcharhinformes such as great hammerhead (*Sphyrna mokarran*), silvertip (*Carcharhinus albimarginatus*), nervous (*C. cautus*), and bignose sharks (*C. altimus*) (Cavanagh et al. 2003). It should be borne in mind that species that are protected or Least Concern at local scales may be threatened by large-scale migrations and vulnerability to fisheries over wider spatial scales (Bonfil et al. 2005; Heithaus et al. 2007).

17.10.4 The Distribution of Threat Is Evolutionarily and Ecologically Nonrandom

The ecological and taxonomic distribution of threat across chondrichthyans also appears to be nonrandom. The most threatened ecological guild of species appears to be the oceanic pelagic sharks, species that are found mainly on the high seas and rarely come within the Exclusive Economic Zones and shelf seas (Compagno 2007; Gilman et al. 2008). Three-quarters of the 21 species of oceanic pelagic sharks have been listed as threatened or near threatened (Dulvy et al. 2008a). Consequently, this group of large oceanic predators may well constitute the most threatened group of animals in the world. They are more threatened, in the sense that a greater proportion of this ecologically distinct group faces an elevated risk of extinction, than maybe even primates or whales or Amazonian frogs or freshwater turtles. These species are threatened because they are caught mainly as bycatch of the exploitation of tunas and billfishes and also because of their high intrinsic sensitivity to exploitation, particularly for lamniform sharks (Garcia, Lucifora, and Myers 2008). Their fins are removed, dried, and sold to southeast Asia to support the demand for shark fin soup. An analysis of the shark fin trade in Hong Kong, the main port of entry for shark fins, has estimated that an average of 38 million (range = 26 to 73 million) sharks are killed each year (Clarke et al. 2006; see Chapter 15). Retrospective analyses of fisheries observer logbooks in the North Atlantic suggests oceanic pelagic sharks have declined rapidly in the last few decades (Baum et al. 2003). While there are some challenges in guaranteeing the taxonomic identity of these observer data (Burgess et al. 2005), these declines appear robust in the face of such uncertainties (Baum, Kehler, and Myers 2005) and appear consistent with the other available evidence such as the rise in estimated catches of pelagic sharks over the past 15 years (Clarke 2008) and a 30% decline in the catch per unit effort over the last 50 years of one of the most productive species, the blue shark (*Prionace glauca*; Aires-da-Silva, Hoey, and Gallucci 2008).

In addition to the high rates of threat in oceanic pelagic sharks and deepwater sharks (Dulvy et al. 2008a; Garcia, Lucifora, and Myers 2008; Kyne and Simpfendorfer 2007), freshwater chondrichthyans are poorly known and face high rates of threat. The distributions of many freshwater species are poorly known, particularly in Australasian regions, such as Indonesia and Papua New Guinea, for example, the Critically Endangered Bizant river shark (*Glyphis glyphis*; previously known as *Glyphis* sp. A) and Northern River shark (*Glyphis garricki*) (Compagno 1997; Thorburn and Morgan 2005; Last and Stevens 2009). The Northern river shark is known to science from only 18 individuals (Thorburn and Morgan 2005). Many species have relatively restricted ranges and tend to suffer from the impacts of habitat degradation and destruction and heavy exploitation. Many of the watersheds these species live in, particularly in Asia, are densely populated. For example, the giant river stingray (*Himantura chaophraya*) inhabits the Chao Phraya river basin that runs through the center of Bangkok.

17.11 Future Threats to Sharks and Rays Due to Climate Change

The widespread scale and intensity of fishing on coastal shelves in deeper waters and across the full extent of the high seas worldwide is increasingly evident (Worm et al. 2005; Morato et al. 2006; Bailey et al. 2009). Fishing is the main cause of marine population extinctions and threat in North American marine fishes (55% to 60%), followed closely by habitat loss (32% to 26%) (Musick et al. 2000; Dulvy, Sadovy, and Reynolds 2003; Reynolds et al. 2005). However there is increasing concern regarding the effect of climate change on marine communities (Brander 2006, 2007; Dulvy et al. 2009), particularly since climate change can interact with the effects of fishing and habitat loss both in temperate and tropical systems (Blanchard et al. 2005; S.K. Wilson et al. 2008). So far only one chondrichthyan has been listed on the IUCN Red List due to the impending threat of climate change, the New Caledonia catshark (*Aulohalaelurus kanakorum*). This species was listed as Vulnerable largely on the basis that it is known from only a single type specimen, within an area that is biologically relatively well known and it is presumed to be endemic to New Caledonia (Fowler and Lisney 2003). Like other species in this genus and family it is likely to be rare within a relatively small geographic range and it is likely to be distributed within a narrow depth band centered on coral reef habitat, which is highly vulnerable to degradation due to the projected increase in the frequency and intensity of coral bleaching (Fowler and Lisney 2003).

Many elasmobranchs are large bodied and feed near the top of food chains (Cortés 1999), and hence one might expect that they are less sensitive to the impacts of climate change. Climate change impacts are readily detectable in the primary producers and near the base of food webs, notably in plankton communities and on coral reefs. Marked impacts have also been noted in predatory species feeding directly on the herbivorous species. Notably fledging success of some North Sea seabirds has been linked to the effect of climate variability on the abundance and food quality of planktivorous fishes (Frederiksen et al. 2006). Similarly in the North Pacific the recent decline of Steller sea lions (*Eumetopias jubatus*) is attributed at least partially to climate change-mediated impacts on the quality of their fish prey (Guénette et al. 2006; Trites et al. 2007). For other higher trophic level fishes, including elasmobranchs, there is concern that the northward movement and deepening of their preferred isotherms will lead to smaller geographic distributions and poleward range shifts (Perry et al. 2005; Dulvy et al. 2008b). It is therefore worth considering the state of scientific understanding of the impact of climate change on sharks and rays. Here we summarize two studies that have considered the impact of climate change on chondrichthyans.

17.11.1 Climate Change, Fishing, and Extinction Risk in the Australian Grey Reef Shark

The grey nurse shark (*Carcharius taurus*) is globally Vulnerable and Critically Endangered in eastern Australia mainly due to recreational and commercial fishing, which is estimated to kill 12 sharks per year, and mortality from beach netting, which kills 2 to 6 sharks per year. A recent study has considered the relative effect of fishing, climate change, and demographic stochasticity on the grey nurse shark (Bradshaw et al. 2008). The current population size has been estimated at between 162 and 766 individuals and there is a 35% chance of quasi-extinction (<50 females) within three generations or 50 years, unless fishing mortality is underreported, in which case there is an almost certain (~100%) chance of quasi-extinction within this timeframe. Presently there are two disjunct east and west Australian populations of grey nurse shark, restricted to areas where winter sea surface temperatures are >14°C for nine or more months a year (Figure 17.6). The most conservative Australian climate projections predict a 1°C sea surface temperature (SST) rise by 2030, which is sufficient to eliminate the cool water separating these populations by 2030. There will be 10+ months each year when SSTs are >14°C throughout the currently unoccupied region south to Victoria and full connectivity and panmixia of east and west populations is likely to occur soon after 2050. Assuming demographic rates are unchanged by climate change, the risk of extinction was reduced by 69% from a 35% to an 11% risk of quasi-extinction within 50 years. This outcome was sensitive to the potential immigration rates,

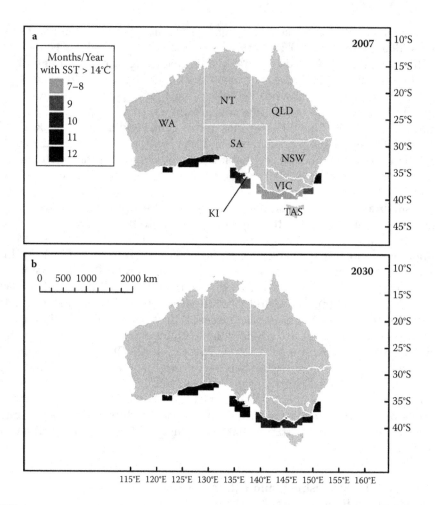

FIGURE 17.6
A color version of this figure follows page 336. Climate change and the increasing distribution of suitable thermal habitat for the grey nurse shark. Present day (A) and predicted to 2030 (B) estimates of the number of months each year where annual minimum monthly sea surface temperature averages are greater than 14°C in 1-degree blocks are along the south Australian coast. Predictions for 2030 are derived from the CSIRO Mk3 model. (Redrawn from Bradshaw CJA, Peddemors VM, Mcauley RB, Harcourt RG (2008) Final Report to the Commonwealth of Australia, Department of the Environment, Water, Heritage and the Arts.)

the relative size of the western Australian population, and the local details of how climate change affects this species, which all remain unknown.

17.11.2 Vulnerability of Australian Sharks and Rays to Climate Change

Vulnerability analyses developed by the social science community have emerged as a promising strategic planning tool with which to evaluate the impact of climate change particularly on data-poor socio-ecological systems (Williams et al. 2008; Allison et al. 2009). Vulnerability is defined as the combination of intrinsic sensitivity to and extrinsic exposure to a threatening process, such as climate change, and the degree to which the potential impact (sensitivity × exposure) can be offset or mitigated against by the adaptive capacity of the system (Figure 17.7). In this study, exposure and sensitivity were defined

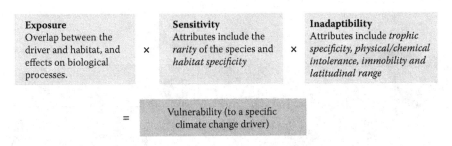

FIGURE 17.7

A conceptual framework to evaluate the vulnerability of Australian sharks and rays to climate change. (Redrawn from Chin A, Kyne PM (2007) In Climate Change and the Great Barrier Reef. Great Barrier Reef Marine Park Authority and Australian Greenhouse Office, Townsville, Australia, pp 393–425.)

as having a negative impact on vulnerability, while adaptive capacity could offset and decrease vulnerability (Chin and Kyne 2007; Chin and Kyne, in press).

The key trick in such an analysis is to develop hypotheses of plausible pathways through which the extrinsic exposure to climate change is likely to affect species as determined by their intrinsic sensitivity to the driver (in this case climate change). Three plausible direct climate impacts on the physiology of Australian chondrichthyans were hypothesized: (1) rising air and sea temperatures; (2) increasing ocean acidification; and (3) increased variability in salinity resulting from greater variability in precipitation and riverine run-off into coastal zones, along with a further nine indirect impacts of climate on Australian sharks (Figure 17.8). The indirect impacts were hypothesized to be mediated through the effects of climate change on habitat distribution and quality and prey availability (Chin and Kyne 2007).

The exposure of six chondrichthyan functional groups to climate change (Figure 17.8) was evaluated and ranked (low, medium, high). Biological attributes were ranked to provide sensitivity and adaptive capacity scores. The overall vulnerability was based on the rankings of all three components. Exposure was defined in terms of: (1) the degree of overlap between species' geographic and depth range and the scale of the climate driver and (2) the extent to which the climate driver was likely to affect the habitats and ecological process upon which the functional group of chondrichthyans depends. Sensitivity was defined based on (1) rarity and (2) habitat specificity, with rare species and species with highest habitat-specificity scoring the highest sensitivity. Ecological adaptive capacity was defined in four terms: (1) trophic specificity, which is the breadth of the diet; (2) physical or chemical tolerance—for example, the bull shark is tolerant to a wide range of salinities; (3) immobility, or the degree to which species are site attached or cannot surmount physical barriers, for example, species on seamounts; and (4) latitudinal range, which was used as a proxy for thermal range.

Temperature change, freshwater input, and oceanic circulation are likely to have higher impacts on elasmobranchs, than, say, ocean acidification, particularly through the effect on prey availability. The freshwater and estuarine and coral reef functional groups were predicted to have the highest vulnerability due to high exposure to the widest range of climate drivers, and the strong direct linkage between climate drivers, such as freshwater flows and sea level rise, and coastal habitat quality. Coral reef species are highly exposed due to the effects of climate change on coral bleaching and, in the longer term, ocean acidification. In contrast shelf, pelagic, and bathyal species were predicted to have low to moderate exposure to projected climate change (Chin and Kyne 2007).

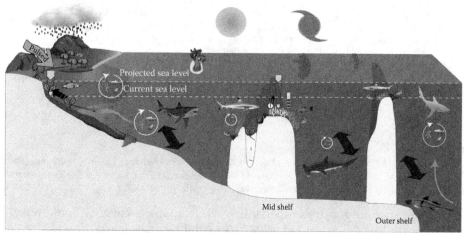

Key climate change factors that affect shark and ray ecological groups

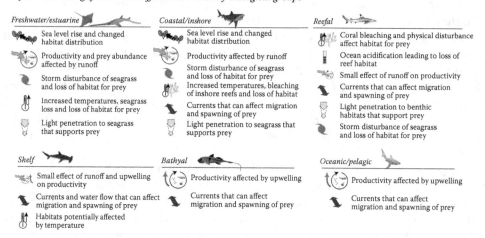

FIGURE 17.8

Six functional groups of sharks and rays and the main climate change drivers that may affect the habitats and biological processes upon which they depend. (Redrawn from Chin A, Kyne PM (2007) In Climate Change and the Great Barrier Reef. Great Barrier Reef Marine Park Authority and Australian Greenhouse Office, Townsville, Australia, pp 393–425.)

17.12 Prioritization for Action: Which Sharks and Rays Require Most Urgent Attention?

Many populations and species of shark, ray, and chimaera have either suffered local and regional extinction or are threatened and face an elevated risk of extinction, as measured by IUCN Red List status. By the time the first IUCN Red List assessment is published (anticipated in 2010) it is likely that more than 300 species will require action to halt and reverse their decline and guarantee their future. However, there is limited scientific capacity to manage and conserve the large number of threatened species! Clearly we need to prioritize our limited scientific and financial capacity (Marris 2007).

There are a wide range of criteria for prioritizing species for conservation and management effort. To gain an insight into priority species one of us (NKD) conducted a straw poll of American Elasmobranch Society members and other scientists involved in marine conservation. A total of 50 people were asked, "If you had the chance to save five species of chondrichthyans, which five would you choose?" No selection criteria were imposed and some respondents freely volunteered their criteria and rationalized their choices. A broad range of criteria was cited, many of which have been considered and subject to debate in the terrestrial conservation literature. For example, one selection criterion, evolutionary distinctness, prioritizes species that represent large amounts of unique evolutionary history (Redding and Mooers 2006; Isaac et al. 2007). All taxonomic orders were represented by the species voted for by the respondents. One approach to examine biases in preference with respect to the taxonomic distribution of species is to compare the proportion of votes cast for each order relative to the proportion of species within each order. The null expectation is that votes will be cast in proportion to the number of species in each order. Unsurprisingly voting was biased. However, the findings are valid and provide insight into the taxonomic distribution in the interests and research capacity of chondrichthyan biologists (Figure 17.8). Four orders comprise almost 90% of all elasmobranchs: Rajiformes, Carcharhiniformes, Squaliformes, and Torpediniformes (Figure 17.9A). The most votes were cast for species in the Carcharhiniformes, Lamniformes, Squaliformes, and Rajiformes (Figure 17.9B). There were a greater proportion of votes for mackerel sharks (Lamniformes) and, to a lesser extent, ground sharks (Carcharhiniformes) than expected, given the proportion of species in these orders (Figure 17.9C). There was considerable underrepresentation of batoids, particularly skates (Rajiformes) and torpedo rays (Torpediniformes) (Figure 17.9C). Clearly, our motivations, values, and scientific capacity are biased toward a few favored groups and these could be the focus of initial conservation efforts. However, the underrepresentation of other taxa, particularly batoids, may suggest that we need to be aware of our potential biases in our interests and scientific capacity when partitioning our limited management and conservation resources.

17.13 Research Required to Manage and Conserve Chondrichthyans

There has been rapid emergence of an awareness of the plight of chondrichthyans among scientists, policy makers, and the public, and a burgeoning of scientific literature on life histories, demography, and population trends over the past decade. We can now make statements on the global and regional status of species and we now have an increased awareness of the wide variety of data from which inferences can be drawn of the former distribution and population trends. The range of data employed is remarkable and includes, but is not restricted to, fisheries observer logbook records (Baum et al. 2003), market surveys (Clarke et al. 2006; Clarke 2008), historical research vessel surveys (Rogers and Ellis 2000), newspaper reports and sightings data (Ferretti et al. 2008; Sims 2008; McPherson and Myers 2009), and taxonomies, museum records, and species lists (Dulvy et al. 2000). The range of demographic and population models that cope with data limitations by using short cuts such as life history invariants (Frisk, Miller, and Fogarty 2001) incorporating uncertainty in parameter estimates (McAllister, Pikitch, and Babcock 2001; Cortés 2002) or are robust to uncertainty in density dependence (Forrest et al. 2008; Forrest and Walters, in press) is remarkable. A new opportunity for assessing the risk of fishing on populations that incorporates less formal data and evidence includes the development of risk evaluation frameworks (Braccini, Gillanders, and Walker 2006b; A.D.M. Smith et

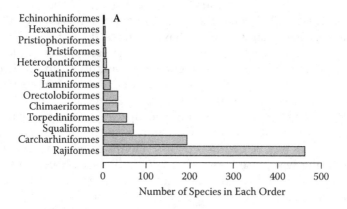

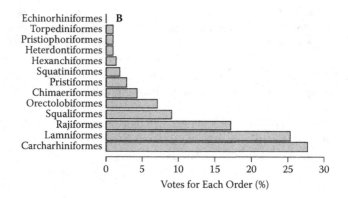

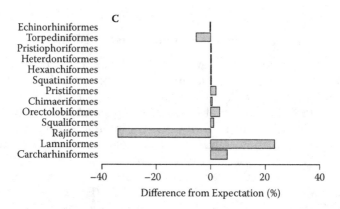

FIGURE 17.9

If you had a choice to save five chondrichthyan species, which five would you choose? A. The distribution of 903 elasmobranch species across the 13 taxonomic orders. B. The proportion of votes for species grouped by taxonomic order. C. The overrepresentation (positive values) or underrepresentation (negative values) of votes compared to the proportion of species in each order.

al. 2007; Pilling et al. 2008). Such frameworks have the capacity to bridge the gap between formal stock assessments and IUCN Red List–style threat assessments and provide prioritization of the species, habitats, and fishing processes for more rigorous assessment and management.

In addition there are clear taxonomic and geographic gaps combined with limited capacity within the scientific community to tackle these gaps. We have little knowledge of the details of the fate and status of coastal and oceanic chondrichthyans in the Indian Ocean and South Atlantic and of the population substructure, movements, and connectivity of deepwater chondrichthyans (Kyne and Simpfendorfer 2007; see Chapter 2). It is unclear whether severe declines in part of the narrow depth range of some species have occurred throughout their geographic range. Increasing evidence of population substructuring suggests some of these apparently widespread species may actually comprise a species complex. We are fortunate that there are a large number of highly active chondrichthyan taxonomists; however, the rate of description of new chondrichthyans is in the region of 20 to 50 species each year (Last 2007). Thus training and support for chondrichthyan taxonomy and systematics would also be a priority.

There are clear biases in the species we value and hence are most likely to study. The charismatic megafauna of the Carcharhiniformes and Lamniformes, rightly or wrongly, attract the greatest attention of the public and the scientific community. Because of the compelling images and elaborate behaviors of these charismatic species, many of us have been attracted to chondrichthyan science. There are hundreds of active white shark and manta ray biologists. Yet for many skates, rays, and freshwater and deepwater chondrichthyans there are many more species than there are active scientists. The point is not that we should reduce white shark or manta ray science, but instead, alongside these efforts, we should encourage scientific capacity and knowledge of the other species occupying the diverse and complex chondrichthyan underworld.

Finally we conclude that there have been numerous losses of populations of chondrichthyans that may represent the permanent loss of some unique, behavioral, morphological, and genetic diversity. There is increasing concern for the Threatened or Near Threatened status of a large proportion of at least three regional faunas and across the globe. However, we have the benefit of a firm theoretical foundation for modeling and predicting the relative risk of extinction of chondrichthyans and fisheries sustainability. The key challenge for the future will be to prioritize species for intervention and to implement effective conservation and fisheries management.

Acknowledgments

We thank Andrew Chin, Mika Diop, Peter Kyne, Cami McCandless, Corey Bradshaw, and Willie Roache for providing reports, preprints, and other invaluable information. Carl Walters, Ken Graham, Tony Pitcher, and Ross Daley were valued advisors during development of RF's research on chondrichthyan population dynamics. We are particularly indebted to the 50 people who responded to NKD's ad hoc questionnaire at the 2008 AES meeting in Montreal: Aleks Maljkovic, Amanda Vincent, Andrea Marshall, Andrew Clark, Andrew Percy, Aaron MacNeill, Bill Duffy, Brooke Flammang, Christina Walker, Christine Bedere, Christine Ward-Paige, Colin Simpfendorfer, Cyrena Riley, Dovi Kacev, Erin Standen, Fiona Hogan, Francesco Ferretti, Heather Marshall, Isabelle Côté, Ivy Baremore,

Jeremy Vando, Jim Ellis, John Carlson, John Foronda, John Froeschke, John Reynolds, Jon Walsh, Julian Dickenson, K. Parsons, Ken Goldman, Lewis Barnett, Luis Lucifora, Lynda Dirk, Marta Calosso, Matt Kolmann, Michelle Heupell, Mike Heithaus, Mike McAllister, Mike Stratton, Neil Ascheilman, Peter Kyne, Simon Brown, Tanya Brunner, Toby Daly-Engel, Tom Kashiwagi, Veronica Garcia, Yannis Papastamatiou, and three other anonymous respondents.

References

Aires-da-Silva AM, Hoey JJ, Gallucci VF (2008) A historical index of abundance for the blue shark (*Prionace glauca*) in the western North Atlantic. Fish Res 92:41–52.

Allison EH, Perry AL, Adger WN, Badjeck M-C, Brown K et al. (2009) Vulnerability of national economies to the impacts of climate change on fisheries. Fish Fish 10:173–196.

Anderson, RC (2002) Elasmobranchs as a recreational resource. In S.L. Fowler, T.M. Reid & F.A. Dipper (eds). Elasmobranch Biodiversity, Conservation and Management. Proceedings of the International Seminar and Workshop. Sabah, East Malaysia. July 1997, pp. 46–51. Occasional Paper of the IUCN Species Survival Commission No 25. IUCN The World Conservation Union: Gland, Switzerland.

Andrew NL, Graham KJ, Hodgson KE, Gordon GNG (1997) Changes after twenty years in relative abundance and size composition of commercial fishes caught during fishery independent surveys on SEF trawl grounds. NSW Fisheries Research Institute, Cronulla, Australia. NSW Fisheries Final Report Series, FRDC Project No. 96/139.

Au DW, Smith SE (1997) A demographic method with population density compensation for estimating productivity and yield per recruit of the leopard shark (*Triakis semifasciata*). Can J Fish Aquat Sci 54:415–420.

Au DW, Smith SE, Show C (2008) Shark productivity and reproductive protection and a comparison with teleosts. In: Camhi MD, Pikitch EK, Babcock EA (eds) Sharks of the open ocean. biology, fisheries and conservation. Blackwell Publishing, Oxford, pp 298–308.

Bailey DM, Collins MA, Gordon JDM, Zuur AF, Priede IG (2009) Long-term changes in deep-water fish populations in the northeast Atlantic: a deeper reaching effect of fisheries? Proc R Soc London B 276:1965–1969.

Baino R, Serena F, Ragonese S, Rey J, Rinelli P (2001) Catch composition and abundance of elasmobranchs based on the MEDITS program. Rapp Comm Int Mer Medit 36:234.

Baum JK, Kehler D, Myers RA (2005) Robust estimates of decline for pelagic shark populations in the northwest Atlantic and Gulf of Mexico. Fisheries 30:27–30.

Baum JK, Myers RA, Kehler DG, Worm B, Harley SJ et al. (2003) Collapse and conservation of shark populations in the northwest Atlantic. Science 299:389–392.

Beverton RJ (1987) Longevity in fish: some ecological and evolutionary considerations. Basic Life Sci 42:161–185.

Beverton RJH (1992) Patterns of reproductive strategy parameters in some marine teleost fishes. J Fish Biol 41:137–160.

Beverton RJH, Holt SJ (1957) On the dynamics of exploited fish populations. Chapman & Hall, London.

Beverton RJH, Holt SJ (1959) A review of the life-spans and mortality rates of fish in nature and their relationship to growth and other physiological characteristics. Ciba Found Colloq Aging 5:142–180.

Blanchard JL, Dulvy NK, Jennings S, Ellis JE, Pinnegar JK et al. (2005) Do climate and fishing influence size-based indicators of Celtic Sea fish community structure? ICES J Mar Sci 62:405–411.

Bonfil R (1994) Overview of world elasmobranch fisheries. Fisheries Technical Report 341. Food and Agriculture Organisation of the United Nations, Rome.

Bonfil R (1996) Elasmobranch fisheries: status, assessment and management. PhD thesis, University of British Columbia, Vancouver.

Bonfil R, Meyer M, Scholl MC, Johnson R, O'Brien S et al. (2005) Transoceanic migration, spatial dynamics, and population linkages of white sharks. Science 310:100–103.

Braccini JM, Gillanders BM, Walker TI (2006a) Determining reproductive parameters for population assessments of chondrichthyan species with asynchronous ovulation and parturition: piked spurdog (*Squalus megalops*) as a case study. Mar Freshw Res 57:105–119.

Braccini JM, Gillanders BM, Walker TI (2006b) Hierarchical approach to the assessment of fishing effects on non-target chondrichthyans: case study of *Squalus megalops* in southeastern Australia. Can J Fish Aqua Sci 63:2456–2466.

Bradshaw CJA, Peddemors VM, Mcauley RB, Harcourt RG (2008) Population viability of eastern Australia grey nurse sharks under fishing mitigation and climate change. Final Report to the Commonwealth of Australia, Department of the Environment, Water, Heritage and the Arts.

Brander K (1981) Disappearance of common skate *Raia batis* from Irish Sea. Nature 290:48–49.

Brander K (2006) Assessment of possible impacts of climate change on fisheries. German Advisory Council on Global Change, Berlin.

Brander K (2007) Global fish production and climate change. Proc Natl Acad Sci USA 104:19709–19714.

Branstetter S (1990) Early life-history implications of selected carcharhinid and lamnoid sharks of the northwest Atlantic. In: Pratt HL, Gruber SH, Taniuchi T (eds) Elasmobranchs as living resources: advances in the biology, ecology, systematics, and the status of the fisheries. NOAA Technical Report NMFS 90. National Marine Fisheries Service, Washington, DC, pp 17–28.

Brewer D, Rawlinson N, Eayrs S, Burridge C (1998) An assessment of bycatch reduction devices in a tropical Australian prawn trawl fishery. Fish Res 36:195–215.

Brook BW, Bradshaw CJA (2006) Strength of evidence for density dependence in abundance time series of 1198 species. Ecology 87:1445–1451.

Brown JH, Gillooly JF, Allen AP, Savage VM, West GB (2004) Toward a metabolic theory of ecology. Ecology 85:1771–1789.

Burgess GH, Beerkircher LR, Cailliet GM, Carlson JK, Cortes E et al. (2005) Is the collapse of shark populations in the northwest Atlantic Ocean and Gulf of Mexico real? Fisheries 30:19–26.

Bush A, Holland K (2002) Food limitation in a nursery area: estimates of daily ration in juvenile scalloped hammerheads, *Sphyrna lewini* (Griffith and Smith, 1834) in Kane'ohe Bay, O'ahu, Hawai'i. J Exp Mar Biol Ecol 278:157–178.

Carlson JK, Osborne J, Schmidt TW (2007) Monitoring the recovery of smalltooth sawfish, *Pristis pectinata*, using standardized relative indices of abundance. Biol Cons 136:195–202.

Casey J, Myers RA (1998) Near extinction of a large, widely distributed fish. Science 281:690–692.

Cavanagh R, Gibson C (2007) Overview of the conservation status of cartilaginous fishes (chondrichthyans) in the Mediterranean Sea. International Union for the Conservation of Nature, Gland, Switzerland.

Cavanagh RD, Kyne PM, Fowler SL, Musick JA, Bennett MB (2003) The conservation status of Australasian chondrichthyans: Report of the IUCN Shark Specialist Group Australia and Oceania Regional Red List Workshop. The University of Queensland, School of Biomedical Sciences, Brisbane, Australia.

Charnov EL (1993) Life history invariants. Oxford University Press, Oxford.

Charvet-Almeida P, Faria V, Furtado M, Cook SF, et al. (2007) *Pristis perotteti*. IUCN Red List of Threatened Species. www.iucnredlist.org. Accessed February 5, 2009.

Chin A, Kyne PM (2007) Vulnerability of chondrichthyan fishes of the Great Barrier Reef to climate change. In: Johnson JE, Marshall PA (eds) Climate change and the Great Barrier Reef. Great Barrier Reef Marine Park Authority and Australian Greenhouse Office, Townsville, Australia, pp 393–425.

Chin A and Kyne PM (in press) A new integrated risk assessment for climate change: analysing the vulnerability of sharks and rays on Australia's Great Barrier Reef. Global Change Biology.

Clarke SC (2008) Use of shark fin trade data to estimate historic total shark removals in the Atlantic Ocean. Aquat Living Resour 21:373–381.

Clarke SC, McAllister MK, Milner-Gulland EJ, Kirkwood GP, Michielsens CGJ et al. (2006) Global estimates of shark catches using trade records from commercial markets. Ecol Lett 9:1115–1126.

Compagno LJV (1990) Alternative life-history styles of cartilaginous fishes in time and space. Environ Biol Fish 28:33–75.

Compagno LJV (1997) Threatened fishes of the world: *Glyphis gangeticus* (Muller & Henle, 1839) (Carcharhinidae). Environ Biol Fish 49:400–400.

Compagno LJV (2007) Checklist of oceanic sharks and rays. In: Camhi MA, Pikitch EK (eds) Sharks of the open ocean: biology, fisheries and conservation. Blackwell Publishing, Oxford.

Cook RM, Sinclair A, Stefansson G (1997) Potential collapse of North Sea cod stocks. Nature 385:521–522.

Cook SF, Compagno LJV (2000) *Pristis pristis*. IUCN Red List of Threatened Species. www.iucnredlist. org. Accessed February 5, 2009.

Cortés E (1998) Demographic analysis as an aid in shark stock assessment and management. Fish Res 39:199–208.

Cortés E (1999) Standardized diet compositions and trophic levels of sharks. ICES J Mar Sci 56:707–717.

Cortés E (2000) Life history patterns and correlations in sharks. Rev Fish Sci 8:299–344.

Cortés E (2002) Incorporating uncertainty into demographic modeling: application to shark populations and their conservation. Conserv Biol 18:1048–1062.

Cortés E (2007) Chondrichthyan demographic modelling: an essay on its use, abuse and future. Mar Freshw Res 58:4–6.

Cortés E (2008) Comparative life history and demography of pelagic sharks. In: Camhi MD, Pikitch EK, Babcock EA (eds) Sharks of the open ocean. biology, fisheries and conservation. Blackwell Publishing, Oxford, pp 309–322.

Daley R, Stevens J, Graham KJ (2002) Catch analysis and productivity of the deepwater dogfish resource in southern Australia. CSIRO, Hobart FRDC Project 1998/108.

Darling DJ, Keogh K (1994) Observations of basking sharks, *Cetorhinus maximus*, in Clayoquot Sound, British Columbia. Can Field-Nat 108:199–210.

da Silva HM (1993) The cause of variability in the stock-recruitment relationship of spiny dogfish, *Squalus acanthias*, in the NW Atlantic. ICES, C. M. 1993/G:52.

Denney NH, Jennings S, Reynolds JD (2002) Life history correlates of maximum population growth rates in marine fishes. Proc R Soc London B 269:2229–2237.

Deprez PP, Volkman JK, Davenport SR (1990) Squalene content and neutral lipid-composition of livers from deep-sea sharks caught in Tasmanian waters. Austr J Mar Freshw Res 41:375–387.

DEWHA (2008) Finalised Priority Assessment List for the Assessment Period Commencing 1 October 2008. http://www.environment.gov.au/biodiversity/threatened/nominations-fpal. html. Accessed March 12, 2009.

Dulvy NK, Reynolds JD (2002) Predicting extinction vulnerability in skates. Conserv Biol 16:440–450.

Dulvy NK, Baum JK, Clarke S, Compagno LVJ, Cortés E et al. (2008a) You can swim but you can't hide: the global status and conservation of oceanic pelagic sharks. Aquat Conserv 18:459–482.

Dulvy NK, Hyde K, Heymans JJ, Chassot E, Pauly D et al. (2009) Climate change, ecosystem variability and fisheries productivity. In: Platt T, Forget M-H, Stuart V (eds) Remote sensing in fisheries and aquaculture: the societal benefits. International Ocean-Colour Coordinating Group, Dartmouth, Canada, pp 11–29.

Dulvy NK, Jennings SJ, Goodwin NB, Grant A, Reynolds JD (2005) Comparison of threat and exploitation status in Northeast Atlantic marine populations. J Appl Ecol 42:883–891.

Dulvy NK, Metcalfe JD, Glanville J, Pawson MG, Reynolds JD (2000) Fishery stability, local extinctions and shifts in community structure in skates. Conserv Biol 14:283–293.

Dulvy NK, Pinnegar JK, Reynolds JD (2009) Holocene extinctions in the sea. In: Turvey ST (ed) Holocene extinctions. Oxford University Press, Oxford, pp 129–150.

Dulvy NK, Reynolds JD (1997) Evolutionary transitions among egg-laying, live-bearing and maternal inputs in sharks and rays. Proc R Soc London B 264:1309–1315.

Dulvy NK, Rogers SI, Jennings S, Stelzenmüller V, Dye SR et al. (2008b) Climate change and deepening of the North Sea fish assemblage: a biotic indicator of regional warming. J Appl Ecol 45:1029–1039.

Dulvy NK, Sadovy Y, Reynolds JD (2003) Extinction vulnerability in marine populations. Fish Fish 4:25–64.

Ellis JR, Dulvy NK, Jennings S, Parker-Humphreys M, Rogers SI (2005) Assessing the status of demersal elasmobranchs in UK waters: a review. J Mar Biol Assoc UK 85:1025–1047.

FAO (2000) Fisheries management. 1. Conservation and management of sharks. Food and Agriculture Organisation of the United Nations, Rome. FAO Technical Guidelines for Responsible Fisheries, 4, suppl. 1.

Ferretti F, Myers RA, Serena F, Lotze HK (2008) Loss of large predatory sharks from the Mediterranean Sea. Conserv Biol 22:952–964.

Fitzmaurice P, Green P (2009) Monkfish migrations. Central Fisheries Board, Ireland. http://www.cfb.ie/fisheries_research/tagging/monkfish.htm. Accessed February 2, 2009.

Forrest RE, Martell SJD, Melnychuk MC, Walters CJ (2008) An age-structured model with leading management parameters, incorporating age-specific selectivity and maturity. Can J Fish Aquat Sci 65:286–296.

Forrest RE, Walters CJ (2009) Estimating thresholds to optimal harvest rate for long-lived, low-fecundity sharks accounting for selectivity and density dependence in recruitment. Can J Fish Aquat Sci 66:2062–2080.

Fowler CW (1981) Density dependence as related to life-history strategy. Ecology 62:602–610.

Fowler SL, Lisney TJ (2003) *Aulohalaelurus kanakorum*. 2008 IUCN Red List of Threatened Species. www.iucnredlist.org. Accessed February 9, 2009.

Frederiksen M, Edwards M, Richardson AJ, Halliday NC, Wanless S (2006) From plankton to top predators: bottom-up control of a marine food web across four trophic levels. J Anim Ecol 75:1259–1268.

Frisk MG, Miller TJ, Dulvy NK (2005) Life histories and vulnerability to exploitation of elasmobranchs: inferences from elasticity, perturbation and phylogenetic analyses. J Northwest Atl Fish Org 35:27–45.

Frisk MG, Miller TJ, Fogarty MJ (2001) Estimation of biological parameters in elasmobranch fishes: a comparative life history study. Can J Fish Aquat Sci 58:969–981.

Frisk MG, Miller TJ, Fogarty MJ (2002) The population dynamics of little skate *Leucoraja erinacea*, winter skate *Leucoraja ocellata*, and barndoor skate *Dipturus laevis*: predicting exploitation limits using matrix analyses. ICES J Mar Sci 59:576–586.

Garcia VB, Lucifora LO, Myers RA (2008) The importance of habitat and life history to extinction risk in sharks, skates, rays and chimaeras. Proc R Soc London B 275:83–89.

Gazey WJ, Gallaway BJ, Cole JG, Fournier DA (2008) Age composition, growth and density-dependent mortality in juvenile red snapper estimated from observer data from the Gulf of Mexico penaeid shrimp fishery. North Am J Fish Manage 28:1828–1842.

Gedamke T, Hoenig JM, DuPaul WD, Musick JA (2008) Total mortality rates of the barndoor skate, *Dipturus laevis*, from the Gulf of Maine and Georges Bank, United States, 1963–2005. Fish Res 89:17–25.

Gedamke T, Hoenig JM, DuPaul WD, Musick JA (2009) Stock-recruitment dynamics and the maximum population growth rate of the barndoor skate on Georges Bank. North Am J Fish Manage 29:512–526.

Gedamke T, Hoenig JM, Musick JA, DuPaul WD, Gruber SH (2007) Using demographic models to determine intrinsic rate of increase and sustainable fishing for elasmobranchs: pitfalls, advances, and applications. North Am J Fish Manage 27:605–618.

Gerber LR, Heppell SS (2004) The use of demographic sensitivity analysis in marine species conservation planning. Biol Cons 120:121–128.

Gerber LR, Heppell SS, Ballantyne F, Sala E (2005) The role of dispersal and demography in determining the efficacy of marine reserves. Can J Fish Aquat Sci 62:863–871.

Gibson C, Valenti SV, Fowler SL, Fordham SV (2008) The conservation status of northeast Atlantic Chondrichthyans. IUCN Shark Specialist Group, Newbury, UK.

Gilman E, Clarke S, Brothers N, Alfaro-Shigueto J, Mandelman J et al. (2008) Shark interactions in pelagic longline fisheries. Mar Policy 32:1–18.

Goodwin NB, Dulvy NK, Reynolds JD (2002) Life history correlates of the evolution of live-bearing in fishes. Philos Trans R Soc London B 356:259–267.

Goodwin NB, Dulvy NK, Reynolds JD (2005) Macroecology of live-bearing in fishes: latitudinal and depth range comparisons with egg-laying relatives. Oikos 110:209–218.

Goodwin NB, Grant A, Perry A, Dulvy NK, Reynolds JD (2006) Life history correlates of density-dependent recruitment in marine fishes. Can J Fish Aquat Sci 63:494–509.

Goodyear CP (1993) Spawning stock biomass per recruit in fisheries management: foundation and current use. In: Smith SJ, Hunt JJ, Rivard D (eds) Risk evaluation and biological reference points for fisheries management, Canadian Special Publication in Fisheries and Aquatic Science, pp 67–81.

Graham KJ, Andrew NL, Hodgson KE (2001) Changes in relative abundance of sharks and rays on Australian South East Fishery trawl grounds after twenty years of fishing. Mar Freshw Res 52:549–561.

Graham KJ, Wood BR, Andrew NL (1997) The 1996–97 survey of upper slope trawling grounds between Sydney and Gabo Island (and comparisons with the 1976–77 survey). Kapala Cruise Report 117. NSW Fisheries Research Institute, Cronulla, Australia.

Grime JP (1974) Vegetation classification by reference to strategies. Nature 250:26–31.

Gruber SH, de Marignac JRC, Hoenig JM (2001) Survival of juvenile lemon sharks at Bimini, Bahamas, estimated by mark-depletion experiments. Trans Am Fish Soc 130:376–384.

Guénette S, Heymans SJJ, Christensen V, Trites AW (2006) Ecosystem models show combined effects of fishing, predation, competition, and ocean productivity on Steller sea lions (*Eumetopias jubatus*) in Alaska. Can J Fish Aquat Sci 63:2495–2517.

Heithaus MR, Wirsing AJ, Dill LM, Heithaus LI (2007) Long-term movements of tiger sharks satellite-tagged in Shark Bay, Western Australia. Mar Biol 151:1455–1461.

Heppell SS, Crowder LB, Menzel TR (1999) Life table analysis of long-lived marine species with implications for conservation and management. In: Musick JA (ed) Life in the slow lane: ecology and conservation of long-lived marine animals. American Fisheries Society, Bethesda, MD, pp 137–147.

Heupel MR, Simpfendorfer CA (2002) Estimation of mortality of juvenile blacktip sharks, *Carcharhinus limbatus*, within a nursery area using telemetry data. Can J Fish Aquat Sci 59:624–632.

Hilborn R, Walters CJ (1992) Quantitative fisheries stock assessment: choice, dynamics and uncertainty. Chapman and Hall, New York.

Hoenig JM, Gruber SH (1990) Life-history patterns in the elasmobranchs: implications for fisheries management. NOAA NMFS Technical Report 90. National Marine Fisheries Service, Washington, DC, pp 1–16.

Holden MJ (1973) Are long-term sustainable fisheries for elasmobranchs possible? Rapp P-v Reun Cons Int Explor Mer 164:360–367.

Holden MJ (1977) Elasmobranchs. In: Gulland JA (ed) Fish population dynamics. John Wiley & Sons, London, pp 117–137.

Holden MJ (1979) The migrations of tope, *Galeorhinus galeus* (L) in the eastern North Atlantic as determined by tagging. J Cons Int Explor Mer 38:314–317.

Hunter E, Berry F, Buckley AA, Stewart C, Metcalfe JD (2006) Seasonal migration of thornback rays and implications for closure management. J Appl Ecol 43:710–720.

ICES. 2006. Report of the ICES Working Group of Fish Ecology 2006. pp. 246, International Council for the Exploration of the Seas, Copenhagen.

Irish Specimen Fish Committee (2009) Irish Trophy Fish. Irish Specimen Fish Committee, Dublin. http://www.irish-trophy-fish.com/trophies/species.htm. Accessed February 4, 2009.

Irvine SB, Stevens JD, Laurenson LJB (2006) Comparing external and internal dorsal-spine bands to interpret the age and growth of the giant lantern shark, *Etmopterus baxteri* (Squaliformes: Etmopteridae). Environ Biol Fish 77:253–264.

Isaac NJB, Turvey ST, Collen B, Waterman C, Baillie JEM (2007) Mammals on the EDGE: conservation priorities based on threat and phylogeny. PLoS ONE 2:e296.

Jennings S, Dulvy NK (2008) Beverton and Holt's insights into life history theory: influence, application and future use. In: Payne AI, Cotter AJR, Potter ECE (eds) Advances in fisheries science: 50 years on from Beverton and Holt. Blackwell Publishing, Oxford.

Jennings S, Mélin F, Blanchard JL, Forster RM, Dulvy NK et al. (2008) Global-scale predictions of community and ecosystem properties from simple ecological theory. Proc R Soc London B 275:1375–1383.

Jennings S, Reynolds JD, Mills SC (1998) Life history correlates of responses to fisheries exploitation. Proc R Soc London B 265:333–339.

Jensen AL (1996) Beverton and Holt life history invariants result from optimal trade-off of reproduction and survival. Can J Fish Aquat Sci 53:820–822.

Kaimmer S, Stoner AW (2008) Field investigation of rare-earth metal as a deterrent to spiny dogfish in the Pacific halibut fishery. Fish Res 94:43–47.

Kinney MJ, Simpfendorfer CA (2009) Reassessing the value of nursery areas to shark conservation and management. Cons Letts 2:53–60.

Kirkwood GP, Beddington JR, Rossouw JA (1994) Harvesting species of different lifespans. In: Edwards PJ, May RM, Webb NR (eds) Large-scale ecology and conservation biology. Blackwell Science, Oxford, pp 199–227.

Koslow JA, Tuck G (2001) The boom and bust of deep-sea fisheries: why haven't we done better? Report Sci. Counc. Res. Doc. NAFO, 01/141. NAFO.

Kulka DW (1999) Barndoor skate on the Grand Banks, Northeast Newfoundland, and Labrador shelves: distribution in relation to temperature and depth based on research survey and commercial fisheries data. Department of Fisheries and Oceans, Canadian Science Advisory Secretariat Research Document, 2002/073.

Kulka DW, Frank KT, Simon JE (2002) Barndoor skate in the northwest Atlantic off Canada: distribution in relation to temperature and depth based on commercial fisheries data. Department of Fisheries and Oceans, Canadian Science Advisory Secretariat Research Document, 2002/073.

Kyne PM, Simpfendorfer CA (2007) A collation and summarization of available data on deepwater chondrichthyans: biodiversity, life history and fisheries. IUCN SSC Shark Specialist Group. International Union for the Conservation of Nature, Gland, Switzerland.

Last PR (2007) The state of chondrichthyan taxonomy and systematics. Marine and Freshwater Research 58:7–9.

Last PR, Stevens JD (1994) Sharks and rays of Australia. CSIRO, Hobart, Australia.

Last PR, Stevens JD (2009) Sharks and rays of Australia. CSIRO, Hobart, Australia.

Last PR, White WT, Pogonoski J (2008a) Descriptions of new Australian chondrichthyans. CSIRO Marine and Atmospheric Research Paper 22. CSIRO, Hobart, Australia.

Last PR, White WT, Pogonoski J (2008b) Descriptions of new dogfishes of the genus *Squalus* (Squaloidea: Squalidae). CSIRO Marine and Atmospheric Research Paper 14. CSIRO, Hobart, Australia.

Last PR, White WT, Pogonoski J, Gledhill DC (2008c) Descriptions of new Australian skates (Batoidea: Rajoidea). CSIRO Marine and Atmospheric Research Paper 21. CSIRO, Hobart, Australia.

Law R (1979) Ecological determinants in the evolution of life histories. In: Anderson RM, Turner BD, Taylor LR (eds) Population dynamics. Blackwell Scientific Publications, Oxford, pp 81–103.

Lessa R, Vooren CM (2007) *Rhinobatos horkelii*. 2008 IUCN Red List of Threatened Species, Cambridge. www.iucnredlist.org. Accessed March 16, 2009.

Liermann M, Hilborn R (1997) Depensation in fish stocks: a hierarchic Bayesian meta-analysis. Canadian Journal of Fisheries and Aquatic Sciences 54:1976–1984.

MacArthur RH, Wilson EO (1967) The theory of island biogeography. Princeton University Press, Princeton, New Jersey.

Mace PM (1994) Relationships between common biological reference points used as thresholds and targets of fisheries management strategies. Can J Fish Aquat Sci 51:110–122.

Mace PM (2001) A new role for MSY in single-species and ecosystem approaches to fisheries stock assessment and management. Fish Fish 2:2–32.

Manire CA, Gruber S (1990) Many sharks may be headed toward extinction. Conserv Biol 4:10–11.

Marris E (2007) Conservation priorities: what to let go. Nature 450:152–155.

Maxwell DL, Jennings S (2005) Power of monitoring programmes to detect decline and recovery of rare and vulnerable fish. J Appl Ecol 42:25–37.

McAllister MK, Pikitch EK, Babcock EA (2001) Using demographic methods to construct Bayesian priors for the intrinsic rate of increase in the Schaefer model and implications for stock rebuilding. Can J Fish Aquat Sci 58:1871–1890.

McPherson JM, Myers RA (2009) How to infer population trends in sparse data: examples with opportunistic sighting records for great white sharks. Divers Distrib 15:880–890.

Mollet HF, Cailliet GM (2002) Comparative population demography of elasmobranchs using life history tables, Leslie matrices and stage-based matrix models. Mar Freshw Res 53:503–516.

Morato T, Watson R, Pitcher TJ, Pauly D (2006) Fishing down the deep. Fish Fish 7:24–34.

Musick JA (1999) Life in the slow lane: ecology and conservation of long-lived marine animals. Am Fish Symp 23:1–10.

Musick JA, Harbin MM, Berkeley SA, Burgess GH, Eklund AM et al. (2000) Marine, estuarine, and diadromous fish stocks at risk of extinction in North America (exclusive of Pacific salmonids). Fisheries 25:6–30.

Myers RA (2001) Stock and recruitment: generalizations about maximum reproductive rate, density dependence, and variability using meta-analytic approaches. ICES J Mar Sci 58:937–951.

Myers RA (2002) Recruitment: understanding density-dependence in fish populations. In: Hart PJB, Reynolds JD (eds) Handbook of fish and fisheries. Blackwell Science, Oxford, pp 123–148.

Myers RA, Baum JK, Shepherd TD, Powers SP, Peterson CH (2007) Cascading effects of the loss of apex predatory sharks from a coastal ocean. Science 315:1846–1850.

Myers RA, Bridson J, Barrowman N (1995) Summary of worldwide stock and recruitment data. Can Tech Rep Fish Aquat Sci 2024:iv + 327.

Myers RA, Brown KG, Barrowman NJ (1999) The maximum reproductive rate of fish at low population sizes. Can J Fish Aquat Sci 56:2404–2419.

Myers RA, Mertz G (1998) The limits of exploitation: a precautionary approach. Ecol Appl 8:S165–S169.

Myers RA, Mertz G, Fowlow PS (1997) Maximum population growth rates and recovery times for Atlantic cod, *Gadus morhua*. Fish Bull 95:762–772.

National Marine Fisheries Service (2000) Status review of smalltooth sawfish (*Pristis pectinata*). National Marine Fisheries Service, Office of Protected Resources, Silver Springs, MD.

Perry AL, Low PJ, Ellis JR, Reynolds JD (2005) Climate change and distribution shifts in marine fishes. Science 308:1912–1915.

Pianka ER (1970) On r & K selection. Am Nat 104:592–597.

Pilling GM, Apostolaki P, Failler P, Floros C, Large PA et al. (2008) Assessment and management of data-poor fisheries. In: Payne AI, Cotter AJ, Potter ECE (eds) Advances in fisheries science: 50 years on from Beverton and Holt. Blackwell, Oxford, p 547.

Pogonoski J, Pollard D (2003) *Centrophorus harrissoni*. 2008 IUCN Red List of Threatened Species. www.iucnredlist.org. Accessed March 2, 2009.

Pratt HL, Casey JG (1990) Shark reproductive strategies as a limiting factor in directed fisheries, with a review of Holden's method of estimating growth parameters. NOAA NMFS Technical Report 90. National Marine Fisheries Service, Washington, DC, pp 97–111.

Prince JD (2005) Gauntlet fisheries for elasmobranchs: the secret of sustainable shark fisheries. J Northwest Atl Fish Org 35:407–416.

Punt AE (2000) Extinction of marine renewable resources: a demographic analysis. Popul Ecol 42:19.

Punt AE, Pribac F, Taylor BL, Walker TI (2005) Harvest strategy evaluation for school and gummy shark. J Northwest Atl Fish Sci 35:387–406.

Punt AE, Smith ADM (2001) The gospel of maximum sustainable yield in fisheries management: birth crucifixion and reincarnation. In: Reynolds JD, Mace GM, Redford KH, Robinson JG (eds) Conservation of exploited species. Cambridge University Press, Cambridge, UK, pp 41–66.

Redding DW, Mooers AO (2006) Incorporating evolutionary measures into conservation prioritization. Conserv Biol 20:1670–1678.

Reynolds JD, Dulvy NK, Goodwin NB, Hutchings JA (2005) Biology of extinction risk in marine fishes. Proc R Soc London B 272:2337–2344.

Reznick D, Bryant MJ, Bashey F (2002) r- and K-selection revisited: the role of population regulation in life-history evolution. Ecology 83:1509–1520.

Ricker WE (1954) Stock and recruitment. J Fish Res Board Can 11:559–623.

Ripley WE (1946) The biology of the soupfin *Galeorhinus zygopterus* and biochemical studies of the liver. Calif Fish Bull 64:1–96.

Robillard M, Seret B (2006) Cultural importance and decline of sawfish (Pristidae) populations in West Africa. Cybium 30:23–30.

Roff DA (1984) The evolution of life history parameters in teleosts. Can J Fish Aquat Sci 41:989–1000.

Rogers SI, Ellis JR (2000) Changes in the demersal fish assemblages of British coastal waters during the 20th century. ICES J Mar Sci 57:866–881.

Rose KA, Cowan JH, Winemiller KO, Myers RA, Hilborn R (2001) Compensatory density dependence in fish populations: importance, controversy, understanding and prognosis. Fish Fish 2:293–327.

Rothwell DR, Haward M (1996) Federal and international perspectives on Australia's maritime claims. Mar Policy 20:29–46.

Rudstam LG, Aneer G and Hilden M (1994) Top-down control in the pelagic Baltic ecosystem. Dana 10:105–129.

Savage VM, Gillooly JF, Brown JH, West GB, Charnov EL (2004) Effects of body size and temperature on population growth. Am Nat 163:429–441.

Schaefer MB (1954) Some aspects of the dynamics of populations important to the management of commercial marine fisheries. IATT Bull 127–156.

Simon JE, Frank KT, Kulka DW (2002) Distribution and abundance of barndoor skate *Dipturus laevis* in the Canadian Atlantic based upon research vessel surveys and industry/science surveys. Department of Fisheries and Oceans, Canadian Science Advisory Secretariat Research Document, 2002/070.

Simpfendorfer CA (2005a) Demographic models: life tables, matrix models and rebound potential. Food and Agriculture Organisation of the United Nations, Rome, pp 187–204.

Simpfendorfer CA (2005b) Threatened fishes of the world: *Pristis pectinata* Latham, 1794 (Pristidae). Environ Biol Fish 73:20.

Simpfendorfer CA (2007) The importance of mangroves as nursery habitat for smalltooth sawfish (*Pristis pectinata*) in South Florida. Bull Mar Sci 80:933–934.

Sims DW (2008) Sieving a living: a review of the biology, ecology and conservation status of the plankton-feeding basking shark *Cetorhinus maximus* Adv Mar Sci 54:171–220.

Sissenwine MP, Shepherd JG (1987) An alternative perspective on recruitment overfishing and biological reference points. Can J Fish Aquat Sci 44:913–918.

Smith ADM, Fulton EJ, Hobday AJ, Smith DC, Shoulder P (2007) Scientific tools to support the practical implementation of ecosystem-based fisheries management. ICES J Mar Sci 64:633–639.

Smith SE, Au DW, Show C (1998) Intrinsic rebound potentials of 26 species of Pacific sharks. Mar Freshw Res 49:663–678.

Smith SE, Au DW, Show C (2008) Intrinsic rates of increase in pelagic elasmobranchs. In: Camhi MD, Pikitch EK, Babcock EA (eds) Sharks of the open ocean: biology, fisheries and conservation. Blackwell Publishing, Oxford, pp 288–297.

Stearns SC (1977) Evolution of life-history traits: critique of theory and a review of data. Annu Rev Ecol Syst 8:145–171.

Stevens JD (1999) Variable resilience to fishing pressure in two sharks: the significance of different ecological and life history parameters. In: Musick JA (ed) Life in the slow lane: ecology and conservation of long-lived marine animals. American Fisheries Society, Bethesda, MD, pp 11–16.

Stevens JD, Bonfil R, Dulvy NK, Walker P (2000) The effects of fishing on sharks, rays and chimaeras (chondrichthyans), and the implications for marine ecosystems. ICES J Mar Sci 57:476–494.

Stobutzki I, Miller M, Brewer D (2001) Sustainability of fishery bycatch: a process for assessing highly diverse and numerous bycatch. Environ Conserv 28:167–181.

Summers G (1987) Squalene: a potential shark byproduct. Catch 14:29.

Sutherland WJ Gill JA (2001) The role of behaviour in studying sustainable exploitation. pp. 259–280. In: Reynolds JD, Mace GM, Redford KH, and Robinson JG (ed.) Conservation of exploited species, Cambridge University Press, Cambridge.

Thorburn DC, Morgan DL (2005) Threatened fishes of the world: *Glyphis* sp C (Carcharhinidae). Environ Biol Fish 73:140.

Thorson TB (1982) The impact of commercial exploitation on sawfish and shark populations in Lake Nicaragua. 7:2–10.

Tilzey RDJ, Rowling KR (2001) History of Australia's South East Fishery: a scientist's perspective. Mar Freshw Res 52:361–375.

Trites AW, Deecke VB, Gregr EJ, Ford JKB, Olesiuk PF (2007) Killer whales, whaling, and sequential megafaunal collapse in the North Pacific: a comparative analysis of the dynamics of marine mammals in Alaska and British Columbia following commercial whaling. Mar Mamm Sci 23:751–765.

van der Elst R (1979) A proliferation of small sharks in the shore-based Natal sport fishery. Environ Biol Fishes 4:349–362.

Walker PA, Heessen HJL (1996) Long-term changes in ray populations in the North Sea. ICES J Mar Sci 53:1085–1093.

Walker TI (1998) Can shark resources be harvested sustainably? A question revisited with a review, of shark fisheries. Mar Freshw Res 49:553–572.

Walker TI, Gason AS (2007) Shark and other chondrichthyan byproduct and bycatch estimation in the Southern and Eastern Scalefish and Shark Fishery. Primary Industries Research, Victoria, Queenscliff, Victoria, Australia. Final report to Fisheries Research and Development Corporation Project.

Wallace S, Gisborne B (2006) Basking sharks: the slaughter of BC's gentle giants. New Star Books, Vancouver.

Walters C, Kitchell JF (2001) Cultivation/depensation effects on juvenile survival and recruitment: implications for the theory of fishing. Canadian Journal of Fisheries and Aquatic Sciences 58:39–50.

Walters CJ, Korman J (1999) Linking recruitment to trophic factors: revisiting the Beverton–Holt recruitment model from a life history and multispecies perspective. Rev Fish Biol Fish 9:187–202.

Walters CJ, Martell SJD (2004) Fisheries ecology and management. Princeton University Press, Princeton, NJ.

White WT, Ebert DA, Compagno LJV (2008) Description of two new species of gulper sharks, genus *Centrophorus* (Chondrichthyes: Squaliformes: Centrophoridae) from Australia. In: Last PR, White WT, Pogonoski JJ (eds) Descriptions of new Australian chondrichthyans. CSIRO Marine and Atmospheric Research Paper 022. CSIRO, Hobart, Australia, pp 1–21.

Williams SE, Shoo LP, Isaac JL, Hoffmann AA, Langham G (2008) Towards an integrated framework for assessing the vulnerability of species to climate change. PLoS Biol 6:e325.

Wilson CD, Seki MP (1994) Biology and population characteristics of *Squalus mitsukurii* from a seamount in the central north Pacific Ocean. Fish Bull 92:851–864.

Wilson SK, Fisher R, Pratchett MS, Graham NAJ, Dulvy NK et al. (2008) Exploitation and habitat degradation as agents of change within coral reef fish communities. Global Change Biol 14:2796–2809.

Winemiller KO (2005) Life history strategies, population regulation, and implications for fisheries management. Can J Fish Aquat Sci 62:872–885.

Winemiller KO, Rose KA (1992) Patterns of life history diversification in North American fishes: implications for population regulation. Can J Fish Aquat Sci 49:2196–2218.

Wood CC, Ketchen KS, Beamish RJ (1979) Population dynamics of the spiny dogfish (*Squalus acanthias*) in British Columbia waters. J Fish Res Board Can 36:647–656.

Worm B, Lotze HK, Myers RA, Sandow M, Oschlies A (2005) Global patterns of predator diversity in the open oceans. Science 309:1365–1369.

Zeeberg J, Corten A, de Graaf E (2006) Bycatch and release of pelagic megafauna in industrial trawler fisheries off Northwest Africa. Fish Res 78:186–195.

Index

Printed in the United States
by Baker & Taylor Publisher Services